Lecture Notes in Computer Science 1573
Edited by G. Goos, J. Hartmanis and J. van Leeuwen

Springer
Berlin
Heidelberg
New York
Barcelona
Hong Kong
London
Milan
Paris
Singapore
Tokyo

José M.L.M. Palma Jack Dongarra
Vicente Hernández (Eds.)

Vector and Parallel Processing – VECPAR'98

Third International Conference
Porto, Portugal, June 21-23, 1998
Selected Papers and Invited Talks

Series Editors

Gerhard Goos, Karlsruhe University, Germany
Juris Hartmanis, Cornell University, NY, USA
Jan van Leeuwen, Utrecht University, The Netherlands

Volume Editors

José M.L.M. Palma
Faculdade de Engenharia da Universidade do Porto
Rua dos Bragas, 4050-123 Porto, Portugal
E-mail: jpalma@fe.up.pt

Jack Dongarra
University of Tennessee, Department of Computer Science
Knoxville, TN 37996-1301, USA
and
Oak Ridge National Laboratory, Mathematical Sciences Section
Oak Ridge, TN 37821-6367, USA
E-mail: dongarra@cs.utk.edu

Vicente Hernández
Universidad Politécnica de Valencia
Departamento de Sistemas Informáticos y Computación
Camino de Vera, s/n, Apartado 22012, E-46020 Valencia, Spain
E-mail: vhernand@dsic.upv.es

Cataloging-in-Publication data applied for

Die Deutsche Bibliothek - CIP-Einheitsaufnahme

Vector and parallel processing : third international conference ; selected papers and invited talks / VECPAR '98, Porto, Portugal, June 21 - 23, 1998. José M. L. M. Palma ... (ed.). - Berlin ; Heidelberg ; New York ; Barcelona ; Hong Kong ; London ; Milan ; Paris ; Singapore ; Tokyo : Springer, 1999
(Lecture notes in computer science ; Vol. 1573)
ISBN 3-540-66228-6

CR Subject Classification (1998): G.1-2, D.1.3, C.2, F.1.2, F.2

ISSN 0302-9743
ISBN 3-540-66228-6 Springer-Verlag Berlin Heidelberg New York

Typesetting: Camera-ready by author
SPIN: 10703040 06/3142 – 5 4 3 2 1 0 Printed on acid-free paper

Preface

This book stands as the visible mark of *VECPAR'98 - 3rd International Meeting on Vector and Parallel Processing*, which was held in Porto (Portugal) from 21 to 23 June 1998. *VECPAR'98* was the third of the *VECPAR* series of conferences initiated in 1993 and organised by *FEUP*, the Faculty of Engineering of the University of Porto.

The conference programme comprised a total of 6 invited talks, 66 contributed papers and 18 posters. The contributed papers and posters were selected from 120 extended abstracts originating from 27 countries.

Outline of the book

The book, with 7 chapters, contains 41 contributed papers and 6 invited talks. The 41 papers included in these proceedings result from the reviewing of all papers presented at the conference.

Each of the first 6 chapters includes 1 of the 6 invited talks of the conference, and related papers. Chapters 1, 2, 4, 5 and 7 are initiated by an introductory text providing the reader with a guide to the chapter contents.

Chapter 1 is on numerical algebra. It begins with an introductory text by Vicente Hernández, followed by the invited talk by Gene Golub, entitled *Some Unusual Eigenvalue Problems.* The remaining 11 contributed articles in this chapter deal either with large scale eigenvalue problems or with linear system problems.

Computational fluid dynamics and crash and structural analysis were brought under the same chapter and that is Chapter 2, which contains the invited talk by Timothy Barth, entitled *Parallel Domain Decomposition Pre-conditioning for Computational Fluid Dynamics*, plus 8 contributed papers. Timothy Barth also authors the introductory text to the chapter.

The invited talk by Peter Welch, entitled *Parallel and Distributed Computing in Education*, alone makes up Chapter 3.

The main article in Chapter 4 is the invited talk, by Jean Vuillemin, entitled *Reconfigurable Systems: Past and Next 10 Years.* José Silva Matos provides the introductory text. Apart from the invited talk, the chapter contains 15 contributed papers, covering subjects such as computer organisation, programming and benchmarking.

Thierry Priol authors the introductory text and also the invited talk in Chapter 5, entitled *High Performance Computing for Image Synthesis.* The chapter includes only one contributed paper.

Database servers are the topic of Chapter 6, which includes the invited talk entitled *The Design of an ODMG Conformant Parallel Object Database Server*, by Paul Watson.

Chapter 7 brings the book to an end with 6 papers on various nonlinear problems. The main features of these articles are discussed by Heather Ruskin and José A.M.S. Duarte in their introductory text.

Readership of this book

The wide range of subjects covered by this book reflects the multi-disciplinarity of the *VECPAR* series of conferences. The book may be of interest to those in scientific areas, as for instance computational methods applied to either engineering or science in general, computer architectures, applied mathematics or computational science.

Student papers

To encourage the participation of students and provide a stimulus that may be of great importance for their careers, *VECPAR'98* included a student paper competition for the first time. There were 10 participants in this competition. The evaluation was performed by a jury of 13 members that attended the presentations, questioned the students and evaluated the written version of the work.

As a result of this competition 2 articles emerged. The first prize was awarded to M. Arenaz (Spain) with the work entitled *High Performance Computing of a New Numerical Algorithm for an Industrial Problem in Tribology*, by M. Arenaz, R. Doallo, G. García and C. Vázquez. An honourable mention was awarded to Carmen Borges (Brazil) for her study entitled *A Parallelisation Strategy for Power Systems Composite Reliability Evaluation*, by C.L.T. Borges and D.M. Falcão.

Paper selection

The extended abstracts and the papers were evaluated on the basis of technical quality, originality, importance to field, and style and clarity of presentation. Each paper was marked in a scale ranging from 1 (poor) to 5 (excellent). No paper with a score lower than 3.5 was included in this book.

One must note that, in our terminology, an *extended abstract* is what some may call conference paper, and a *paper*, as included in the book and presented at the conference, is what some may call a journal article. The extended abstracts had a maximum of 2000 words (about 7 printed pages) and the articles could run to 14 printed pages, or 30 printed pages in the case of the invited talks. We believe that authors should have no major space limitations that could prevent them from presenting their work in full detail.

It is our hope that the efforts put into the selection process have contributed to the prestige of the *VECPAR* series of conferences and the quality of this book, making it a valuable source of future reference.

January 1999

José M.L.M. Palma,
Jack Dongarra
Vicente Hernández

Acknowledgments

This book, a 700 page document, contains the results of the organisational works performed over 2 years, since the 2nd edition of *VECPAR* in 1996.

First, one must acknowledge the excellent work of the Organising Committee. Lígia Ribeiro's and Augusto de Sousa's attention to details contributed to a general feeling of satisfaction among the participants of the conference.

The Local Advisory Organising Committee was much more than its name may suggest. It was its responsibility, jointly with the Organising Committee and a restricted number of members of the Scientific Committee, to identify the topics of the conference, to select those for which an invited speaker should be designated, and also to identify the Invited Speaker. To honour the multidisciplinarity of the *VECPAR* conferences, the help of colleagues was required (too many to mention here), to whom we are grateful for providing us with information and suggestions on who should be invited.

The selection of the articles was much facilitated by the excellent work of the Scientific Committee. Our colleagues were extremely generous in providing part of their precious time for careful reading of the many documents that we asked them to review. The depth and detail of their comments were essential: firstly to us, when making the final decision on which manuscript to accept or reject; and secondly to authors, providing them with suggestions that contributed to further improvement of their work.

Our word of gratitude also to all authors and participants at *VECPAR'98*. They were those who made the conference possible and also those for whom the conference was intended.

The financial support of all sponsoring organisations is also acknowledged.

Committees

Organising Committee

Lígia Maria Ribeiro (Chair)
A. Augusto de Sousa (Co-Chair)

Local Advisory Organising Committee

F. Nunes Ferreira
J. Carlos Lopes
J. Silva Matos
J. César Sá
J. Marques dos Santos
R. Moreira Vidal

Scientific Committee

J. Palma (Chair)	Universidade do Porto, Portugal
J. Dongarra (Co-Chair)	University of Tennessee and Oak Ridge National Laboratory, USA
V. Hernández (Co-Chair)	Universidad Politécnica de Valencia, Spain
P. Amestoy	ENSEEIHT-IRIT, France
E. Aurell	Stockholm University, Sweden
A. Chalmers	University of Bristol, England
A. Coutinho	Universidade Federal do Rio de Janeiro, Brazil
J.C. Cunha	Universidade Nova de Lisboa, Portugal
F. d'Almeida	Universidade do Porto, Portugal
M. Daydé	ENSEEIHT-IRIT, France
J. Dekeyser	Université des Sciences et Technologies de Lille, France
R. Delgado	Universidade do Porto, Portugal
J. Duarte	Universidade do Porto, Portugal
I. Duff	Rutherford Appleton Laboratory, England, and CERFACS, France
D. Falcão	Universidade Federal do Rio de Janeiro, Brazil
S. Gama	Universidade do Porto, Portugal
L. Giraud	CERFACS, France
S. Hammarling	Numerical Algorithms Group Ltd (NAG), England
D. Heermann	Universität Heidelberg, Germany
W. Janke	Universität Leipzig, Germany

D. Knight	Rutgers – The State University of New Jersey, USA
V. Kumar	University of Minnesota, USA
R. Lins	Universidade Federal de Pernambuco em Recife, Brazil
J. Long	Centro de Supercomputacion de Galicia, Spain
J.P. Lopes	Universidade do Porto, Portugal
E. Luque	Universidad Autònoma de Barcelona, Spain
P. Marquet	Université des Sciences et Technologies de Lille, France
P. de Miguel	Universidad Politécnica de Madrid, Spain
F. Moura	Universidade do Minho, Portugal
K. Nagel	Los Alamos National Laboratory, USA
M. Novotny	Florida State University, USA
P. Oliveira	Universidade do Minho, Portugal
E. Oñate	Universitat Politécnica de Catalunya, Spain
A. Padilha	Universidade do Porto, Portugal
R. Pandey	University of Southern Mississipi, USA
J. Pereira	INESC, Portugal
M. Perić	Technische Universität Hamburg-Harburg, Germany
H. Pina	Universidade Técnica de Lisboa, Portugal
A. Proença	Universidade do Minho, Portugal
R. Ralha	Universidade do Minho, Portugal
Y. Robert	École Normale Supérieure de Lyon, France
A. Ruano	Universidade do Algarve, Portugal
D. Ruiz	ENSEEIHT-IRIT, France
H. Ruskin	Dublin City University, Ireland
F. Silva	Universidade do Porto, Portugal
J.G. Silva	Universidade de Coimbra, Portugal
F. Tirado	Universidad Complutense, Spain
B. Tourancheau	École Normale Supérieure de Lyon, France
M. Valero	Universitat Politécnica de Catalunya, Spain
V. Venkatakrishnan	Boeing Commercial Airplane Group, USA
P. Veríssimo	Universidade de Lisboa, Portugal
J.-S. Wang	National University of Singapore, Singapore
E. Zapata	Universidad de Malaga, Spain

Other Reviewers

T. Barth	NASA Ames Research Center, USA
G. Golub	Stanford University, USA
P. Hernandez	Universidad Autónoma de Barcelona, Spain
M. Joshi	University of Minnesota, USA
D. Kranzlmueller	Johannes Kepler Universität Linz, Austria
J. Lopez	Universidad Autònoma de Barcelona, Spain
J. Macedo	Universidade do Porto, Portugal
T. Margalef	Universidad Autònoma de Barcelona, Spain
R. Moreno	Universidad Complutense Madrid, Spain
T. Priol	IRISA/INRIA, France
VCV. Rao	University of Minnesota, USA
Y. Robert	École Normale Supérieure de Lyon, France
C. Sá	Universidade do Porto, Portugal
A. Sousa	Universidade do Porto, Portugal
K. Schloegel	University of Minnesota, USA
P. Watson	University of Newcastle, England
P. Welch	The University of Kent at Canterbury, England

Best Student Paper Award (Jury)

Filomena d'Almeida	Universidade do Porto, Portugal
Michel Daydé	ENSEEIHT-IRIT, France
Jack Dongarra	University of Tennessee and Oak Ridge National Laboratory, USA
José Duarte	Universidade do Porto, Portugal
José Fortes	University of Purdue, USA
Sílvio Gama	Universidade do Porto, Portugal
Gene Golub	Stanford University, USA
Vicente Hernández	Universidad Politécnica de Valencia, Spain
João Peças Lopes	Universidade do Porto, Portugal
José Laginha Palma	Universidade do Porto, Portugal
Alberto Proença	Universidade do Minho, Portugal
Rui Ralha	Universidade do Minho, Portugal
Heather Ruskin	Dublin City University, Ireland

Sponsoring Organisations

FEUP - Faculdade de Engenharia da Universidade do Porto
FCT - Fundação para a Ciência e Tecnologia
EOARD - European Office of Aerospace Research and Development
Fundação Dr. António Cupertino de Miranda
FLAD - Fundação Luso Americana para o Desenvolvimento
Reitoria da Universidade do Porto
FCCN - Fundação para a Computação Científica Nacional
Fundação Calouste Gulbenkian
AEFEUP - Associação de Estudantes da Faculdade de Engenharia da Universidade do Porto
Câmara Municipal do Porto
Digital Equipment Portugal
Bull Portugal
ICL
SGI - Silicon Graphics Inc.
Porto Convention Bureau
PT - Portugal Telecom
Sicnet - Sistemas Integrados de Comunicações, Novos Equipamentos e Tecnologias, Lda

Invited Lecturers

- Timothy Barth
  NASA Ames Research Center, Moffett Field, USA
- Gene Golub
  Stanford University, Stanford, USA
- Thierry Priol
  IRISA/INRIA, Rennes, France
- Jean Vuillemin
  École Normale Supérieure, Paris, France
- Paul Watson
  University of Newcastle-upon-Tyne, England
- Peter Welch
  University of Kent at Canterbury, England

Table of Contents

Chapter 1: Eigenvalue Problems and Solution of Linear Systems

Chapter 2: Computational Fluid Dynamics, Structural Analysis and Mesh Partioning Techniques

Chapter 3: Computing in Education

Chapter 4: Computer Organisation, Programming and Benchmarking

Chapter 5: Image, Analysis and Synthesis

Chapter 6: Parallel Database Servers

Chapter 7: Nonlinear Problems

Chapter 1: Eigenvalue Problems and Solution of Linear Systems

Introduction

Vicente Hernández

Universidad Politécnica de Valencia
Camino de Vera, s/n, 46071 Valencia, Spain
vhernand@dsic.upv.es

In this chapter, one invited talk and 11 articles are included dealing with the solution of linear systems and eigenvalue problems. These problems arise in many engineering and scientific areas, and the articles selected show applications in fields such as nuclear engineering, simulation of chemical plants, simulation of electronic circuits, quantum chemistry, structural mechanics, etc. Issues covered include, among others, algorithm modifications for particular architectures, use of parallel preconditioning techniques, direct and iterative linear solvers and eigensolvers. What follows is a general overview of the articles in this chapter.

The first article to be presented is the invited talk, *Some Unusual Eigenvalue Problems*, by Bai and Golub. The article is centred on some eigenvalue problems that are not presented in a standard form. These problems are converted to the problem of computing a quadratic form, and then numerical algorithms based on the Gauss-type quadrature rules and Lanczos process are applied. Since the algorithms reference the matrix in question only through a matrix-vector product operation, they are suitable for large-scale sparse problems. Some numerical examples are presented to illustrate the performance of the approach.

In *Multi-Sweep Algorithms for the Symmetric Eigenproblem*, Gansterer, Kvasnicka and Ueberhuber are concerned with the solution of the symmetric eigenvalue problem by the reduction of the matrix to tridiagonal form. The algorithm presented uses multi-sweep in order to improve data locality and allow the use of Level 3 BLAS instead of Level 2. The eigenvectors are computed from an intermediate band matrix, instead of a tridiagonal matrix. The new algorithm improves the floating-point performance with respect to algorithms currently implemented in LAPACK.

A different approach that can be taken to solve the symmetric eigenvalue problem is the use of the Jacobi method, which offers a high level of parallelism. This is the case studied in *A Unified Approach to Parallel Block-Jacobi Methods for the Symmetric Eigenvalue Problem* by Giménez, Hernández and Vidal. This approach is based on two steps: first, a logical algorithm is designed by dividing the matrices into square blocks and considering each block as a process; then an algorithm for a particular topology is obtained by grouping processes and assigning them to processors.

J. Palma, J. Dongarra, and V. Hernández (Eds.): VECPAR'98, LNCS 1573, pp. 1–3, 1999.

An application of the eigenvalue problem is shown in *Calculation of Lambda Modes of a Nuclear Reactor: A Parallel Implementation Using the Implicitly Restarted Arnoldi Method.* Hernández, Román, A. Vidal and V. Vidal show how the calculation of the dominant lambda-modes of a nuclear power reactor leads to a generalized eigenvalue problem, which can be reduced to a standard one. The Implicitly Restarted Arnoldi method is used by means of a parallel approach adequate for the large dimension of the arising problems. The algorithm shown involves the parallel iterative solution of a linear system.

Two articles study the Jacobi-Davidson method for the solution of eigenvalue problems. *Parallel Jacobi-Davidson for Solving Generalized Eigenvalue Problems*, by Nool and van der Ploeg, describes two Jacobi-Davidson variants, one using standard Ritz values and one harmonic Ritz values. In both cases, a process is performed of complete LU decomposition, with reordering strategies based on block cyclic reduction and domain decomposition. Results are presented for different Magneto-Hydrodynamics problems. In *Parallel Preconditioned Solvers for Large Sparse Hermitian Eigenproblems*, Basermann studies a pre-conditioning technique for the Jacobi-Davidson method, using the Quasi-Minimal Residual iteration. In order to parallelise the solvers, matrix and vector partitioning are investigated and sparsity of the matrix exploited. Results are provided from problems of quantum chemistry and structural mechanics.

Low cost and availability are important advantages that make networks of processors an increasingly attractive choice as parallel platform. Thus, it is important to study the performance of algorithms on networks of clusters. The paper, *Solving Eigenvalue Problems on Networks of Processors* by Giménez, Jiménez, Majado, Marín, and Martín, studies different eigenvalue solvers on networks of processors. In particular, the Power method, deflation, Givens algorithm, Davidson methods and Jacobi methods are analysed. As a conclusion, the article points out the interest in the use of networks of processors, although the high cost of communications gives rise to small modifications in the algorithms.

Solving Large-Scale Eigenvalue Problems on Vector Parallel Processors, by Harrar II and Osborne, considers the development of eigensolvers on distributed memory parallel arrays of vector processors, platforms in which it is important to achieve high levels of both vectorisation and parallelisation. It is shown that this fact leads both to novel implementations of known methods and to the development of completely new algorithms. The paper covers both symmetric and non-symmetric matrices, and tridiagonal matrices. Results are shown for eigenvalue problems arising in a variety of applications.

Friedrich Grund studies the solution of linear systems of equations involving un-symmetric sparse matrices, by means of direct methods in *Direct Linear Solvers for Vector and Parallel Computers.* Special attention is paid to the case when several linear systems with the same pattern structure have to be solved. The approach used consists in the generation of a pseudo-code containing the required operations for the factorisation and the solution of the triangular systems. Then this pseudo-code can be interpreted repeatedly to compute the solution of these systems. Numerical results are provided for matrices arising from different scientific and technical problems, and the application of the method to problems of dynamic simulation of chemical plants is also shown.

When the solution of linear systems is computed by means of iterative methods, looking for efficient parallel pre-conditioners is an important topic in current research. García-Loureiro, Pena, López-González and Viñas present parallel versions of two pre-conditioners in *Parallel Preconditioners for Solving Nonsymmetric Linear Systems.* The first one is based on a block form of the ILU and the second one is based on the SPAI (Sparse Approximate Inverse) method. Both methods are analysed using the Bi-CGSTAB algorithm to solve general sparse, non-symmetric systems. Results are presented for matrices arising from the simulation of hetero-junction bipolar transistors.

Parallel algorithms for solving almost linear systems are studied in *Synchronous and Asynchronous Parallel Algorithms with Overlap for Almost Linear Systems*, by Arnal, Migallón and Penadés. The article presents a non-stationary parallel algorithm based on the multi-splitting technique and its extension to an asynchronous model are considered. Convergence properties of these methods are studied, and results are shown for problems arising from partial differential equations.

Finally, Husbands and Isbell present in *The Parallel Problems Server: A Client-Server Model for Interactive Large Scale Scientific Computation*, an interesting approach for the application of fast scientific computing algorithms to large problems. A novel architecture is used based on a client-server model for interactive linear algebra computation. The advantages of the new approach with respect to other standard approaches, such as linear algebra libraries and interactive prototyping systems, are presented.

Some Unusual Eigenvalue Problems
Invited Talk

Zhaojun Bai[1] and Gene H. Golub[2]

[1] University of Kentucky
Lexington, KY 40506, USA,
bai@ms.uky.edu
[2] Stanford University
Stanford, CA 94305, USA
golub@sccm.stanford.edu

Abstract. We survey some unusual eigenvalue problems arising in different applications. We show that all these problems can be cast as problems of estimating quadratic forms. Numerical algorithms based on the well-known Gauss-type quadrature rules and Lanczos process are reviewed for computing these quadratic forms. These algorithms reference the matrix in question only through a matrix-vector product operation. Hence it is well suited for large sparse problems. Some selected numerical examples are presented to illustrate the efficiency of such an approach.

1 Introduction

Matrix eigenvalue problems play a significant role in many areas of computational science and engineering. It often happens that many eigenvalue problems arising in applications may not appear in a standard form that we usually learn from a textbook and find in software packages for solving eigenvalue problems. In this paper, we described some unusual eigenvalue problems we have encountered. Some of those problems have been studied in literature and some are new. We are particularly interested in solving those associated with large sparse problems. Many existing techniques are only suitable for dense matrix computations and becomes inadequate for large sparse problems.

We will show that all these unusal eigenvalue problems can be converted to the problem of computing a quadratic form $u^T f(A)u$, for a properly defined matrix A, a vector u and a function f. Numerical techniques for computing the quadratic form to be discussed in this paper will based on the work initially proposed in [6] and further developed in [11, 12, 2]. In this technique, we first transfer the problem of computing the quadratic form to a Riemann-Stieltjes integral problem, and then use Gauss-type quadrature rules to approximate the integral, which then brings the orthogonal polynomial theory and the underlying Lanczos procedure into the scene. This approach is well suitable for large sparse problems, since it references the matrix A through a user provided subroutine to form the matrix-vector product Ax.

J. Palma, J. Dongarra, and V. Hernández (Eds.): VECPAR'98, LNCS 1573, pp. 4–19, 1999.

The basic time-consuming kernels for computing quadratic forms using parallelism are vector inner products, vector updates and matrix-vector products; this is similar to most iterative methods in linear algebra. Vector inner products and updates can be easily parallelized: each processor computes the vector-vector operations of corresponding segments of vectors (local vector operations (LVOs)), and if necessary, the results of LVOs have to sent to other processors to be combined for the global vector-vector operations. For the matrix-vector product, the user can either explore the particular structure of the matrix in question for parallelism, or split the matrix into strips corresponding to the vector segments. Each process then computes the matrix-vector product of one strip. Furthermore, the iterative loop of algorithms can be designed to overlap communication and computation and eliminating some of the synchronization points. The reader may see [8, 4] and references therein for further details.

The rest of the paper is organized as follows. Section 2 describes some unusual eigenvalue problems and shows that these problems can be converted to the problem of computing a quadratic form. Section 3 reviews numerical methods for computing a quadratic form. Section 4 shows that how these numerical methods can be applied to those problems described in section 2. Some selected numerical examples are presented in section 5. Concluding remarks are in section 5.

2 Some Unusual Matrix Eigenvalue Problems

2.1 Constrained eigenvalue problem

Let A be a real symmetric matrix of order N, and c a given N vector with $c^T c = 1$. We are interested in the following optimization problem

$$\max_x \quad x^T A x \tag{1}$$

subject to the constraints

$$x^T x = 1 \tag{2}$$

and

$$c^T x = 0. \tag{3}$$

Let

$$\phi(x, \lambda, \mu) = x^T A x - \lambda(x^T x - 1) + 2\mu x^T c, \tag{4}$$

where λ, μ are Lagrange multipliers. Differentiating (4) with respect to x, we are led to the equation

$$Ax - \lambda x + \mu c = 0.$$

Then

$$x = -\mu(A - \lambda I)^{-1} c.$$

Using the constraint (3), we have

$$c^T (A - \lambda I)^{-1} c = 0. \tag{5}$$

An equation of such type is referred as a *secular equation.* Now the problem becomes finding the largest λ of the above secular equation.

We note that in [10], the problem is cast as computing the largest eigenvalue of the matrix PAP, where P is a project matrix $P = I - cc^T$.

2.2 Modified eigenvalue problem

Let us consider solving the following eigenvalue problems

$$Ax = \lambda x$$

and

$$(A + cc^T)\bar{x} = \bar{\lambda}\bar{x}$$

where A is a symmetric matrix and c is a vector and without loss of generality, we assume $c^T c = 1$. The second eigenvalue problem can be regarded as a modifed or perturbed eigenvalue problem of the first one. We are interested in obtaining some, not all, of the eigenvalues of both problems. Such computation task often appears in structural dynamic (re-)analysis and other applications [5].

By simple algebraic derivation, it is easy to show that the eigenvalues $\bar{\lambda}$ of the second problem satisfy the following secular equation

$$1 + c^T(A - \bar{\lambda}I)^{-1}c = 0. \tag{6}$$

2.3 Constraint quadratic optimization

Let A be a symmetric positive definite matrix of order N and c a given N vector. The quadratic optimization problem is stated as the following:

$$\min_x \quad x^T Ax - 2c^T x \tag{7}$$

with the constraint

$$x^T x = \alpha^2, \tag{8}$$

where α is a given scalar. Now let

$$\phi(x, \lambda) = x^T Ax - 2c^T x + \lambda(x^T x - \alpha^2) \tag{9}$$

where λ is the Lagrange multiplier. Differentiating (9) with respect to x, we are led to the equation

$$(A + \lambda I)x - c = 0$$

By the constraint (8), we are led to the problem of determining $\lambda > 0$ such that

$$c^T(A + \lambda I)^{-2}c = \alpha^2. \tag{10}$$

Furthermore, one can show the existence of a unique positive λ^* for which the above equation is satisfied. The solution of the original problem (7) and (8) is then $x^* = (A + \lambda^* I)^{-1}c$.

2.4 Trace and determinant

The trace and determinant problems are simply to estimate the quantities

$$\text{tr}(A^{-1}) = \sum_{i=1}^{n} e_i^T A^{-1} e_i$$

and

$$\det(A)$$

for a given matrix A. For the determinant problem, it can be easily verified that for a symmetric positive definite matrix A:

$$\ln(\det(A)) = \text{tr}(\ln(A)) = \sum_{i=1}^{n} e_i^T (\ln(A)) e_i. \tag{11}$$

Therefore, the problem of estimating the determinant is essentially to estimate the trace of the matrix natural logarithm function $\ln(A)$.

2.5 Partial eigenvalue sum

The partial eigenvalue sum problem is to compute the sum of all eigenvalues less than a prescribed value α of the generalized eigenvalue problem

$$Ax = \lambda Bx, \tag{12}$$

where A and B are real $N \times N$ symmetric matrices with B positive definite. Specifically, let $\{\lambda_i\}$ be the eigenvalues; one wants to compute the quantity

$$\tau_\alpha = \sum_{\lambda_i < \alpha} \lambda_i$$

for a given scalar α.

Let $B = LL^T$ be Cholesky decomposition of B, the problem (12) is then equivalent to

$$(L^{-1}AL^{-T})L^T x = \lambda L^T x.$$

Therefore the partial eigenvalue sum of the matrix pair (A, B) is equal to the partial eigenvalue sum of the matrix $L^{-1}AL^{-T}$, which, in practice, does not need to be formed explicitly.

A number of approaches might be found in literature to solve such problem. Our approach will based on constructing a function f such that the trace of $f(L^{-1}AL^{-T})$ approximates the desired sum τ_α. Specifically, one wants to construct a function f such that

$$f(\lambda_i) = \begin{cases} \lambda_i, & \text{if } \lambda_i < \alpha \\ 0, & \text{if } \lambda_i > \alpha, \end{cases} \tag{13}$$

for $i = 1, 2, \ldots, N$. Then $\mathrm{tr}(f(L^{-1}AL^{-T}))$ is the desired sum τ_α. One of choices is to have the f of the form

$$f(\zeta) = \zeta g(\zeta) \tag{14}$$

where

$$g(\zeta) = \frac{1}{1 + \exp\left(\frac{\zeta - \alpha}{\kappa}\right)},$$

where κ is a constant. This function, among other names, is known as the Fermi-Dirac distribution function [15, p. 347]. In the context of a physical system, the usage of this distribution function is motivated by thermodynamics. It directly represents thermal occupancy of electronic states. κ is proportional to the temperature of the system, and α is the chemical potential (the highest energy for occupied states).

It is easily seen that $0 < g(\zeta) < 1$ for all ζ with horizontal asymptotes 0 and 1. $(\alpha, \frac{1}{2})$ is the inflection point of g and the sign of κ determines whether g is decreasing ($\kappa > 0$) or increasing ($\kappa < 0$). For our application, we want the sum of all eigenvalues less than α, so we use $\kappa > 0$. The magnitude of κ determines how "close" the function g maps $\zeta < \alpha$ to 1 and $\zeta > \alpha$ to 0. As $\kappa \to 0^+$, the function $g(\zeta)$ rapidly converges to the step function $h(\zeta)$.

$$h(\zeta) = \begin{cases} 1 & \text{if } \zeta < \alpha \\ 0 & \text{if } \zeta > \alpha. \end{cases}$$

The graphs of the function $g(\zeta)$ for $\alpha = 0$ and different values of the parameter κ are plotted in Figure 1.

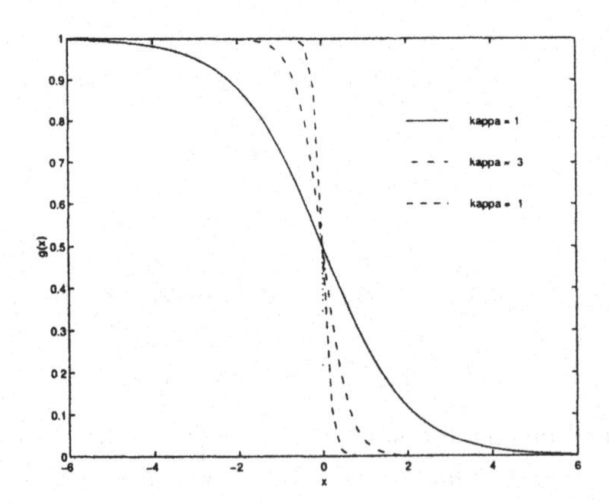

Fig. 1. Graphs of $g(\zeta)$ for different values of κ where $\alpha = 0$.

With this choice of $f(\zeta)$, we have

$$\tau_\alpha = \sum_{\lambda_i < \alpha} \lambda_i \approx \mathrm{tr}(f(L^{-1}AL^{-T})) = \sum_{i=1}^{n} e_i^T f(L^{-1}AL^{-T}) e_i. \tag{15}$$

In summary, the problem of computing partial eigenvalue sum becomes computing the trace of $f(L^{-1}AL^{-T})$.

3 Quadratic Form Computing

As we have seen, all those unusual eigenvalue problems presented in section 2 can be summarized as the problem of computing the quadratic form $u^T f(A)u$, where A is a $N \times N$ real matrix, and u is a vector, and f is a proper defined function. One needs to find an approximate of the quantity $u^T f(A)u$, or give a lower bound ℓ and/or an upper bound ν of it. Without loss of generality, one may assume $u^T u = 1$.

The quadratic form computing problem is first proposed in [6] for bounding the error of CG method for solving linear system of equations. It has been further developed in [11, 12, 2] and extended to other applications. The main idea is to first transform the problem of the quadratic form computing to a Riemann-Stieltjes integral problem, and then use Gauss-type quadrature rules to approximate the integral, which then brings the orthogonal polynomial theory and the underlying Lanczos procedure into the picture.

Let us go through the main idea. Since A is symmetric, the eigen-decomposition of A is given by $A = Q^T \Lambda Q$, where Q is an orthogonal matrix and Λ is a diagonal matrix with increasingly ordered diagonal elements λ_i. Then we have

$$u^T f(A)u = u^T Q^T f(\Lambda)Qu = \tilde{u}^T f(\Lambda)\tilde{u} = \sum_{i=1}^{N} f(\lambda_i)\tilde{u}_i^2,$$

where $\tilde{u} = (\tilde{u}_i) \equiv Qu$. The last sum can be considered as a Riemann-Stieltjes integral

$$u^T f(A)u = \int_a^b f(\lambda)d\mu(\lambda),$$

where the measure $\mu(\lambda)$ is a piecewise constant function and defined by

$$\mu(\lambda) = \begin{cases} 0, & \text{if } \lambda < a \leq \lambda_1, \\ \sum_{j=1}^{i} \tilde{u}_j^2, & \text{if } \lambda_i \leq \lambda < \lambda_{i+1} \\ \sum_{j=1}^{N} \tilde{u}_j^2 = 1, & \text{if } b \leq \lambda_N \leq \lambda \end{cases}$$

and a and b are the lower and upper bounds of the eigenvalues λ_i.

To obtain an estimate for the Riemann-Stieltjes integral, one can use the Gauss-type quadrature rule [9, 7]. The general quadrature formula is of the form

$$I[f] = \sum_{j=1}^{n} \omega_j f(\theta_j) + \sum_{k=1}^{m} \rho_k f(\tau_k), \tag{16}$$

where the weights $\{\omega_j\}$ and $\{\rho_k\}$ and the nodes $\{\theta_j\}$ are unknown and to be determined. The nodes $\{\tau_k\}$ are prescribed. If $m = 0$, then it is the well-known

Gauss rule. If $m = 1$ and $\tau_1 = a$ or $\tau_1 = b$, it is the Gauss-Radau rule. The Gauss-Lobatto rule is for $m = 2$ and $\tau_1 = a$ and $\tau_2 = b$.

The accuracy of the Gauss-type quadrature rules may be obtained by an estimation of the remainder $R[f]$:

$$R[f] = \int_a^b f(\lambda) d\mu(\lambda) - I[f].$$

For example, for the Gauss quadrature rule,

$$R[f] = \frac{f^{(2n)}(\eta)}{(2n)!} \int_a^b \left[\prod_{i=1}^n (\lambda - \theta_i) \right]^2 d\mu(\lambda),$$

where $a < \eta < b$. Similar formulas exist for Gauss-Radau and Gauss-Lobatto rules. If the sign of $R[f]$ is determined, then the quadrature formula $I[f]$ is a lower bound (if $R[f] > 0$) or an upper lower bound (if $R[f] < 0$) of the quantity $u^T f(A) u$.

Let us briefly recall how the weights and the nodes in the quadrature formula are obtained. First, we know that a sequence of polynomials $p_0(\lambda), p_1(\lambda), p_2(\lambda), \ldots$ can be defined such that they are orthonormal with respect to the measure $\mu(\lambda)$:

$$\int_a^b p_i(\lambda) p_j(\lambda) d\mu(\lambda) = \begin{cases} 1 & \text{if } i = j \\ 0 & \text{if } i \neq j \end{cases}$$

where it is assumed that the normalization condition $\int d\mu = 1$ (i.e., $u^T u = 1$). The sequence of orthonormal polynomials $\pi_j(\lambda)$ satisfies a three-term recurrence

$$\gamma_j p_j(\lambda) = (\lambda - \alpha_j) p_{j-1}(\lambda) - \gamma_{j-1} p_{j-2}(\lambda),$$

for $j = 1, 2, \ldots, n$ with $p_{-1}(\lambda) \equiv 0$ and $p_0(\lambda) \equiv 1$. Writing the recurrence in matrix form, we have

$$\lambda p(\lambda) = T_n p(\lambda) + \gamma_n p_n(\lambda) e_n$$

where

$$p(\lambda)^T = [p_0(\lambda), p_1(\lambda), \ldots, p_{n-1}(\lambda)], \qquad e_n^T = [0, 0, \ldots, 1]$$

and

$$T_n = \begin{pmatrix} \alpha_1 & \gamma_1 & & & \\ \gamma_1 & \alpha_2 & \gamma_2 & & \\ & \gamma_2 & \alpha_3 & \ddots & \\ & & \ddots & \ddots & \ddots \\ & & & \ddots & \ddots & \gamma_{n-1} \\ & & & & \gamma_{n-1} & \alpha_n \end{pmatrix}.$$

Then for the Gauss quadrature rule, the eigenvalues of T_n (which are the zeros of $p_n(\lambda)$) are the nodes θ_j. The weights ω_j are the squares of the first elements of the normalized (i.e., unit norm) eigenvectors of T_n.

For the Gauss-Radau and Gauss-Lobatto rules, the nodes $\{\theta_j\}$, $\{\tau_k\}$ and weights $\{\omega_j\}$, $\{\rho_j\}$ come from eigenvalues and the squares of the first elements of the normalized eigenvectors of an adjusted tridiagonal matrices of $\tilde{T}_{n+1}$, which has the prescribed eigenvalues a and/or b.

To this end, we recall that the classical Lanczos procedure is an elegant way to compute the orthonormal polynomials $\{p_j(\lambda)\}$ [16, 11]. We have the following algorithm in summary form. We refer it as the Gauss-Lanczos (GL) algorithm.

GL algorithm: Let A be a $N \times N$ real symmetric matrix, u a real N vector with $u^T u = 1$. f is a given smooth function. Then the following algorithm computes an estimation I_n of the quantity $u^T f(A) u$ by using the Gauss rule with n nodes.

- Let $x_0 = u$, and $x_{-1} = 0$ and $\gamma_0 = 0$
- For $j = 1, 2, \ldots, n$,
  1. $\alpha_j = x_{j-1}^T A x_{j-1}$
  2. $v_j = A x_{j-1} - \alpha_j x_{j-1} - \gamma_{j-1} x_{j-2}$
  3. $\gamma_j = \|v_j\|_2$
  4. $x_j = r_j / \gamma_j$
- Compute eigenvalues θ_k and the first elements ω_k of eigenvectors of T_n
- Compute $I_n = \sum_{k=1}^{n} \omega_k^2 f(\theta_k)$

We note that the "For" loop in the above algorithm is an iteration step of the standard symmetric Lanczos procedure [16]. The matrix A in question is only referenced here in the form of the matrix-vector product. The Lanczos procedure can be implemented with only 3 n-vectors in the fast memory. This is the major storage requirement for the algorithm and is an attractive feature for large scale problems.

On the return of the algorithm, from the expression of $R[f]$, we may estimate the error of the approximation I_n. For example, if $f^{(2n)}(\eta) > 0$ for any n and η, $a < \eta < b$, then I_n is a lower bound ℓ of the quantity $u^T f(A) u$.

Gauss-Radau-Lanczos (GRL) algorithm: To implement the Gauss-Radau rule with the prescribed node $\tau_1 = a$ or $\tau_1 = b$, the above GL algorithm just needs to be slightly modified. For example, with $\tau_1 = a$, we need to extend the matrix T_n to

$$\tilde{T}_{n+1} = \begin{bmatrix} T_n & \gamma_n e_n \\ \gamma_n e_n^T & \phi \end{bmatrix}.$$

Here the parameter ϕ is chosen such that $\tau_1 = a$ is an eigenvalue of $\tilde{T}_{n+1}$. From [10], it is known that

$$\phi = a + \delta_n,$$

where δ_n is the last component of the solution δ of the tridiagonal system

$$(T_n - aI)\delta = \gamma_n^2 e_n.$$

Then the eigenvalues and the first components of eigenvectors of $\tilde{T}_{n+1}$ gives the nodes and weight of the Gauss-Radau rule to compute an estimation $\tilde{I}_n$ of $u^T f(A)u$.

Furthermore, if $f^{(2n+1)}(\eta) < 0$ for any n and η, $a < \eta < b$, then $\tilde{I}_n$ (with b as a prescribed eigenvalue of $\tilde{T}_{n+1}$) is a lower bound ℓ of the quantity $u^T f(A)u$. $\tilde{I}_n$ (with a as a prescribed eigenvalue of $\tilde{T}_{n+1}$) is an upper bound ν.

Gauss-Lobatto-Lanczos (GLL) algorithm: To implement the Gauss-Lobatto rule, T_n computed in the GL algorithm is updated to

$$\hat{T}_{n+1} = \begin{bmatrix} T_n & \psi e_n \\ \psi e_n^T & \phi \end{bmatrix}.$$

Here the parameters ϕ and ψ are chosen so that a and b are eigenvalues of $\hat{T}_{n+1}$. Again, from [10], it is known that

$$\phi = \frac{\delta_n b - \mu_n a}{\delta_n - \mu_n} \qquad \text{and} \qquad \psi^2 = \frac{b + a}{\delta_n - \mu_n},$$

where δ_n and μ_n are the last components of the solutions δ and μ of the tridiagonal systems

$$(T_n - aI)\delta = e_n \qquad \text{and} \qquad (T_n - bI)\mu = e_n.$$

The eigenvalues and the first components of eigenvectors of $\hat{T}_{n+1}$ gives the nodes and weight of the Gauss-Lobatto rule to compute an estimation $\hat{I}_n$ of $u^T f(A)u$. Moreover, if $f^{(2n)}(\eta) > 0$ for any η, $a < \eta < b$, then $\hat{I}_n$ is an upper bound ν of the quantity $u^T f(A)u$.

Finally, we note that we need not always compute the eigenvalues and the first components of eigenvectors of the tridiagonal matrix T_n or its modifications $\tilde{T}_{n+1}$ or $\hat{T}_{n+1}$ for obtaining the estimation I_n or $\tilde{I}_n$, $\hat{I}_n$. We have following proposition.

Proposition 1. *For Gaussian rule:*

$$I_n = \sum_{k=1}^{n} \omega_k^2 f(\theta_k) = e_1^T f(T_n) e_1. \tag{17}$$

For Gauss-Radau rule:

$$\tilde{I}_n = \sum_{k=1}^{n} \omega_k^2 f(\theta_k) + \rho_1 f(\tau_1) = e_1^T f(\tilde{T}_{n+1}) e_1. \tag{18}$$

For Gauss-Lobatto rule:

$$\hat{I}_n = \sum_{k=1}^{n} \omega_k^2 f(\theta_k) + \rho_1 f(\tau_1) + \rho_2 f(\tau_2) = e_1^T f(\hat{T}_{n+1}) e_1. \tag{19}$$

Therefore, if the (1,1) entry of $f(T_n)$, $f(\tilde{T}_{n+1})$ or $f(\hat{T}_{n+1})$ can be easily computed, for example, $f(\lambda) = 1/\lambda$, we do not need to compute the eigenvalues and eigenvectors.

4 Solving the UEPs by Quadratic Form Computing

In this section, we use the GL, GRL and GLL algorithms for solving those unusual eigenvalue problems discussed in section 2.

Constraint eigenvalue problem Using the GL algorithm with the matrix A and the vector c, we have

$$c_1^T(A-\lambda I)^{-1}c = e_1^T(T_n-\lambda I)^{-1}e_1 + R,$$

where R is the remainder. Now we may solve reduced-order secular equation

$$e_1^T(T_n-\lambda I)^{-1}e_1 = 0$$

to find the largest λ as the approximate solution of the problem. This secular equation can be solved using the method discussed in [17] and its implementation available in LAPACK [1].

Modified eigenvalue problem Again, using the GL algorithm with the matrix A and the vector c, we have

$$1 + c^T(A-\bar{\lambda} I)^{-1}c = 1 + e_1^T(T_n-\bar{\lambda} I)^{-1}e_1 + R,$$

where R is the remainder. Then we may solve the eigenvalue problem of T_n to approximate some eigenvalues of A, and then solve reduced-order secular equation

$$1 + e_1^T(T_n-\bar{\lambda} I)^{-1}e_1 = 0$$

for $\bar{\lambda}$ to find some approximate eigenvalues of the modified eigenvalue problem.

Constraint quadratic programming By using the GRL algorithm with the prescribed node $\tau_1 = b$ for the matrix A and vector c, it can be shown that

$$c^T(A+\lambda I)^{-2}c \geq e_1^T(\tilde{T}_{n+1}+\lambda I)^{-2}e_1.$$

for all $\lambda > 0$. Then by solving the reduced-order secular equation

$$e_1^T(\tilde{T}_{n+1}+\lambda I)^{-2}e_1 = \alpha^2$$

for λ, we obtain $\underline{\lambda}_n$, which is a lower bound of the solution λ^*: $\underline{\lambda}_n \leq \lambda^*$

On the other hand, using the GRL algorithm with the prescribed node $\tau_1 = a$, we have

$$c^T(A+\lambda I)^{-2}c \leq e_1^T(\tilde{T}_{n+1}+\lambda I)^{-2}e_1$$

for all $\lambda > 0$. Then by solving the reduced-order secular equation

$$e_1^T(\tilde{T}_{n+1}+\lambda I)^{-2}e_1 = \alpha^2$$

for λ. We have an upper bound $\bar{\lambda}_n$ of the solution λ^*: $\bar{\lambda}_n \geq \lambda^*$.

Using such two-sided approximation as illustrated in Figure 2, the iteration can be adaptively proceeded until the estimations $\underline{\lambda}_n$ and $\bar{\lambda}_n$ are sufficiently close, we then obtain an approximation

$$\lambda^* \approx \frac{1}{2}(\underline{\lambda}_n + \bar{\lambda}_n)$$

of the desired solution λ^*.

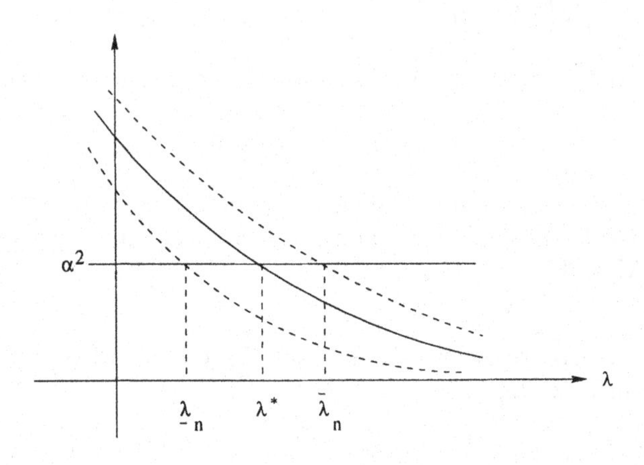

Fig. 2. Two-sided approximation approximation of the solution λ^* for the constraint quadratic programming problem (7) and (8).

Trace, determinant and partial eigenvalue sum As shown in sections 2.4 and 2.5, the problems of computing trace of the inverse of a matrix A, determinant of a matrix A and partial eigenvalue sum of a symmetric positive definite pair (A, B) can be summarized as the problem of computing the trace of a corresponding matrix function $f(H)$, where $H = A$ or $H = L^{-1}AL^{-T}$ and $f(\lambda) = 1/\lambda$, $\ln(\lambda)$ or $\lambda/\left(1 + \exp(\frac{\lambda-\alpha}{\kappa})\right)$. To efficiently compute the trace of $f(H)$, instead of applying GR algorithm or its variations N times for each diagonal element of $f(H)$, we may use a Monte Carlo approach which only applies the GR algorithm m times to obtain an unbiased estimation of $\text{tr}(f(H))$. For practical purposes, m can be chosen much smaller than N. The saving in computational costs could be significant. Such a Monte Carlo approach is based on the following lemma due to Hutchinson [14].

Proposition 2. *Let $C = (c_{ij})$ be an $N \times N$ symmetric matrix with $tr(C) \neq 0$. Let $\mathcal{V}$ be the discrete random variable which takes the values* 1 *and* -1 *each with probability 0.5 and let z be a vector of n independent samples from $\mathcal{V}$. Then z^TCz is an unbiased estimator of $tr(C)$, i.e.,*

$$E(z^TCz) = tr(C),$$

and

$$var(z^T Cz) = 2\sum_{i\neq j} c_{ij}^2.$$

To use the above proposition in practice, one takes m such sample vectors z_i, and then uses GR algorithm or its variations to obtain an estimation $I_n^{(i)}$, a lower bound $\ell_n^{(i)}$ and/or an upper bound $\nu_n^{(i)}$ of the quantity $z_i^T f(H) z_i$:

$$\ell_n^{(i)} \leq z_i^T f(H) z_i \leq \nu_n^{(i)}.$$

Then by taking the mean of the m computed estimation $I_n^{(i)}$ or lower and upper bounds $\ell_n^{(i)}$ and $\nu_n^{(i)}$, we have

$$\mathrm{tr}(f(H)) \approx \frac{1}{m}\sum_{i=1}^{m} I_n^{(i)}$$

or

$$\frac{1}{m}\sum_{i=1}^{m} \ell_n^{(i)} \leq \frac{1}{m}\sum_{i=1}^{m} z_i^T f(H) z_i \leq \frac{1}{m}\sum_{i=1}^{m} \nu_n^{(i)}.$$

It is natural to expect that with a suitable sample size m, the mean of the computed bounds yields a satisfactory estimation of the quantity $\mathrm{tr}(f(H))$. To assess the quality of such estimation, one can also obtain probabilistic bounds of the approximate value [2].

5 Numerical Examples

In this section, we present some numerical examples to illustrate our quadratic form based algorithms for solving some of the unusual eigenvalue problems discussed in section 2.

Table 1. Numerical Results of estimating $\mathrm{tr}(A^{-1})$

Matrix	N	"Exact"	Iter	Estimated	Rel.err
Poisson	900	$5.126e+02$	30–50	$5.020e+02$	2.0%
VFH	625	$5.383e+02$	12–21	$5.366e+02$	0.3%
Wathen	481	$2.681e+01$	33–58	$2.667e+01$	0.5%
Lehmer	200	$2.000e+04$	38–70	$2.017e+04$	0.8%

Table 2. Numerical results of estimating $\ln(\det(A)) = \mathrm{tr}(\ln A)$

Matrix	N	"Exact"	Iter	Estimated	Rel.err
Poisson	900	$1.065e+03$	11–29	$1.060e+03$	0.4%
VFH	625	$3.677e+02$	10–14	$3.661e+02$	0.4%
Heat Flow	900	$5.643e+01$	4	$5.669e+01$	0.4%
Pei	300	$5.707e+00$	2–3	$5.240e+00$	8.2%

5.1 Trace and determinant

Numerical results for a set of test matrices presented in Tables 1 and 2 are first reported in [2]. Some of these test matrices are model problems and some are from practical applications. For example, VFH matrix is from the analysis of transverse vibration of a Vicsek fractal. These numerical experiments are carried out on an Sun Sparc workstation. The so-called "exact" value is computed by using the standard methods for dense matrices. The numbers in the "Iter"-column are the number of iterations n required for the estimation $I_n^{(i)}$ to reach stationary value within the given tolerance value $tol = 10^{-4}$, namely,

$$|I_n - I_{n-1}| \leq tol * |I_n|.$$

The number of random sample vector z_i is $m = 20$. For those test matrices, the relative accuracy of the new approach within 0.3% to 8.2% may be sufficient for practical purposes.

5.2 Partial eigenvalue sum

Here we present a numerical example from the computation of the total energy of an electronic structure. Total energy calculation of a solid state system is necessary in simulating real materials of technological importance [18]. Figure 3 shows a carbon cluster that forms part of a "knee" structure connecting nanotubes of different diameters and the distribution of eigenvalues such carbon structure with 240 atoms. One is interested in computing the sum of all these eigenvalues less than zero. Comparing the performance of our method with dense methods, namely symmetric QR algorithm and bisection method in LAPACK, our method achieved up to a factor of 20 speedup for large system on an Convex Exemplar SPP-1200 (see Table 3). Because of large memory requirements, we were not able to use LAPACK divide-and-conquer symmetric eigenroutines. Furthermore, algorithms for solving large-sparse eigenvalue problems, such as Lanczos method or implicitly restarted methods for computing some eigenvalues are found inadequate due to large number of eigenvalues required. Since the

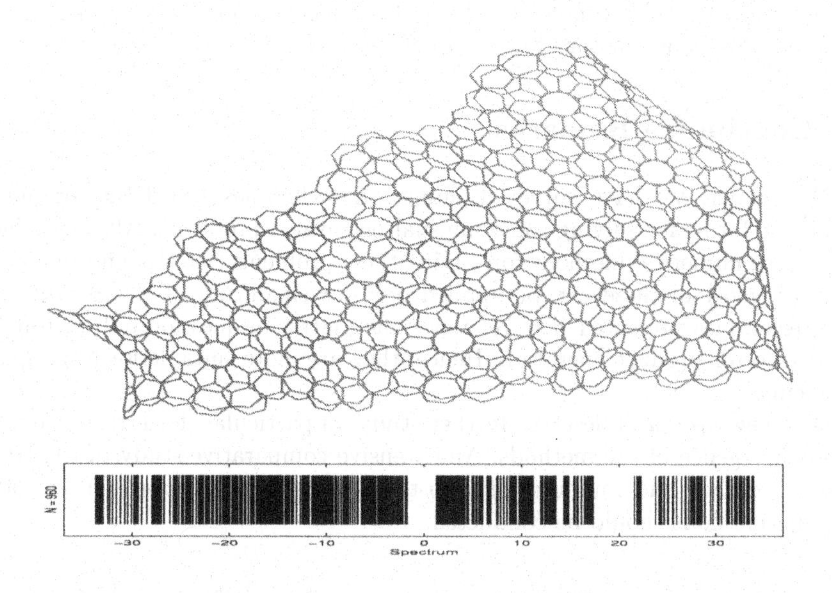

Fig. 3. A carbon cluster that forms part of a "knee" structure, and the corresponding spectrum

Table 3. Performance of our method vs. dense methods on Convex Exemplar SPP-1200. Here, 10 Monte Carlo samples were used to obtain estimates for each systems size.

		Dense methods			GR Algorithm		% Relative
n	m	Partial Sum	QR Time	BI Time	Estimate	Time	Error
480	349	-4849.8	7.4	7.6	-4850.2	2.8	0.01
960	648	-9497.6	61.9	51.8	-9569.6	18.5	0.7
1000	675	-9893.3	80.1	58.6	-10114.1	22.4	2.2
1500	987	-14733.1	253.6	185.6	-14791.8	46.4	0.4
1920	1249	-18798.5	548.3	387.7	-19070.8	72.6	1.4
2000	1299	-19572.9	616.9	431.8	-19434.7	78.5	0.7
2500	1660	-24607.6	1182.2	844.6	-24739.6	117.2	0.5
3000	1976	-29471.3	1966.4	1499.7	-29750.9	143.5	0.9
3500	2276	-34259.5	3205.9	2317.4	-33738.5	294.0	1.5
4000	2571	-39028.9	4944.3	3553.2	-39318.0	306.0	0.7
4244	2701	-41299.2	5915.4	4188.0	-41389.8	339.8	0.2

problem is required to be solved repeatly, we are now able to solve previously intractable large scale problems. The relative accuracy of new approach within 0.4% to 1.5% is satisfactory for the application [3].

6 Concluding Remarks

In this paper, we have surveyed numerical techniques based on computing quadratic forms for solving some unusual eigenvalue problems. Although there exist some numerical methods for solving these problems (see [13] and references therein), most of these can be applied only for small and/or dense problems. The techniques presented here reference the matrix in question only through a matrix-vector product operation. Hence, they are more suitable for large sparse problems.

The new approach deserves further study; in particular, for error estimation and convergence of the methods. An extensive comparative study of the tradeoffs in accuracy and computational costs between the new approach and other existing methods should be conducted.

Acknowledgement Z. B. was supported in part by an NSF grant ASC-9313958, an DOE grant DE-FG03-94ER25219.

References

1. Anderson, E., Bai, Z., Bischof, C., Demmel, J., Dongarra, J., Du Croz, J., Greenbaum, A., Hammarling, S., McKenney, A., Ostrouchov, S., Sorensen, D.:, LAPACK Users' Guide (second edition), SIAM, Philadelphia, 1995.
2. Bai, Z., Fahey, M., Golub, G.: Some large-scale matrix computation problems. J. Comp. Appl. Math. **74** (1996) 71–89.
3. Bai, Z., Fahey, M., Golub, G., Menon, M., Richter, E.: Computing partial eigenvalue sum in electronic structure calculations, Scientific Computing and Computational Mathematics Program, Computer Science Dept., Stanford University, SCCM-98-03, 1998.
4. Barrett. R., Berry. M., Chan. F., Demmel. J., Donato. J., Dongarra. J., Eijkhout. V., Pozo. R., Romine. C., van der Vorst., H.: Templates for the solution of linear systems: Building blocks for iterative methods. SIAM, Philadelphia, 1994.
5. Carey, C., Golub, G., Law, K.: A Lanczos-based method for structural dynamic reanalysis problems. Inter. J. Numer. Methods in Engineer., 37 (1994) 2857–2883.
6. Dahlquist, G., Eisenstat, S., Golub, G.: Bounds for the error of linear systems of equations using the theory of moments. J. Math. Anal. Appl. 37 (1972) 151–166.
7. Davis. P., Rabinowitz. P.: Methods of Numerical Integration. Academic Press, New York, 1984.
8. Demmel, J., Heath. M., van der Vorst., H.: Parallel linear algebra, in Acta Numerica, Vol.2, Cambridge Press, New York, 1993
9. Gautschi. W., A survey of Gauss-Christoffel quadrature formulae. In P. L Bultzer and F. Feher, editors, *E. B. Christoffel – the Influence of His Work on on Mathematics and the Physical Sciences*, pages 73–157. Birkhauser, Boston, 1981.

10. Golub, G.: Some modified matrix eigenvalue problems. SIAM Review, 15 (1973) 318–334.
11. Golub, G., Meurant, G.: Matrices, moments and quadrature, in Proceedings of the 15th Dundee Conference, June 1993, D. F. Griffiths and G. A. Watson, eds., Longman Scientific & Technical, 1994.
12. Golub, G., Strakoš, Z.: Estimates in quadratic formulas, Numerical Algorithms, 8 (1994) 241–268.
13. Golub. G., Van Loan. C.: Matrix Computations. Johns Hopkins University Press, Baltimore, MD, third edition, 1996.
14. Hutchinson, M.: A stochastic estimator of the trace of the influence matrix for Laplacian smoothing splines, Commun. Statist. Simula., 18 (1989) 1059–1076.
15. Kerstin. K., Dorman. K. R.: A Course in Statistical Thermodynamics, Academic Press, New York, 1971.
16. Lanczos, C.: An iteration method for the solution of the eigenvalue problem of linear differential and integral operators, J. Res. Natl. Bur. Stand, 45 (1950) 225-280.
17. Li., R.-C.: Solving secular equations stably and efficiently, Computer Science Division, Department of EECS, University of California at Berkeley. Technical Report UCB//CSD-94-851,1994
18. Menon. M., Richter. E., Subbaswamy. K. R.: Structural and vibrational properties of fullerenes and nanotubes in a nonorthogonal tight-binding scheme. J. Chem. Phys., 104 (1996) 5875–5882.

Multi-sweep Algorithms for the Symmetric Eigenproblem

Wilfried N. Gansterer[1], Dieter F. Kvasnicka[2], and Christoph W. Ueberhuber[3]

[1] Institute for Applied and Numerical Mathematics,
University of Technology, Vienna
ganst@aurora.tuwien.ac.at
[2] Institute for Technical Electrochemistry,
University of Technology, Vienna
dieter@titania.tuwien.ac.at
[3] Institute for Applied and Numerical Mathematics,
University of Technology, Vienna
christof@uranus.tuwien.ac.at

Abstract. This paper shows how the symmetric eigenproblem, which is the computationally most demanding part of numerous scientific and industrial applications, can be solved much more efficiently than by using algorithms currently implemented in LAPACK routines.
The main techniques used in the algorithm presented in this paper are (i) sophisticated blocking in the tridiagonalization, which leads to a two-sweep algorithm; and (ii) the computation of the eigenvectors of a band matrix instead of a tridiagonal matrix.
This new algorithm improves the locality of data references and leads to a significant improvement of the floating-point performance of symmetric eigensolvers on modern computer systems. Speedup factors of up to four (depending on the computer architecture and the matrix size) have been observed.

Keywords: Numerical Linear Algebra, Symmetric Eigenproblem, Tridiagonalization, Performance Oriented Numerical Algorithm, Blocked Algorithm

1 Introduction

Reducing a dense symmetric matrix A to tridiagonal form T is an important preprocessing step in the solution of the symmetric eigenproblem. LAPACK (Anderson et al. [1]) provides a blocked tridiagonalization routine whose memory reference patterns are not optimal on modern computer architectures. In this LAPACK routine a significant part of the computation is performed by calls to Level 2 BLAS. Unfortunately, the ratio of floating-point operations to data movement of Level 2 BLAS is not high enough to enable efficient reuse of data that

The work described in this paper was supported by the Special Research Program SFB F011 "AURORA" of the Austrian Science Fund.

J. Palma, J. Dongarra, and V. Hernández (Eds.): VECPAR'98, LNCS 1573, pp. 20–28, 1999.

reside in cache or local memory (see Table 1). Thus, software construction based on calls to Level 2 BLAS routines is not well suited to computers with a memory hierarchy and multiprocessor machines.

Table 1. Ratio of floating-point operations to data movement for three closely related operations from the Level 1, 2, and 3 BLAS (Dongarra et al. [6])

BLAS	Routine	Memory Accesses	Flops	Flops per Memory Access
Level 1	**daxpy**	$3n$	$2n$	$2/3$
Level 2	**dgemv**	n^2	$2n^2$	2
Level 3	**dgemm**	$4n^2$	$2n^3$	$n/2$

Bischof et al. [4, 5] recently developed a general framework for reducing the bandwidth of symmetric matrices in a multi-sweep manner, which improves data locality and allows for the use of Level 3 BLAS instead of Level 2 BLAS. Their multi-sweep routines improve efficiency of symmetric eigensolvers if only eigenvalues have to be computed. Accumulating the transformation information for the eigenvectors incurs an overhead which can outweigh the benefits of a multi-sweep reduction.

In this paper we introduce an important modification of this framework in case eigenvectors have to be computed, too: We compute the required eigenvectors directly from the intermediate band matrix. Analyses and experimental results show that our approach to improving memory access patterns is superior to established algorithms in many cases.

2 Eigensolver with Improved Memory Access Patterns

A real symmetric $n \times n$ matrix A can be factorized as

$$A = VBV^T = QTQ^T = Z\Lambda Z^T \tag{1}$$

where V, Q, and Z are orthogonal matrices. B is a symmetric band matrix with band width $2b + 1$, T is a symmetric tridiagonal matrix ($b = 1$), and Λ is a diagonal matrix whose diagonal elements are the eigenvalues of the (similar) matrices A, B and T. The column vectors of Z are the eigenvectors of A.

LAPACK reduces a given matrix A to tridiagonal form T by applying Householder similarity transformations. A bisection algorithm is used to compute selected[1] eigenvalues of T, which are also eigenvalues of A. The eigenvectors of T are found by inverse iteration on T. These eigenvectors have to be transformed into the eigenvectors of A using the transformation matrix Q.

[1] If *all* eigenvalues are required, LAPACK provides other algorithms.

Following the multi-sweep framework of Bischof et. al [4,5], the new method proposed does not compute the tridiagonal matrix T from A directly, but derives a band matrix B as an intermediate result. This band reduction can be organised with good data locality, which is critical for high performance on modern computer architectures. Using a block size b in the first reduction sweep results in a banded matrix B with a semi-bandwidth of at least b. This relationship leads to a tradeoff:

- If smaller values of b are chosen, then more elements of A are eliminated in the first reduction sweep. This decrease in the number of non-zero elements leads to a smaller amount of data to be processed in later reduction sweeps.
- If larger values of b are chosen, then the Level 3 BLAS routines operate on larger matrices and thus better performance improvements are obtained in the first reduction sweep.

Using appropriate values of b speedups of up to ten can be achieved with the two-sweep reduction as compared with the LAPACK tridiagonal reduction (see Gansterer et al. [7,8]).

Eigenvectors can be computed by inverse iteration either from A, B, or T. In numerical experiments inverse iteration on B turned out to be the most effective. The eigenvectors of A have to be computed from the eigenvectors of B using the orthogonal transformation matrix V.

The new algorithm includes two special cases:

1. If $V = Q$ then $B = T$, and the inverse iteration is performed on the tridiagonal matrix T. This variant coincides with the LAPACK algorithm.
2. If $V = I$ then $B = A$, and the inverse iteration is performed on the original matrix A. This variant is to be preferred if only a few eigenvectors have to be computed and the corresponding eigenvalues are known.

3 The New Level 3 Eigensolver

The blocked LAPACK approach puts the main emphasis on accumulating several elimination steps. b rank-2 updates (each of them a Level 2 BLAS operation) are aggregated to form one rank-$2b$ update and hence one Level 3 BLAS operation. However, this approach does not take into account memory access patterns in the updating process. We tried to introduce blocking in a much stricter sense, namely by localising data access patterns.

Partitioning matrix A into blocks consisting of b columns, and further partitioning each column block into $b \times b$ sub-matrices makes it possible to process the block-columns and their sub-matrices the same way that the single columns and their elements are processed in the original unblocked method: all *blocks*[2] below the sub-diagonal *block* are eliminated, which leaves a *block* tridiagonal matrix, i.e., a *band* matrix (when considered elementwise). At first sight this matrix has bandwidth $4b - 1$. Further examination shows that the first elimination sweep

[2] Replace *block* by *element* (or *elementwise*) to get the original, unblocked, algorithm.

can be organised such that the sub-diagonal blocks in the block tridiagonal matrix are of upper triangular form. Hence the band width of B can be reduced to $2b + 1$.

Two features distinguish the newly developed Level 3 algorithm from standard algorithms, as, for example, implemented in LAPACK (see Fig. 1):

Tridiagonalization in two sweeps: The first sweep reduces the matrix A to a band matrix B. The second sweep reduces B to a tridiagonal matrix T. No fill-in occurs in the first reduction sweep. In the second sweep, however, additional operations are needed to remove fill-in (see Bischof et al. [4]).

Inverse iteration on the band matrix: Calculating the eigenvectors of B avoids the large overhead entailed by the back-transformation of the eigenvectors of T. It is crucial that the overhead caused by inverse iteration on B instead of T does not outweigh the benefits of this approach.

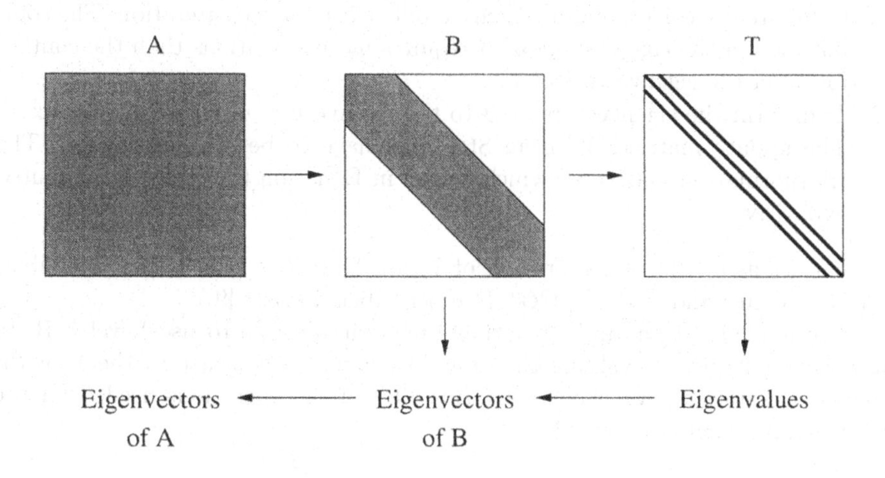

Fig. 1. Basic concept of the new eigensolver

4 Implementation Details

The resulting algorithm has five steps.

1. Reduce the matrix A to the band matrix B.
   (a) Compute the transformation

$$V_i A V_i^T = (I - u_b u_b^T) \cdots (I - u_1 u_1^T) A (I - u_1 u_1^T) \cdots (I - u_b u_b^T)$$

   for a properly chosen sub-block of A. Because each Householder vector u_i eliminates only elements below sub-diagonal b in the current column

i, the corresponding update from the right does not affect the current column block. Therefore, the Householder vectors for the transformation of the columns of the current column block do not depend on each other. This independence enables a new way of blocking in Step 1c.
   (b) Collect the Householder vectors as column vectors in an $n \times b$ matrix Y and compute the $n \times b$ matrix W such that the transformation matrix V_i is represented as $I - WY^T$ (see Bischof, Van Loan [3]). Matrix Y can be stored in those parts of A which have just been eliminated. Matrix W requires separate storage of order $O(n^2)$; therefore it is overwritten and has to be computed anew in the back-transformation step.
   (c) Perform a rank-$2b$ update using the update matrix $(I - WY^T)$.
   (d) Iterate Steps 1a to 1c over the entire matrix A.
2. Reduce the matrix B to the tridiagonal matrix T. A significant portion of the flops in this step is required to remove fill-in.
3. Compute the desired eigenvalues of the tridiagonal matrix T.
4. Compute the corresponding eigenvectors of B by inverse iteration. The computation of the eigenvectors of B requires more operations than the computation of the eigenvectors of T.
5. Transform the eigenvectors of B to the eigenvectors of the input matrix A. The update matrices W from Step 1(b) have to be computed anew. The transformation matrix V, which occurs in Equation (1), is never computed explicitly.

Variations of the representation of V_i are $V_i = I - YUY^T$ (see Schreiber, Van Loan [10]) and $V_i = I - GG^T$ (see Schreiber, Parlett [9]).

The new algorithm and its variants make it possible to use Level 3 BLAS in all computations involving the original matrix, in contrast to the LAPACK routine `LAPACK/dsyevx`, which performs 50 % of the operations needed for the tridiagonalization in Level 2 BLAS.

5 Complexity

The total execution time T of our algorithm consists of five parts:

$T_1 \sim c_1(b)n^3$ for reducing the symmetric matrix A to a band matrix B,
$T_2 \sim c_2(b)n^2$ for reducing the band matrix B to a tridiagonal matrix T,
$T_3 \sim c_3kn$ for computing k eigenvalues of T,
$T_4 \sim c_4kb^2n$ for computing the corresponding k eigenvectors[3] of B ($n \gg b$), and
$T_5 \sim c_5kn^2$ for transforming the eigenvectors of B into the eigenvectors of A.

The parameters c_1 and c_2 depend on the semi-bandwidth b. The parameter c_1 *decreases* in b, whereas c_2 *increases* in b due to an increasing number of operations to be performed. c_3, c_4, and c_5 are independent of the problem size.

[3] If eigenvalues are clustered and reorthogonalization is required, an additional term $\hat{c}_4k^2n$ dominates.

The first reduction step is the only part of the algorithm requiring an $O(n^3)$ effort. Thus, a large semi-bandwidth b of B seems to be desirable. However, b should be chosen appropriately not only to speed up the first band reduction (T_1), but also to make the tridiagonalization (T_2) and the eigenvector computation (T_4) as efficient as possible.

For example, on an SGI Power Challenge the calculation of $k = 200$ eigenvalues and eigenvectors of a symmetric 2000×2000 matrix requires a total execution time $T = 90$ s. T_1 is 57 s, i. e., 63 % of the total time. With increasing matrix size this percentage increases.

6 Results

A first implementation of our algorithm uses routines from the symmetric band reduction toolbox (SBR; see Bischof et al. [2, 4, 5]), EISPACK routines (Smith et al. [11]), LAPACK routines (Anderson et al. [1]), and some of our own routines.

In numerical experiments we compare the well established LAPACK routine `LAPACK/dsyevx` with the new algorithm. On an SGI Power Challenge (with a MIPS R8000 processor running with 90 MHz), speedup factors of up to 4 were observed (see Table 2 and Fig. 5).

Table 2. Execution times (in seconds) on an SGI Power Challenge. $k = n/10$ of the n eigenvalues and eigenvectors were computed

n	k	b	LAPACK dsyevx	New Method	Speedup
500	50	6	1.5 s	2.1 s	0.7
1000	100	6	16.6 s	12.6 s	1.3
1500	150	6	74.4 s	39.2 s	1.9
2000	200	10	239.2 s	89.7 s	2.7
3000	300	12	945.5 s	286.4 s	3.3
4000	400	12	2432.4 s	660.5 s	3.7

Fig. 2 shows the normalised computing time $T(n)/n^3$. The significant speedup of the new algorithm when applied to large problems is striking. This speedup has nothing to do with complexity, which is nearly identical for both algorithms (see Fig. 3). The reason for the good performance of the new algorithm is its significantly improved utilisation of the computer's potential peak performance (see Fig. 4). The deteriorating efficiency of the LAPACK routine (due to cache effects) causes the $O(n^4)$ behaviour of its computation time between $n = 500$ and $n = 2000$ (see Fig. 2).

The new algorithm shows significant speedups compared to existing algorithms for a small subset of eigenvalues and eigenvectors ($k = n/10$) of large matrices ($n \geq 1500$) on all computers at our disposal, including workstations of DEC, HP, IBM, and SGI.

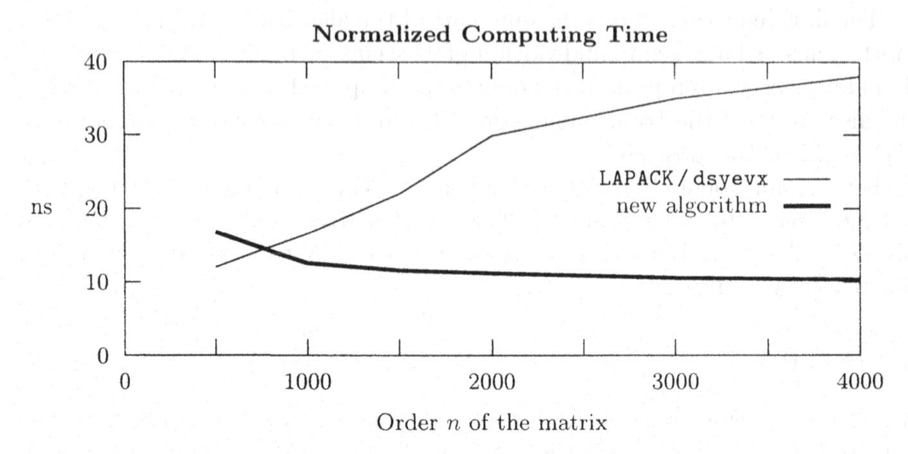

Fig. 2. Normalised computing time $T(n)/n^3$ in nanoseconds

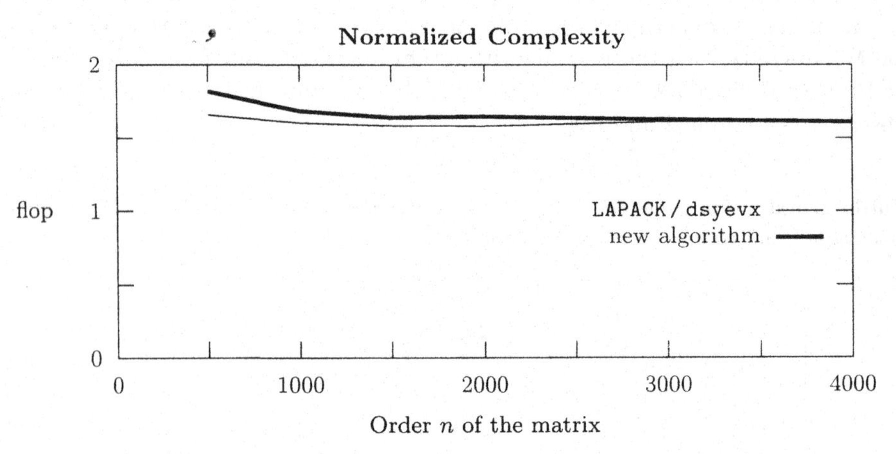

Fig. 3. Normalised number $\mathrm{op}(n)/n^3$ of floating-point operations

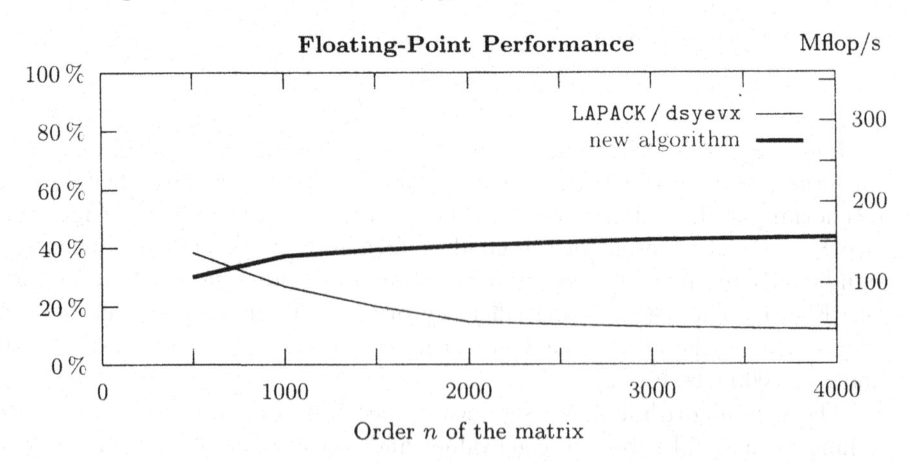

Fig. 4. Floating-point performance (MFlop/s) and efficiency (%)

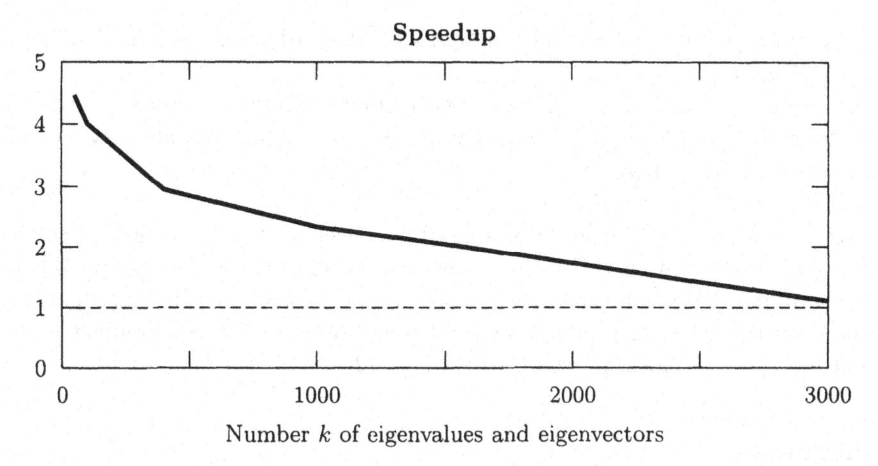

Fig. 5. Speedup of the new algorithm as compared with the LAPACK routine dsyevx. Performance improvements are all the better if k is only a small percentage of the $n = 3000$ eigenvalues and eigenvectors

Choosing the block size b. Experiments show that the optimum block size b, which equals the smallest possible band width, increases slightly with larger matrix sizes (see Table 2). A matrix of order 500 requires a b of only 4 or 6 for optimum performance, depending on the architecture (and the size of cache lines). Band widths of 12 or 16 are optimum on most architectures when the matrix order is larger than 2000. A hierarchically blocked version of the new eigensolver which is currently under development allows increasing the block size of the update without increasing the band width and therefore leads to even higher performance.

7 Conclusion

We presented an algorithm for the computation of selected eigenpairs of symmetric matrices which is significantly faster than existing algorithms as, for example, implemented in LAPACK. This speedup is achieved by improved blocking in the tridiagonalization process, which significantly improves data locality.

LAPACK algorithms for the symmetric eigenproblem spend up to 80 % of their execution time in Level 2 BLAS, which do not perform well on cache-based and multiprocessor computers. In our algorithm all performance relevant steps make use of Level 3 BLAS.

The price that has to be paid for the improved performance in the tridiagonalization process is that the eigenvectors cannot be computed from the tridiagonal matrix because of prohibitive overhead in the back-transformation. They have to be computed from an intermediate band matrix.

When choosing the block size, a compromise has to be made: Larger block sizes improve the performance of the band reduction. Smaller block sizes reduce the band width and, therefore, speed up the inverse iteration on the band matrix.

The new algorithm is particularly attractive (on most architectures) if not all eigenvectors are required.

If the gap between processor speed and memory bandwidth further increases, our algorithm will be highly competitive also for solving problems where all eigenvectors are required.

Future Development. Routines dominated by Level 3 BLAS operations, like the eigensolver presented in this paper, have the potential of speeding up almost linearly on parallel machines. That is why we are currently developing a parallel version of the new method. Another promising possibility of development is the use of hierarchical blocking (see Ueberhuber [12]).

References

1. E. Anderson et al., LAPACK *Users' Guide, 2nd ed.*, SIAM Press, Philadelphia, 1995.
2. C. H. Bischof, B. Lang, X. Sun, *Parallel Tridiagonalization through Two-Step Band Reduction*, Proceedings of the Scalable High-Performance Computing Conference, IEEE Press, Washington, D. C., 1994, pp. 23–27.
3. C. H. Bischof, C. F. Van Loan, *The WY Representation for Products of Householder Matrices*, SIAM J. Sci. Comput. 8 (1987), pp. s2–s13.
4. C. H. Bischof, B. Lang, X. Sun, *A Framework for Symmetric Band Reduction*, Technical Report ANL/MCS-P586-0496, Argonne National Laboratory, 1996.
5. C. H. Bischof, B. Lang, X. Sun, *The* SBR *Toolbox—Software for Successive Band Reduction*, Technical Report ANL/MCS-P587-0496, Argonne National Laboratory, 1996.
6. J. J. Dongarra, I. S. Duff, D. C. Sorensen, H. A. van der Vorst, *Linear Algebra and Matrix Theory*, SIAM Press, Philadelphia, 1998.
7. W. Gansterer, D. Kvasnicka, C. Ueberhuber, *High Performance Computing in Material Sciences. Higher Level* BLAS *in Symmetric Eigensolvers*, Technical Report AURORA TR1998-18, Vienna University of Technology, 1998.
8. W. Gansterer, D. Kvasnicka, C. Ueberhuber, *High Performance Computing in Material Sciences. Numerical Experiments with Symmetric Eigensolvers*, Technical Report AURORA TR1998-19, Vienna University of Technology, 1998.
9. R. Schreiber, B. Parlett, *Block Reflectors: Theory and Computation*, SIAM J. Numer. Anal. 25 (1988), pp. 189–205.
10. R. Schreiber, C. Van Loan, *A Storage-Efficient WY Representation for Products of Householder Transformations*, SIAM J. Sci. Stat. Comput. 10-1 (1989), pp. 53–57.
11. B. T. Smith et al., *Matrix Eigensystem Routines*—EISPACK *Guide*, Lecture Notes in Computer Science, Vol. 6, Springer-Verlag, Berlin Heidelberg New York Tokyo, 1976.
12. C. W. Ueberhuber, *Numerical Computation*, Springer-Verlag, Berlin Heidelberg New York Tokyo, 1997.

A Unified Approach to Parallel Block-Jacobi Methods for the Symmetric Eigenvalue Problem*

D. Giménez[1,**], Vicente Hernández[2,***], and Antonio M. Vidal[2]

[1] Departamento de Informática, Lenguajes y Sistemas Informáticos. Univ de Murcia. Aptdo 4021. 30001 Murcia, Spain.
domingo@dif.um.es

[2] Dpto. de Sistemas Informáticos y Computación. Univ Politécnica de Valencia Aptdo 22012. 46071 Valencia. Spain.
{vhernand,avidal}@dsic.upv.es

Abstract. In this paper we present a unified approach to the design of different parallel block-Jacobi methods for solving the Symmetric Eigenvalue Problem. The problem can be solved designing a logical algorithm by considering the matrices divided into square blocks, and considering each block as a process. Finally, the processes of the logical algorithm are mapped on the processors to obtain an algorithm for a particular system. Algorithms designed in this way for ring, square mesh and triangular mesh topologies are theoretically compared.

1 Introduction

The Symmetric Eigenvalue Problem appears in many applications in science and engineering, and in some cases the problems are of large dimension with high computational cost, therefore it might be better to solve in parallel.

Different approaches can be utilised to solve the Symmetric Eigenvalue Problem on multicomputers:

- The initial matrix can be reduced to condensed form (tridiagonal) and then the reduced problem solved. This is the approach in ScaLAPACK [1].
- A Jacobi method can be used taking advantage of the high level of parallelism of the method to obtain high performance on multicomputers. In addition, the design of block methods allows us to reduce the communications and to use the memory hierarchy better. Different block-Jacobi methods have been

* The experiments have been performed on the 512 node Paragon on the CSCC parallel computer system operated by Caltech on behalf of the Concurrent Supercomputing Consortium (access to this facility was provided by the PRISM project).

** Partially supported by Comisión Interministerial de Ciencia y Tecnología, project TIC96-1062-C03-02, and Consejería de Cultura y Educación de Murcia, Dirección General de Universidades, project COM-18/96 MAT.

*** Partially supported by Comisión Interministerial de Ciencia y Tecnología, project TIC96-1062-C03-01.

J. Palma, J. Dongarra, and V. Hernández (Eds.): VECPAR'98, LNCS 1573, pp. 29–42, 1999.

designed to solve the Symmetric Eigenvalue Problem or related problems on multicomputers [2–5].

- There are other type of methods in which high performance is obtained because most of the computation is in matrix-matrix multiplications, which can be optimised both in shared or distributed memory multiprocessors. Methods of that type are those based in spectral division [6–8] or the Yau-Lu method [9].

In this paper a unified approach to the design of parallel block-Jacobi methods is analized.

2 A sequential block-Jacobi method

Jacobi methods work by constructing a matrix sequence $\{A_l\}$ by means of $A_{l+1} = Q_l A_l Q_l^t$, $l = 1, 2, \ldots$, where $A_1 = A$, and Q_l is a plane-rotation that annihilates a pair of non-diagonal elements of matrix A_l. A cyclic method works by making successive sweeps until some convergence criterion is fulfilled. A sweep consists of successively nullifying the $n(n-1)/2$ non-diagonal elements in the lower-triangular part of the matrix. The different ways of choosing pairs (i, j) have given rise to different versions of the method. The odd-even order will be used, because it simplifies a block based implementation of the sequential algorithm, and allows parallelisation. With $n = 8$, numbering indices from 1 to 8, and initially grouping the indices in pairs $\{(1,2), (3,4), (5,6), (7,8)\}$, the sets of pairs of indices are obtained as follows:

$$
\begin{array}{ll}
k=1 & \{(1,2),(3,4),(5,6),(7,8)\} \\
k=2 & \{2,(1,4),(3,6),(5,8),7\} \\
k=3 & \{(2,4),(1,6),(3,8),(5,7)\} \\
k=4 & \{4,(2,6),(1,8),(3,7),5\} \\
k=5 & \{(4,6),(2,8),(1,7),(3,5)\} \\
k=6 & \{6,(4,8),(2,7),(1,5),3\} \\
k=7 & \{(6,8),(4,7),(2,5),(1,3)\} \\
k=8 & \{8,(6,7),(4,5),(2,3),1\}
\end{array}
$$

When the method converges we have $D = Q_k Q_{k-1} \ldots Q_1 A Q_1^t \ldots Q_{k-1}^t Q_k^t$ and the eigenvalues are the diagonal elements of matrix D and the eigenvectors are the rows of the product $Q_k Q_{k-1} \ldots Q_1$.

The method works over the matrix A and a matrix V where the rotations are accumulated. Matrix V is initially the identity matrix. To obtain an algorithm working by blocks both matrices A and V are divided into columns and rows of square blocks of size $s \times s$. These blocks are grouped to obtain bigger blocks of size $2sk \times 2sk$.

The scheme of an algorithm by blocks is shown in figure 1.

In each block the algorithm works by making a sweep over the elements in the block. Blocks corresponding to the first Jacobi set are considered to have size $2s \times 2s$, adding to each block the two adjacent diagonal blocks. A sweep is performed covering all elements in these blocks and accumulating the rotations

```
WHILE convergence not reached DO
      FOR every pair (i,j) of indices in a sweep DO
            perform a sweep on the block of size 2s × 2s formed
              by the blocks of size s × s, A_ii, A_ij and A_jj,
              accumulating the rotations on a matrix Q of size 2s × 2s
            update matrices A and V performing matrix-matrix
              multiplications
      ENDFOR
ENDWHILE
```

Fig. 1. Basic block-Jacobi iteration.

to form a matrix Q of size $2s \times 2s$. Finally, the corresponding columns and rows of blocks of size $2s \times 2s$ of matrix A and the rows of blocks of matrix V are updated using Q.

After completing a set of blocked rotations, a swap of column and row blocks is performed. This brings the next blocks of size $s \times s$ to be zeroed to the sub-diagonal, and the process continues nullifying elements on the sub-diagonal blocks.

Because the sweeps over each block are performed using level-1 BLAS, and matrices A and V can be updated using level-3 BLAS, the cost of the algorithm is:

$$8k_3n^3 + (12k_1 - 16k_3)\,n^2s + 8k_3ns^2 \quad flops, \tag{1}$$

when computing eigenvalues and eigenvectors. In this formula k_1 and k_3 represent the execution time to perform a floating point operation using level-1 or level-3 BLAS, respectively.

3 A logical parallel block-Jacobi method

To design a parallel algorithm, what we must do first is to decide the distribution of data to the processors. This distribution and the movement of data in the matrices determine the necessities of memory and data transference on a distributed system. We begin analysing these necessities considering processes but not processors, obtaining a logical parallel algorithm.

Each one of the blocks of size $2sk \times 2sk$ is considered as a process and will have a particular necessity of memory. At least it needs memory to store the initial blocks, but some additional memory is necessary to store data in successive steps of the execution.

A scheme of the method is shown in figure 2. The method using this scheme is briefly explained below.

```
On each process:
WHILE convergence not reached DO
      FOR every Jacobi set in a sweep DO
            perform sweeps on the blocks of size 2s × 2s
              corresponding to indices associated to the processor,
              accumulating the rotations
            broadcast the rotation matrices
            update the part of matrices A and V associated
              to the process, performing matrix-matrix multiplications
            transfer rows and columns of blocks of A and rows
              of blocks of V
      ENDFOR
ENDWHILE
```

Fig. 2. Basic parallel block-Jacobi iteration.

Each sweep is divided into a number of steps corresponding each step to a Jacobi set.

For each Jacobi set the rotations matrices can be computed in parallel, but working only processes associated to blocks $2sk \times 2sk$ of the main diagonal of A. On these processes a sweep is performed on each one of the blocks it contains corresponding to the Jacobi set in use, and the rotations on each block are accumulated on a rotations matrix of size $2s \times 2s$.

After the computation of the rotations matrices, they are sent to the other processes corresponding to blocks in the same row and column in the matrix A, and the same row in the matrix V. And then the processes can update the part of A or V they contain.

In order to obtain the new grouping of data according to the next Jacobi set it is necessary to perform a movement of data in the matrix, and that implies a data transference between processes and additional necessities of memory. It is illustrated in figure 3, where $\frac{n}{s} = 16$, the rows and columns of blocks are numbered from 0 to 15, and the occupation of memory on an odd and an even step is shown. In the figure blocks containing data are marked with $\times$.

Thus, to store a block of matrix A it is necessary to reserve a memory of size $(2sk + s) \times (2sk + s)$, and to store a block of V is necessary a memory of size $(2sk + s) \times 2sk$.

4 Parallel block-Jacobi methods

To obtain parallel algorithms it is necessary to assign each logical process to a processor, and this assignation must be in such a way that the work is balanced. The most costly part of the algorithm is the updating of matrices A and V (that

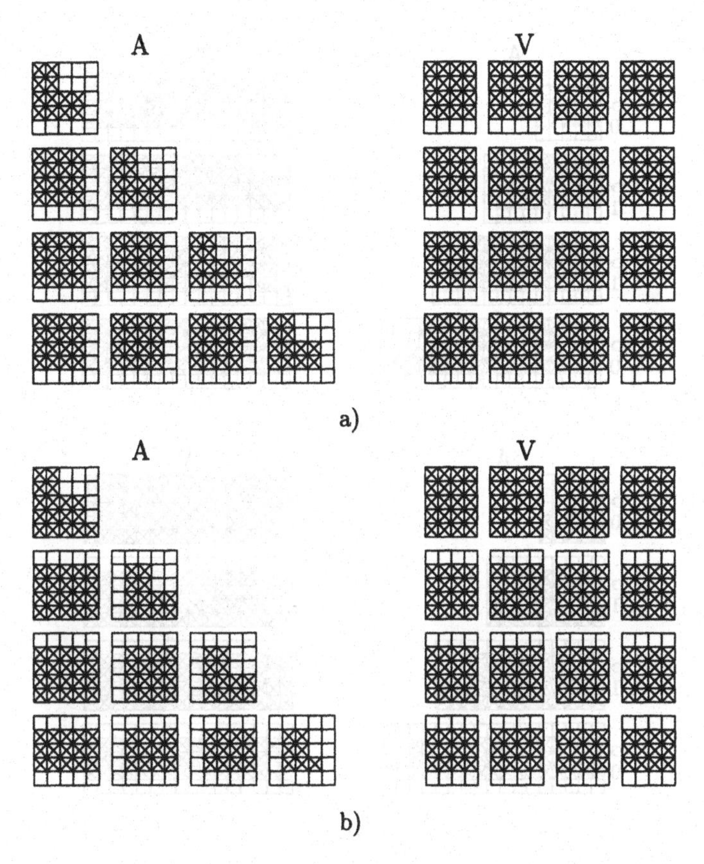

Fig. 3. Storage of data: a) on an odd step, b) on an even step.

produces the cost of order $O\left(n^3\right)$). To assign the data in a balanced way it suffices to balance the updating of the matrices only. In the updating of non-diagonal blocks of matrix A, $\frac{n}{2sk} \times \frac{n}{2sk}$ data are updated pre- and post-multiplying by rotations matrices. In the updating of diagonal blocks only elements in the lower triangular part of the matrix are updated pre- and post-multiplying by rotations matrices. And in the updating of matrix V, $\frac{n}{2sk} \times \frac{n}{2sk}$ data are updated but only pre-multiplying by rotations matrices. So, we can see the volume of computation on the updating of matrices on processes corresponding to a non-diagonal block of matrix A is twice that of processes corresponding to a block of V or a diagonal block of A. This must be had in mind when designing parallel algorithms.

An algorithm for a ring. Considering $q = \frac{n}{2sk}$ and a ring with $p = \frac{q}{2}$ processors, P_0, P_1, ..., P_{p-1}, a balanced algorithm can be obtained assigning to each processor P_i, rows i and $q-1-i$ of matrices A and V. So, each processor P_i contains blocks A_{ij}, with $0 \leq j \leq i$, $A_{q-1-i,j}$, with $0 \leq j \leq q-1-i$, and V_{ij} and $V_{q-1-i,j}$, with $0 \leq j < q$.

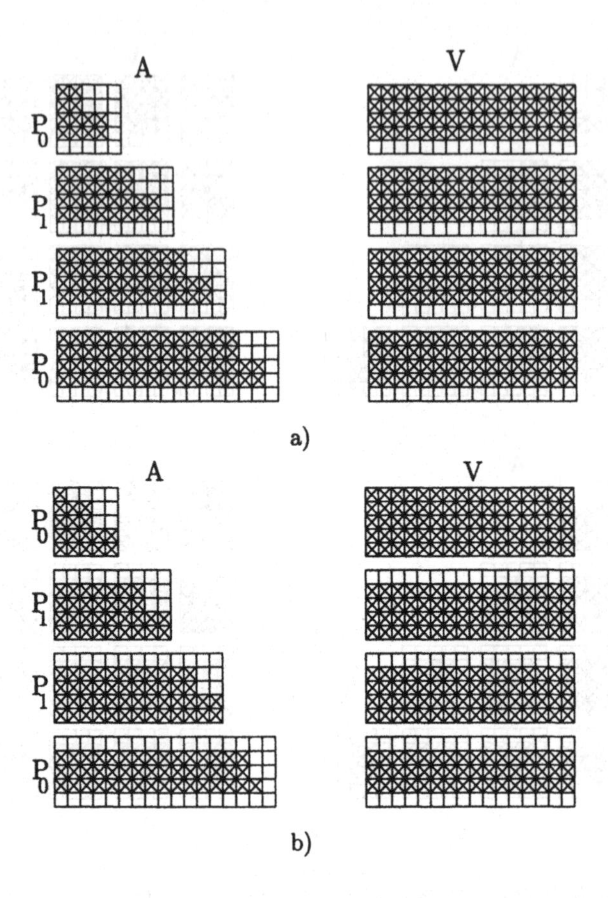

Fig. 4. Storage of data: a) on an odd step, b) on an even step. Algorithm for ring topology.

To save memory and to improve computation some of the processes in the logical method can be grouped, obtaining on each processor four logical processes, corresponding to the four rows of blocks of matrices A and V contained in the processor. In figure 4 the distribution of data and the memory reserved are shown, for $p = 2$ and $\frac{n}{s} = 16$. In that way, $(2sk + s)(3n + 2sk + 2s)$ possitions of memory are reserved on each processor.

The arithmetic cost per sweep when computing eigenvalues and eigenvectors is:

$$8k_3\frac{n^3}{p} + (12k_1 - 8k_3)\frac{n^2s}{p} + 12k_1\frac{ns^2}{p} \quad flops. \tag{2}$$

And the cost per sweep of the communications is:

$$\beta(p+3)\frac{n}{s} + \tau\left(8n^2 + 2ns - \frac{2n^2}{p}\right) , \tag{3}$$

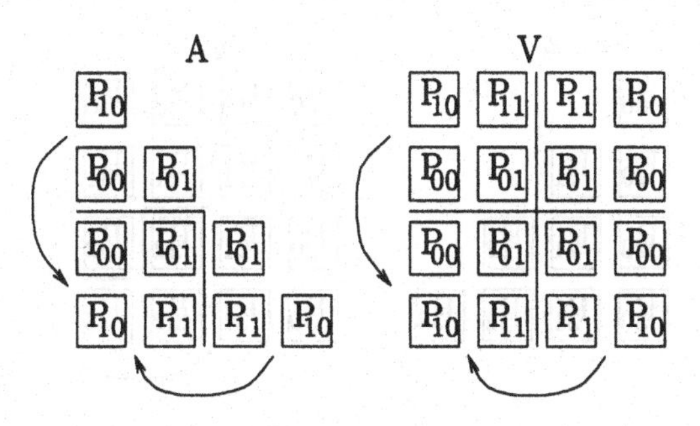

Fig. 5. Folding the matrices in a square mesh topology.

where β represents the start-up time, and τ the time to send a double precision number.

An algorithm for a square mesh. In a square mesh, a way to obtain a balanced distribution of the work is to fold the matrices A and V in the system of processors, such as is shown in figure 5, where a square mesh with four processors is considered.

To processor P_{ij} are assigned the next blocks: from matrix A block $A_{\overline{p}+i,j}$, if $i \leq \overline{p}-1-j$ block $A_{\overline{p}-1-i,j}$, and if $i \geq \overline{p}-1-j$ block $A_{\overline{p}+i,2\overline{p}-1-j}$; and from matrix V blocks $V_{\overline{p}+i,j}$, $V_{\overline{p}-1-i,j}$, $V_{\overline{p}+i,2\overline{p}-1-j}$ and $V_{\overline{p}-1-i,2\overline{p}-1-j}$. The memory reserved on each processor in the main anti-diagonal is $(2sk+s)(14sk+3s)$, and in each one of the other processors $(2sk+s)(12sk+2s)$.

This data distribution produces an imbalance in the computation of the rotations matrices, because only processors in the main anti-diagonal of processors work in the sweeps over blocks in the diagonal of matrix A. On the other hand, this imbalance allows us to overlap computations and communications.

The arithmetic cost per sweep when computing eigenvalues and eigenvectors is:

$$8k_3\frac{n^3}{p} + (12k_1 + 2k_3)\frac{n^2 s}{\overline{p}} + 12k_1\frac{ns^2}{\overline{p}} \quad flops. \tag{4}$$

And the cost per sweep of the communications is:

$$\beta(\overline{p} + 4)\frac{n}{s} + \tau n^2\left(2 + \frac{7}{\overline{p}}\right) \quad . \tag{5}$$

Comparing equations 4 and 2 we can see the arithmetic cost is lower in the algorithm for a ring, but only in the terms of lower order. Furthermore, communications and computations can be overlapped in some parts of the algorithm for a mesh.

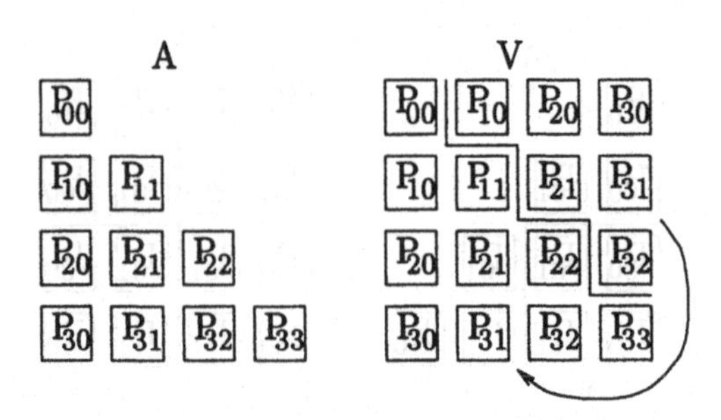

Fig. 6. The storage of matrices in a triangular mesh topology.

An algorithm for a triangular mesh. In a triangular mesh, matrix A can be assigned to the processors in an obvious way, and matrix V can be assigned folding the upper triangular part of the matrix over the lower triangular part (figure 6).

To processor P_{ij} ($i \geq j$) are assigned the blocks A_{ij}, V_{ij} and V_{ji}. The memory reserved on each processor in the main diagonal is $(2sk + s)(4sk + s)$, and in each one of the other processors $(2sk + s)(6sk + s)$.

Only processors in the main diagonal work in the sweeps over blocks in the diagonal of matrix A. In this case, as happens in a square mesh, the imbalance allows us to overlap computations and communications.

If we call r to the number of rows and columns in the processors system, r and p are related by the formula $r = \frac{-1+\sqrt{1+8p}}{2}$, and the arithmetic cost per sweep of the algorithm when computing eigenvalues and eigenvectors is:

$$16k_3\frac{n^3}{r^2} + (12k_1 + 2k_3)\frac{n^2 s}{r} + 12k_1\frac{ns^2}{r} \quad flops. \tag{6}$$

And the cost per sweep of the communications is:

$$\beta(r+7)\frac{n}{s} + \tau\left(2n^2 + \frac{6n^2}{r} + 4ns\right) . \tag{7}$$

The value of r is a little less than $\sqrt{2p}$. Thus, the arithmetic cost of this algorithm is worse than that of the algorithm for a square mesh, and the same happens with the cost of communications. But when p increases the arithmetic costs tend to be equal, and the algorithm for triangular mesh is better than the algorithm for a square mesh, due to a more regular distribution of data.

5 Comparison

Comparing the theoretical costs of the algorithms studied it is possible to conclude the algorithm for a ring is the best and the algorithm for a triangular mesh is the worst. This can be true when using a small number of processors, but it

Table 1. Theoretical costs per sweep of the different parts of the algorithms.

	update matrices	*comp. rotations*	*broadcast*	*trans. data*
ring	$8\frac{n^3}{p} - 8\frac{n^2 s}{p}$	$12\frac{n^2 s}{p} + 12\frac{ns^2}{p}$	$(p-1)\frac{n}{s}\beta$	$4\frac{n}{s}\beta$
			$\left(2n^2 - 2\frac{n^2}{p}\right)\tau$	$\left(6n^2 + 2ns\right)\tau$
sq. mesh	$8\frac{n^3}{p} + 2\frac{n^2 s}{\sqrt{p}}$	$12\frac{n^2 s}{\sqrt{p}} + 12\frac{ns^2}{\sqrt{p}}$	$\sqrt{p}\frac{n}{s}\beta$	$4\frac{n}{s}\beta$
			$2n^2\tau$	$7\frac{n^2}{\sqrt{p}}\tau$
tr. mesh	$16\frac{n^3}{r^2} + 2\frac{n^2 s}{r}$	$12\frac{n^2 s}{r} + 12\frac{ns^2}{r}$	$r\frac{n}{s}\beta$	$7\frac{n}{s}\beta$
			$2n^2\tau$	$\left(6\frac{n^2}{r} + 4ns\right)\tau$

is just the opposite when the number of processors and the matrix size increase, due to the overlapping of computations and communications on the algorithms for a mesh.

Some attention has been paid to the optimisation of parallel Jacobi methods by overlapping communication and computation [10, 11], and in the mesh algorithms here analysed the imbalance in the computation of rotation matrices makes possible this overlapping. Adding the arithmetic and the communication costs in equations 2 and 3, 4 and 5, and 6 and 7, the total cost per sweep of the algorithms for ring, square mesh and triangular mesh, respectively, can be estimated; but these times have been obtained without regard to the overlapping of computation and communication. In the algorithm for a ring there is practically no overlapping because the computation of rotations is balanced, and after the computation of the rotation matrices each processor is involved in the broadcast, and the updating of matrices begins only after the broadcast finishes. But in the algorithms for mesh the computation of rotations is performed only by the processors in the main diagonal or anti-diagonal in the system of processors, and this makes the overlapping possible.

To compare in more detail the three methods, in table 1 the costs per sweep of each part of the algorithms are shown.

The three algorithms have an iso-efficiency function $f(n) = p$, but the algorithms for mesh are more scalable in practice. The value of the iso-efficiency function appears from the term corresponding to rotations broadcast, which has a cost $O\left(n^2\right)$, but in the algorithm for a ring this is the real cost, because the matrices A and V can not be updated before the rotations have been broadcast. It is different in the algorithms for a mesh topology, where the execution times obtained are upper-bounds. In these algorithms the rotations broadcast can be overlapped with the updating of the matrices (as shown in [12] for systolic arrays) and when the size of the matrices increases the total cost can be better approximated by adding the costs of table 1 but without the cost of broadcast, which is overlapped with the updating of the matrices. In this way, the iso-efficiency function of the algorithms for mesh topology is $f(n) = \sqrt{p}$, and these methods are more scalable.

In addition, few processors can be utilised efficiently in the algorithm for a ring, for example, with $n = 1024$, if $p = 64$ the block size must be lower or equal to 8, but when using 64 processors on the algorithm for a square mesh the block size must be lower or equal to 64.

The overlapping of communication and computation in the algorithm for triangular mesh is illustrated in figure 7. In this figure matrices A and V are shown, and a triangular mesh with 21 processors is considered. The first steps of the computation are represented writing into each block of the matrices which part of the execution is carried out: R represents computation of rotations, B broadcast of rotation matrices, U matrix updating, and D transference of data. The numbers represent which Jacobi set is involved, and an arrow indicates a movement of data between blocks in the matrices and the corresponding communication of data between processors. We will briefly explain some of the aspects in the figure:

- Step 1: Rotation matrices are computed by the processors in the main diagonal of processors.
- Step 2: The broadcast of rotation matrices to the other processors in the same row and column of processors begins.
- Step 3: Computation and communication are overlapped. All the steps in the figure have not the same cost (the first step has a cost $24k_1\frac{ns^2}{p}$ and the second $2\beta + 4\frac{ns}{p}\tau$), but if a large size of the matrices is assumed the cost of the computational parts is much bigger than that of the communication parts, therefore communication in this step finishes before computation.
- Step 4: More processors begin to compute and the work is more balanced.
- Step 5: Update of matrices has finished in processors in the main diagonal and the sub-diagonal of processors, and the movement of rows and columns of blocks begins in order to obtain the data distribution needed to perform the work corresponding to the second Jacobi set.
- Step 6: The computation of the second set of rotation matrices begins in the diagonal before the updating of the matrices have finished. All the processors are involved in this step, and the work is more balanced than in the previous steps. If the cost of computation is much bigger than the cost of communication, the broadcast could have finished and all the processors could be computing.
- Step 7: Only four diagonals of processors can be involved at the same time in communications. Then, if the number of processors increases the cost of communication becomes less important.
- Step 8: First and second updating are performed at the same time.
- Step 9: After this step all the data have been moved to the positions corresponding to the second Jacobi set.
- Step 10: When the step finishes the third set of rotations can be computed.

As we can see, there is an inbalance at the beginning of the execution, but that is compensated by the overlapping of communication and computation, which makes it possible to overlook the cost of broadcast to analyse the scalability of the algorithm. The same happens with the algorithm for square mesh.

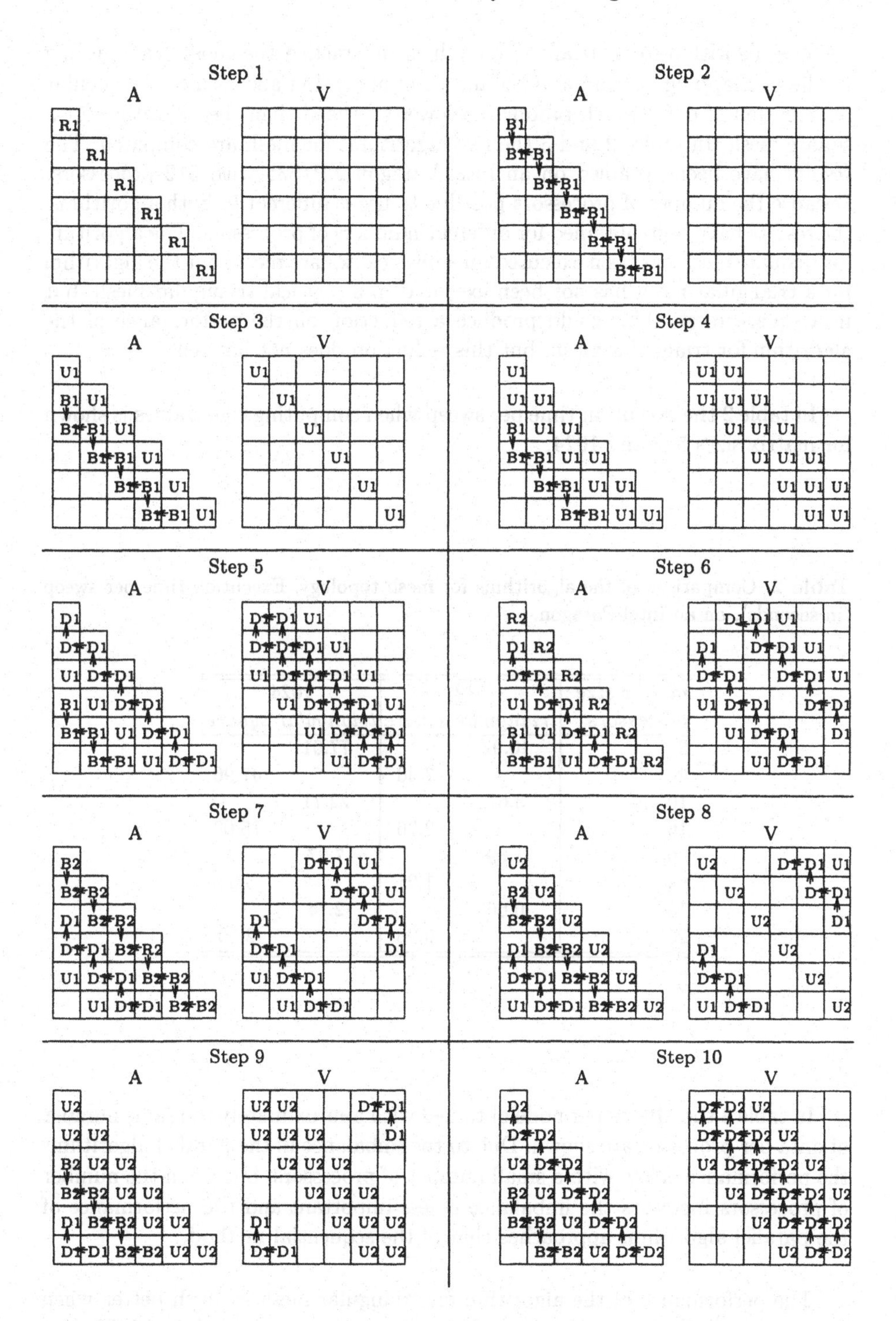

Fig. 7. Overlapping communication and computation on the algorithm for triangular mesh.

The algorithm for a triangular mesh is in practice the most scalable due to the overlapping of computation and communication and also to the regular distribution of data, which produces a lower overhead than the algorithm for a square mesh. In tables 2 and 3 the two algorithms for mesh are compared. The results have been obtained on an Intel Paragon XP/S35 with 512 processors. Because the number of processors possible to use is different in both algorithms, the results have been obtained for different numbers of processors. The algorithm for a square mesh has been executed on a physical square mesh, but the algorithm for a triangular mesh has not been executed in a physical triangular mesh, but in a square mesh. This could produce a reduction on the performance of the algorithm for triangular mesh, but this reduction does not happen.

In table 2 the execution time per sweep when computing eigenvalues is shown for matrix sizes 512 and 1024.

Table 2. Comparison of the algorithms for mesh topology. Execution time per sweep (in seconds), on an Intel Paragon.

matrix size :	512		1024	
processors	*triangular*	*square*	*triangular*	*square*
3	10.92		77.61	
4		7.40		47.96
10	3.67		22.71	
16		2.76		15.02
36	1.39		7.27	
64		1.21		5.53
136	0.66		2.74	
256		0.96		2.50

In table 3 the Mflops per node obtained with approximately the same number of data per processor are shown. Due to the imbalance in the parallel algorithms the performance is low with a small number of processors, but when the number of processors increases the imbalance is less important and the performance of the parallel algorithms approaches that of the sequential method.

The performance of the algorithm for triangular mesh is much better when the number of processors increases. For example, for a matrix size of 1408, in a square mesh of 484 processors 2.99 Gflops were obtained, while in a triangular mesh of 465 processors 4.23 Gflops were obtained.

Table 3. Comparison of the algorithms for mesh topologies. Mflops per node with approximately the same number of data per processor, on an Intel Paragon XP/S35.

data/pro. :	4068		8192		18432		32768	
seq. :	13.10		14.71		18.63		18.39	
proc.	*trian–*	*square*	*trian–*	*square*	*trian–*	*square*	*trian–*	*square*
	gular		*gular*	*square*	*gular*	*square*	*gular*	*square*
4		10.52		9.34		15.73		18.14
6	10.95		13.63		14.41		17.00	
15	11.46		12.78		15.53		18.30	
16		10.36		12.16		15.62		17.87
36	10.73	10.21	14.38	12.13	17.04	15.41	18.93	
64		10.15		12.14		15.46		17.65

6 Conclusions

We have shown how parallel block-Jacobi algorithms can be designed in two steps: first associating one process to each block in the matrices, and then obtaining algorithms for a topology by grouping processes and assigning them to processors.

Scalable algorithms have been obtained for mesh topologies and the more scalable in practice is the algorithm for a triangular mesh.

References

1. J. Demmel and K. Stanley. The Performance of Finding Eigenvalues and Eigenvectors of Dense Symmetric Matrices on Distributed Memory Computers. In David H. Bailey, Petter E. Bjørstad, John R. Gilbert, Michael V. Mascagni, Robert S. Schreiber, Horst D. Simon, Virginia J. Torczon and Layne T. Watson, editor, *Proceedings of the Seventh SIAM Conference on Parallel Processing for Scientific Computing*, pages 528–533. SIAM, 1995.
2. Robert Schreiber. Solving eigenvalue and singular value problems on an undersized systolic array. *SIAM J. Sci. Stat. Comput.*, 7(2):441–451, 1986.
3. Gautam Schroff and Robert Schreiber. On the convergence of the cyclic Jacobi method for parallel block orderings. *SIAM J. Matrix Anal. Appl.*, 10(3):326–346, 1989.
4. Christian H. Bischof. Computing the singular value decomposition on a distributed system of vector processors. *Parallel Computing*, 11:171–186, 1989.
5. D. Giménez, V. Hernández, R. van de Geijn and A. M. Vidal. A block Jacobi method on a mesh of processors. *Concurrency: Practice and Experience*, 9(5):391–411, May 1997.
6. L. Auslander and A. Tsao. On parallelizable eigensolvers. *Ad. App. Math.*, 13:253–261, 1992.
7. S. Huss-Lederman, A. Tsao and G. Zhang. A parallel implementation of the invariant subspace decomposition algorithm for dense symmetric matrices. In *Proceedings Sixth SIAM Conf. on Parallel Processing for Scientific Computing.* SIAM, 1993.

8. Xiaobai Sun. Parallel Algorithms for Dense Eigenvalue Problems. In *Whorkshop on High Performance Computing and Gigabit Local Area Networks, Essen, Germany, 1996*, pages 202–212. Springer-Verlag, 1997.
9. Stéphane Domas and Françoise Tisseur. Parallel Implementation of a Symmetric Eigensolver Based on the Yau and Lu Method. In José M. L. M. Palma and Jack Dongarra, editor, *Vector and Parallel Processing- VECPAR'96*, pages 140–153. Springer-Verlag, 1997.
10. Makan Pourzandi and Bernard Tourancheau. A Parallel Performance Study of Jacobi-like Eigenvalue Solution. Technical report, 1994.
11. El Mostafa Daoudi and Abdelhak Lakhouaja. Exploiting the symmetry in the parallelization of the Jacobi method. *Parallel Computing*, 23:137–151, 1997.
12. Richard P. Brent and Franklin T. Luk. The solution of singular-value and symmetric eigenvalue problems on multiprocessor arrays. *SIAM J. Sci. Stat. Comput.*, 6(1):69–84, 1985.

Calculation of Lambda Modes of a Nuclear Reactor: A Parallel Implementation Using the Implicitly Restarted Arnoldi Method *

Vicente Hernández, José E. Román, Antonio M. Vidal, and Vicent Vidal

Dept. Sistemas Informáticos y Computación, Universidad Politécnica de Valencia, Camino de Vera, s/n, 46071 Valencia (Spain)
{vhernand,jroman,avidal,vvidal}@dsic.upv.es

Abstract. The objective of this work is to obtain the dominant λ-modes of a nuclear power reactor. This is a real generalized eigenvalue problem, which can be reduced to a standard one. The method used to solve it has been the Implicitly Restarted Arnoldi (IRA) method. Due to the dimensions of the matrices, a parallel approach has been proposed, implemented and ported to several platforms. This includes the development of a parallel iterative linear system solver. To obtain the best performance, care must be taken to exploit the structure of the matrices.

Keywords: Parallel computing, eigenproblems, lambda modes

1 Introduction

The generalised algebraic eigenvalue problem is a standard problem that frequently arises in many fields of science and engineering. In particular, it appears in approximations of differential operators eigenproblems. The application presented here is taken from nuclear engineering.

The analysis of the lambda modes are of great interest for reactor safety and modal analysis of neutron dynamical processes. In order to study the steady state neutron flux distribution inside a nuclear power reactor and the sub-critical modes responsible for the regional instabilities produced in the reactors, it is necessary to obtain the dominant λ-modes and their corresponding eigenfunctions.

The discretisation of the problem leads to an algebraic eigensystem which can reach considerable sizes in real cases. The main aim of this work is to solve this problem by using appropriate numerical methods and introducing High Performance Computing techniques so that response time can be reduced to the minimum. In addition to this, other benefits can be achieve d, as well. For example, larger problems can be faced and a better precision in the results can be attained.

This contribution is organised as follows. Section 2 is devoted to present how the algebraic generalised eigenvalue problem is derived from the neutron diffusion equation. In section 3, a short description of the matrices which arise

* This work has been developed under the support of the Spanish (TIC96-1062-C03-01) and Valencian (FPI-97-CM-05-0422-G) governments.

J. Palma, J. Dongarra, and V. Hernández (Eds.): VECPAR'98, LNCS 1573, pp. 43–57, 1999.

in this problem is done. Section 4 reviews the method used for the solution of the eigenproblem. Section 5 describes some im plementation issues, whereas in section 6 the results are summarised. Finally, the main conclusions are exposed in section 7.

2 The Neutron Diffusion Equation

Reactor calculations are usually based on the multigroup neutron diffusion equation [12]. If this equation is modeled with two energy groups, then the problem we have to deal with is to find the eigenvalues and eigenfunctions of

$$\mathcal{L}\phi_i = \frac{1}{\lambda_i}\mathcal{M}\phi_i \ , \tag{1}$$

where

$$\mathcal{L} = \begin{bmatrix} -\nabla\left(D_1\nabla\right) + \Sigma_{a1} + \Sigma_{12} & 0 \\ -\Sigma_{12} & -\nabla\left(D_2\nabla\right) + \Sigma_{a2} \end{bmatrix} ,$$

$$\mathcal{M} = \begin{bmatrix} \nu_1\Sigma_{f1} & \nu_2\Sigma_{f2} \\ 0 & 0 \end{bmatrix} \quad \text{and} \quad \phi_i = \begin{bmatrix} \phi_{f_i} \\ \phi_{t_i} \end{bmatrix} ,$$

with the boundary conditions $\phi_i|_\Gamma = 0$, where Γ is the reactor border.

For a numerical treatment, this equation must be discretized in space. Nodal methods are extensively used in this case. These methods are based on approximations of the solution in each node in terms of an adequate base of functions, for example, Legendre polynomials [9]. It is assumed that the nuclear propert ies are constant in every cell. Finally, appropriate continuity conditions for fluxes and currents are imposed.

This process allows to transform the original system of partial differential equations into an algebraic large sparse generalised eigenvalue problem

$$L\psi_i = \frac{1}{\lambda_i}M\psi_i \ ,$$

where L and M are matrices of order $2N$ with the following N-dimensional block structure

$$\begin{bmatrix} L_{11} & 0 \\ -L_{21} & L_{22} \end{bmatrix} \begin{bmatrix} \psi_{1_i} \\ \psi_{2_i} \end{bmatrix} = \frac{1}{\lambda_i} \begin{bmatrix} M_{11} & M_{12} \\ 0 & 0 \end{bmatrix} \begin{bmatrix} \psi_{1_i} \\ \psi_{2_i} \end{bmatrix} \tag{2}$$

being L_{11} and L_{22} nonsingular sparse symmetric matrices, and M_{11}, M_{12} and L_{21} diagonal matrices. By eliminating ψ_{2_i}, we obtain the following N-dimensional non-symmetric standard eigenproblem

$$A\psi_{1_i} = \lambda_i\psi_{1_i} \ ,$$

where the matrix A is given by

$$A = L_{11}^{-1}\left(M_{11} + M_{12}L_{22}^{-1}L_{21}\right) \ . \tag{3}$$

All the eigenvalues of this equation are real. We are only interested in calculating a few dominant ones with the corresponding eigenvectors.

This problem has been solved with numerical methods such as Subspace Iteration, [10], [11]. Here we present a more effective strategy.

3 Matrix Features

In order to validate the correctness of the implemented programs, two reactors have been chosen as test cases.

The first benchmark is the BIBLIS reactor [2], which is a pressure water reactor (PWR). Due to its characteristics, this reactor has been modelled in a bidimensional fashion with 1/4 symmetry. The nodalization scheme is shown in figure 1(a). The darkest cells represent the 23.1226 cm wide reflector whereas the other cells correspond to the reactor kernel, with a distance between nodes of 23.1226 cm as well. The kernel cells can be of up to 7 different materials.

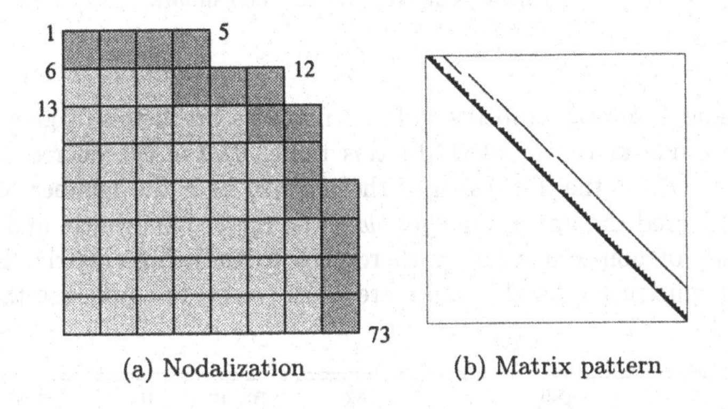

(a) Nodalization (b) Matrix pattern

Fig. 1. The BIBLIS benchmark.

The numbers of the nodes follow a left-right top-down ordering, as shown in the figure. This leads to a staircase-like matrix pattern which can be seen in figure 1(b). Note that only the upper triangular part is stored. This pattern is ide ntical for both L_{11} and L_{22}. In particular, the matrix pattern depicted in the figure corresponds to a space discretisation using Legendre polynomials of 5th degree. This degree is directly related to the number of rows and columns associated to every node in the mesh. The dimensions of the matrices are shown in table 1.

The other reference case is the RINGHALS reactor [3], a real boiling water reactor (BWR). This reactor has been discretized three-dimensionally in 27 axial planes (25 for the fuel and 2 for the reflector). In its turn, each axial plane is di vided in 15.275 cm × 15.275 cm cells distributed as shown in figure 2(a). In this case, it is not possible to simplify the problem because the reactor does not have any symmetry by planes. Each of the 15600 cells has different neu tronic properties. As expected, matrices arising from this nodalization scheme are much larger and have a much more regular structure (figure 2(b)).

Apart from symmetry and sparsity, the most remarkable feature of the matrices is bandedness. This aspect has been exploited in the implementation.

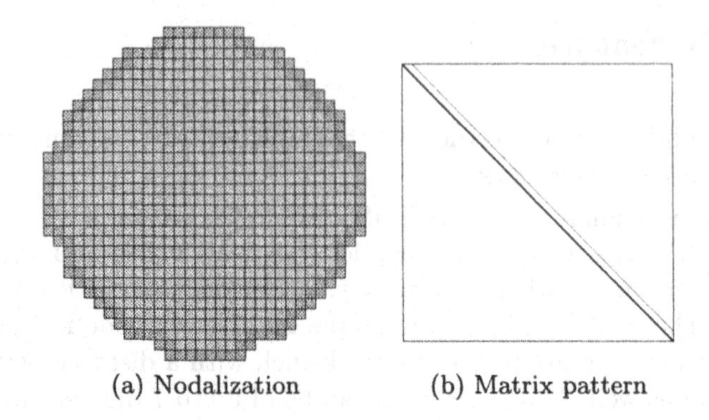

(a) Nodalization (b) Matrix pattern

Fig. 2. The RINGHALS benchmark.

In table 1 several properties of the matrices are listed to give an idea of the magnitude of the problem. In this table, *dpol* is the degree of Legendre polynomials, n is the dimension of the matrix, nz is the number of non-z ero elements stored the upper triangle, *bw* is the upper bandwidth, and *disp* is the percentage of non-zero values with respect to the whole matrix. Finally, the storage requirements for the values are given (*mem*) to emphasise this issue.

	dpol	n	nz	mem	bw	disp
	1	73	201	1.6 Kb	10	0.062
	2	219	1005	7.9 Kb	29	0.037
Biblis	3	438	2814	22 Kb	57	0.027
	4	730	6030	47 Kb	94	0.021
	5	1095	11055	86 Kb	140	0.017
	1	20844	80849	0.6 Mb	773	0.00032
	2	83376	505938	3.9 Mb	3090	0.00013
Ringhals	3	208440	1721200	13 Mb	7723	0.000074
	4	416880	4355110	33 Mb	15444	0.000048
	5	729540	9218685	70 Mb	27025	0.000033

Table 1. Several properties of the matrices.

4 Implicitly Restarted Arnoldi Method

As we only want to determine the dominant eigenvalues which define the reactor behaviour, we approach this partial eigenproblem with the Arnoldi method.

The Arnoldi method is a *Krylov subspace* or *orthogonal projection* method for extracting spectral information. We call

$$\mathcal{K}_k(A, v_0) = \mathrm{span}\left\{v_0, Av_0, A^2v_0, \cdots, A^{k-1}v_0\right\}$$

the k-th Krylov subspace corresponding to $A \in C^{n\times n}$ and $v_0 \in C^n$. The idea is to construct approximate eigenvectors in this subspace.

We define a k-step Arnoldi factorization of A as a relationship of the form

$$AV = VH + fe_k^T$$

where $V \in C^{n\times k}$ has orthonormal columns, $V^H f = 0$, and $H \in C^{k\times k}$ is upper Hessenberg with a non-negative sub-diagonal. The central idea behind this factorisation is to construct eigenpairs of the large matrix A from t he eigenpairs of the small matrix H.

In general, we would like the starting vector v_0 to be rich in the directions of the desired eigenvectors. In some sense, as we get a better idea of what the desired eigenvectors are, we would like to adaptively refine v_0 to be a linear comb ination of the approximate eigenvectors and restart the Arnoldi factorisation with this new vector instead. A convenient and stable way to do this without explicitly computing a new Arnoldi factorisation is given by the Implicitly Restarted Arnoldi (IRA) method, based on the implicitly shifted QR factorisation [8].

The idea of the IRA method is to extend a k-step Arnoldi factorisation

$$AV_k = V_k H_k + f_k e_k^T$$

to a $(k+p)$-step Arnoldi factorisation

$$AV_{k+p} = V_{k+p} H_{k+p} + f_{k+p} e_{k+p}^T \,.$$

Then p implicit shifts are applied to the factorisation, resulting in the new factorisation

$$AV_+ = V_+ H_+ + f_{k+p} e_{k+p}^T Q \,,$$

where $V_+ = V_{k+p}Q$, $H_+ = Q^H H_{k+p} Q$, and $Q = Q_1 Q_2 \cdots Q_p$, where Q_i is associated with factoring $(H - \sigma_i I) = Q_i R_i$. It turns out that the first $k-1$ entries of $e_{k+p}Q$ are zero, so that a new k-step Arnoldi factorisation can be obtai ned by equating the first k columns on each side:

$$AV_k^+ = V_k^+ H_k^+ + f_k^+ e_k^T \,.$$

We can iterate the process of extending this new k-step factorisation to a $(k+p)$-step factorisation, applying shifts, and condensing. The payoff is that every iteration implicitly applies a pth degree polynomial in A to the initial vector v_0 . The roots of the polynomial are the p shifts that were applied to the factorisation. Therefore, if we choose as the shifts σ_i eigenvalues that are "unwanted", we can effectively filter the starting vector v_0 so that it is rich in the dire ction of the "wanted" eigenvectors.

5 Implementation

This section describes some details of the implemented codes for a distributed memory environment. The Message Passing Interface (MPI), [6], has been used as the message passing layer so that portability is guaranteed.

5.1 Eigensolver Iteration

The Implicitly Restarted Arnoldi (IRA) method which has been used in this work is that implemented in the ARPACK [4] software package. This package contains a suit of codes for the solution of several types of eigenvalue related prob lems, including standard and generalised eigenproblems and singular value decompositions for both real and complex matrices. It implements the IRA method for non-symmetric matrices and the analogous Lanczos method for symmetric matrices.

In particular, the programs implemented for the calculation of the lambda modes make use of the parallel version of ARPACK [5], which is oriented to a SPMD/MIMD programming paradigm. This package uses a distribution of all t he vectors involved in the algorithms by blocks among the available processors. The size of the block can be established by the user, thus allowing more flexibility for load balancing.

As well as in many other iterative methods packages, ARPACK subroutines are organised in a way that they offer a reverse communication interface to the user [1]. The primary aim of this scheme is to isolate the matrix-vector operat ions. Whenever the iterative method needs the result of an operation such as a matrix-vector product, it returns control to the user's subroutine that called it. After performing this operation, the user invokes the iterative method subroutine again.

The flexibility of this scheme gives the possibility of using various matrix storage formats as well as obtaining the eigenvalues and eigenvectors of a matrix for which an explicit form is not available. Indeed, the problem we are presenting here is a non-symmetric standard partial eigenproblem of an operator given by the expression (3), where A is not calculated explicitly.

The explicit construction of the inverses would imply the loss of sparsity properties, thus making the storage needs prohibitive. For this reason, the matrix-vector product needed in the Arnoldi process has to be calculated by performing the operations which appear in (3) one at a time. The necessary steps to compute $y = Ax$ are the following:

1. Calculate $w_1 = M_{11}x$.
2. Calculate $w_2 = L_{21}x$.
3. Solve the system $L_{22}w_3 = w_2$ for w_3.
4. Calculate $w_4 = w_1 + M_{12}w_3$.
5. Solve the system $L_{11}y = w_4$ for y.

It has to be noted that the above matrix-vector products involve only diagonal matrices. Therefore, the most costly operations are the solution of linear systems of equations (steps 3 and 5).

5.2 Linear Systems of Equations

The resolution of the linear systems can be approached with iterative methods, such as the Conjugate Gradient [7]. These methods, in their turn, typically use the aforementioned reverse communication scheme.

For the parallel implementation, two basic operations have to be provided, namely the matrix-vector product and the dot product.

In this case, the matrix-vector product subroutine deals with sparse symmetric matrices (L_{ii}). The parallel implementation of this operation is described later. For the distributed dot product function, each processor can perform a dot produc t on the sub-vectors it has, and then perform a summation of all the partial results.

In order to accelerate the convergence, a Jacobi preconditioning scheme was used because of its good results and also because its parallelisation is straightforward.

5.3 Parallel Matrix-Vector Product

The matrices involved in the systems of equations (L_{ii}) are symmetric, sparse and with their nonzero elements within a narrow band. This structure must be exploited for an optimal result. The storage scheme used has been Compressed Sparse Row containing only the upper triangle elements including the main diagonal.

A multiplication algorithm which exploits the symmetry can view the product $L_{ii}x = y$ as $(U + (L_{ii} - U))x = y$, where U is the stored part. Thus, the product can be written as the addition of two partial products,

$$\underbrace{Ux}_{y_1} + \underbrace{(L_{ii} - U)x}_{y_2} = y \, .$$

The algorithm for the first product can be row-oriented ($y_i = U^{(i)}x$) whereas the other must be column-oriented ($y = 0,\ y = y + x_i(A - U)_i = y + x_iU^{(i)T}$).

The parallel matrix-vector product has to be highly optimised in order to achieve good overall performance, since it is the most time-consuming operation. The matrices have been partitioned by blocks of rows conforming with the partitioning of the vectors. It has been implemented in three stages, one initial communication stage, parallel computation of intermediate results and finally another communication operation. This last stage is needed because only the upper triangular part of the symmetric matrices is stored. In the communication stages, only the minimum amount of information is exchanged and this is done synchronously by all the processors.

Figure 3 shows a scheme of the parallel matrix-vector product for the case of 4 processors. The shadowed part of the matrix corresponds to the elements stored in processor 1. Before one processor can carry out its partial product, a fragment of the vector operand owned by the neighbour processor is needed. Similarly, after the partial product is complete, part of the solution vector must be exchanged.

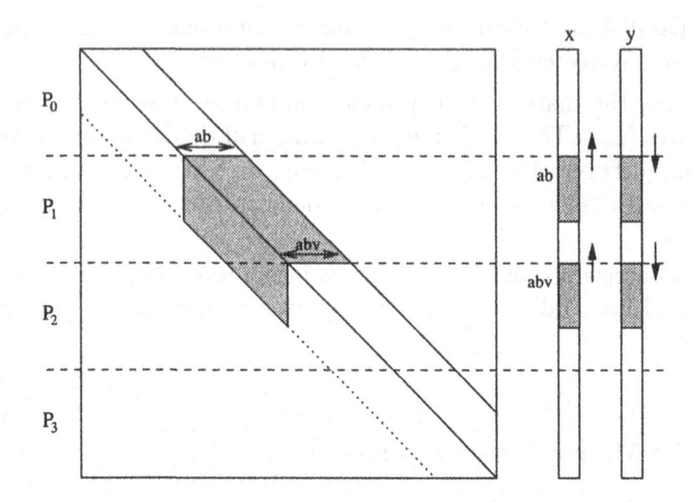

Fig. 3. Message passing scheme for the matrix-vector product.

The detailed sequence of operations in processor i is the following:

1. Receive from processor $i+1$ the necessary components of x. Also send to processor $i-1$ the corresponding ones.
2. Compute the partial result of y.
3. Send to processor $i+1$ the fragment of y assigned to it. Get from processor $i-1$ the part which must be stored in the local processor, as well.
4. Add the received block of y to the block calculated locally, in the appropriate positions.

The communication steps 1 and 3 are carried out synchronously by all the processors. The corresponding MPI primitives have been used in order to minimise problems such as network access contention.

6 Results

The platforms on which the code has been tested are a Sun Ultra Enterprise 4000 multiprocessor and a cluster of Pentium II processors at 300 MHz with 128 Mb of memory each connected with a Fast Ethernet network. The code has been ported to several other platforms as well.

Several experiments have shown that the most appropriate method for the solution of the linear systems in this particular problem is the Conjugate Gradient (CG) with Jacobi preconditioning. Other iterative methods have been tested, including BGC, BiCGStab, TFQMR and GMRES. In all the cases, the performance of these solvers has turned out to be worse than that of CG. Table 2 compares the average number of $L_{ii}x_k$ products and the time (in seconds) spent by the programs in the solution of the Ringhals benchmark ($dpol = 1$) for each of the tested methods. Apart from the response time, it can be observed also in this table that CG is the solver which requires less memory. The time corresponding to eight processors is also included to show that the efficiency (E_p) is modified slightly.

Method	$L_{ii}x_k$	Time (p=1)	Time (p=8)	E_p (%)	Memory
CG	12	102.04	15.12	84	$5n$
BCG	22	174.32	24.84	88	$7n$
BiCGStab	15	129.99	19.03	85	$8n$
TFQMR	13	155.59	22.75	86	$11n$
GMRES(3)	16	210.79	30.49	86	$\sim 5n$

Table 2. Comparison between several iterative methods.

6.1 Adjustment of Parameters

When applied to this particular problem, the IRA method is a convergent process in all the studied cases. However, it is not an approximate method in the sense that the precision of the obtained approximate solution depends to a great extent on th e tolerance demanded in the iterative process for the solution of linear systems of equations. Table 3 reflects the influence of the precision required in the Conjugate Gradient process ($\|L_{ii}\tilde{x} - b\|_2 < tol_{CG}$) in values such as average number of matrix-vector products ($L_{ii}x_k$), number of Arnoldi iterations (IR A) and precision of the obtained eigenvalue. It can be observed that the number of significant digits in the approximate solution λ_1 matches the required precision tol_{CG} and that, after a certain threshold (10^{-4}), the greater the tolera nce, the worse the approximate solution is.

tol_{CG}	$L_{ii}x_k$	IRA	Ax	Time	λ_1	$\|Ax - \lambda x\|/\|\lambda\|$
10^{-9}	33	19	60	157.59	1.012189	0.000096
10^{-8}	29	19	60	139.33	1.012189	0.000096
10^{-7}	25	19	60	120.48	1.012189	0.000096
10^{-6}	21	19	60	102.16	1.012189	0.000097
10^{-5}	17	19	60	85.71	1.012189	0.000101
0.0001	13	19	60	67.16	1.012170	0.000354
0.001	9	23	72	57.47	1.011175	0.004809
0.01	5	38	117	58.14	1.003579	0.033075
0.1	2	14	44	12.04	0.801034	0.297213

Table 3. Influence of the CG tolerance in the precision of the final result.

In conclusion, the tolerance to be demanded in both the IRA and CG methods should be of the same magnitude. In our case, it is sufficient to choose 10^{-4}

and 10^{-5}, respectively, since input data are perturbed by inherent errors of the order of 10^{-6}.

Another parameter to be considered is the number of columns of V (ncv), that is, the maximum dimension of the Arnoldi basis before restart. This value has a great influence in the effectiveness of the IRA method, either in its computational co st as well as in the memory requirements. If ncv takes a great value, the computational cost per iteration and the storage requirements can be prohibitive. However, a small value can imply that the constructed Krylov subspace contains too few informatio n and, consequently, the process needs too many iterations until convergence is achieved. Some experiments reveal that, in this particular application, a value of 2 or 3 times the number of desired eigenvalues gives good results.

Figure 4 shows the number of matrix-vector operations (Ax_k) necessary to achieve convergence for different values of ncv. The experiment has been repeated for a value of desired eigenvalues (nev) ranging from 1 to 6. In the case of only one eigenvalue, it is clear that a small subspace dimension makes the convergence much slower. On the other hand, incrementing ncv beyond some value is useless with respect to convergence while increasing the memory requirements.

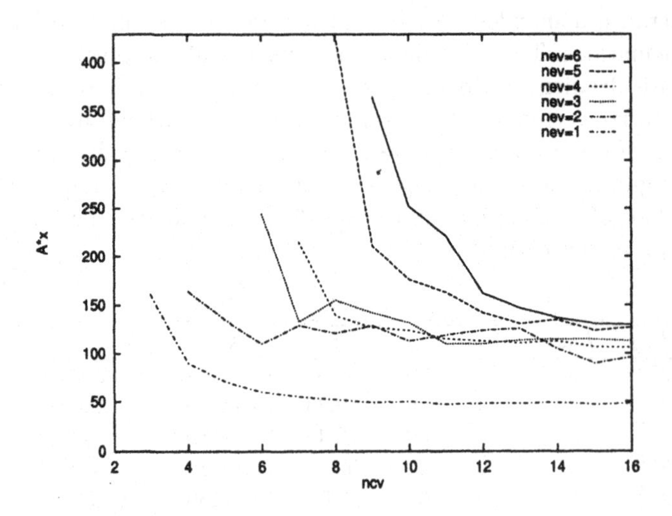

Fig. 4. Influence of ncv in the convergence of the algorithm.

In the other five cases, one can observe a similar behaviour. However, the lines are mixed, mainly in the next three eigenvalues. This is caused by the known fact that clusters of eigenvalues affect the convergence. Figure 5 shows the convergence history of the first 6 eigenvalues. In this graphic it can be appreciated that eigenvalues 2, 3 and 4 converge nearly at the same time, because they are very close to each other (see table 4).

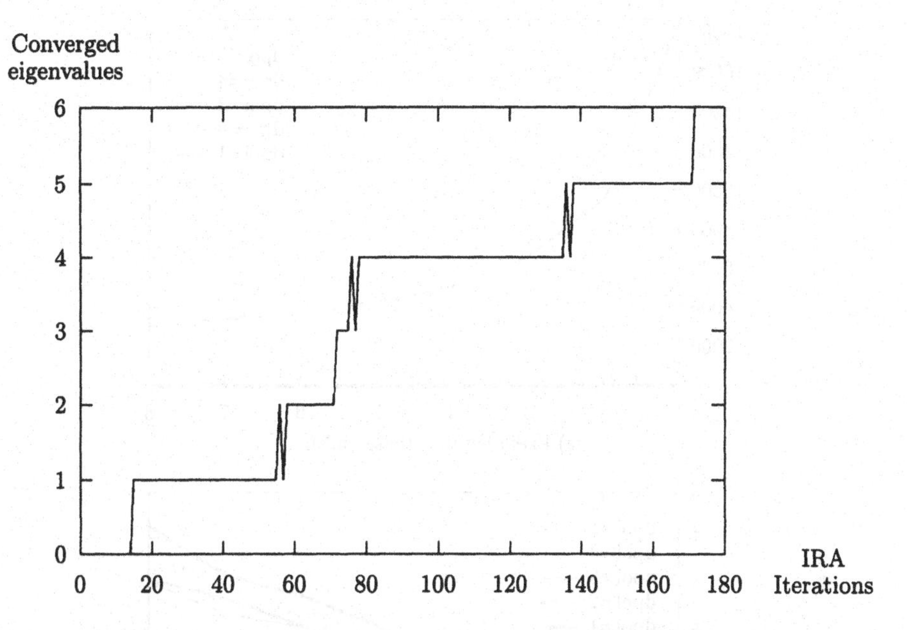

Fig. 5. Convergence history of the first 6 dominant eigenvalues of the Ringhals case (dpol=2).

	λ_i	$\|Ax - \lambda_i x\|/\|\lambda_i\|$
λ_1	1.012190	0.000038
λ_2	1.003791	0.000042
λ_3	1.003164	0.000043
λ_4	1.000775	0.000036
λ_5	0.995716	0.000049
λ_6	0.993402	0.000096

Table 4. Dominant eigenvalues of the Ringhals benchmark (dpol=2).

6.2 Speed-up and Efficiency

In figure 6 we give the performance of the parallel eigensolver measured in the Sun multiprocessor. The graphics show the execution time (in seconds), speed-up and efficiency with up to eight processors for each of the five matric es corresponding to the Ringhals benchmark. The Biblis reactor has not been considered for measurement of parallel performance because of its small size.

In these graphics, it can be seen that the speed-up increases almost linearly with the number of processors. The efficiency does not fall below 75% in any case. We can expect these scalability properties to maintain with more processors.

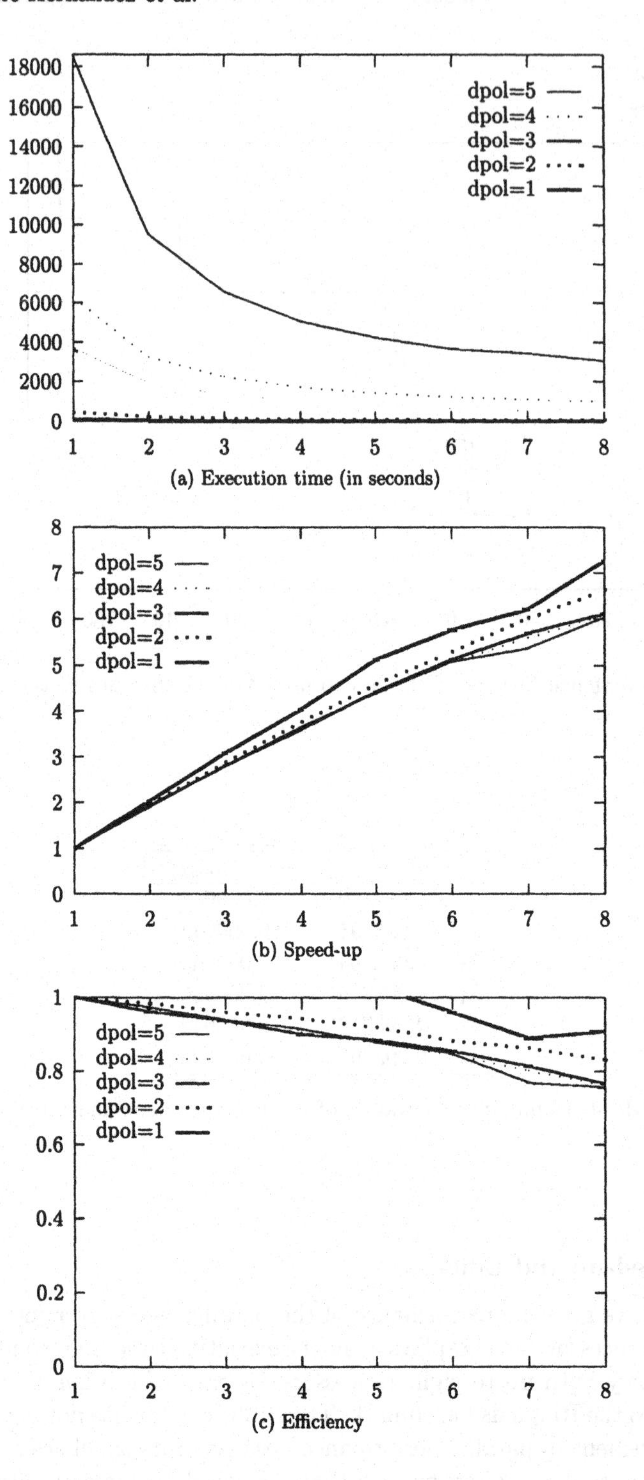

Fig. 6. Graphics of performance of the eigensolver.

The discretisation with polynomials of 2nd degree gives a reasonably accurate approximation for the Ringhals benchmark. In this case, the response time with 8 processors is about 70 seconds.

In the case of the cluster of Pentiums, the efficiency reduces considerably as a consequence of a much slower communication system. However, the results are quite acceptable. Figure 7 shows a plot of the efficiency attained for t he five cases of the Ringhals benchmark. It can be observed that for the biggest matrix, the efficiency is always greater than 50% even when using 8 computers, corresponding to a speed-up of 4.

Another additional advantage in the case of personal computers must be emphasised: the utilisation of the memory resources. The most complex cases (Ringhals with *dpol* = 4 and *dpol* = 5) can not be run on a single computer because of the memory requ irements. Note that the memory size of the running process can be more than 250 Mb. The parallel approach gives the possibility to share the memory of several computers to solve a single problem.

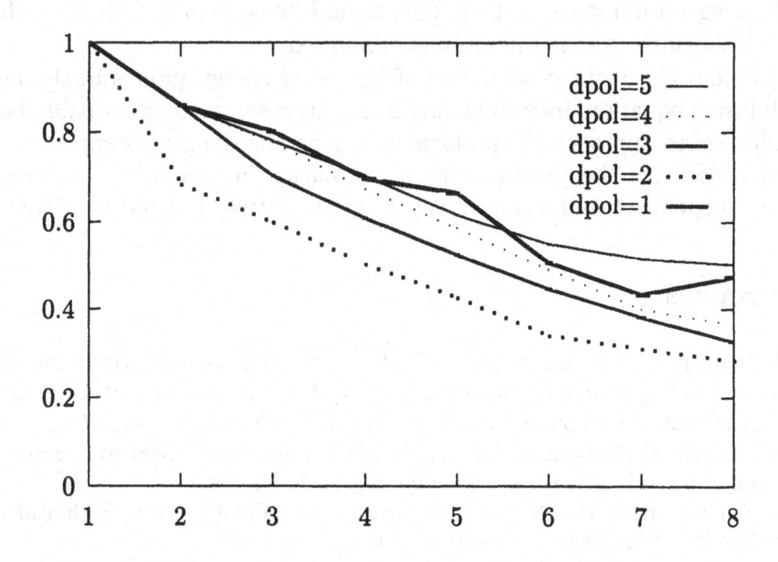

Fig. 7. Efficiency obtained in a cluster of PC's.

7 Conclusions

In this work, we have developed a parallel implementation of the Implicitly Restarted Arnoldi method for the problem of obtaining the λ-modes of a nuclear reactor. This parallel implementation can reduce the response time significantly, sp ecially in complex realistic cases. The nature of the problem has forced to implement a parallel iterative linear system solver as a part of the solution process. With regard to scalability, the experiments with up to 8 processors have shown a good behaviour of the eigensolver.

Apart from the gain in computing time, another advantage of the parallel implementation is the possibility to cope with larger problems using inexpensive platforms with limited physical memory such as networks of personal computers. With this appr oach, the total demanded memory is evenly distributed among all the available processors.

The following working lines are being considered for further refinement of the programs:

- Store the solution vectors of the linear systems to use them as initial solution estimations in subsequent IRA iterations.
- Begin the Arnoldi process taking as initial estimated solution an extrapolation of the solution of the previous order problem.
- Use a more appropriate numbering scheme for the nodes of the grid in order to reduce the bandwidth of the matrices and, consequently, reduce the size of the messages exchanged between processors.
- Implement a parallel direct solver for the linear systems of equations, instead of using an iterative method. This should be combined with the reduction of bandwidth so that the fill-in is minimised.
- Consider the more general case of having G energy groups in the neutron diffusion equation, instead of only 2. In this case, it would probably be more effective to approach the problem as a generalised eigensystem.
- Consider also the general dynamic problem. In this case, the solution is time-dependent and a convenient way of updating it would be of interest.

References

1. J. Dongarra, V. Eijkhout, and A. Kalhan. Reverse communication interface for linear algebra templates for iterative methods. Technical Report UT-CS-95-291, Department of Computer Science, University of Tennessee, May 1995.
2. A. Hébert. Application of the Hermite Method for Finite Element Reactor Calculations. *Nuclear Science and Engineering*, 91:34–58, 1985.
3. T. Lefvert. RINGHALS1 Stability Benchmark - Final Report. Technical Report NEA/NSC/DOC(96)22, November 1996.
4. R. B. Lehoucq, D. C. Sorensen, and C. Yang. *ARPACK USERS GUIDE: Solution of Large Scale Eigenvalue Problems by Implicitly Restarted Arnoldi Methods*. SIAM, Philadelphia, PA, 1998.
5. K. J. Maschhoff and D. C. Sorensen. PARPACK: An Efficient Portable Large Scale Eigenvalue Package for Distributed Memory Parallel Architectures. *Lecture Notes in Computer Science*, 1184:478–486, 1996.
6. MPI Forum. MPI: a message-passing interface standard. *International Journal of Supercomputer Applications and High Performance Computing*, 8(3/4):159–416, Fall-Winter 1994.
7. Y. Saad. SPARSKIT: a basic tool kit for sparse matrix computation. RIACS Technical Report 90.20, NASA Ames Research Center, 1990.
8. Danny C. Sorensen. Implicitly Restarted Arnoldi/Lanczos Methods For Large Scale Eigenvalue Calculations. Technical Report TR-96-40, Institute for Computer Applications in Science and Engineering, May 1996.

9. G. Verdú, J. L. Muñoz-Cobo, C. Pereira, and D. Ginestar. Lambda Modes of the Neutron-Diffusion Equation. Application to BWRs Out-of-Phase Instabilities. *Ann. Nucl. Energy*, 7(20):477–501, 1993.
10. V. Vidal, J. Garayoa, G. Verdú, and D. Ginestar. Optimization of the Subspace Iteration Method for the Lambda Modes Determination of a Nuclear Power Reactor. *Journal of Nuclear Science and Technology*, 34(9):929–947, September 1997.
11. V. Vidal, G. Verdú, D. Ginestar, and J. L. Muñoz-Cobo. Variational Acceleration for Subspace Iteration Method. Application to Nuclear Power Reactors. *International Journal for Numerical Methods in Engineering*, 41:391–407, 1998.
12. J. R. Weston and M. Stacey. *Space-Time Nuclear Reactor Kinetics*. Academic Press, New York, 1969.

Parallel Jacobi-Davidson for Solving Generalized Eigenvalue Problems

Margreet Nool[1] and Auke van der Ploeg[2]

[1] CWI, P.O. Box 94079, 1090 GB Amsterdam, The Netherlands
Margreet.Nool@cwi.nl
[2] MARIN, P.O. Box 28, 6700 AA Wageningen, The Netherlands
A.v.d.Ploeg@marin.nl

Abstract. We study the Jacobi-Davidson method for the solution of large generalised eigenproblems as they arise in MagnetoHydroDynamics. We have combined Jacobi-Davidson (using standard Ritz values) with a shift and invert technique. We apply a complete LU decomposition in which reordering strategies based on a combination of block cyclic reduction and domain decomposition result in a well-parallelisable algorithm. Moreover, we describe a variant of Jacobi-Davidson in which harmonic Ritz values are used. In this variant the same parallel LU decomposition is used, but this time as a preconditioner to solve the 'correction' equation.
The size of the relatively small projected eigenproblems which have to be solved in the Jacobi-Davidson method is controlled by several parameters. The influence of these parameters on both the parallel performance and convergence behaviour will be studied. Numerical results of Jacobi-Davidson obtained with standard and harmonic Ritz values will be shown. Executions have been performed on a Cray T3E.

1 Introduction

Consider the generalised eigenvalue problem

$$Ax = \lambda Bx, \qquad A, B \in \mathcal{C}^{N_t \times N_t}, \tag{1}$$

in which A and B are complex block tridiagonal N_t-by-N_t matrices and B is Hermitian positive definite. The number of diagonal blocks is denoted by N and the blocks are n-by-n, so $N_t = N \times n$. In close cooperation with the FOM Institute for Plasma Physics "Rijnhuizen" in Nieuwegein, where one is interested in such generalised eigenvalue problems, we have developed a parallel code to solve (1). In particular, the physicists like to have accurate approximations of certain interior eigenvalues, called the *Alfvén spectrum*. A promising method for computing these eigenvalues is the Jacobi-Davidson (JD) method [3, 4]. With this method it is possible to find several interior eigenvalues in the neighbourhood of a given target σ and their associated eigenvectors.

J. Palma, J. Dongarra, and V. Hernández (Eds.): VECPAR'98, LNCS 1573, pp. 58–70, 1999.

In general, the sub-blocks of A are dense, those of B are rather sparse ($\approx 20\%$ nonzero elements) and N_t can be very large (realistic values are $N = 500$ and $n = 800$), so computer storage demands are very high. Therefore, we study the feasibility of parallel computers with a large distributed memory for solving (1).

In [2], Jacobi-Davidson has been combined with a parallel method to compute the action of the inverse of the block tridiagonal matrix $A - \sigma B$. In this approach, called DDCR, a block-reordering based on a combination of Domain Decomposition and Cyclic Reduction is combined with a complete block LU decomposition of $A - \sigma B$. Due to the special construction of L and U, the solution process parallelises well.

In this paper we describe two Jacobi-Davidson variants, one using standard Ritz values and one harmonic Ritz values. The first variant uses DDCR to transform the generalised eigenvalue problem into a standard eigenvalue problem. In the second one DDCR has been applied as a preconditioner to solve approximately the 'correction' equation. This approach results also into a projected standard eigenvalue problem with eigenvalues in the dominant part of the spectrum. In Section 2 both approaches are described. To avoid that the projected system becomes too large, we make use of a restarting technique. Numerical results, based on this technique, are analysed in Section 3. We end up with some conclusions and remarks in Section 4.

2 Parallel Jacobi-Davidson

2.1 Standard Ritz Values

The availability of a complete LU decomposition of the matrix $A - \sigma B$ gives us the opportunity to apply Jacobi-Davidson to a *standard* eigenvalue problem instead of a *generalised* eigenvalue problem. To that end, we rewrite (1) as

$$(A - \sigma B)x = (\lambda - \sigma)Bx. \tag{2}$$

If we define $Q := (A - \sigma B)^{-1}B$ then (2) can be written as

$$Qx = \mu x, \qquad \text{with } \mu = \frac{1}{\lambda - \sigma} \Leftrightarrow \lambda = \sigma + \frac{1}{\mu}. \tag{3}$$

The eigenvalues we are interested in form the dominant part of the spectrum of Q, which makes them relatively easy to find. The action of the operator Q consists of a matrix-vector multiplication with B, a perfectly scalable parallel operation, combined with two triangular solves with L and U.

At the k-th step of Jacobi-Davidson, an eigenvector x is approximated by a linear combination of k search vectors v_j, $j = 1, 2, \cdots, k$, where k is very small compared with N_t. Consider the N_t-by-k matrix V_k, whose columns are given by v_j. The approximation to the eigenvector can be written as $V_k s$, for some k-vector s. The search directions v_j are made orthonormal to each other, using Modified Gram-Schmidt (MGS), hence $V_k^* V_k = I$.

Let θ denote an approximation of an eigenvalue associated with the Ritz vector $u = V_k s$. The vector s and the scalar θ are constructed in such a way that the residual vector $r = QV_k s - \theta V_k s$ is orthogonal to the k search directions. From this Rayleigh-Ritz requirement it follows that

$$V_k^* Q V_k s = \theta V_k^* V_k s \Longleftrightarrow V_k^* Q V_k s = \theta s. \qquad (4)$$

The order of the matrix $V_k^* Q V_k$ is k. By using a proper restart technique k stays so small that this 'projected' eigenvalue problem can be solved by a sequential method.

In order to obtain a new search direction, Jacobi-Davidson requires the solution of a system of linear equations, called the 'correction equation'. Numerical experiments show that fast convergence to selected eigenvalues can be obtained by solving the correction equation to some *modest accuracy* only, by some steps of an inner iterative method, e.g. GMRES.

Below we show the Jacobi-Davidson steps used for computing several eigenpairs of (3) using standard Ritz values.

step 0: initialize
Choose an initial vector v_1 with $\|v_1\|_2 = 1$; set $V_1 = [v_1]$;
$W_1 = [Qv_1]$; $k = 1$; $it = 1$; $n_{ev} = 0$
step 1: update the projected system
Compute the last column and row of $H_k := V_k^* W_k$
step 2: solve and choose approximate eigensolution of projected system
Compute the eigenvalues $\theta_1, \cdots, \theta_k$ of H_k and choose $\theta := \theta_j$ with $|\theta_j|$ maximal and $\theta_j \neq \mu_i$, for $i = 1, \cdots, n_{ev}$; compute associated eigenvector s with $\|s\|_2 = 1$
step 3: compute Ritz vector and check accuracy
Let u be the Ritz vector $V_k s$; compute the residual vector $r := W_k s - \theta u$;
if $\|r\|_2 < tol_{sJD}.|\theta|$ **then**
$n_{ev} := n_{ev} + 1$; $\mu_{n_{ev}} := \theta$; **if** $n_{ev} = N_{ev}$ **stop; goto 2**
else if $it = iter$ **stop**
end if
step 4: solve correction equation approximately with it_{SOL} steps of GMRES
Determine an approximate solution $\tilde{z}$ of z in
$(I - uu^*)(Q - \theta I)(I - uu^*)z = -r \;\wedge\; u^* z = 0$
step 5: restart if projected system has reached its maximum order
if $k = m$ **then**
5a: Set $k = k_{min} + n_{ev}$. Construct $C \in \mathcal{C}^{m \times k} \subset H_m$;
Orthonormalize columns of C; compute $H_k := C^* H_m C$
5b: Compute $V_k := V_m C$; $W_k := W_m C$
end if
step 6: add new search direction
$k := k + 1$; $it := it + 1$; call $MGS\ [V_{k-1}, \tilde{z}]$; set $V_k = [V_{k-1}, \tilde{z}]$; $W_k = [W_{k-1}, Q\tilde{z}]$;
goto 1

Steps 2 and 5a deal with the small projected system (4). Those sequential steps are performed by all processors in order to avoid communication. The basic ingredients of the other steps are matrix-vector products, vector updates and inner products. Since, for our applications, N_t is much larger than the number of processors, those steps parallelise well.

2.2 Harmonic Ritz Values

For the introduction of harmonic Ritz values we return to the original generalised eigenvalue problem (1). Assume $(\theta, V_k s)$ approximates an eigenpair (λ, x), then the residual vector r is given by

$$r = AV_k s - \theta BV_k s.$$

In case of standard Ritz values, the correction vector r has to be orthogonal to V_k; the harmonic Ritz values approach asks for vectors r to be orthogonal to $(A - \sigma B)V_k$. Let W_k denote $(A - \sigma B)V_k$, then we have

$$\begin{aligned} r &= AV_k s - \theta BV_k s \\ &= (A - \sigma B)V_k s - (\theta - \sigma)B(A - \sigma B)^{-1}(A - \sigma B)V_k s \\ &= W_k s - (\theta - \sigma)B(A - \sigma B)^{-1}W_k s. \end{aligned} \tag{5}$$

Obviously, $\nu = \frac{1}{(\theta - \sigma)}$ is a Ritz value of the matrix $B(A - \sigma B)^{-1}$ with respect to W_k. To obtain eigenvalues in the neighbourhood of σ, ν must lie in the dominant spectrum of $B(A - \sigma B)^{-1}$. The orthogonalisation requirement leads to

$$\nu W_k^* W_k s = W_k^* BV_k s. \tag{6}$$

To obtain a standard eigenvalue problem we require $W_k^* W_k = I$. By introducing $C := (A - \sigma B)^*(A - \sigma B)$ this requirement gives

$$W_k^* W_k = V_k^*(A - \sigma B)^*(A - \sigma B)V_k = V_k^* C V_k = I \tag{7}$$

and we call V_k a C-orthonormal matrix.

The new search direction $\tilde{v}_k$ must be C-orthonormal to V_{k-1}, which implies that

$$V_{k-1}^* v_k = 0 \text{ and } \tilde{v}_k = \frac{v_k}{\|v_k\|_C} = \frac{v_k}{\|w_k\|_2}, \tag{8}$$

where $w_k = (A - \sigma B)v_k$.

To move from standard to harmonic Ritz values, the adjustments in the algorithm are not radical. In comparison to the original implementation, the harmonic case requires two extra matrix-vector multiplications and in addition extra memory to store an N_t-by-k matrix. The main difference is that the LU decomposition of $A - \sigma B$ is used as a preconditioner and not as a shift and invert technique.

3 Numerical Results

In this section, we show some results obtained on both an 80 processor Cray T3E situated at the HPαC centre in Delft, The Netherlands and a 512 processor Cray T3E at Cray Research, Eagan, MN, USA. The local memory per processor is at least 128 Mbytes. On these machines, the best results were obtained by a `MESSAGE PASSING` implementation using Cray intrinsic `SHMEM` routines for data transfer and communication. For more details, we refer to [2].

3.1 Problems

We have timed five MHD problems of the form (1). The *Alfvén* spectra of Problems **1**, **2** and **3**, on the one hand, and Problems **4** and **5**, on the other hand, do not correspond because different MHD equilibria have been used. For more details we refer to CASTOR [1]. The choices of the acceptance criteria will be explained in the next section.

1 A small problem of $N = 64$ diagonal blocks of size $n = 48$. We look for eigenvalues in the neighbourhood of $\sigma = (-0.08, 0.60)$, and stop after 10 eigenpairs have been found with $tol_{sJD} = 10^{-8}$ and $tol_{hJD} = 10^{-6}$. The experiments have been performed on $p = 8$ processors.

2 The size of this problem is four times as big as that of the previous problem; $N = 128$ and $n = 96$. Again, we look for eigenvalues in the neighbourhood of $\sigma = (-0.08, 0.60)$, and stop after 10 eigenpairs have been found with $tol_{sJD} = 10^{-8}$ and $tol_{hJD} = 10^{-6}$. The experiments have been performed on $p = 8$ processors.

3 The same as Problem **2**, but performed on $p = 32$ processors.

4 The size of this large problem is: $N = 256$ and $n = 256$. We took $\sigma = (-0.15, 0.15)$ and look for $N_{ev} = 12$ eigenpairs with $tol_{sJD} = 10^{-8}$ and $tol_{hJD} = 10^{-5}$. The experiments are performed on $p = 128$ processors.

5 The size of this very large problem is: $N = 4096$ and $n = 64$, we took $\sigma = (-0.10, 0.23)$ leading to another branch in the *Alfvén spectrum.* Now, we look for $N_{ev} = 20$ eigenpairs with $tol_{sJD} = 10^{-8}$ and $\tilde{tol}_{hJD} = 10^{-5}$. For this problem a slightly different acceptance criterion has been applied:

$$\|r\|_2 < \tilde{tol}_{hJD}.|\sigma + \frac{1}{\nu}|.\|u\|_2. \tag{9}$$

For the harmonic case, the 2-norm of u can be very large, about 10^6, so the results can be compared with $tol_{hJD} = 10^{-6}$. At present, we prefer to control the residue as described in Section 3.2. Figure 1 shows the distribution of 20 eigenvalues in the neighbourhood of $\sigma = (-0.10, 0.23)$.

3.2 Acceptance Criterion

For the standard approach we accept an eigenpair $(\sigma + \frac{1}{\nu}, u)$ if the residual vector satisfies:

$$\|r\|_2 = \|(Q - \nu I)u\|_2 < tol_{sJD}.|\nu|, \text{ with } \|u\|_2 = 1 \tag{10}$$

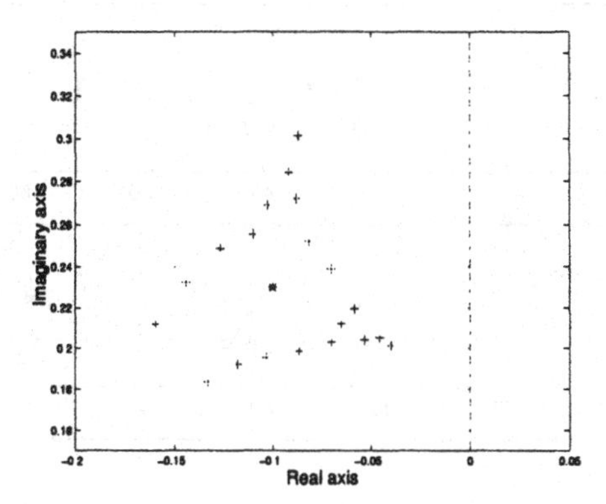

Fig. 1. The eigenvalue distribution of problem 5

and for the harmonic approach we require:

$$\|r\|_2 = \|(A - (\sigma + \frac{1}{\nu}))Bu\|_2 < tol_{hJD}.|\sigma + \frac{1}{\nu}|, \text{ with } \|u\|_C = 1. \tag{11}$$

To compare both eigenvalue solvers it is not advisable to choose the tolerance parameters tol_{sJD} equal to tol_{hJD} in (10) and (11), respectively. There are two reasons to take different values: firstly, within the same number of iterations the standard approach will result into more eigenpair solutions that satisfy (10) than into solutions that satisfy (11). Secondly, if we compute for each *accepted* eigenpair (λ, u) the true normalised residue γ defined by

$$\gamma := \frac{\|(A - \lambda B)u\|_2}{|\lambda|.\|u\|_2}, \tag{12}$$

then we see that the harmonic approach leads to much smaller γ values.

In Figure 2, the convergence behaviour of both the standard and harmonic approach is displayed, with and without restarts. A $\circ$ indicates that the eigenpair satisfies (10) or (11), a $\times$ denotes the γ value. We observe that the accuracy for the eigenpairs achieved by means of harmonic Ritz values is better than suggested by tol_{hJD}. On the other hand, tol_{sJD} seems to be too optimistic about the accuracy compared to the γ values shown in Figure 2. In our experiments we took $tol_{sJD} = 10^{-8}$ and $tol_{hJD} = 10^{-6}$ and $tol_{hJD} = 10^{-5}$. It is not yet clear to us how these parameters depend on the problem size or the choice of the target.

3.3 Restarting Strategy

The algorithm has two parameters that control the size of the projected system: k_{min} and m. During each restart, the k_{min} eigenvalues with maximal norm and not included in the set of accepted eigenvalues, that correspond to the k_{min} most promising search directions are maintained. Moreover, since an implicit deflation technique is applied in our implementation, the n_{ev} eigenpairs found so far are

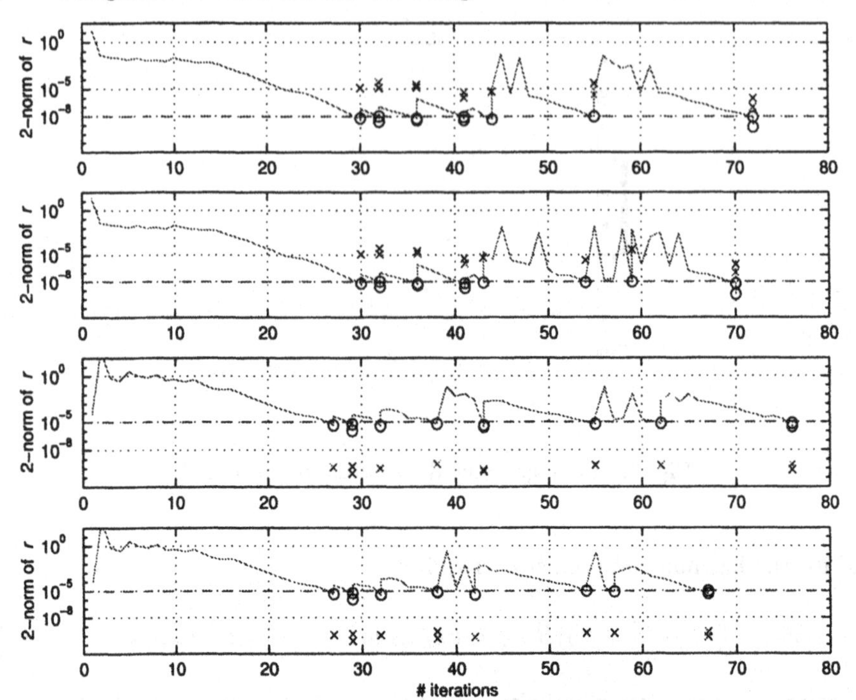

Fig. 2. The two upper plots result on problem 4 using standard Ritz values, the lower two on the same problem but using harmonic Ritz values. The first and third one show the convergence behaviour of Jacobi-Davidson restarting each time when the size of the projected system reaches $m = 37$, where $k_{min} = 25$ and $k_{min} = 20$, respectively. The second and fourth plots demonstrate the convergence in case of no restarts. The process ended when $N_{ev} = 12$ eigenvalues were found. It may happen that two eigenvalues are found within the same iteration step.

kept in the system too. The maximum size m should be larger than $k_{min} + N_{ev}$, where N_{ev} denotes the number of eigenvalues we are looking for. The influence of several (k_{min}, m) parameter combinations on both the parallel performance and convergence behaviour is studied.

3.4 Timing Results of (k_{min}, m) Parameter Combinations

For each experiment we take m constant and for k_{min} we choose the values $5, 10, \cdots, m - N_{ev}$. In Figures 4, 5, 6 and 7, the results of a single m value have been connected by a dashed or dotted line. Experiments with several m values have been performed. In the plots we only show the most interesting m values; m reaches its maximum if N_{ev} eigenpairs were found without using a restart. In the pictures this is indicated by a solid horizontal line, which is of course independent of k_{min}. If the number of iterations equals 80 and besides less than N_{ev} eigenpairs have been found, we consider the result as negative. This implies that, although the execution time is low, this experiment cannot be a candidate for the best (k_{min}, m) combination.

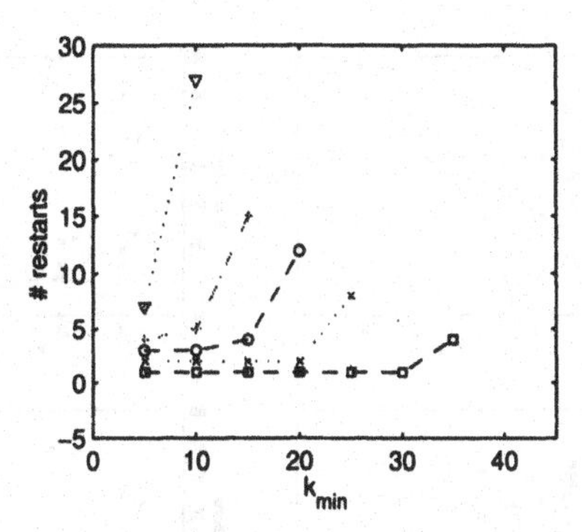

Fig. 3. The number of restarts needed to compute N_{ev} eigenvalues of Problem **2**. Results are shown for different m values: $m = 20$ ($\triangledown \cdots$), $m = 25$ ($+ - \cdot$ line), $m = 30$ ($\circ - -$ line), $m = 35$ ($\times \cdots$ line), $m = 40$ ($\triangleright - \cdot$ line), $m = 45$ ($\square - -$ line).

Before we describe the experiments illustrated by Figures 4, 5, 6 and 7 we make some general remarks:

- We observed that if a (k_{min}, m) parameter combination is optimal on p processors, it is optimal on q processors too, with $p \neq q$.
- For k_{min} small, for instance $k_{min} = 5$ or 10, probably too much information is thrown away, leading to a considerable increase of iteration steps.
- For k_{min} large the number of restarts will be large at the end of the process; suppose that in the extreme case, $k_{min} = m - N_{ev}$, already $N_{ev} - 1$ eigenpairs have been found, then after a restart k becomes $k_{min} + N_{ev} - 1 = m - 1$. In other words, each step will require a restart. In Figure 3, the number of restarts is displayed corresponding to the results of Problem **2** obtained with harmonic Ritz values.
- The number of iterations is almost independent of the number of processors involved; it may happen that an increase of the number of processors causes a decrease by one or two iterations under the same conditions, because the LU decomposition becomes more accurate if the number of cyclic reduction steps increases at the cost of the domain decomposition part.

The first example (Figure 4) explicitly shows that the restarting technique can help to reduce the wall clock time for both the standard and harmonic method. The minimum number of iterations to compute 10 eigenvalues in the neighbourhood of σ is achieved in case of no restarts, viz, 53 for the standard case, 51 for the harmonic case. The least time to compute 10 eigenvalues is attained for $k_{min} = 15$ and $m = 30, 35$, but also for $k_{min} = 10$ and $m = 30, 35$ and $m = 40$ and $k_{min} = 15, 20, 25$ leads to a reduction in wall clock time of about 15 %. The harmonic approach leads to comparable results: for $(k_{min}, m) = (15, 30 : 35)$, but also $(k_{min}, m) = (10, 30 : 35)$ and $(k_{min}, m) = (15 : 25, 40)$ a reasonable reduction in time is achieved. The score for $k_{min} = 5$ in combination with $m = 35$ is striking, the unexpected small number of iterations in combination with a small k_{min} results into a fast time.

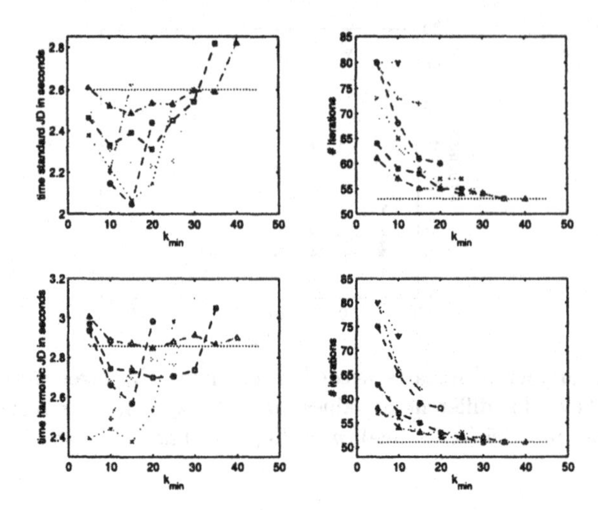

Fig. 4. The upper pictures result on problem 1 using standard Ritz values. The lower pictures result on the same problem with harmonic Ritz values. Results are shown for different m values: $m = 20$ ($\nabla \cdots$), $m = 25$ ($+ - \cdot$ line), $m = 30$ ($\circ - -$ line), $m = 35$ ($\times \cdots$ line), $m = 40$ ($\triangleright - \cdot$ line), $m = 45$ ($\square - -$ line), $m = 50$ ($\triangle - \cdot$ line). The solid lines give the value for no restart.

The plots in Figure 5 with the timing results for the Jacobi-Davidson process for Problem **2** give a totally different view. There is no doubt of benefit from restarting, although the numbers of iterations pretty well correspond with those of Problem **1**. This can be explained as follows: the size of the projected system k is proportionally much smaller compared to N_t/p than in case of Problem **1**; both the block size and the number of diagonal blocks is twice as big. For Problem **1** the sequential part amounts 45% and 36% of the total wall clock time, respectively, for the standard and harmonic Ritz values. For Problem **2** these values are 10.5% and 8%, respectively. These percentages hold for the most expensive sequential case of no restarts. The increase of JD iterations due to several restarts can not be compensated by a reduction of serial time by keeping the projected system small.

When we increase the number of active processors by a factor 4, as is done in Problem **4** (see Figure 6), we observe that again a reduction in wall clock time can be achieved by using a well-chosen (k_{min}, m) combination. The number of iterations slightly differ from those given in Figure 5, but the pictures with the Jacobi-Davidson times look similar to those in Figure 5. If we should have enlarged N by a factor of 4 and left the block size unchanged, we may expect execution times as in Figure 5.

For Problem **4**, the limit of 80 iterations seems to be very critical. The right-hand plots of Figure 7 demonstrate that the number of iterations does not decrease monotonically when k_{min} increases for a fixed value m as holds for the previous problems. Moreover, it may happen that for some (k_{min}, m) combination, the limit of JD iterations is too strictly, while for both a smaller and larger

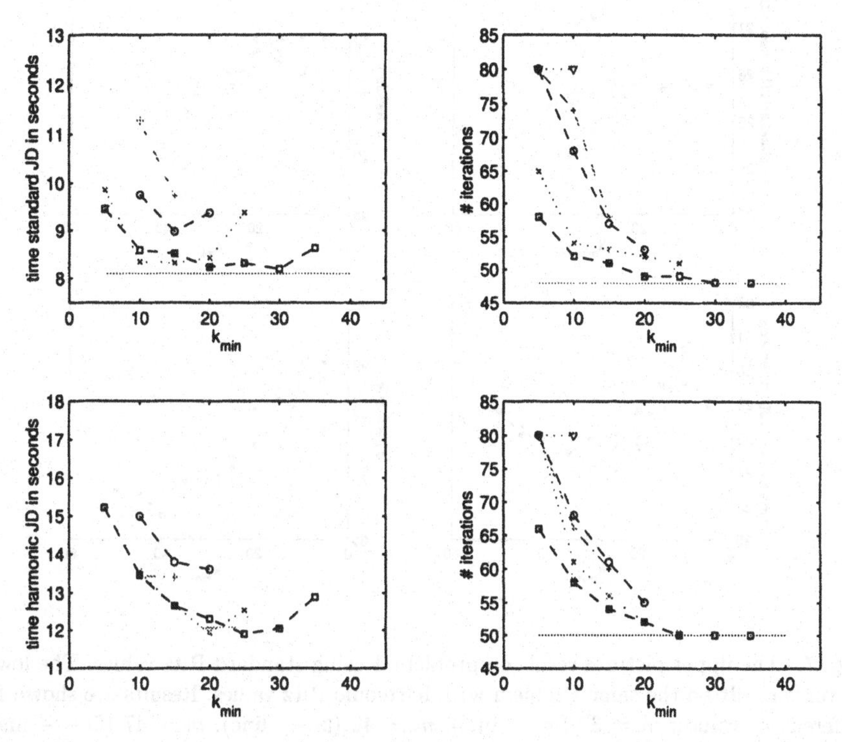

Fig. 5. The upper pictures result on problem **2** using standard Ritz values. The lower pictures result on the same problem with harmonic Ritz values. Results are shown for different m values: $m = 20$ ($\nabla \cdots$), $m = 25$ ($+ - \cdot$ line), $m = 30$ ($\circ - -$ line), $m = 35$ ($\times \cdots$ line), $m = 40$ ($\triangleright - \cdot$ line), $m = 45$ ($\square - -$ line). The solid lines give the value for no restart.

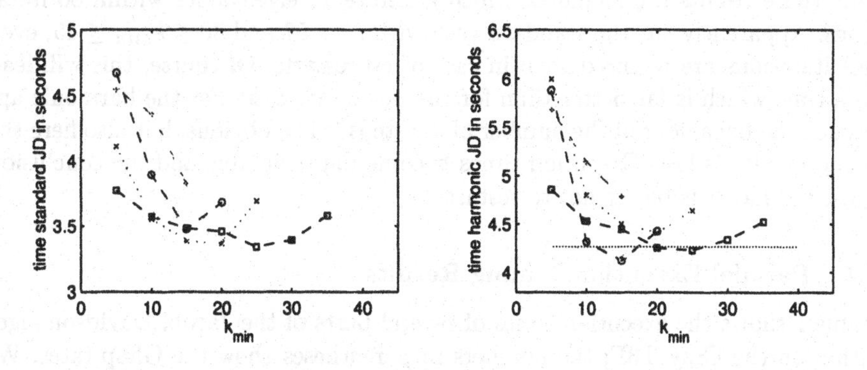

Fig. 6. The left pictures results on problem **3** using standard Ritz values. The right pictures result on the same problem with harmonic Ritz values. Results are shown for different m values: $m = 20$ ($\nabla \cdots$), $m = 25$ ($+ - \cdot$ line), $m = 30$ ($\circ - -$ line), $m = 35$ ($\times \cdots$ line), $m = 40$ ($\triangleright - \cdot$ line), $m = 45$ ($\square - -$ line). The solid lines give the value for no restart.

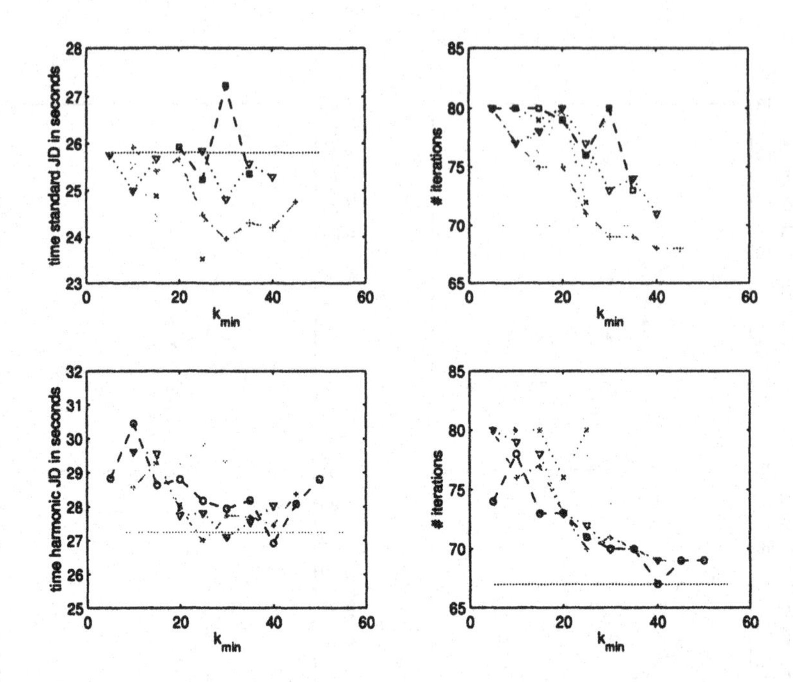

Fig. 7. The upper pictures result on problem 4 using standard Ritz values. The lower pictures result on the same Problem with harmonic Ritz values. Results are shown for different m values: $m = 37$ (× ··· line), $m = 42$ (▷ – · line), $m = 47$ (□ – – line), $m = 52$ (▽ – · line), $m = 57$ (+ – · line), $m = 62$ (○ – – line). The solid lines give the value for no restart.

k_{min} value the desired N_{ev} eigenpairs were easily found. In the left-hand plots only those results are included, which generate 12 eigenvalues within 80 iterations. Apparently, for the standard case with $m = 57$ and $30 \leq k_{min} \leq 45$, even less iterations are required than in case of no restarts. Of course, this will lead to a time which is far better than for the no-restart case. For the harmonic approach the behaviour of the number of JD steps is less obvious, but also here the monotonicity is lost. Execution times become unpredictable and the conclusion must be that it is better not to restart.

3.5 Parallel Execution Timing Results

Table 1 shows the execution times of several parts of the Jacobi-Davidson algorithm on the Cray T3E; the numbers in parentheses show the Gflop-rates. We took

$$N_{ev} = 20;\ tol_{sJD} = 10^{-8};\ tol_{hJD} = 10^{-5};\ k_{min} = 10;\ m = 30{+}N_{ev};\ it_{SOL} = 0.$$

The number of eigenvalues found slightly depends on the number of processors involved: about 11 for the standard and 13 for the harmonic approach within 80 iterations.

Table 1. Wall clock times in seconds for the standard and harmonic Ritz approach. $N = 4096$, $n = 64$.

p	Preprocessing	Time *standard* JD	Time *harmonic* JD	Triangular solves
32	7.90 (6.75)	64.59	88.61	25.56 (2.08)
64	4.08 (13.21)	31.70	43.78	13.28 (4.02)
128	2.19 (24.78)	15.07	21.33	7.28 (7.36)
256	1.27 (42.69)	8.55	11.48	4.36 (12.29)
512	0.84 (64.65)	5.64	7.02	3.01 (17.81)

The construction of L and U is a very time-consuming part of the algorithm. However, with a well-chosen target σ ten up to twenty eigenvalues can be found within 80 iterations. Hence, the life-time of a (L, U) pair is about 80 iterations. On account of the cyclic reduction part of the LU factorisation, a process that starts on all processors, while at each step half of the active processors becomes idle, we may not expect linear speed-up. The fact that the parallel performance of DDCR is quite good is caused by the domain decomposition part of the LU. For more details we refer to [2, 5].

About 40% of the execution time is spent by the computation of the LU factorisation (in Table 1 'Preprocessing'), which does not depend on the number of processors. The storage demands for Problem **5** are so large that at least the memories of 32 processors are necessary. DDCR is an order $\mathcal{O}(Nn^3)$ process performed by Level 3 BLAS and it needs less communication: only sub- and super diagonal blocks of size n-by-n must be transfered. As a consequence, for the construction of L and U, the communication time can be neglected also due to the fast communication between processors on the Cray T3E. The Gflop-rates attained for the construction of the LU are impressively high just like its parallel speed-up.

The application of L and U, consisting of two triangular solves, is the most expensive component of the JD process after preprocessing. It parallelises well, but its speed is much lower, because it is built up of Level 2 BLAS operations. The wall clock times for *standard* and *harmonic* JD are given including the time spent on the triangular solves. Obviously, a harmonic iteration step is more expensive than a standard step, but the overhead becomes less when more processors are used, because the extra operations parallelise very well.

4 Conclusions

We have examined the convergence behaviour of two Jacobi-Davidson variants, one using standard Ritz values, the other one harmonic Ritz values. For the kind of eigenvalue problems we are interested in, arising from Magneto-Hydrodynamics, both methods converge very fast and parallelise pretty well. With $tol_{sJD} = 10^{-8}$ and $tol_{hJD} = 10^{-5}$ in the acceptance criteria (10) and

(11), respectively, both variants give about the same amount of eigenpairs. The harmonic variant is about 20% more expensive, but results in more accurate eigenpairs. With a well-chosen target ten up to twenty eigenvalues can be found. Even for very large problems, $N_t = 65,536$ and $N_t = 262,144$, we obtain more than 10 sufficiently accurate eigenpairs in a few seconds.

Special attention has been paid to a restarting technique. The (k_{min}, m) parameter combination prescribes the amount of information that remains in the system after a restart and the maximum size of the projected system. In this paper we have demonstrated that k_{min} may not be too small. because then too much information gets lost. On the other hand, too large k_{min} values lead to many restarts and become expensive in execution time. In general, the number of iterations decreases when m increases. It depends on the N_t/p value, as we have shown, whether restarts lead to a reduction in the wall clock time for the Jacobi-Davidson process.

Acknowledgements

The authors wish to thank Herman te Riele for many stimulating discussions and suggestions for improving the presentation of the paper. They gratefully acknowledge HPαC (Delft, The Netherlands) for their technical support, and Cray Research for a sponsored account on the Cray T3E (Eagan, MN, USA), and the Dutch National Computing Facilities Foundation NCF for the provision of computer time on the Cray C90 and the Cray T3E.

References

1. W. Kerner, S. Poedts, J.P. Goedbloed, G.T.A. Huysmans, B. Keegan, and E. Schwartz. Computing the damping and destabilization of global Alfvén waves in tokamaks. In P. Bachman and D.C. Robinson, editors, *Proceedings of 18th Conference on Controlled Fusion and Plasma Physics*. EPS: Berlin, 1991. IV.89-IV.92.
2. Margreet Nool and Auke van der Ploeg. A Parallel Jacobi-Davidson Method for solving Generalized Eigenvalue Problems in linear Magnetohydrodynamics. Technical Report NM-R9733, CWI, Amsterdam, December 1997.
3. G.L.G. Sleijpen, J.G.L. Booten, D.R. Fokkema, and H.A. van der Vorst. Jacobi-Davidson Type Methods for Generalized Eigenproblems and Polynomial Eigenproblems. *BIT*, 36:595–633, 1996.
4. G.L.G Sleijpen and H.A. van der Vorst. A Jacobi-Davidson iteration method for linear eigenvalue problems. *SIAM J. Matrix Anal. Appl.*, 17(2):401–425, april 1996.
5. A. van der Ploeg. Reordering Strategies and LU-decomposition of Block Tridiagonal Matrices for Parallel Processing. Technical Report NM-R9618, CWI, Amsterdam, October 1996.

Parallel Preconditioned Solvers for Large Sparse Hermitian Eigenproblems

Achim Basermann

C&C Research Laboratories, NEC Europe Ltd.
Rathausallee 10, 53757 Sankt Augustin, Germany
basermann@ccrl-nece.technopark.gmd.de
http://www.ccrl-nece.technopark.gmd.de/~baserman/

Abstract. Parallel preconditioned solvers are presented to compute a few extreme eigenvalues and -vectors of large sparse Hermitian matrices based on the Jacobi-Davidson (JD) method by G.L.G. Sleijpen and H.A. van der Vorst. For preconditioning, an adaptive approach is applied using the QMR (Quasi-Minimal Residual) iteration. Special QMR versions have been developed for the real symmetric and the complex Hermitian case. To parallelise the solvers, matrix and vector partitioning is investigated with a data distribution and a communication scheme exploiting the sparsity of the matrix. Synchronization overhead is reduced by grouping inner products and norm computations within the QMR and the JD iteration. The efficiency of these strategies is demonstrated on the massively parallel systems NEC Cenju-3 and Cray T3E.

1 Introduction

The simulation of quantum chemistry and structural mechanics problems is a source of computationally challenging, large sparse real symmetric or complex Hermitian eigenvalue problems. For the solution of such problems, parallel preconditioned solvers are presented to determine a few eigenvalues and -vectors based on the Jacobi-Davidson (JD) method [9].

For preconditioning, an adaptive approach using the QMR (Quasi-Minimal Residual) iteration [2, 5, 7] is applied, i.e., the preconditioning system of linear equations within the JD iteration is solved iteratively and adaptively by checking the residual norm within the QMR iteration [3, 4]. Special QMR versions have been developed for the real symmetric and the complex Hermitian case.

The matrices A considered are *generalised sparse*, i.e., the computation of a matrix-vector multiplication $A \cdot v$ takes considerably less than n^2 operations. This covers ordinary sparse matrices as well as dense matrices from quantum chemistry built up additively from a diagonal matrix, a few outer products, and an FFT. In order to exploit the advantages of such structures with respect to operational complexity and memory requirements when solving systems of linear equations or eigenvalue problems, it is natural to apply iterative methods.

To parallelise the solvers, matrix and vector partitioning is investigated with a data distribution and a communication scheme exploiting the sparsity of the

J. Palma, J. Dongarra, and V. Hernández (Eds.): VECPAR'98, LNCS 1573, pp. 71–84, 1999.

matrix. Synchronization overhead is reduced by grouping inner products and norm computations within the QMR and the JD iteration. Moreover, in the complex Hermitian case, communication coupling of QMR's two independent matrix-vector multiplications is investigated.

2 Jacobi-Davidson Method

To solve large sparse Hermitian eigenvalue problems numerically, variants of a method proposed by Davidson [8] are frequently applied. These solvers use a succession of subspaces where the update of the subspace exploits approximate inverses of the problem matrix, A. For A, $A = A^H$ or $A^* = A^T$ holds where A^* denotes A with complex conjugate elements and $A^H = (A^T)^*$ (transposed and complex conjugate).

The basic idea is: Let $\mathbf{V}^k$ be a subspace of $\mathbb{R}^n$ with an orthonormal basis $w_1^k, \ldots, w_m^k$ and W the matrix with columns w_j^k, $S := W^H A W$, $\bar{\lambda}_j^k$ the eigenvalues of S, and T a matrix with the eigenvectors of S as columns. The columns x_j^k of WT are approximations to eigenvectors of A with Ritz values $\bar{\lambda}_j^k = (x_j^k)^H A x_j^k$ that approximate eigenvalues of A. Let us assume that $\bar{\lambda}_{j_s}^k, \ldots, \bar{\lambda}_{j_{s+l-1}}^k \in [\lambda_{\text{lower}}, \lambda_{\text{upper}}]$. For $j \in j_s, \ldots, j_{s+l-1}$ define

$$q_j^k = (A - \bar{\lambda}_j^k I)\, x_j^k\,, \qquad r_j^k = (\bar{A} - \bar{\lambda}_j^k I)^{-1} q_j^k\,, \tag{1}$$

and $\mathbf{V}^{k+1} = \text{span}(\mathbf{V}^k \cup r_{j_s}^k \cup \ldots \cup r_{j_{s+l-1}}^k)$ where $\bar{A}$ is an easy to invert approximation to A ($\bar{A} = \text{diag}(A)$ in [8]). Then $\mathbf{V}^{k+1}$ is an $(m+l)$-dimensional subspace of $\mathbb{R}^n$, and the repetition of the procedure above gives in general improved approximations to eigenvalues and -vectors. Restarting may increase efficiency.

For good convergence, $\mathbf{V}^k$ has to contain crude approximations to all eigenvectors of A with eigenvalues smaller than λ_{lower} [8]. The approximate inverse must not be too accurate, otherwise the method stalls. The reason for this was investigated in [9] and leads to the Jacobi-Davidson (JD) method with an improved definition of r_j^k:

$$[(I - x_j^k\,(x_j^k)^H)\,(\bar{A} - \bar{\lambda}_j^k I)\,(I - x_j^k\,(x_j^k)^H)]\, r_j^k = q_j^k\,. \tag{2}$$

The projection $(I - x_j^k\,(x_j^k)^H)$ in (2) is not easy to incorporate into the matrix, but there is no need to do so, and solving (2) is only slightly more expensive than solving (1).

The method converges quadratically for $\bar{A} = A$.

3 Preconditioning

The character of the JD method is determined by the approximation $\bar{A}$ to A. For obtaining an approximate solution of the preconditioning system (2), we may try an iterative approach [3, 4, 9]. Here, a real symmetric or a complex Hermitian version of the QMR algorithm are used [2, 5, 7] that are directly applied to the projected system (2) with $\bar{A} = A$. The control of the QMR iteration is as follows. Iteration is stopped when the current residual norm is smaller than the residual norm of QMR in the previous inner JD iteration. By controlling the QMR residual norms, we achieve that the preconditioning system (2) is solved in low accuracy in the beginning and in increasing accuracy in the course of the JD iteration. For a block version of JD, the residual norms of each preconditioning system (2) are separately controlled for each eigenvector to approximate since some eigenvector approximations are more difficult to obtain than others. This adapts the control to the properties of the matrix's spectrum.

Algorithm 1 shows the QMR iteration used to precondition JD for complex Hermitian matrices. The method is derived from the QMR variant described in [5]. Within JD, the matrix B in Algorithm 1 corresponds to the matrix $[(I - x_j^k\,(x_j^k)^H)\,(A - \bar{\lambda}_j^k I)\,(I - x_j^k\,(x_j^k)^H)]$ of the preconditioning system (2).

Per QMR iteration, two matrix-vector operations with B and B^* (marked by frames in Algorithm 1) are performed since QMR bases on the non-Hermitian Lanczos algorithm that requires operations with B and $B^T = B^*$ but not with B^H [7]. For real symmetric problems, only one matrix-vector operation per QMR iteration is necessary since then $q^i = Bp^i$ and thus $v^{i+1} = q^i - (\tau^i/\gamma^i)v^i$ hold. The only matrix-vector multiplication to compute per iteration is then Bw^{i+1}.

Naturally, B is not computed element-wise from $[(I - x_j^k\,(x_j^k)^H)\,(A - \bar{\lambda}_j^k I)\,(I - x_j^k\,(x_j^k)^H)]$; the operation Bp^i, e.g., is splitted into vector-vector operations and one matrix-vector operation with A.

Note that the framed matrix-vector operations in the complex Hermitian QMR iteration are independent from each other. This can be exploited for a parallel implementation (see 5.2). Moreover, all vector reductions in Algorithm 1 (marked by bullets) are grouped. This in addition makes the QMR variant well suited for a parallel implementation (see 5.3).

4 Storage Scheme

Efficient storage schemes for large sparse matrices depend on the sparsity pattern of the matrix, the considered algorithm, and the architecture of the computer system used [1]. Here, the CRS format (Compressed Row Storage) is applied. This format is often used in FE programs and is suited for matrices with regular as well as irregular structure. The principle of the scheme is illustrated in Fig. 1 for a matrix A with non-zeros $a_{i,j}$.

$$A = \begin{pmatrix} a_{1,1} & 0 & 0 & 0 & 0 & 0 & 0 & 0 \\ 0 & a_{2,2} & a_{2,3} & 0 & 0 & 0 & 0 & 0 \\ 0 & a_{3,2} & a_{3,3} & a_{3,4} & 0 & 0 & 0 & 0 \\ 0 & 0 & a_{4,3} & a_{4,4} & a_{4,5} & a_{4,6} & a_{4,7} & a_{4,8} \\ 0 & 0 & 0 & a_{5,4} & a_{5,5} & 0 & a_{5,7} & 0 \\ 0 & 0 & 0 & a_{6,4} & 0 & a_{6,6} & a_{6,7} & 0 \\ 0 & 0 & 0 & a_{7,4} & a_{7,5} & a_{7,6} & a_{7,7} & 0 \\ 0 & 0 & 0 & a_{8,4} & 0 & 0 & 0 & a_{8,8} \end{pmatrix}$$

Algorithm 1. Complex Hermitian QMR

$p^0 = q^0 = d^0 = s^0 = 0,\ \nu^1 = 1,\ \kappa^0 = -1,\ w^1 = v^1 = r^0 = b - Bx^0$
$\gamma^1 = \|v^1\|,\ \xi^1 = \gamma^1,\ \rho^1 = (w^1)^T v^1,\ \epsilon^1 = (B^* w^1)^T v^1,\ \mu^1 = 0, \tau^1 = \frac{\epsilon^1}{\rho^1}$

$i = 1, 2, \ldots$

$$p^i = \frac{1}{\gamma^i} v^i - \mu^i p^{i-1}$$

$$q^i = \frac{1}{\xi^i} B^* w^i - \frac{\gamma^i \mu^i}{\xi^i} q^{i-1}$$

$$v^{i+1} = \boxed{Bp^i} - \frac{\tau^i}{\gamma^i} v^i$$

$$w^{i+1} = q^i - \frac{\tau^i}{\xi^i} w^i$$

- *if* ($\|r^{i-1}\| <$ *tolerance*) *then STOP*
- $\gamma^{i+1} = \|v^{i+1}\|$
- $\xi^{i+1} = \|w^{i+1}\|$
- $\rho^{i+1} = (w^{i+1})^T v^{i+1}$
- $\epsilon^{i+1} = (\boxed{B^* w^{i+1}})^T v^{i+1}$

$$\mu^{i+1} = \frac{\gamma^i \xi^i \rho^{i+1}}{\gamma^{i+1} \tau^i \rho^i}$$

$$\tau^{i+1} = \frac{\epsilon^{i+1}}{\rho^{i+1}} - \gamma^{i+1} \mu^{i+1}$$

$$\theta^i = \frac{|\tau^i|^2 (1 - \nu^i)}{\nu^i |\tau^i|^2 + |\gamma^{i+1}|^2}$$

$$\kappa^i = \frac{-\gamma^i (\tau^i)^* \kappa^{i-1}}{\nu^i |\tau^i|^2 + |\gamma^{i+1}|^2}$$

$$\nu^{i+1} = \frac{\nu^i |\tau^i|^2}{\nu^i |\tau^i|^2 + |\gamma^{i+1}|^2}$$

$$d^i = \theta^i d^{i-1} + \kappa^i p^i$$

$$s^i = \theta^i s^{i-1} + \kappa^i B p^i$$

$$x^i = x^{i-1} + d^i$$

$$r^i = r^{i-1} - s^i$$

value:

$a_{1,1}$	$a_{2,3}$	$a_{2,2}$	$a_{3,4}$	$a_{3,2}$	$a_{3,3}$	$a_{4,3}$	$a_{4,4}$	$a_{4,8}$	$a_{4,6}$	$a_{4,7}$	$a_{4,5}$
1	2	3	4	5	6	7	8	9	10	11	12
$a_{5,4}$	$a_{5,5}$	$a_{5,7}$	$a_{6,7}$	$a_{6,4}$	$a_{6,6}$	$a_{7,4}$	$a_{7,5}$	$a_{7,7}$	$a_{7,6}$	$a_{8,4}$	$a_{8,8}$
13	14	15	16	17	18	19	20	21	22	23	24

col_ind:

1	3	2	4	2	3	3	4	8	6	7	5
1	2	3	4	5	6	7	8	9	10	11	12
4	5	7	7	4	6	4	5	7	6	4	8
13	14	15	16	17	18	19	20	21	22	23	24

row_ptr:

1	2	4	7	13	16	19	23	25

Fig. 1. CRS storage scheme

The non-zeros of matrix A are stored row-wise in three one-dimensional arrays. `value` contains the values of the non-zeros, `col_ind` the corresponding column indices. The elements of `row_ptr` point to the position of the beginning of each row in `value` and `col_ind`.

5 Parallelisation Strategies

5.1 Data Distribution

The data distribution scheme considered here balances both matrix-vector and vector-vector operations for irregularly structured sparse matrices on distributed memory systems (see also [2]). The scheme results in a row-wise distribution of the matrix arrays `value` and `col_ind` (see 4); the rows of each processor succeed one another. The distribution of the vector arrays corresponds component-wise to the row distribution of the matrix arrays. In the following, n_k denotes the number of rows of processor k, $k = 0, \ldots, p-1$; n is the total number. g_k is the index of the first row of processor k, and z_i is the number of non-zeros of row i. For these quantities, the following equations hold: $n = \sum_{k=0}^{p-1} n_k$ and $g_k = 1 + \sum_{i=0}^{k-1} n_i$.

In each iteration of an iterative method like JD or QMR, s sparse matrix-vector multiplications and c vector-vector operations are performed. Scalar operations are neglected here. With the data distribution considered, the load generated by row i is proportional to

$$l_i = z_i \cdot s \cdot \zeta + c .$$

The parameter ζ is hardware dependent since it considers the ratio of the costs for a regular vector-vector operation and an irregular matrix-vector operation. However, different matrix patterns could result in different memory access costs, e.g., different caching behaviour. Therefore, the parameter ζ is determined at run-time by timings for a row block of the current matrix within the symmetric or Hermitian QMR solver used. The measurement is performed once on one processor with a predefined number of QMR iterations before the data are distributed. With approximating ζ at run-time for the current matrix, the slight dependence of ζ on the matrix pattern is considered in addition.

For computational load balance, each processor has to perform the p-th fraction of the total number of operations. Hence, the rows of the matrix and the vector components are distributed according to (3).

$$n_k = \begin{cases} \min\limits_{1 \le t \le n - g_k + 1} \left\{ t \,\middle|\, \sum_{i=1}^{t} l_{i+g_k-1} \ge \frac{1}{p} \sum_{i=1}^{n} l_i \right\} & \text{for } k = 0, 1, \ldots, q \\ n - \sum_{i=0}^{q} n_i & \text{for } k = q+1 \\ 0 & \text{for } k = q+2, \ldots, p-1 \end{cases} \tag{3}$$

For large sparse matrices and $p \ll n$, usually $q = p - 1$ or $q + 1 = p - 1$ hold. It should be noted that for $\zeta \to 0$ each processor gets nearly the same number of rows and for $\zeta \to \infty$ nearly the same number of non-zeros.

Fig. 2 illustrates the distribution of `col_ind` from Fig. 1 as well as the distribution of the vectors x and y of the matrix-vector multiplication $y = Ax$ to four processors for $\zeta = 5$, $s = 2$, and $c = 13$.

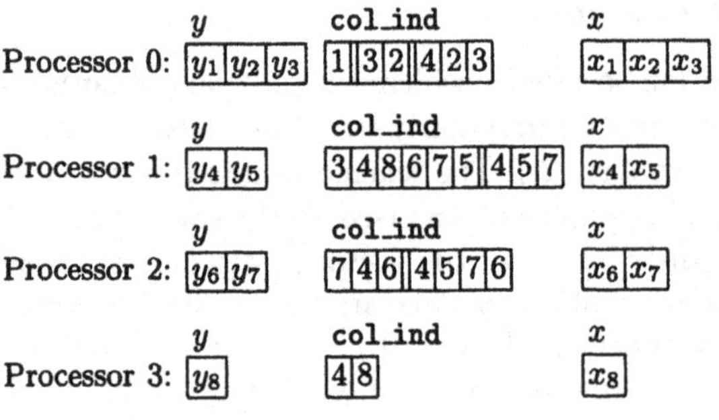

	y	col_ind	x
Processor 0:	y_1 y_2 y_3	1 | 3 2 | 4 2 3	x_1 x_2 x_3
Processor 1:	y_4 y_5	3 4 8 6 7 5 | 4 5 7	x_4 x_5
Processor 2:	y_6 y_7	7 4 6 | 4 5 7 6	x_6 x_7
Processor 3:	y_8	4 8	x_8

Fig. 2. Data distribution for $\zeta = 5$, $s = 2$, and $c = 13$

In case of an heterogeneous computing environment, e.g., workstation clusters with fast network connections or high-speed connected parallel computers, the data distribution criterion (3) can easily be adapted to different per processor performance or memory resources by predefining weights ω_k per processor k. Only the fraction $1/p$ in (3) has then to be replaced by $\omega_k / \sum_{i=0}^{p-1} \omega_i$.

5.2 Communication Scheme

On a distributed memory system, the computation of the matrix-vector multiplications requires communication because each processor owns only a partial vector. For the efficient computation of the matrix-vector multiplications, it is necessary to develop a suitable communication scheme (see also [2]). The goal of the scheme is to enable the overlapped execution of computations and data transfers to reduce waiting times based on a parallel matrix pattern analysis and, subsequently, a block rearranging of the matrix data.

First, the arrays `col_ind` (see 4 and 5.1) are analysed on each processor k to determine which elements result in access to non-local data. Then, the processors exchange information to decide which local data must be sent to which processors. If the matrix-vector multiplications are performed row-wise, components of the vector x of $y = Ax$ are communicated. After the analysis, `col_ind` and `value` are rearranged in such a way that the data that results in access to processor h is collected in block h. The elements of block h succeed one another row-wise with increasing column index per row. Block k is the first block in the arrays `col_ind` and `value` of processor k. Its elements result in access to local data; therefore, in the following, it is called the local block. The goal of this rearranging is to perform computation and communication overlapped. Fig. 3 shows the rearranging for the array `col_ind` of processor 1 from Fig. 2.

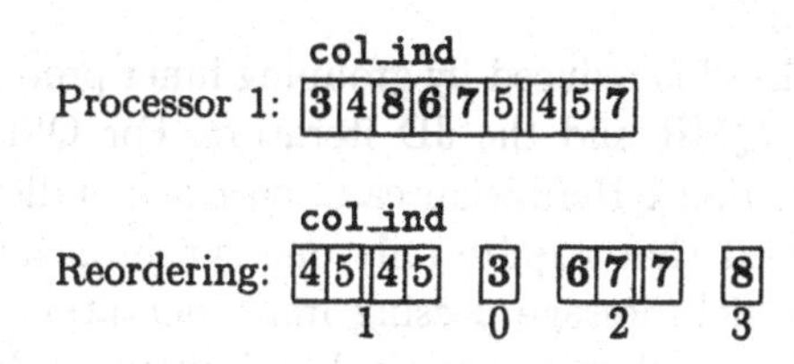

Fig. 3. Rearranging into blocks

The elements of block 1, the local block, result in access to the local components 4 and 5 of x during the row-wise matrix-vector multiplication, whereas operations with the elements of the blocks 0, 2, and 3 require communication with the processors 0, 2, and 3, respectively. For parallel matrix-vector multiplications, each processor first executes asynchronous receive-routines to receive necessary non-local data. Then all components of x that are needed on other processors are sent asynchronously. While the required data is on the network, each processor k performs operations with block k. After that, as soon as non-local data from processor h arrives, processor k continues the matrix-vector multiplication by accessing the elements of block h. This is repeated until the matrix-vector multiplication is complete. Computation and communication are performed overlapped so that waiting times are reduced.

The block structure of the matrix data and the data structures for communications have been optimised for both the real and the complex case to reduce memory requirements and to save unnecessary operations. In addition, cache exploitation is improved by these structures. All message buffers and the block row pointers of the matrix structure are stored in a modified compressed row format. Thus memory requirements per processor almost proportionally decrease with increasing processor number even if the number of messages per processor markedly rises due to a very irregular matrix pattern.

A parallel pre analysis phase to determine the sizes of all data structures precedes the detailed communication analysis and the matrix rearranging. This enables dynamic memory allocation and results in a further reduction of memory requirements since memory not needed any more, e.g., after the analysis phases, can be deallocated. Another advantage is that the same executable can be used for problems of any structure and size.

For complex Hermitian problems, two independent matrix-vector products with B and B^* have to be computed per QMR iteration (see the framed operations in Algorithm 1). Communications for both operations — they possess the same communication scheme — are coupled to reduce communication overhead and waiting times.

The data distribution and the communication scheme do not require any knowledge about a specific discretisation mesh; the schemes are determined automatically by the analysis of the indices of the non-zero matrix elements.

5.3 Synchronization

Synchronization overhead is reduced by grouping inner products and norm computations within the QMR and the JD iteration. For QMR in both the real symmetric and the complex Hermitian case, special parallel variants based on [5] have been developed that require only one synchronization point per iteration step. For a parallel message passing implementation of Algorithm 1, all local values of the vector reductions marked by bullets can be included into one global communication to determine the global values.

6 Results

All parts of the algorithms have been investigated with various application problems on the massively parallel systems NEC Cenju-3 with up to 128 processors (64 Mbytes main memory per processor) and Cray T3E with up to 512 processors (128 Mbytes main memory per processor). The codes have been written in FORTRAN 77 and C; MPI is used for message passing.

6.1 Numerical Test Cases

Numerical and performance tests of the JD implementation have been carried out with the large sparse real symmetric matrices **Episym1** to **Episym6** and the large sparse complex Hermitian matrices **Epiherm1** and **Epiherm2** stemming from the simulation of electron/phonon interaction [10], with the real symmetric matrices **Struct1** to **Struct3** from structural mechanics problems (finite element discretisation), and with the dense complex Hermitian matrix **Thinfilms** from the simulation of thin films with defects. The smaller real symmetric test matrices **Laplace**, **GregCar**, **CulWil**, and **RNet** originate from finite difference discretisation problems. Table 1 gives a survey of all matrices considered.

Table 1. Numerical data of the considered large sparse matrices

Matrix	Properties	Order	Number of non-zeros
Episym1	Real symmetric	98,800	966,254
Episym2	Real symmetric	126,126	1,823,812
Episym3	Real symmetric	342,200	3,394,614
Episym4	Real symmetric	1,009,008	14,770,746
Episym5	Real symmetric	5,513,508	81,477,386
Episym6	Real symmetric	11,639,628	172,688,506
Epiherm1	Complex Hermitian	126,126	1,823,812
Epiherm2	Complex Hermitian	1,009,008	14,770,746
Thinfilms	Complex Hermitian	1,413	1,996,569
Struct1	Real symmetric	835	13,317
Struct2	Real symmetric	2,839	299,991
Struct3	Real symmetric	25,222	3,856,386
Laplace	Real symmetric	900	7,744
GregCar	Real symmetric	1,000	2,998
CulWil	Real symmetric	1,000	3,996
RNet	Real symmetric	1,000	6,400

6.2 Effect of Preconditioning

For the following investigation, the JD iteration was stopped if the residual norms divided by the initial norms are less than 10^{-5}.

In Fig. 4, times for computing the four smallest eigenvalues and -vectors of the two real symmetric matrices **Episym2** and **Struct3** on 64 NEC Cenju-3 processors are compared for different preconditioners.

The best results are gained for JD with adaptive QMR preconditioning and a few preceding, diagonally preconditioned outer JD steps (4 or 1). Compared with pure diagonal preconditioning, the number of matrix-vector multiplications required decreases from 6,683 to 953 for the matrix **Episym2** from electron/phonon interaction. Note that the Lanczos algorithm used in the application code requires about double the number of matrix-vector multiplications as QMR preconditioned JD for this problem. For the matrix **Struct3** from structural mechanics, the diagonally preconditioned method did not converge in 100 minutes. Note that in this case the adaptive approach is markedly superior to preconditioning with a fixed number of 10 QMR iterations; the number of matrix-vector multiplications decreases from 55,422 to 11,743.

6.3 JD versus Lanczos Method

In Table 2, the sequential execution times on an SGI O^2 workstation (128 MHz, 128 Mbytes main memory) of a common implementation of the symmetric Lanc-

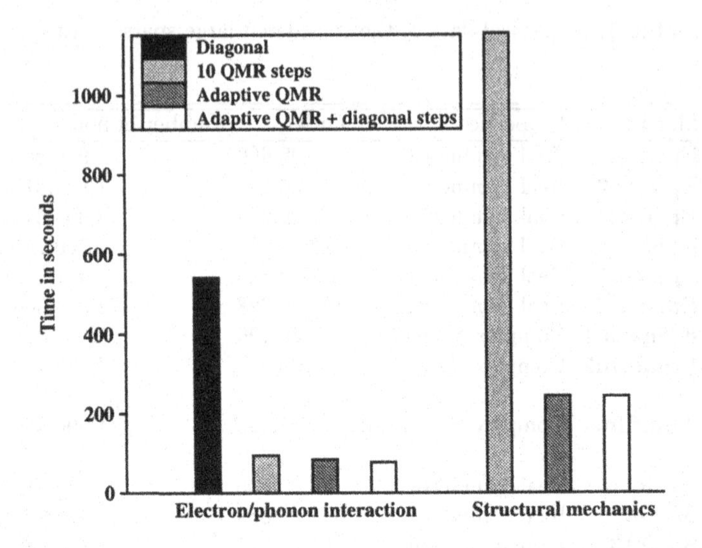

Fig. 4. Different preconditioners. Real symmetric matrices **Episym2** and **Struct3**. NEC Cenju-3, 64 processors

zos algorithm [6] and adaptively QMR preconditioned JD are compared for computing the four smallest eigenvalues and -vectors. In both cases, the matrices are stored in CRS format. The required accuracy of the results was set close to machine precision. Except for matrix **CulWil** — the Lanczos algorithm did not converge within 120 minutes since the smallest eigenvalues of the matrix are very close to each other — both methods gave the same eigenvalues and -vectors within the required accuracy. Since the Lanczos method applied stores all Lanczos vectors to compute the eigenvectors of A only the smallest matrices from Table 1 could be used for the comparison.

Table 2. Comparison of JD and the symmetric Lanczos algorithm. Sequential execution times. SGI O^2 workstation

Matrix	Lanczos	JD	Ratio
Struct1	79.3 s	34.4 s	2.3
Struct2	5633.4 s	899.7 s	6.3
Laplace	97.4 s	2.7 s	36.1
GregCar	95.9 s	15.7 s	6.1
CulWil	—	16.1 s	—
RNet	197.9 s	41.8 s	4.7

Table 2 shows that the JD method is markedly superior to the Lanczos algorithm for the problems tested. Moreover, the results for the matrices **Laplace** and **CulWil**— some of the smallest eigenvalues are very close to each other — appear to indicate that JD can handle the problem of close eigenvalues much better than the frequently used Lanczos method from [6]. It should be remarked that there are many variants of the well-known Lanczos algorithm which are not considered in this investigation. In particular, block Lanczos methods may show a more advantageous behaviour in the case of very close eigenvalues than the algorithm from [6].

6.4 Parallel Performance

In all following investigations, the JD iteration was stopped if the residual norms divided by the initial norms are less than 10^{-5} (10^{-10} for the Cray T3E results).

Fig. 5 shows the scaling of QMR preconditioned JD for computing the four smallest eigenpairs of the large real symmetric electron/phonon interaction matrices **Episym1**, **Episym3**, and **Episym4** on NEC Cenju-3. On 128 processors, speedups of 45.5, 105.7, and 122.3 are achieved for the matrices with increasing order; the corresponding execution times are 8.3 s, 23.3 s, and 168.3 s.

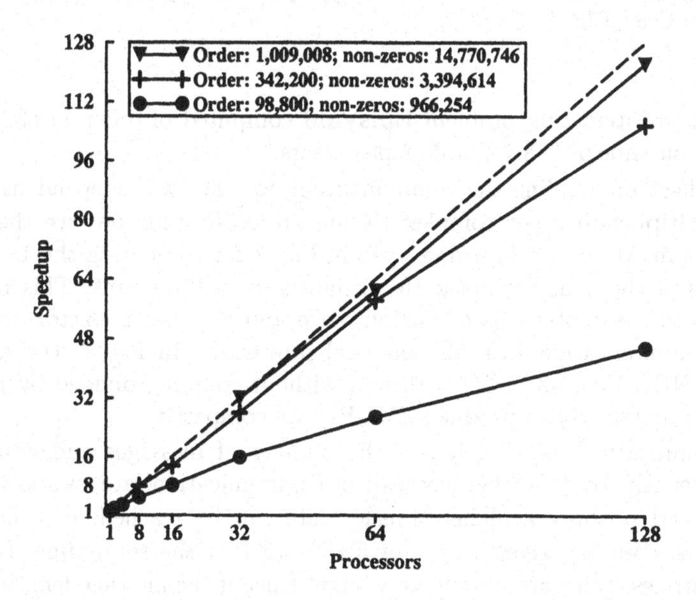

Fig. 5. Speedups. Real symmetric matrices **Episym1**, **Episym3**, and **Episym4**, electron/phonon interaction. NEC Cenju-3

In Fig. 6, speedups of QMR preconditioned JD for computing the four smallest eigenpairs of the two large real symmetric electron/phonon interaction matrices **Episym4** and **Episym5** on Cray T3E are displayed. The problems **Episym4** and **Episym5** of order 1,009,008 and 5,513,508 result in execution times of 15.2 s and 129.2 s, respectively, on 512 processors. The largest real symmetric elec-

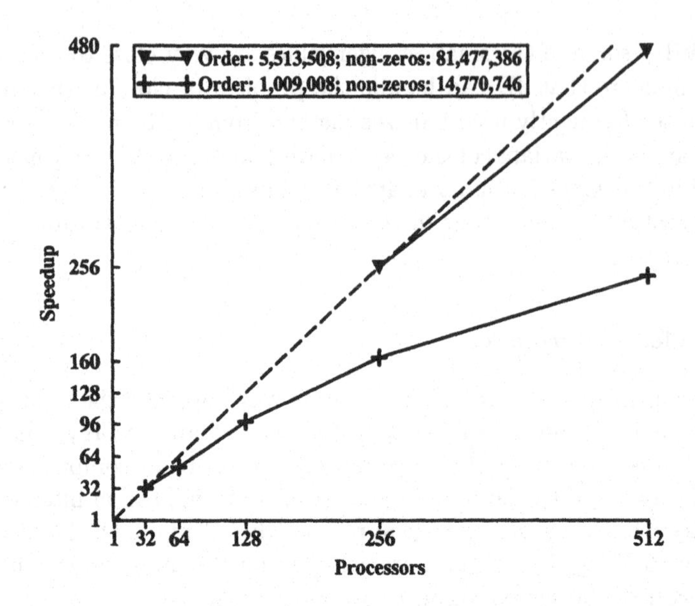

Fig. 6. Speedups. Real symmetric matrices **Episym4** and **Episym5**, electron/phonon interaction. Cray T3E

tron/phonon interaction problem **Episym6** computed of order 11,639,628 has an execution time of 278.7 s on 512 processors.

The effect of coupling the communication for the two independent matrix-vector multiplications per complex Hermitian QMR iteration (see the framed operations in Algorithm 1) is displayed in Fig. 7 for computing the 18 smallest eigenpairs of the dense complex Hermitian matrix **Thinfilms**. This matrix is chosen since the problem is of medium size and the matrix-vector operations require communication with all non-local processors. In Fig. 7, the execution times on NEC Cenju-3 of JD with and without coupling divided by the total number of matrix-vector products (MVPs) are compared.

Communication coupling halves the number of messages and doubles the message length. By this, the overhead of communication latencies is markedly reduced, and possibly a higher transfer rate can be reached. For the matrix **Thinfilms**, coupling gives a gain of 5% to 15% of the total time. For much larger matrices, gains are usually very slight since if the message lengths are big latency is almost negligible and higher transfer rates cannot be reached. For the matrix **Episym2**, e.g., corresponding timings on 128 processors give 64.5 ms without coupling and 64.4 ms with coupling.

Fig. 8 shows the scaling of the complex Hermitian version of QMR preconditioned JD for computing the four smallest eigenpairs of the two large complex Hermitian electron/phonon interaction matrices **Epiherm1** and **Epiherm2** on NEC Cenju-3. On 128 processors, speedups of 57.5 (execution time 38.8 s) and 122.5 (execution time 320.2 s) are achieved for the matrices **Epiherm1** and **Epiherm2** of order 126,126 and 1,009,008, respectively.

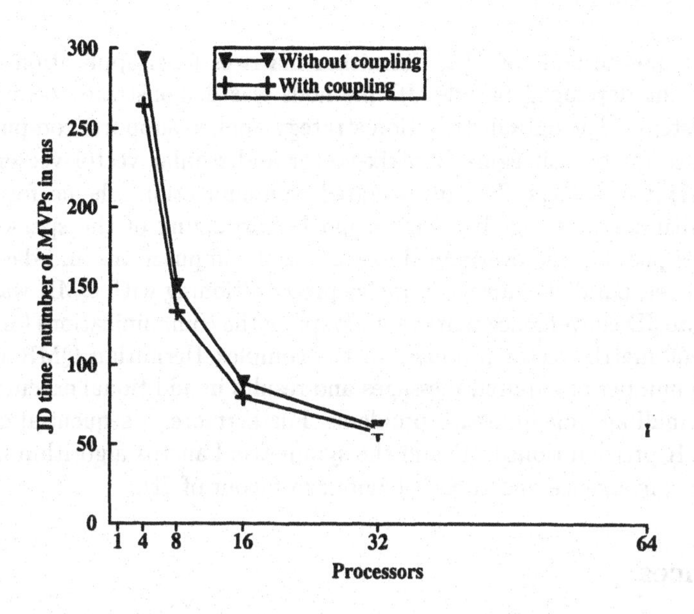

Fig. 7. Communication coupling. Complex Hermitian matrix **Thinfilms**. NEC Cenju-3

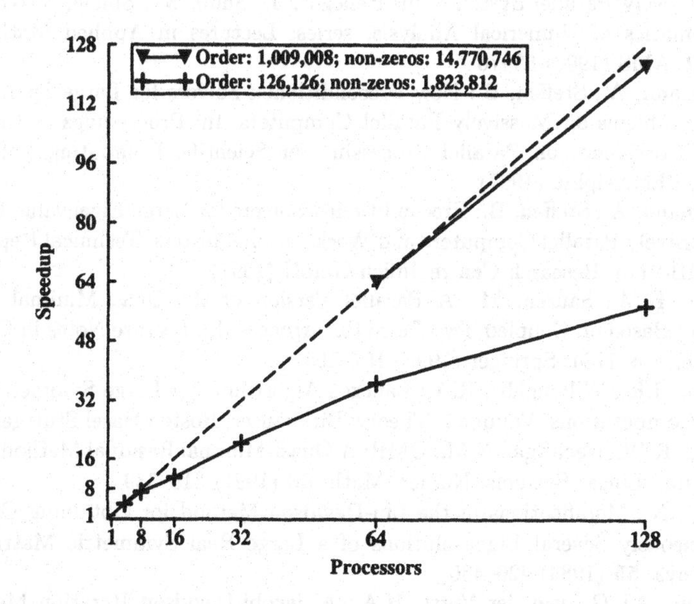

Fig. 8. Speedups. Complex Hermitian matrices **Epiherm1** and **Epiherm2**, electron/phonon interaction. NEC Cenju-3

7 Conclusions

By real symmetric and complex Hermitian matrices from applications, the efficiency of the developed parallel JD methods was demonstrated on massively parallel systems. The data distribution strategy applied supports computational load balance for both irregular matrix-vector and regular vector-vector operations in iterative solvers. The investigated communication scheme for matrix-vector multiplications together with a block rearranging of the sparse matrix data makes possible the overlapped execution of computations and data transfers. Moreover, parallel adaptive iterative preconditioning with QMR was shown to accelerate JD convergence markedly. Coupling the communications for the two independent matrix-vector products in the complex Hermitian QMR iteration halves the number of required messages and results in additional execution time gains for small and medium size problems. Furthermore, a sequential comparison of QMR preconditioned JD and the symmetric Lanczos algorithm indicates a superior convergence and time behaviour in favour of JD.

References

1. Barrett, R., Berry, M., Chan, T., Demmel, J., Donato, J., Dongarra, J., Eijkhout, V., Pozo, R., Romine, C., van der Vorst, H.: Templates for the Solution of Linear Systems: Building Blocks for Iterative Methods. SIAM, Philadelphia (1993)
2. Basermann, A.: QMR and TFQMR Methods for Sparse Nonsymmetric Problems on Massively Parallel Systems. In: Renegar, J., Shub, M., Smale, S. (eds.): The Mathematics of Numerical Analysis, series: Lectures in Applied Mathematics, Vol. 32. AMS (1996) 59–76
3. Basermann, A., Steffen, B.: New Preconditioned Solvers for Large Sparse Eigenvalue Problems on Massively Parallel Computers. In: Proceedings of the Eighth SIAM Conference on Parallel Processing for Scientific Computing (CD-ROM). SIAM, Philadelphia (1997)
4. Basermann, A., Steffen, B.: Preconditioned Solvers for Large Eigenvalue Problems on Massively Parallel Computers and Workstation Clusters. Technical Report FZJ-ZAM-IB-9713. Research Centre Jülich GmbH (1997)
5. Bücker, H.M., Sauren, M.: A Parallel Version of the Quasi-Minimal Residual Method Based on Coupled Two-Term Recurrences. In: Lecture Notes in Computer Science, Vol. 1184. Springer (1996) 157–165
6. Cullum, J.K., Willoughby, R.A.: Lanczos Algorithms for Large Symmetric Eigenvalue Computations, Volume I: Theory. Birkhäuser, Boston Basel Stuttgart (1985)
7. Freund, R.W., Nachtigal, N.M.: QMR: A Quasi-Minimal Residual Method for Non-Hermitian Linear Systems. Numer. Math. **60** (1991) 315–339
8. Kosugi, N.: Modifications of the Liu-Davidson Method for Obtaining One or Simultaneously Several Eigensolutions of a Large Real Symmetric Matrix. Comput. Phys. **55** (1984) 426–436
9. Sleijpen, G.L.G., van der Vorst, H.A.: A Jacobi-Davidson Iteration Method for Linear Eigenvalue Problems. SIAM J. Matrix Anal. Appl. **17** (1996) 401–425
10. Wellein, G., Röder, H., Fehske, H.: Polarons and Bipolarons in Strongly Interacting Electron-Phonon Systems. Phys. Rev. B **53** (1996) 9666–9675

Solving Eigenvalue Problems on Networks of Processors*

D. Giménez, C. Jiménez, M. J. Majado, N. Marín, and A. Martín

Departamento de Informática, Lenguajes y Sistemas Informáticos.
Univ de Murcia. Aptdo 4021. 30001 Murcia, Spain.
{domingo,mmajado,nmarin}@dif.um.es

Abstract. In recent times the work on networks of processors has become very important, due to the low cost and the availability of these systems. This is why it is interesting to study algorithms on networks of processors. In this paper we study on networks of processors different Eigenvalue Solvers. In particular, the Power method, deflation, Givens algorithm, Davidson methods and Jacobi methods are analized using PVM and MPI. The conclusion is that the solution of Eigenvalue Problems can be accelerated by using networks of processors and typical parallel algorithms, but the high cost of communications in these systems gives rise to small modifications in the algorithms to achieve good performance.

1 Introduction

Within the different platforms that can be used to develop parallel algorithms, in recent years special attention is being paid to networks of processors. The main reasons for their use are the lesser cost of the connected equipment, the greater availability and the additional utility as a usual means of work. On the other hand, communications, the heterogeneity of the equipment, their shared use and, generally, the small number of processors used are the negative factors.

The biggest difference between multicomputers and networks of processors is the high cost of communications in the networks, due to the small bandwidth and the shared bus which allows us to send only one message at a time. This characteristic of networks makes it a difficult task to obtain acceptable efficiencies, and also lets one think of the design of algorithms with good performances on a small number of processors, more than of the design of scalable algorithms.

So, despite not being as efficient as supercomputers, networks of processors come up as a new environment to the development of parallel algorithms with a good ratio cost/efficiency. Some of the problems that the networks have can be overlooked using faster networks, better algorithms and new environments appropriate to the network features.

* Partially supported by Comisión Interministerial de Ciencia y Tecnología, project TIC96-1062-C03-02, and Consejería de Cultura y Educación de Murcia, Dirección General de Universidades, project COM-18/96 MAT.

J. Palma, J. Dongarra, and V. Hernández (Eds.): VECPAR'98, LNCS 1573, pp. 85–99, 1999.

The most used matricial libraries (BLAS, LAPACK [1], ScaLAPACK [2]) are not implemented for those environments, and it seems useful to programme these linear algebra libraries over networks of processors [3–6]. The implementation of these algorithms can be done over programming environments like PVM [7] or MPI [8], which makes the work easier, although they do not take advantage of all the power of the equipment. The communications libraries utilised have been the free access libraries PVM version 3.4 and MPICH [9] (which is an implementation of MPI) version 1.0.11.

The results obtained using other systems or libraries could be different, but we are interested in the general behaviour of network of processors, and more particularly Local Area Networks (LANs), when solving Linear Algebra Problems. Our intention is to design a library of parallel linear algebra routines for LANs (we could call this library LANLAPACK), and we are working in Linear System Solvers [10] and Eigenvalue Solvers. In this paper some preliminary studies of Eigenvalue Solvers are shown. These problems are of great interest in different fields in science and engineering, and it is possibly better to solve them with parallel programming due to the high cost of computation [11]. The Eigenvalue Problem is still open in parallel computing, where it is necessary to know the eigenvalues efficiently and exactly. For that, we have carried out a study on the development of five methods to calculate eigenvalues over two different environments: PVM and MPI, and using networks with both Ethernet and Fast-Ethernet connections. Experiments have been performed in four different systems:

- A network of 5 SUN Ultra 1 140 with Ethernet connections and 32 Mb of memory on each processor.
- A network of 7 SUN Sparcstation with Ethernet connections and 8 Mb of memory on each processor.
- A network of 13 PC 486, with Ethernet connections, a memory of 8 Mb on each processor, and using Linux.
- A network of 6 Pentiums, with Fast-Ethernet connections, a memory of 32 Mb on each processor, and using Linux.

In this paper we will call these systems SUNUltra, SUNSparc, PC486 and Pentium, respectively.

The approximated cost of floating point operations working with double precision numbers and the theoretical cost of communicating a double precision number, in the four systems, is shown in table 1. The arithmetic cost has been obtained with medium sized matrices (stored in main memory). Also the quotient of the arithmetic cost with respect to the communication cost is shown. These approximations are presented to show the main characteristics of the four systems, but they will not be used to predict execution times because in networks of processors many factors, which are difficult to measure, influence the execution time: collisions when accessing the bus, other users in the system, assignation of processes to processors ... The four systems have also different utilisation characteristics: SUNUltra can be isolated to obtain results, but the other three are shared and it is more difficult to obtain representative results.

Table 1. Comparison of arithmetic and communication costs in the four systems utilised.

	floating point cost	*word − sending time*	*quotient*
SUNUltra	0.025 μs	0.8 μs	32
SUNSparc	0.35 μs	0.8 μs	2.28
PC486	0.17 μs	0.8 μs	4.7
Pentium	0.062 μs	0.08 μs	1.29

```
given v_0
FOR i = 1, 2, ...
        r_i = A v_{i-1}
        β_i = || r_i ||_∞
        v_i = r_i / β_i
ENDFOR
```

Fig. 1. Scheme of the sequential Power method.

2 Eigenvalue Solvers

Methods of partial resolution (the calculation of some eigenvalues and/or eigenvectors) of Eigenvalue Problems are studied: the Power method, deflation technique, Givens algorithm and Davidson method; and also the Jacobi method to compute the complete spectrum. We are interested in the parallelisation of the methods on networks of processors. Mathematical details can be found in many books ([12–15]).

Power method. The Power method is a very simple method to compute the eigenvalue of biggest absolute value and the associated eigenvector. Some variations of the method allow us to compute the eigenvalue of lowest absolute value or the eigenvalue nearest to a given number. This method is too slow to be considered as a good method in general, but in some cases it can be useful. In spite of the bad behaviour of the method, it is very simple and will allow us to begin to analyse Eigenvalue Solvers on networks of processors.

A scheme of the algorithm is shown in figure 1. The algorithm works by generating a succession of vectors v_i convergent to an eigenvector q_1 associated to the eigenvalue λ_1, as well as another succession of values β_i convergent to the eigenvalue λ_1. The speed of convergency is proportional to $\frac{\lambda_2}{\lambda_1}$.

Each iteration in the algorithm has three parts. The most expensive is the multiplication matrix-vector, and in order to parallelise the method the attention must be concentrated on that operation. In the parallel implementation, a master-slave scheme has been carried out. Matrix A is considered distributed

```
master:
        given v_0
        FOR i = 1,2,...
                broadcast v_{i-1} to the slaves
                receive r_i^(k) from the slaves, and form r_i
                β_i = || r_i ||_∞
                compute norm
                broadcast norm to the slaves
                IF convergence not reached
                        v_i = r_i / β_i
                ENDIF
        ENDFOR

slave k, with k = 0,1,...,p-1:
        FOR i = 1,2,...
                receive v_{i-1} from master
                r_i^(k) = A^(k) v_{i-1}
                send r_i^(k) to master
                receive norm from master
        ENDFOR
```

Fig. 2. Scheme of the parallel Power method.

between the slave processes in a block striped partition by rows [16]. A possible scheme of the parallel algorithm is shown in figure 2. The multiplication matrix-vector is performed by the slave processes, but the master obtains β_i and forms v_i. These two operations of cost $O(n)$ could be done in parallel in the slaves, but it would generate more communications, which are very expensive operations in networks of processors.

The arithmetic cost of the parallel method is: $\frac{2n^2}{p} + 3n$ flops, and the theoretical efficiency is 100%.

The cost of communications varies with the way in which they are performed. When the distribution of vector v_i and the norm is performed using a broadcast operation, the cost of communications per iteration is: $2\tau_b(p) + \beta_b(p)(n+1) + \tau + \beta n$, where τ and β represent the start-up and the word-sending time, respectively, and $\tau_b(p)$ and $\beta_b(p)$ the start-up and the word-sending time when using a broadcast operation on a system with p slaves. If the broadcasts are replaced by point to point communications the cost of communications per iteration is $\tau(2p+1) + \beta(pn+n+p)$. Which method of communication is preferable depends on the characteristics of the environment (the communication library and the network of processors) we are using.

The parallel algorithm is theoretically optimum if we only consider efficiency, but when studying scalability, the isoefficiency function is $n \propto p^2$, and the scalability is not very good. This bad scalability is caused by the use of a shared bus which avoids sending data at the same time. Also the study of scalability is not useful in this type of system due to the reduced number of processors.

We will experimentally analyse the algorithm in the most and the least adequate systems for parallel processing (Pentium and SUNUltra, respectively). In table 2 the execution time of the sequential and the parallel algorithms on the two systems is shown, varying the number of slaves and the matrix size. Times have been obtained for random matrices, and using PVM and the routine `pvm_mcast` to perform the broadcast in figure 2. The results greatly differ in the two systems due to the big difference in the proportional cost of arithmetic and communication operations. Some conclusions can be obtained:

- Comparing the execution time of the sequential and the parallel algorithms using one slave, we can see the very high penalty of communications in these systems, especially in SUNUltra.
- The best execution times are obtained with a reduced number of processors because of the high cost of communications, this number being bigger in Pentium. The best execution time for each system and matrix size is marked in table 2.
- The availability of more potential memory is an additional benefit of parallel processing, because it allows us to solve bigger problems without swapping. For example, the parallel algorithm in SUNUltra is quicker than the sequential algorithm only when the matrix size is big and the sequential execution produces swapping.
- The use of more processes than processors produces in Pentium with big matrices better results than one process per processor, and this is because communications and computations are better overlapped.

The basic parallel algorithm (figure 2) is very simple but it is not optimized for a network of processors. We can try to improve communications in at least two ways:

- The broadcast routine is not optimized for networks, and it would be better if we replace this routine by point to point communications.
- The diffusion of the norm and the vector can be assembled in only one communication. In that way more data are transferred because the last vector need not be sent, but less communications are performed.

In table 3 the execution time of the basic version (version 1), the version with point to point communications (version 2), and the version where the diffusion of norm and vector are assembled (version 3) are compared. Versions 2 and 3 reduce the execution time in both systems, and the reduction is clearer in SUNUltra because of the bigger cost of communication in this system.

Until now, the results shown have been those obtained with PVM, but the use of MPI produces better results (obviously it depends on the versions we are using). In table 4 the execution time obtained with the basic version of the programme using MPI on SUNUltra is shown. Comparing this table with table 2

Table 2. Execution time (in seconds) of the Power method using PVM, varying the number of processors and the matrix size.

	sequential	1 slave	2 slaves	3 slaves	4 slaves	5 slaves	6 slaves	7 slaves	8 slaves
					SUNUltra				
300	**0.069**	0.149	0.255	0.289	0.381	0.547	0.526	0.552	0.638
600	**0.292**	0.497	0.450	0.491	0.629	0.864	0.838	0.849	0.954
900	**0.599**	0.882	0.679	0.735	1.065	1.491	1.457	1.284	1.603
1200	4.211	7.582	1.277	**1.062**	1.426	1.722	1.940	2.004	
1500	23.613	54.464	**1.481**	1.796	1.901	2.233	2.324	2.421	
					Pentium				
300	**0.172**	0.281	0.250	0.188	0.292	0.323	0.407	0.405	0.436
600	0.592	1.153	0.639	**0.569**	0.608	0.682	0.599	0.662	0.656
900	1.302	1.762	1.903	1.272	0.908	0.892	**0.842**	0.891	1.171
1200	2.138	8.141	1.544	1.750	1.568	1.224	1.275	1.231	**1.169**
1500	3.368	254.42	2.776	1.904	4.017	2.431	1.911	1.750	**1.680**

we can see the programme with MPI works better when the number of processors increases.

Deflation technique. Deflation technique is used to compute the next eigenvalue and its associated eigenvector starting from a previously known one. This technique is used to obtain some eigenvalues and it is based on the transformation of the initial matrix to another one that has got the same eigenvalues, replacing λ_1 by zero.

To compute `numEV` eigenvalues of the matrix A, the deflation technique can be used performing `numEV` steps (figure 3), computing in each step, using the Power method, the biggest eigenvalue (λ_i) and the corresponding eigenvector ($q_i^{(i)}$) of a matrix A_i, with $A_1 = A$. Each matrix A_{i+1} is obtained from matrix A_i using $q_i^{(i)}$, which is utilized to form matrix B_{i+1}:

$$B_{i+1} = \begin{bmatrix} 1\,0\cdots 0 & -q_1 & 0\cdots 0 \\ 0\,1\cdots 0 & -q_2 & 0\cdots 0 \\ \vdots\;\vdots\;\ddots\;\vdots & \vdots & \vdots\;\ddots\;\vdots \\ 0\,0\cdots 1 & -q_{k-1} & 0\cdots 0 \\ 0\,0\cdots 0 & 0 & 0\cdots 0 \\ 0\,0\cdots 0 & -q_{k+1} & 1\cdots 0 \\ \vdots\;\vdots\;\ddots\;\vdots & \vdots & \vdots\;\ddots\;\vdots \\ 0\,0\cdots 0 & -q_n & 0\cdots 1 \end{bmatrix},$$

where $q_i^{(i)^t} = (q_1, \ldots, q_{k-1}, 1, q_{k+1}, \ldots, q_n)$.

Table 3. Comparison of the execution time of the Power method using PVM, with different communication strategy.

	2 slaves			3 slaves			4 slaves		
	ver 1	ver 2	ver 3	ver 1	ver 2	ver 3	ver 1	ver 2	ver 3
SUNUltra									
300	0.255	0.187	**0.168**	0.289	0.262	**0.187**	0.381	0.339	**0.203**
600	0.450	0.418	**0.363**	0.491	0.430	**0.386**	0.629	0.605	**0.341**
900	0.679	0.570	**0.566**	0.735	0.666	**0.567**	1.065	0.688	**0.542**
1200	1.277	1.089	**0.877**	1.062	1.045	**0.854**	1.426	0.949	**0.911**
Pentium									
300	0.250	**0.157**	0.271	0.188	0.188	**0.141**	0.292	0.294	**0.192**
600	0.639	**0.574**	0.585	0.569	**0.516**	0.599	0.608	0.603	**0.557**
900	1.903	**1.879**	2.714	1.272	**0.983**	1.090	0.908	**0.813**	0.985

Table 4. Execution time (in seconds) of the Power method using MPI, varying the number of processors and the matrix size, on SUNUltra.

	1 slave	2 slaves	3 slaves	4 slaves	5 slaves	6 slaves
300	0.15	**0.14**	0.22	0.28	0.27	0.38
600	0.39	**0.31**	**0.31**	0.37	0.55	0.62
900	0.73	0.52	**0.47**	0.56	0.61	0.66
1200	3.58	0.76	**0.70**	0.71	0.79	0.93
1500	22.16	1.37	**0.97**	0.99	1.06	0.94

The eigenvalue λ_i is the biggest eigenvalue in absolute value of matrix A_i, and it is also the i-th eigenvalue in absolute value of matrix A. The eigenvector q_i associated to the eigenvalue λ_i in matrix A is computed from the eigenvector $q_i^{(i)}$ associated to λ_i in the matrix A_i. This computation is performed repeatedly applying the formula $q_i^{(w)} = q_i^{(w+1)} + \frac{a_k^{(w)^t} q_i^{(w+1)}}{\lambda_i - \lambda_w} q_w^{(w)}$, where $a_k^{(w)^t}$ is the k-th column of matrix A_w, with k the index where $q_w^{(w)} = 1$.

The most costly part of the algorithm is the application of the Power method, which has a cost of order $O\left(n^2\right)$ per iteration. Therefore, the previously explained Power method can be applied using a scheme master-slave. The cost of the deflation part (update matrix) is $2n^2$ flops, and the cost of the computation of q_i depends on the step and is $5n(i-1)$ flops. These two parts can be performed simultaneously in the parallel algorithm: the master process computes q_i while the slaves processes update matrix A_i. The deflation part of the parallel algorithm (update matrix and compute q_i) is not scalable if a large number of eigenvalues are computed. As we have previously mentioned, scalability is not very important in these types of systems. In addition, this method is used

```
A_1 = A
FOR i = 1, 2, ..., numEV
        compute by the Power method λ_i and q_i^(i)
        update matrix A: A_{i+1} = B_{i+1} A_i
        compute q_i from q_i^(i)
ENDFOR
```

Fig. 3. Scheme of the deflation technique.

Table 5. Execution time (in seconds) of the deflation method using PVM, varying the number of processors and the matrix size, when computing 5% of the eigenvalues.

	sequential	1 slave	2 slaves	3 slaves	4 slaves	5 slaves
			SUNUltra			
300	**11.4**	29.6	25.2	28.5	32.4	33.7
600	**184.7**	308.3	239.7	252.1	244.2	273.2
900	914.4	1258.3	862.5	826.6	**765.7**	831.0
			Pentium			
300	25.0	32.4	29.9	23.0	22.4	**22.2**
600	407.0	462.4	288.9	239.5	195.8	**194.6**
900	2119.0	2316.5	1460.3	993.7	823.2	**720.7**

to compute only a reduced number of eigenvalues, due to its high cost when computing a large number of them.

In table 5 the execution time of the sequential and parallel algorithms are shown on SUNUltra and Pentium, using PVM and the basic version of the Power method. Compared with table 2, we can see the behaviour of the parallel Power and deflation algorithms is similar, but that of the deflation technique is better, due to the high amount of computation, the work of the master processor in the deflation part of the execution and the distribution of data between processes in the parallel algorithm.

Davidson method. The Power method is a very simple but not very useful method to compute the biggest eigenvalue of a matrix. Other more efficient methods, as for example Davidson methods [17] or Conjugate Gradient methods [18, 19], can be used.

The Davidson algorithm lets us compute the highest eigenvalue (in absolute value) of a matrix, though it is especially suitable for large, sparse and symmetric matrices. The method is valid for real as well as complex matrices. It works by building a sequence of search subspaces that contain, in the limit, the desired eigenvector. At the same time these subspaces are built, so approximations to the desired eigenvector in the current subspace are also built.

```
V_0 = [ ]
given v_1
k = 1
WHILE convergence not reached
        V_k = [V_{k-1}|v_k]
        orthogonalize V_k using modified Gram-Schmith
        compute H_k = V_k^c A V_k
        compute the highest eigenpair (θ_k, y_k) of H_k
        obtain the Ritz vector u_k = V_k y_k
        compute the residual r_k = A u_k - θ_k u_k
        IF k = k_max
                reinitialize
        ENDIF
        obtain v_{k+1} = (θ_k I - D)^{-1} r_k
        k = k + 1
ENDWHILE
```

Fig. 4. Scheme of the sequential Davidson method.

Figure 4 shows a scheme of a sequential Davidson method. In successive steps a matrix V_k with k orthogonal column vectors is formed. After that, matrix $H_k = V_k^c AV_k$ is formed and the biggest eigenvalue in absolute value (θ_k) and its associated eigenvector (y_k) are computed. This can be done using the Power method, because matrix H_k is of size $k \times k$ and k can be kept small using some reinitialization strategy (when $k = k_{max}$ the process is reinitialized). The new vector v_{k+1} to be added to the matrix V_k to form V_{k+1} can be obtained with the succession of operations: $u_k = V_k y_k$, $r_k = Au_k - \theta_k u_k$ and $v_{k+1} = (\theta_k I - D)^{-1} r_k$, with D the diagonal matrix which has in the diagonal the diagonal elements of matrix A.

To obtain a parallel algorithm the cost of the different parts in the sequential algorithm can be analysed. The only operations with cost $O(n^2)$ are two matrix-vector multiplications: Av_k in the computation of H_k and Au_k in the computation of the residual. AV_k can be accomplished in order $O(n^2)$ because it can be decomposed as $[AV_{k-1}|Av_k]$, and AV_{k-1} was computed in the previous step. The optimum value of k_{max} varies with the matrix size and the type of the matrix, but it is small and it is not worthwhile to parallelize the other parts of the sequential algorithm. Therefore, the parallelization of the Davidson method is done basically in the same way as the Power method: parallelizing matrix-vector multiplications.

```
master:
        V_0 = [ ]
        given v_1
        k = 1
        WHILE convergence not reached
                V_k = [V_{k-1}|v_k]
                orthogonalize V_k using modified Gram-Schmith
                send v_k to slaves
                in parallel compute Av_k and accumulate in the master
                compute H_k = V_k^c (AV_k)
                compute the highest eigenpair (θ_k, y_k) of H_k
                obtain the Ritz vector u_k = V_k y_k
                send u_k to slaves
                in parallel compute Au_k and accumulate in the master
                compute the residual r_k = Au_k - θ_k u_k
                IF k = k_max
                        reinitialize
                ENDIF
                obtain v_{k+1} = (θ_k I - D)^{-1} r_k
                k = k + 1
        ENDWHILE

slave k, with k = 1,...,p-1:
        WHILE convergence not reached
                receive v_k from master
                in parallel compute Av_k and accumulate in the master
                receive u_k from master
                in parallel compute Au_k and accumulate in the master
                IF k = k_max
                        reinitialize
                ENDIF
                k = k + 1
        ENDWHILE
```

Fig. 5. Scheme of the parallel Davidson method.

Table 6. Execution time of the Davidson method using MPI, varying the number of processors and the matrix size, on PC486.

	sequ	p=2	p=3	p=4	p=5	p=6	p=7	p=8
300	0.048	0.051	0.048	0.045	0.043	**0.041**	**0.041**	0.042
600		0.129	0.097	0.083	0.076	0.072	0.065	**0.064**
900				0.165	0.131	0.118	0.119	**0.112**

This method has been parallelized using a master-slave scheme (figure 5), working all the processes in the two parallelized matrix-vector multiplications, and performing the master process non parallelized operations. In that way, matrix A is distributed between the processes and the result of the local matrix-vector multiplications is accumulated in the master and distributed from the master to the other processes.

Because the operations parallelized are matrix-vector multiplications, as in the Power method, the behaviour of the parallel algorithms must be similar, but better results are obtained with the Davidson method because in this case the master process works in the multiplications.

Table 6 shows the execution time of 50 iterations of the Davidson method for symmetric complex matrices on PC486 and using MPI, varying the number of processors and the matrix size.

Givens algorithm or bisection. As we have seen in previous paragraphs, on parallel Eigenvalue Solvers whose cost is of order $O\left(n^2\right)$ only a small reduction in the execution time can be achieved in networks of processors in some cases: when the matrices are big or the quotient between the cost of communication and computation is small.

In some other Eigenvalue Solvers the behaviour is slightly better. For example, the bisection method is an iterative method to compute eigenvalues in an interval or the k biggest eigenvalues. It is applicable to symmetric tridiagonal matrices. This method is especially suitable to be parallelized, due to the slight communication between processes, which factor increases the total time consumed in a network. When computing the eigenvalues in an interval, the interval is divided in subintervals and each process works in the computation of the eigenvalues in a subinterval. When computing the k biggest eigenvalues, each slave knows the number of eigenvalues it must compute. After that, communications are not necessary but inbalance is produced by the distribution of the spectrum. More details on the parallel bisection method are found in [20].

The eigenvalues are computed by the processes performing successive iterations, and each iteration has a cost of order $O(n)$. Despite the low computational cost good performance is achieved because communications are not necessary after the subintervals are broadcast. Table 7 shows the efficiency obtained using this method to calculate all the eigenvalues or only 20% of them, on SUNUltra

Table 7. Efficiency of the Givens method using PVM, varying the number of processors, on SUNUltra and SUNSparc, with matrix size 100.

	$p=2$	$p=3$	$p=4$	$p=5$	$p=2$	$p=3$	$p=4$	$p=5$
	all the eigenvalues				*20% of the eigenvalues*			
SUNUltra	0.72	0.63	0.55	0.49	0.52	0.38	0.26	0.19
SUNSparc	1.17	0.94	0.60	0.64	0.81	0.61	0.36	0.37

and SUNSparc for matrix size 100. The efficiencies are clearly better than in the previous algorithms, even with small matrices and execution time.

Jacobi method. The Jacobi method for solving the Symmetric Eigenvalue Problem works by performing successive sweeps, nullifying once on each sweep the $n(n-1)/2$ nondiagonal elements in the lower-triangular part of the matrix.

It is possible to design a Jacobi method considering the matrix A of size $n \times n$, dividing it into blocks of size $t \times t$ and doing a sweep on these blocks, regarding each block as an element. The blocks of size $t \times t$ are grouped into blocks of size $2kt \times 2kt$ and these blocks are assigned to the processors in such a way that the load is balanced. Parallel block Jacobi methods are explained in more detail in [21].

In order to obtain a distribution of data to the processors, an algorithm for a logical triangular mesh can be used. The blocks of size $2kt \times 2kt$ must be assigned to the processors in such a way that the work is balanced. Because the most costly part of the execution is the updating of the matrix, and nondiagonal blocks contain twice more elements to be nullified than diagonal blocks, the load of nondiagonal blocks can be considered twice the load of diagonal blocks.

Table 8. Theoretical speed-up of the Jacobi method, varying the number of processes and processors.

	$p=2$	$p=3$	$p=4$	$p=5$	$p=6$	$p=7$	$p=8$	$p=9$	$p=10$
3	2	2	2	2	2	2	2	2	2
6		3	3	4.5	4.5	4.5	4.5	4.5	4.5
10			4	4	5.3	5.3	8	8	8

Table 8 shows the theoretical speed-up of the method when logical meshes of 3, 6 or 10 processes are assigned to a network, varying the number of processors in the network from 2 to 10. Higher theoretical speed-up is obtained increasing the number of processes. This produces better balancing, but also more communications and not always a reduction of the execution time. It can be seen in table 9, where the execution time per sweep for matrices of size 384 and 768 is shown, varying the number of processors and processes. The shown results have been obtained in PC486 and SUNUltra using MPI, and the good behaviour of the parallel algorithm is observed because a relatively large number of processors can be used reducing the execution time.

Table 9. Execution time per sweep of the Jacobi method on PC486 and SUNUltra using MPI.

	p=1	p=2	p=3	p=4	p=5	p=6	p=7	p=8	p=9	p=10
SUNUltra: 384 (non swapping)										
1	6.25									
3		4.78	**4.55**							
6			7.08	7.31	8.26					
SUNUltra: 768 (non swapping)										
1	50.78									
3		28.37	**25.50**							
6			39.51	33.11	32.18					
PC486: 384 (non swapping)										
1	43.26									
3		30.32	26.81							
6			28.01	23.15	19.81	18.53				
10				42.20	27.15	30.79	21.46	19.82	**15.49**	17.44
PC486: 768 (swapping)										
1	698.8									
3		217.4	195.6							
6			187.2	142.1	111.6	104.6				
10				158.9	145.6	106.5	96.9	83.5	76.5	**72.5**

3 Conclusions

Our goal is to develop a library of linear algebra routines for LANs. In this paper some previous results on Eigenvalue Solvers are shown. The characteristics of the environment propitiate small modifications in the algorithms to adapt them to the system. In these environments we do not generally have many processors, and also, when the number of processors goes up, efficiency unavoidably goes down. For these reasons, when designing algorithms for networks of processors it is preferable to think on good algorithms for a small number of processors, and not on scalable algorithms. Because of the great influence of the cost of communications, a good use of the available environments -MPI or PVM- is essential.

References

1. E. Anderson, Z. Bai, C. Bischof, J. Demmel, J. Dongarra, J. Du Croz, A. Greenbaum, S. Hammarling, A. McKenney, S. Ostrouchov and D. Sorensen. *LAPACK Users' Guide.* SIAM, 1995.
2. L. S. Blackford, J. Choi, A. Cleary, E. D'Azevedo, J. Demmel, I. Dhillon, J. Dongarra, S. Hammarling, G. Henry, A. Petitet, K. Stanley, D. Walker and R. C. Whaley. ScaLAPACK User's Guide. SIAM, 1997.
3. J. Demmel and K. Stanley. The Performance of Finding Eigenvalues and Eigenvectors of Dense Symmetric Matrices on Distributed Memory Computers. In D. H. Bailey, P. E. Bjørstad, J. R. Gilbert, M. V. Mascagni, R. S. Schreiber, H. D. Simon, V. J. Torczon and L. T. Watson, editor, *Proceedings of the Seventh SIAM Conference on Parallel Processing for Scientific Computing*, pages 528–533. SIAM, 1995.
4. D. Giménez, M. J. Majado and I. Verdú. Solving the Symmetric Eigenvalue Problem on Distributed Memory Systems. In H. R. Arabnia, editor, *Proceedings of the International Conference on Parallel and Distributed Processing Techniques and Applications. PDPTA'97*, pages 744–747, 1997.
5. Kuo-Chan Huang, Feng-Jian Wang and Pei-Chi Wu. Parallelizing a Level 3 BLAS Library for LAN-Connected Workstations. *Journal of Parallel and Distributed Computing*, 38:28–36, 1996.
6. Gen-Ching Lo and Yousef Saad. Iterative solution of general sparse linear systems on clusters of workstations. May 1996.
7. A. Geist, A. Begelin, J. Dongarra, W. Jiang, R. Manchek and V. Sunderam. Parallel Virtual Machine. A User's Guide and Tutorial for Networked Parallel Computing.. The MIT Press, 1995.
8. Message Passing Interface Forum. A Message-Passing Interface Standard. *International Journal of Supercomputer Applications*, (3), 1994.
9. Users guide to mpich. preprint.
10. F. J. García and D. Giménez. Resolución de sistemas triangulares de ecuaciones lineales en redes de ordenadores. Facultad de Informática. Universidad de Murcia. 1997.
11. A. Edelman. Large dense linear algebra in 1993: The parallel computing influence. *The International Journal of Supercomputer Applications*, 7(2):113–128, 1993.
12. G. H. Golub and C. F. Van Loan. *Matrix Computations.* The Johns Hopkins University Press, 1989. Segunda Edición.
13. L. N. Trefethen and D. Bau III. *Numerical Linear Algebra.* SIAM, 1997.
14. David S. Watkins. *Matrix Computations.* John Wiley & Sons, 1991.
15. J. H. Wilkinson. *The Algebraic Eigenvalue Problem.* Clarendon Press, 1965.
16. V. Kumar, A. Grama, A. Gupta and G. Karypis. *Introduction to Parallel Computing. Design and Analysis of Algorithms.* The Benjamin Cummings Publishing Company, 1994.
17. E. R. Davidson. The Iterative Calculation of a Few of the Lowest Eigenvalues and Corresponding Eigenvectors of Large Real-Symmetric Matrices. *Journal of Computational Physics*, 17:87–94, 1975.
18. W. W. Bradbury and R. Fletcher. New Iterative Methods for Solution of the Eigenproblem. *Numerische Mathematik*, 9:259–267, 1966.
19. A. Edelman and S. T. Smith. On conjugate gradient-like methods for eigenvalue-like problems. *BIT*, 36(3):494–508, 1996.

20. J. M. Badía and A. M. Vidal. Exploiting the Parallel Divide-and-Conquer Method to Solve the Symmetric Tridiagonal Eigenproblem. In *Proceedings of the Sixth Euromicro Workshop on Parallel and Distributed Processing, Madrid, January 21-23*, 1998.
21. D. Giménez, V. Hernández and A. M. Vidal. A Unified Approach to Parallel Block-Jacobi Methods for the Symmetric Eigenvalue Problem. In *Proceedings of VECPAR'98*, 1998.

Solving Large-Scale Eigenvalue Problems on Vector Parallel Processors

David L. Harrar II and Michael R. Osborne

Centre for Mathematics and its Applications, School of Mathematical Sciences,
Australian National University, Canberra ACT 0200, Australia
{David.Harrar,Michael.Osborne}@anu.edu.au
{http://wwwmaths.anu.edu.au/}~dlh, ~mike

Abstract. We consider the development and implementation of eigensolvers on distributed memory parallel arrays of vector processors and show that the concomitant requirements for vectorisation and parallelisation lead both to novel algorithms and novel implementation techniques. Performance results are given for several large-scale applications and some performance comparisons made with LAPACK and ScaLAPACK.

1 Introduction

Eigenvalue problems (EVPs) arise ubiquitously in the numerical simulations performed on today's high performance computers (HPCs), and often their solution comprises the most computationally expensive algorithmic component. It is imperative that efficient solution techniques and high-quality software be developed for the solution of EVPs on high performance computers.

Generally, specific attributes of an HPC architecture play a significant, if not deterministic, role in terms of choosing/designing appropriate algorithms from which to construct HPC software. There are many ways to differentiate today's HPC architectures, one of the coarsest being *vector* or *parallel*, and the respective algorithmic priorities can differ considerably. However, on some recent HPC architectures – those comprising a (distributed-memory) parallel array of powerful vector processors, e.g., the Fujitsu VPP300 (see Section 5) – it is important, not to focus exclusively upon one or the other, but to strive for high levels of *both* vectorisation *and* parallelisation.

In this paper we consider the solution of large-scale eigenvalue problems,

$$Au = \lambda u, \tag{1}$$

and discuss how the concomitant requirements for vectorisation and parallelisation on vector parallel processors has lead both to novel implementations of known methods and to the development of completely new algorithms. These include techniques for both symmetric (or Hermitian) and nonsymmetric A and for matrices with special structure, for example tridiagonal, narrow-banded, etc. Performance results are presented for various large-scale problems solved on a Fujitsu (Vector Parallel Processor) VPP300, and some comparisons are made with analogous routines from the LAPACK [1] and ScaLAPACK [4] libraries.

J. Palma, J. Dongarra, and V. Hernández (Eds.): VECPAR'98, LNCS 1573, pp. 100–113, 1999.

2 Tridiagonal Eigenvalue Problems

Consider (1) with A replaced by a symmetric, irreducible, tridiagonal matrix T:

$$Tu = \lambda u, \quad \text{where } T = \text{tridiag}(\beta_i, \alpha_i, \beta_{i+1}), \quad \text{with } \beta_i \neq 0, \quad i = 2, \cdots, n. \tag{2}$$

One of the most popular methods for computing (selected) eigenvalues of T is *bisection* based on the *Sturm sign count*. The widely used LAPACK/Scalapack libraries, for example, each contain routines for (2) which use bisection. Since the techniques are standard, they are not reviewed in detail here (see, e.g., [26]). Instead, we focus on the main computational component: Evaluation of the Sturm sign count, defined by the recursion

$$\sigma_1(\lambda) = \alpha_1 - \lambda, \quad \sigma_i(\lambda) = \left(\alpha_i - \frac{\beta_i^2}{\sigma_{i-1}(\lambda)}\right) - \lambda, \; i = 1, 2, \cdots, n. \tag{3}$$

Zeros of $\sigma_n(\lambda)$ are eigenvalues of T, and the number of times $\sigma_i(\lambda) < 0$, $i = 1, \cdots, n$, is equal to the number of eigenvalues less than the approximation λ.

Vectorisation via multisection: In terms of vectorisation, the pertinent observation is that the recurrence (3) does not vectorise over the index i. However, if it is evaluated over a sequence of m estimates, λ_j, it is trivial implementationally to interchange i- and j-loops and vectorise over j. This is the basic idea behind *multisection*: An interval containing eigenvalues is split into more than the two subintervals used with bisection. The hope is that the efficiency obtained via vectorisation offsets the spurious computation entailed in sign count evaluation for subintervals containing no eigenvalues. For $r > 1$ eigenvalues, another way to vectorise is to bisect r eigenvalue intervals at a time, i.e. *multi-bisection*.[1]

On scalar processors bisection is optimal.[2] On vector processors, however, this is not the case, as shown in an aptly-titled paper [37]: "Bisection is not optimal on vector processors".[3] The non-optimality of bisection on a single VPP300 PE ("processing element") is illustrated in Figure 1 (left), where we plot the time required to compute one eigenvalue of a tridiagonal matrix ($n = 1000$) as a function of m, the number of multisection points (i.e. vector length). The tolerance is $\epsilon = 3 \times 10^{-16}$. For all plots in this section, times are averages from 25 runs.

---

[1] Nomenclaturally, multisecting r intervals might consistently be termed "multi-multisection"; we make no distinction and refer to this also as "multisection".

[2] This is probably not true for most *super*scalar processors; these should be able to take advantage of the chaining/pipelining inherent in multisection.

[3] This may not be true for vector processors of the near future, nor even perhaps all of today's, specifically those with large $n_{1/2}$, compared with, e.g., those in [37]. Additionally, many of today's vector PEs have a separate scalar unit so it is not justifiable to model bisection performance by assuming vector processing with vector length one – bisection is performed by the scalar unit. See the arguments in [10].

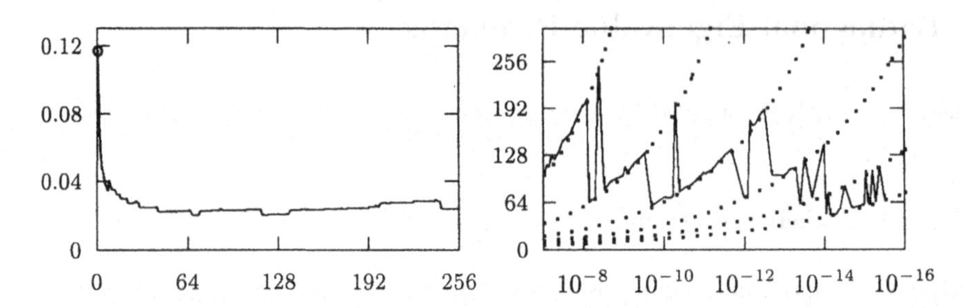

Fig. 1. (Left) Time vs. number of multisection points $m = 1, 256$. (Right) Optimal (i.e., time-minimizing) m vs. accuracy ϵ; dotted lines show $m_{\min}(\epsilon)$ for fixed numbers of multisection steps $\nu = 2, \ldots, 7$ (from left to right).

Clearly, the assertion in [37] holds: Bisection is not optimal. In fact, multisection using up to 3400 points is superior. The minimum time, obtained using 70 points, is roughly 17% the bisection time, i.e. that obtained using LAPACK.

Once it is decided to use multisection, there still remains a critical question: What is the optimal number of multisection points (equivalently, vector length)? The answer, which depends on r, is discussed in detail in [9]; here we highlight only a few key observations and limit discussion to the case $r = 1$.

Although not noted in [37], the optimal number of points, m_{opt}, depends on the desired accuracy ϵ, as illustrated in Figure 1 (right), where we plot m_{opt} vs. $\epsilon \in [10^{-7}, 3 \times 10^{-16}]$. Note that m_{opt} varies between about 50 and 250.[4] Two effects explain this apparently erratic behavior – one obvious, one not so.

To reach a desired accuracy of ϵ using m multisection points requires

$$\nu = -\lceil \log \epsilon / \log(m + 1) \rceil \tag{4}$$

multisection steps. Generally, $m_{\text{opt}}(\epsilon)$ corresponds to some $m_{\min}(\nu, \epsilon)$ at which the ceiling function in (4) causes a change in ν; that is, a minimal number of points effecting convergence to an accuracy of ϵ in ν steps. The dotted lines in Figure 1 (right) indicate $m_{\text{opt}}(\nu, \epsilon)$ for fixed ν. As ϵ is decreased (i.e., moving to the right) $m_{\min}(\nu, \epsilon)$ increases until for some $\epsilon_{\text{crit}}(\nu)$ it entails too much spurious computation in comparison with the (smaller) vector length, $m_{\min}(\nu+1, \epsilon)$, which is now large "enough" to enable adequate vectorisation. The optimal now follows along the curve $m_{\min}(\nu + 1, \epsilon)$ until reaching $\epsilon_{\text{crit}}(\nu + 1)$, etc. This explains the occasional transitions from one ν-curve to a $\nu + 1$-curve, but not the oscillatory switching. This is an artifact of a specific performance anomaly associated with the VPP300 processor, as we now elucidate.

In Figure 2 we extend the range of the plot on the left in Figure 1 to include vector lengths up to 2048. The plot "wraps around" in that the bottom line ($i = 0$) is for $m = 1, \ldots, 256$, the second ($i = 1$) for $m = 257, \ldots, 512$, etc. Accuracy is now $\epsilon = 3 \times 10^{-10}$. Note that there is a 10-20% increase in time when the vector length increases from $64i + 8$ to $64i + 9$ (dotted vertical lines)

[4] These values of m_{opt} are roughly five to twenty times those determined in [37], manifesting the effect of the significantly larger $n_{1/2}$ of the VPP300.

throughout the entire range of vector lengths ($i = 0, \cdots, 31$), though the effect lessens as i increases. This anomalous behaviour fosters the erraticity in Figure 1 (right). Whenever $m_{\min}(\nu, \epsilon) = 73, 137, 201, \ldots$ (or one or two more), the time is decreased using $\hat{m}_{\min} = m_{\min}(\hat{\nu}, \epsilon)$ points, where $\hat{\nu} \neq \nu$ is such that $\hat{m}_{\min}$ is not an anomalous vector length.

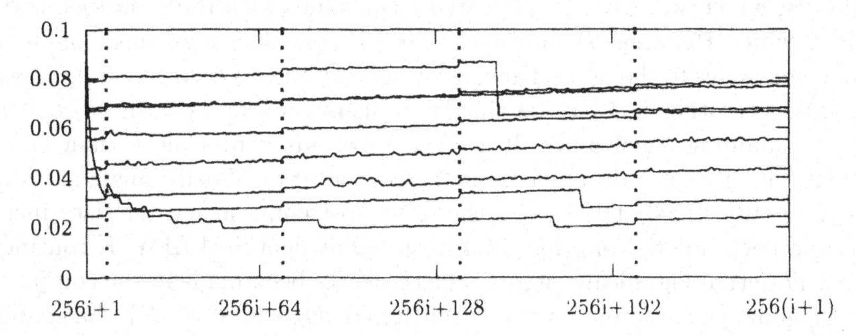

Fig. 2. Time vs. vector length $m = 1, \ldots, 2048$. Dotted lines at $m = 64i + 9$.

This performance anomaly is apparent also for standard vector operations (addition, multiplication, etc.) on the VPP300 [9]; we do not believe it has been noted before, and we are also currently unable to explain the phenomenon. The savings is of course only 10-20% but it is still probably worthwhile to avoid these anomalous vector lengths, in general.

When computing $r > 1$ eigenvalues, the optimal number of points *per eigenvalue interval*, m_{opt}, tends to decrease as r increases, until for some $r = r_b$ multi-bisection is preferable to multisection; this occurs when r is large enough that multi-bisection entails a satisfactory vector length (in relation to $n_{1/2}$). The value of r_b at which this occurs depends on the desired accuracy ϵ. For more details see [9].

Parallelization – Invariant Subspaces for Clustered Eigenvalues: Eigenvectors are calculated using the standard technique of inverse iteration (see, e.g., [14, 40]). Letting λ_i denote a converged eigenvalue, choose u^0 and iterate

$$(T - \lambda_i I)u^k = u^{k-1}, \quad k = 1, 2, \cdots.$$

Generally, one step suffices. Solution of these tridiagonal linear systems is efficiently vectorised via "wrap-around partitioning", discussed in Section 4.

The computation of distinct eigenpairs is communication free. However, computed invariant subspaces corresponding to multiple or tightly clustered eigenvalues are likely to require reorthogonalization. If these eigenvalues reside on different PEs, orthogonalisation entails significant communication. Hence, clustered eigenvalues should reside on individual PEs. This can be accomplished in a straightforward manner if complete spectral data is available to all PEs – for example, if the eigenvalue computation is performed redundantly on each PE, or if all-to-all communication is initiated after a distributed computation – in which case redistribution decisions can be made concurrently. However, we insist on a distributed eigenvalue calculation and opine that all-to-all communication is too expensive.

We detect clustering during the refinement process and effect the redistribution in an implicit manner. Once subinterval widths reach a user-defined "cluster tolerance", the smallest and largest sign counts on each PE are communicated to the PE which was initially allocated eigenvalues with those indices, and it is decided which PE keeps the cluster. This PE continues refinement of the clustered eigenvalues to the desired accuracy and computes a corresponding invariant subspace; the other ignores the clustered eigenvalues and continues refinement of any remaining eigenvalues. If a cluster extends across more than two PEs intermediate PEs are dropped from the computation. Load-imbalance is likely (and generally unavoidable) with the effect worsening as cluster sizes increase. This approach differs from that used in the equivalent ScaLAPACK routines [5].

We note that significant progress has recently been made in the computation of orthogonal eigenvectors for tightly clustered eigenvalues [7, 27], the goal being to obtain these vectors without the need for reorthogonalization. These ideas have not yet been used here, though they may be incorporated into later versions of our routines; it is expected they will be incorporated into future releases of LAPACK and ScaLAPACK [6].

3 Symmetric Eigenvalue Problems

The methods we use for symmetric EVPs entail the solution of tridiagonal EVPs, and this is accomplished using the procedures just described; overall, the techniques are relatively standard and are not discussed in detail.

The first is based on the usual Householder reduction to tridiagonal form, parallelised using a panel-wrapped storage scheme [8]; there is also a version for Hermitian matrices. For sparse matrices we use a Lanczos method [16]. Performance depends primarily on efficient matrix-vector multiplication; the routine uses a diagonal storage format, loop unrolling, and column-block matrix distribution for parallelisation. The tridiagonal EVPs that arise are solved redundantly using a single-PE version of the tridiagonal eigensolver. No performance data are presented for the Lanczos solver, but comparisons of the Householder-based routine with those in LAPACK/ScaLAPACK are included in Section 5.

4 Nonsymmetric Eigenvalue Problems

Arnoldi Methods: Arnoldi's method [2] was originally developed to reduce a matrix to upper Hessenberg form; its practicability as a Krylov subspace projection method for EVPs was established in [31]. Letting $V_m = [v_1|\cdots|v_m]$ denote the matrix whose columns are the basis vectors (orthonormalized via, e.g., modified Gram-Schmidt) for the m-dimensional Krylov subspace, we obtain the projected EVP

$$Hy = V_m^* A V_m y = \lambda y, \tag{5}$$

where H is upper Hessenberg and of size $m \ll n$. This much smaller eigenproblem is solved using, e.g., a QR method. Dominant eigenvalues of H approximate those of A with the accuracy increasing with m.

A plethora of modifications can be made to the basic Arnoldi method to increase efficiency, robustness, etc. These include: restarting [31], including the relatively recent implicit techniques [18, 38]; deflation, implicit or explicit, when computing $r > 1$ eigenvalues; preconditioning/acceleration techniques based on, e.g., Chebyshev polynomials [32], least-squares [33, 34], etc.; spectral transformations for computing non-extremal eigenvalues, e.g., shift-invert [28], Cayley [17, 22], etc.; and of course block versions [35, 36]. For a broad overview see [34].

Our current code is still at a rudimentary stage of development, but we have incorporated a basic restart procedure, shift-invert, and an implicit deflation scheme similar to that outlined in [34] and closely related to that used in [30]. Although it is possible to avoid most complex arithmetic even in the case of a complex shift [28], our routine is currently restricted to real shifts.

In order to be better able to compute multiple or clustered eigenvalues we are also developing a block version. Here matrix-vector multiplication is replaced by matrix-matrix multiplication, leading to another potential advantage, particularly in the context of high performance computing: They enable the use of level-3 BLAS. On some machines this can result in block methods being preferable even in the case of computing only a single eigenvalue [36].

Parallelization opportunities seem to be limited to the reduction phase of the algorithm. Parallelising, e.g., QR is possible and has been considered by various authors (see, e.g., [3, 13] and the references therein); however, since the projected systems are generally small, it is probably not worthwhile parallelising their eigensolution. This is the approach taken with P_ARPACK [21], an implementation of ARPACK [19] for distributed memory parallel architectures; although these packages are based on the implicitly restarted Arnoldi method [18, 38], parallelization issues are, for the most part, identical to those for the standard methods. It is probably more worthwhile to limit the maximum Hessenberg dimension to one that is viably solved redundantly on each processor and focus instead on increasing the efficiency of the restarting and deflation procedures and to add some form of preconditioning/acceleration; however, the choices for these strategies should, on vector parallel processors, be predicated on their amenability to vectorisation. As mentioned, our code is relatively nascent, and it has not yet been parallelised, nor efficiently vectorised.

Newton-Based Methods: Let $K : C^n \to C^n$ and consider the eigenvalue problem

$$K(\lambda)u = 0, \qquad K(\lambda) \equiv (A - \lambda I). \tag{6}$$

(For generalized EVPs, $Au = \lambda Bu$, define $K(\lambda) \equiv (A - \lambda B)$.) Reasonable smoothness of $K(\lambda)$ is assumed but it need not be linear in λ. The basic idea behind using Newton's method to solve EVPs is to replace (6) by the problem of finding zeros of a nonlinear function. Embed (6) in the more general family

$$K(\lambda)u = \beta(\lambda)x, \qquad s^* u = \kappa. \tag{7}$$

As λ approaches an eigenvalue $(K - \lambda I)$ becomes singular so the solution u of the first equation in (7) becomes unbounded for almost all $\beta(\lambda)x$. Hence, the second equation – a scaling condition – can only be satisfied if $\beta(\lambda) \to 0$ as λ

approaches an eigenvalue. The vectors s and x can be chosen dynamically as the iteration proceeds; this freedom results in the possibility of exceeding the second-order convergence rate characteristic of Newton-based procedures [25].

Differentiating equations (7) with respect to λ gives

$$K\frac{du}{d\lambda} + \frac{dK}{d\lambda}u = \frac{d\beta}{d\lambda}x, \qquad s^*\frac{du}{d\lambda} = 0.$$

Solving for the Newton correction, the Newton iteration takes the form

$$\lambda \longleftarrow \lambda - \Delta\lambda, \qquad \Delta\lambda = \frac{\beta(\lambda)}{\frac{d\beta}{d\lambda}} = \frac{s^*u}{s^*K^{-1}\frac{dK}{d\lambda}u}.$$

Note that for the non-generalized problem (1), $dK/d\lambda = -I$. The main computational component is essentially inverse iteration with the matrix K; this is effected using a linear solver highly tuned for the VPP300 [23].

Convergence rates, including conditions under which third-order convergence is possible, are discussed in [24]. A much more recent reference is [25] in which the development is completely in terms of generalised EVPs.

Deflation for k converged eigenvalues can be effected by replacing $\beta(\lambda)$ with

$$\phi_k(\lambda) = \frac{\beta(\lambda)}{\prod_{i=1}^{k}(\lambda - \lambda_i)}.$$

However, it is likely to be more beneficial to use as much of the existing spectral information as possible (i.e., not only the eigenvalues). Weilandt deflation (see, e.g., [34]) requires knowledge of left and right eigenvectors, hence involves matrix transposition which is highly inefficient on distributed memory architectures. Hence, we opt for a form of Schur-Weilandt deflation; see [25] for details.

A separate version of the routines for the Newton-based procedures has been developed specifically for block bidiagonal matrices. This algorithm exhibits an impressive convergence rate of 3.56, and uses a multiplicative form of Wielandt deflation so as to preserve matrix structure. Implementationally, inversion of block bidiagonal matrices is required, and this is efficiently vectorised by the technique of *wrap-around partitioning*, which we now describe.

Vectorisation – Wrap-Around Partitioning for Banded Systems: Wrap-around partitioning [12] is a technique enabling vectorisation of the elimination process in the solution of systems of linear equations. Unknowns are reordered into q blocks of p unknowns each, thereby highlighting groups of unknowns which can be eliminated independently of one another. The natural formulation is for matrices with block bidiagonal (BBD) structure, shown below on the left, but the technique is also applicable to narrow-banded matrices of sufficiently regular structure, as illustrated by reordering a tridiagonal matrix to have BBD form,

$$\begin{bmatrix} A_1 & 0 & 0 & \cdots & B_1 \\ B_2 & A_2 & 0 & \cdots & 0 \\ 0 & B_3 & A_3 & \cdots & 0 \\ \vdots & \vdots & \ddots & \ddots & \vdots \\ 0 & 0 & \cdots & B_n & A_n \end{bmatrix}; \ \text{tridiag}(\beta_i, \alpha_i, \beta_{i+1}) \longrightarrow \left[\begin{array}{cc|cc|cc|cc} 0 & 0 & 0 & 0 & \cdots & \cdots & \beta_n & \alpha_n \\ \alpha_1 & \beta_2 & 0 & 0 & \cdots & \cdots & 0 & 0 \\ \hline \beta_2 & \alpha_2 & \beta_3 & 0 & \cdots & \cdots & 0 & 0 \\ 0 & \beta_3 & \alpha_3 & \beta_4 & \cdots & \cdots & 0 & 0 \\ \hline \vdots & \vdots & \ddots & \ddots & \ddots & \ddots & \vdots & \vdots \\ \hline 0 & 0 & \cdots & \cdots & \beta_{n-2} & \alpha_{n-2} & \beta_{n-1} & 0 \\ 0 & 0 & \cdots & \cdots & 0 & \beta_{n-1} & \alpha_{n-1} & \beta_n \end{array}\right].$$

Significant speed-ups – of roughly a factor of 20 over scalar speed – are easily attainable for matrices of size $n > 1000$ in the case that the subblock dimension, m, is small ($m = 2$ for tridiagonal matrices). The case $q = 2$ corresponds to cyclic reduction, but with wrap-around partitioning p and q need not be exact factors of n; this means that stride-two memory access, which may result in bank conflicts on some computers (e.g., the VPP300), can be avoided. However, q should be relatively small; stride-3 has proven effective on the VPP300. Orthogonal factorisation is used rather than Gaussian elimination with partial pivoting since the latter is known to have less favourable stability properties for BBD matrices [41]. Stable factorisation preserves BBD structure so that wrap-around partitioning can be applied recursively.

An example is illustrative. Consider a BBD matrix of size $n = 11$. With inexact factors $p = 4$ and $q = 2$ the reordered matrix has the form shown below on the left. In the first stage the blocks B_i, $i = 2, 4, 6, 8$ are eliminated; since each is independent of the others this elimination can be vectorised. Fill-in occurs whether orthogonal factorisation or Gaussian elimination with pivoting is used; this is shown at the right, where blocks remaining nonzero are indicated by *, deleted blocks by 0, and blocks becoming nonzero by a 1 indicating fill-in occurred in that position during the first stage.

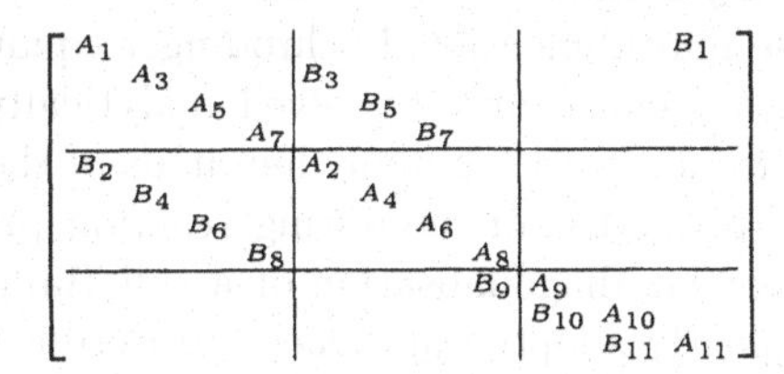

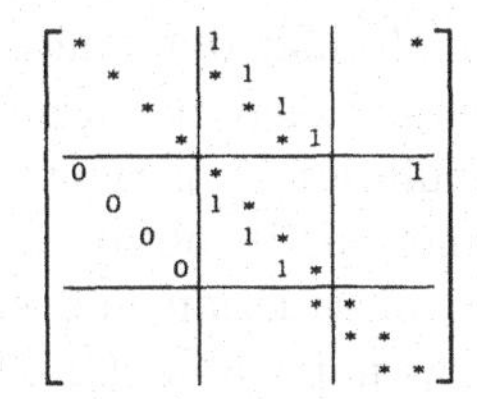

The potential for recursive application of the reordering is evinced by noting that the trailing block 2×2 submatrix – the one now requiring elimination – is again block bidiagonal. Recursion terminates when the final submatrix is small enough to be solved sequentially.

Arnoldi-Newton Methods: Although the Newton-based procedures ultimately result in convergence rates of up to 3.56, they suffer when good initial data are unavailable; unfortunately this is often the case when dealing with large-scale EVPs. Conversely, Arnoldi methods seem nearly always to move in the right direction at the outset, but may stall or breakdown as the iteration continues, for instance if the maximal Krylov dimension is chosen too small. Heuristics are required to develop efficient, robust Arnoldi eigensolvers. In a sense then, these methods may be viewed as having orthogonal difficulties: Newton methods suffer at the outset, but ultimately perform very well, and Arnoldi methods start off well but perhaps run into difficulties as the iteration proceeds. For this reason we have considered a composition of the two methods: The Arnoldi method is used to get good initial estimates to the eigenvalues of interest and their corresponding Schur vectors for use with the Newton-based procedures. These methods are in the early stages of development.

5 Performance Results on Applications

We now investigate the performance of some of our routines on eigenvalue problems arising in a variety of applications and make some comparisons with corresponding routines from LAPACK and ScaLAPACK. More details and extended performance results on these applications can be found in [11].

Fujitsu VPP300: All performance experiments were performed on the thirteen processor Fujitsu VPP300 located at the Australian National University. The Fujitsu VPP300 is a distributed-memory parallel array of powerful vector PEs, each with a peak rate of 2.2 Gflops and 512 MB or 1.92 GB of memory – the ANU configuration has a peak rate of about 29 Gflops and roughly 14 GB of memory. The PEs consist of a scalar unit and a vector unit with one each of load, store, add, multiply and divide pipes. The network is connected via a full crossbar switch so all processors are equidistant; peak bandwidth is 570MB/s bi-directional. Single-processor routines are written in Fortran 90 and parallel routines in VPP Fortran, a Fortran 90-based language with compiler directives for data layout specification, communication, etc.

Tridiagonal EVP – Molecular Dynamics: First we consider an application arising in molecular dynamics. The multidimensional Schrödinger equation describing the motion of polyatomic fragments cannot be solved analytically. In numerical simulations it is typically necessary to use many hundreds of thousands of basis functions to obtain accurate models of interesting reaction processes; hence, the construction and subsequent diagonalisation of a full Hamiltonian matrix, H, is not viable. One frequently adopted approach is to use a Lanczos method, but the Krylov dimension – the size of the resulting tridiagonal matrix, T – often exceeds the size of H. Thus, computation of the eigenvalues of T becomes the dominant computational component.

To test the performance of the tridiagonal eigensolver described in Section 2 we compute some eigenvalues and eigenvectors for a tridiagonal matrix of order $n = 620,000$ arising in a molecular dynamics calculation [29]. In Table 1 we present results obtained using our routines (denoted SSL2VP since they have been incorporated into Fujitsu's Scientific Subroutine Library for Vector Processors) and the corresponding ones from LAPACK on a single VPP300 processor for the computation of one eigenpair and 100 eigenpairs of the 11411 which are of interest for this particular matrix. For further comparison we also include times for the complete eigendecomposition of matrices of size 1000, 3000, and 5000. Values are computed to full machine precision.

# (λ, u)	1	100	1000	3000	5000
SSL2VP	23.06	162.7	3.208	23.90	63.43
LAPACK	25.11	768.1	7.783 (25.46)	132.4 (395.5)	581.1 (1717.)

Table 1. Tridiagonal eigensolver: Molecular dynamics application.

For large numbers of eigenvalues the optimal form of multisection is multi-bisection so that there are no significant performance differences between the two eigenvalue routines. However, the eigenvector routine using wrap-around partitioning is significantly faster than the LAPACK implementation, resulting in considerably reduced times when large numbers of eigenvectors are required. We note that the LAPACK routine uses a tridiagonal QR method when *all* eigenvalues are requested. If $r \leq n-1$ are requested, bisection with inverse iteration is used; the parenthetical times - a factor of three larger - are those obtained computing $n-1$ eigenpairs and serve further to illustrate the efficiency of our implementation. Effective parallelisation is evident from the performance results of the symmetric eigensolver, which we next address.

Symmetric EVP – Quantum Chemistry: An application arising in computational quantum chemistry is that of modelling electron interaction in protein molecules. The eigenvalue problem again arises from Schrödinger's equation, $\mathcal{H}\Psi = E\Psi$, where $\mathcal{H}$ is the Hamiltonian operator, E is the total energy of the system, and Ψ is the wavefunction. Using a semi-empirical, as opposed to *ab initio*, formulation we arrive at an eigenvalue problem,

$$F\psi = \epsilon\psi,$$

where ϵ is the energy of an electron orbital and ψ the corresponding wavefunction. The software package MNDO94 [39] is used to generate matrices for three protein structures: (1) single helix of pheromone protein from Euplotes Raikovi (310 atoms, $n = 802$), (2) third domain of turkey ovomucoid inhibitor (814 atoms, $n = 2068$), and (3) bovine pancreatic ribonuclease A (1856 atoms, $n = 4709$). In Table 2 we give the times required to compute all eigenpairs of the resulting symmetric EVPs using our routines based on Householder reduction and the tridiagonal eigensolver; SSL2VP indicates the single processor version and SSL2VPP the parallel version included in Fujitsu's Scientific Subroutine Library for Vector Parallel Processors. Also given are times obtained using the analogous routines from LAPACK and ScaLAPACK, respectively. As noted above, when all eigenvalues are requested these routines use QR on the resulting tridiagonal system and this is faster than their routines based on bisection and inverse iteration. Thus, if fewer than n eigenvalues are requested the performance differences between those routines and ours are amplified considerably.

n	802			2068			4709		
# PEs	1	2	4	1	2	4	1	2	4
SSL2V(P)P	4.130	3.157	1.783	35.55	24.76	12.69	373.1	217.5	113.1
(Sca)LAPACK	5.884	7.362	6.437	68.22	59.62	46.03	694.2	467.8	314.6

Table 2. Symmetric eigensolver: Quantum chemistry application.

The reduction and eigenvector recovery algorithmic components are apparently more efficiently implemented in ScaLAPACK, but the efficiency of our tridiagonal eigensolvers results in superior performance for the SSL2V(P)P routines. Good scalability of our parallel implementation is also evident.

Nonsymmetric EVP – Optical Physics: Now we consider an application from optical physics, namely the design of dielectric waveguides, e.g., optical fibers. We solve the vector wave equation for general dielectric waveguides which, for the magnetic field components, takes the form of two coupled PDEs

$$\frac{\partial^2 H_x}{\partial x^2} + \frac{\partial^2 H_x}{\partial y^2} - 2\frac{\partial \ln(n)}{\partial y}\left(\frac{\partial H_y}{\partial x} - \frac{\partial H_x}{\partial y}\right) + (n^2k^2 - \beta^2)H_x = 0$$

$$\frac{\partial^2 H_y}{\partial x^2} + \frac{\partial^2 H_y}{\partial y^2} - 2\frac{\partial \ln(n)}{\partial x}\left(\frac{\partial H_y}{\partial x} - \frac{\partial H_x}{\partial y}\right) + (n^2k^2 - \beta^2)H_y = 0.$$

Following [20], a Galerkin procedure is applied to this system, resulting in a coupled system of algebraic equations. Given an optical fiber with indices of refraction n_o and n_i for the cladding and core regions, respectively, eigenvalues of interest correspond to "guided modes" and are given by $\lambda = (\beta/k)^2$, where $\beta/k \in [n_o, n_i)$. The matrix is full, real, and nonsymmetric; despite the lack of symmetry all eigenvalues are real.

Since the Arnoldi-based procedures have not yet been efficiently vectorised/parallelised we do not compare performance against, e.g., ARPACK or P_ARPACK. Instead we illustrate the considerable reduction in time obtained using the Arnoldi method to acquire good initial data for the Newton-based procedures – that is, the Arnoldi-Newton method. This comparison is somewhat contrived since the Newton codes use full complex arithmetic and the matrix for this problem is real. However, it serves to elucidate the effectiveness of combining the two procedures. Considering a fiber with indices of refraction $n_o = 1.265$ and $n_i = 1.415$, we use shift-invert Arnoldi (in real arithmetic) with a shift of $\sigma = 1.79 \in [1.265^2, 1.415^2)$. Once Ritz estimates are $O(1 \times 10^{-9})$, eigenvalues and Schur vectors are passed to the Newton-based routine. Generally, only one Newton step is then necessary to reach machine precision. Times for complex LU factorisation are shown, and the number of factorisations required with the Newton and Arnoldi-Newton methods appears in parentheses.

n	cmplx LU	Newton	Arnoldi-Newton
1250	4.105	1414. (108)	328.4 (26)
3200	62.92	20904 (103)	3597. (31)

Table 3. Nonsymmetric eigensolver: Optical physics application.

Matrices are built using the (C++) software library NPL [15] which also includes an eigensolver; our routine consistently finds eigenvalues which NPL's fails to locate – in this example, NPL located ten eigenvalues of interest while our routines find fourteen. The techniques used here are robust and effective. Clearly the use of Arnoldi's method to obtain initial eigendata results in a significant reduction in time. More development work is needed for these methods and for their implementations, in particular the (block) Arnoldi routine which is not nearly fully optimised.

Complex Nonsymmetric Block Bidiagonal EVP – CFD: Our final application is from hydrodynamic stability; we consider flow between infinite parallel plates. The initial behaviour of a small disturbance to the flow is modelled by the Orr-Sommerfeld equation

$$\frac{i}{\alpha R}\left(\frac{d^2}{dz^2}-\alpha^2\right)^2\phi+(U(z)-\lambda)\left(\frac{d^2}{dz^2}-\alpha^2\right)\phi-\frac{d^2U(z)}{dz^2}\phi=0,$$

where α is the wave number, R is the Reynolds' number, and $U(z)$ is the velocity profile of the basic flow; we assume plane Poiseuille flow for which $U(z)=1-z^2$. Boundary conditions are $\phi = d\phi/dz = 0$ at solid boundaries and, for boundary layer flows, $\phi \sim 1$ as $z \to \infty$. The differential equation is written as a system of four first-order equations which, when integrated using the trapezoidal rule, yields a generalised EVP

$$K(\alpha,R)u = s^T K(\alpha,R), \quad K(\alpha,R)=A(\alpha,R)-\lambda B,$$

where $A(\alpha,R)$ is complex and block bidiagonal with 4×4 blocks. Further description of the problem formulation can be found in [11].

We compute the neutral curve, i.e. the locus of points in the (α,R)-plane for which $\mathrm{Im}\{c(\alpha,R)\}=0$, using the Newton-based procedures of the last section. The resulting algorithm has convergence rate 3.56 [25]. We use a grid with 5000 points for which A is of order $n = 20000$. A portion of the neutral curve is plotted in Figure 3.

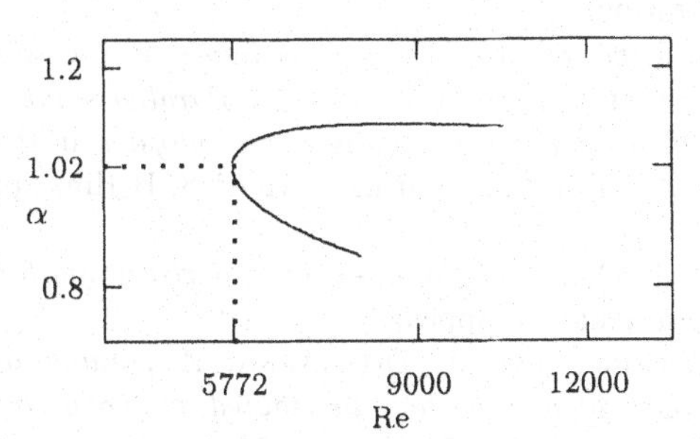

Fig. 3. Complex banded nonsymmetric eigensolver: Neutral stability diagram for Poiseuille flow.

The significant degree of vectorisation obtained using wrap-around partitioning enables us to consider highly refined discretizations. Additional results, including consideration of a Blasius velocity profile, can be found in [11].

Acknowledgements

The authors thank the following for assistance with the section on applications: Anthony Rasmussen, Sean Smith, Andrey Bliznyuk, Margaret Kahn, Francois Ladouceur, and David Singleton. This work was supported as part of the Fujitsu-ANU Parallel Mathematical Subroutine Library Project.

References

1. E. Anderson, Z. Bai, C. Bischof, J. Demmel, J. Dongarra, J. Du Croz, A. Greenbaum, S. Hammarling, A. McKenny, S. Ostrouchov, and D. Sorensen, *LAPACK: Linear Algebra PACKage*. software available from `http://www.netlib.org` under directory "lapack".
2. W. Arnoldi, *The principle of minimized iterations in the solution of the matrix eigenvalue problem*, Quarterly of Appl. Math., 9 (1951), pp. 17–29.
3. Z. Bai and J. Demmel, *Design of a parallel nonsymmetric eigenroutine toolbox, Part I*, Tech. Rep. Computer Science Division Report UCB/CSD-92-718, University of California at Berkeley, 1992.
4. L. Blackford, J. Choi, A. Cleary, E. D'Azevedo, J. Demmel, I. Dhillon, J. Dongarra, S. Hammarling, G. Henry, A. Petitet, K. Stanley, D. Walker, and R. Whaley, *ScaLAPACK: Scalable Linear Algebra PACKage*. software available from `http://www.netlib.org` under directory "scalapack".
5. J. Demmel, I. Dhillon, and H. Ren, *On the correctness of some bisection-like parallel eigenvalue algorithms in floating point arithmetic*, Electronic Trans. Num. Anal. (ETNA), 3 (1996), pp. 116–149.
6. I. Dhillon, 1997. Private communication.
7. I. Dhillon, G. Fann, and B. Parlett, *Application of a new algorithm for the symmetric eigenproblem to computational quantum chemisty*, in Proc. of the Eight SIAM Conf. on Par. Proc. for Sci. Comput., SIAM, 1997.
8. J. Dongarra and R. Van de Geijn, *Reduction to condensed form for the eigenvalue problem on distributed memory architectures*, Parallel Computing, 18 (1992), pp. 973–982.
9. D. Harrar II, *Determining optimal vector lengths for multisection on vector processors*. In preparation.
10. ———, *Multisection vs. bisection on vector processors*. In preparation.
11. D. Harrar II, M. Kahn, and M. Osborne, *Parallel solution of some large-scale eigenvalue problems arising in chemistry and physics*, in Proc. of Fourth Int. Workshop on Applied Parallel Computing: PARA'98, Berlin, Springer-Verlag. To appear.
12. M. Hegland and M. Osborne, *Wrap-around partitioning for block bidiagonal systems*, IMA J. Num. Anal. to appear.
13. G. Henry, D. Watkins, and J. Dongarra, *A parallel implemenations of the nonsymmetric QR algorithm for distributed memory architectures*, Tech. Rep. Computer Science Technical Report CS-97-355, University of Tennessee at Knoxville, 1997.
14. I. Ipsen, *Computing an eigenvector with inverse iteration*, SIAM Review, 39 (1997), pp. 254–291.
15. F. Ladouceur, 1997. Numerical Photonics Library, version 1.0.
16. C. Lanczos, *An iteration method for the solution of the eigenvalue problem of linear differential and integral operators*, J. Res. Nat. Bur. Standards, 45 (1950), pp. 255–282.
17. R. Lehoucq and K. Meerbergen, *Using generalized Cayley transformations within an inexact rational Krylov sequence method*, SIAM J. Mat. Anal. and Appl. To appear.
18. R. Lehoucq and D. Sorensen, *Deflation techniques for an implicitly restarted Arnoldi iteration*, SIAM J. Mat. Anal. and Appl., 8 (1996), pp. 789–821.

19. R. LEHOUCQ, D. SORENSEN, AND P. VU, *ARPACK: An implementation of the Implicitly Restarted Arnoldi Iteration that computes some of the eigenvalues and eigenvectors of a large sparse matrix*, 1995.
20. D. MARCUSE, *Solution of the vector wave equation for general dielectric waveguides by the Galerkin method*, IEEE J. Quantum Elec., 28(2) (1992), pp. 459–465.
21. K. MASCHOFF AND D. SORENSEN, *P_ARPACK: An efficient portable large scale eigenvalue package for distributed memory parallel architectures*, in Proc. of the Third Int. Workshop on Appl. Parallel Comp. (PARA96), Denmark, 1996.
22. K. MEERBERGEN, A. SPENCE, AND D. ROOSE, *Shift-invert and Cayley transforms for detection of rightmost eigenvalues of nonsymmetric matrices*, BIT, 34 (1995), pp. 409–423.
23. M. NAKANISHI, H. INA, AND K. MIURA, *A high performance linear equation solver on the VPP500 parallel supercomputer*, in Proc. Supercomput. '94, 1994.
24. M. OSBORNE, *Inverse iteration, Newton's method, and nonlinear eigenvalue problems*, in The Contributions of J.H. Wilkinson to Numerical Analysis, Symposium Proc. Series No. 19, The Inst. for Math. and its Appl., 1979.
25. M. OSBORNE AND D. HARRAR II, *Inverse iteration and deflation in general eigenvalue problems*, Tech. Rep. Mathematics Research Report No. MRR 012-97, Australian National University. submitted.
26. B. PARLETT, *The Symmetric Eigenvalue Problem*, Prentice Hall, Englewood Cliffs, 1980.
27. B. PARLETT AND I. DHILLON, *Fernando's solution to Wilkinson's problem: an application of double factorization*, Lin. Alg. Appl., 267 (1997), pp. 247–279.
28. B. PARLETT AND Y. SAAD, *Complex shift and invert stategies for real matrices*, Lin. Alg. Appl., 88/89 (1987), pp. 575–595.
29. A. RASMUSSEN AND S. SMITH, 1998. Private communication.
30. A. RUHE, *The rational Krylov algorithm for nonsymmetric eigenvalue problems, III: Complex shifts for real matrices*, BIT, 34 (1994), pp. 165–176.
31. Y. SAAD, *Variations on Arnoldi's method for computing eigenelements of large unsymmetric matrices*, Lin. Alg. Appl., 34 (1980), pp. 269–295.
32. ———, *Chebyshev acceleration techniques for solving nonsymmetric eigenvalue problems*, Math. Comp., 42(166) (1984), pp. 567–588.
33. ———, *Least squares polynomials in the complex plane and their use for solving parse nonsymmetric linear systems*, SIAM J. Numer. Anal., 24 (1987), pp. 155-169.
34. ———, *Numerical Methods for Large Eigenvalue Problems*, Manchester University Press (Series in Algorithms and Architectures for Advanced Scientific Computing), Manchester, 1992.
35. M. SADKANE, *A block Arnoldi-Chebyshev method for computing the leading eigenpairs of large sparse unsymmetric matrices*, Numer. Math., 64 (1993), pp. 181–193.
36. J. SCOTT, *An Arnoldi code for computing selected eigenvalues of sparse real unsymmetric matrices*, ACM Trans. on Math. Soft., 21 (1995), pp. 432–475.
37. H. SIMON, *Bisection is not optimal on vector processors*, SIAM J. Sci. Stat. Comput., 10 (1989), pp. 205–209.
38. D. SORENSEN, *Implicit application of polynomial filters in a k-step Arnoldi method*, SIAM J. Mat. Anal. and Appl., 13 (1992), pp. 357–385.
39. W. THIEL, 1994. Program MNDO94, version 4.1.
40. J. WILKINSON, *The Algebraic Eigenvalue Problem*, Clarendon Press, Oxford, 1965.
41. S. WRIGHT, *A collection of problems for which Gaussian elimination with partial pivoting is unstable*, SIAM J. Sci. Stat. Comput., 14 (1993), pp. 231–238.

Direct Linear Solvers for Vector and Parallel Computers*

Friedrich Grund

Weierstrass Institute for Applied Analysis and Stochastics
Mohrenstrasse 39, 10117 Berlin, Germany
grund@wias-berlin.de
http://www.wias-berlin.de/~grund

Abstract. We consider direct methods for the numerical solution of linear systems with unsymmetric sparse matrices. Different strategies for the determination of the pivots are studied. For solving several linear systems with the same pattern structure we generate a pseudo code, that can be interpreted repeatedly to compute the solutions of these systems. The pseudo code can be advantageously adapted to vector and parallel computers. For that we have to find out the instructions of the pseudo code which are independent of each other. Based on this information, one can determine vector instructions for the pseudo code operations (vectorisation) or spread the operations among different processors (parallelisation). The methods are successfully used on vector and parallel computers for the circuit simulation of VLSI circuits as well as for the dynamic process simulation of complex chemical production plants.

1 Introduction

For solving systems of linear equations

$$Ax = b, \quad A \in \mathbb{R}^{n \times n}, \quad x, b \in \mathbb{R}^n \tag{1}$$

with non singular, unsymmetric and sparse matrices A, we use the Gaussian elimination method. Only the nonzero elements of the matrices are stored for computation. In general, we need to establish a suitable control for the numerical stability and for the fill-in of the Gaussian elimination method.

For the time domain simulation in many industrial applications structural properties are used for a modular modelling. Thus electronic circuits usually consist of identical subcircuits as inverter chains or adders. Analogously, complex chemical plants consist of process units as pumps, reboilers or trays of distillation columns. A mathematical model is assigned to each subcircuit or unit and they are coupled. This approach leads to initial value problems for large systems of differential–algebraic equations. For solving such problems we use backward differentiation formulas and the resulting systems of nonlinear equations are

* This work was supported by the Federal Ministry of Education, Science, Research and Technology, Bonn, Germany under grants GA7FVB-3.0M370 and GR7FV1.

J. Palma, J. Dongarra, and V. Hernández (Eds.): VECPAR'98, LNCS 1573, pp. 114–127, 1999.

solved with Newton methods. The Jacobian matrices are sparse and maintain their sparsity structure during the integration over many time steps. In general, the Gaussian elimination method can be used with the same ordering of the pivots for these steps. A pseudo code is generated to perform the factorisations of the matrices and the solving of the systems with triangular matrices efficiently. This code contains only the required operations for the factorisation and for solving the triangular systems. It is defined independently of a computer and can be adapted to vector and parallel computers.

The solver has been proven successfully for the dynamic process simulation of large real life chemical production plants and for the electric circuit simulation as well. Computing times for complete dynamic simulation runs of industrial applications are given. For different linear systems with matrices arising from scientific and technical problems the computing times for several linear solvers are compared.

2 The Method

The Gaussian elimination method

$$PAQ = LU, \tag{2}$$

$$Ly = Pb, \quad UQ^{-1}x = y \tag{3}$$

is used for solving the linear systems (1). The nonzero elements of the matrix A are stored in compressed sparse row format, also known as sparse row wise format. L is a lower triangular and U an upper triangular matrix. The row permutation matrix P is used to provide numerical stability and the column permutation matrix Q is used to control sparsity. In the following, we consider two approaches for the determination of the matrices P and Q.

In the first approach, at each step of the elimination, the algorithm searches for the 1st column with a minimal number of nonzero elements. This column becomes the pivot column [6] and the columns are reordered (dynamic ordering). This approach can be considered as a variant of a pivotal strategy described in [13]. For keeping the method numerically stable at stage k of the elimination, the pivot $a_{i,j}$ is selected among those candidates satisfying the numerical threshold criterion

$$|a_{i,j}| \geq \beta \max_{l \geq k} |a_{l,j}|$$

with a given threshold parameter $\beta \in (0, 1]$. This process is called partial pivoting. In our applications we usually choose $\beta = 0.01$ or $\beta = 0.001$.

In the second approach, we determine in a first step the permutation matrix Q by minimum degree ordering of $A^T A$ or of $A^T + A$, using the routine *get_perm_c.c* from SuperLU [8]. Then the columns are reordered and in a separate step the permutation matrix P is determined by using partial pivoting.

3 Pseudo Code

It is sometimes possible to use the Gaussian elimination method with the same pivot ordering to solve several linear systems with the same pattern structure of the coefficient matrix. To do this, we generate a pseudo code to perform the factorisation of the matrix as well as to solve the triangular systems (forward and back substitution).

For the generation of the pseudo code, the factorisation of the Gaussian elimination method is used as shown in Fig. 1.

```
for i = 2, n do
    a_{i-1,i-1} = 1/a_{i-1,i-1}
    for j = i, n do
        a_{j,i-1} = (a_{j,i-1} - Σ_{k=1}^{i-2} a_{j,k} a_{k,i-1}) a_{i-1,i-1}
    enddo
    for j = i, n do
        a_{i,j} = a_{i,j} - Σ_{k=1}^{i-1} a_{i,k} a_{k,j}
    enddo
enddo
a_{n,n} = 1/a_{n,n}.
```

Fig. 1. Gaussian elimination method (variant of Crout algorithm)

The algorithm needs n divisions. Six different types of pseudo code instructions are sufficient for the factorisation of the matrix, four instructions for the computation of the elements of the upper triangular matrix and two of the lower triangular matrix. For computing the elements of the upper triangular matrix one has to distinguish between the cases that the element is a pivot or not and that it exists or that it is generated by fill-in. For the determination of the elements of the lower triangular matrix one has only to distinguish that the element exists or that it is generated by fill-in.

Let l, with $1 \leq l \leq 6$, denote the type of the pseudo code instruction, n the number of elements of the scalar product and $k, m, i_\kappa, j_\kappa, \kappa = 1, 2, \ldots, n$ the indices of matrix elements. Then, the instruction of the pseudo code to compute an element of the lower triangular matrix

$$a(k) = \left(a(k) - \sum_{\kappa=1}^{n} a(i_\kappa) a(j_\kappa) \right) a(m)$$

is coded in the following form

1	n	i_1	j_1	...	i_n	j_n	k	m

.

The integer numbers $l, n, i_\kappa, j_\kappa, k$ and m are stored in integer array elements. For l and n only one array element is used.

The structure of the other pseudo code instructions is analogous.

Let μ denote the number of multiplications and divisions for the factorisation of the matrix and ν the number of nonzero elements of the upper and lower triangular matrices. Then one can estimate the number of integer array elements that are necessary to store the pseudo code with

$$\gamma(\mu + \nu).$$

For many linear systems arising from chemical process and electric circuit simulation $\gamma \approx 2.2$ was found to be sufficient for large systems with more than thousand equations while one has to choose $\gamma \approx 4$ for smaller systems.

4 Vectorisation and Parallelisation

The pseudo code instructions are used for the vectorisation and the parallelisation as well. For the factorisation in (2) and for solving the triangular systems in (3), elements have to be found that can be computed independently of each other.

In the case of the factorisation, a matrix

$$M = (m_{i,j}), \quad m_{i,j} \in \mathbb{N} \cup \{0, 1, 2, \ldots, n^2\}$$

is assigned to the matrix

$$LU = PAQ,$$

where $m_{i,j}$ denotes the level of independence.

In the case of solving the triangular systems, vectors

$$p = (p_i) \quad and \quad q = (q_i), \quad p_i, q_i \in \{0, 1, \ldots, n\}$$

are assigned analogously to the vectors x and y from

$$Ly = Pb \quad and \quad UQ^{-1}x = y.$$

Here the levels of independence are denoted by p_i and q_i.

The elements with the assigned level zero do not need any operations. Now, all elements with the same level in the factorised matrix (2) as well as in the vectors x and y from (3) can be computed independently. First all elements with level one are computed, then all elements with level two and so on.

The levels of independence for the matrix elements in (2) and for the vector elements in (3) can be computed with the algorithm of Yamamoto and Takahashi [10]. The algorithm for the determination of the levels of independence $m_{i,j}$ is shown in Fig. 2. The corresponding algorithm for the determination of the elements of the vectors p and q is analogous to it.

```
M = 0
for i = 1, n - 1 do
    for all {j : a_{j,i} ≠ 0 & j > i} do
        m_{j,i} = 1 + max(m_{j,i}, m_{i,i})
        for all {k : a_{i,k} ≠ 0 & k > i} do
            m_{j,k} = 1 + max(m_{j,k}, m_{j,i}, m_{i,k})
        enddo
    enddo
enddo.
```

Fig. 2. Algorithm of Yamamoto and Takahashi

For a vector computer, we have to find vector instructions at the different levels of independence [2, 6]. Let $a(i)$ denote the nonzero elements in LU. The vector instructions, shown in Fig. 3, have been proven to be successful in the case of factorisation. The difficulty is that the array elements are addressed indirectly. But adequate vector instructions exist for many vector computers. The Cray vector computers, for example, have explicit calls to gather/scatter routines for the indirect addressing.

$$s = \sum_{\kappa} a(i_\kappa) * a(j_\kappa)$$

$$a(i_k) = 1/a(i_k)$$

$$a(i_k) = a(i_k) * a(i_l)$$

$$a(i_k) = (a(i_l) * a(i_m) + a(i_p) * a(i_q)) * a(i_k)$$

Fig. 3. Types of vector instructions for factorisation

For parallelisation, it needs to distinguish between parallel computers with shared memory and with distributed memory.

In the case of parallel computers with shared memory and p processors, we assign the pseudo code for each level of independence in parts of approximately same size to the processors. After the processors have executed their part of the pseudo code instructions of a level concurrently, a synchronization among the processors is needed. Then the execution of the next level can be started. If the processors are vector processors then this property is also used. The moderate parallel computer Cray J90 with a maximum number of 32 processors is an example for such a computer.

In the case of parallel computers with distributed memory and q processors, the pseudo code for each level of independence is again partitioned into q parts of approximately same size. But in this case, the parts of the pseudo code are moved to the memory of each individual processor. The transfer of parts of the code to the memories of the individual processors is done only once. A synchronization is carried out analogous to the shared memory case. The partitioning and the storage of the matrix as well as of the vectors is implemented in the following way. For small problems the elements of the matrix, right hand side and solution vector are located in the memory of one processor, while for large problems, they have to be distributed over the memories of several processors. We assume that the data communication between the processors for the exchange of data concerning elements of the matrix, right hand side and solution vector is supported by a work sharing programming model. The massive parallel computers Cray T3D and T3E are examples for such computers.

Now, we consider a small example to illustrate our approach. For a matrix

$$A = \begin{pmatrix} 9 & & & 2 & 1 \\ 1 & & 3 & & 5 \\ & 2 & & 4 & \\ 1 & & 7 & & 8 \\ & 5 & & 7 & 9 \end{pmatrix} \tag{4}$$

the determination the permutation matrices P and Q gives

$$PAQ = \begin{pmatrix} 2 & 4 & & & \\ 5 & 7 & & & 9 \\ & 2 & 9 & & 1 \\ & & 1 & 7 & 8 \\ & & 1 & 3 & 5 \end{pmatrix}. \tag{5}$$

The nonzero elements of the matrix A are stored in sparse row format in the vector a. Let $\boxed{\text{i}}$ denote the index of the i-th element in the vector a, then the elements of the matrix PAQ are stored in the following way

$$\begin{pmatrix} \boxed{7} & \boxed{8} & & & \\ \boxed{12} & \boxed{13} & & & \boxed{14} \\ & \boxed{2} & \boxed{1} & & \boxed{3} \\ & & \boxed{9} & \boxed{10} & \boxed{11} \\ & & \boxed{4} & \boxed{5} & \boxed{6} \end{pmatrix}. \tag{6}$$

The matrix M assigned to the matrix PAQ is found to be

$$M = \begin{pmatrix} 0 & 0 & & & \\ 1 & 2 & & & 0 \\ & 3 & 0 & & 4 \\ & & 1 & 0 & 5 \\ & & 1 & 1 & 6 \end{pmatrix}. \tag{7}$$

From (7), we can see, that six independent levels exist for the factorisation.

Table 1. Instructions for the factorisation

Level	Instructions
1	a(12) = a(12)/a(7)
	a(9) = a(9)/a(1)
	a(4) = a(4)/a(1)
	a(5) = a(5)/a(10)
2	a(13) = a(13) – a(12) ⋆ a(8)
3	a(2) = a(2)/a(13)
4	a(3) = a(3) – a(2) ⋆ a(14)
5	a(11) = a(11) – a(5) ⋆ a(3)
6	a(6) = a(6) – a(4) ⋆ a(3) – a(5) ⋆ a(11)

The instructions for the factorisation of the matrix A resulting from (4) – (7) are shown in Table 1. Now, we consider, for example, the instructions of level one in Table 1 only. One vector instruction of the length four can be generated (see Fig.3) on a vector computer.

On a parallel computer with distributed memory and two processors, the allocation of the instructions of level one to the processors is shown in Table 2. The transfer of the instructions to the local memory of the processors is done during the analyse step of the algorithm. The data transfer is carried out by the operating system. On a parallel computer with shared memory the approach is analogous. The processors have to be synchronised after the execution of the instructions of each level.

Table 2. Allocation of instructions to processors

	processor one	processor two
computation of	a(12), a(9)	a(4), a(5)
	synchronization	

From our experiments with many different matrices arising from the process simulation of chemical plants and the circuit simulation respectively, it was found that the number of levels of independence is small. The number of instructions in the first two levels is very large, in the next four to six levels it is large and finally it becomes smaller and smaller.

5 Numerical Results

The developed numerical methods are realized in the program package GSPAR. GSPAR is implemented on workstations (Digital AlphaStation, IBM RS/6000, SGI, Sun UltraSparc 1 and 2), vector computers (Cray J90, C90), parallel computers with shared memory (Cray J90, C90, SGI Origin2000, Digital AlphaServer) and parallel computers with distributed memory (Cray T3D).

The considered systems of linear equations result from real life problems in the dynamic process simulation of chemical plants, in the electric circuit simulation and in the account of capital links (political sciences) [1]. The $n \times n$ matrices A with $|A|$ nonzero elements are described in Table 3. In Table 4 results for the

Table 3. Test matrices

name	discipline	n	$\|A\|$
bayer01	chemical	57 735	277 774
b_dyn	engineering	1 089	4 264
bayer02		13 935	63 679
bayer03		6 747	56 196
bayer04		20 545	159 082
bayer05		3 268	27 836
bayer06		3 008	27 576
bayer09		3 083	21 216
bayer10		13 436	94 926
advice3388	circuit	33 88	40 545
advice3776	simulation	3 776	27 590
cod2655_tr		2 655	24 925
meg1		2 904	58 142
meg4		5 960	46 842
rlxADC_dc		5 355	24 775
rlxADC_tr		5 355	32 251
zy3315		3 315	15 985
poli	account of	4 008	8 188
poli_large	capital links	15 575	33 074

matrices in Table 3 are shown using the method GSPAR on a DEC AlphaServer with an alpha EV5.6 (21164A) processor. Here, *# op LU* is the number of operations (only multiplications and divisions) and *fill-in* is the number of fill-ins during the factorisation. The cpu time (in seconds) for the first factorisation,

[1] Some matrices, which are given in Harwell–Boeing format and interesting details about the matrices, can be found in Tim Davis, University of Florida Sparse Matrix Collection, http://www.cise.ufl.edu/~davis/sparse/

presented in *strat*, includes the times for the analysis as well as for the numerical factorisation. The cpu time for the generation of the pseudo code is given in *code*. At the one hand, a dynamic ordering of the columns can be applied during the pivoting. At the other hand, a minimum degree ordering of A^TA (upper index *) or of $A^T + A$ (upper index $^+$) can be used before the partial pivoting.

Table 4. GSPAR first factorisation and generation pseudo code

	dynamic ordering				minimum degree ordering			
name	# op LU	fill-in	strat.	code	# op LU	fill-in	strat.	code
bayer01	10 032 621	643 898	35.18	12.72	13 860 173	812 505	5.75	9.95 *
b_dyn	15 902	2 909	0.02	0	21 556	8 231	0.02	0.02 *
bayer02	2 095 207	134 546	2.28	1.30	2 030 130	165 357	1.03	2.20 *
bayer03	1 000 325	64 130	0.68	0.47	625 272	53 991	0.25	0.35 *
bayer04	5 954 718	268 006	5.33	3.93	6 340 579	290 021	1.95	2.77 *
bayer05	119 740	11 024	0.15	0.03	474 273	33 797	0.18	0.17 *
bayer06	3 042 620	73 773	0.85	1.00	5 008 097	129 278	1.42	1.52 *
bayer09	364 731	23 145	0.18	0.15	287 947	22 022	0.12	0.12 *
bayer10	5 992 500	227 675	3.05	2.55	3 953 687	203 633	1.28	1.40 *
advice3388	310 348	9 297	0.38	0.65	396 965	9 818	0.75	0.95 +
advice3776	355 465	25 656	0.35	0.75	382 224	26 074	0.62	0.98 +
cod2655_tr	3 331 105	113 640	0.90	1.00	4 839 771	144 875	1.50	1.40 +
meg1	796 797	40 436	0.32	0.40	1 245 847	59 558	0.48	0.78 +
meg4	420 799	38 784	0.68	0.62	376 324	35 008	0.30	0.48 +
rlxADC_dc	73 612	5 404	0.38	0.13	63 227	2 906	0.08	0.08 +
rlxADC_tr	988 759	47 366	0.85	1.13	1 049 623	48 888	0.72	1.13 +
zy3315	47 326	8 218	0.12	0.03	49 263	8 202	0.03	0.02 +
poli	4 620	206	0.15	0	6 094	41	0.02	0 *
poli_large	43 310	10 318	2.38	0.25	34 115	588	0.08	0.03 +

The results in Table 4 show the following characteristics. For linear systems arising from the process simulation of chemical plants, the analyse step with the minimum degree ordering is in most cases, particularly for large systems, faster than the dynamic ordering, but the fill-in and the number of operations for the factorisation are larger. On the other hand, for systems arising from the circuit simulation the factorisation with the dynamic ordering is in most cases faster then the minimum degree ordering. The factorisation with the minimum degree ordering of A^TA is favourable for systems arising from chemical process simulation, while using an ordering of $A^T + A$ is recommendable for systems arising from the circuit simulation. The opposite cases of the minimum degree ordering are unfavourable because the number of operations and the amount of fill-in are very large.

In Table 5, cpu times (in seconds) for a second factorisation using the existing pivot sequence are shown for the linear solvers UMFPACK [4], SuperLU with minimum degree ordering of $A^T A$ (upper index *) or of $A^T + A$ (upper index$^+$) [5], Sparse [7] and GSPAR with dynamical column ordering, using a DEC AlphaStation with an alpha EV4.5 (21064) processor. In many applications, mainly in the numerical simulation of physical and chemical problems, the analysis step including ordering and first factorisation is performed only a few times, but the second factorisation is performed often. Therefore the cpu time for the second factorisation is essential for the overall simulation time.

Table 5. Cpu times for second factorisation

name	UMFPACK	SuperLU	Sparse	GSPAR
bayer01	5.02	6.70 *	7.78	3.20
b_dyn	0.05	0.05 *	0.07	0.00
bayer02	1.13	1.47 *	10.433	0.55
bayer03	0.72	0.70 *	17.467	0.27
bayer04	3.37	2.77 *	187.88	1.70
bayer05	0.13	0.75 *	0.08	0.05
bayer06	0.83	0.90 *	54.33	0.82
bayer09	0.23	0.23 *	3.57	0.10
bayer10	1.60	1.57 *	379.75	1.65
advice3388	0.25	0.28 $^+$	0.15	0.10
advice3776	0.30	0.42 $^+$	0.20	0.10
cod2655_tr	0.30	0.55 $^+$	0.27	0.10
megl	0.58	1.43 $^+$	13.95	0.22
meg4	0.37	0.75 $^+$	0.25	0.13
rlxADC_dc	0.15	0.18 $^+$	0.04	0.03
rlxADC_tr	0.40	0.90 $^+$	0.72	0.30
zy3315	0.15	0.18 $^+$	0.03	0.02
poli	0.03	0.07 $^+$	0.00	0.00
poli_large	0.13	0.27 $^+$	0.04	0.03

GSPAR achieves a fast second factorisation for all linear systems in Table 5. For linear systems with a large number of equations GSPAR is at least two times faster then UMFPACK, SuperLU and Sparse respectively.

The cpu times for solving the triangular matrices are one order of magnitude smaller then the cpu times for the factorisation. The proportions between the different solvers are comparable to the results in Table 5.

The vector version of GSPAR has been compared with the frontal method FAMP [11] on a vector computer Cray Y–MP8E using one processor. The used version of FAMP is the routine from the commercial chemical process simulator

SPEEDUP [2] [1]. The cpu times (in seconds) for the second factorisation are shown in Table 6.

Table 6. Cpu times for second factorisation

name	FAMP	GSPAR
b_dyn	0.034	0.011
bayer09	0.162	0.082
bayer03	0.404	0.221
bayer02	0.683	0.421
bayer10	1.290	0.738
bayer04	2.209	0.983

GSPAR is at least two times faster then FAMP for these examples. The proportions for solving the triangular systems are again the same.

For two large examples the number of levels of independence are given in Table 7, using GSPAR with two different ordering for pivoting. The algorithm for lower triangular systems is called forward substitution and the analogous algorithm for upper triangular systems is called back substitution.

Table 7. Number of levels of independence

example		dynamical ordering	minimum degree ordering
	factorisation	3 077	3 688
bayer01	forward sub.	1 357	1 562
	back substit.	1 728	2 476
	factorisation	876	820
bayer04	forward sub.	399	338
	back substit.	556	495

In Table 8, wall–clock times (in seconds) are shown for the second factorisation, using GSPAR with different pivoting on a DEC AlphaServer with four alpha EV5.6 (21164A) processors. The parallelisation technique is based on OpenMP [9]. The wall–clock times have been determined with the system routine *gettimeofday*.

[2] Used under licence 95122131717 for free academic use from Aspen Technology, Cambridge, MA, USA; Release 5.5–5

Table 8. Wall–clock times for second factorisation

processors	dynamical ordering		minimum degree ordering	
	bayer01	bayer04	bayer01	bayer04
1	0.71	0.39	1.08	0.43
2	0.54	0.27	0.75	0.29
3	0.45	0.23	0.63	0.25
4	0.49	0.24	0.70	0.30

In Table 9, the cpu times (in seconds) on a Cray T3D are given for the second factorisation, using GSPAR with dynamic ordering for pivoting. The linear systems can not be solved with less then four or sixteen processors respectively, because the processors of the T3D do have not enough local memory for the storage of the pseudo code in this cases. The speedup factors are set equal to one for four or sixteen processors respectively.

Table 9. Cpu times for second factorisation on Cray T3D

example	processors	cpu time	speedup factor
	4	1.59	1.00
	8	0.99	1.60
bayer04	16	0.60	2.65
	32	0.37	4.30
	64	0.24	6.63
	16	2.36	1.00
bayer01	32	1.45	1.63
	64	0.95	2.47

6 Applications

Problems of the dynamic process simulation of chemical plants can be modelled by initial value problems for systems of differential–algebraic equations. The numerical solution of these systems [3] involves the solution of large scale systems of nonlinear equations, which can be solved with modified Newton methods. The Newton corrections are found by solving large unsymmetric sparse systems of linear equations. The overall computing time of the simulation problems is often dominated by the time needed to solve the linear systems. In industrial applications, the solution of sparse linear systems requires often more then 70 %

of the total simulation time. Thus a reduction of the linear system solution time usually results into a significant reduction of the overall simulation time [12].

Table 10 shows three large scale industrial problems of the Bayer AG Leverkusen. The number of differential–algebraic equations as well as an estimate for the condition number of the matrices of the linear systems are given. The condition numbers are very large, what is typical for industrial applications in this field.

Table 10. Large scale industrial problems

name	chemical plants	equations	condition numbers
bayer04	nitration plant	3 268	2.95E+26, 1.4E+27
bayer10	distillation column	13 436	1.4E+15
bayer01	five coupled distillation columns	57 735	6.0E+18 6.96E+18

The problems have been solved on a vector computer Cray C90 using the chemical process simulator SPEEDUP [1]. In SPEEDUP the vector versions of the linear solvers FAMP and GSPAR have been used alternatively. The cpu time (in seconds) for complete dynamic simulation runs are shown in Table 11.

Table 11. Cpu time for complete dynamic simulation

name	FAMP	GSPAR	in %
bayer04	451.7	283.7	62.8
bayer10	380.9	254.7	66.9

For the large plant bayer01 benchmark tests have been performed on a dedicated computer Cray J90, using the simulator SPEEDUP with the solvers FAMP and GSPAR alternatively. The results are given in Table 12. The simulation of

Table 12. Bench mark tests

time	FAMP	GSPAR	in %
cpu time	6 066.4	5 565.8	91.7
wall–clock time	6 697.9	5 797.1	86.5

plant bayer01 has been performed also on a vector computer Cray C90 connected with a parallel computer Cray T3D, using SPEEDUP and the parallel version of GSPAR. Here, the linear systems have been solved on the parallel computer while the other parts of the algorithms of SPEEDUP have been performed on the vector computer. GSPAR needs 1 440.5 seconds cpu time on a T3D with 64 used processors. When executed on the Cray C90 only, 2 490 seconds are needed for the total simulation.

Acknowledgements. The author thanks his coworkers J. Borchardt and D. Horn for useful discussions. The valuable assistance and the technical support from the Bayer AG Leverkusen, the Cray Research Munich and Aspen Technology, Inc., Cambridge, MA, USA are gratefully acknowledged.

References

1. AspenTech: SPEEDUP, User Manual, Library Manual. Aspen Technology, Inc., Cambridge, Massachusetts, USA (1995)
2. Borchardt, J., Grund, F., Horn, D.: Parallelized numerical methods for large systems of differential–algebraic equations in industrial applications. Preprint No. 382, WIAS Berlin (1997). Surv. Math. Ind. (to appear)
3. Brenan, K.E., Campbell, S.L., Petzold, L.R.: Numerical solution of initial–value problems in differential–algebraic equations. North–Holland, New York (1997)
4. Davis, T.A., Duff, I.S.: An unsymmetric–pattern multifrontal method for sparse LU factorization. Tech. Report TR–94–038, CIS Dept., Univ. of Florida, Gainsville, FL (1994)
5. Demmel, J.W., Gilbert, J.R. Li, X.S.: SuperLU Users' Guide. Computer Science Division, U.C. Berkeley (1997)
6. Grund, F., Michael, T., Brüll, L., Hubbuch, F., Zeller, R., Borchardt, J., Horn, D., Sandmann, H.: Numerische Lösung großer strukturierter DAE–Systeme der chemischen Prozeßsimulation. In Mathematik Schlüsseltechnologie für die Zukunft, K.-H. Hoffmann, W. Jäger, T. Lohmann, H. Schunk, eds., Springer–Verlag Berlin Heidelberg (1997) 91–103
7. Kundert, K.S., Sangiovanni-Vincentelli, A.: Sparse User's Guide, A Sparse Linear Equation Solver. Dep. of Electr. Engin. and Comp. Sc., U.C. Berkeley (1988)
8. Li, Xiaoye S.: Sparse Gaussian elimination on high performance computers. Technical Reports UCB//CSD-96-919, Computer Science Division, U.C. Berkeley (1996), Ph.D. dissertation
9. OpenMP: A proposed standard API for shared memory programming. White paper, http://www.openmp.org (1997)
10. Yamamoto, F., Takahashi, S.: Vectorized LU decomposition algorithms for large–scale nonlinear circuits in the time domain. IEEE Trans. on Computer–Aided Design **CAD–4** (1985) 232–239
11. Zitney, S.E., Stadtherr, M.A.: Frontal algorithms for equation–based chemical process flowsheeting on vector and parallel computers. Computers chem. Engng. **17** (1993) 319–338
12. Zitney, S.E., Brüll, L., Lang, L., Zeller, R.: Plantwide dynamic simulation on supercomputers: Modelling a Bayer distillation process. AIChE Symp. Ser. **91** (1995) 313–316
13. Zlatev, Z.: On some pivotal strategies in Gaussian elimination by sparse technique. SIAM J. Numer. Anal. **17** (1980) 18–30

Parallel Preconditioners for Solving Nonsymmetric Linear Systems

Antonio J. García-Loureiro[1], Tomás F. Pena[1],
J.M. López-González[2], and Ll. Prat Viñas[2]

[1] Dept. Electronics and Computer Science. Univ. Santiago de Compostela
Campus Sur. 15706 Santiago de Compostela. Spain.
{antonio,tomas}@dec.usc.es

[2] Dept. of Electronic Engineering. Univ. Politécnica de Catalunya
Modulo 4. Campus Norte.c) Jordi Girona 1 y 3. 08034 Barcelona. Spain
jmlopezg@eel.upc.es

Abstract. *In this work we present a parallel version of two preconditioners. The first one, is based on a partially decoupled block form of the ILU. We call it Block–ILU(fill,τ,overlap), because it permits the control of both, the block fill and the block overlap. The second one, is based on the SPAI (SParse Approximate Inverse) method. Both methods are analysed and compared to the ILU preconditioner using the Bi–CGSTAB to solve general sparse, nonsymmetric systems. Results have been obtained for different matrices. The preconditioners have been compared in terms of robustness, speedup and time of execution, to determine which is the best one in each situation. These solvers have been implemented for distributed memory multicomputers, making use of the MPI message passing standard library.*

1 Introduction

In the development of simulation programs in different research fields, from fluid mechanics to semiconductor devices, the solution of the systems of equations which arise from the discretisation of partial differential equations, is the most CPU consuming part [12]. In general, the matrices are very large, sparse, nonsymmetric and are not diagonal dominant [3, 11]. So, using an effective method to solve the system is essential.

We are going to consider a linear system of equations such as:

$$Ax = b, \qquad A \in \mathbb{R}^{n\times n}, \quad x, b \in \mathbb{R}^n \tag{1}$$

where A is a sparse, nonsymmetric matrix.

Direct methods, such as Gaussian elimination, LU factorisation or Cholesky factorisation may be excessively costly in terms of computational time and memory, specially when n is large. Due to these problems, iterative methods [1, 13] are generally preferred for the solution of large sparse systems. In this work we have chosen a non stationary iterative solver, the Bi–Conjugate Gradient Stabilised [18]. Bi–CGSTAB is one of the methods that obtains better results in the

J. Palma, J. Dongarra, and V. Hernández (Eds.): VECPAR'98, LNCS 1573, pp. 128–141, 1999.

solution of non–symmetric linear systems, and its attractive convergence behaviour has been confirmed in many numerical experiments in different fields [7].

In order to reduce the number of iterations needed in the Bi–CGSTAB process, it is convenient to precondition the matrices. That is, transform the linear system into an equivalent one, in the sense that it has the same solution, but which has more favourable spectral properties.

Looking for efficient parallel preconditioners is a very important topic in current research in the field of scientific computing. A broad class of preconditioners are based on incomplete factorisations (incomplete Cholesky or ILU) of the coefficient matrix. One important problem associated with these preconditioners is their inherently sequential character. This implies that they are very hard to parallelise, and only a modest account of parallelism can be attained, with complicated implementations. So, it is important to find alternative forms of preconditioners that are more suitable for parallel architectures.

The first preconditioner we present is based on a partially decoupled block form of the ILU [2]. This new version, called Block–ILU(*fill*,τ,*overlap*), permits the control of its effectiveness through a dropping parameter τ and a block fill–in parameter. Moreover, it permits the control of the overlap between the blocks. We have verified that the fill–in control is very important for getting the most out of this preconditioner. Its main advantage is that it presents a very efficient parallel execution, because it avoids the data dependence of sequential ILU, obtaining high performance and scalability. As a disadvantage is that it is less robust than complete ILU, due to the loss of information, and this can be a problem in very bad conditioned systems.

The second preconditioner we present is an implementation of preconditioner SPAI (*SParse Approximate Inverse*) [5, 8]. This alternative has been proposed in the last few years as an alternative to ILU, in situations where the last obtain very poor results (situations which often arise when the matrices are indefinite or have large nonsymmetric parts). These methods are based on finding a matrix M which is a direct approximation to the inverse of A, so that $AM \approx I$.

This paper presents a parallel version of these preconditioners. Section 2 presents the iterative methods we have used. Section 3 introduces the characteristics of the Block–ILU and the SPAI preconditioners. Section 4 indicates the numerical experiment we have studied. The conclusions are given in Section 5.

2 Iterative Methods

The *iterative methods* are a wide range of techniques that use successive approximations to obtain more accurate solutions to linear systems at each step. There are two types of iterative methods. Stationary methods, like Jacobi, Gauss–Seidel, SOR, etc., are older, simpler to understand and implement, but usually not very effective. Nonstationary methods, like Conjugate Gradient, Minimum Residual, QMR, Bi–CGSTAB, etc., are a relatively recent development and can be highly effective. These methods are based on the idea of sequences of orthogonal vectors.

In recent years the Conjugate Gradient–Squared (CGS) method [1] has been recognised as an attractive variant of the Bi–Conjugate Gradient (Bi–CG) for the solution of certain classes of nonsymmetric linear systems. Recent studies indicate that the method is often competitive with other well established methods, such as GMRES [14]. The CG–S method has tended to be used in the solution of two or tree–dimensional problems, despite its irregular convergence pattern, because when it works -which is most of the time- it works quite well. Recently, van der Vorst [18] has presented a new variant of Bi–CG, called Bi–CGSTAB, which combines the efficiency of CGS with the more regular convergence pattern of Bi–CG.

In this work we have chosen the Bi–Conjugate Gradient Stabilised [1, 18], because of its attractive convergence behaviour. This method was developed to solve nonsymmetric linear systems while avoiding irregular convergence patterns of the Conjugate Gradient Squared methods. Bi–CGSTAB requires two matrix–vector products and four inner products per iteration.

In order to reduce the number of iterations needed in the Bi–CGSTAB process, it is convenient to precondition the matrices. The preconditioning can be applied in two ways: either we solve the explicitly preconditioned system using the normal algorithm, or we introduce the preconditioning process in the iterations of the Bi–CGSTAB. This last method is usually preferred.

3 Preconditioners

The rate at which an iterative method converges depends greatly on the spectrum of the coefficient matrix. Hence iterative methods usually involve a second matrix that transforms the coefficient matrix into one with a more favourable spectrum. A preconditioner is a matrix that affects such a transformation.

We are going to consider a linear system of equations such as:

$$Ax = b, \qquad A \in \mathbb{R}^{n\times n}, \quad x, b \in \mathbb{R}^n \tag{2}$$

where A is a large, sparse, nonsymmetric matrix.

If a matrix M right–preconditions coefficient matrix A in some way, we can transform the original system as follows:

$$Ax = b \quad \rightarrow \quad AM^{-1}(Mx) = b \tag{3}$$

Similarly, a left–preconditioner can be defined by:

$$Ax = b \quad \rightarrow \quad M^{-1}Ax = M^{-1}b \tag{4}$$

Another way of deriving the preconditioner would be to split M as $M = M_1 M_2$, where the matrices M_1 and M_2 are called the left and right preconditioners, and to transform the system as

$$Ax = b \quad \rightarrow \quad M_1^{-1}AM_2^{-1}(M_2 x) = M_1^{-1}b \tag{5}$$

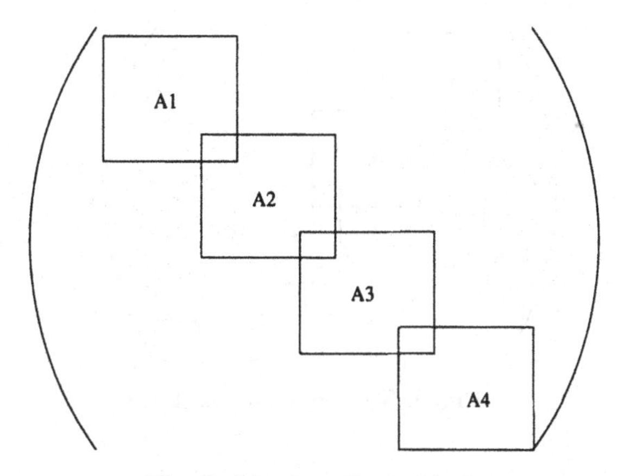

Fig. 1. Matrix splits in blocks

In this section we present a parallel version of two preconditioners. The first one, is based on a partially decoupled block form of the ILU. We call it Block–ILU(fill,τ,overlap), because it permits the control of both the block fill and the block overlap. The second one is based on the SPAI (SParse Approximate Inverse) method [13]. Both methods are analysed and compared to the ILU preconditioner using the Bi–CGSTAB to solve general sparse, nonsymmetric systems.

3.1 Parallel Block–ILU preconditioner

In this section we present a new version of a preconditioner based on a partially decoupled block form of the ILU [2]. This new version, called Block–ILU(*fill*,τ,*overlap*), permits the control of its effectiveness through a dropping parameter τ and a block fill–in parameter. Moreover, it permits the control of the overlap between the blocks. We have verified that the fill–in control is very important for getting the most out of this preconditioner. The original matrix is subdivided into a number of overlapping blocks, and each block is assigned to a processor. This setup produces a partitioning effect represented in Figure 1, for the case of 4 processors, where the ILU factorisation for all the blocks is computed in parallel, obtaining $A_i = L_iU_i$, $1 \leqslant i \leqslant p$, where p is the number of blocks. Due to the characteristics of this preconditioner, there is a certain loss of information. This means that the number of iterations will increase as the number of blocks increases (as a direct consequence of increasing the number of processors). This loss can be compensated to a certain extent by the information provided by the overlapping zones.

To create the preconditioner the rows of each block indicated for the parameter *overlap* are interchanged between the processors. These rows correspond to regions A and C of figure 2. After, the factorisation is carried out. Within the loop of solution algorithm it is necessary to carry out the operation of preconditioning

$$L_iU_iv = w \tag{6}$$

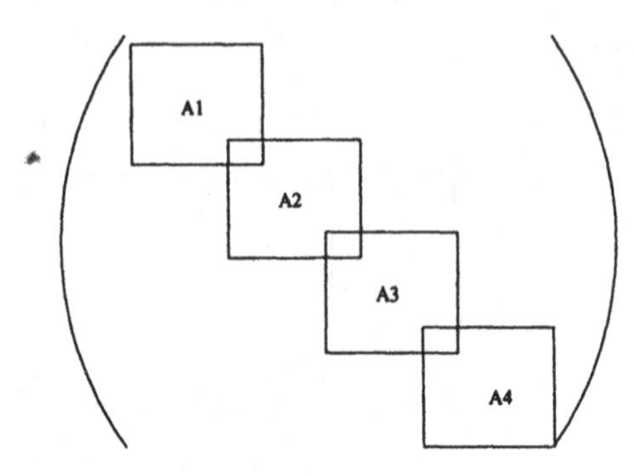

Fig. 2. Scheme of one block

To reduce the number of operations of the algorithm, each processor only works with its local rows. The first operation is to extend vector w's information to the neighbouring processors. Later we carry out in each processor the solution of the superior and inferior triangular system to calculate vector v. As regions A and C have also been calculated by other processors, the value that we obtain will vary in different processors. In order to avoid this and improve the convergency of the algorithm it is necessary to interchange these data and calculate the average of these values.

The main advantage of this method is that it presents a very efficient parallel execution, because it avoids the data dependence of sequential ILU, thereby obtaining high performance and scalability. A disadvantage is that it is less robust than complete ILU, due to the loss of information, and this can be a problem in very bad conditioned systems, as we will show in section 4.

3.2 Parallel SPAI preconditioner

One of the main drawback of ILU preconditioner is the low parallelism it implies. A natural way to achieve parallelism is to compute an approximate inverse M of A, such that $M \cdot A \simeq I$ in some sense. A simple technique for finding approximate inverses of arbitrary sparse matrices is to attempt to find a sparse matrix M which minimises the Frobenius norm of the residual matrix $AM - I$,

$$F(M) = \|AM - I\|_F^2 \tag{7}$$

A matrix M whose value $F(M)$ is small would be a right–approximate inverse of A. Similarly, a left–approximate inverse can be defined by using the objective function

$$\|MA - I\|_F^2 \tag{8}$$

These cases are very similar. The objective function 7 decouples into the sum of the squares of the 2–norms of the individual columns of the residual matrix $AM - I$,

$$F(M) = \|AM - I\|_F^2 = \sum_{j=1}^{n} \|Am_j - e_j\|_2^2 \tag{9}$$

in which e_j and m_j are the j–th columns of the identity matrix and of the matrix M. There are two different ways to proceed in order to minimise 9. The first one consists of in minimising it globally as a function of the matrix M, e.g., by a gradient–type method. Alternatively, in the second way the individual functions

$$f_j(m) = \|Am_j - e_j\|_F^2, j = 1, \ldots, n \tag{10}$$

can be minimised. This second approach is attractive for parallel computers, and it is the one we have used in this paper. A good, inherently parallel solution would be to compute the columns k of M, m_k, in an independent way from each other, resulting:

$$\|AM - I\|_F^2 = \sum_{k=1}^{n} \|(AM - I)e_k\|_2^2 \tag{11}$$

The solution of 11 can be organised into n independent systems,

$$\min_{m_k} \|Am_k - e_k\|_2, \quad k = 1, ..., n, \quad e_k = (0, ..., 0, 1, 0, ..., 0)^T \tag{12}$$

We have to solve n systems of equations. If these linear systems were solved without taking advantage of sparsity, the cost of constructing the preconditioner would be of order n^2. This is because each of the n columns would require $O(n)$ operations. Such a cost would become unacceptable for large linear systems. To avoid this, the iterations must be performed in *sparse–sparse mode.* As A is sparse, we could work with systems of much lower dimension. Let $L(k)$ be the set of indices j such that $m_k(j) \neq 0$. We denote the reduced vector of unknowns as $m_k(L)$ by $\hat{m}_k(L)$ and the resulting sub-matrix $A(L, L)$ as $\hat{A}$. Similarly, we define $\hat{e}_k = e_k(L)$. Now, solving 12 is transformed into solving:

$$\min_{m_k} \|\hat{A}\hat{m}_k - \hat{e}_k\|_2 \tag{13}$$

Due to the sparsity of A and M, the dimension of systems 13 is very small. To solve these systems we have chosen direct methods. We are using these methods instead of an iterative one, mainly because the systems 13 are very small and almost dense. Of the different alternatives we have concentrated on QR and LU methods [16].

The QR factorisation of matrix $A \in \mathbb{R}^{m \times n}$ is given by $A = QR$ where R is an m–by–n upper triangular matrix and Q is an m–by–m unitary matrix. This factorisation is better than LU because it can be used for the case of non squared matrices, and also works in some cases in which LU fails due to problems with too small pivots. The cost of this factorisation is $O(\frac{4}{3}n^3)$. The other direct method we have tested is LU. This factorisation and the closely related Gaussian elimination algorithm are widely used in the solution of linear systems of equations. LU factorisation expresses the coefficient matrix, A, as the product of a lower triangular matrix, L, and an upper triangular matrix, U. After factorisation, the original system of equations can be written as a pair of triangular systems,

$$Ax = b \tag{14}$$

$$Ly = b \quad Ux = y \tag{15}$$

The first of the systems can be solved by forward reduction, and then back substitution can be used to solve the second system to give x. The advantage of this factorisation is that its cost is $O(\frac{2}{3}n^3)$, lower than that of QR. We have implemented the two solvers in our code, specifically, the QR and the LU decomposition with pivoting. An efficient implementation consists of selecting the QR method if the matrix is not squared. In the case that it is squared, we will solve the system using LU, as this is faster than QR. Moreover, there is also the possibility of using QR if some error is produced in the construction of the factorisation LU.

In the next section we have compared the results we have obtained with these methods. In this code the SPAI parameter $\mathfrak{L}$ indicates the number of neighbours of each point we use to reduce the system. The main drawback of preconditioners based on the SPAI idea is that they need more computations than the rest. So, in the simplest situations and when the number of processors is small, they may be slower than ILU based preconditioners.

4 Numerical experiments

4.1 Test problem specification

The matrices we have tested were obtained by applying the method of finite elements to hetero-junction bipolar devices, indeed for transistors of InP/InGaAs [6]. These matrices are rank $N = 25000$, highly sparse, not symmetric and, in general, not diagonal dominant. We have calculated an estimate in 1–norm of the condition number of these matrices and we have obtained 5.0 ($\mathbb{A}$), 23.0 ($\mathbb{B}$) and $1.08\ 10^5$ ($\mathbb{C}$).

The basic equations of the semiconductor devices are Poisson's eq. and electron and hole continuity, in a stationary state [9, 10]:

$$div(\varepsilon\nabla\psi) = q(p - n + N_D^+ - N_A^-) \qquad (16)$$

$$div(J_n) = qR \qquad (17)$$

$$div(J_p) = -qR \qquad (18)$$

where ψ is the electrostatic potential, q is the electronic charge, ε is the dielectric constant of the material, n and p are the electron and hole densities, N_D^+ and N_A^+ are the doping effective concentration and J_n and J_p are the electron and hole current densities, respectively. The term R represents the volume recombination term, taking into account Schokley–Read–Hall, Auger and band–to–band recombination mechanisms [19].

For this type of semiconductors it is usual to apply at first a Gummel type solution method [15], which uncouples the three equations and allows us to obtain an initial solution for the system coupled with the three equations. For the semiconductors we use we have to solve the three equations simultaneously. The pattern of these matrices is similar to those in other fields such as applications of CFD [3, 4].

Table 1. Time (sec) for Block-ILU

Proc	2	4	6	8	10
FILL=0/Overlap=1	0.86	0.47	0.29	0.21	0.18
FILL=0/Overlap=3	0.83	0.41	0.29	0.22	0.18
FILL=0/Overlap=6	0.83	0.43	0.29	0.21	0.17
FILL=2/Overlap=1	0.38	0.18	0.12	0.094	0.077
FILL=4/Overlap=1	0.38	0.19	0.12	0.095	0.077

We have partitioned the matrix in rows and group of rows have been assigned to processors in order to get a good load-balance.

All the results have been obtained in a CRAY T3E multicomputer [17]. We have programmed it using the SPMD paradigm, with the MPI library, and we have obtained results with several matrices of different characteristics.

4.2 Parallel Block–ILU preconditioner

We have carried out different tests to study how the parameters of *fill–in* and *overlap* affect the time of calculation and speedup for the solution of a system of equations. In tables 1 and 2 we show the times of execution and speedup for matrix $\mathbb{B}$. Time is measured from two processors onwards, because we have memory problems trying to run the code in a single processor. So the speedup is computed as:

$$speedup_p = \frac{T_p}{\frac{p}{2}T_2} \tag{19}$$

where T_p is the time of execution with two processors.

With respect to the results shown in table 1 note that, if we maintain constant the value of the *fill–in*, when the value of the *overlap* is increased the time of execution hardly varies. This is because the only variation is in the size of the message to be transmitted, whereas the size of the *overlap* zone in comparison to the total is minimum. Therefore the increase in the computations is small. However, if we maintain constant the value of the *overlap* and increase the *fill–in* a significant variation is observed. This is because the number of iterations decreases considerably

As regards the values of speedup in table 2, the values obtained are significantly better in all cases, although the algorithm obtains slightly better results when the level of *fill–in* is increased for a constant level of *overlap*. However, for a constant level of *fill–in* the speedup decreases very smoothly as the level of overlap increases. This is because it is necessary to carry out a large number of operations and the cost of communications is also a little higher. From the results obtained it is possible to conclude that the best option is to choose the lowest value of overlap with which we can assure convergency with an average value of *fill–in*.

Table 2. Speedup for Block-ILU

Proc	2	4	6	8	10
FILL=0/Overlap=1	1.0	1.82	2.96	4.09	4.77
FILL=0/Overlap=3	1.0	2.02	2.86	3.77	4.66
FILL=0/Overlap=6	1.0	1.91	2.84	3.92	4.53
FILL=2/Overlap=1	1.0	2.10	3.16	4.04	4.93
FILL=4/Overlap=1	1.0	2.00	3.16	4.0	4.93

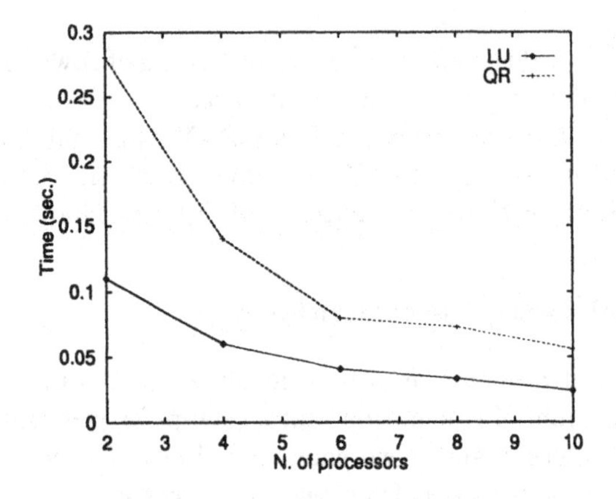

(a) Time versus number of processors

Fig. 3. QR versus LU on the CRAY T3E for matrix $\mathbb{B}$

4.3 Parallel SPAI preconditioner

First we are going to compare the results we have obtained with the two direct solvers we have implemented in section 3.2. For matrix $\mathbb{B}$ we have obtained the results shown in figure 3. These data refer to the cost of generating the matrix for each node with an overlap level 1. In this case resulting sub-matrices are of rank 3. Note that the cost of QR factorisation is significantly higher than that of LU. This difference is much larger for higher values of the overlap level.

Table 3 shows the time used to solve a matrix $\mathbb{B}$, as well as the number of iterations of the Bi-CGSTAB solver. Note that as the value of parameter $\mathfrak{L}$ increases, the number of iterations decreases because the preconditioner is more exact. Considering speedup, in all the cases values close to optimum are obtained, and in some cases even surpassed due to phenomena of super-linearity. For this class of matrices the optimum value of parameter $\mathfrak{L}$ would be 3 or 4. From the rest of results we can conclude that the more diagonally dominant the matrix, the smaller is the optimum value of this parameter, and inversely, for worse conditioned matrices we will need higher values of $\mathfrak{L}$ to assure the convergency.

Table 3. Time (sec) for SPAI with LU

Proc	2	4	6	8	10	Iter.
$\mathcal{L}$=0	2.19	1.10	0.76	0.54	0.46	47
$\mathcal{L}$=1	1.93	0.97	0.66	0.48	0.41	35
$\mathcal{L}$=2	1.51	0.86	0.52	0.38	0.32	22
$\mathcal{L}$=3	1.41	0.73	0.49	0.35	0.32	17
$\mathcal{L}$=4	1.46	0.73	0.49	0.37	0.31	13
$\mathcal{L}$=5	1.60	0.79	0.51	0.39	0.33	11

4.4 Parallel Block–ILU versus Parallel SPAI

In order to test the effectiveness of the parallel implementation of Block–ILU and SPAI, we have compared them to a parallel version of the ILU(*fill*,τ) preconditioner.

In Figure 4, results are shown for the complete solution of a system of equations for matrix $\mathbb{A}$. Again time is measured from two processors, because we have memory problems trying to run the code in a single processor. It can be seen, in Figure 4(a), that the parallel SPAI method obtains the best speedup, and that parallel Block–ILU(0,0,1) obtains very similar results. However, the ILU(0,0) preconditioner obtains very bad results. This is because of the bottleneck implied in the solution of the upper and lower sparse triangular systems. On the other hand, parallel SPAI is slower (Figure 4(b)) when the number of processors is small, because of the high number of operations it implies.

In Figure 5, results are shown for matrix $\mathbb{B}$. Again (Figure 5(a)) parallel SPAI and Block–ILU(0,0,1) obtain very similar speedup results. The ILU(0,0) preconditioner obtains the worst results. And again, parallel SPAI is the slower solution when the number of processors is small (Figure 5(b)).

From the point of view of scalability, parallel Block–ILU is worse than parallel SPAI. This is due to the fact that Block–ILU suffers a loss of information with respect to the sequential algorithm when the number of processors increases. This means that, with some matrices, the number of iterations, and, therefore, the total time for the BI–CGSTAB to converge, grows when the number of processors increases, thereby degrading the effectiveness of the preconditioner.

Figure 6 shows the results for matrix $\mathbb{C}$. In this case, the system converges with the three preconditioners, but a significant difference is noted between the SPAI and the incomplete factorisations. Note that with preconditioner SPAI we obtain a nearly ideal value of speedup, whereas in the other cases this hardly reaches 1, irrespective of the number of processors. However, if we examine the measures of time, it can be established that the fastest preconditioner is the ILU(3,0), together with the Block–ILU(3,0,3), although this time hardly varies with different numbers of processors. On the other hand, the SPAI is much slower than the other two. The motive for this behaviour is that, on the one hand, Block–ILU increases considerably the number of iterations as the number

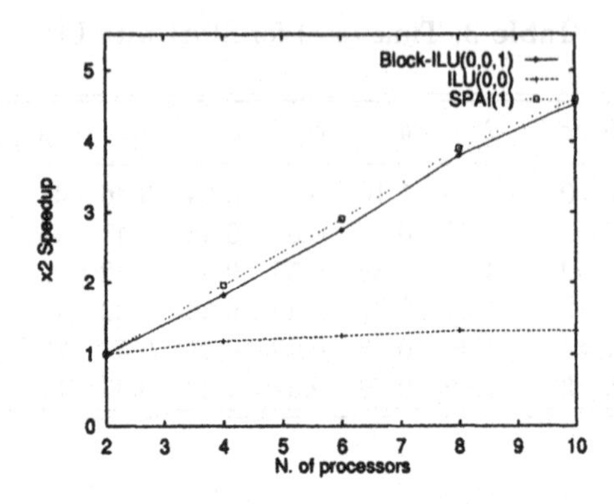

(a) Speedup versus number of processors

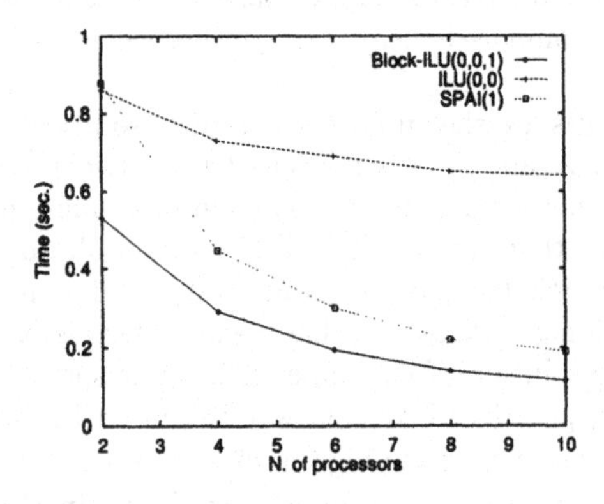

(b) Time versus number of processors

Fig. 4. Results on the CRAY T3E for matrix $\mathbb{A}$

of processors is increased, due to the loss of information that this method implies. This increase compensates the reduction in the cost for iteration, which means that the speedup does not increase. On the other hand, to guarantee convergency we must use SPAI with high values of $\mathfrak{L}$ (with small values $\mathfrak{L}$'s the SPAI does not converge), which supposes a high cost of each iteration. However, the number of iterations does not grow as the number of processors increases, and thereby we obtain a high level of speedup. With a large number of processors, Parallel SPAI probably overcomes ILU based preconditioners.

5 Conclusions

Choosing the best preconditioner is going to be conditioned by the characteristics of the system we have to solve. When it is not a very badly conditioned

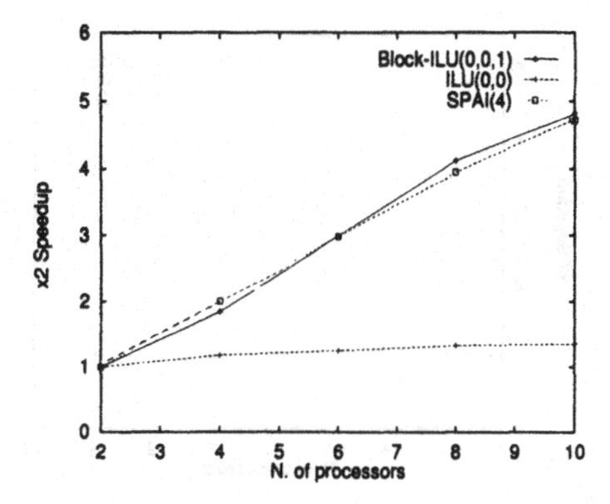

(a) Speedup versus number of processors

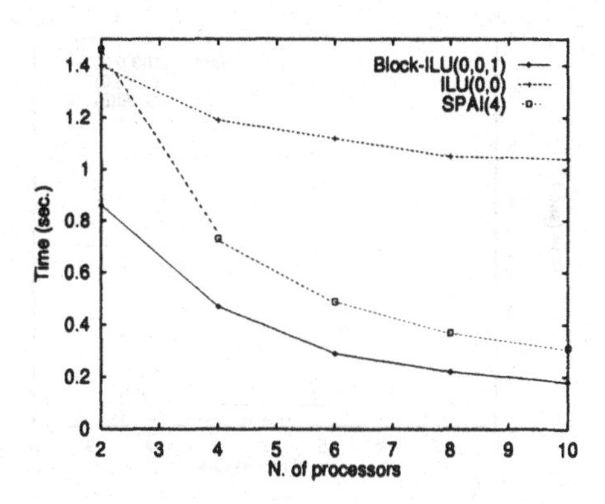

(b) Time versus number of processors

Fig. 5. Results on the CRAY T3E for matrix $\mathbb{B}$

system, parallel Block–ILU appears to be the best solution, because of both the high level of speedup it achieves and the reduced time it requires to obtain the final solution. The Parallel SPAI preconditioner obtains very good results in scalability, so it could be the best choice when the number of processors grows. Moreover, we have verified that it achieves convergence in some situations where ILU based preconditioners fail. Finally, the direct parallel implementations of ILU obtain very poor results.

Acknowledgements

The work described in this paper was supported in part by the Ministry of Education and Science (CICYT) of Spain under projects TIC96–1125–C03 and TIC96–1058. We want to thank CIEMAT (Madrid) for providing us access to the Cray T3E multicomputer.

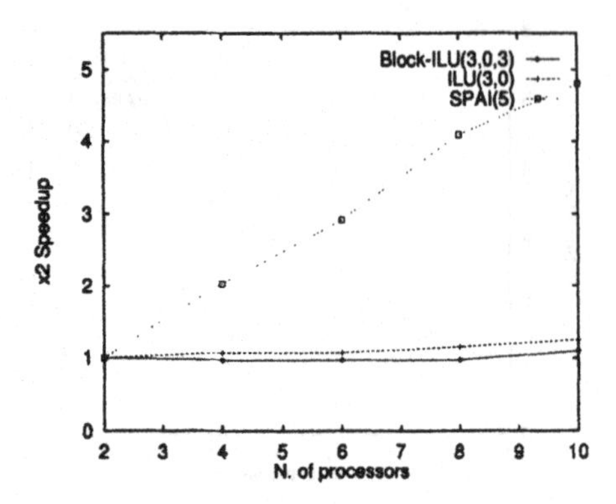

(a) Speedup versus number of processors

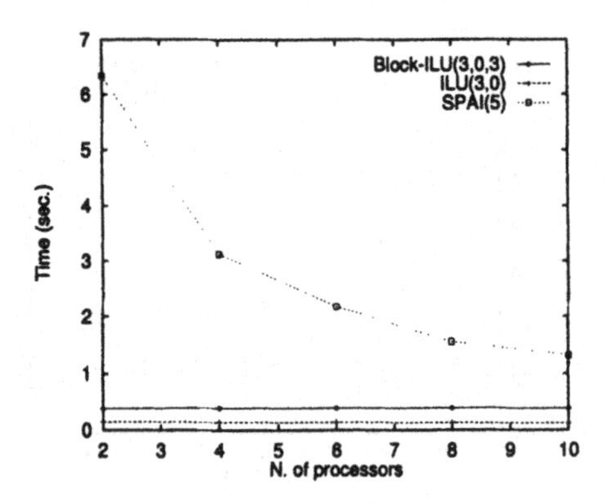

(b) Time versus number of processors

Fig. 6. Results on the CRAY T3E for matrix C

References

[1] R. Barrett, M. Berry, et al. *Templates for the Solution of Linear Systems: Building Blocks for Iterative Methods.* SIAM, 1994.

[2] G. Radicati di Brozolo and Y. Robert. Parallel Conjugate Gradient-like algorithms for solving sparse nonsymmetric linear systems on a vector multiprocessor. *Parallel Computing*, 11:223–239, 1989.

[3] A. Chapman, Y. Saad, and L. Wigton. High order ILU preconditioners for CFD problems. Technical report, Minnesota Supercomputer Institute. Univ. of Minnesota, 1996.

[4] Filomena D. d'Almeida and Paulo B. Vasconcelos. Preconditioners for nonsymmetric linear systems in domain decomposition applied to a coupled discretisation of Navier-Stokes equations. In *Vector and Parallel Processing – VECPAR'96*, pages 295–312. Springer-Verlag, 1996.

[5] V. Deshpande, M. Grote, P. Messmer, and W. Sawyer. Parallel implementation of a sparse approximate inverse preconditioner. In Springer-Verlag, editor, *Proceedings of Irregular'96*, pages 63–74, August 1996.
[6] A.J. García-Loureiro, J.M. López-González, T. F. Pena, and Ll. Prat. Numerical analysis of abrupt heterojunction bipolar transistors. *International Journal of Numerical Modelling: Electronic Networks, Devices and Fields*, (11):221–229, 8 1998.
[7] A.J. García-Loureiro, T. F. Pena, J.M. López-González, and Ll. Prat. Preconditioners and nonstationary iterative methods for semiconductor device simulation. In *Conferencia de Dispositivos Electrónicos (CDE-97)*, pages 403–409. Universitat Politecnica de Catalunya, Barcelona, February 1997.
[8] Marcus J. Grote and Thomas Huckle. Parallel preconditioning with sparse approximate inverses. *Siam J. Sci. Comput.*, 18(3):838–853, May 1997.
[9] K. Horio and H. Yanai. Numerical modeling of heterojunctions including the heterojunction interface. *IEEE Trans. on ED*, 37(4):1093–1098, April 1990.
[10] J. M. Lopez-Gonzalez and Lluis Prat. Numerical modelling of abrupt InP/InGaAs HBTs. *Solid-St. Electron*, 39(4):523–527, 1996.
[11] T.F. Pena, J.D. Bruguera, and E.L. Zapata. Finite element resolution of the 3D stationary semiconductor device equations on multiprocessors. *J. Integrated Computer-Aided Engineering*, 4(1):66–77, 1997.
[12] C.S. Rafferty, M.R. Pinto, and R.W. Dutton. Iterative methods in semiconductors device simulation. *IEEE trans on Computer-Aided Design*, 4(4):462–471, October 1985.
[13] Y. Saad. *Iterative Methods for Sparse Linear Systems.* PWS Publishing Co., 1996.
[14] Y. Saad and M.H. Schultz. GMRES: a generalized minimal residual algorithm for solving nonsymmetric linear systems. *SIAM J. Sci. Statist. Comput.*, 7:856–869, 1986.
[15] D.L. Scharfetter and H.K. Gummel. Large-signal analysis of a silicon read diode oscillator. *IEEE Trans. on ED*, pages 64–77, 1969.
[16] H. R. Schwarz. *Numerical Analysis.* John Wiley & Sons, 1989.
[17] S. L. Scott. Synchronization and communication in the T3E multiprocessor. Technical report, Inc. Cray Research, 1996.
[18] A. Van der Vorst. Bi-CGSTAB: A fast and smoothly converging variant of Bi-CG for the solution of nonsymmetric linear systems. *SIAM J. Sci. Statist. Comput.*, 13:631–644, 1992.
[19] C.M. Wolfe, N. Holonyak, and G.E. Stillman. *Physical Properties of Semiconductors*, chapter 8. Ed. Prentice Hall, 1989.

Synchronous and Asynchronous Parallel Algorithms with Overlap for Almost Linear Systems

Josep Arnal, Violeta Migallón, and José Penadés

Departamento de Ciencia de la Computación e Inteligencia Artificial,
Universidad de Alicante,
E-03071 Alicante, Spain
{arnal,violeta,jpenades}@dtic.ua.es

Abstract. Parallel algorithms for solving almost linear systems are studied. A non-stationary parallel algorithm based on the multi-splitting technique and its extension to an asynchronous model are considered. Convergence properties of these methods are studied for M-matrices and H-matrices. We implemented these algorithms on two distributed memory multiprocessors, where we studied their performance in relation to overlapping of the splittings at each iteration.

1 Introduction

We are interested in the parallel solution of almost linear systems of the form

$$Ax + \Phi(x) = b \,, \tag{1}$$

where $A = (a_{ij})$ is a real $n \times n$ matrix, x and b are n-vectors and $\Phi : \mathbb{R}^n \to \mathbb{R}^n$ is a nonlinear diagonal mapping (i.e., the ith component Φ_i of Φ is a function only of x_i).

These systems appear in practice from the discretisation of differential equations, which arise in many fields of applications such as trajectory calculation or the study of oscillatory systems; see e.g., [3], [5] for some examples.

Considering that system (1) has in fact a unique solution, White [18] introduced the parallel nonlinear Gauss-Seidel algorithm, based on both the classical nonlinear Gauss-Seidel method (see [13]) and the multi-splitting technique (see [12]). Until then, the multi-splitting technique had only been used for linear problems. Recently, in the context of relaxed methods, Bai [1] has presented a class of algorithms, called parallel nonlinear AOR methods, for solving system (1). These methods are a generalisation of the parallel nonlinear Gauss-Seidel algorithm [18].

In order to get a good performance of all processors and a good load balance among processors, in this paper we extend the idea of the non-stationary methods to the problem of solving the almost linear system (1). This technique was introduced in [6] for solving linear systems, (see also [8], [11]). In a formal way,

J. Palma, J. Dongarra, and V. Hernández (Eds.): VECPAR'98, LNCS 1573, pp. 142–155, 1999.

let us consider a collection of splittings $A = (D - L_{\ell,k,m}) - U_{\ell,k,m}$, $\ell = 1, 2, \ldots$, $k = 1, 2, \ldots, \alpha$, $m = 1, 2, \ldots, q(\ell, k)$, such that $D = \text{diag}(A)$ is nonsingular and $L_{\ell,k,m}$ are strictly lower triangular matrices. Note that matrices $U_{\ell,k,m}$ are not generally upper triangular. Let E_k be nonnegative diagonal matrices such that $\sum_{k=1}^{\alpha} E_k = I$.

Let us define $r_i : \mathbb{R} \to \mathbb{R}$, $1 \le i \le n$, such that

$$r_i(t) = a_{ii}t + \Phi_i(t), \quad t \in \mathbb{R}\,, \tag{2}$$

and suppose that there exists the inverse function of each r_i, denoted by r_i^{-1}.

Let us consider the operators $P_{\ell,k,m} : \mathbb{R}^n \to \mathbb{R}^n$ such that each of them maps x into y in the following way

$$\begin{cases} y_i = \omega \hat{y}_i + (1-\omega)x_i, \ 1 \le i \le n, \ \omega \in \mathbb{R}, \ \omega \neq 0\,, \\ \text{and } \hat{y}_i = r_i^{-1}(z_i), \ \text{with} \\ z = \mu L_{\ell,k,m}\hat{y} + (1-\mu)L_{\ell,k,m}x + U_{\ell,k,m}x + b, \ \mu \in \mathbb{R}\,. \end{cases} \tag{3}$$

With this notation, the following algorithm describes a non-stationary parallel nonlinear method to solve system (1). This algorithm is based on the AOR-type methods. It is assumed that processors update their local approximation as many times as the *non-stationary* parameters $q(\ell, k)$ indicate.

Algorithm 1 (NON-STATIONARY PARALLEL NONLINEAR ALG.).
Given the initial vector $x^{(0)}$, and a sequence of numbers of local iterations $q(\ell, k)$, $\ell = 1, 2, \ldots,$ $k = 1, 2, \ldots, \alpha$

For $\ell = 1, 2, \ldots,$ until convergence
 In processor k, $k = 1$ to α
  $x^{\ell,k,0} = x^{(\ell-1)}$
  For $m = 1$ to $q(\ell, k)$
   $x^{\ell,k,m} = P_{\ell,k,m}(x^{\ell,k,m-1})$

$$x^{(\ell)} = \sum_{k=1}^{\alpha} E_k x^{\ell,k,q(\ell,k)}\,.$$

We note that Algorithm 1 extends the nonlinear algorithms introduced in [1] and [18]. Moreover, Algorithm 1 reduces to Algorithm 2 in [8], when $\Phi(x) = 0$ and for all $\ell = 1, 2,$ $ldots$, $m = 1, 2, \ldots, q(\ell, k)$, $L_{\ell,k,m} = L_k$ and $U_{\ell,k,m} = U_k$, $k = 1, 2, \ldots, \alpha$. Here, the formulation of Algorithm 1 allows us to use different splittings not only in each processor but at each global iteration ℓ and/or at each local iteration m. Furthermore, the overlap is allowed as well.

In this algorithm all processors complete their local iterations before updating the global approximation $x^{(\ell)}$. Thus, this algorithm is synchronous.

To construct an asynchronous version of Algorithm 1 we consider an iterative scheme on $\mathbb{R}^{\alpha n}$. More precisely, we consider that, at the ℓth iteration, processor k performs the calculations corresponding to its $q(\ell, k)$ splittings, saving the update vector in $x_k^{(\ell)}$, $k = 1, 2, \ldots, \alpha$. Moreover, at each step, processors make use of the most recent vectors computed by the other processors, which are previously weighted with the matrices E_k, $k = 1, 2, \ldots, \alpha$.

In a formal way, let us define the sets $J_\ell \subseteq \{1, 2, \ldots, \alpha\}$, $\ell = 1, 2, \ldots$, as $k \in J_\ell$ if the kth part of the iteration vector is computed at the ℓth step. The superscripts $r(\ell, k)$ denote the iteration number in which the processor k computed the vector used at the beginning of the ℓth iteration.

As it is customary in the description and analysis of asynchronous algorithms (see e.g., [2], [4]), we always assume that the superscripts $r(\ell, k)$ and the sets J_ℓ satisfy the following conditions

$$r(\ell, k) < \ell \text{ for all } k = 1, 2, \ldots, \alpha, \ \ell = 1, 2, \ldots. \tag{4}$$

$$\lim_{\ell \to \infty} r(\ell, k) = \infty \text{ for all } k = 1, 2, \ldots, \alpha \ . \tag{5}$$

$$\text{The set } \{\ell \mid k \in J_\ell\} \text{ is unbounded for all } k = 1, 2, \ldots, \alpha \ . \tag{6}$$

Let us consider the operators $P_{\ell,k,m}$ used in Algorithm 1. With this notation, the asynchronous counterpart of that algorithm corresponds to the following algorithm.

Algorithm 2 (ASYNC. NON-STATIONARY PARALLEL NONLINEAR ALG.). Given the initial vectors $x_k^{(0)}$, $k = 1, 2, \ldots, \alpha$, and a sequence of numbers of local iterations $q(\ell, k)$, $\ell = 1, 2, \ldots$, $k = 1, 2, \ldots, \alpha$

For $\ell = 1, 2, \ldots$, until convergence

$$x_k^{(\ell)} = \begin{cases} x_k^{(\ell-1)} & \text{if } k \notin J_\ell \\ P_{\ell,k,q(\ell,k)} \cdot \ldots \cdot P_{\ell,k,2} \cdot P_{\ell,k,1} \left(\sum_{j=1}^{\alpha} E_j x_j^{(r(\ell,j))} \right) & \text{if } k \in J_\ell \ . \end{cases} \tag{7}$$

Note that Algorithm 2 computes iterate vectors of size αn, while it only uses n-vectors to perform the updates. For that reason, from the experimental point of view, we can consider that the sequence of iterate vectors is made up of that n-vectors, that is, $\sum_{j=1}^{\alpha} E_j x_j^{(r(\ell,j))}$, $\ell = 1, 2, \ldots$. Another consequence of what has been mentioned above is that only components of the vectors $x_k^{(\ell)}$ corresponding to nonzero diagonal entries of the matrix E_k need to be computed. Then, the local storage is of order n and not αn.

In order to rewrite the asynchronous iteration (7) more clearly, we define the operators $G^{(\ell)} = (G_1^{(\ell)}, \ldots, G_\alpha^{(\ell)})$, with $G_k^{(\ell)} : \mathbb{R}^{\alpha n} \to \mathbb{R}^n$ such that, if $\tilde{y} \in \mathbb{R}^{\alpha n}$

$$G_k^{(\ell)}(\tilde{y}) = P_{\ell,k,q(\ell,k)} \cdot \ldots \cdot P_{\ell,k,2} \cdot P_{\ell,k,1}(Q\tilde{y}), \quad k = 1, 2, \ldots, \alpha \ ,$$

where

$$Q = [E_1, \ldots, E_k, \ldots, E_\alpha] \in \mathbb{R}^{n \times \alpha n} \ . \tag{8}$$

Then, iteration (7) can be rewritten as the following iteration

$$x_k^{(\ell)} = \begin{cases} x_k^{(\ell-1)} & \text{if } k \notin J_\ell \\ G_k^{(\ell)}\left(x_1^{(r(\ell,1))}, \ldots, x_k^{(r(\ell,k))}, \ldots, x_\alpha^{(r(\ell,\alpha))}\right) & \text{if } k \in J_\ell\,. \end{cases} \tag{9}$$

In Section 2, we study the convergence properties of the above algorithms when the matrix in question is either M-matrix or H-matrix. The last section contains computational results which illustrate the behaviour of these algorithms on two distributed multiprocessors. In the rest of this section we introduce some notation, definitions and preliminary results.

We say that a vector $x \in \mathbb{R}^n$ is nonnegative (positive), denoted $x \geq 0$ ($x > 0$), if all its entries are nonnegative (positive). Similarly, if $x, y \in \mathbb{R}^n$, $x \geq y$ ($x > y$) means that $x - y \geq 0$ ($x - y > 0$). Given a vector $x \in \mathbb{R}^n$, $|x|$ denotes the vector whose components are the absolute values of the corresponding components of x. These definitions carry over immediately to matrices.

A nonsingular matrix A is said to be an M-matrix if it has non-positive off-diagonal entries and it is monotone, i.e., $A^{-1} \geq O$; see e.g., Berman and Plemmons [3] or Varga [17]. Given a matrix $A = (a_{ij}) \in \mathbb{R}^{n\times n}$, its comparison matrix is defined by $\langle A\rangle = (\alpha_{ij})$, $\alpha_{ii} = |a_{ii}|$, $\alpha_{ij} = -|a_{ij}|$, $i \neq j$. A is said to be an H-matrix if $\langle A\rangle$ is a nonsingular M-matrix.

Lemma 1. *Let $H^{(1)}, H^{(2)}, \ldots, H^{(\ell)}, \ldots$ be a sequence of nonnegative matrices in $\mathbb{R}^{n\times n}$. If there exists a real number $0 \leq \theta < 1$, and a vector $v > 0$ in $\mathbb{R}^n$, such that*

$$H^{(\ell)}v \leq \theta v,\ \ell = 1, 2, \ldots,$$

then $\rho(K_\ell) \leq \theta^\ell < 1$, where $K_\ell = H^{(\ell)}H^{(\ell-1)}\cdots H^{(1)}$, and therefore $\lim_{\ell\to\infty} K_\ell = O$.

Proof. The proof of this lemma can be found, e.g., in [15].

Lemma 2. *Let $A = (a_{ij}) \in \mathbb{R}^{n\times n}$ be an H-matrix and let $\Phi : \mathbb{R}^n \to \mathbb{R}^n$ be a continuous and diagonal mapping. If $sign(a_{ii})\,(t-s)\,(\Phi_i(t) - \Phi_i(s)) \geq 0$, $i = 1, 2, \ldots, n$, for all $t, s \in \mathbb{R}$, then the almost linear system (1) has a unique solution.*

Proof. It is essentially the proof of [1, Lemma 2].

2 Convergence

In order to analyse the convergence of Algorithm 1 we rewrite it as the following iteration scheme

$$x^{(\ell)} = \sum_{k=1}^{\alpha} E_k x^{\ell,k,q(\ell,k)},\ \ \ell = 1, 2, \ldots, \tag{10}$$

where $x^{\ell,k,q(\ell,k)}$ is computed according to the iteration

$$
\begin{aligned}
&x^{\ell,k,0} = x^{(\ell-1)} \\
&\text{For } m = 1 \text{ to } q(\ell,k) \\
&\quad x_i^{\ell,k,m} = \omega \hat{x}_i^{\ell,k,m} + (1-\omega) x_i^{\ell,k,m-1}, \ 1 \le i \le n, \ \omega \in \mathbb{R}, \ \omega \ne 0
\end{aligned} \tag{11}
$$

and $\hat{x}_i^{\ell,k,m}$ is determined by

$$
\hat{x}_i^{\ell,k,m} = r_i^{-1}(z_i^{\ell,k,m}) , \tag{12}
$$

where r_i is defined in (2) and

$$
z^{\ell,k,m} = \mu L_{\ell,k,m} \hat{x}^{\ell,k,m} + (1-\mu) L_{\ell,k,m} x^{\ell,k,m-1} + U_{\ell,k,m} x^{\ell,k,m-1} + b, \ \mu \in \mathbb{R} .
$$

The following theorem ensures the existence of a unique solution of system (1) and shows the convergence of scheme (10) (or Algorithm 1) when A is an H-matrix and $0 \le \mu \le \omega < \frac{2}{1+\rho(|D|^{-1}|B|)}$, with $\omega \ne 0$, where $D = \text{diag}(A)$ and $A = D - B$. Note that, from [17, Theorem 3.10], $|D|$ is a nonsingular matrix and $\rho(|D|^{-1}|B|) < 1$.

Theorem 1. *Let* $A = D - L_{\ell,k,m} - U_{\ell,k,m} = D - B$, $\ell = 1,2,\ldots$, $k = 1,2,\ldots,\alpha$, $m = 1,2,\ldots,q(\ell,k)$, *be an H-matrix, where* $D = \text{diag}(A)$ *and* $L_{\ell,k,m}$ *are strictly lower triangular matrices. Assume that* $|B| = |L_{\ell,k,m}| + |U_{\ell,k,m}|$. *Let* Φ *be a continuous and diagonal mapping satisfying*

$$
sign(a_{ii})\,(t-s)\,(\Phi_i(t) - \Phi_i(s)) \ge 0, \quad i = 1,2,\ldots,n, \quad \text{for all } t,s \in \mathbb{R} . \tag{13}
$$

If $0 \le \mu \le \omega < \frac{2}{1+\rho}$, *with* $\omega \ne 0$, *where* $\rho = \rho(|D|^{-1}|B|)$, *and* $q(\ell,k) \ge 1$, $\ell = 1,2,\ldots$, $k = 1,2,\ldots,\alpha$, *then the iteration* (10) *is well-defined and converges to the unique solution of the almost linear system* (1), *for every initial vector* $x^{(0)} \in \mathbb{R}^n$.

Proof. Since Φ_i, $1 \le i \le n$, are continuous mappings satisfying (13), it follows that each r_i given in (2) is one-to-one and maps $\mathbb{R}$ onto $\mathbb{R}$. Hence, each r_i has an inverse function defined in all of $\mathbb{R}$ and thus iteration (10) is well-defined. On the other hand, by Lemma 2, system (1) has a unique solution, denoted $x^\star$. Let $\varepsilon^{(\ell)} = x^{(\ell)} - x^\star$ be the error vector at the ℓth iteration of scheme (10). Then, $|\varepsilon^{(\ell)}| \le \sum_{k=1}^{\alpha} E_k \left| x^{\ell,k,q(\ell,k)} - x^* \right|$. Using (13) and reasoning in a similar way as in the proof of [13, Theorem 13.13], it is easy to prove that $|a_{ii}|\,|y - \bar{y}| \le |r_i(y) - r_i(\bar{y})|$, for all $y, \bar{y} \in \mathbb{R}$, where r_i is defined in (2). Therefore, we obtain that $|a_{ii}|\,\left|r_i^{-1}(z) - r_i^{-1}(\bar{z})\right| \le |z - \bar{z}|$, for all $z, \bar{z} \in \mathbb{R}$. Then, from (12) and using the fact that $x_i^\star = r_i^{-1}([\mu L_{\ell,k,m} x^\star + (1-\mu)\, L_{\ell,k,m} x^\star + U_{\ell,k,m} x^\star + b]_i)$, we obtain, for each $i = 1,2,\ldots,n$

$$
\begin{aligned}
|a_{ii}| \left| \hat{x}_i^{\ell,k,m} - x_i^\star \right| &= |a_{ii}| \left| r_i^{-1}(z_i^{\ell,k,m}) - r_i^{-1}(z_i^k) \right| \le \left| z_i^{\ell,k,m} - z_i^k \right| \\
&= \left| \left[\mu L_{\ell,k,m} \left(\hat{x}^{\ell,k,m} - x^* \right) + (1-\mu)\, L_{\ell,k,m} \left(x^{\ell,k,m-1} - x^* \right) \right.\right. \\
&\quad \left.\left. + U_{\ell,k,m} \left(x^{\ell,k,m-1} - x^* \right) \right]_i \right| .
\end{aligned}
$$

Since these inequalities are true for all $i = 1, 2, \ldots, n$, we can write

$$|D|\left|\hat{x}^{\ell,k,m} - x^*\right| \le \left|\mu L_{\ell,k,m}\left(\hat{x}^{\ell,k,m} - x^*\right) + (1-\mu)\, L_{\ell,k,m}\left(x^{\ell,k,m-1} - x^*\right) \right.$$
$$\left. + U_{\ell,k,m}\left(x^{\ell,k,m-1} - x^*\right)\right| .$$

Since $(|D| - \mu|L_{\ell,k,m}|)^{-1} \ge O$, making use of (11) we obtain, after some algebraic manipulations, that

$$\left|x^{\ell,k,m} - x^*\right| \le \left(|D| - \mu\,|L_{\ell,k,m}|\right)^{-1}\left((\omega-\mu)\,|L_{\ell,k,m}| + \omega\,|U_{\ell,k,m}|\right.$$
$$\left. + |1-\omega|\,|D|\right)\left|x^{\ell,k,m-1} - x^*\right|, \quad m = 1, 2, \ldots, q(\ell,k) .$$

Therefore, $\left|x^{\ell,k,q(\ell,k)} - x^*\right| \le H_k^{(\ell)}\left|x^{(\ell-1)} - x^*\right|$, where $H_k^{(\ell)} = H_{\ell,k,q(\ell,k)} \cdot \ldots \cdot H_{\ell,k,2} \cdot H_{\ell,k,1}$, and

$$H_{\ell,k,m} = \left(|D| - \mu\,|L_{\ell,k,m}|\right)^{-1}\left((\omega-\mu)\,|L_{\ell,k,m}| + \omega\,|U_{\ell,k,m}| + |1-\omega|\,|D|\right) . \tag{14}$$

Then $\left|\varepsilon^{(\ell)}\right| \le H^{(\ell)}|\varepsilon^{(\ell-1)}| \le \cdots \le H^{(\ell)} \cdots H^{(1)}|\varepsilon^{(0)}|$, where $H^{(\ell)} = \sum_{k=1}^{\alpha} E_k H_k^{(\ell)}$.

Since A is an H-matrix, following the proof of [8, Theorem 4.1] we conclude that for $0 \le \mu \le \omega < \frac{2}{1+\rho}$, with $\omega \ne 0$, there exist real constants $0 \le \theta_k < 1$ and a positive vector v such that $H_k^{(\ell)} v \le \theta_k v$. Hence, setting $\theta = \max_{k=1,2,\ldots,\alpha} \theta_k$, it obtains $H^{(\ell)} v \le \theta v$. Then, from Lemma 1 the product $H^{(\ell)}H^{(\ell-1)} \cdots H^{(1)}$ tends to the null matrix as $\ell \to \infty$ and thus $\lim_{\ell\to\infty} \varepsilon^{(\ell)} = 0$. Therefore, the proof is done.

Next we show the convergence of the asynchronous Algorithm 2 under similar hypotheses as in the synchronous case.

Theorem 2. *Let $A = D - L_{\ell,k,m} - U_{\ell,k,m} = D - B$, $\ell = 1, 2, \ldots$, $k = 1, 2, \ldots, \alpha$, $m = 1, 2, \ldots, q(\ell,k)$, be an H-matrix, where $D = \mathrm{diag}(A)$ and $L_{\ell,k,m}$ are strictly lower triangular matrices. Assume that $|B| = |L_{\ell,k,m}| + |U_{\ell,k,m}|$. Let Φ be a continuous and diagonal mapping satisfying for all $t, s \in \mathbb{R}$*

$$sign(a_{ii})\,(t-s)\,(\Phi_i(t) - \Phi_i(s)) \ge 0, \quad i = 1, 2, \ldots, n .$$

Assume further that the sequence $r(\ell,k)$ and the sets J_ℓ, $k = 1, 2, \ldots, \alpha$, $l = 1, 2, \ldots$, satisfy conditions (4–6). If $0 \le \mu \le \omega < \frac{2}{1+\rho}$, with $\omega \ne 0$, where $\rho = \rho(|D|^{-1}|B|)$, and $q(\ell,k) \ge 1$, $\ell = 1, 2, \ldots$, $k = 1, 2, \ldots, \alpha$, then the asynchronous Algorithm 2 is well-defined and converges to $\tilde{x}^\star = (x^{\star T}, \ldots, x^{\star T})^T \in \mathbb{R}^{\alpha n}$, where $x^\star$ is the unique solution of the almost linear system (1), for all initial vectors $x_k^{(0)} \in \mathbb{R}^n$, $k = 1, 2, \ldots, \alpha$.

Proof. By Lemma 2, the existence and uniqueness of a solution of system (1) is guaranteed. From the proof of Theorem 1 it follows that Algorithm 2 is well-defined. Moreover, there exists a positive vector v and a constant $0 \le \theta < 1$ such that

$$H_k^{(\ell)} v \le \theta v, \; k = 1, 2, \ldots, \alpha, \; \ell = 1, 2, \ldots . \tag{15}$$

Let us consider $\tilde{v} = (v^T, \ldots, v^T)^T \in \mathbb{R}^{\alpha n}$. As $G_k^{(\ell)}(\tilde{x}^\star) = x^\star$, then $\tilde{x}^\star$ is a fixed point of $G^{(\ell)}$, $\ell = 1, 2, \ldots$. Following the proof of Theorem 1 it easy to prove that

$$|G_k^{(\ell)}(\tilde{y}) - G_k^{(\ell)}(\tilde{z})| \leq H_k^{(\ell)} Q |\tilde{y} - \tilde{z}|, \text{ for all } \tilde{y}, \tilde{z} \in \mathbb{R}^{\alpha n} ,$$

where Q is defined in (8). Then,

$$|G^{(\ell)}(\tilde{y}) - G^{(\ell)}(\tilde{z})| \leq T^{(\ell)} |\tilde{y} - \tilde{z}|, \text{ for all } \tilde{y}, \tilde{z} \in \mathbb{R}^{\alpha n} ,$$

where

$$T^{(\ell)} = \begin{bmatrix} H_1^{(\ell)} Q \\ \vdots \\ H_\alpha^{(\ell)} Q \end{bmatrix} \in \mathbb{R}^{\alpha n \times \alpha n} . \tag{16}$$

From equations (15) and (16) it follows that

$$T^{(\ell)} \tilde{v} \leq \theta \tilde{v}, \ \ell = 1, 2, \ldots . \tag{17}$$

Due to the uniformity assumption (17), we can apply [2, Theorem 1] to our case in which the operators change with the iteration superscript. Then, the convergence is shown.

Note that, in the particular case in which A is an M-matrix, condition (13) is reduced to state that the mapping Φ is nondecreasing. Moreover, condition $|B| = |L_{\ell,k,m}| + |U_{\ell,k,m}|$ is equivalent to assume that $L_{\ell,k,m}$ and $U_{\ell,k,m}$ are nonnegative matrices.

3 Numerical Experiments

We have implemented the above algorithms on two distributed multiprocessors. The first platform is an IBM RS/6000 SP with 8 nodes. These nodes are 120 MHz Power2 Super Chip (Thin SC) and they are connected through a high performance switch with latency time of 40 microseconds and a bandwidth of 30 to 35 Mbytes per second. The second platform is an Ethernet network of five 120 MHz Pentiums. The peak performance of this network is 100 Mbytes per second with a bandwidth around 6.5 Mbytes per second. In order to manage the parallel environment we have used the PVMe library of parallel routines for the IBM RS/6000 SP and the PVM library for the cluster of Pentiums [9], [10].

In order to illustrate the behaviour of the above algorithms, we have considered the following semi-linear elliptic partial differential equation (see e.g., [7], [16], [18])

$$\begin{aligned} -(K^1 u_x)_x - (K^2 u_y)_y &= -g e^u \quad (x,y) \in \Omega , \\ u &= x^2 + y^2 \ (x,y) \in \partial\Omega , \end{aligned} \tag{18}$$

where

$$\begin{aligned} K^1 &= K^1(x,y) = 1 + x^2 + y^2 , \\ K^2 &= K^2(x,y) = 1 + e^x + e^y , \\ g &= g(x,y) = 2(2 + 3x^2 + y^2 + e^x + (1+y)e^y)e^{-x^2-y^2} , \\ \Omega &= (0,1) \times (0,1) . \end{aligned}$$

It is well known that this problem has the unique solution $u(x,y) = x^2 + y^2$. To solve equation (18) using the finite difference method, we consider a grid in Ω of d^2 nodes equally spaced by $h = \Delta x = \Delta y = \frac{1}{d+1}$. This discretisation yields an almost linear system $Ax + \Phi(x) = b$, where A is a block tridiagonal symmetric matrix $A = (D_{i-1}, T_i, D_i)_{i=1}^d$, where T_i are tridiagonal matrices of size $d \times d$, $i = 1, 2, \ldots, d$, and D_i are $d \times d$ diagonal matrices, $i = 1 \ldots, d-1$; see e.g., [7].

Let $S = \{1, 2, \ldots, n\}$ and let n_k, $k = 1, 2, \ldots, \alpha$, be positive integers which add n. Consider $S_{k,m}$, $k = 1, 2, \ldots, \alpha$, $m = 1, 2, \ldots, q(\ell, k)$, subsets of S defined as

$$S_{k,m} = \{s^1_{k,m}, s^1_{k,m} + 1, \ldots, s^2_{k,m}\}\ , \tag{19}$$

where

$$\begin{cases} s^1_{k,m} = \max\{1, 1 + \sum_{i<k} n_i - bd - (m-1)d\}, \text{ and} \\ s^2_{k,m} = \min\{n, \sum_{i\leq k} n_i + bd + (m-1)d\}\ , \end{cases} \tag{20}$$

with b being a nonnegative integer. Note $S = \bigcup_{k=1}^{\alpha} S_{k,m}$, $m = 1, 2, \ldots, q(\ell, k)$.

Let us further consider multi-splittings of the form

$$\{D - L_{\ell,k,m}, U_{\ell,k,m}, E_k\}_{k=1}^{\alpha}, \text{ where } L_{\ell,k,m} = L_{k,m} \equiv \begin{cases} -a_{ij}, j < i, \ i, j \in S_{k,m} \\ \quad 0 \text{ otherwise,} \end{cases} \tag{21}$$

and $U_{\ell,k,m} = U_{k,m}$, for all $\ell = 1, 2, \ldots$. The $n \times n$ nonnegative diagonal matrices E_k, $1 \leq k \leq \alpha$, are defined such that their ith diagonal entry $(E_k)_{ii}$ is calculated as follows

$$(E_k)_{ii} = \begin{cases} 1 & \text{if } i \in S_{k,1} \text{ and } i \notin S_{j,1},\ j \neq k\ , \\ 0.5 & \text{if } i \in S_{k,1} \cap S_{k-1,1} \text{ or } i \in S_{k,1} \cap S_{k+1,1}\ , \\ 0 & \text{if } i \notin S_{k,1}\ . \end{cases}$$

Experiments were performed with almost linear systems of different orders. The conclusions were similar for all tested problems. Here we discuss the results obtained with $d = 64$ and $d = 200$, that originate almost linear systems of sizes 4096 and 40000 respectively. In this paper all the times obtained for the parallel algorithms correspond to REAL times; moreover, they are reported in seconds. The initial vector used was $x^{(0)} = (1, \ldots, 1)^T$. The stopping criterion used for the almost linear system of size 4096 was $\|x^{(\ell)} - v\|_2 \leq h^2$, where $\|\cdot\|_2$ is the Euclidean norm and v is the vector which entries are the values of the exact solution of (18) on the nodes (ih, jh), $i, j = 1, \ldots, d$. However, for the almost linear system of size 40000 the convergence criterion was changed to $\|x^{(\ell)} - x^{(\ell-1)}\|_1 < 10^{-5}$. This change was done because for the last problem the norm $\|x^\star - v\|_2$, where $x^\star$ is the exact solution of the almost linear system of size 40000, is very close to h^2 but greater than h^2. Thus, with the first stopping criterion, convergence cannot be achieved. Moreover, we have used $\|x^{(\ell)} - x^{(\ell-1)}\|_1 < 10^{-5}$ because this is a possible stopping criterion in practice when the actual solution is not known.

On the other hand, to solve the one-dimensional nonlinear equation (3) the Newton method is used. The best results were obtained performing only one iteration of this method.

Table 1 shows some of these results for the almost linear system of size 4096, setting $\omega = \mu = 1$ in Algorithm 1 and using different multi-splittings depending on the number of processors used (α) and on the choice of the values n_k, $1 \leq k \leq \alpha$. Moreover, in this table, the case in which the splittings do not change with the local iterations (i.e., $L_{k,m} = L_{k,1}$, $m = 1, 2, \ldots, q(\ell, k)$) is analysed together with the case in which the splittings change according to (19) and (21). No overlapping is considered, that is, the integer b in (20) is taken as zero. It is observed that when the splittings change, the number of global iterations decreases. Therefore, the communications among processors are reduced and less execution time is observed.

		Without varying the splittings			Varying the splittings		
α / n_k	$q(\ell,k)$	It.	Time Cluster	Time SP2	It.	Time Cluster	Time SP2
2	1,1	4292	70.46	11.82	4292	70.46	11.82
2048	2,2	2161	64.21	10.53	2144	59.42	10.29
2048	4,4	1091	58.11	10.14	1065	53.79	9.54
	8,8	552	58.74	10.36	532	53.69	9.85
	3,2	1802	70.45	11.81	1786	68.72	11.89
2	1,1	4266	111.4	15.95	4266	111.4	15.95
1216	3,3	1435	85.84	14.11	1416	83.57	13.52
2880	8,8	545	81.55	14.16	530	70.71	12.83
	10,9	476	80.04	13.98	464	69.35	12.60
4	1,1,1,1	4391	57.82	7.73	4391	57.82	7.73
1024	3,3,3,3	1499	39.50	5.98	1447	37.33	5.92
1024	4,4,4,4	1113	34.94	6.01	1081	35.12	5.94
1024	3,4,4,3	1206	38.79	6.46	1152	36.14	6.30
1024	8,8,8,8	580	40.08	6.82	536	36.57	6.64
4	1,1,1,1	4418	62.84	7.77	4418	62.84	7.77
1216	3,3,3,3	1513	44.27	6.55	1453	41.89	6.42
832	4,4,4,4	1145	42.38	6.51	1085	39.49	6.30
832	3,4,4,3	1256	36.94	5.81	1195	34.52	5.62
1216	4,5,5,4	989	36.69	5.89	930	33.89	5.61

Table 1. Non-stationary synchronous models without overlap. Size of the almost linear system: 4096.

One can observe that the number of iterations of the non-stationary algorithms decreases when the parameters $q(\ell, k)$ are increased. Furthermore, if the decrease in the number of global iterations balances the realization of more local updates then, less execution time is observed. Note that when $q(\ell, k) = 1$ and the splittings do not change with the local iterations, the method reduces to the parallel nonlinear Gauss-Seidel method (see [18]) and as it can be appreciated the non-stationary parallel methods are generally better than the parallel nonlinear Gauss-Seidel method.

On the other hand, it is interesting to compare the parallel results of Table 1 with the results of the well-known one-step Gauss-Seidel Newton method [13]. The latter performs 4196 iterations and the CPU time in the IBM RS/6000 SP computer was 20.09 seconds. So, we calculated the speed-up setting as sequential reference algorithm that method, that is, we have considered Speed-up= $\frac{\text{CPU time of one-step GS Newton algorithm}}{\text{REAL time of parallel algorithm}}$. In this context, it is observed that we obtain parallel non-stationary algorithms such that processors can achieve between 84 % and 105 % of efficiency ($\frac{\text{Speed-up}}{\text{processors's number}}$) when it uses two processors and between 62 % and 89 % of efficiency using four processors.

α b / n_k	$q(\ell,k)$	Without varying the splittings: It.	Without varying the splittings: Time SP2	Varying the splittings: It.	Varying the splittings: Time SP2
2 1	1,1	4253	12.43	4292	12.43
2048	2,2	2134	10.95	2124	10.70
2048	4,4	1072	10.42	1059	9.99
	8,8	541	10.55	530	10.22
	3,2	1777	12.35	1769	12.42
2 2	1,1	4246	12.89	4246	12.89
2048	2,2	2129	11.31	2121	11.10
2048	4,4	1068	10.67	1058	10.32
	8,8	538	10.81	529	10.49
	3,2	1769	12.67	1761	12.81
2 3	1,1	4244	13.27	4244	13.27
2048	2,2	2127	11.67	2120	11.47
2048	4,4	1067	11.00	1057	10.64
	8,8	537	11.17	529	10.78
	3,2	1763	12.99	1755	13.12
4 1	1,1,1,1	4327	9.05	4327	9.05
1216	3,3,3,3	1461	7.04	1432	6.89
832	4,4,4,4	1102	6.87	1071	6.76
832	3,4,4,3	1210	6.36	1179	6.29
1216	4,5,5,4	951	6.31	920	6.23
4 2	1,1,1,1	4311	9.84	4311	9.84
1216	3,3,3,3	1452	7.37	1427	7.35
832	4,4,4,4	1093	7.19	1068	7.12
832	3,4,4,3	1200	7.01	1175	6.96
1216	4,5,5,4	942	6.95	917	6.92

Table 2. Non-stationary synchronous models with overlap. Size of the almost linear system: 4096.

The above conclusions are independent of the computer used. However, the network of the cluster is very slow compared to the network of the other computing platform. Hence, while the REAL time and the CPU time are similar in the IBM RS/6000 SP computer, there is a significant difference between these two times in the cluster. In the rest of this section all the numerical experiments have been run in the IBM RS/6000 SP multiprocessor.

Table 2 illustrates the influence of the overlap according to different choices of the overlapping level $b = 1, 2, 3$; see (20). Note that the parameter b indicates that the splitting assigned to a processor k, $k = 1, 2, \ldots, \alpha$, has an overlap of $2b$ blocks (each one of size d) with the splittings assigned to the processors

$k-1$ and $k+1$. The conclusions are simil ar to those presented in Table 1. However, it is observed that while the number of iterations decreases when the overlap increases, this decrease does not get less execution time. This is due to the increase of the number of operations performed at each processor and the increase of the communications among processors.

α	b / n_k	$q(\ell,k)$	Without varying the splittings It.	Time SP2	Error	Varying the splittings It.	Time SP2	Error
4	0	1,1,1,1	40983	818.18	0.00025	40983	818.18	0.00025
	$n_k=10000$	3,3,3,3	14961	611.29	0.00012	14799	604.55	0.00012
	$1\leq k\leq 4$	4,4,4,4	11484	598.31	0.00010	11318	585.85	0.00010
4	1	1,1,1,1	40757	845.37	0.00025	40757	845.37	0.00025
	$n_k=10000$	3,3,3,3	14820	622.18	0.00012	14741	616.90	0.00012
	$1\leq k\leq 4$	4,4,4,4	11365	607.84	0.00010	11280	606.88	0.00010
6	0	1,1,1,1,1,1	41266	608.52	0.00025	41266	608.52	0.00025
	$n_k=6800, k=1,6$	3,3,3,3,3,3	15120	446.33	0.00012	14874	437.73	0.00012
	$n_k=6600$	5,5,5,5,5,5	9474	433.37	0.000095	9223	415.58	0.000094
	$2\leq k\leq 5$	4,5,5,5,5,4	9622	430.46	0.000096	9369	413.59	0.000095
6	1	1,1,1,1,1,1	40925	636.56	0.00025	40925	636.56	0.00025
	$n_k=6800, k=1,6$	3,3,3,3,3,3	14906	463.43	0.00012	14786	453.39	0.00015
	$n_k=6600$	5,5,5,5,5,5	9318	440.76	0.000094	9183	434.38	0.000094
	$2\leq k\leq 5$	3,4,4,4,4,3	11679	449.56	0.00010	11548	445.59	0.00010

Table 3. Non-stationary synchronous models without and with overlap. Size of the almost linear system: 40000.

Now, we report in Table 3 results of non-stationary methods for the almost linear system of size 40000. Moreover, we have calculated for each method, the error $\|x^{(\ell)}-v\|_2$. As it can be appreciated when the non-stationary paramet ers $q(\ell,k)$ increase the error is reduced and therefore the approximation to the solution of the semi-linear elliptic partial differential equation (18) is better. The remaining conclusions are similar to those obtained for the almost linear system of size 4096.

Figure 1 illustrates the influence of the parameters $\omega=\mu\neq 1$ for different overlapping levels, $b=0,1,2$ when Algorithm 1, varying the splittings, is used. These results correspond to the problem of size 4096 when we use four processors and the multi-splitting is defined by $n_k = 1024$, $k = 1,2,3,4$. As it can be observed, in a neighbourhood of the optimum relaxation parameter ω the models with overlap are better than the corresponding non overlapped one. We note that similar results were obtained without varying the splittings.

To finish this section we consider two different implementations of asynchronous Algorithm 2. In the first implementation we consider α processors connected to a host processor, where α is the number of splittings. Thus, we use $\alpha+1$ processors. The role of the host processor is to receive, in an asynchronous way, the approximation computed by other processors, to update the global approximation and to send it to the corresponding processor.

In the second implementation we use α processors, as many as splittings. Then, one of the processors, we assume the first one, has to compute the ap-

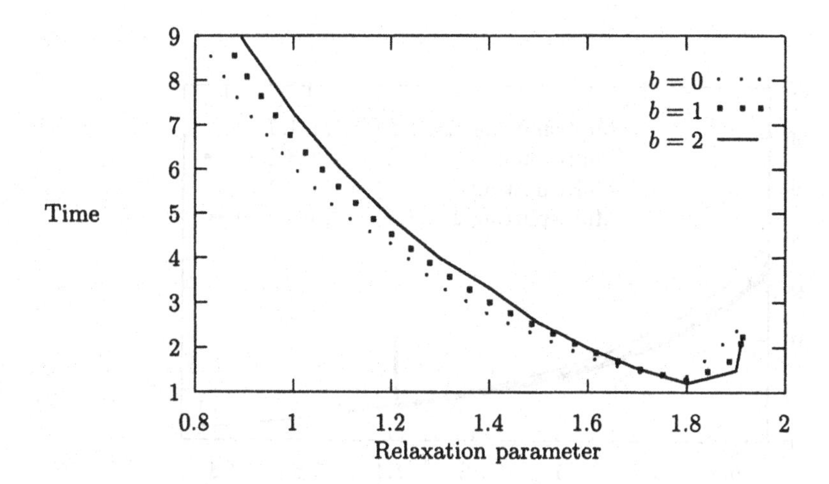

Fig. 1. Comparison synchronous parallel models for different overlapping levels. Non-stationary parameters $q(\ell, k) = 4,\ \ell = 1, 2, \ldots, k = 1, 2, 3, 4$.

proximation corresponding to one of the splittings and, moreover, it has to take the role of host processor. For this purpose, in the process executed by this processor we have added some PVMe calls between the sentences which compute its approximation. This allows us to check if the approximations of some of the other processors have arrived. In this case the host processor executes the following tasks,

1. it stops the calculation of its approximation and it receives the approximation or approximations sent by other processors,
2. it updates the global approximation,
3. if the stopping criterion is not satisfied, then it sends the updated approximation to the corresponding processors, and finally
4. it continues computing its approximation.

Note that in this implementation there are some waiting times.

Figures 2 and 3 illustrate, respectively, the behavior of the above asynchronous implementations of Algorithm 2. In these figures, the multi-splitting 1 makes reference to the one obtained from $n_k = 1024,\ k = 1, 2, 3, 4$, while the multi-splitting 2 corresponds to the values $n_1 = n_4 = 1216,\ n_2 = n_3 = 832$. Overlap and variation of the splittings are not considered in these figures. The conclusions were similar to those of the synchronous models whether the overlap and variation of the splittings were considered or not. However, these asynchronous implementations have not accelerated the convergence. This is due to the fact that in the asynchronous implementations of our example the communications increase compared to the synchronous ones, while the number of operations

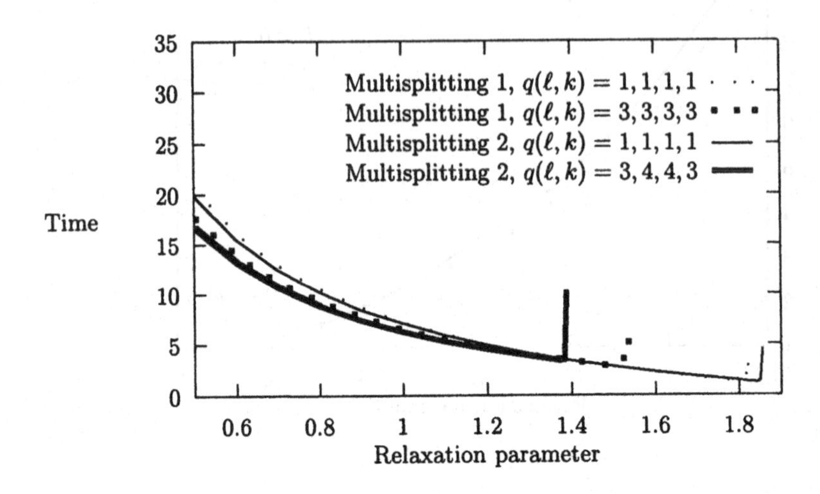

Fig. 2. Comparison asynchronous parallel models without varying the splittings and without overlap (first implementation). Size of almost linear system: 4096.

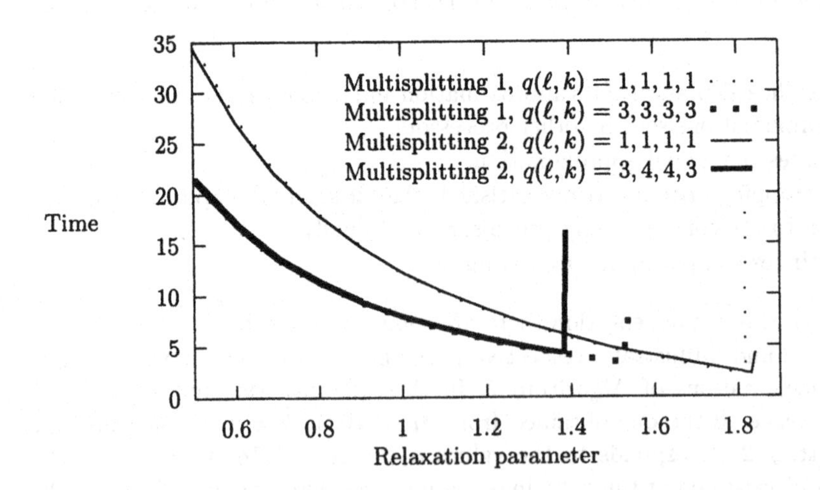

Fig. 3. Comparison asynchronous parallel models without varying the splittings and without overlap (second implementation). Size of almost linear system: 4096.

performed remains of the same order. This is specially problematic when a distributed memory multiprocessor is used.

References

1. Bai, Z.: Parallel nonlinear AOR method and its convergence. Computers and Mathematics with Applications, **31**(2) (1996) 21–31
2. Baudet, G. M.: Asynchronous iterative methods for multiprocessors. Journal of the Association for Computing Machinery, **25**(2) (1978) 226–244
3. Berman A., Plemmons R. J.: Nonnegative Matrices in the Mathematical Sciences. Academic Press, New York, third edition (1979). Reprinted by SIAM, Philadelphia (1994)
4. Bertsekas, D. P., Tsitsiklis, J. N.: Parallel and Distributed Computation. Prentice-Hall, Englewood Cliffs, New Jersey (1989)
5. Birkhoff, G.: Numerical Solution of Elliptic Equations. Vol. 1 of CBMS Regional Conference Series in Applied Mathematics. Society for Industrial and Applied Mathematics, Philadelphia (1970)
6. Bru R., Elsner L., Neumann M.: Models of parallel chaotic iteration methods. Linear Algebra and its Applications, **103** (1988) 175–192
7. Frommer, A.: Parallel nonlinear multi-splitting methods. Numerische Mathematik, **56** (1989) 269–282
8. Fuster, R., Migallón, V., Penadés, J.: Non-stationary parallel multisplitting AOR methods. Electronic Transactions on Numerical Analysis, **4** (1996) 1–13
9. Geist, A., Beguelin, A., Dongarra, J., Jiang, W., Manchek, R., Sunderam, V.: PVM 3 User's Guide and Reference Manual. Technical Report ORNL/TM-12187. Oak Ridge National Laboratory, Tennessee (1994)
10. IBM Corporation: IBM PVMe for AIX User's Guide and Subroutine Reference. Technical Report GC23-3884-00, IBM Corp. Poughkeepsie, New York (1995)
11. Mas, J., Migallón, V., Penadés, J., Szyld, D. B.: Non-stationary parallel relaxed multisplitting methods. Linear Algebra and its Applications, **241/243** (1996) 733–748
12. O'Leary, D. P., White, R. E.: Multi-splittings of matrices and parallel solution of linear systems. SIAM Journal on Algebraic Discrete Methods, **6** (1985) 630–640
13. Ortega, J. M., Rheinboldt, W. C.: Iterative Solution of Nonlinear Equations in Several Variables. Academic Press, San Diego (1970)
14. Ostrowski, A. M.: Über die determinanten mit überwiegender hauptdiagonale. Commentarii Mathematici Helvetici, **10** (1937) 69–96
15. Robert, F., Charnay, M., Musy, F.: Itérations chaotiques série-parallèle pour des équations non-linéaires de point fixe. Aplikace Matematiky, **20** (1975) 1–38
16. Sherman, A.: On Newton-iterative methods for the solution of systems of nonlinear equations. SIAM Journal on Numerical Analysis, **15** (1978) 755–771
17. Varga, R. S.: Matrix Iterative Analysis. Prentice Hall, (1962)
18. White, R. E.: Parallel algorithms for nonlinear problems. SIAM Journal on Algebraic Discrete Methods, **7** (1986) 137–149

The Parallel Problems Server: A Client-Server Model for Interactive Large Scale Scientific Computation

Parry Husbands[1] and Charles Isbell[2]

[1] Laboratory for Computer Science NE43-218,
MIT, Cambridge, MA 02139 USA
parry@supertech.lcs.mit.edu
[2] AT&T Labs/Research Shannon Laboratories,
180 Park Avenue Room B282, Florham Park, NJ 07932-0971 USA
isbell@research.att.com

Abstract. Applying fast scientific computing algorithms to large problems presents a difficult engineering problem. We describe a novel architecture for addressing this problem that uses a robust client-server model for interactive large-scale linear algebra computation.
We discuss competing approaches and demonstrate the relative strengths of our approach. By way of example, we describe MITMatlab, a powerful transparent client interface to the linear algebra server. With MITMatlab, it is now straightforward to implement full-blown algorithms intended to work on very large problems while still using the powerful interactive and visualisation tools that MATLAB provides. We also examine the efficiency of our model by timing selected operations and comparing them to commonly used approaches.

1 Introduction

We describe a novel architecture for a "linear algebra server" that operates on very large matrices. Matrices are created by the server and distributed across many machines or processors. Operations take place automatically in parallel. The server includes a general communication interface to clients and is extensible via a robust package system.

We are motivated by three observations. First, many widely-used algorithms in machine learning, differential equations, simulation, etc. can be realized as operations on matrices. Second, it is vital to be able to test new ideas quickly in an interactive setting. Finally, algorithms that appear promising on small data sets can fail on large problems and it would be helpful to have a tool that easily enables experimentation on large problems.

Common approaches suffer from several difficulties. Interactive prototyping environments such as Mathematica, Maple, Octave, and MATLAB exist; however, they often fail to work well on large problems. Linear algebra libraries designed to work on large problems abound; however, they involve steep learning curves. Further they are typically not interactive, requiring that applications

J. Palma, J. Dongarra, and V. Hernández (Eds.): VECPAR'98, LNCS 1573, pp. 156–169, 1999.

be written in a compiled language, such as C++ or Fortran. This is a burden both for users who simply want a library's functionality and for programmers who wish to extend it.

We address these problems directly. Like standard libraries, our system encapsulates basic functionality; however, by modelling the system as a server, we allow for on-the-fly interaction with arbitrary user interfaces. Further, the server is a self-contained application, so we are able to extend it at run-time.

In this paper, we show that our model opens several key possibilities. We briefly describe standard approaches in Sect. 2 before describing the Parallel Problems Server itself in Sect. 3. We detail its architecture, focusing on its extensibility. Section 4 describes MITMatlab, a system that enables users to compute interactively with very large data sets directly from within MATLAB. We then report on the results of some performance experiments in Sect. 5. Finally, we conclude, discussing further extensions to the system.

2 Standard Approaches

2.1 Linear Algebra Libraries

For many compute-intensive tasks, the best way to maximise performance is to use a library. For example, optimised versions of LAPACK [1] exist that outperform similar code written in a high-level programming language (thanks primarily to native implementations of the BLAS). For distributed memory architectures, vendor-optimised libraries (e.g. Sun's S3L and IBM's ESSL) coexist with public domain offerings such as ScaLAPACK [5], PARPACK [11] and PETSc [4,9].

Each of these libraries has its own idiosyncratic interface and assumptions about the types and distributions of data allowed. It is often a major programming effort to incorporate library routines into an application.

2.2 Interactive Systems

The power of prototyping systems like Maple, MATLAB, Mathematica, and Octave is that they are interactive. It is straightforward for both seasoned programmers and relatively naive users to develop algorithms and to visualise results from such algorithms. Unfortunately, while these tools work well for small problems, they are often inadequate for production-level data.

There have been many attempts to extend prototyping tools in order to make them work in parallel with large data sets. Here, we focus on systems that add parallel features to MATLAB, a widely-used scientific computing tool.

Both MultiMATLAB from Cornell University [13] and the Parallel Toolbox for MATLAB from Wake Forest University [10], make it possible to manage MATLAB processes on different machines. MATLAB is extended to include *send*, *receive* and *collective* operations so that separate MATLAB processes can communicate. In short, these approaches implement traditional message passing with MATLAB as the implementation language.

Compilers for MATLAB are also an active area. Both the CONLAB system from the University of Umeå [7] and the FALCON environment from the University of Illinois at Urbana-Champaign [3,12] translate MATLAB-like languages into intermediate languages for which high performance compilers exist. For example, FALCON compiles MATLAB to Fortran 90 and pC++. Sophisticated analyses of the MATLAB source are performed so that efficient target code is generated.

Both of these approaches have merits; however, it is our claim that they do not adequately address the issues we have raised. The former approach is too involved for the naive user and the latter approach sacrifices direct interaction with the computation and includes an edit-compile-run cycle that may increase development time.

3 The Parallel Problems Server

The Parallel Problems Server (PPServer) combines many aspects of the approaches we have described so far. Like standard linear algebra packages, the PPServer neatly encapsulates basic functionality; however, because it is a server with a general communication protocol, interaction with arbitrary programs (with their own user interfaces) is possible. Also, the server implements a robust protocol for accessing compiled libraries. Thus, extending the functionality of the PPServer is a simple, modular task.

3.1 The Client-Server Model

The client-server model is ubiquitous. There are HTTP servers that allow access to data via the World Wide Web and database servers that admit access to specially indexed data. Because these servers implement robust protocols for communicating the information they provide, it is possible to build useful clients, such as web browsers.

We believe that this model is also a useful one for scientific computation. First, there is no need to force a client to operate in parallel by endowing it with communication primitives; rather, such communication should remain implicit. As a result, the user is not responsible for managing data among various processes. The user simply issues the client's standard commands; these are then transparently executed on multiple machines.

Secondly, there is no need to use the client as the computational engine. While this has the possible short-term disadvantage of the server's functionality being different than the client's, we gain extremely high performance. We are free to use the fastest distributed memory implementations of the algorithms that we need. Furthermore, we are not required to use the client's data representation. For example, MATLAB uses double precision numbers. For the very large operations that concern us, it often preferable to use single precision, gaining significant time and space advantages when accuracy is not a concern.

A high-level view of our implementation of the PPServer is shown in Fig. 1. Clients make requests of the server. Data is created in a distributed fashion and managed among worker processes, which may live on different machines. Currently we support row and column distributed dense arrays, column distributed sparse arrays, and replicated arrays in single precision. Communication and synchronisation among the workers is accomplished using the MPI [8] message passing library. This is a standard library available on a wide range of platforms; it is currently the most portable way to develop applications on distributed memory computers.

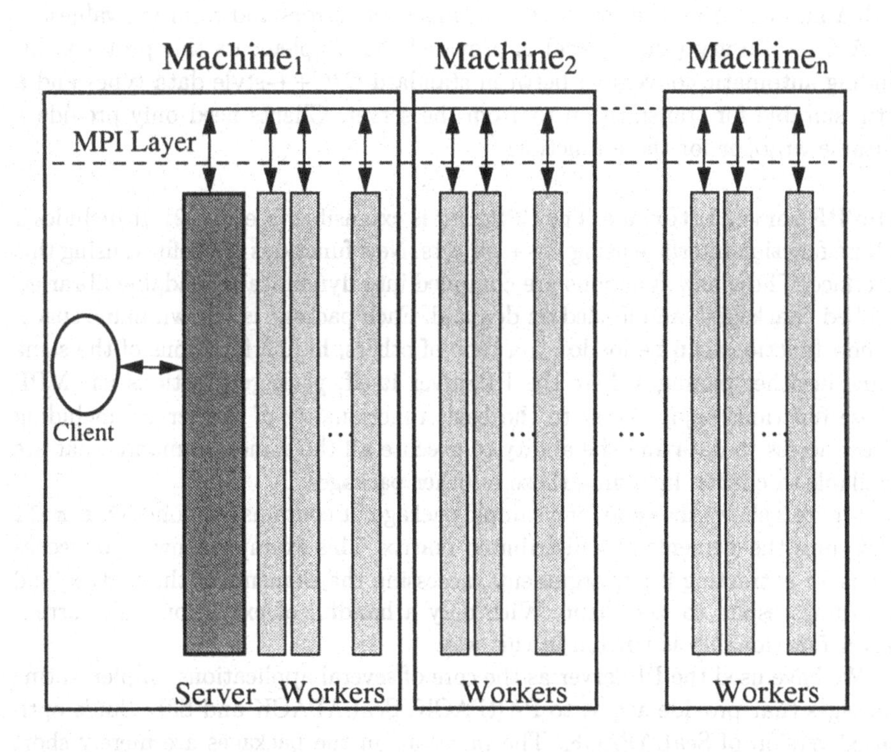

Fig. 1. The General Organisation of the Parallel Problems Server. The server process provides an interface to any client that implements its communication protocol.

3.2 Communication and Extensibility

We use the client-server model in two ways. First, there is a protocol for communicating with clients. Just as importantly, there is a separate plug-in architecture that allows for straightforward run-time extensibility of the PPServer.

The Client Interface While we believe that servers are crucial, they remain only academic oddities without useful clients. HTTP servers are useful but they are much more useful when powerful browsers exist. Therefore, it is important that the client interface be simple to use but powerful enough to allow for arbitrary operations.

The PPServer uses standard Unix sockets for client communication. The protocol is straightforward. A client sends a request, consisting of a command and arguments. A command is a string, naming a function. Functions may request data or the loading, saving, or creating of data. Furthermore, they may require that specific operations be performed on already existing data or that library extensions to be included with the server. Arguments are lists of characters, integers and real numbers. Once a command has been completed, it is acknowledged with a message from the server that includes any errors and returned values.

A C++ library (and source) is provided that implements this protocol, including automatic conversion between standard C/C++-style data types and a form suitable for transmission to/from the server. Clients need only provide a suitable wrapper for these functions.

The PPServer Interface The PPServer is extensible (see Fig. 2). It includes a robust function interface using C++ objects. New functions are defined using this interface. These new functions are compiled into dynamically loadable libraries, dubbed "packages" and loaded on demand. Each package is its own name space, so new functions can be loaded "on top" of others, hiding functions of the same name in other packages. Like the PPServer itself, package functions use MPI. These functions enjoy access to the basic functionality of the server, including direct access to data and the ability to execute all the same commands that are available to clients, including those in other packages.

Figure 3 shows the code for a sample package. It contains one function `sumall` that sums the elements of a distributed matrix. This example shows the mechanisms for extracting input arguments, accessing the elements of the matrix, and returning results to the client. With only a handful of exceptions, all current server functionality is written in this way.

We have used the PPServer as the core of several applications, implementing packages that provide access to PARPACK, ScaLAPACK and S3L, Sun's optimised version of ScaLAPACK. The functions in the packages are merely short wrappers for the underlying functions provided by the libraries.

Portability The use of standard C++ and MPI has allowed us to develop a system that is highly portable. Although the PPServer was originally developed on a network of symmetric multiprocessors from Sun Microsystems, we have been able to port it to a cluster of SMPs from Digital Equipment Corporation with minimal effort. A similar effort resulted in a port to Pentium-driven Linux systems.

3.3 Other Client-Server Models

There have been previous library systems that implement a similar model. Both RCS [2] and Netsolve [6] act as fast back-ends for slower clients. In their model,

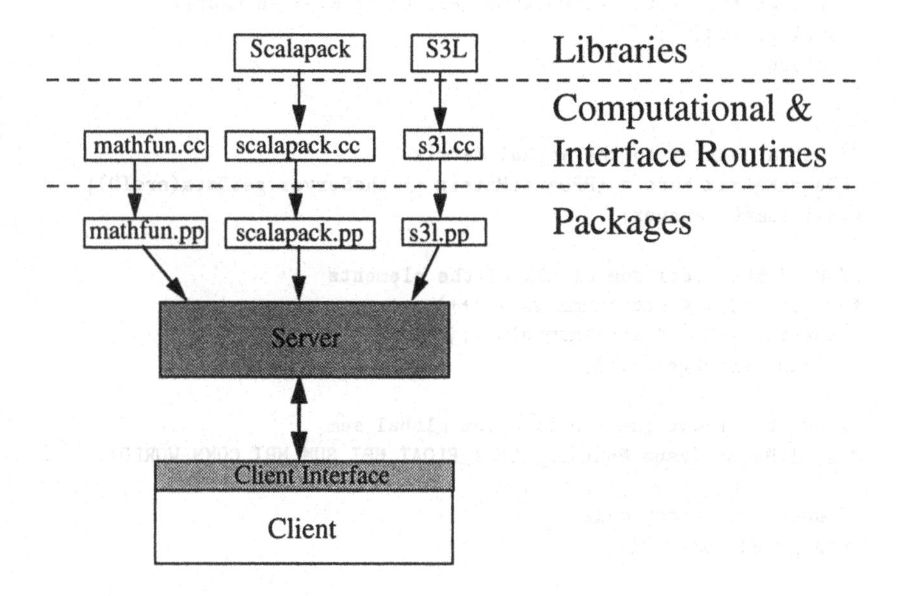

Fig. 2. Extending the PPServer. A client communicates with the PPServer using a simple command-argument protocol. The Server itself uses a "package" mechanism to implement all but its most basic functions. New functionality can be added to the PPServer and managed in a reasonable way. (S3L is Sun's optimised version of some ScaLAPACK routines)

```
void sumall(PPServer &theServer, PPArgList &inArgs, PPArgList &outArgs)
{
  // Get the matrix identifier that was passed in
  PPMatrixID srcID=*(inArgs[0]);

  // Make sure that we're passing in a dense matrix
  if(!theServer.isDense(srcID)) {
    // Return the corresponding error
    outArgs.addError(BADINPUTARGS,"Expecting a Dense Matrix");
    outArgs.add(0);
    return;
  }

  // Get a pointer to the actual matrix
  PPDenseMatrix *src = (PPDenseMatrix *) theServer.getData(srcID);
  float sum=0, answer;

  // Find the local sum of all of the elements
  for(int i=0;i < src->numRows();i++)
    for(int j=0;j < src->numCols();j++)
      sum+=src->get(i,j);

  // Add the local sums to find the global sum
  MPI_AllReduce(&sum,&answer,1,MPI_FLOAT,MPI_SUM,MPI_COMM_WORLD);

  // Return an error code
  outArgs.addNoError();

  // Return the result to the client
  outArgs.add(answer);
}

// Register this function to the server
extern "C" PPError ppinitialize(PPServer &theServer);
PPError ppinitialize(PPServer &theServer)
{
  theServer.addPPFunction("sumall",sumall);
  return(NOERR);
}
```

Fig. 3. A Sample Server Extension. This code is essentially complete other than a few header files.

clients issue requests, arguments are communicated to the remote machine and results sent back. Clients have been developed for Netsolve using both MATLAB and Java.

Our approach to this problem is different in many respects. Our clients are not responsible for storing the data to be computed on. Generally, data is created and stored on the server itself; clients receive only a "handle" to this data (see Fig. 4 for an example). This means that there is no cost for sending and receiving large datasets to and from the computational server. Further, this approach allows computation on data sets too large for the client itself to even store.

We also support transparent access to server data from clients. As we shall see below, given a sufficiently powerful client, PPServer variables can be created remotely but still be treated like local variables.

RCS assumes that the routines that perform needed computation have already been written. Through our package system we support on-the-fly creation of parallel functions. Thus, the server is a meeting place for both data and algorithms.

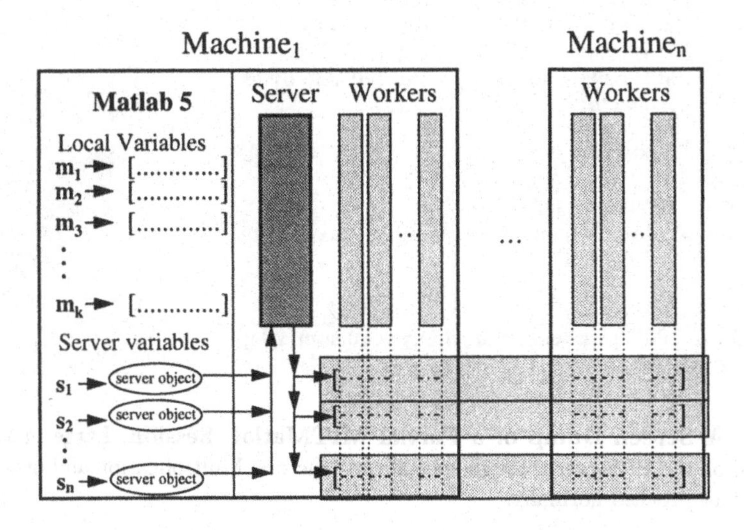

Fig. 4. MITMatlab Variables. Use of the PPServer by MATLAB is almost completely transparent. PPServer variables remain tied to the server itself while MATLAB receives "handles" to the data. Using MATLAB scripts and MATLAB's object and typing mechanisms, functions using PPServer variables invoke PPServer commands implicitly.

4 MITMatlab

Using the client interface, we have implemented a MATLAB front end, called MITMatlab. At present, we can process gigabyte-sized sparse and dense matrices "within" MATLAB, admitting many of MATLAB's operations transparently (see Fig. 5). By using a client as the user interface, we take advantage of whatever interactive mechanisms are available to it. In MATLAB's case, we inherit a host of parsing capabilities, a scripting language, and a host of powerful visualisation tools.

For example, we have implemented BRAZIL, a text retrieval system for large databases. BRAZIL can process queries on a million documents comprised of hundreds of thousands of different words. Because of MATLAB's scripting capabilities, little functionality had to be added to the server directly; rather, most of BRAZIL was "written" in MATLAB.

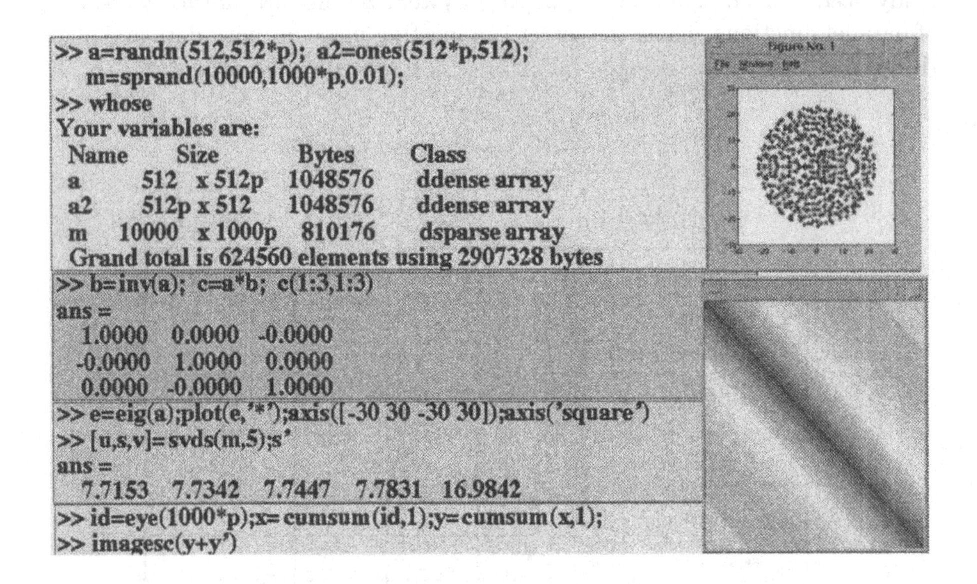

Fig. 5. A Screen Dump of a Partial MITMatlab Session. Large matrices are created on the PPServer through special constructors. Multiplication and other matrix operations proceed normally.

5 Performance

In this section we present results demonstrating the performance of the PPServer. We begin with experiments comparing the efficiency of individual operations in MATLAB with the same operations using MITMatlab. We conclude with a case study of a computation that requires more than a single individual operation. We compare the performance impact of implementing a short program in MATLAB, directly on the PPServer, and using optimised Fortran.

5.1 Individual Operations

We categorise individual operations into two broad classes, according to the amount of computation that is performed relative to the overhead involved in communicating with the PPServer. For *fine grained* operations, most of the time is spent communicating with the server. A typical fine grained task would involve accessing or setting an individual element of a matrix. *Coarse grained* operations include functions such as matrix multiplication, singular value decompositions, and eigenvalue computations where the majority of the time is spent computing instead of communicating input and output arguments with the server.

Below we assess MITMatlab's performance on both kinds of operations. Experiments were performed on a network of Digital AlphaServer 4/4100s connected with Memory Channel.

Fine Grained Operations These operations are understandably slow. For example, in order to access an individual element, the client sends a message to the server specifying the matrix and location, the server locates the desired element among its worker processes, finally sending the result back to the client.

MITMatlab cannot compete with the local function calls that MATLAB uses for these operations. For example, accessing an element in MATLAB only takes 139 microseconds on average, while on a request from the server such can take 2.8 milliseconds. This result can be entirely explained by the overhead involved in communicating with the server; a simple "ping" operation where MITMatlab asks the PPServer for nothing more than an empty reply takes 2 milliseconds.

Coarse Grained Operations For coarse grained operations, the overhead of client/server communication is only a small fraction of the computation to be performed.

Table 1 shows the performance of dense matrix multiplication using MATLAB and MITMatlab. Large performance gains result from the parallelism obtained by using the server; however, even in the case where the server is only using a single processor, it gains significantly over MATLAB. This is due in part because the PPServer can use an optimised version of the BLAS. This illustrates one of the advantages of our model. We can use the fastest operations available on a given platform.

Using PARPACK, MITMatlab also shows superior performance in computing singular value decompositions on sparse matrices (see Table 2).

It is worth noting that MATLAB's operations were performed in double precision while the PPServer's used single precision. While this clearly has an effect on performance, we do not believe that it can account for the great performance difference between the two systems.

Table 1. Matrix multiplication performance of the MITMatlab on p processors. Times are in seconds. Here "$p = 3 + 3$" means 6 processors divided between two machines.

	Matrix Size N		
	1Kx1K	2Kx2K	4Kx4K
MATLAB	41.1	267.1	2814.9
MITMatlab			
with $p = 1$	5.5	45.1	357.9
$p = 2$	2.8	21.5	175.6
$p = 4$	3.9	12.9	94.7
$p = 3 + 3$	1.4	14.4	64.5

Table 2. SVD performance of MITMatlab on p processors using PARPACK. These tests found the first 5 singular triplets of a random 10K by 10K sparse matrix with approximately 1, 2, and 4 million nonzero elements. MATLAB failed to complete the computation in a reasonable amount of time. Times are in seconds.

Processors	Nonzeros		
used	1M	2M	4M
2	136.8	169.2	433.5
4	88.8	91.9	241.0
$3 + 3$	75.2	78.8	168.6

Discussion These results make it clear what types of tasks are best performed on the server. Computations that can be described as a series of coarse grained operations on large matrices fare very well. By contrast, those that use many fine grained operations may be slower than MATLAB. Such tasks should be recoded to use coarse grained operations if possible, or incorporated directly into the server via the package system. Note that on many tasks that involve computation on large matrices, fine grained operations occupy a very small amount of time and so the advantages that we gain using the server are not lost.

5.2 Executing Programs

Figure 6 shows the MATLAB function that we used for this experiment. It performs a matrix-vector multiplication and a vector addition in a loop. Table 3 shows the results when the function is executed: 1) in MATLAB, 2) in MATLAB with server operations, 3) directly on the server through a package, and 4) in Fortran. Experiments were performed using a Sun E5000 with 8 processors. The Fortran code used Sun's optimised version of LAPACK.

The native Fortran version is the fastest; however, the PPServer package version did not incur a substantial performance penalty. The interpreted MITMatlab version, while still faster than the pure MATLAB version, was predictably slower than the two compiled versions. It had to manage the temporary variables that were created in the loop and incurred a little overhead for every server function called. We believe that this small cost is well worth the advantages we obtain in ease of implementation (a simple MATLAB script) and interactivity.

```
A=rand(3000,3000);
x0=rand(3000,1);
Q=rand(3000,9);
n=10;

function X=testfun(A,x0,Q,n)

X(:,1)=x0;
for i=1:n-1
  X(:,i+1)=A*X(:,i)+Q(:,i);
end
```

Fig. 6. MATLAB code for the program test. The MATLAB version that used server operations included some garbage collection primitives in the loop.

Table 3. The performance of the various implementations of the program test. Although MATLAB takes some advantage of multiple processors in the SMP we list it in the $p = 1$ row.

Processors Used	Time (sec)			
	Fortran	Server Package	MATLAB with Server	MATLAB
1	3.07			49.93
2	1.61	1.92	2.43	
4	0.90	1.02	1.49	
6	0.62	0.78	1.26	
8	0.55	0.67	1.84	

6 Conclusions

Applying fast scientific computing algorithms to large everyday problems represents a major engineering effort. We believe that a client-server architecture provides a robust approach that makes this problem much more manageable.

We have shown that we can create tools that allow easy interactive access to large matrices. With MITMatlab, researchers can use MATLAB as more than a prototyping engine restricted to toy problems. It is now possible to implement full-blown algorithms intended to work on very large problems without sacrificing interactive power. MITMatlab has been used successfully in a graduate course in parallel scientific computing. Students have implemented algorithms from areas including genetic algorithms and computer graphics. Packages encapsulating various machine learning techniques, including gradient-based search methods, have been incorporated as well.

Work on the PPServer continues. Naturally, we intend to incorporate more standard libraries as packages. We also intend to implement out-of-core algorithms for extremely large problems, as well as implement interfaces to other clients, such as Java-enabled browsers. Finally, we wish to use the PPServer as real tool for understanding the role of interactivity in supercomputing.

Acknowledgements Parry Husbands is supported by a fellowship from Sun Microsystems. When doing much of this work, Charles Isbell was supported by a fellowship from AT&T Labs/Research. Most of this research was performed on clusters of SMPs provided by Sun Microsystems and Digital Corp.

References

1. E. Anderson, Z. Bai, C. Bischof, J. Demmel, J. Dongarra, J. Du Criz, A. Greenbaum, S. Hammarling, A. McKenney, S. Ostrouchov, and D. Sorensen. *LAPACK Users' Guide.* Siam Publications, Philadelphia, 1995.
2. P. Arbenz, W. Gander, and M. Oettli. The Remote Computation System. Technical Report 245, ETH Zurich, 1996.
3. FALCON Group at the University of Illinois at Urbana-Champaign. The FALCON Project. http://www.csrd.uiuc.edu/falcon/falcon.html.
4. S. Balay, W. D. Gropp, L. C. McInnes, and B. F. Smith. *Efficient Management of Parallelism in Object-Oriented Numerical Software Libraries.* Birkhauser Press, 1997.
5. L. S. Blackford, J. Choi, A. Cleary, E. D'Azevedo, J. Demmel, I. Dhilon, J. Dongarra, S. Hammarling, G. Henry, A. Petitet, K. Stanley, D. Walker, and R.C. Whaley. ScaLAPACK Users' Guide. http://www.netlib.org/scalapack/slug/scalapack_slug.html, May 1997.
6. Henri Casanova and Jack Dongarra. Netsolve: A Network Server for Solving Computational Science Problems. In *Proceedings of SuperComputing 1996*, 1996.
7. Peter Drakenberg, Peter Jacobson, and Bo Kågström. A CONLAB Compiler for a Distributed Memory Multicomputer. In *Proceedings of the Sixth SIAM Conference on Parallel Processing from Scientific Computing*, volume 2, pages 814–821. Society for Industrial and Applied Mathematics, 1993.
8. William Gropp, Ewing Lusk, and Anthong Skjellum. *Using MPI: Portable Parallel Programming with the Message-Passing Interface.* The MIT Press, 1994.
9. PETSc Group. PETSc - the Portable, Extensible Toolkit for Scientific Computation. http://www.mcs.anl.gov/home/gropp/petsc.html.
10. J. Hollingsworth, K. Liu, and P. Pauca. *Parallel Toolbox for MATLAB PT v. 1.00: Manual and Reference Pages.* Wake Forest University, 1996.
11. K. J. Maschhoff and D. C. Sorensen. A Portable Implementation of ARPACK for Distributed Memory Parallel Computers. In *Preliminary Proceedings of the Copper Mountain Conference on Iterative Methods*, 1996.

12. L. De Rose, K. Gallivan, E. Gallopoulos, B. Marsolf, and D. Padua. FALCON: An Environment for the Development of Scientific Libraries and Applications. In *Proceedings of KBUP'95 - First International Workshop on Knowledge-Based Systems for the (re)Use of Program Libraries*, November 1995.
13. Anne E. Trefethen, Vijay S. Menon, Chi-Chao Chang, Gregorz J. Czajkowski, Chris Myers, and Lloyd N. Trefethen. MultiMATLAB: MATLAB on Multiple Processors. http://www.cs.cornell.edu/Info/People/lnt/multimatlab.html, 1996.

Chapter 2:

Computational Fluid Dynamics, Structural Analysis and Mesh Partioning Techniques

Introduction

Timothy J. Barth

NASA Ames Research Center
Information Sciences Directorate
NAS Division
Mail Stop T27A-1
Moffett Field, CA 94035, USA
barth@nas.nasa.gov

A Brief Introduction to the Design and Implementation of Parallel Numerical Simulation Codes

Large scale scientific computing problems associated with the numerical simulation of physical processes such as those occurring in fluid flow and structural dynamics have greatly stimulated the recent evolution of numerical algorithms suitable for parallel computing architectures. The development of efficient parallel numerical simulation algorithms is particularly difficult since computer software and hardware are also evolving at a relatively rapid pace to satisfy the ever-increasing demand for floating point arithmetic performance. Even though a diverse (almost combinatorial) range of architectural possibilities must be carefully weighed by the software architect, the papers contained in this chapter discuss a number of important reoccuring themes in the development of efficient numerical simulation codes:

1. **Code portability.** Owing to the overall complexity of numerical simulation codes, portability and re-usability of software has become an increasing issue for software providers. For large commercial simulation software packages, it is simply not feasible to provide a large number of code modifications to gain compatibility with each new computer architecture. Fortunately, a number of standardised libraries are now routinely used for explicit message passing (e.g. PVM and MPI) and for shared memory access (e.g. SHMEM). In addition, standardised languages such as FORTRAN-90 and C as well as object oriented languages such as C++ provide stable development environments together with parallel scientific libraries such as ScaLAPACK for parallel linear algebra tasks and many others. The performance of these libraries is extensively evaluated in subsequent papers.

J. Palma, J. Dongarra, and V. Hernández (Eds.): VECPAR'98, LNCS 1573, pp. 171–175, 1999.

2. **Minimised communication, data locality and load balancing.** When compared to local (on-board) communication, the relative cost of communication between remote (off-board) processors can be quite high. This cost of communication is strongly related to the bandwidth and latency of sending messages or cache lines between processors. Consequently, an essential ingredient in obtaining high computational efficiency is the mapping of data to memory and/or processors of the computer. This is true to a variable degree across a broad range of architectures utilising distributed memory access via message passing protocol as well as shared memory access (local or global) via cache line protocol. In addition, the work tasked to each processor should be balanced with a minimum number of synchronization points so that processors do not wait idly for other processors to catch up. Several papers contained in this chapter address this problem by explicitly partitioning the domain into a number of smaller (possibly overlapping) sub-domains using a variety of mesh partitioning techniques. These partitioning techniques divide the computational domain among processors so that the work load is approximately balanced and interprocessor communication is approximately minimised. Inside each sub-domain, the data may be reordered in memory to further improve data locality so that cache memory misses are reduced.
3. **Speedup and Scalability.** Common threads contained in the papers of this chapter are the basic motivations for using a parallel computer in large scale scientific computations. These motivations embody the expectation that by introducing more processors into the computation it will be become possible to (1) solve a problem with fixed number of unknowns in a shorter amount of wall clock time (speedup), (2) solve a problem that exceeds the capacity of a single processor in terms of memory and/or computer time, and (3) to control the overall algorithmic complexity in terms of arithmetic operations as the number of solution unknowns is increased (scalability) by employing multiple meshes and processors. This latter motivation is the divide-and-conquer effect of certain domain decomposition methods and the multi-level effect of multi-grid methods. A number of techniques for obtaining motivations (1-3) are discussed in subsequent papers based on non-overlapping domain decomposition, additive Schwarz on overlapping meshes, and multi-grid. Each has relative merits depending on the relative cost of communication and computation, granularity of parallel computations, differential equation discretized, and discretisation technique used. Note that for some applications that require only a few processors, parallel speedup may be the most important issue. In other applications requiring thousands of processors, scalability may become a dominate issue. Each technique must be evaluated against the desired class of computations.

The broad range of computing architectures used in actual parallel computations suggests that the standardisation of libraries for interprocessor communication, parallel linear algebra, etc. has produced an environment suitable for the development of large scale numerical simulation codes. Even so, the demands for tera- and peta-flop computing capability needed in complex simulations suggests that the evolution of parallel computing must continue at the current unrelenting rate.

A Roadmap to Articles in This Chapter

The contents of this chapter consists of one invited article and 8 contributed articles. The first five articles consider parallel algorithms applied to problems in computational fluid dynamics (CFD), the next two articles consider parallel algorithms and load balancing applied to structural analysis. The remaining two articles address the areas of object-oriented library development and mesh partitioning with specific relevance to finite element and finite difference solvers. Further consideration of this latter category is given in Chapter 4.

Parallel Domain Decomposition Preconditioning for Computational Fluid Dynamics by Timothy J. Barth, Tony Chan and Wei-Pai Tang, NASA Ames Research Center (USA). This article considers domain decomposition solution procedures for discretized scalar advection-diffusion equations as well as the equations of computational fluid dynamics. Particular emphasis is given to the non-overlapping (Schur complement) domain decomposition technique. Several techniques are presented for simplifying the formation of the Schur complement and improving the overall performance of implementations on parallel computers. Calculations on the IBM SP2 parallel computer show scalability of the non-overlapping technique but reveal some sensitivity to load balancing of mesh interfaces between adjacent sub-domains.

Influence of the Discretisation Scheme on the Parallel Efficiency of a Code for the Modelling of a Utility Boiler by P.J. Coelho. Using the Navier-Stokes equations to simulate fluid flow inside a utility boiler, the author compares the parallel efficiency of the baseline computer code when using a hybrid (central/upwind) difference scheme or a MUSCL discretisation scheme for the convective transport terms. Benchmark computations were executed on a Cray T3D using 1 to 128 processors in multiples of 2. Although the amount of computation required for the MUSCL scheme is larger than the hybrid scheme, the MUSCL scheme displays a superior speedup versus number of processors when compared to the hybrid scheme.

Parallel 3D Air Flow Simulation on Workstation Cluster by Jean-Baptiste Vicaire, Loic Prylli, Georges Perrot and Bernard Tourancheau. The authors discuss the development of a panel method for 3D air flow using the portable BLAS, BLACS, and ScaLAPACK linear algebra packages together with PVM or MPI communication protocol. Particular attention is given to code portability issues. Benchmark calculations are presented using a variety of interconnection networks and computer platforms. The authors successfully demonstrate that good parallel performance can be achieved using relatively low cost hardware.

Parallel Turbulence Simulation: Resolving the Inertial Subrange of Kolmogorov Spectra by Thomas Gerz and Martin Strietzel. The authors evaluate the parallel efficiency of a direct numerical simulation (DNS) code for simulating stratified turbulence. The code is based on a second-order finite difference algorithm with pseudo-spectral approximation in the x-direction. The paper shows informative comparisons of various parallel computer architectures (IBM SP2, T3D, SGI Power Challenge and Cray J916). For example, a speedup of 7.4 is achieved on an eight-processor SGI Power Challenge computer. Numerical results include the simulation of a stratified flow (Richardson number, 0.13) with a Reynolds number of 600 exhibiting an inertial subrange in the energy spectrum.

The Study of a Parallel Algorithm Using the Laminar Backward-Facing Step Flow as a Test Case by P.M. Areal and J.M.L.M. Palma. Using the laminar backward-facing step flow as a benchmark test case, the authors present a study of the overlapping domain decomposition solution for incompressible flow. The convergence characteristics of the numerical method are studied as the number of sub-domains, mesh size, and Reynolds number are varied. The results show that the convergence characteristics are sensitive to the Reynolds number and the flow pattern. The results of the parallel code using 4 or fewer sub-domains show very little deterioration when compared with the results obtained using the serial code.

A Low Cost Distributed System for FEM Parallel Structural Analysis by C.O. Moretti, T.N. Bittencourt and L.F. Martham. A distributed computational system for finite element structural analysis is described. The authors perform calculations using a low cost system consisting of a 100 Mbit Fast-Ethernet network cluster of 8 Pentium (200 MHz) processors together with the increasingly popular LINUX operating system. The article discusses the various components of the software package (pre-processing, partioning, analysis and post-processing). Different implementation aspects concerning scalability and performance speedup are discussed. Significant improvement is observed using nodal reordering and direct access data files.

Dynamic Load Balancing in Crashworthiness Simulation by H.G. Galbas and O. Kolp. The authors consider the parallelisation of a commercial software code for crashworthiness analysis. Due to the relatively large amount of computational resources used in typical crash simulations, parallelisation of the simulation is extremely beneficial. Unfortunately, high parallel efficiency of the simulation hinges critically on load balancing of work in the contact and non-contact zones. The article presents a new dynamic load balancing strategy for crash simulations that keeps the contact and non-contact zones separately loaded balanced. Significant improvements are shown and future improvements are suggested.

Some Concepts of the Software Package FEAST by Ch. Becker, S. Kilian, S. Turek and the FEAST Group. This article describes a new object-oriented finite element software package called FEAST. The authors describe the numerous algorithms and data structures chosen for use in FEAST so that high-performance simulations are obtained. FEAST implements local mesh refinement strategies with the framework of multi-grid accelerated FEM solvers. Computational examples demonstrate the high efficiency of FEAST when compared to other existing methods.

Multilevel Mesh Partitioning for Optimising Aspect Ratio by C. Walshaw, M. Cross, R. Diekmann and F. Shlimbach. Multilevel graph partitioning algorithms are a class of optimisation techniques which address the mesh partitioning problem by approximately minimising the associated graph cut-weight. The graph theoretic nature of these algorithms precludes the overall control of the shape of sub-domains. For certain classes of iterative solution methods for solving discretized differential equations on partitioned meshes, it is known that the shape of sub-domains can influence the convergence properties of the iterative method. For this reason, the authors consider modifications to a multilevel graph partitioning algorithm which permit the control of sub-domain aspect ratio via cost (penalty) function. Using this technique, the authors demonstrate the improvement in sub-domain aspect ratios for meshes similar to those used in numerical simulations.

Parallel Domain-Decomposition Preconditioning for Computational Fluid Dynamics

Invited Talk

Timothy J. Barth[1], Tony F. Chan[2*], and Wei-Pai Tang[3**]

[1] NASA Ames Research Center,
Mail Stop T27A-1,Moffett Field, CA 94035, USA
barth@nas.nasa.gov

[2] UCLA Department of Mathematics
Los Angeles, CA 90095-1555, USA
chan@math.ucla.edu

[3] University of Waterloo Department of Computer Science,
Waterloo, Ontario N2L 3G1, Canada
wptang@bz.uwaterloo.ca

Abstract. Algebraic preconditioning algorithms suitable for computational fluid dynamics (CFD) based on overlapping and non-overlapping domain decomposition (DD) are considered. Specific distinction is given to techniques well-suited for time-dependent and steady-state computations of fluid flow. For time-dependent flow calculations, the overlapping Schwarz algorithm suggested by Wu et al. [28] together with stabilised (upwind) spatial discretisation shows acceptable scalability and parallel performance without requiring a coarse space correction. For steady-state flow computations, a family of non-overlapping Schur complement DD techniques are developed. In the Schur complement DD technique, the triangulation is first partitioned into a number of non-overlapping subdomains and interfaces. The permutation of the mesh vertices based on subdomains and interfaces induces a natural 2×2 block partitioning of the discretisation matrix. Exact LU factorisation of this block system introduces a Schur complement matrix which couples subdomains and the interface together. A family of simplifying techniques for constructing the Schur complement and applying the 2×2 block system as a DD preconditioner are developed. Sample fluid flow calculations are presented to demonstrate performance characteristics of the simplified preconditioners.

* The second author was partially supported by the National Science Foundation grant ASC-9720257, by NASA under contract NAS 2-96027 between NASA and the Universities Space Research Association (USRA).

** The third author was partially supported by NASA under contract NAS 2-96027 between NASA and the Universities Space Research Association (USRA), by a Natural Sciences and Engineering Research Council of Canada and by the Information Technology Research Centre which is funded by the Province of Ontario.

J. Palma, J. Dongarra, and V. Hernández (Eds.): VECPAR'98, LNCS 1573, pp. 176–202, 1999.

1 Overview

The efficient numerical simulation of compressible fluid flow about complex geometries continues to be a challenging problem in large scale computing. Many computational problems of interest in combustion, turbulence, aerodynamic performance analysis and optimisation will require orders of magnitude increases in mesh resolution and/or solution degrees of freedom (dofs) to adequately resolve relevant fluid flow features. In solving these large problems, issues such as algorithmic scalability [1] and efficiency become fundamentally important. Furthermore, current computer hardware projections suggest that the needed computational resources can only be achieved via parallel computing architectures. Under this scenario, two algorithmic solution strategies hold particular promise in computational fluid dynamics (CFD) in terms of complexity and implementation on parallel computers: (1) multi-grid (MG) and (2) domain decomposition (DD). Both are known to possess essentially optimal solution complexity for model discretised elliptic equations. Algorithms such as DD are particularly well-suited to distributed memory parallel computing architectures with high off-processor memory latency since these algorithms maintain a high degree of on-processor data locality. Unfortunately, it remains an open challenge to obtain similar optimal complexity results using DD and/or MG algorithms for the hyperbolic-elliptic and hyperbolic-parabolic equations modelling compressible fluid flow. In the remainder of this paper, we report on promising domain decomposition strategies suitable for the equations of CFD. In doing so, it is important to distinguish between two types of flow calculations:

1. *Steady-state computation of fluid flow.* The spatially hyperbolic-elliptic nature of the equations places special requirements on the solution algorithm. In the elliptic-dominated limit, global propagation of decaying error information is needed for optimality. This is usually achieved using either a coarse space operator (multi-grid and overlapping DD methods) or a global interface operator (non-overlapping DD methods). In the hyperbolic-dominated limit, error components are propagated along characteristics of the flow. This suggests specialised coarse space operators (MG and overlapping DD methods) or special interface operators (non-overlapping DD methods). In later sections, both overlapping and non-overlapping DD methods are considered in further detail.
2. *Time-dependent computation of fluid flow.* The hyperbolic-parabolic nature of the equations is more forgiving. Observe that the introduction of numerical time integration implies that error information can only propagate over relatively small distances during a given time step interval. In the context of overlapping DD methods with backward Euler time integration, it becomes mathematically possible to show that scalability is retained without a coarse space correction by choosing the time step small enough and the subdomain overlap sufficiently large enough, cf. Wu et al. [28]. Section 5.3

[1] the arithmetic complexity of algorithms with increasing number of degrees of freedom

reviews the relevant theory and examines the practical merits by performing time-dependent Euler equation computations using overlapping DD with no coarse space correction.

2 Scalability and Preconditioning

To understand algorithmic scalability and the role of preconditioning, we think of the partial differential equation (PDE) discretisation process as producing linear or linearised systems of equations of the form

$$Ax - b = 0 \tag{1}$$

where A is some large (usually sparse) matrix, b is a given right-hand-side vector, and x is the desired solution. For many practical problems, the amount of arithmetic computation required to solve (1) by iterative methods can be estimated in terms of the condition number of the system $\kappa(A)$. If A is symmetric positive definite (SPD), the well-known conjugate gradient method converges at a constant rate which depends on κ. After n iterations of the conjugate gradient method, the error ϵ satisfies

$$\frac{\|\epsilon^n\|_2}{\|\epsilon^0\|_2} \leq \left(\frac{\sqrt{\kappa(A)} - 1}{\sqrt{\kappa(A)} + 1} \right)^n . \tag{2}$$

For most applications of interest in computational fluid dynamics, the condition number associated with A depends on computational parameters such as the mesh spacing h, added stabilisation terms, and/or artificial viscosity coefficients. In addition, $\kappa(A)$ can depend on physical parameters such as the Peclet number and flow direction as well as the underlying stability and well-posedness of the PDE and boundary conditions. Of particular interest in algorithm design and implementation is the parallel scalability experiment whereby a mesh discretisation of the PDE is successively refined while keeping a fixed physical domain so that the mesh spacing h uniformly approaches zero. In this setting, the matrix A usually becomes increasingly ill-conditioned because of the dependence of $\kappa(A)$ on h. A standard technique to overcome this ill-conditioning is to solve the prototype linear system in right (or left) preconditioned form

$$(AP^{-1})Px - b = 0 . \tag{3}$$

The solution is unchanged but the convergence rate of iterative methods now depends on properties of AP^{-1}. Ideally, one seeks preconditioning matrices P which are easily solved and in some sense nearby A, e.g. $\kappa(AP^{-1}) = O(1)$ when A is SPD. The situation changes considerably for advection dominated problems. The matrix A ceases to be SPD so that the performance of iterative methods is not directly linked to the condition number behaviour of A. Moreover, the convergence properties associated with A can depend on nonlocal properties of the PDE. To see this, consider the advection and advection-diffusion problems shown in Fig. 1. The entrance/exit flow shown in Fig. 1(a) transports the solution and any error components along 45° characteristics which eventually exit the domain. This is contrasted with the recirculation flow shown in Fig. 1(b) which has circular characteristics in the advection dominated limit. In this (singular) limit, any radially symmetric error components persist for all time. More generally, these recirculation error components are removed by the physical *cross-wind* diffusion terms present in the PDE or the artificial cross-wind diffusion terms introduced by the numerical discretisation. When the advection speed is large and the cross-wind diffusion small, the problem becomes ill-conditioned. Brandt and Yavneh [7] have studied both entrance/exit and recirculation flow within the context of multi grid acceleration. The behaviour of multi-grid (or most

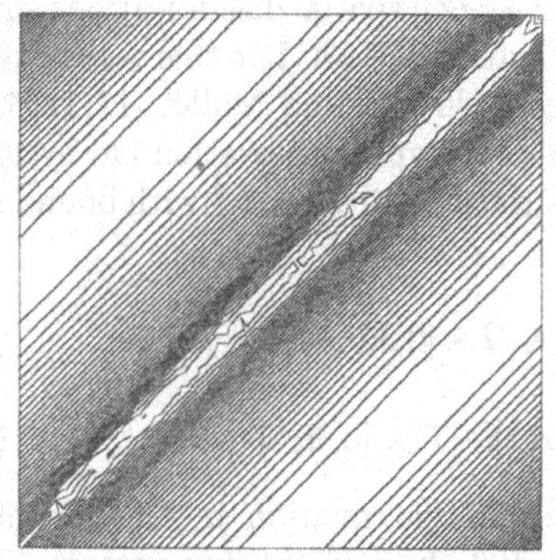

(a) 45° error component transport for entrance/exit flow.

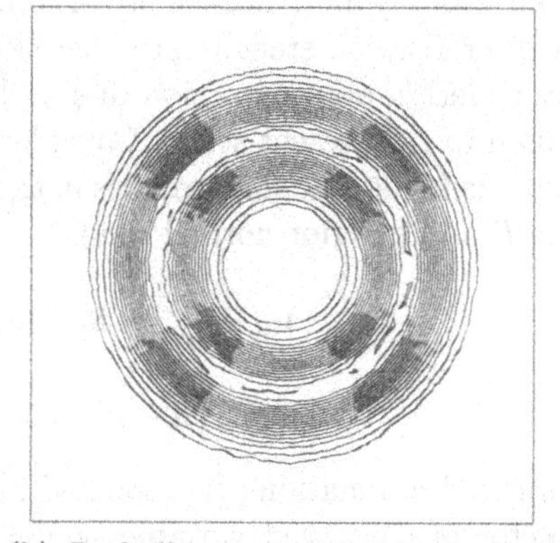

(b) Radially symmetric error component for recirculating flow.

Fig. 1. Two model advection flows: (a) entrance/exit flow $u_x + u_y = 0$, (b) recirculating flow $yu_x - xu_y = \lim_{\epsilon \downarrow 0} \epsilon \Delta u$.

other iterative methods) for these two flow problems is notably different. For example, Fig. 2 graphs the convergence history of ILU(0)-preconditioned GMRES in solving Cuthill-McKee ordered matrix problems for entrance/exit flow and recirculation flow discretised using the Galerkin least-squares (GLS) procedure described in Sect. 3. The entrance/exit flow matrix problem is solved to a 10^{-8} accuracy tolerance in approximately 20 ILU(0)-GMRES iterations. The recirculation flow problem with $\epsilon = 10^{-3}$ requires 45 ILU(0)-GMRES iterations to reach the 10^{-8} tolerance and approximately 100 ILU(0)-GMRES iterations with $\epsilon = 0$. This difference in the number of iterations required for each problem increases dramatically as the mesh is refined. Using the non-overlapping DD method described in Sect. 5.5, we can remove the ill-conditioning observed in the recirculating flow problem. Let V_H denote the set of vertices along a nearly horizontal line from the centre of the domain to the right boundary and V_S the set of remaining vertices in the mesh, see Fig. 3. Next, permute the discretisation matrix so that solution unknowns corresponding to V_H are ordered last. The remaining mesh vertices have a natural ordering along characteristics of the advection operator which renders the discretisation matrix associated with V_S nearly lower triangular. Using the technique of Sect. 5.5 together with exact factorisation of the small $|V_H| \times |V_H|$ Schur complement, acceptable convergence rates for ILU(0)-preconditioned GMRES are once again obtainable as shown in Fig. 3. These promising results have strengthened our keen interest in DD for fluid flow problems.

3 Stabilised Numerical Discretisation

Non-overlapping domain-decomposition procedures such as those developed in Sect. 5.5 strongly motivate the use of compact-stencil spatial discretisations since larger discretisation stencils produce larger interface sizes. For this reason, the Petrov-Galerkin approximation due to Hughes, Franca and Mallet [17,18] has been used in the present study. Consider the prototype conservation law system in m coupled independent variables in the spatial domain $\Omega \subset \mathbb{R}^d$ with boundary surface Γ and exterior normal $\mathbf{n}(x)$

$$\mathbf{u}_{,t} + \mathbf{f}^i_{,x_i} = 0, \quad (x,t) \in \Omega \times [0, \mathbb{R}^+] \tag{4}$$

$$(n_i \mathbf{f}^i_{,\mathbf{u}})^- (\mathbf{u} - \mathbf{g}) = 0, \quad (x,t) \in \Gamma \times [0, \mathbb{R}^+] \tag{5}$$

with implied summation over repeated indices. In this equation, $\mathbf{u} \in \mathbb{R}^m$ denotes the vector of conserved variables and $\mathbf{f}^i \in \mathbb{R}^m$ the inviscid flux vectors. The vector $\mathbf{g}$ can be suitably chosen to impose characteristic data or surface flow tangency using reflection principles. The conservation law system (4) is assumed to possess a generalised entropy pair so that the change of variables $\mathbf{u}(\mathbf{v}) : \mathbb{R}^m \mapsto \mathbb{R}^m$ symmetrises the system in quasi-linear form

$$\mathbf{u}_{,\mathbf{v}} \mathbf{v}_{,t} + \mathbf{f}^i_{,\mathbf{v}} \mathbf{v}_{,x_i} = 0 \tag{6}$$

with $\mathbf{u}_{,\mathbf{v}}$ symmetric positive definite and $\mathbf{f}^i_{,\mathbf{v}}$ symmetric. The computational domain Ω is composed of non-overlapping simplicial elements T_i, $\Omega = \cup T_i$, $T_i \cap T_j = \emptyset$, $i \neq j$. For purposes of the present study, our attention is restricted to steady-state calculations. Time derivatives are retained in the Galerkin integral so that a pseudo-time marching strategy can be used for obtaining steady-state solutions. The Galerkin least-squares method due to Hughes, Franca and Mallet [17] can be defined via the following variational problem with time derivatives omitted from the least-squares bilinear form: Let $\mathcal{V}^h$ denote the finite element space $\mathcal{V}^h = \left\{ \mathbf{w}^h | \mathbf{w}^h \in \left(C^0(\Omega)\right)^m, \mathbf{w}^h|_T \in \left(\mathcal{P}_k(T)\right)^m \right\}$.

Find $\mathbf{v}^h \in \mathcal{V}^h$ such that for all $\mathbf{w}^h \in \mathcal{V}^h$

$$B(\mathbf{v}^h, \mathbf{w}^h)_{gal} + B(\mathbf{v}^h, \mathbf{w}^h)_{ls} + B(\mathbf{v}^h, \mathbf{w}^h)_{bc} = 0 \tag{7}$$

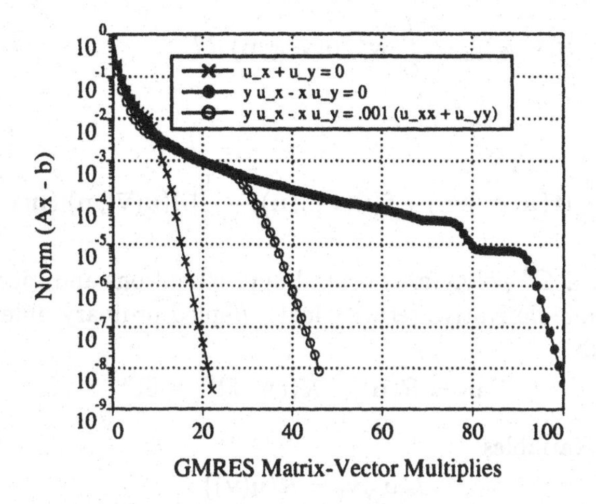

Fig. 2. Convergence behaviour of ILU(0)-preconditioned GMRES for entrance/exit and recirculation flow problems using GLS discretisation in a triangulated square (1600 dofs).

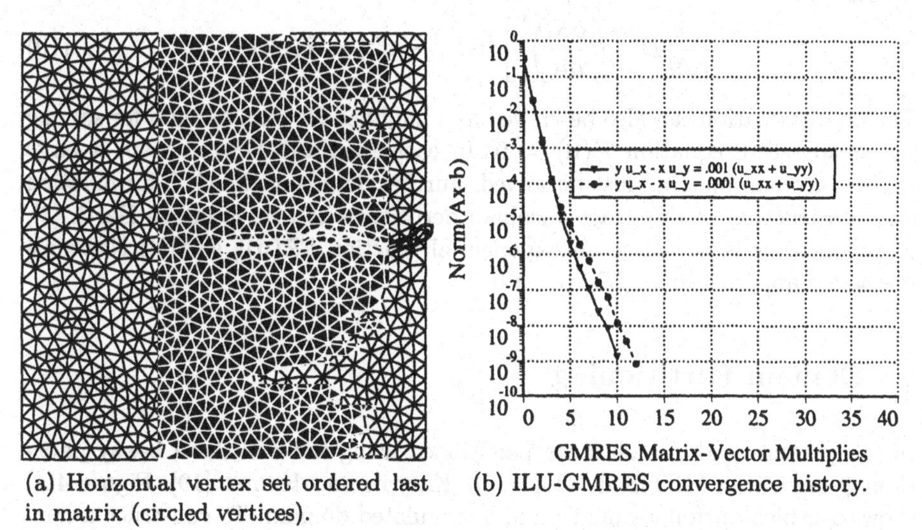

(a) Horizontal vertex set ordered last in matrix (circled vertices).

(b) ILU-GMRES convergence history.

Fig. 3. Sample mesh and ILU-GMRES convergence history using the non-overlapping DD technique of Sect. 5.5.

with

$$B(\mathbf{v},\mathbf{w})_{gal} = \int_{\Omega} (\mathbf{w}^T \mathbf{u}(\mathbf{v})_{,t} - \mathbf{w}^T_{,x_i} \mathbf{f}^i(\mathbf{v}))\, d\Omega$$

$$B(\mathbf{v},\mathbf{w})_{ls} = \sum_{T\in\Omega} \int_T \left(\mathbf{f}^i_{,\mathbf{v}} \mathbf{w}_{,x_i}\right)^T \boldsymbol{\tau} \left(\mathbf{f}^i_{,\mathbf{v}} \mathbf{v}_{,x_i}\right)\, d\Omega$$

$$B(\mathbf{v},\mathbf{w})_{bc} = \int_\Gamma \mathbf{w}^T\, \mathbf{h}(\mathbf{v},\mathbf{g};\mathbf{n})\, d\Gamma$$

where

$$\mathbf{h}(\mathbf{v}_-,\mathbf{v}_+,\mathbf{n}) = \frac{1}{2}\left(\mathbf{f}(\mathbf{u}(\mathbf{v}_-);\mathbf{n}) + \mathbf{f}(\mathbf{u}(\mathbf{v}_+);\mathbf{n})\right) - \frac{1}{2}|A(\mathbf{u}(\overline{\mathbf{v}});\mathbf{n})|(\mathbf{u}(\mathbf{v}_+) - \mathbf{u}(\mathbf{v}_-)).$$

Inserting standard C^0 polynomial spatial approximations and mass-lumping of the remaining time derivative terms, yields coupled ordinary differential equations of the form:

$$D\,\mathbf{u}_t = \mathcal{R}(\mathbf{u}), \quad \mathcal{R}(\mathbf{u}) : \mathbb{R}^n \to \mathbb{R}^n \tag{8}$$

or in symmetric variables

$$D\,\mathbf{u}_{,\mathbf{v}}\mathbf{v}_t = \mathcal{R}(\mathbf{u}(\mathbf{v})) \tag{9}$$

where D represents the (diagonal) lumped mass matrix. In the present study, backward Euler time integration with local time linearisation is applied to Eqn. (8) yielding:

$$\left[\frac{1}{\Delta t}D - \left(\frac{\partial \mathcal{R}}{\partial \mathbf{u}}\right)^n\right](\mathbf{u}^{n+1} - \mathbf{u}^n) = \mathcal{R}(\mathbf{u}^n)\ . \tag{10}$$

The above equation can also be viewed as a modified Newton method for solving the steady-state equation $\mathcal{R}(\mathbf{u}) = 0$. In each modified Newton step, a large sparse Jacobian matrix must be solved. Further details of the GLS formulation and construction of the exact Jacobian derivatives can be found in Barth [3]. In practice Δt is varied as an exponential function $\|\mathcal{R}(\mathbf{u})\|$ so that Newton's method is approached as $\|\mathcal{R}(\mathbf{u})\| \to 0$.

4 Domain Partitioning

In the present study, meshes are partitioned using the multilevel k-way partitioning algorithm METIS developed by Karypis and Kumar [19]. Figure 4(a) shows a typical airfoil geometry and triangulated domain. To construct a non-overlapping partitioning, a dual triangulation graph has been provided to the METIS partitioning software. Figure 4(b) shows partition boundaries and sample solution contours using the spatial discretisation technique described in the previous section. By partitioning the dual graph of the triangulation, the number of elements in each subdomain is automatically balanced by the METIS software. Unfortunately, a large percentage of computation in our domain-decomposition algorithm is proportional to the interface size associated with each subdomain. On general meshes containing non-uniform element densities, balancing subdomain sizes does not imply a balance of interface sizes. In fact, results shown in Sect. 6 show increased imbalance of interface sizes as meshes are partitioned into larger numbers of subdomains. This ultimately leads to poor load balancing of the parallel computation. This topic will be revisited in Sect. 6.

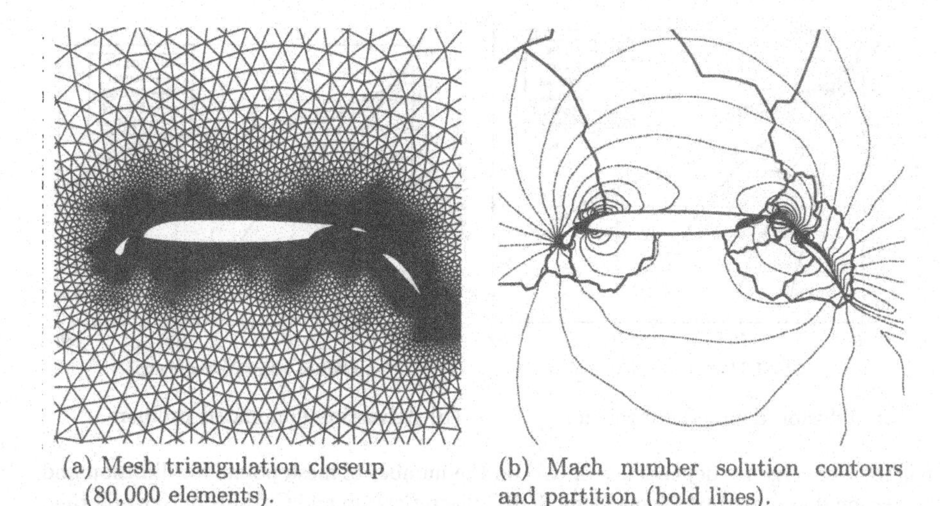

(a) Mesh triangulation closeup (80,000 elements).

(b) Mach number solution contours and partition (bold lines).

Fig. 4. Multiple component airfoil geometry with 16 subdomain partitioning and sample solution contours ($M_\infty = .20, \alpha = 10°$).

5 Preconditioning Algorithms for CFD

In this section, we consider several candidate preconditioning techniques based on overlapping and non-overlapping domain decomposition.

5.1 ILU Factorisation

A common preconditioning choice is incomplete lower-upper factorisation with arbitrary fill level k, ILU[k]. Early application and analysis of ILU preconditioning is given in Evans [15], Stone [27] and Meijerink and van der Vorst [21]. Although the technique is algebraic and well-suited to sparse matrices, ILU-preconditioned systems are not generally scalable. For example, Dupont et al. [14] have shown that ILU[0] preconditioning does not asymptotically change the $O(h^{-2})$ condition number of the 5-point difference approximation to Laplace's equation. Figure 5 shows the convergence of ILU-preconditioned GMRES for Cuthill-McKee ordered matrix problems obtained from diffusion and advection dominated problems discretised using Galerkin and Galerkin least-squares techniques respectively with linear elements. Both problems show pronounced convergence deterioration as the number of solution unknowns (degrees of freedom) increases. Note that matrix orderings exist for discretised scalar advection equations that are vastly superior to Cuthill-McKee ordering. Unfortunately, these orderings do not generalise naturally to coupled systems of equations which do not have a single characteristic direction. Some ILU matrix ordering experiments are given in [10]. Keep in mind that ILU *does* recluster eigenvalues of the preconditioned matrix so that for small enough problems a noticeable improvement

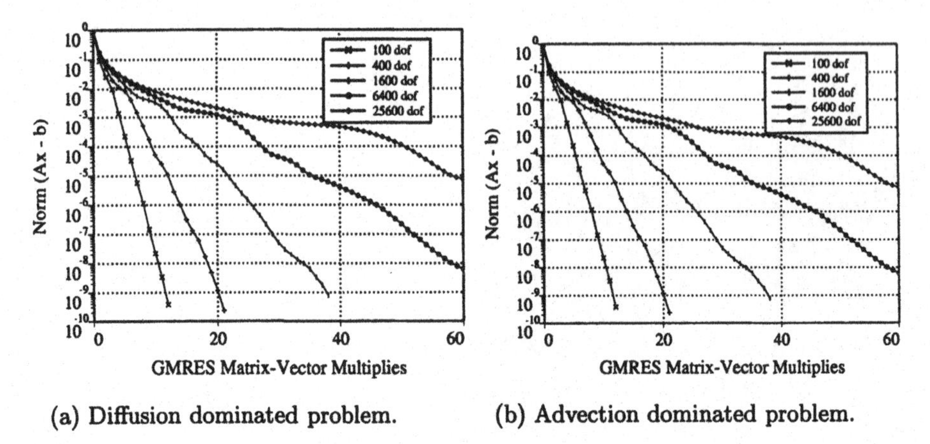

(a) Diffusion dominated problem. (b) Advection dominated problem.

Fig. 5. Convergence dependence of ILU on the number of mesh points for diffusion and advection dominated problems using SUPG discretisation and Cuthill-McKee ordering.

can often be observed when ILU preconditioning is combined with a Krylov projection sequence.

5.2 Additive Overlapping Schwarz Methods

Let V denote the triangulation vertex set. Assume the triangulation has been partitioned into N overlapping subdomains with vertex sets $V_i, i = 1, \ldots, N$ such that

$$V = \cup_{i=1}^{N} V_i \ .$$

Let R_i denote the rectangular restriction matrix that returns the vector of coefficients in the subdomain Ω_i, i.e.

$$x_{\Omega_i} = R_i x \ .$$

Note that $A_i = R_i \, A \, R_i^T$ is the subdomain discretisation matrix in Ω_i. The additive Schwarz preconditioner P^{-1} from (3) is then written as

$$P^{-1} = \sum_{i=1}^{N} R_i A_i^{-1} R_i^T \ .$$

The additive Schwarz algorithm [24] is appealing since each subdomain solve can be performed in parallel. Unfortunately the performance of the algorithm deteriorates as the number of subdomains increases. Let H denote the characteristic size of each subdomain, δ the overlap distance, and h the mesh spacing. Dryja and Widlund [12, 13] give the following condition number bound for the method when used as a preconditioner for elliptic discretisations

$$\kappa(AP^{-1}) \leq CH^{-2} \left(1 + (H/\delta)^2\right) \tag{11}$$

where C is a constant independent of H and h. This result describes the deterioration as the number of subdomains increases (and H decreases). With some additional work this deterioration can be removed by the introduction of a global coarse subspace with restriction matrix R_0 with scale H so that

$$P^{-1} = R_0 A R_0^T + \sum_{i=1}^{N} R_i A_i^{-1} R_i^T \ .$$

Under the assumption of "generous overlap" the condition number bound [12, 13, 8] can be improved to

$$\kappa(AP^{-1}) \leq C\,(1 + (H/\delta)) \ . \tag{12}$$

The addition of a coarse space approximation introduces implementation problems similar to those found in multi-grid methods described below. Once again,

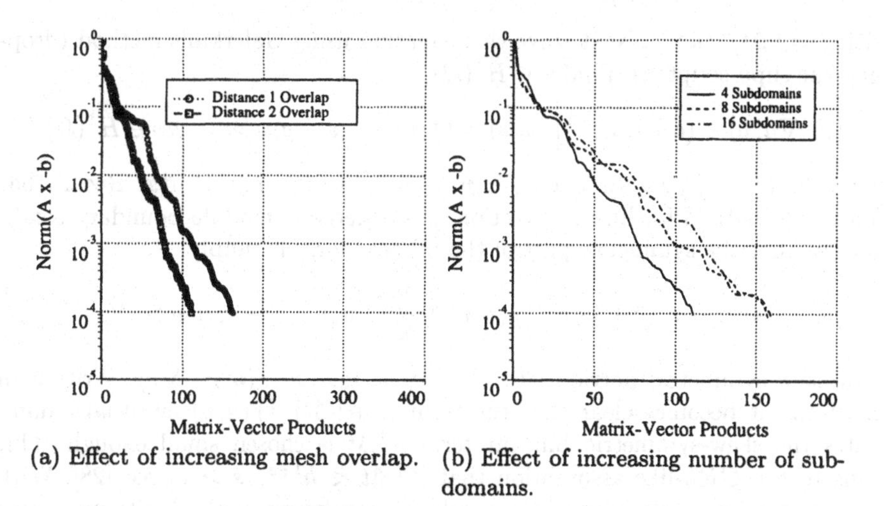

(a) Effect of increasing mesh overlap. (b) Effect of increasing number of subdomains.

Fig. 6. Performance of GMRES with additive overlapping Schwarz preconditioning.

the theory associated with additive Schwarz methods for hyperbolic PDE systems is not well-developed. Practical applications of the additive Schwarz method for the steady-state calculation of hyperbolic PDE systems show similar deterioration of the method when the coarse space is omitted. Figure 6 shows the performance of the additive Schwarz algorithm used as a preconditioner for GMRES. The test matrix was taken from one step of Newton's method applied to an upwind finite volume discretisation of the Euler equations at low Mach number ($M_\infty = .2$), see Barth [1] for further details. These calculations were performed without coarse mesh correction. As expected, the graphs show a degradation in quality with decreasing overlap and increasing number of mesh partitions.

5.3 Additive Overlapping Schwarz Methods for Time-Dependent Fluid Flow

We begin by giving a brief sketch of the analysis given in Wu et al. [28] which shows that for small enough time step and large enough overlap, the additive Schwarz preconditioner for hyperbolic problems behaves optimally without requiring a coarse space correction.

Consider the model scalar hyperbolic equation for the spatial domain $\Omega \subset \mathbb{R}^d$ with characteristic boundary data g weakly imposed on Γ

$$\begin{aligned} \frac{\partial u}{\partial t} + \beta \cdot \nabla u + c\, u = 0, \quad & (x,t) \in \Omega \times [0,T] \\ (\beta \cdot n(x))^- (u - g) = 0, \quad & x \in \Gamma \end{aligned} \tag{13}$$

with $\beta \in \mathbb{R}^d$, $c > 0$, and suitable initial data. Suppose that backward Euler time integration is employed ($u^n(x) = u(x, n\,\Delta t)$), so that (13) can then be written as

$$\beta \cdot \nabla u^n + \left(c + (\Delta t)^{-1}\right) u^n = f$$

with $f = (\Delta t)^{-1} u^{n-1}$. Next solve this equation using Galerkin's method (dropping the superscript n): Find $u \in H^1(\Omega)$

$$(\beta \cdot \nabla u, v) + \left(c + (\Delta t)^{-1}\right)(u,v) = (f,v) + < u - g, v >_- \quad \forall v \in H^1(\Omega)$$

where $(u,v) = \int_\Omega u\, v\; dx$ and $< u, v >_\pm = \int_\Gamma u\, v (\beta \cdot n(x))^\pm \; dx$. Recall that Galerkin's method for linear advection is iso-energetic modulo boundary conditions so that the symmetric part of the bilinear form is simply

$$A(u,v) = \left(c + (\Delta t)^{-1}\right)(u,v) + \frac{1}{2}(< u, v >_+ - < u, v >_-)$$

with skew-symmetric portion $S(u,v) = \frac{1}{2} < u, v > -(u, \beta \cdot \nabla v)$. Written in this form, it becomes clear that the term $\left(c + (\Delta t)^{-1}\right)(u,v)$ eventually dominates the skew-symmetric bilinear term if Δt is chosen small enough. This leads to the CFL-like assumption that $|\beta|\, \Delta t < h^{1+s}$, $s \geq 0$, see [28]. With this assumption, scalability of the overlapping Schwarz method without coarse space correction can be shown. Unfortunately, the assumed CFL-like restriction makes the method impractical since more efficient explicit time advancement strategies could be used which obviate the need for mesh overlap or implicit subdomain solves. The situation changes considerably if a Petrov-Galerkin discretisation strategy if used such as described in Sect. 3. For a the scalar model equation (13) this amounts to added the symmetric bilinear stabilisation term $B_{ls}(u,v) = (\beta \cdot \nabla u, \tau \beta \cdot \nabla v)$ to the previous Galerkin formulation: Find $u \in H^1(\Omega)$

$$(\beta \cdot \nabla u, v) + (\beta \cdot \nabla u, \tau \beta \cdot \nabla v) + \left(c + (\Delta t)^{-1}\right)(u,v) = (f,v) + < u - g, v >_-$$

$\forall v \in H^1(\Omega)$ with $\tau = h/(2|\beta|)$. This strategy is considered in a sequel paper to [28] which has yet to appear in the open literature. Practical CFD calculations show surprising good performance of overlapping Schwarz preconditioning when combined with Galerkin least-squares discretisation of hyperbolic systems as discussed in Sect. 3. Figure 8 shows iso-density contours for Mach 3 flow over a forward-facing step geometry using a triangulated mesh containing 22000 mesh vertices which has been partitioned into 1, 4, 16, and 32 subdomains for evaluation purposes. Owing to the nonlinearity of the strong shock-wave profiles, the solution must be evolved in time at a relatively small Courant number < 20 to prevent nonlinear instability in the numerical method. On the other hand, the solution eventually reaches an equilibrium state. (Note that on finer resolution meshes, the fluid contact surface emanating from the Mach triple point eventually makes the flow field unsteady.) This test problem provides an ideal candidate scenario for the overlapping Schwarz method since time accuracy is not essential to reaching the correct steady-state solution. Computations were

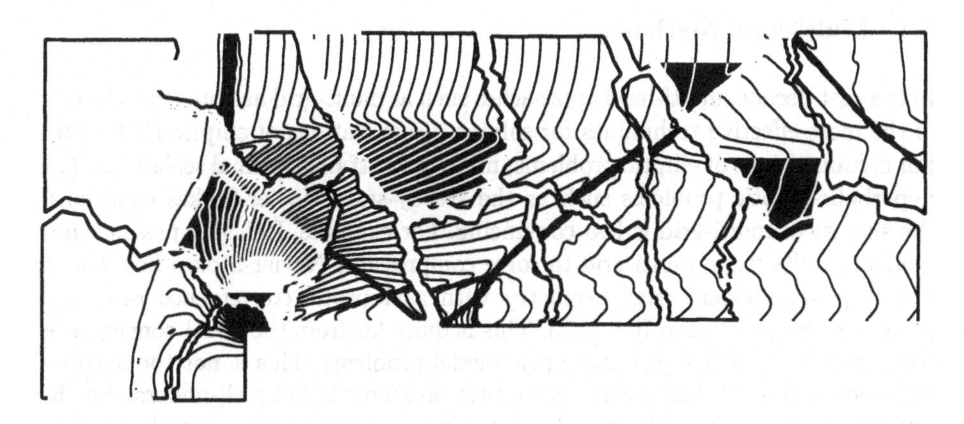

Fig. 7. Iso-density contours for Mach 3 inviscid Euler flow over a backward-facing step with exploded view of 16 subdomain partitioning.

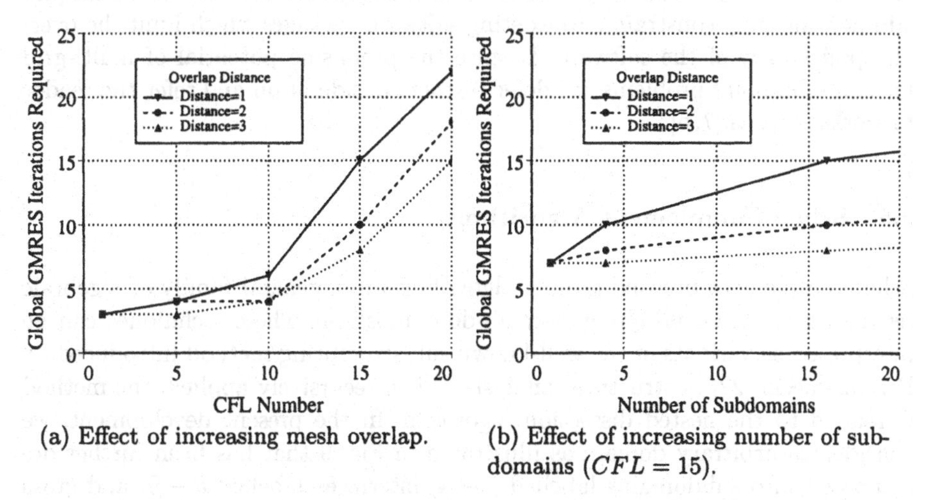

(a) Effect of increasing mesh overlap.

(b) Effect of increasing number of subdomains ($CFL = 15$).

Fig. 8. Number of ILU(0)-GMRES iterations required to reduce $\|Ax - b\| < 10^{-5}$.

performed on a fixed size mesh with 1, 4, 16, and 32 subdomains while also varying the overlap (graph) distances values and the CFL number. Figure 8(a) shows the effect of increasing CFL number and subdomain mesh overlap distance on the number of global GMRES iterations required to solve the global matrix problem to an accuracy of less than 10^{-5} using additive Schwarz-like ILU(0) on overlapped subdomain meshes. For CFL numbers less than about 10, the number of GMRES iterations is relatively insensitive to the amount of overlap. Figure 8(b) shows the effect of increased mesh partitioning on the number of GMRES iterations required (assuming a fixed $CFL = 15$). For overlap distance ≥ 2, the iterative method is relatively insensitive to the number of subdomains. By lowering the CFL number to 10, the results become even less sensitive to the number of subdomains.

5.4 Multi-level Methods

In the past decade, multi-level approaches such as multi-grid has proven to be one of the most effective techniques for solving discretisations of elliptic PDEs [29]. For certain classes of elliptic problems, multi-grid attains optimal scalability. For hyperbolic-elliptic problems such as the steady-state Navier-Stokes equations, the success of multi-grid is less convincing. For example, Ref. [20] presents numerical results using multi-grid to solve compressible Navier-Stokes flow about a multiple-component wing geometry with asymptotic convergence rates approaching .98 (Fig. 12 in Ref. [20]). This is quite far from the usual convergence rates quoted for multi-grid on elliptic model problems. This is not too surprising since multi-grid for hyperbolic-elliptic problems is not well-understood. In addition, some multi-grid algorithms require operations such as mesh coarsening which are poorly defined for general meshes (especially in 3-D) or place unattainable shape-regularity demands on mesh generation. Other techniques add new meshing constraints to existing software packages which limit the overall applicability of the software. Despite the promising potential of multi-grid for non-selfadjoint problems, we defer further consideration and refer the reader to works such as [7,6].

5.5 Schur Complement Algorithms

Schur complement preconditioning algorithms are a general family of algebraic techniques in non-overlapping domain-decomposition. These techniques can be interpreted as variants of the well-known sub-structuring method introduced by Przemieniecki [22] in structural analysis. When recursively applied, the method is related to the nested dissection algorithm. In the present development, we consider an arbitrary domain as illustrated in Fig. 9 that has been further decomposed into subdomains labelled $1-4$, interfaces labelled $5-9$, and cross points x. A natural 2×2 partitioning of the system is induced by permuting rows and columns of the discretisation matrix so that subdomain unknowns are ordered first, interface unknowns second, and cross points ordered last

$$Ax = \begin{bmatrix} A_{\mathcal{DD}} & A_{\mathcal{DI}} \\ A_{\mathcal{ID}} & A_{\mathcal{II}} \end{bmatrix} \begin{pmatrix} x_{\mathcal{D}} \\ x_{\mathcal{I}} \end{pmatrix} = \begin{pmatrix} f_{\mathcal{D}} \\ f_{\mathcal{I}} \end{pmatrix} \quad (14)$$

where $x_{\mathcal{D}}, x_{\mathcal{I}}$ denote the subdomain and interface variables respectively. The

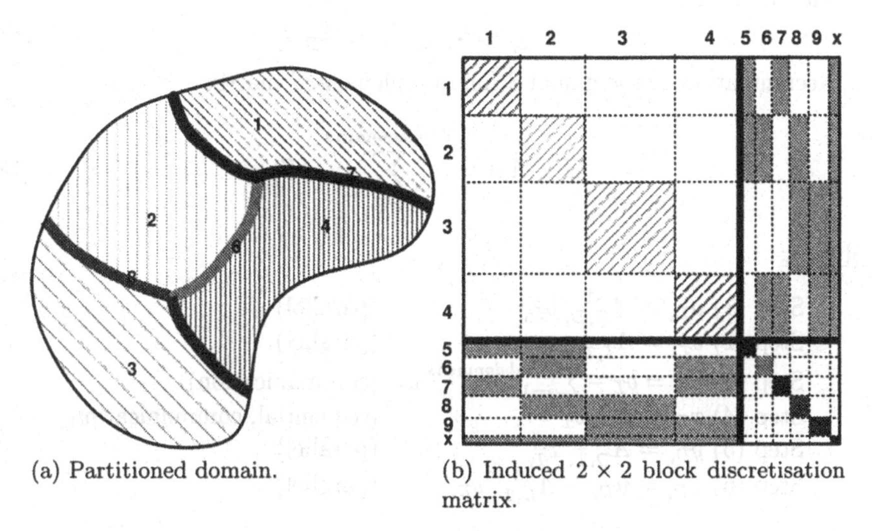

(a) Partitioned domain. (b) Induced 2 × 2 block discretisation matrix.

Fig. 9. Domain decomposition and the corresponding block matrix.

block LU factorisation of A is then given by

$$A = LU = \begin{bmatrix} A_{\mathcal{DD}} & 0 \\ A_{\mathcal{ID}} & I \end{bmatrix} \begin{bmatrix} I & A_{\mathcal{DD}}^{-1} A_{\mathcal{DI}} \\ 0 & S \end{bmatrix}, \quad (15)$$

where

$$S = A_{\mathcal{II}} - A_{\mathcal{ID}} A_{\mathcal{DD}}^{-1} A_{\mathcal{DI}} \quad (16)$$

is the Schur complement for the system. Note that $A_{\mathcal{DD}}$ is block diagonal with each block associated with a subdomain matrix. Subdomains are decoupled from each other and only coupled to the interface. The subdomain decoupling property is exploited heavily in parallel implementations.

In the next section, we outline a naive parallel implementation of the "exact" factorisation. This will serve as the basis for a number of simplifying approximations that will be discussed in later sections.

5.6 "Exact" Factorisation

Given the domain partitioning illustrated in Fig. 9, a straightforward (but naive) parallel implementation would assign a processor to each subdomain and a single processor to the Schur complement. Let $\overline{\mathcal{I}}_i$ denote the union of interfaces surrounding $\mathcal{D}_i$. The entire solution process would then consist of the following steps:

Parallel Preprocessing:

1. Parallel computation of subdomain $A_{\mathcal{D}_i\mathcal{D}_i}$ matrix LU factors.
2. Parallel computation of Schur complement block entries associated with each subdomain $\mathcal{D}_i$
$$\Delta S_{\overline{\mathcal{I}}_i} = A_{\overline{\mathcal{I}}_i\mathcal{D}_i} A^{-1}_{\mathcal{D}_i\mathcal{D}_i} A_{\mathcal{D}_i\overline{\mathcal{I}}_i} \ . \tag{17}$$
3. Accumulation of the global Schur complement S matrix
$$S = A_{\mathcal{I}\mathcal{I}} - \sum_{i=1}^{\#subdomains} \Delta S_{\overline{\mathcal{I}}_i} \ . \tag{18}$$

Solution:

Step (1) $u_{\mathcal{D}_i} = A^{-1}_{\mathcal{D}_i\mathcal{D}_i} b_{\mathcal{D}_i}$ (parallel)
Step (2) $v_{\overline{\mathcal{I}}_i} = A_{\overline{\mathcal{I}}_i\mathcal{D}_i} u_{\mathcal{D}_i}$ (parallel)
Step (3) $w_{\mathcal{I}} = b_{\mathcal{I}} - \sum_{i=1}^{\#subdomains} v_{\overline{\mathcal{I}}_i}$ (communication)
Step (4) $x_{\mathcal{I}} = S^{-1} w_{\mathcal{I}}$ (sequential, communication)
Step (5) $y_{\mathcal{D}_i} = A_{\mathcal{D}_i\overline{\mathcal{I}}_i} x_{\overline{\mathcal{I}}_i}$ (parallel)
Step (6) $x_{\mathcal{D}_i} = u_{\mathcal{D}_i} - A^{-1}_{\mathcal{D}_i\mathcal{D}_i} y_{\mathcal{D}_i}$ (parallel)

This algorithm has several deficiencies. Steps 3 and 4 of the solution process are sequential and require communication between the Schur complement and subdomains. More generally, the algorithm is not scalable since the growth in size of the Schur complement with increasing number of subdomains eventually overwhelms the calculation in terms of memory, computation, and communication.

5.7 Iterative Schur Complement Algorithms

A number of approximations have been investigated in Barth et al. [2] which simplify the exact factorisation algorithm and address the growth in size of the Schur complement. During this investigation, our goal has been to develop algebraic techniques which can be applied to both elliptic and hyperbolic partial differential equations. These approximations include iterative (Krylov projection) subdomain and Schur complement solves, element dropping and other sparsity control strategies, localised subdomain solves in the formation of the Schur complement, and partitioning of the interface and parallel distribution of the Schur complement matrix. Before describing each approximation and technique, we can make several observations:

Observation 1. (Ill-conditioning of Subproblems) For model elliptic problem discretisations, it is known in the two subdomain case that $\kappa(A_{\mathcal{D}_i\mathcal{D}_i}) = O((L/h)^2)$ and $\kappa(S) = O(L/h)$ where L denotes the domain size. From this perspective, both subproblems are ill-conditioned since the condition number depends on the mesh spacing parameter h. If one considers the scalability experiment, the situation changes in a subtle way. In the scalability experiment, the number of mesh points and the number of subdomains is increased such that the ratio of subdomain size to mesh spacing size H/h is held constant. The subdomain matrices for elliptic problem discretisations now exhibit a $O((H/h)^2)$ condition number so the cost associated with iteratively solving them (with or without preconditioning) is approximately constant as the problem size is increased. Therefore, this portion of the algorithm is scalable. Even so, it may be desirable to precondition the subdomain problems to reduce the overall cost. The Schur complement matrix retains (at best) the $O(L/h)$ condition number and becomes increasingly ill-conditioned as the mesh size is increased. Thus in the scalability experiment, it is ill-conditioning of the Schur complement matrix that must be controlled by adequate preconditioning, see for example Dryja, Smith and Widlund [11].

Observation 2. (Non-stationary Preconditioning) The use of Krylov projection methods to solve the local subdomain and Schur complement subproblems renders the global preconditioner non-stationary. Consequently, Krylov projection methods designed for non-stationary preconditioners should be used for the global problem. For this reason, FGMRES [23], a variant of GMRES designed for non-stationary preconditioning, has been used in the present work.

Observation 3. (Algebraic Coarse Space) The Schur complement serves as an algebraic coarse space operator since the system

$$Sx_{\mathcal{I}} = b_{\mathcal{I}} - A_{\mathcal{I}\mathcal{D}}A_{\mathcal{D}\mathcal{D}}^{-1}b_{\mathcal{D}} \tag{19}$$

globally couples solution unknowns on the entire interface. The rapid propagation of information to large distances is a crucial component of optimal algorithms.

5.8 ILU-GMRES Subdomain and Schur complement Solves

The first natural approximation is to replace exact inverses of the subdomain and Schur complement subproblems with an iterative Krylov projection method such as GMRES (or stabilized biconjugate gradient).

Iterative Subdomain Solves Recall from the exact factorisation algorithm that a subdomain solve is required once in the preprocessing step and twice in the solution step. This suggests replacing these three inverses with m_1, m_2, and m_3 steps of GMRES respectively. As mentioned in Observation 1, although the condition number of subdomain problems remains roughly constant in the scalability experiment, it is beneficial to precondition subdomain problems to improve the overall efficiency of the global preconditioner. By preconditioning subdomain problems, the parameters m_1, m_2, m_3 can be kept small. This will be exploited in later approximations. Since the subdomain matrices are assumed given, it is straightforward to precondition subdomains using ILU[k]. For the GLS spatial discretisation, satisfactory performance is achieved using ILU[2].

Iterative Schur complement Solves It is possible to avoid explicitly computing the Schur complement matrix for use in Krylov projection methods by alternatively computing the action of S on a given vector p, i.e.

$$S\,p = A_{\mathcal{I}\mathcal{I}}\,p - A_{\mathcal{I}\mathcal{D}}A_{\mathcal{D}\mathcal{D}}^{-1}A_{\mathcal{D}\mathcal{I}}\,p\ . \tag{20}$$

Unfortunately S is ill-conditioned, thus some form of interface preconditioning is needed. For elliptic problems, the rapid decay of elements away from the diagonal in the Schur complement matrix [16] permits simple preconditioning techniques. Bramble, Pasciak, and Schatz [5] have shown that even the simple block Jacobi preconditioner yields a substantial improvement in condition number

$$\kappa(SP_S^{-1}) \le CH^{-2}\left(1+\log^2(H/h)\right) \tag{21}$$

for C independent of h and H. For a small number of subdomains, this technique is very effective. To avoid the explicit formation of the diagonal blocks, a number of simplified approximations have been introduced over the last several years, see for examples Bjorstad [4] or Smith et al. [26]. By introducing a further coarse space coupling of cross points to the interface, the condition number is further improved

$$\kappa(SP_S^{-1}) \le C\left(1+\log^2(H/h)\right)\ . \tag{22}$$

Unfortunately, the Schur complement associated with advection dominated discretisations may not exhibit the rapid element decay found in the elliptic case. This can occur when characteristic trajectories of the advection equation traverse a subdomain from one interface edge to another. Consequently, the Schur complement is not well-preconditioned by elliptic-like preconditioners that use the action of local problems. A more basic strategy has been developed in the present work whereby elements of the Schur complement are *explicitly computed.* Once the elements have been computed, ILU factorisation is used to precondition the Schur complement iterative solution. In principle, ILU factorisation with a suitable reordering of unknowns can compute the long distance interactions associated with simple advection fields corresponding to entrance/exit-like flows. For general advection fields, it remains a topic of current research to find reordering algorithms suitable for ILU factorisation. The situation is further complicated for coupled systems of hyperbolic equations (even in two independent variables) where multiple characteristic directions and/or Cauchy-Riemann systems can be produced. At the present time, Cuthill-McKee ordering has been used on all matrices although improved reordering algorithms are currently under development.

In the present implementation, each subdomain processor computes (in parallel) and stores portions of the Schur complement matrix

$$\Delta S_{\overline{\mathcal{I}}_i} = A_{\overline{\mathcal{I}}_i\mathcal{D}_i}A_{\mathcal{D}_i\mathcal{D}_i}^{-1}A_{\mathcal{D}_i\overline{\mathcal{I}}_i}\ . \tag{23}$$

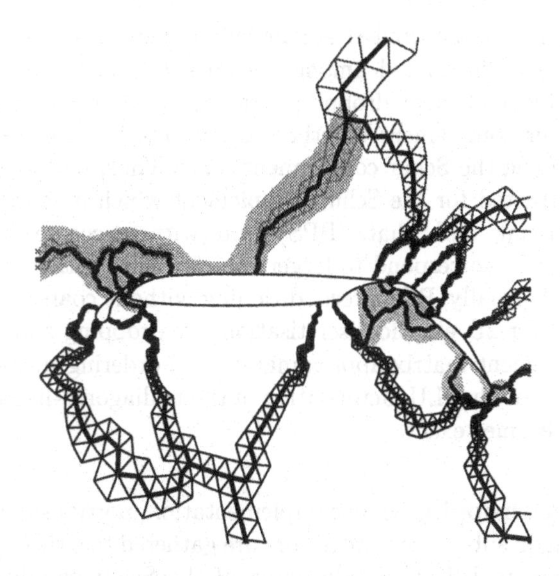

Fig. 10. Interface (bold lines) decomposed into 4 sub-interfaces indicated by alternating shaded regions.

To gain improved parallel scalability, the interface edges and cross points are partitioned into a smaller number of generic "sub-interfaces". This sub-interface partitioning is accomplished by assigning a supernode to each interface edge separating two subdomains, forming the graph of the Schur complement matrix in terms of these supernodes, and applying the METIS partitioning software to this graph. Let $\bar{\bar{\mathcal{I}}}_j$ denote the j-th sub-interface such that $\mathcal{I} = \cup_j \bar{\bar{\mathcal{I}}}_j$. Computation of the action of the Schur complement matrix on a vector p needed in Schur complement solves now takes the (highly parallel) form

$$Sp = \sum_{j=1}^{\#subinterfaces} A_{\bar{\bar{\mathcal{I}}}_j \bar{\bar{\mathcal{I}}}_j} \, p(\bar{\bar{\mathcal{I}}}_j) - \sum_{i=1}^{\#subdomains} \Delta S_{\bar{\mathcal{I}}_i} \, p(\bar{\mathcal{I}}_i) \ . \tag{24}$$

Using this formula it is straightforward to compute the action of S on a vector p to any required accuracy by choosing the subdomain iteration parameter m_i large enough. Figure 10 shows an interface and the immediate neighbouring mesh that has been decomposed into 4 smaller sub-interface partitions for a 32 subdomain partitioning. By choosing the number of sub-interface partitions proportional to the square root of the number of 2-D subdomains and assigning a processor to each, the number of solution unknowns associated with each sub-interface is held approximately constant in the scalability experiment. Note that the use of iterative subdomain solves renders both Eqns. (20) and (24) approximate.

In our investigation, the Schur complement is preconditioned using ILU factorisation. This is not a straightforward task for two reasons: (1) portions of the

Schur complement are distributed among subdomain processors, (2) the interface itself has been distributed among several sub-interface processors. In the next section, a block element dropping strategy is proposed for gathering portions of the Schur complement together on sub-interface processors for use in ILU preconditioning the Schur complement solve. Thus, a block Jacobi preconditioner is constructed for the Schur complement which is more powerful than the Bramble, Pasciak, and Schatz (BPS) form (without coarse space correction) since the blocks now correspond to larger sub-interfaces rather than the smaller interface edges. Formally, BPS preconditioning without coarse space correction can be obtained for 2D elliptic discretisations by dropping additional terms in our Schur complement matrix approximation and ordering unknowns along interface edges so that the ILU factorisation of the tridiagonal-like system for each interface edge becomes exact.

Block Element Dropping In our implementation, portions of the Schur complement residing on subdomain processors are gathered together on sub-interface processors for use in ILU preconditioning of the Schur complement solve. In assembling a Schur complement matrix approximation on each sub-interface processor, certain matrix elements are neglected:

1. All elements that couple sub-interfaces are ignored. This yields a block Jacobi approximation for sub-interfaces.
2. All elements with matrix entry location that exceeds a user specified graph distance from the diagonal as measured on the triangulation graph are ignored. Recall that the Schur complement matrix can be very dense. The graph distance criteria is motivated by the rapid decay of elements away from the matrix diagonal for elliptic problems. In all subsequent calculations, a graph distance threshold of 2 has been chosen for block element dropping.

Figures 11(a) and 11(b) show calculations performed with the present non-overlapping domain-decomposition preconditioner for diffusion and advection problems. These figures graph the number of global FGMRES iterations needed to solve the discretisation matrix problem to 10^{-6} accuracy tolerance as a function of the number of subproblem iterations. In this example, all the subproblem iteration parameters have been set equal to each other ($m_1 = m_2 = m_3$). The horizontal lines show poor scalability of single domain ILU-FGMRES on meshes containing 2500, 10000, and 40000 solution unknowns. The remaining curves show the behaviour of the Schur complement preconditioned FGMRES on 4, 16, and 64 subdomain meshes. Satisfactory scalability for very small values (5 or 6) of the subproblem iteration parameter m_i is clearly observed.

Wireframe Approximation A major cost in the explicit construction of the Schur complement is the matrix-matrix product

$$A^{-1}_{\mathcal{D}_i\mathcal{D}_i} A_{\mathcal{D}_i\mathcal{I}_i} \; . \tag{25}$$

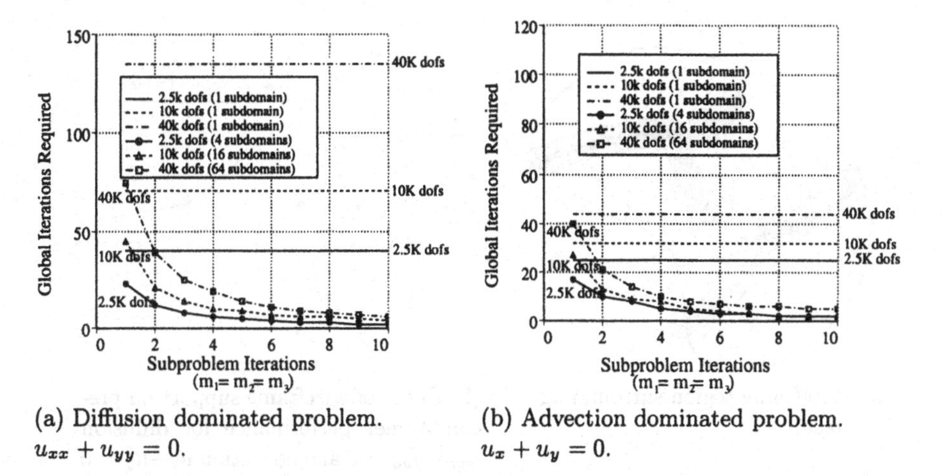

(a) Diffusion dominated problem. $u_{xx} + u_{yy} = 0$.

(b) Advection dominated problem. $u_x + u_y = 0$.

Fig. 11. Effect of the subproblem iteration parameters m_i on the global FGMRES convergence, $m_1 = m_2 = m_3$ for meshes containing 2500, 10000, and 40000 solution unknowns.

Since the subdomain inverse is computed iteratively using ILU-GMRES iteration, forming (25) is equivalent to solving a multiple right-hand sides system with each right-hand side vector corresponding to a column of $A_{\mathcal{D}_i \mathcal{I}_i}$. The number of columns of $A_{\mathcal{D}_i \mathcal{I}_i}$ is precisely the number of solution unknowns located on the interface surrounding a subdomain. This computational cost can be quite large. Numerical experiments with Krylov projection methods designed for multiple right-hand side systems [25] showed only marginal improvement owing to the fact that the columns are essentially independent. In the following paragraphs, "wireframe" and "supersparse" approximations are introduced to reduce the cost in forming the Schur complement matrix.

The wireframe approximation idea [9] is motivated from standard elliptic domain-decomposition theory by the rapid decay of elements in S with graph distance from the diagonal. Consider constructing a relatively thin *wireframe* region surrounding the interface as shown in Fig. 12(a). In forming the Eqn. (25) expression, subdomain solves are performed using the much smaller wireframe subdomains. In matrix terms, a principal submatrix of A, corresponding to the variables within the wireframe, is used to compute the (approximate) Schur complement of the interface variables. It is known from domain-decomposition theory that the exact Schur complement of the wireframe region is spectrally equivalent to the Schur complement of the whole domain. This wireframe approximation leads to a substantial savings in the computation of the Schur complement matrix. Note that the full subdomain matrices are used everywhere else in the Schur complement algorithm. The wireframe technique introduces a new adjustable parameter into the preconditioner which represents the width of the wireframe. For simplicity, this width is specified in terms of graph distance on the mesh triangulation. Figure 12(b) demonstrates the performance of this approx-

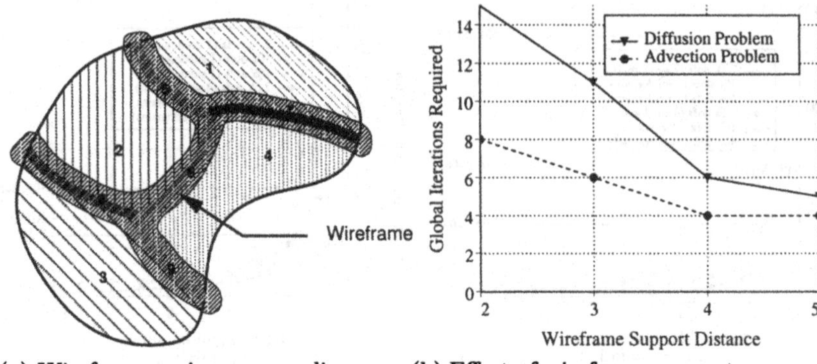

(a) Wireframe region surrounding interface.

(b) Effect of wireframe support on preconditioner performance for diffusion $u_{xx}+u_{yy}=0$ and advection $u_x+u_y=0$ problems.

Fig. 12. Wireframe region surrounding interface and preconditioner performance results for a fixed mesh size (1600 vertices) and 16 subdomain partitioning.

imation by graphing the total number of preconditioned FGMRES iterations required to solve the global matrix problem to a 10^{-6} accuracy tolerance while varying the width of the wireframe. As expected, the quality of the preconditioner improves rapidly with increasing wireframe width with full subdomain-like results obtained using modest wireframe widths. As a consequence of the wireframe construction, the time taken form the Schur complement has dropped by approximately 50%.

Supersparse Matrix-Vector Operations It is possible to introduce further approximations which improve upon the overall efficiency in forming the Schur complement matrix. One simple idea is to exploit the extreme sparsity in columns of $A_{\mathcal{D}_i\mathcal{I}_i}$ or equivalently the sparsity in the right-hand sides produced from $A^{-1}_{\mathcal{D}_i\mathcal{D}_i}A_{\mathcal{D}_i\mathcal{I}_i}$ needed in the formation of the Schur complement. Observe that m steps of GMRES generates a small sequence of Krylov subspace vectors $[p, A\,p, A^2\,p, \ldots, A^m\,p]$ where p is a right-hand side vector. Consequently for small m, if both A and p are sparse then the sequence of matrix-vector products will be relatively sparse. Standard sparse matrix-vector product subroutines utilize the matrix in sparse storage format and the vector in dense storage format. In the present application, the vectors contain only a few non-zero entries so that standard sparse matrix-vector products waste many arithmetic operations. For this reason, a "supersparse" software library have been developed to take advantage of the sparsity in matrices as well as in vectors by storing both in compressed form. Unfortunately, when GMRES is preconditioned using ILU factorization, the Krylov sequence becomes $[p, A\,P^{-1}\,p, (A\,P^{-1})^2\,p, \ldots, (A\,P^{-1})^m\,p]$. Since the inverse of the ILU approximate factors $\widetilde{L}$ and $\widetilde{U}$ can be dense, the first application of ILU preconditioning produces a dense Krylov vector result. All subsequent Krylov vectors can become dense as well. To prevent this densification of vectors using ILU preconditioning, a fill-level-like strategy has been incorporated into the ILU *backsolve* step. Consider the ILU preconditioning problem, $\widetilde{L}\,\widetilde{U}\,r = b$. This system is conventionally solved by a lower triangular backsolve, $w = \widetilde{L}^{-1}b$, followed by a upper triangular backsolve $r = \widetilde{U}^{-1}w$. In our supersparse strategy, sparsity is controlled by imposing a non-zero fill pattern for the vectors w and r during lower and upper backsolves. The backsolve fill patterns are most easily specified in terms fill-level distance, i.e. graph distance from existing nonzeros of the right-hand side vector in which new fill in the resultant vector is allowed to occur. This idea is motivated from the element decay phenomena observed for elliptic problems. Table 1 shows the performance benefits of using supersparse computations together with backsolve fill-level specification for a 2-D test problem consisting of Euler flow past a multi-element airfoil geometry partitioned into 4 subdomains with 1600 mesh vertices in each subdomain. Computations

Table 1. Performance of the Schur complement preconditioner with supersparse arithmetic for a 2-D test problem consisting of Euler flow past a multi-element airfoil geometry partitioned into 4 subdomains with 1600 mesh vertices in each subdomain.

Backsolve Fill-Level Distance k	Global GMRES Iterations	Time(k)/Time(∞)
0	26	0.325
1	22	0.313
2	21	0.337
3	20	0.362
4	20	0.392
∞	20	1.000

were performed on the IBM SP2 parallel computer using MPI message passing protocol. Various values of backsolve fill-level distance were chosen while monitoring the number of global GMRES iterations needed to solve the matrix problem and the time taken to form the Schur complement preconditioner. Results for this problem indicate preconditioning performance comparable to exact ILU backsolves using backsolve fill-level distances of only 2 or 3 with a 60-70% reduction in cost.

6 Numerical Results on the IBM SP2

In the remaining paragraphs, we assess the performance of the Schur complement preconditioned FGMRES in solving linear matrix problems associated an approximate Newton method for the nonlinear discretised compressible Euler equations. All calculations were performed on an IBM SP2 parallel computer using MPI message passing protocol. A scalability experiment was performed on meshes containing 4/1, 16/2, and 64/4 subdomains/subinterfaces with each subdomain containing 5000 mesh elements. Figures 13(a) and 13(b) show mesh

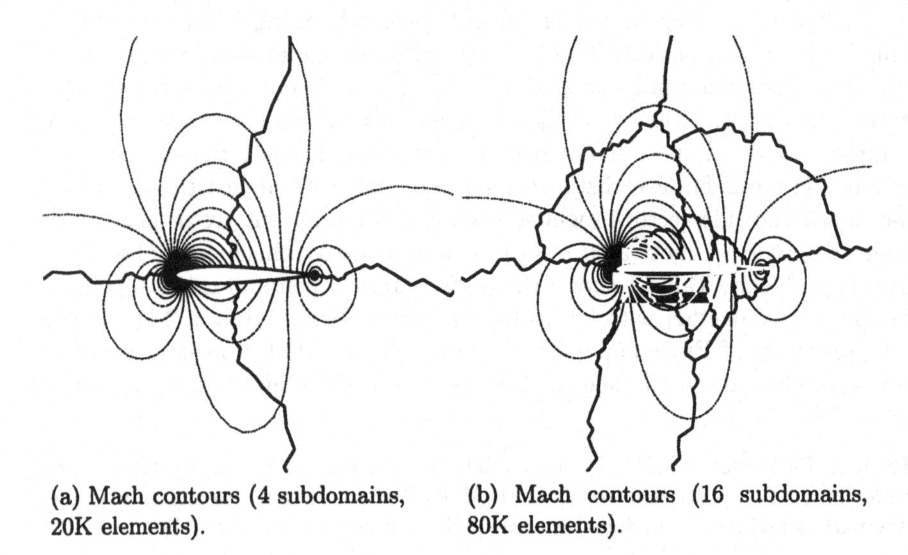

(a) Mach contours (4 subdomains, 20K elements).

(b) Mach contours (16 subdomains, 80K elements).

Fig. 13. Mach number contours and mesh partition boundaries for NACA0012 airfoil geometry.

partitionings and sample Mach number solution contours for subsonic ($M_\infty = .20, \alpha = 2.0°$) flow over the airfoil geometry. The flow field was computed using the stabilised GLS discretisation and approximate Newton method described in Sect. 3. Figure 14 graphs the convergence of the approximate Newton method for the 16 subdomain test problem. Each approximate Newton iterate shown in Fig. 14 requires the solution of a linear matrix system which has been solved using the Schur complement preconditioned FGMRES algorithm. Figure 15 graphs the convergence of the FGMRES algorithm for each matrix from the 4 and 16 subdomain test problems. These calculations were performed using ILU[2] and $m_1 = m_2 = m_3 = 5$ iterations on subproblems with supersparse distance equal to 5. The 4 subdomain mesh with 20000 total elements produces matrices that are easily solved in 9-17 global FGMRES iterations. Calculations corresponding to the largest CFL numbers are close approximations to exact Newton iterates. As is typically observed by these methods, the final few Newton iterates are solved more easily than matrices produced during earlier iterates. The most difficult matrix problem required 17 FGMRES iterations and the final Newton iterate required only 12 FGMRES iterations. The 16 subdomain mesh containing 80000 total elements produces matrices that are solved in 12-32 global FGMRES. Due to the nonlinearity in the spatial discretisation, several approximate Newton iterates were relatively difficult to solve, requiring over 30 FGMRES iterations. As nonlinear convergence is obtained the matrix problems become less

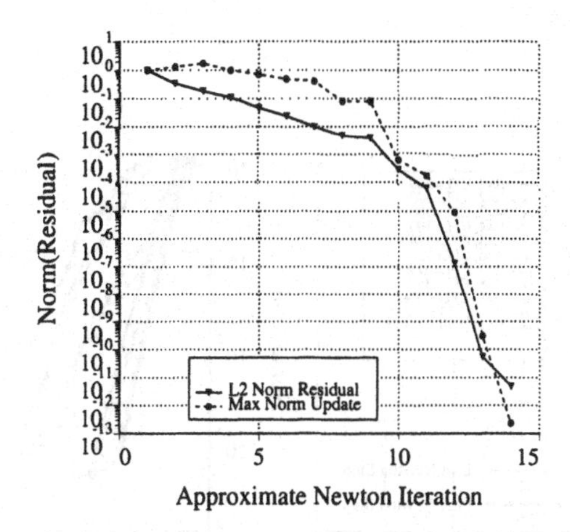

Fig. 14. Nonlinear convergence behaviour of the approximate Newton method for subsonic airfoil flow.

demanding. In this case, the final Newton iterate matrix required 22 FGMRES iterations. This iteration degradation from the 4 subdomain case can be reduced by increasing the subproblem iteration parameters m_1, m_2, m_3 but the overall computation time is increased. In the remaining timing graphs, we have sampled timings from 15 FGMRES iterations taken from the final Newton iterate on each mesh. For example, Fig. 16(a) gives a raw timing breakdown for several of the major calculations in the overall solver: calculation of the Schur complement matrix, preconditioning FGMRES with the Schur complement algorithm, matrix element computation and assembly, and FGMRES solve. Results are plotted on each of the meshes containing 4, 16, and 64 subdomains with 5000 elements per subdomain. Since the number of elements in each subdomain is held constant, the time taken to assemble the matrix is also constant. Observe that in our implementation the time to form and apply the Schur complement preconditioner currently dominates the calculation. Although the growth observed in these timings with increasing numbers of subdomains comes from several sources, the dominate effect comes from a very simple source: *the maximum interface size growth associated with subdomains.* This has a devastating impact on the parallel performance since at the Schur complement synchronization point all processors must wait for subdomains working on the largest interfaces to finish. Figure 16(b) plots this growth in maximum interface size as a function of number of subdomains in our scalability experiment. Although the number of elements in each subdomain has been held constant in this experiment, the largest interface associated with any subdomain has more than doubled. This essentially translates into a doubling in time to form the Schur complement matrix. This doubling in time is clearly observed in the raw timing breakdown in Fig. 16(a). At this point in time, we known of no partitioning method that actively addresses controlling the maximum interface size associated with subdomains. We suspect that other non-overlapping methods are sensitive to this effect as well.

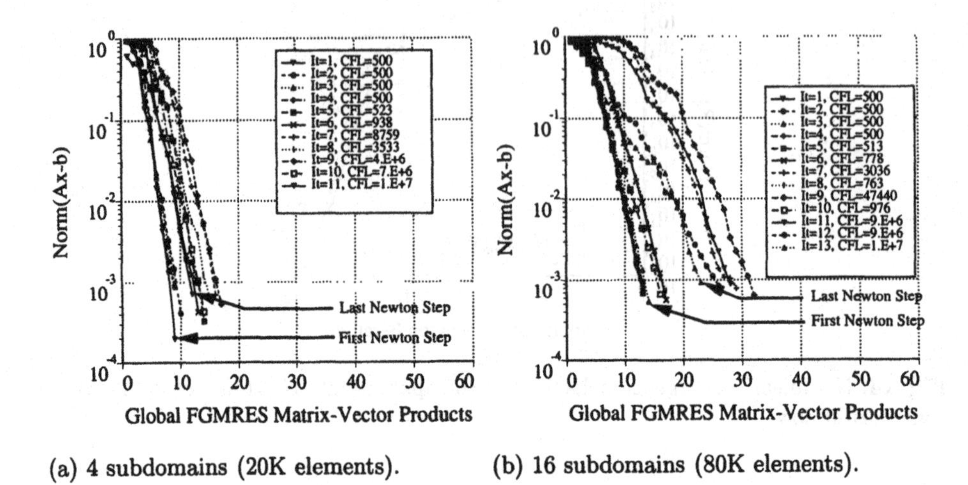

(a) 4 subdomains (20K elements). (b) 16 subdomains (80K elements).

Fig. 15. FGMRES convergence history for each Newton step.

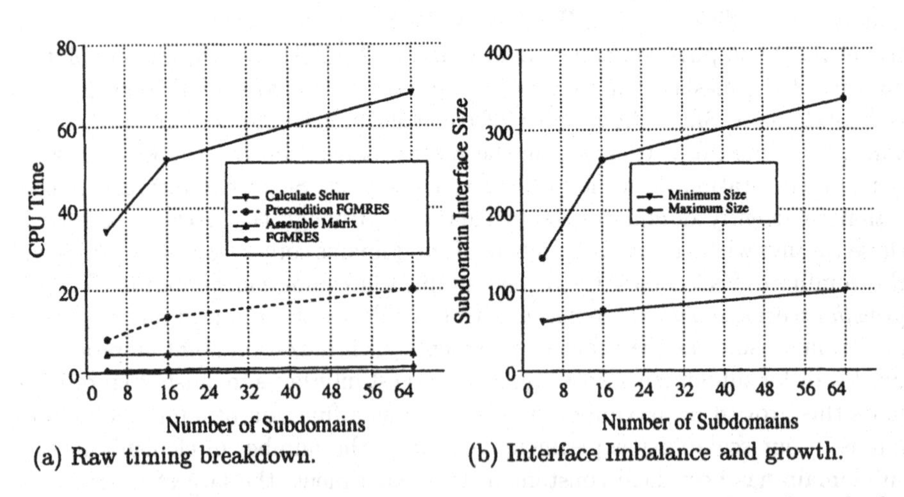

(a) Raw timing breakdown. (b) Interface Imbalance and growth.

Fig. 16. Raw IBM SP2 timing breakdown and the effect of increased number of subdomains on smallest and largest interface sizes.

7 Concluding Remarks

Experience with our non-overlapping domain-decomposition method with an algebraically generated coarse problem shows that we can successfully trade off some of the robustness of the exact Schur complement method for increased efficiency by making appropriately designed approximations. In particular, the localised wireframe approximation and the supersparse matrix-vector operations together result in reduced cost without significantly degrading the overall convergence rate.

It remains an outstanding problem to partition domains such that the maximum interface size does grow with increased number of subdomains and mesh size. In addition, it may be cost effective to combine this technique with multigrid or multiple-grid techniques to improve the robustness of Newton's method.

References

1. T. J. Barth, *Parallel CFD algorithms on unstructured meshes*, Tech. Report AGARD R-907, Advisory Group for Aerospace Research and Development, 1995.
2. T. J. Barth, T. F. Chan, and W.-P. Tang, *A parallel non-overlapping domain-decomposition algorithm for compressible fluid flow problems on triangulated domains*, AMS Cont. Math. **218** (1998).
3. T.J. Barth, *Numerical methods for gas-dynamic systems on unstructured meshes*, An Introduction to Recent Developments in Theory and Numerics for Conservation Laws (Kröner, Ohlberger and Rohde eds.), Lecture Notes in Computational Science and Engineering, vol. 5, Springer-Verlag, Heidelberg, 1998, pp. 195–285.
4. P. Bjorstad and O. B. Widlund, *Solving elliptic problems on regions partitioned into substructures*, SIAM J. Numer. Anal. **23** (1986), no. 6, 1093–1120.
5. J. H. Bramble, J. E. Pasciak, and A. H. Schatz, *The construction of preconditioners for elliptic problems by substructuring, I*, Math. Comp. **47** (1986), no. 6, 103–134.
6. J. H. Bramble, J. E. Pasciak, and J. Xu, *The analysis of multigrid algorithms for nonsymmetric and indefinite elliptic problems*, Math. Comp. **51** (1988), 289–414.
7. A. Brandt and I. Yavneh, *Accelerated multigrid convergence for high-Reynolds recirculating flow*, SIAM J. Sci. Comp. **14** (1993), 607–626.
8. T. Chan and J. Zou, *Additive Schwarz domain decomposition methods for elliptic problems on unstructured meshes*, Tech. Report CAM 93-40, UCLA Department of Mathematics, December 1993.
9. T. F. Chan and T. Mathew, *Domain decomposition algorithms*, Acta Numerica (1994), 61–143.
10. D. F. D'Azevedo, P. A. Forsyth, and W.-P. Tang, *Toward a cost effective ilu preconditioner with high level fill*, BIT **32** (1992), 442–463.
11. M. Dryja, B. F. Smith, and O. B. Widlund, *Schwarz analysis of iterative substructuring algorithms for elliptic problems in three dimensions*, SIAM J. Numer. Anal. **31** (1994), 1662–1694.
12. M. Dryja and O.B. Widlund, *Some domain decomposition algorithms for elliptic problems*, Iterative Methods for Large Linear Systems (L. Hayes and D. Kincaid, eds.), 1989, pp. 273–291.

13. M. Dryja and O.B. Widlund, *Additive Schwarz methods for elliptic finite element problems in three dimensions*, Fifth Conference on Domain Decomposition Methods for Partial Differential Equations (T. F. Chan, D.E. Keyes, G.A. Meurant, J.S. Scroggs, and R.G. Voit, eds.), 1992.
14. T. Dupont, R. Kendall, and H. Rachford, *An approximate factorization procedure for solving self-adjoint elliptic difference equations*, SIAM J. Numer. Anal. **5** (1968), 558–573.
15. D. J. Evans, *The use of pre-conditioning in iterative methods for solving linear equations with symmetric positive definite matrices*, J. Inst. Maths. Appl. **4** (1968), 295–314.
16. G. Golub and D. Mayers, *The use of preconditioning over irregular regions*, Comput. Meth. Appl. Mech. Engrg. **6** (1984), 223–234.
17. T. J. R. Hughes, L. P. Franca, and M. Mallet, *A new finite element formulation for CFD: I. symmetric forms of the compressible Euler and Navier-Stokes equations and the second law of thermodynamics*, Comp. Meth. Appl. Mech. Engrg. **54** (1986), 223–234.
18. T. J. R. Hughes and M. Mallet, *A new finite element formulation for CFD: III. the generalized streamline operator for multidimensional advective-diffusive systems*, Comp. Meth. Appl. Mech. Engrg. **58** (1986), 305–328.
19. G. Karypis and V. Kumar, *Multilevel k-way partitioning scheme for irregular graphs*, Tech. Report 95-064, U. of Minn. Computer Science Department, 1995.
20. D. J. Mavriplis, *A three dimensional multigrid Reynolds-averaged Navier-Stokes solver for unstructured meshes*, Tech. Report 94-29, ICASE, NASA Langley R.C., 1994.
21. J. A. Meijerink and H. A. van der Vorst, *An iterative solution method for linear systems of which the coefficient matrix is a symmetric M-matrix*, Math. Comp. **34** (1977), 148–162.
22. J. S. Przemieniecki, *Matrix structural analysis of substructures*, Am. Inst. Aero. Astro. J. **1** (1963), 138–147.
23. Y. Saad, *A flexible inner-outer preconditioned GMRES algorithm*, SIAM J. Sci. Stat. Comp. **14** (1993), no. 2, 461–469.
24. H. A. Schwarz, *Uber einige abbildungensaufgaben*, J. Reine Angew. Math. **70** (1869), 105–120.
25. V. Simoncini and E. Gallopoulos, *An iterative method for nonsymmetric systems with multiple right-hand sides*, SIAM J. Sci. Comp. **16** (1995), no. 4, 917–933.
26. B. Smith, P. Bjorstad, and W. Gropp, *Domain decomposition: parallel multilevel methods for elliptic partial differential equations*, Cambridge University Press, 1996.
27. H. Stone, *Iterative solution of implicit approximations of multidimensional partial differential equations*, SIAM J. Numer. Anal. **5** (1968), 530–558.
28. Y. Wu, X.-C. Cai, and D. E. Keyes, *Additive Schwarz methods for hyperbolic equations*, AMS Cont. Math. **218** (1998).
29. J. Xu, *An introduction to multilevel methods*, Lecture notes: VIIth EPSRC numerical analysis summer school, 1997.

Influence of the Discretization Scheme on the Parallel Efficiency of a Code for the Modelling of a Utility Boiler

Pedro Jorge Coelho

Instituto Superior Técnico, Technical University of Lisbon
Mechanical Engineering Department
Av. Rovisco Pais, 1096 Lisboa Codex, Portugal

Abstract. A code for the simulation of the turbulent reactive flow with heat transfer in a utility boiler has been parallelized using MPI. This paper reports a comparison of the parallel efficiency of the code using the hybrid central differences/upwind and the MUSCL schemes for the discretization of the convective terms of the governing equations. The results were obtained using a Cray T3D and a number of processors in the range 1 - 128. It is shown that higher efficiencies are obtained using the MUSCL scheme and that the least efficient tasks are the solution of the pressure correction equation and the radiative heat transfer calculation.

Keywords: Parallel Computing; Discretization Schemes; Computational Fluid Dynamics; Radiation; Boilers

1 Introduction

The numerical simulation of the physical phenomena that take place in the combustion chamber of a utility boiler is a difficult task due to the complexity of those phenomena (turbulence, combustion, radiation) and to the range of geometrical length scales which spans fours or five orders of magnitude [1]. As a consequence, such a simulation is quite demanding as far as the computational resources are concerned. Therefore, parallel computing can be very useful in this field. The mathematical modelling of a utility boiler is often based on the numerical solution of the equations governing conservation of mass, momentum and energy, and transport equations for scalar quantities describing turbulence and combustion. These equations are solved in an Eulerian framework and their numerical discretization yields convective terms which express the flux of a dependent variable across the faces of the control volumes over which the discretization is carried out. Many dicretization schemes for the convective terms have been proposed along the years and this issue has been one of the most important topics in computational fluid dynamics research.

The hybrid central differences/upwind scheme has been one the most popular ones, especially in incompressible flows. However, it reverts to the first order upwind scheme whenever the absolute value of the local Péclet number is higher

J. Palma, J. Dongarra, and V. Hernández (Eds.): VECPAR'98, LNCS 1573, pp. 203–214, 1999.

than two, which may be the case in most of the flow regions. This yields poor accuracy and numerical diffusion errors. These can only be overcome using a fine grid which enables a reduction of the local Peclet number, and will ultimately revert the scheme to the second order accurate central differences scheme. However, this often requires a grid too fine, and there is nowadays general consensus that the hybrid scheme should not be used (see, e.g., [2]). Moreover, some leading journals presently request that solution methods must be at least second order accurate in space. Alternative discretization schemes, such as the skew upwind, second order upwind and QUICK, are more accurate but may have stability and/or boundedness problems. Remedies to overcome these limitations have been proposed more recently and there are presently several schemes available which are stable, bounded and at least second order accurate (see, e.g., [3]-[9]).

Several high resolution schemes have been incorporated in the code presented in [10] for the calculation of laminar or turbulent incompressible fluid flows in two or three-dimensional geometries. Several modules were coupled to this code enabling the modelling of combustion, radiation and pollutants formation. In this work, the code was applied to the simulation of a utility boiler, and a comparison of the efficiency obtained using the hybrid and the MUSCL ([11]) schemes is presented. The mathematical model and the parallel implementation are described in the next two sections. Then, the results are presented and discussed, and the conclusions are summarized in the last section.

2 The Mathematical Model

2.1 Main features of the model

The mathematical model is based on the numerical solution of the density weighted averaged form of the equations governing conservation of mass, momentum and energy, and transport equations for scalar quantities. Only a brief description of the reactive fluid flow model is given below. Further details may be found in [12].

The Reynolds stresses and the turbulent scalar fluxes are determined by means of the k-ε eddy viscosity/diffusivity model which comprises transport equations for the turbulent kinetic energy and its dissipation rate. Standard values are assigned to all the constants of the model.

Combustion modelling is based on the conserved scalar/probability density function approach. A chemical equilibrium code is used to obtain the relationship between instantaneous values of the mixture fraction and the density and chemical species concentrations. The calculation of the mean values of these quantities requires an integration of the instantaneous values weighted by the assumed probability density function over the mixture fraction range. These calculations are performed a *priori* and stored in tabular form.

The discrete ordinates method [13] is used to calculate the radiative heat transfer in the combustion chamber. The S_4 approximation, the level symmetric quadrature satisfying sequential odd moments [14] and the step scheme are employed. The radiant superheaters which are suspended from the top of the combustion chamber are simulated as baffles without thickness as reported in [15]. The radiative properties of the medium are calculated using the weighted sum of grey gases model.

The governing equations are discretized over a Cartesian, non-staggered grid using a finite volume/finite difference method. The convective terms are discretized using either the hybrid or the MUSCL schemes. The solution algorithm is based on the SIMPLE method. The algebraic sets of discretized equations are solved using the Gauss-Seidel line-by-line iterative procedure, except the pressure correction equation which is solved using a preconditioned conjugate gradient method.

2.2 Discretization of the convective terms

The discretized equation for a dependent variable ϕ at grid node P may be written in the following compact form:

$$a_P \phi_P = \sum_i a_i \phi_i + b \tag{1}$$

where the coeficients a i denote combined convection/diffusion coefficients and b is a source term. The summation runs over all the neighbours of grid node P (east, west, north, south, front and back). Derivation of this discretized equation may be found, e.g., in [16]. If the convective terms are computed by means of the hybrid upwind/central differences method, then the system of equations (1) is diagonally dominant and can be solved using any conventional iterative solution technique. If a higher order discretization scheme is used, the system of equations may still have a diagonally dominant matrix of coefficients provided that the terms are rearranged using a deferred correction technique [17]. In this case, equation (1) is written as:

$$a_P^U \phi_P = \sum_i a_i^U \phi_i + b + \sum_j C_j (\phi_j^U - \phi_j) \tag{2}$$

where the superscript U means that the upwind scheme is used to compute the corresponding variable or coefficient, and Cj is the convective flux at cell face j. The last term on the right hand side of the equation is the contribution to the source term due to the deferred correction procedure.

The high order schemes were incorporated in the code using the normalized variable and space formulation methodology [18]. According to this formulation, denoting by U, C, and D the upstream, central and downstream grid nodes surrounding the control volume face f (see Figure 1), the following normalized variables are defined:

$$\tilde{\phi} = \frac{\phi - \phi_U}{\phi_D - \phi_U} \tag{3}$$

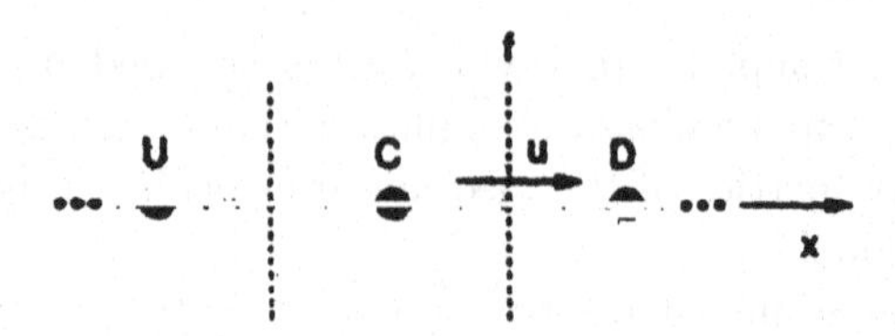

Fig. 1. Interpolation grid nodes involved in the calculation of ϕ_f.

$$\tilde{x} = \frac{x - x_U}{x_D - x_I} \tag{4}$$

where x is the coordinate along the direction of these nodes. The upwind scheme yields:

$$\tilde{\phi}_f = \tilde{\phi}_C \tag{5}$$

while the MUSCL scheme is given by:

$$\tilde{\phi}_f = (2\tilde{x}_f - \tilde{x}_C)\tilde{\phi}_C/\tilde{x}_C \quad \text{if} \quad 0 < \tilde{\phi}_C < \tilde{x}_C/2 \tag{6a}$$

$$\tilde{\phi}_f = \tilde{x}_f - \tilde{x}_C) + \tilde{\phi}_C \quad \text{if} \quad \tilde{x}_c/2 < \tilde{\phi}_C < 1 + +\tilde{x}_C - \tilde{x}_f \tag{6b}$$

$$\tilde{\phi}_f = 1 \quad \text{if} \quad 1 + \tilde{x}_C - \tilde{x}_f < \tilde{\phi}_C < 1 \tag{6c}$$

$$\tilde{\phi}_f = \tilde{\phi}_C \quad \text{elsewhere} \tag{6d}$$

3 Parallel Implementation

The parallel implementation is based on a domain decomposition strategy and the communications among the processors are accomplished using MPI. This standard is now widely available and the code is therefore easily portable across hardware ranging from workstation clusters, through shared memory modestly parallel servers to massively parallel systems. Within the domain decomposition approach the computational domain is split up into non-overlapping subdomains, and each subdomain is assigned to a processor. Each processor deals with a subdomain and communicates and/ synchronizes its actions with those of other processors by exchanging messages.

The calculation of the coefficients of the discretized equations in a control volume requires the knowledge of the values of the dependent variables at one or two neighbouring control volumes along each direction. Only one neighbour is involved if the hybrid scheme is used, while two neighbours are involved if the MUSCL scheme is employed. In the case of control volumes on the boundary of a subdomain, the neighbours lie on a subdomain assigned to a different processor. In distributed memory computers a processor has only in its memory the data of the subdomain assigned to that processor. This implies that it must exchange data with the neighbouring subdomains.

Data transfer between neighbouring subdomains is simplified by the use of a buffer of halo points around the rectangular subdomain assigned to every processor. Hence, two planes of auxiliary points are added to each subdomain boundary, which store data calculated in neighbouring subdomains. These data is periodically exchanged between neighbouring subdomains to ensure the correct coupling of the local solutions into the global solution. This halo data transfer between neighbouring processors is achieved by a pair-wise exchange of data. This transfer proceeds in parallel and it will be referred to as local communications. Local communication of a dependent variable governed by a conservation or transport equation is performed just before the end of each sweep (inner iteration) of the Gauss-Seidel procedure for that variable. Local communication of the mean temperature, density, specific heat and effective viscosity is performed after the update of these quantities, i.e., once per outer iteration. Besides the local communications, the processors need to communicate global data such as the values of the residuals which need to be accumulated, or maximum or minimum values determined, or values broadcast. These data exchange are referred to as global communications and are available in standard message passing interfaces.

While the parallelization of the fluid flow equations solver has been widely addressed in the literature, the parallelization of the radiation model has received little attention. The method employed here is described in detail in [19],[20] and uses the spatial domain decomposition method for the parallelization of the discrete ordinates method. It has been found that this method is not as efficient as the angular decomposition method, since the convergence rate of the radiative calculations is adversely influenced by the decomposition of the domain, dropping fast as the number of processors increases. However, the compatibility with the domain decomposition technique used in parallel computational fluid dynamics (CFD) codes favours the use of the spatial domain decomposition method for the radiation in the case of coupled fluid flow/heat transfer problems.

4 Results and Discussion

The code was applied to the simulation of the physical phenomena taking place in the combustion chamber of a power station boiler of the Portuguese Electricity Utility. It is a natural circulation drum fuel-oil fired boiler with a pressurized combustion chamber, parallel passages by the convection chamber and preheating. The boiler is fired from three levels of four burners each, placed on the front wall. Vaporization of the fuel is assumed to occur instantaneously. At maximum capacity (771 ton/h at 167 bar and 545°C) the output power is 250 MWe. This boiler has been extensively investigated in the past, both experimentally and numerically (see, e.g., [1], [21] - [23]), and therefore no predictions are shown in this paper which is concentrated only on the parallel performance of the code.

The calculations were performed using the Cray T3D of the University of Edinburgh in U.K. It comprises 256 nodes each with two processing units. Each processing element consists of a DEC Alpha 21064 processor running at 150MHz and delivering 150 64-bit Mflop/s. The peak performance of the 512 processing elements is 76.8 Gflop/s.

Table 1. Parallel performance of the code for the first 80 iterations using the hybrid discretization scheme.

n_p	1	2	4	8	16	32	64	128
Partition	1x1x1	1x2x1	1x4x1	1x8x1	2x4x2	2x4x4	2x4x8	2x8x8
t_{total} (s)	1666.6	850.9	439.5	231.5	123.5	67.8	36.8	21.5
S	1	1.96	3.79	7.20	13.5	24.6	45.2	77.4
ε (%)	—	97.9	94.8	90.0	84.3	76.8	70.7	60.5
ε_{vel} (%)	—	99.5	97.9	94.9	89.5	84.3	79.6	71.7
ε_p (%)	—	97.9	91.9	83.2	73.5	63.9	56.0	39.2
$\varepsilon_{k,\varepsilon}$(%)	—	96.7	92.5	86.8	82.8	75.6	69.4	60.9
$\varepsilon_{scalars}$ (%)	—	97.1	94.3	88.9	84.7	77.2	70.5	61.5
ε_{prop} (%)	—	97.6	94.9	93.6	87.1	76.5	73.5	71.7

Jobs running in the computer used in the present calculations and using less than 64 processors are restricted to a maximum of 30 minutes. Therefore, to allow a comparison between runs with different number of processors the initial calculations, summarized in tables 1 and 2, were carried out for a fixed number of iterations (30 for the MUSCL discretization scheme and 80 for the hybrid scheme). They were obtained using a grid with 64,000 grid nodes (20x40x80). Radiation is not accounted for in this case. For a given number of processors different partitions of the computational domain were tried, yielding slightly different results. The influence of the partition on the attained efficiency is discussed in [24]. Only the results for the best partitions, as far as the efficiency is concerned, are shown in tables 1 and 2.

The parallel performance of the code is examined by means of the speedup, S, defined as the ratio of the execution time of the parallel code on one processor to the execution time on n_p processors (t_total), and the efficiency, ε, defined as the ratio of the speedup to the number of processors. The results obtained show that the highest speedups are obtained when the MUSCL discretization scheme is employed. For example when 128 processors are used speedups of 95.1 and 77.4 are achieved using the MUSCL and the hybrid discretization schemes, respectively. As the number of processors increases, so does the speedup of the MUSCL calculations compared to the hybrid calculations. For n_p = 2 we have S(MUSCL)/S(hybrid) = 1.04 while for n_p = 128 that ratio is 1.23. There are two opposite trends responsible for this behaviour. In fact, there is more data to communicate among the processors when the MUSCL scheme is used, because there are two planes of grid nodes in the halo region compared to only one when the hybrid scheme is employed. However, the calculation time is significantly higher for the MUSCL scheme since the computation of the coefficients of the discretized equations is more involved. Overall, the ratio of the communications to the total time is larger for the hybrid scheme, yielding smaller speedups.

Table 2. Parallel performance of the code for the first 30 iterations using the MUSCL discretization scheme.

n_p	1	2	4	8	16	32	64	128
Partition	1x1x1	1x2x1	1x4x1	1x8x1	2x4x2	2x4x4	2x4x8	2x8x8
t_{total} (s)	1692.0	832.2	420.8	215.9	116.1	61.0	31.9	17.8
S	1	2.03	4.02	7.84	14.6	27.7	53.1	95.1
ε (%)		101.7	100.5	98.0	91.1	86.7	82.9	74.3
ε_{vel} (%)		97.7	97.8	97.1	91.3	88.7	86.7	82.1
ε_p (%)	—	95.3	84.1	70.7	59.2	49.9	41.9	28.6
$\varepsilon_{k,\varepsilon}$(%)	—	103.0	102.2	100.0	93.2	89.4	86.0	78.5
$\varepsilon_{scalars}$ (%)	—	106.8	106.4	104.5	98.2	94.2	90.9	84.5
ε_{prop} (%)	—	97.9	95.7	93.6	86.8	80.6	76.9	71.8

The calculations were divided into five tasks, for analysis purposes, and their partial efficiency is shown in tables 1 and 2. These tasks are: i) the solution of the momentum equations, including the calculation of the convective, diffusive and source terms, the incorporation of the boundary conditions, the calculation of the residuals, the solution of the algebraic sets of equations and the associated communications among processors; ii) the solution of the pressure correction equation and the correction of the velocity and pressure fields according to the SIMPLE algorithm; iii) the solution of the transport equations for the turbulent kinetic energy and its dissipation rate; iv) the solution of other scalar transport equations, namely the enthalpy, mixture fraction and mixture fraction variance equations; v) the calculation of the mean properties, namely the turbulent viscosity and the mean values of density, temperature and chemical species concentrations. The efficiencies of these five tasks are referred to as ε_{vel}, ε_p, $\varepsilon_{k,\varepsilon}$, $\varepsilon_{scalars}$ and ε_{prop}, respectively.

It can be seen that the efficiency of the pressure correction task is the lowest one, and decreases much faster than the efficiencies of the other tasks when the number of processors increases. The reason for this behaviour is that the amount of data to be communicated associated with this task is quite large, as discussed in [24],[25]. Therefore, the corresponding efficiency is strongly affected by the number of processors. The computational load of this task is independent of the discretization scheme of the convective terms because the convective fluxes across the faces of the control volumes are determined by means of the interpolation procedure of Rhie and Chow [26] and there is no transport variable to be computed at those faces as in the other transport equations. However, the communication time is larger for the MUSCL scheme, due to the larger halo region. Therefore, the task efficiency is smaller for the MUSCL scheme, in contrast to the overall efficiency.

The efficiency of the properties calculation is slightly higher for the hybrid scheme if $n_p = 16$, and equal or slightly lower in the other cases, but it does not differ much from one scheme to the other. This is a little more difficult to interpret since the computational load of this task is also independent of the discretization scheme and the communication time is larger for the MUSCL scheme. Hence, it would be expected a smaller efficiency in the case of the MUSCL scheme, exactly as observed for the pressure task. But the results do not confirm this expectation. It is believed that the reason for his behaviour is the following.

If the turbulent fluctuations are small, the mean values of the properties (e.g., density and temperature) are directly obtained from the mean mixture fraction, neglecting those fluctuations. If they are significant, typically when the mixture fraction variance exceeds $10^{-}4$, then the mean values are obtained from interpolation of the data stored in tabular form. This data is obtained *a priori* accounting for the turbulent fluctuations for a range of mixture fraction and mixture fraction variance values. Although the interpolation of the stored data is relatively fast, it is still more time consuming than the determination of the properties in the case of negligible fluctuations. Therefore, when the number of grid nodes with significant turbulent fluctuations increases, the computational load increases too. Since the calculations start from a mixture fraction variance field uniform and equal to zero, a few iterations are needed to increase the mixture fraction variance values above the limit of $10^{-}4$. The results given in tables 1 and 2 were obtained using a different number of iterations, 30 for the MUSCL scheme and 80 for the hybrid scheme. So, it is expected that in the former case the role of the turbulent fluctuations is still limited compared to the last case. This means that the computational load per iteration will be actually higher for the hybrid scheme, rather than identical in both cases as initially assumed. This would explain the similar task efficiency observed for the two schemes.

The three remaining tasks, i), iii) and iv), exhibit a similar behaviour, the efficiency being higher for the calculations using the MUSCL scheme. This is explained exactly by the same reasons given for the overall efficiency. In both cases, the efficiency of these tasks is higher than the overall efficiency, compensating the smaller efficiency of the pressure task. For a small number of processors the efficiency of these tasks slightly exceeds 100%. This has also been found by other researchers and is certainly due to a better use of cache memory.

Tables 3 and 4 summarize the results obtained for a complete run, i.e., for a converged solution, using 32, 64 and 128 processors. Convergence is faster if the hybrid scheme is employed, as expected. Regardless of the discretization scheme, there is a small influence of the number of processors on the convergence rate, and although this rate tends to decrease for a large number of processors, it does not change monotonically. The complex interaction between different phenomena and the non- linearity of the governing equations may be responsible for the non-monotonic behaviour which has also found in other studies. Since the smaller number of processors used in these calculations was 32, the efficiency and the speedup were computed taking the run with 32 processors as a reference. This

Table 3. Parallel performance of the code using the hybrid discretization scheme

n_p	32	64	128
Partition	2x4x4	2x4x8	2x8x8
n_{iter}	1758	1749	1808
t_{total} (s)	1982	1107	701
S_{rel}	1	1.79	2.83
ε_{rel} (%)	—	89.5	70.7
$\varepsilon_{rel,vel}$ (%)	—	95.0	83.0
$\varepsilon_{rel,p}$ (%)	—	86.3	58.9
$\varepsilon_{rel,k,\epsilon}$ (%)	—	91.6	78.1
$\varepsilon_{rel,scalars}$ (%)	—	90.9	77.5
$\varepsilon_{rel,prop}$ (%)	—	99.1	92.6
$\varepsilon_{rel,radiation}$ (%)	—	82.7	58.1

means that the values presented in tables 3 and 4, denoted by the subscript rel, are relative efficiencies and speedups. The relative speedup is higher when the MUSCL scheme is employed, in agreement with the trend observed in tables 1 and 2 for the first few iterations.

Table 4. Parallel performance of the code using the MUSCL scheme

n_p	32	64	128
Partition	2x4x4	2x4x8	2x8x8
n_{iter}	3244	3215	3257
t_{total} (s)	7530	4119	2524
S_{rel}	1	1.83	2.98
ε_{rel} (%)	—	91.4	74.6
$\varepsilon_{rel,vel}$ (%)	—	98.1	90.7
$\varepsilon_{rel,p}$ (%)	—	82.8	54.5
$\varepsilon_{rel,k,\epsilon}$ (%)	—	96.5	86.4
$\varepsilon_{rel,scalars}$ (%)	—	97.3	88.3
$\varepsilon_{rel,prop}$ (%)	—	96.1	87.9
$\varepsilon_{rel,radiation}$ (%)	—	75.6	49.0

There are two tasks that exhibit a much smaller efficiency than the others: the solution of the pressure correction equation and the calculation of the radiative heat transfer. The low efficiency of the radiative heat transfer calculations is due to the decrease of the convergence rate with the increase of the number of processors [19],[20]. The radiation subroutine is called with a certain frequency, typically every 10 iterations of the main loop of the flow solver (SIMPLE algorithm). The radiative transfer equation is solved iteratively and a maximum number of iterations, 10 in the present work, is allowed. If the number of processors is small, convergence is achieved in a small number of iterations. But when the number of processors is large, the limit of 10 iterations is achieved, and a number of iterations smaller than this maximum is sufficient for convergence only when a quasi-converged solution has been obtained. Both the pressure and the radiation tasks have a lower partial efficiency if the MUSCL scheme is used. In fact, the computational effort of these tasks is independent of the discretization scheme, and the communication time is higher for the MUSCL scheme. The same is true, at least after the first few iterations, for the properties task. The other tasks (momentum, turbulent quantities and scalars) involve the solution of transport equations and their computational load strongly depends on how the convective terms are discretized. Hence, their efficiencies are higher than the overall efficiency, the highest efficiencies being achieved for the MUSCL scheme.

5 Conclusions

The combustion chamber of a power station boiler was simulated using a Cray T3D and a number of processors ranging from 1 to 128. The convective terms of the governing equations were discretized using either the hybrid central differences/upwind or the MUSCL schemes, and a comparison of the parallel efficiencies attained in both cases was presented. The MUSCL scheme is more computationally demanding, and requires more data to be exchanged among the processors, but it yields higher speedups than the hybrid scheme. An examination of the computational load of different tasks of the code shows that two of them are controlling the speedup. These are the solution of the pressure correction equation, which requires a lot of communications among processors, and the calculation of the radiative heat transfer, whose convergence rate is strongly dependent on the number of processors. The efficiency of these tasks, as well as the efficiency of the properties calculation task, is higher for the hybrid than for the MUSCL schemes. On the contrary, the efficiency of the tasks that involve the solution of transport equations is higher for the MUSCL than for the hybrid scheme, and it is also higher than the overall efficiency.

6 Acknowledgements

The code used in this work was developed within the framework of the project Esprit No. 8114 — HP-PIPES sponsored by the European Commission. The calculations performed in the Cray T3D at the University of Edinburgh were supported by the TRACS programme, coordinated by the Edinburgh Parallel Computing Centre and funded by the Training and Mobility of Researchers Programme of the European Union.

References

1. Coelho, P.J. and Carvalho, M.G.: Application of a Domain Decomposition Technique to the Mathematical Modelling of a Utility Boiler. International Journal for Numerical Methods in Engineering, 36 (1993) 3401-3419.
2. Leonard, B.P. and Drummond, J.E.: Why You Should Not Use "Hybrid" Powerlaw, or Related Exponential Schemes for Convective Modelling — Are Much Better Alternatives. Int. J. Num. Meth. Fluids, 20 (1995) 421-442.
3. Harter, A., Engquist, B., Osher S., and Chakravarthy S. : Uniformly High Order Essentialy Non-Oscillatory Schemes, III. J. Comput Phys., 71 (1987) 231-303.
4. Gaskell, P.H. and Lau, A.K.C.: Curvature-compensated Convective Transport: SMART, a New Boundedness-transport Algorithm. Int. J. Num. Meth. Fluids, 8 (1988) 617-641.
5. Zhu, J.: On the Higher-order Bounded Discretization Schemes for Finite Volume Computations of Incompressible Flows. Computer Methods Appl. Mech. Engrg., 98 (1992) 345-360.
6. Darwish, M.S.: A New High-Resolution Scheme Based on the Normalized Variable Formulation. Numerical Heat Transfer, Part B, 24 (1993) 353-371.
7. Lien, F.S. and Leschziner, M.A.: Upstream Monotonic Interpolation for Scalar Transport with Application to Complex Turbulent Flows. Int. J. Num. Meth. Fluids, 19 (1994) 527-548.
8. Choi, S.K., Nam, H.Y., Cho, M.: A Comparison of Higher-Order Bounded Convection Schemes. Computer Methods Appl. Mech. Engrg., 121 (1995) 281-301.
9. Kobayashi, M.H., and Pereira, J.C.F.: A Comparison of Second Order Convection Discretization Schemes for Incompressible Fluid Flow. Communications in Numerical Methods in Engineering, 12 (1996) 395-411.
10. Blake, R., Carter, J., Coelho, P.J., Cokljat, D. and Novo, P.: Scalability and Efficiency in Parallel Calculation of a Turbulent Incompressible Fluid Flow in a Pipe. Proc. 2nd Int. Meeting on Vector and Parallel Processing (Systems and Applications), Porto, 25-25 June (1996).
11. Van Leer, B.: Towards the Ultimate Conservative Difference Scheme. V. A Second-Order Sequel to Godunov's Method. J. Comput. Physics, 32 (1979) 101-136.
12. Libby, P.A. and Williams, F.A.: Turbulent Reacting Flows. Springer-Verlag, Berlin (1980).
13. Fiveland, W.A.: Discrete-ordinates Solutions of the Radiative Transport Equation for Rectangular Enclosures. J. Heat Transfer, 106 (1984) 699-706.
14. Fiveland, W.A.: The Selection of Discrete Ordinate Quadrature Sets for Anistropic Scattering. HTD-160, ASME (1991) 89-96.
15. Coelho, P.J., Gonalves, J.M., Carvalho, M.G. and Trivic, D.N.: Modelling of Radiative Heat Transfer in Enclosures with Obstacles. International Journal of Heat and Mass Transfer, 41 (1998) 745-756.
16. Patankar, S.V.: Numerical Heat Transfer and Fluid Flow. Hemisphere Publishing Corporation (1980).
17. Khosla, P.K. and Rubin, S.G.: A Diagonally Dominant Second-order Accurate Implicit Scheme. Computers & Fluids, 1 (1974) 207-209.
18. Darwish, M.S. and Moukalled, F.H.: Normalized Variable and Space Formulation Methodology for High-Resolution Schemes. Numerical Heat Transfer, Part B, 26 (1994) 79-96.
19. Coelho, P.J., Gonalves, J. and Novo, P.: Parallelization of the Discrete Ordinates Method: Two Different Approaches In Palma, J., Dongarra, J. (eds.): Vector

and Parallel Processing - VECPAR'96. Lecture Notes in Computer Science, 1215. Springer-Verlag (1997) 22-235.
20. Gonalves, J. and Coelho, P.J.: Parallelization of the Discrete Ordinates Method. Numerical Heat Transfer, Part B: Fundamentals, 32 (1997) 151-173.
21. Cassiano, J., Heitor, M.V., Moreira, A.L.N. and Silva, T.F.: Temperature, Species and Heat Transfer Characteristics of a 250 MWe Utility Boiler. Combustion Science and Technology, 98 (1994) 199-215.
22. Carvalho, M.G., Coelho, P.J., Moreira, A.L.N., Silva, A.M.C. and Silva, T.F.: Comparison of Measurements and Predictions of Wall Heat Flux and Gas Composition in an Oil-fired Utility Boiler. 25th Symposium (Int.) on Combustion, The Combustion Institute (1994) 227-234.
23. Coelho, P.J. and Carvalho, M.G.: Evaluation of a Three-Dimensional Mathematical Model of a Power Station Boiler. ASME J. Engineering for Gas Turbines and Power, 118 (1996) 887-895.
24. Coelho, P.J.: Parallel Simulation of Flow, Combustion and Heat Transfer in a Power Station Boiler. 4th ECCOMAS Computational Fluid Dynamics Conference, Athens, Greece, 7-11 September (1998).
25. Coelho, P.J., Novo, P.A. and Carvalho, M.G.: Modelling of a Utility Boiler using Parallel Computing. 4th Int. Conference on Technologies and Combustion for a Clean Environment, 7-10 July (1997).
26. Rhie, C.M., and Chow, W.L.: Numerical Study of the Turbulent Flow past an Airfoil with Trailing Edge Separation. AIAA J., 21 (1983) 1525-1532.

Parallel 3D Airflow Simulation on Workstation Clusters

Jean-Baptiste Vicaire[1], Loic Prylli[1], Georges Perrot[1], and Bernard Tourancheau[2]*

[1] LHPC & INRIA REMAP, laboratoire LIP, ENS-Lyon 699364 Lyon - France
Loic.Prylli@ens-lyon.fr,
http://www.ens-lyon.fr/LIP
[2] LHPC & INRIA REMAP, laboratoire LIGIM bat710, UCB-Lyon
69622 Villeurbanne - France
Bernard.Tourancheau@inria.fr,
http://lhpca.univ-lyon1.fr

Abstract. *Thesee* is a 3D panel method code, which calculates the characteristic of a wing in an inviscid, incompressible, irrotational, and steady airflow, in order to design new paragliders and sails.
In this paper, we present the parallelisation of *Thesee* for low cost workstation/PC clusters. *Thesee* has been parallelised using the ScaLAPACK library routines in a systematic manner that lead to a low cost development. The code written in C is thus very portable since it uses only high level libraries. This design was very efficient in term of manpower and gave good performance results. The code performances were measured on 3 clusters of computers connected by different LANs : an Ethernet LAN of SUN SPARCstation, an ATM LAN of SUN SPARCstation and a Myrinet LAN of PCs. The last one was the less expensive and gave the best timing results and super-linear speedup.

1 Introduction

The aim of this work is to compare the performance of various parallel platforms on a public domain aeronautical engineering simulation software similar to those routinely used in the aeronautical industry where the same numerical solver is used, with a l ess user-friendly interface, which results in a more portable code (smaller size, no graphic library).

Parallel *Thesee* is written in C with the ScaLAPACK[3] library routines and can be run on top of MPI[14] or PVM[6], thus the application is portable on a wide range of distributed memory platforms.

We introduce in the following the libraries that are necessary to understand the ScaLAPACK package. Then we give an insight of the parallelisation of the code. Tests are presented before the conclusion.

* This work was supported by EUREKA contract EUROTOPS, LHPC (Matra MSI, CNRS, ENS-Lyon, INRIA, Région Rhône-Alpes), INRIA Rhône-Alpes project REMAP, CNRS PICS program, CEE KIT contract

J. Palma, J. Dongarra, and V. Hernández (Eds.): VECPAR'98, LNCS 1573, pp. 215–226, 1999.

2 Software Libraries

2.1 LAPACK, BLAS and BLACS Libraries

The BLAS (Basic Linear Algebra Subprograms) are high quality "building block" routines for performing basic vector and matrix operations. Level 1 BLAS do vector-vector operations, Level 2 BLAS do matrix-vector operations, and Level 3 BLAS do matrix-matrix operations.

LAPACK[1] provides routines for solving systems of simultaneous linear equations, least-squares solutions of linear systems of equations, eigenvalue problems, and singular value problems. The associated matrix factorisations (LU, Cholesky, QR, SVD, Schur, generalised Schur) are also provided. The LAPACK implementation used as much as possible the BLAS building block to ensure efficiency, reliability and portability.

The BLACS (Basic Linear Algebra Communication Subprograms) are dedicated to communication operations used for the parallelisation of the level 3 BLAS or the ScaLAPACK libraries. They can of course be used for other applications that need matrix communication inside a network. BLACS are not a multi-usage library for every parallel application but an efficient library for matrix computation.

In the BLACS, processes are grouped in one or two dimension grids. BLACS provide point to point synchronous receive, broadcast and combine. There is also routines to build, modify or to consult a grid. Processes can be enclosed in multiple overlapping or disjoint grids, each one identified by a context. Different release of BLACS are available on top of PVM,MPI and others. In this project, BLACS are used on top of PVM or MPI.

2.2 The ScaLAPACK Library

ScaLAPACK is a library of high-performance linear algebra routines for distributed memory message passing MIMD computers and networks of workstations supporting PVM and/or MPI . It is a continuation of the LAPACK project, which designed and produced analogous software for workstations, vector supercomputers, and shared-memory parallel computers. Both libraries contain routines for solving systems of linear equations, least squares problems, and eigenvalue problems. The goals of both projects are efficiency (to run as fast as possible), scalability (as the problem size and number of processors grow), reliability (including error bounds), portability (across all important parallel machines), flexibility (so users can construct new routines from well-designed parts), and ease of use (by making the interface to LAPACK and ScaLAPACK look as similar as possible). Many of these goals, particularly portability, are aided by developing and promoting standards , especially for low-level communication and computation routines. LAPACK will run on any machine where the BLAS are available, and ScaLAPACK will run on any machine where both the BLAS and the BLACS are available.

3 Implementation

The *Thesee* sequential code [8] uses the 3D panel method (also called the singularity element method), which is a fast method to calculate a 3D low speed airflow[9]. It can be separated in 3 parts :

Part P1 is the fill-in of the element influence matrix from the 3D mesh. Its complexity is $O(n^2)$ with n the mesh size (number of nodes). Each matrix element gives the contribution of a double layer (source + vortex) singularity distribution o n facet i at the centre of facet j.

Part P2 is the LU decomposition of the element influence matrix and the resolution of the associated linear system $(O(n^3))$, in order to calculate the strength of each element singularity distribution.

Part P3 is the speed field computation. Its complexity is $(O(n^2))$, because the contribution of every nodes has to be taken into account for the speed calculation at each node. Pressure is then obtained using the Bernoulli equation

Each of these parts are parallelised independently and are linked together by the redistribution of the matrix data. For each part, the data distribution is chosen to insure the best possible efficiency of the parallel computation.

The rest of the computation, is the acquisition of the initial data and the presentation of the results. Software tools are uses to build the initial wing shape and to modify it as the results of the simulation gives insight that are valuable. The results are presented using a classical viewer showing the pressure field with different colours and with a display of the raw data in a window.

3.1 Fill in of the Influence Matrix

The 3D meshing of the wing is defined by an airfoils file, i.e. by points around the section of the wing that is parallel to the airflow, in order to define the shape of the wing in this section, as in the NACA tables[1]. Thus two airfoils delimit one strip of the wing, i.e. the narrow surface between the airfoils. The wing is then divided in strips and each strip is divided in facets by perpendicular divisions joining two similar points of the airfoils. This meshing is presented in Figure 1. Notice that the points are not regularly distributed in order to have more facets, thus more precision in the computation, in areas where the pressure gradient is greater.

During the computation, for each facet the whole wing must be examined (every other facet). Since the wing is symmetrical, the computation is made with a half-wing, thus as the dimension of the equation system is $2 \times n$, its size is divided by four. The results for the whole wing are then deduced from those of the half-wing. Let K be the number of strip, and N be the number of facets per strip. The sequential code for the fill in of the influence matrix is made of nested loops as described in Figure 2.

[1] http://www.larc.nasa.gov/naca/

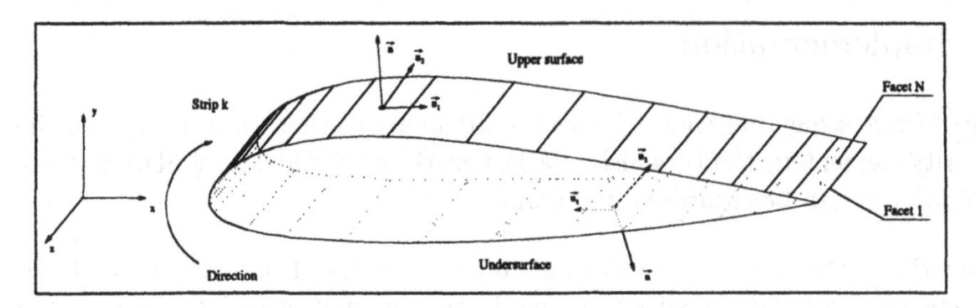

Fig. 1. Strip of the meshing

```
/* computation of the influence coefficient (source)*/
 for (i1=1;i1<=K/2;i1++) {      /*for each strip */
    cste1=(i1-1)*(N+2)+1;
    for (j1=cste1;j1<=(cste1+N-1);j1++) {
      /* 2 loops to examine each point of the half wing */
      for (i2=1;i2<=K/2;i2++) {
         cste2=(i2-1)*(N+2)+1;
         for(j2=cste2;j2<=(cste2+N);j2++)
         {
           ............................
           M[ind++]=....................
         }
```

Fig. 2. Sequential code for the influence matrix fill in.

The computation of the influence coefficient $M[i]$ for one facet is independent of the other coefficients. The external loop can be split up straightforwardly in order to parallelise the computation on different processors. Each processor will then compute the facets of a given number of strips. The strips are assigned to a processor according to their number. For instance, with 4 processors and 20 strips to compute, each processor will compute five strips; strips 1 to 5 on processor 1, strips 6 to 10 on processor 2, ...

Hence the external loop becomes like in Figure 3.

```
for (i1=ceil((MyPRow)*(K/2)/((NPRow)))+1;
     i1<=ceil((MyPRow+1)*(K/2)/(NPRow));
     i1 ++)
```

Fig. 3. The parallel version of the external loop corresponding to the given strips assignment.

The data needed for these computations (the initial meshing of the wing) are distributed to each processor using the PDGEMR2D routine from the ScaLAPACK parallel library [3], the results is then gathered with the same routine. This routine provides the distribution of data between virtual grids of processors with any kind of block cyclic data distribution. We used it from the initial "grid" of size 1×1 which contains the matrix to compute, to the computation virtual grid of processors with 2 to 4 processors arranged in a 1×2 or 1×4 shape with, for instance, a full block data distribution. Notice that this data distribution introduces no other communication cost in this part because it is embarrassingly parallel.

The second inner step of this part, the computation of doublet influence coefficient, is realized with the same method, splitting the outer loop and using the *PDGEMR2D* routine for the data repartition.

3.2 Influence Matrix Resolution

This part mainly consists in the resolution of a linear algebra system (Part P2). In the sequential version of *Thesee* it is carried out with a simple call of the *DGESV* routine from LAPACK [1]. DGESV solves a linear equation system with a LU factorisation and then a back-solve substitution.

```
DGESV_(&n, &nrhs, &M[1], &lda, &ipiv[1], &B[1], &ldb, &info);
```

In this procedure DGESV solves the $M * X = B$ system of equations and stores the results in the `B` vector.

The parallel solve routine *PDGESV* from ScaLAPACK provides the same system resolution with a parallel LU decomposition on block cyclically distributed data on a virtual grid of processors. A redistribution of the data on each processor is then needed to use the parallelised resolution.

The main idea is that each processor involved in the resolution holds a submatrix of the matrix `M` to solve. The processors of the parallel machine with P processors are presented to the user as a linear array of process IDs, labelled 0 through (P-1). It is often more convenient while doing matrix computations to map this 1-D array of P processes into a logical two dimensional process mesh, or grid while doing matrix computation. This grid will have R processor rows and C processor columns, where $R * C = G <= P$. A processor can now be referenced by its coordinates within the grid (indicated by the notation i, j, where $0 <= i < R$, and $0 <= j < C$). An example of such a mapping is shown in Figure 4.

A processor can be a member of several overlapping or disjoint virtual grids during the computation, each one identified by a *context*.

The ScaLAPACK library uses a block cyclic data distribution on a virtual grid of processors in order to reach a good load-balance, good computation efficiency on arrays, and an equal memory usage between processors. The load-balance is insure by the cyclic distribution that gives to each processor matrix elements that are coming from "different" locations of the matrix (compared to

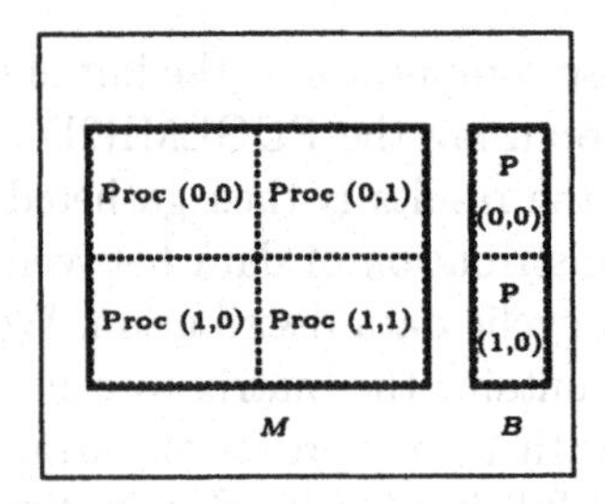

Fig. 4. Example of redistribution with a 2 × 2 grid

a classical full block decomposition). The communication efficiency is obtained because in a cyclic distribution, the row and column shape of the matrix is preserved, so most of communication of 1D arrays can happen without a complex index computation (see [4, 5, 13, 2] among others). We ran tests in order to choose the best grid shape and the best block size of the data distribution for our problem on each of the platforms.

Matrices and arrays are then wrapped by blocks in all dimensions corresponding to the processor grid using the PDGEMR2D routine. Figure 5 illustrates the organisation of the block cyclic distribution of a 2D array on a 2D grid of processors and Figure 6 shows how the main matrix and vector are distributed from processor 0 to processors 0 to N.

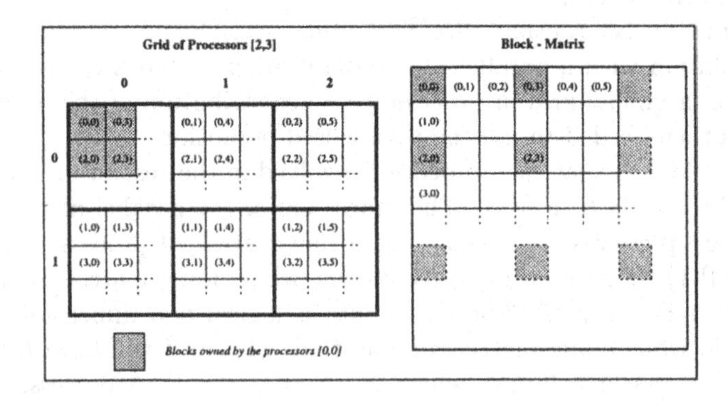

Fig. 5. The block cyclic data distribution of a 2D array on a 2 × 3 grid of processors.

In the parallel version of *Thesee* , we used the following parameters for the data distribution of the LU factorisation: the block size is 32 × 32 and the processor grid shape is a 1D-grid that gave the best overall computation timings (for further information see[10]).

In the first parts of parallel *Thesee* , the global system matrix (`M`) is hold in 1 × 1 grid by the processor 0 (`Context1x1`). Then, it is distributed over a

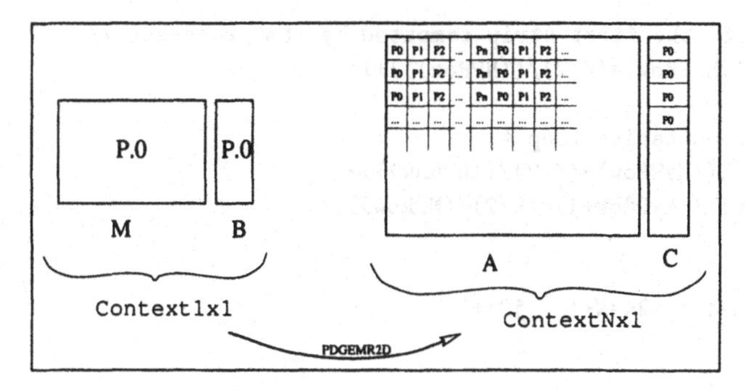

Fig. 6. Data redistribution inside parallel Thesee

$1 \times NbProc$ grid (`Context1xN`) with the *PDGEMR2D*[2]. The same operation is also performed for the `B` vector. Then each processor calls the PDGESV routine from ScaLAPACK instead of DGESV from LAPACK. After this, the solution vector is distributed on the local sub-matrix of `B`. A new call to PDGEMR2D is needed to gather the solution sub-vectors from the `Context1xN` grid to the `Context1x1` grid.

3.3 Speed Computation

The speed computation procedure (Part P3) is made of loops to compute the speed array (`S`) and the potential array (`Pot`).

The sequential code uses two nested loops for the speed array computation:

```
for(i2=1 ; i2<=K/2 ; i2++) {
   for(j2=1 ; j2<=N+1 ; j2++) {
      ........
      S[(i2-1)*(N+1)+j2]=...;
   }
}
```

Fig. 7. Sequential code of the speed computation

The computation of an element of the speed array is independent from all the others, giving us another nice embarrassingly parallel problem. The external loop is thus split like in Part P1 and each processor is assigned a given number of strips to compute. The computation code of the speed array is modified as shown in Figure 8.

The computation of the potential array is done with the same method.

[2] Parallel Double GEneral Matrix Redistribution (from ScaLAPACK)

```
/* number of  the first strip computed by the processor */
 ideb=ceil((MyPRow)*(K/2)/((NPRow)))+1;

 /* speed computation loop */
 for(i2=ceil((MyPRow)*(K/2)/((NPRow)))+1 ;
     i2<=ceil((MyPRow+1)*(K/2)/(NPRow));
     i2++)
   {
     for(j2=1 ; j2<=N+1 ; j2++)
       {
        .........
        S[(i2-ideb)*(N+1)+j2]=.......
       }
   }
```

Fig. 8. Parallelization of the external loop of the Figure 7

4 Performance Tests

The tests were run on 2 different platforms with 3 different networks: an Ethernet and ATM network of SUN Sparc5 (85MHz) with Solaris and an Ethernet and Myrinet network of Pentium-Pro (200MHz) running Linux. On both systems, the MPI message passing interface was used with either the LAM implementation on top of IP or MPI-BIP user level implementation. The lower level computation libraries BLAS were an optimised version on the SUNs and a compiled version on the Pentium-Pros.

The efficiency of the fill in of the influence matrix and of computation of the speed and potential arrays are roughly the same on every configuration. The code for these parts is "embarrassingly parallel" and thus the speed-up is almost equal to the number of processor. Whereas the LU factorisation which involves a lot of communications is highly dependent of the network software and hardware.

4.1 Optimal Block Size

There is only slight differences between the different block sizes performances on such small platforms with our problem sizes. However a block size of 32×32 was the optimal on every configuration (i.e. Ethernet, ATM, Myrinet).

We present in Table 1 the timing for the LU resolution with a system 1722×1722 which correspond to our production problem size.

The results show the advantage of the new Pentium-Pro (200MHz) generation over the rather old Sparc5 (85MHz) and gives the better block size for each configuration.

Block size	Timings for a system 1722 × 1722 (seconds)		
	IP/Ethernet (on SUN Sparc)	IP/ATM (on SUN Sparc)	IP/Myrinet (on Pentium Pro)
8 × 8	71.3	67.7	30.1
16 × 16	68.9	61.3	29.5
32 × 32	68.3	61.1	28.1
64 × 64	74.4	66.8	34.1

Table 1. Timings of the LU resolution for different block sizes.

4.2 Comparison between Ethernet and ATM

We present here the timing results of the whole computation using the different platforms. First, we compare the networks with the same processor kind (SUN Sparc 5) over PVM.

	System size	Timings(seconds) [speedups]	
		IP/Ethernet (on SUN Sparc)	IP/ATM (on SUN Sparc)
4 Proc	902 × 902	16.3 [1.24]	13.5 [1.50]
2 Proc	902 × 902	18.8 [1.07]	16.2 [1.25]
Sequential	902 × 902	20.3	
4 Proc	1722 × 1722	68.3 [1.92]	61.1 [2.15]
2 Proc	1722 × 1722	108.9 [1.21]	95.9 [1.37]
Sequential	1722 × 1722	131.8	

Table 2. Timings and speedups of the whole computation (using the best block size).

Both networks do not provide efficient enough communications for this kind of fine grained parallel computation. Moreover, although it provides more throughput, the gain obtained with ATM is small because the startup time of this two network is similar to the one on Ethernet. This is the big part of the communication delay.

Note that the low speedup is mostly due to the high elementary communication/computation ratio. For these matrix sizes and this number of processors, there is not more load balancing issues thanks to the use of block cyclic data distribution.

However, the speedup obtained is not negligible while using the code in production because the response time is critical.

4.3 Comparison between Ethernet and Myrinet

We present here the timings obtained on the PC platform. The Myrinet network is driven by the BIP[11, 12] (Basic Interface for Parallelism) software. BIP is a new protocol that provides a small parallel API implemented on the Myrinet network. Other protocol layers are implemented for the classical interfaces. BIP delivers to the application the maximal performance achievable by the hardware using a low latency zero copy mechanism. An IP stack has been build on top of BIP. As well, a port of MPI-CH was realized with MPI-BIP[15, 12, 7]. The results of parallel *Thesee* have been measured with PVM over IP/Ethernet and PVM over IP/Myrinet and MPI over BIP/Myrinet. This gives an idea of the portability of our code that uses library calls either from IP, PVM or MPI.

	Timings (seconds) [speedups]		
System size	IP/Ethernet (on SUN Sparc)	IP/Myrinet (on Pentium Pro)	MPI-BIP/Myrinet (on Pentium Pro)
4 Proc 902 × 902	10.2 [0.97]	4.7 [2.10]	3.1 [3.24]
2 Proc 902 × 902	9.6 [1.03]	6.7 [1.49]	5.0 [1.97]
Sequential 902 × 902		10.0	
4 Proc 1722 × 1722	45.9 [1.90]	28.1 [3.10]	21.3 [4.09]
2 Proc 1722 × 1722	56.4 [1.54]	44.3 [1.97]	38.1 [2.29]
Sequential 1722 × 1722		87.4	

Table 3. Timings and speedups with Myrinet (using the best block size).

First notice the advantage of the Pentium-Pro speed against the Sparc on sequential numbers.

Not surprisingly the best results are achieved with MPI-BIP that provides a very low $9\mu s$ latency for the basic send communication. The gain on large problem size and platform is more than 50% over the Ethernet run, leading to a super-linear speed-up (in bold in Table 3).

This can be explained by a better cache hit ratio in the parallel version of the code. As the matrix is distributed cyclically on the processors, the computation occurs on blocked data that fits better in the cache during the LU decomposition, leading to a better use of the processor's pipeline units. Moreover, an overlap of the communications is done in the parallel LU decomposition, this overlap can be (relatively) increased because the total amount of the communication cost is greatly reduced with the Myrinet+BIP platform.

These outstanding results show that low-level access to high-speed network is essential to achieve the best possible performances while doing parallel computation.

4.4 Industrial Use of the Code

We ran the code on an industrial version of the PC-Myrinet cluster, the POPC (Pile of PCs) machine designed by Matra. This architecture is a little bit slower than the original test-bed (thus does not achieve super-linear speedups) but gives interesting results on Table 4 about the scalability of the code up to 12 processors. When using the compiled BLAS kernels, depending of the data size of the problem, there is little interest in going further than 6 processors. When using an optimised version of the BLAS designed for the Pentium processors, the timings dropped down by a factor of more than 2 for the small wing and more than 3 for the big one and there is little gain going for more than 4 processors with the small wing and more than 8 processors for the big one. From a user point of view, the elapse time was decreased by a factor of 25 (from more than a minute to 5 second). This will drastically improve the production iterative process of the new wing shapes for the cost of a few PCs, the use of a good BLAS kernel, and a very simple parallelisation method.

System size	Number of processors					
	1	2	4	6	8	10
wing 902	8.97	5.11	3.11	2.36	2.12	1.79
wing 902 (optimized BLAS)	4.199	2.71	1.88	1.59	1.39	1.31
wing 1722	75.72	38.17	21.07	14.11	12.63	
wing 1722 (optimised BLAS)	25.337	14.6	8.09	6.32	5.41	5.10

Table 4. Timings (seconds) of the whole execution on the POPC Pile of PC machine

These outstanding results show that low-level access to high-speed network is essential to achieve the best possible performances while doing parallel computation.

5 Conclusion

We described our work on the parallelisation of an air flow 3D simulation that use the singularity method that is well suited for low speed airflows.

We presented a very easy and clean way to parallelise such numerical code, using only parallel library routines (this requires the sequential code to be written with sequential library routines too), loop splitting and calls to a data redistribution routi ne. The parallel code is thus portable (we ran it on 3 different , IP, PVM, MPI) and efficient (super-linear speedup over Myrinet).

We demonstrate that low cost parallel hardware and good software can lead to significant improvement for production codes, starting from a more than 2 minutes delay and going to 21s on a four PCs platform with Myrinet.

Our future work will consist in the automation of this parallelisation process of numerical code using library routines with a software tool.

References

1. Anderson, Bai, Bischof, Demmel, Dongarra, Du Croz, Greenbaum, Hammarling, McKenney, Ostrouchov, and Sorensen. *LAPACK Users' Guide.* SIAM, 1994. http://www.netlib.org/lapack/.

2. E. Anderson, A. Benzoni, J. Dongarra, S. Moulton, S. Ostrouchov, B. Tourancheau, and R. Van de geijn. LAPACK for distributed memory architecture. In *Fifth SIAM Conference on Parallel Processing for Scientific Computing*, USA, 1991.
3. Blackford, Choi, Cleary, d'Azevedo, Demmel, Dhillon, Dongarra, Hammarling, Henry, Petitet, Stanley, Walker, and Whaley. *ScaLAPACK Users' Guide*. SIAM, 1997. http://www.netlib.org/scalapack/.
4. F. Desprez, J.J. Dongarra, and B. Tourancheau. Performance Study of LU Factorization with Low Communication Overhead on Multiprocessors. *Parallel Processing Letters*, 5(2):157–169, 1995.
5. F. Desprez and B. Tourancheau. LOCCS: Low Overhead Communication and Computation Subroutines. *Future Generation Computer Systems*, 10:279–284, 1994.
6. Al Geist, Beguelin, Dongarra, Jiang, Manchek, and Sunderam. *PVM, a users' guide and tutorial*. MIT Press, 1994. http://www.netlib.org/pvm/.
7. Marc Herbert, Frederic Naquin, Loïc Prylli, Bernard Tourancheau, and Roland Westrelin. Protocole pour le gbit/s en reseau local: l'experience myrinet. *Calculateurs Paralleles, Reseaux et Systemes Repartis*, 1998.
8. L. Giraudeau Georges Perrot S. Petit and B. Tourancheau. 3-d air flow simulation software for paragliders. Technical Report 96-35, LIP-ENS Lyon, 69364 Lyon, France, 1996.
9. J. Katz & A. Plotkin. *Low-speed Aerodynamics From Wing Theory to Panel Methods*. MCGraw-Hill, Inc., 1991.
10. L. Prylli and B. Tourancheau. Efficient block cyclic array redistribution. *Journal of Parallel and Distributed Computing*, (45):63–72, 1997.
11. Loïc Prylli and Bernard Tourancheau. Protocol design for high performance networkin g: Myrinet experience. Technical Report 97-22, LIP-ENS Lyon, 69364 Lyon, France, 1997.
12. Loïc Prylli and Bernard Tourancheau. Bip: a new protocol designed for high performance networking on myrinet. In *Workshop PC-NOW, IPPS/SPDP98*, Orlando, USA, 1998.
13. Y. Robert, B. Tourancheau, and G. Villard. Data allocation strategies for the gauss and jordan algorithms on a ring of processors. *Information Processing Letters*, 31:21–29, 1989.
14. M. Snir, S.W. Otto, S. Huss-Lederman, D.W. Walker, and J.J. Dongarra. Mpi: The complete reference, 1996. http://www.netlib.org/mpi/.
15. Roland Westrelin. Réseaux haut débit et calcul parallèle: étude de myrinet. Master's thesis, LHPC, CPE-Lyon, 1997.

Parallel Turbulence Simulation: Resolving the Inertial Subrange of the Kolmogorov Spectrum

Martin Strietzel[1] and Thomas Gerz[2]

[1] German Aerospace Center (DLR), Simulation and Softwaretechnology, D-51170 Köln, Germany
[2] German Aerospace Center (DLR), Institute of Atmospheric Physics, Oberpfaffenhofen
D-82230 Weßling, Germany
{Martin.Strietzel, Thomas.Gerz}@dlr.de
http://www.dlr.de

Abstract. A parallel implementation for direct and large–eddy simulation of turbulent flow (PARDISTUF) is described. The code is based on the three-dimensional incompressible Navier–Stokes equation. Benchmark results on a set of European supercomputers under the message-passing platform MPI are presented. Using this programme on a 48 node IBM-SP2 with a domain resolution of 480^3 grid cells, the inertial subrange of the kinetic energy spectrum of homogeneous turbulence in a sheared and stratified fluid could be resolved over three wavenumber decades.

1 Introduction

The Institute of Atmospheric Physics at the German Aerospace Center (DLR) in Oberpfaffenhofen is investigating the physics of turbulent fluids. The studies are motivated by the need to understand and predict the diffusion of species concentrations in atmospheric flows which often are turbulent, stably stratified and sheared. One special point of interest is the concern that exhaust gases from aircraft may influence the global climate. The aim of the parallelisation activities is to tackle the particular problem of the diffusion properties at small scales.
During the last ten years, a programme for DIrect numerical Simulation of TUrbulent Fluids (DISTUF) under the influence of background shear and stable thermal stratification was developed [?], extended to the large–eddy simulation technique to allow for higher Reynolds numbers [5], and optimised for daily use on vector computers such as the Cray Y-MP. For a simulation run on 128^3 grid points with 3000 time steps DISTUF requires about 44 MWords (64bit) of memory and eight hours of CPU time on one processor.
To resolve the inertial subrange of the turbulent kinetic energy spectrum the resolution has to be increased to the order of about 500^3 grid points in the computational domain. This is not feasible on nowadays vector computers. At this point we decided to make our code suitable for state-of-the-art massively parallel systems to access more computing power and more memory. The parallelisation

J. Palma, J. Dongarra, and V. Hernández (Eds.): VECPAR'98, LNCS 1573, pp. 227–237, 1999.

is based on the concepts of message-passing and domain decomposition. We use MPI to achieve a high portability.
The successful parallelisation is demonstrated by a flow simulation on the IBM–SP-2 computer of DLR using 48 nodes. In this run the inertial subrange of Kolmogorov's universal energy spectrum could be resolved in a homogeneously and stationary turbulent flow with constant background shear and stable thermal stratification. The background shear, dU_0/dz, which enhances turbulence, and the stable thermal stratification, dT_0/dz, which damps turbulence, were selected such that the ratio, the so–called gradient Richardson number

$$Ri := \frac{\beta g \, dT_0/dz}{(dU_0/dz)^2}, \tag{1}$$

was 0.13 which enabled the evolution of statistically stationary turbulence [3]. In eq. (1), β and g are the constant isobaric volumetric expansion coefficient and the gravitational acceleration, respectively.

2 Method of Simulation

We integrate the time-dependent, incompressible, three-dimensional Navier - Stokes and temperature concentration equations in a rectangular domain and in time. The methods of large-eddy simulation (LES) or direct numerical simulation (DNS) are selectable.
We consider a rectangular domain with coordinates x, y, z or x_i ($i \in \{1,2,3\}$),

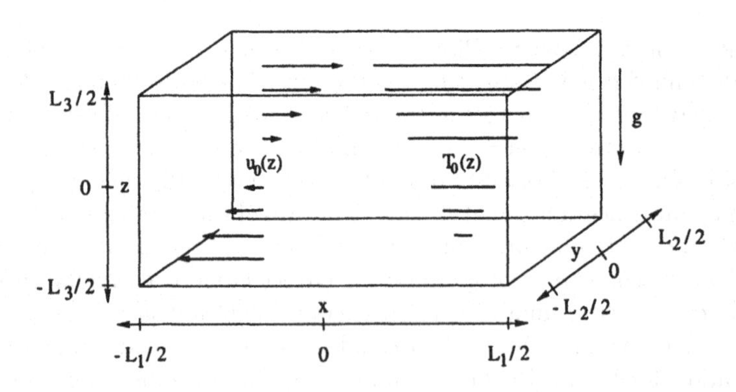

Fig. 1. (a) Simulation domain and (b) mean profiles of velocity and temperature

and side-length L_i. The mean horizontal velocity $U_0(z)$ and mean (reference) temperature $T_0(z)$ possess uniform and constant gradients relative to the vertical coordinate z and are constant in the other directions. The fluid is assumed to have constant molecular diffusivities for momentum and heat. All

fields are expressed non-dimensionally using $L := L_3$, $\Delta U = \|dU_0/dx_3\|L$ and $\Delta T = \|dT_0/dx_3\|L$ as reference scales for length, velocity and temperature. The turbulent fluctuations relative to these mean values are u_i ($i \in \{1,2,3\}$) for velocity, T for temperature, and p for pressure.
The normalised Navier-Stokes-equation, the heat balance, and the continuity equation read

$$\frac{\partial u_i}{\partial t} + \frac{\partial}{\partial x_j}(u_j u_i) + Sx_3\frac{\partial u_i}{\partial x_1} + Su_3\delta_{i1} = -\frac{\partial \tau_{ij}}{\partial x_j} - \frac{\partial p}{\partial x_i} + Ri\frac{S^2}{s}T\delta_{i3} \quad (2)$$

$$\frac{\partial T}{\partial t} + \frac{\partial}{\partial x_j}(u_j T) + Sx_3\frac{\partial T}{\partial x_1} + su_3 = -\frac{\partial \tau_{Tj}}{\partial x_j} \quad (3)$$

$$\frac{\partial u_j}{\partial x_j} = 0, \quad (4)$$

where $S = (L/\Delta U)(dU_0/dx_3) \in \{0,1\}$ is the non-dimensional shear parameter, $s = (L/\Delta T)(dT_0/dx_3) \in \{-1,0,1\}$ is the stratification parameter, and δ_{ij} is the Kronecker-Delta. τ_{ij} and τ_{Tj} denote the diffusive fluxes of momentum and heat. Additionally the distribution of three passive scalars can be calculated by solving transport equations for each one.
Periodic–periodic–shear-periodic boundaries are used in the x, y, and z directions, respectively.
The equations are discretized in an equidistant Eulerian framework using a second-order finite-difference technique on a staggered grid for all the terms in the equations except the mean advection, where pseudo-spectral (Fourier) approximation in x-direction is used. The Adams-Bashforth scheme is employed for time integration of the acceleration terms. The pressure p^{n+1} at the new time-level $n+1$ is obtained by solving the Poisson equation in finite difference form (5), with δ_i as the common finite difference operator and $\tilde{u}_i$ denoting the velocity terms resulting from the Adams-Bashforth scheme.

$$\delta_i \delta_i p^{n+1} = \frac{1}{\Delta t}\delta_i \tilde{u}_i. \quad (5)$$

The solution of (5) is obtained using a fast Poisson solver, which includes the shear-periodic boundary condition at time t^{n+1} and applies a combination of fast Fourier transforms and Gaussian elimination. Finally, the velocities are updated by the new pressure terms.
As initial conditions we use isotropic variance spectra with prescribed intensity and spectral shape for the velocity fluctuations [3], such that the turbulent flow fulfils the continuity equation (4). Initial temperature fluctuations are set to zero. The flow evolution is only controlled by the gradient Richardson number Ri which is set to 0.13.

3 Parallel Approach

Our main aim was to develop a parallel code which is not only efficient, scalable, and numerically correct, but also portable on a wide range of supercomputers.

For this reason we decided to take advantage of the MPI message passing standard. The first experiences with the MPI implementations show that this was the right decision. We were able to port the programme to a wide set of computers very fast and without any changes concerning the message passing calls.

3.1 Domain Decomposition

Analysing the sequential code we found out that under the aspect of parallelism the one-, two- and three-dimensional fast Fourier transforms (FFT) are the most critical parts of the programme. The one-dimensional FFT in x-direction is used in every time step in order to consider the shear flow in the horizontal velocity components. The two-dimensional FFT in x- and y-direction is a part of the Poisson solver for equation (5). For statistic evaluations in the Fourier space, which can be done at user defined intervals, we need a parallel three-dimensional Fourier transform, too.
The best way of implementing two- and three-dimensional FFT on parallel computers is to treat them as a sequence of one-dimensional transforms, which are computed independently on the processors. (cf. [1],[10],[2]). This is no problem at all on shared memory computers and can be implemented on distributed memory machines by using efficient algorithms for data transposition. For this reason we partition the three-dimensional grid into sub-domains allowing to perform a one-dimensional Fourier transform in x-direction without communication. That means that the domain is decomposed in horizontal bars in x-direction (two-dimensional decomposition) or in horizontal (or vertical) planes in x/y (or x/z) direction, what we call one-dimensional decomposition. According to this, the process topology is a one-dimensional or a two-dimensional grid (fig. 2). With the described domain partitioning we can perform all necessary Fourier transforms in y and z direction after data transposition, which can be implemented very efficiently using MPI calls (cf. [8]).

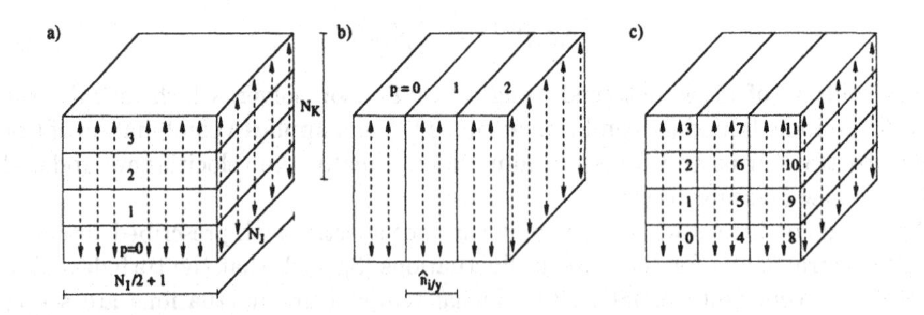

Fig. 2. Data decomposition: One-dimensional (a, b) and two-dimensional (c). The arrows show the distribution of the tridiagonal systems (6)

In the Poisson solver we have to solve tridiagonal systems of equations distributed in z-direction. This is another point where we found an efficient parallel algorithm. Our approach is described in chapter 3.2. All other parts of the algorithm can be calculated independently on all processors, if we ensure an overlap of one row or column of grid points in each direction, which has to be updated after each time step.

3.2 Parallel Poisson Solver

For pressure correction we have to solve the Poisson equation

$$\frac{\partial \tilde{u}_i(x)}{\partial x_i} = \Delta t \frac{\partial^2 p}{\partial x_i^2}$$

for the pressure p in each time step. This is done by two-dimensional Fourier transforms of the left hand-sides in both horizontal directions, which results in $(N_I/2 + 1) * N_J$ ($N_{I,J,K}$ = number of grid points in x, y, and z direction) tridiagonal systems of equations of rank N_K with shear-periodic boundaries (ω is a complex shear factor):

$$\begin{aligned} \hat{p}_{i-1} - 2\hat{p}_i + \hat{p}_{i+1} &= \hat{u}_i\,, \qquad i = 1, 2, \ldots, N_K \quad \hat{p}_i, \hat{u}_i \in C \\ \hat{p}_0 &= \omega \hat{p}_{N_K}\,, \qquad \hat{p}_{N_K+1} = \omega^{-1}\hat{p}_1 \end{aligned} \tag{6}$$

The components $\hat{p}_i$, $\hat{u}_i$ of each of these systems are distributed in z-direction on the grid. The systems themselves are distributed in x- and y-direction (fig. 2). After solving the equations the results are transformed back, and we finally get the pressure terms p for the next time step.

For parallelisation we must distinguish between the phase of Fourier transforms and the algorithm for solving the tridiagonal systems. The Fourier transforms can be done simultaneously on the distributed datasets. In case of the two-dimensional decomposition we have to include a data transposition step.

The distributed equations can be solved independently by the subsets of processes with the same x and y coordinates. Each of the in z-direction distributed systems (6) is solved by an improved divide & conquer method (cf. [9]) based on an algorithm from *Mehrmann* (cf. [7]).

4 Experiences on Parallel Systems

We had the opportunity to run the parallel turbulence simulation code on a set of parallel computers, including shared memory and distributed memory systems. The machines are summarised in table 4.

The parallel test runs are compared with a run on one processor of the CRAY J916 with 256 MB of memory and a peak performance of 200 MFlops. This is the machine the original code was designed and highly optimised for.

In order to evaluate performance data on all machines we run simulations with shear, stratification, and three passive scalars about 64 time steps on 128^3 grid points. Here we discuss the timings for one time-step in this run.

Table 1. Technical data of the parallel systems

System (manufacturer)	Processor (manufacturer)	Max. number of proc.	Memory[1] p. proc. (MB)	Performance per proc. peak (MFlop/s)	SPEC fp92[2]	cache (KB)
Massively-parallel systems with distributed memory:						
GC/PowerPlus (Parsytec)	601+ (PowerPC)	192 (96)	32 (64)	80	125	32
SP2 (IBM)	Power2 (IBM)	58 (+8)	256 (128)	267 (133)	244.6 (202.1)	256 (128)
T3D (Cray)	alpha 21064 (DEC)	512	64	150	200	8
Systems with shared memory:						
Power Challenge XL (SGI/SNI)	R8000 (MIPS)	16	500	300	311	4000
Ultra Enterprise 4000 (SUN)	Ultra I (SPARC)	8	128	unknown to the author	386	512
Parallel vector machines:						
J916 (Cray)		16	256	200		

On the shared memory systems we get a speed-up of 7.44 on eight SGI processors and one of 6.56 on eight SUN CPUs. On more than 8 processors of the Power Challenge this value could not be increased. This restriction depends on the fixed bandwidth of the underlying communication system, the SGI data bus. A totally different picture is the speed-up on the distributed systems. The timing results are presented in figure 3. Only on the SP2 the 128^3 grid fits on one processor and here we get a speed-up of 62.95 for 64 processors.

The best timing results for one time-step has the SP2, which with 16 processors is 3.5 times faster than the Parsytec GC or the Cray T3D. The T3D with 128 CPUs is still 2 times slower than 64 SP2 processors, but 2 times faster than the Parsytec GC.

On the IBM SP2 and the SUN Enterprise we need at least three processors to get the same performance than one CRAY J90 vector processor. On the SGI Power Challenge two CPUs already are faster than the J90.

The results show that our parallel code has a good efficiency on medium sized shared memory and larger distributed memory machines. But not only the performance is important. Running the programme on 48 fat nodes of the SP2 allows us to use 12 Gbyte memory. Therfore we are able to compute grids with 480^3 grid points, this is more than 8 times larger than on vector machines.

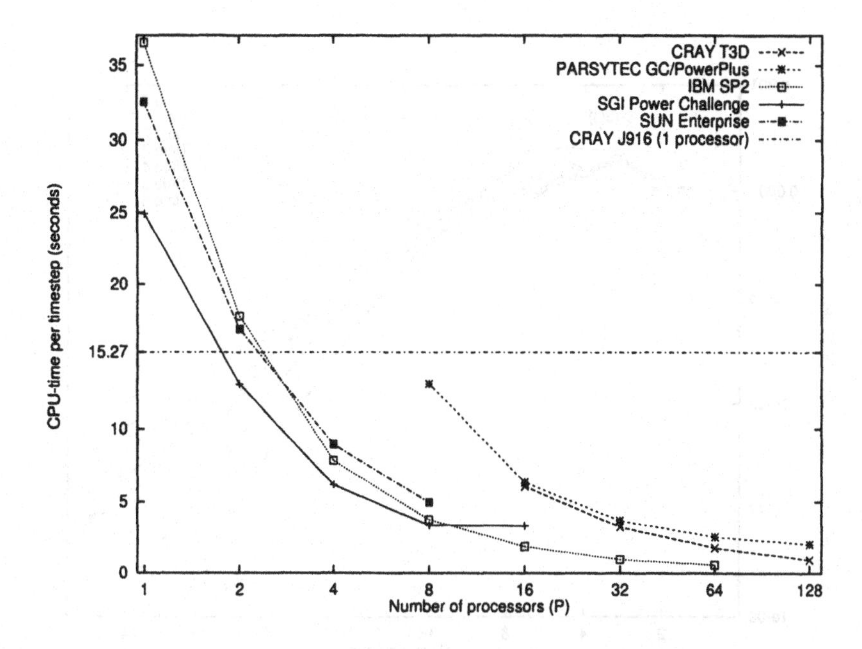

Fig. 3. Runtimes for one time-step on 128^3 grid points

5 Resolving the Inertial Subrange

The Kolmogorov spectrum [6] exhibits the turbulence kinetic energy $E(k)$ in the flow versus the magnitude of the three-dimensional wavenumber vector k. The wavenumber is reciprocal to a length scale such that $E(k)$ describes the energy content of an turbulent eddy of size $\sim 1/k$. This spectrum can be divided into three subranges: production, inertia, and dissipation. In the inertial subrange the flow has universal properties which do not depend on the geometry or other external parameters. The inertial subrange exists if the Reynolds number Re of the flow is large. Then, the spectrum of the inertial subrange obeys the universal formula

$$E(k) = \alpha\, \epsilon^{2/3}\, k^{-5/3}, \tag{7}$$

where α is the Kolmogorov constant of 1.6 ± 0.1.
In a direct numerical simulation, the domain resolution is proportional to Re, $M > Re^{3/4}$, where M^3 is the number of grid cells in a cubic domain. On the SP2 we could perform one run with $M = 480$, achieving a Reynolds number of 600, based on the rms velocity value and the integral length scale of turbulence. By setting the external flow parameters shear and stratification such that $Ri = 0.13$ (see eq. 1), we force the flow to evolve into a stationary state [3] which means that its turbulence kinetic energy is constant in time after an initial phase of adjustment.

a)

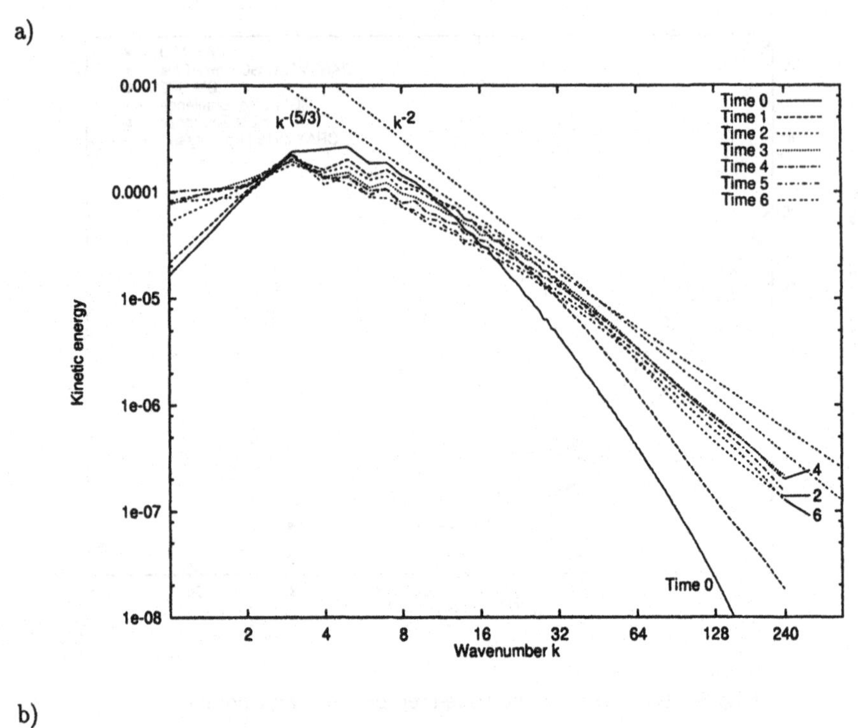

b)

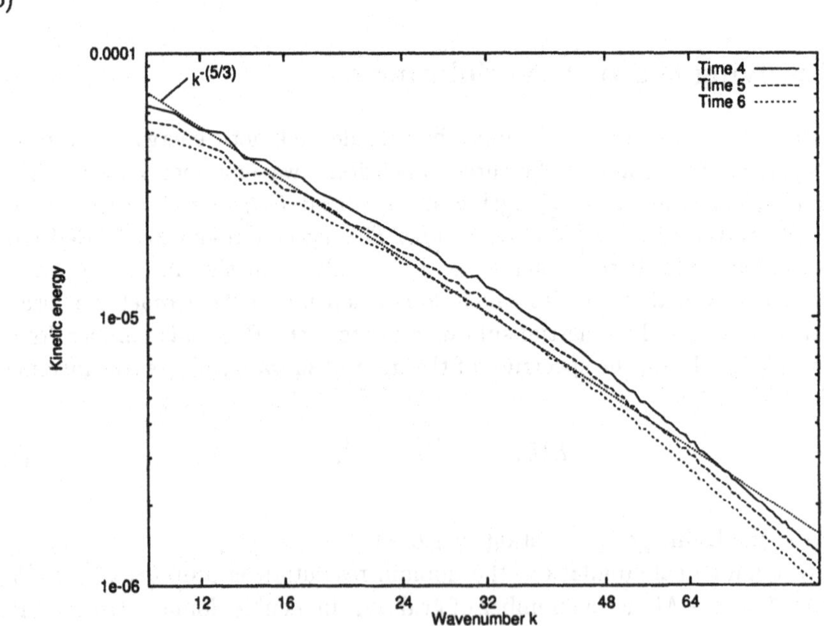

Fig. 4. Spectra of the kinetic energy of the simulation with 480^3 grid points, Reynolds number 600 (based on velocity fluctuation and integral scale):
a) Complete spectra for time 0 to 6;
b) Inertial subrange with spectra parallel to $k^{-(5/3)}$

For that flow, figure 4 displays $E(k)$ as it evolves in time and becomes self-similar and approximately stationary after a normalised time of $t\,dU_0/dz = 2$. The reader may further observe that the flow establishes an inertial subrange for some decades of wavenumber k (fig. 4 b). Since the dissipation rate ϵ of turbulence energy scales like $\epsilon \sim k^2\ E(k)$, a wavenumber dependence of $\epsilon \sim k^{1/3}$ should be observed in the inertial subrange. This is indeed demonstrated in figure 5.

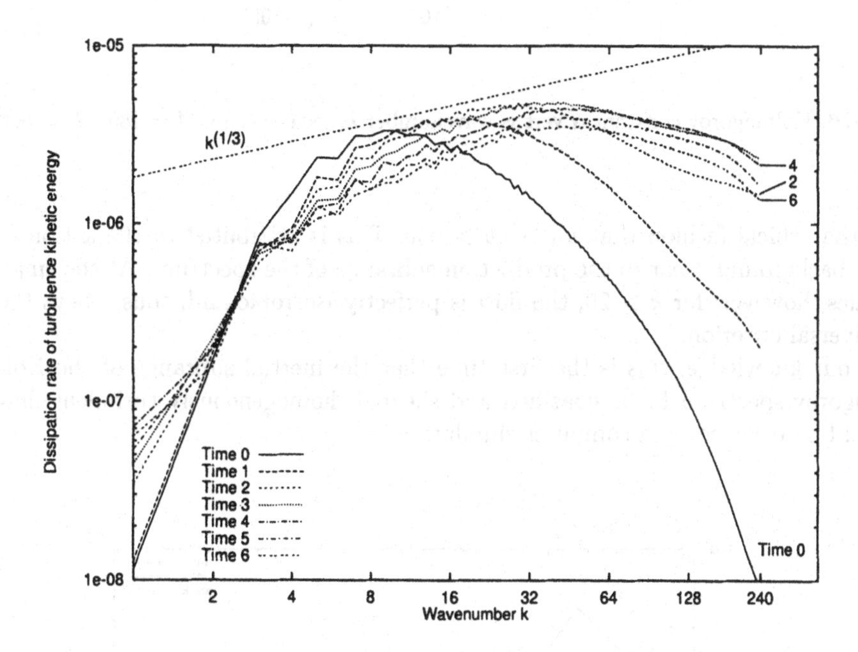

Fig. 5. Spectra of the dissipation rate of turbulence kinetic energy for $M = 480$, $Re = 600$

A more demanding since quantitative evaluation of our flow data in terms of inertial subrange resolution can be performed by plotting the Kolmogorov constant α of eq. (7) versus wavenumber. The result is shown in figure 6.

Now, we can precisely define the range of wave-numbers where the simulated flow behaves according to Kolmogorov's universal theory: For wave-numbers $20 < k < 50$, the Kolmogorov constant of our simulation data lies in the very well established range of $\alpha = 1.6 \pm 0.1$.

A further requirement for an universal turbulent flow behaviour is the flow isotropy at the respective (small) scales. That is that the variances of the three velocity components u, v, and w have the same values at large wave-numbers. Figure 7 elucidates that the simulated sheared and stratified turbulent flow at $Ri = 0.13$ is strongly anisotropic at the largest and energy containing scales

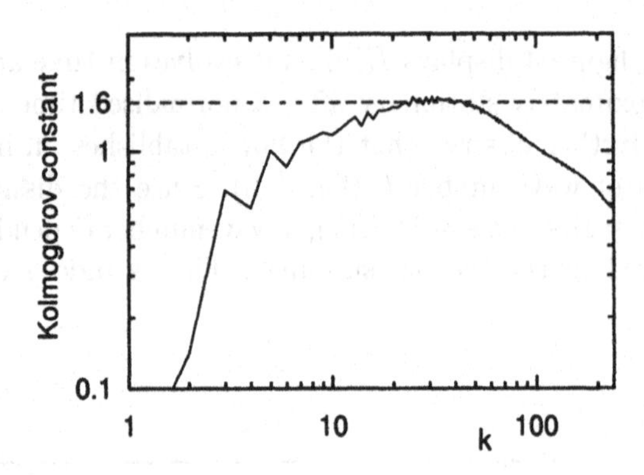

Fig. 6. Kolmogorov constant α versus wavenumber k, see eq. 7, for $M = 480$, $Re = 600$

in the typical fashion that $uu > vv > ww$. This is attributed to the action of the background shear in the production subrange of the spectrum. At the small scales, however, for $k > 20$, the flow is perfectly isotropic and, thus, obeys the universal criterion.

To our knowledge, this is the first time that the inertial subrange of the Kolmogorov spectrum for a stratified and sheared, homogeneously turbulent flow could be resolved by a computer simulation.

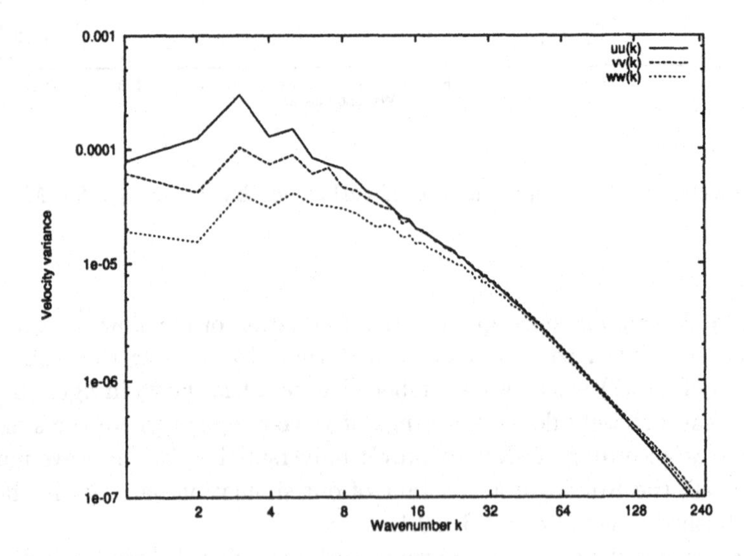

Fig. 7. Velocity variance spectra versus wavenumber for $M = 480$, $Re = 600$

6 Conclusion

A fully parallelised version of an incompressible turbulence simulation code has been presented. The parallel code achieves 2.446 GFlop/s on 64 SP2 processors, this is 26.3 times faster than on one J90 vector processor. Due to the message passing standard MPI we achieved a perfect portability. The developed code is now fitted for the use on state-of-the-art parallel computers.
We have demonstrated that the concept on message passing is well suited for distributed memory machines as well as for shared memory machines. More over, since MPI takes advantage from the fast communication possibilities on shared memory machines with a few processors, these systems can be better suited for parallel computing than some other dedicated message passing machines.
With the parallelised code PARDISTUF we were able to already produce new physical results, i.e. the direct numerical simulation of the inertial subrange of the kinetic energy spectrum of a homogeneously turbulent flow under the influence of background shear and stable thermal stratification. This elucidates the way of how parallel computing may open the door to new fundamental research in fluid dynamics.

References

1. Martin Bücker. Zweidimensionale Schnelle Fourier–Transformation auf massiv parallelen Rechnern. Technical Report Jül-2833, ZAM, Forschungszentrum Jülich, D-52425 Jülich, November 1993.
2. Clare Yung-Lei Chu. *The fast Fourier transform on hypercube parallel computers.* PhD thesis, Cornell University, 1988.
3. Thomas Gerz and Ulrich Schumann. Direct simulation of homogeneous turbulence and gravity waves in sheared and unsheared stratified flows. In Durst et al., editors, *Turbulent Shear Flows 7*, pages 27–45. Springer-Verlag Berlin Heidelberg, 1989.
4. Thomas Gerz, Ulrich Schumann, and S. E. Elghobashi. Direct numerical simulation of stratified homogeneous turbulent shear flows. *J. Fluid Mech.*, 200:563–594, 1989.
5. H.-J. Kaltenbach, Thomas Gerz, and Ulrich Schumann. Large eddy simulation of homogeneous turbulence and diffusion in stably stratified shear flow. *J. Fluid Mech.*, 280:1–40, 1994.
6. A. N. Kolmogorov. The local structure of turbulence in incompressible viscous fluid for very large Reynolds number. *C. R. Acad. Nauk SSSR*, 30:301–303, 1941. Reprinted in Proc. Soc. Lond. A, 434, 9-13(1991).
7. Volker Mehrmann. Divide and conquer methods for block tridiagonal systems. *Parallel Comput.*, 19:257–279, 1992.
8. Message Passing Interface Forum. *MPI: A message-passing interface standard*, June 1995.
9. Ulrich Schumann and Martin Strietzel. Parallel solution of tridiagonal systems for the Poisson equation. *J. Sci. Comput.*, 10(2):181–190, Juni 1995.
10. Paul N. Swarztrauber. Multiprocessor FFTs. *Parallel Comput.*, 5:197–210, 1987.

The Study of a Parallel Algorithm Using the Laminar Backward-Facing Step Flow as a Test Case

P.M. Areal[1] and J.M.L.M. Palma[2]

[1] Instituto Superior de Engenharia do Porto
Rua de São Tomé
4200 Porto, Portugal
pareal@sfc6.fe.up.pt

[2] Faculdade de Engenharia da Universidade do Porto
Rua dos Bragas
4099 Porto Codex, Portugal
jpalma@fe.up.pt

Abstract. The current study discusses the results of parallelisation of a computer program based on the SIMPLE algorithm and applied to the prediction of the laminar backward-facing step flow. The domain of integration has been split into subdomains, as if the flow were made up of physically distinct domains of integration.
The convergence characteristics of the parallel algorithm have been studied as a function of grid size, number of subdomains and flow Reynolds number. The results showed that the difficulties of convergence increase with the complexity of the flow, as the Reynolds number increases and extra recirculation regions appear.

1 Introduction

Parallel computing has become more and more common, and has developed from a subject of specialised scientific meetings and journals into an affordable technology, to a point where we feel that no references are needed to support this statement.

Fluid flow algorithms are complex, given the number of equations involved, the elliptic nature, non-linearity and peculiarities of the pressure-velocity coupling for incompressible flows. All these features are familiar to those in the fluid dynamics community and make this a formidable set of equations intractable by theoretical approaches. The parallelisation makes things even worse and our option was to study parallel fluid algorithm through applications to a series of flow geometries and conditions.

In the present study we discuss the numerical behaviour of a parallel version of the SIMPLE algorithm [15]. We show how the convergence was influenced by the Reynolds number, grid size, number of subdomains and overlapping between the subdomains. A previous work [3] has been followed and new findings and conclusions were added as a result of a new set of calculations using different flow geometry and conditions, i.e. the laminar backward-facing step flow.

J. Palma, J. Dongarra, and V. Hernández (Eds.): VECPAR'98, LNCS 1573, pp. 238–249, 1999.

2 Mathematical Model and Strategy of Parallelisation

The fluid flow equations, assuming steady-state, incompressible and Newtonian fluid, were solved on a Cartesian coordinate system,

$$\frac{\partial u_i}{\partial x_i} = 0 \tag{1}$$

$$\rho \frac{\partial u_i u_j}{\partial x_j} = -\frac{\partial p}{\partial x_i} + \frac{\partial}{\partial x_j}\left[\mu\left(\frac{\partial u_i}{\partial x_j} + \frac{\partial u_j}{\partial x_i}\right)\right] \tag{2}$$

where u_i is the velocity along the direction x_i, p is the pressure, and ρ and μ are the density and fluid dynamic viscosity, respectively.

These equations were discretized in a numerical grid, with all variables being defined at the same location (the collocated grid [16]) and following the finite volume approach [7]. The hybrid and central finite differencing schemes were used for discretisation of the convective and diffusive terms, respectively.

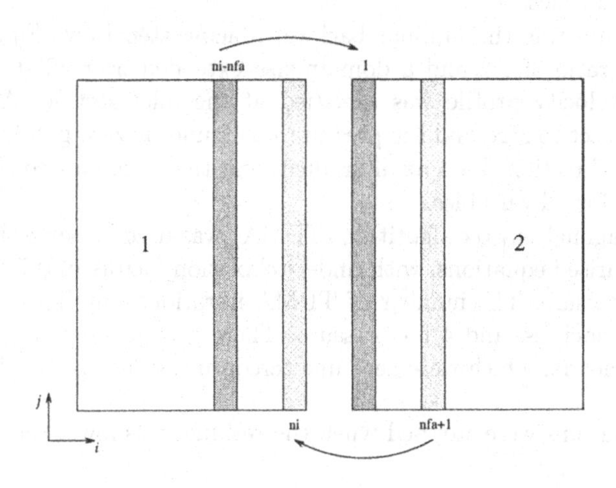

Fig. 1. Overlapping and data exchange between subdomains

The algorithm entitled SIMPLE [15], was used. An approach where, after rewriting the continuity equation (1) as a function of a pressure correction variable, mass and momentum conservation are alternately enforced. Equations (1) and (2) are solved as if they were independent (or segregated) systems of equations, until a prescribed criterion of convergence can be satisfied.

For parallelisation of the algorithm, the integration domain was split along the horizontal direction into a variable number of subdomains, with overlapping (Figure 1). This approach is identical to the well known Schwarz's method (e.g. [19]). Within each subdomain, the SIMPLE algorithm was used, followed by exchange of data (pressure gradient, and both u and v velocities) between subdomains. This was one iteration; equivalent to one iteration of the sequential version of the SIMPLE algorithm, comprising the full domain.

The simplest case, with 2 subdomains only, will be used as an example (Figure 1). The first column of subdomain 2 (the west boundary condition) was taken as one of the columns interior to subdomain 1. The last column of subdomain 1 (the east boundary condition) was taken as one of the columns interior to subdomain 2. The level of overlapping between the subdomains (`nfa`) depended on where inside the subdomains the interior columns were located. The amount of transferred data did not depend on the overlapping: that was a function of the number of grid nodes along the vertical only. The nodes within the overlapping region are calculated twice (in case of 2 subdomains); however, a minimum level of overlapping is required to secure convergence. There is here a compromise which is also a drawback of the Schwarz's method [19].

Other strategies of parallelisation have been suggested involving parallelisation of the equation solver only, either in case of a segregated approach (e.g.: [12]) or in cases where all the fluid equations are solved simultaneously as part of a single system (e.g.: [6]). Either of these two alternatives, when compared with the present parallelisation strategy, is more robust at a cost of increased communication times.

The test case was the laminar backward-facing step flow (Figure 2), with an expansion ratio of 1:2 and a domain size extended over 30 step height, h. A parabolic velocity profile was specified at the inlet section. At the walls, velocities were set to zero and the pressure was found by zero gradient along the perpendicular direction. Zero axial gradient was the boundary condition at the outlet section for all variables.

The tri-diagonal matrix algorithm (TDMA) was used to solve the system of algebraic linearised equations, with under-relaxation factors of 0.7 for velocities and 0.3 for pressure. The number of TDMA iterations was fixed and set at 2 for u and v velocities, and 4 for pressure. There was no previous optimisation of these parameters, which were kept unaltered during the course of the current study.

The calculations were stopped when the residual was lower than 10^{-4}.

3 Results

Results were obtained for numerical grids ranging from 100×64 up to 250×160, and Reynolds number between 10^2 and 10^3 in steps of 10^2. The Reynolds number was defined by $Re=\rho U 2h/\mu$, where U is the average axial velocity at the inlet section. The flow regime remains laminar for Reynolds number lower than 1200 (cf. [2]).

The calculations were performed on a shared memory computer architecture with a 1.2 G_2B/s system bus bandwidth (SGI Power Challenge, with 4 processors R8000/75 MHz and a total of 512 M_2B of RAM; operating system IRIX 6.1 and fortran compiler MIPSPro Power Fortran77 6.1). The communication between processors was performed via version 3.3 of PVM message passing protocol [8].

The available hardware restricted our studies of parallelisation to a maximum of 4 subdomains.

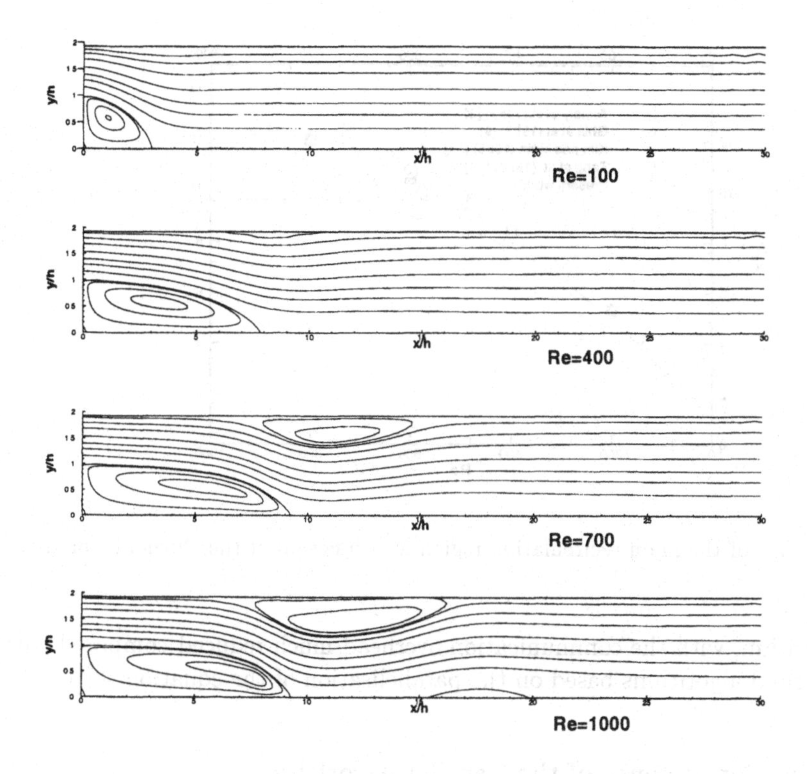

Fig. 2. Streamline pattern as a function of the Reynolds number

3.1 The Flow Pattern

For a prior assessment of the computer program, the streamline pattern (Figure 2) was analysed. For all Reynolds number being tested, downstream of the step, there was a main recirculation region, whose length increased with the Reynolds number (Figure 3). The results follow the trend as observed in experiments (cf. [2], [17]) and previous calculations (eg. [11], [20], [4]). The deviation from the experimental curve at *Re*=500 has been attributed to three-dimensional effects, not accounted for by our calculations.

For a Reynolds number around 400, attached to the top wall of the channel, a second recirculation appears (Figure 2), reaching a maximum length of about $10h$ at *Re*=1000 (the highest Reynolds number being used). This is a flow pattern that has been observed experimentally (cf. [2], [17]) and is shown here to confirm the quality of our calculations.

Around 50% of the computing time was spent within the routine for solution of the linear system of equations. The assembling of the coefficients of the three differential equations of pressure, and u and v velocities required 48%. The communication time was estimated to be 2% of the total computing time. This was a consequence of the computer architecture, but also a consequence of

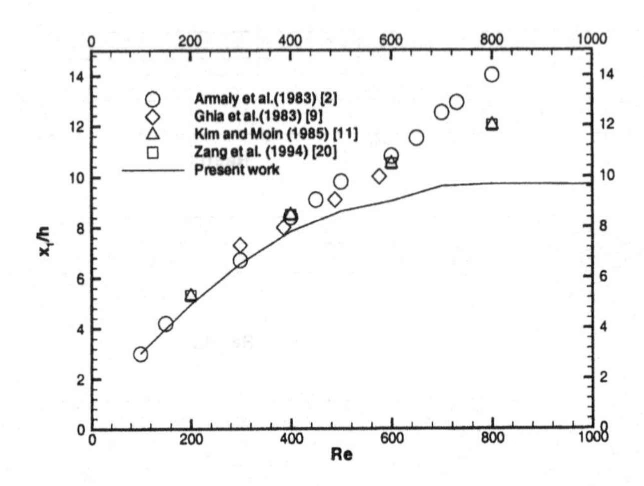

Fig. 3. Size of the main recirculation region as a function of the Reynolds number

the algorithm, with the communication overhead much reduced compared with parallel fluid algorithms based on the parallelisation of the equation solver.

3.2 The Convergence of the Parallel Algorithm

We were interested in studying the effect of parallelisation on the convergence of the algorithm and Figure 4 shows the number of iterations as a function of Reynolds number for four grid sizes. In case of the parallel version the results have been plotted in terms of global iteration and the equivalent grid size of the sequential version. There are two regions in this Figure.

In region I the number of iterations decreased to a minimum, obtained at Re=400 for grids 200×128 and 250×160, and Re=500 for grids 100×64 and 150×96. This is related with the recirculation region attached to the top wall and can be confirmed by joint observation of Figures 4 and 2.

For Reynolds number higher than 400 (or 500 for grids 100×64 and 150×96), in region II, the number of iterations increased with the Reynolds number. The convergence becomes more difficult as the recirculation in the top wall increases (see also Figure 2). The convergence is tightly coupled with the flow pattern.

There was a minimum value of the Reynolds number (region I), depending on the grid size, for which the residual did not fall below 10^{-4} and the convergence criterion was not satisfied. For instance, for Re=100 after 9000, 6000 and 3550 iterations, for grids 250×160, 200×128 and 150×96, the residual became constant at 3.28×10^{-4}, 1.97×10^{-4} and 1.12×10^{-4}, respectively. For coarser grids, the constant residual was lower and could be reached at a lower number of iterations. However, the flow pattern was as expected and we concluded that the criterion of convergence was too restrictive. Nevertheless, and for consistency with the

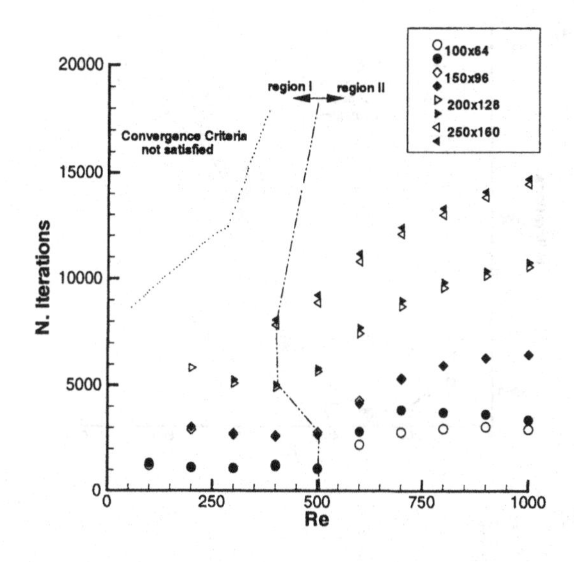

Fig. 4. Number of iterations as a function of the Reynolds number and grid size (filled symbols: parallel version (2 subdomains); open symbols: sequential version)

remaining cases the criterion of convergence was not changed and we considered these as non-converged solutions. To some extent, this is related with the numerical technique used to solve the linear system of equations. For instance, calculations performed using the SIP (*Strong Implicit Procedure* [18]), solver were able to satisfy the convergence criterion for grid 150×96, *Re*=100.

The results of the parallel version with 2 subdomains (filled symbols) are also included in Figure 4. In most of the cases, the number of iterations (global iterations) of the parallel was identical to the number of iterations of the sequential version. However, there were cases where the parallel version converged faster (e.g. grid 150×96 and *Re*=600), whereas in the most unfavourable situation (grid 100×64 and *Re*=700), the parallel required 38% more iterations than the sequential version.

The convergence characteristics of grids 100×64 and 200×128 are shown in Figure 5 as a function of the Reynolds number. In case of grid 100×64 (Figure 5a) for *Re* lower than 500 the sequential requires more iterations to converge than the parallel version, whereas for *Re* higher than 500 the situation is reversed. In general, and for *Re* higher than 500 the number of iterations increases with the number of subdomains. In case of *Re*=1000 the calculation with 3 subdomains requires 3100 iterations compared with 2800 of the sequential version.

The behaviour for grid 200×128 (Figure 5b), apart from also an increased number of iterations with the Reynolds number (as in the case of grid 100×64), evidences other details worth referring to. For instance, convergence could not be achieved for Reynolds number lower than 300, even in case of the sequential version. There was no optimization of the under-relaxation factors. This exer-

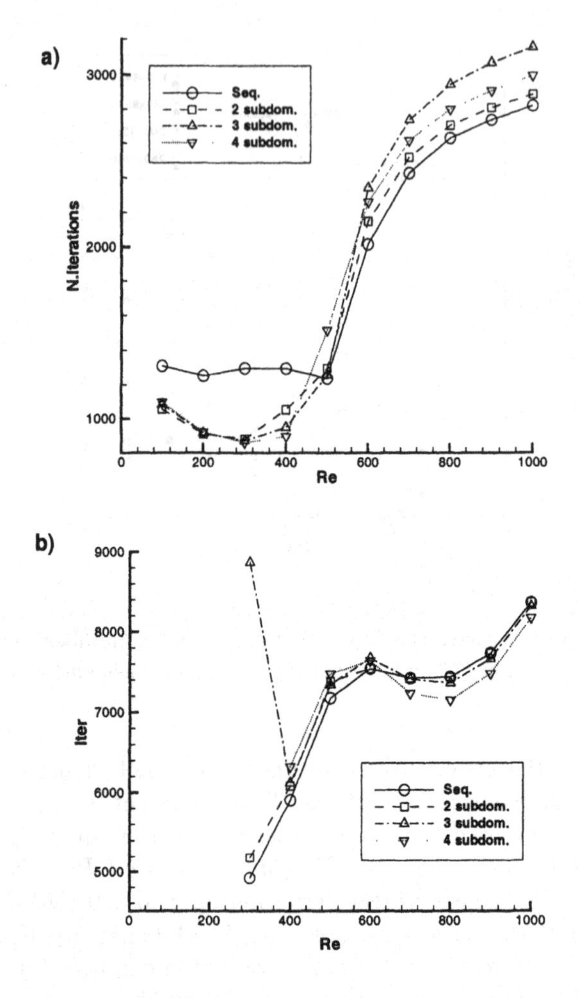

Fig. 5. Number of iterations as a function of Reynolds number and number of subdomains for two numerical grids. a) 100×64; b) 200×128.

cise was beyond the scope of the present work. It is our experience [13] that the SIMPLE algorithm requires lower under-relaxation factors for finer grids. The plateau shown for 500 $<Re<$800 by grid 200×128 may be associated with the occurrence of the recirculation region at the bottom wall around x/h=17. A feature to which the coarse grid could not be sensitive because of lack of resolution. This is a statement that we found difficult to confirm.

Figure 5 shows that the number of subdomains did not alter the general pattern of the convergence characteristics compared with the sequential version. This is valid throughout the current study and led us to the conclusion that the domain splitting, even when the interface between the subdomains divides the recirculation region, does not affect the convergence.

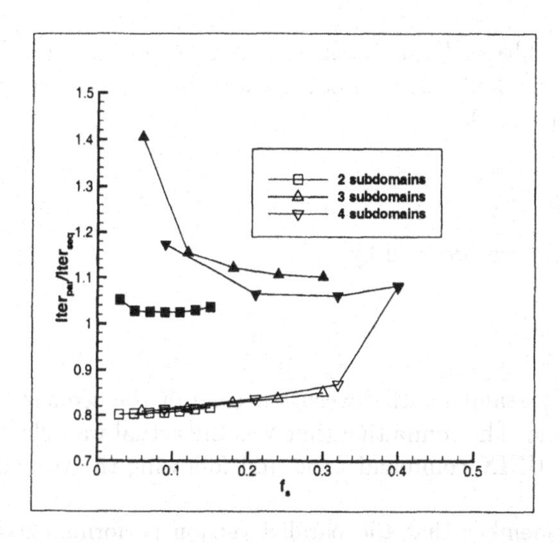

Fig. 6. Number of iterations as a function of subdomain overlapping and number of subdomains for grid 100×64 and two Reynolds number. *Re*=100 (empty symbols); *Re*=1000 (filled symbols)

3.3 Domain Overlapping

The effect of overlapping between subdomains was studied [1] for *Re*=100, 500 and 1000 on the grid 100×64. Results are shown here (Figure 6) for *Re*=100 and 1000 as a function of f_s, defined as the ratio between the number of nodes in the overlapping regions and the total number of grid nodes. The parameter f_s was the measure of domain overlapping; here we preferred f_s as an alternative to nfa (in Figure 1), because it did not depend on the grid size.

The number of nodes was identical for each subdomain, to avoid load balancing problems. This, however, restricted the overlapping (i.e. f_s) to a discrete set of values, depending on the grid nodes and on the number of subdomains.

The results at *Re*=100 (Figure 6, empty symbols) are markedly different from those at *Re*=1000 (Figure 6, filled symbols). At *Re*=100 the parallel requires less iterations than the sequential version and is not sensitive to the number of subdomains. The number of iterations increases slightly with the overlapping, with the exception of f_s=0.4.

At *Re*=1000 for each case, with 2, 3 or 4 subdomains, there is a value of f_s beyond which there is no reduction of the number of iterations. Furthermore, for all cases, the number of iterations of the parallel exceeds the number of iterations of the sequential version.

These results reinforce the point that we made earlier, that the flow pattern influences the convergence pattern. A stronger coupling between the subdomains, a larger f_s, is required if the interfaces between the subdomains are crossing recirculation regions, which does not occur in case of *Re*=100.

Of the two results in Figure 6, those corresponding to Re=1000 are typical of the Schwarz's method (see section 2), where the convergence improves as the overlapping is increased.

3.4 Speed-up

The speed-up (S_n) was defined by

$$S_n = \frac{T_S}{T_p} \tag{3}$$

where T_S and T_p stand for the execution time of the sequential and parallel version of the code. The computing time was the actual wall-clock elapsed time, as given by the UNIX command `time` and following the recommendations of ref. [10].

One must remember that the parallel version performs extra calculations, because the grid nodes within the overlapping region are calculated twice. Taking that into account, the expected speed-up values are 1.8, 2.5 and 3.2, for 2, 3 and 4 subdomains, respectively. These values are shown by horizontal lines in Figure 7 and were obtained assuming identical number of iterations for both sequential and parallel calculations, time per iteration directly proportional to the number of nodes and negligible communication overhead.

Figure 7a shows that for grid 100×64 and Reynolds number lower than 500 the speed-up exceeded the expected value. The domain splitting was favourable to the convergence of the algorithm, and the number of iterations was reduced compared with the sequential version. For Reynolds numbers higher than 500, the speed-up was below the ideal value, in particular for cases with 3 and 4 subdomains. Nevertheless in real time, in case of 4 subdomains a converged solution can be obtained in at least 2.6× faster compared with a sequential calculation.

Figure 7b shows a fairly uniform trend and all values were close to the maximum speed-up.

For each set of calculations (number of subdomains) in Figure 7 we used only one value of f_s. This value, which may not be constant, was chosen (based on the results in Figure 6) in order to keep both the overlapping (f_s) and the number of iterations to the minimum.

4 Conclusions

Results were shown of the simulation of the laminar backward-facing step flow. A parallelised version of the SIMPLE algorithm was used, based on the partitioning of the domain. The main conclusions were:

1. The minimum number of iterations was obtained for Reynolds number about 400 or 500, depending on the grid size. The convergence pattern was tightly coupled with the flow pattern, and the number of iterations increased as

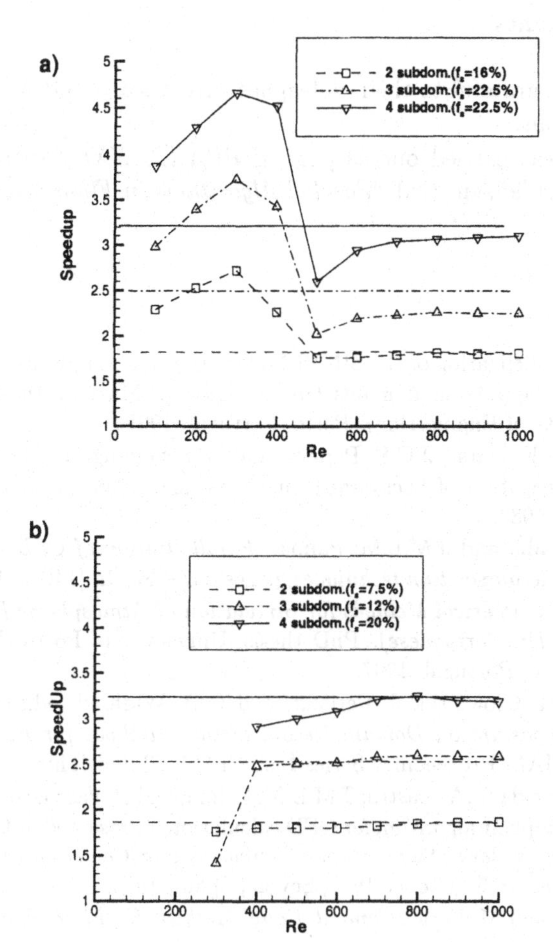

Fig. 7. Speed-up as a function of the Reynolds number, in case of 2, 3 and 4 subdomains for two numerical grids. a) 100×64; b) 200×128.

soon as a second recirculation region attached to the top wall appeared. The consistency observed throughout our series of tests led us to believe that this conclusion may be extended to other cases, i.e different flows and geometries.

2. The comparison between the sequential and the parallel versions of the algorithm showed that the number of iterations was in many cases identical in both versions. The communication overhead was 2% of the total computing time.
3. The number of subdomains did not alter the general pattern of the convergence characteristics compared with the sequential version.
4. The convergence was not affected by domain splitting, even when the interface between the subdomains divided the recirculation region.

Acknowledgments

The authors are grateful to their colleagues F.A. Castro and A. Silva Lopes for helpful discussions.

This work was carried out as part of *PRAXIS XXI* research contract N. 3/3.1/CEG/2511/95, entitled *"Parallel Algorithms in Fluid Mechanics"*.

References

1. P. Areal. Parallelization of the SIMPLE algorithm for computational fluid dynamics using overlapping subdomains (*in portuguese*). Master's thesis, University of Porto (Faculty of Engineering), Porto, Portugal, 1998.
2. B.F. Armaly, F. Durst, J.C.F. Pereira, and B. Schönung. Experimental and theoretical investigation of backward-facing step flow. *Journal of Fluid Mechanics*, 127:473–496, 1983.
3. L.M.R. Carvalho and J.M.L.M. Palma. *Parallelization of CFD code using PVM and domain decomposition techniques*, pages 247–257. In [14], 1997.
4. F.A. Castro. *Numerical Methods for Simulation of Atmospheric Flows over Complex Terrain (In Portuguese).* PhD thesis, University of Porto (Faculty of Engineering), Porto, Portugal, 1997.
5. T.F. Chan, R. Glowinski, J. Periaux, and O.B. Widlund, editors. *Third International Symposium on Domain Decomposition Methods for Partial Differential Equations.* SIAM Proceedings Series List. SIAM, Philadelphia, 1989.
6. F. Dias d'Almeida, F.A. Castro, J.M.L.M. Palma, and P. Vasconcelos. Development of a parallel implicit algorithm for CFD calculations. Presented at the AGARD *77th Fluid Dynamics Panel Symposium on Progress and Challenges in CFD Methods and Algorithms*, 2–5 October 1995, Seville, Spain, 1995.
7. J.H. Ferziger and M. Perić. *Computational Methods for Fluid Dynamics.* Springer, Berlin, 1996.
8. G.A. Geist, A. Beguelin, J.J. Dongarra, W. Jiang, R. Manchek, and V.S. Sunderam. *PVM Parallel Virtual Machine. A Users' Guide and Tutorial for Networked Parallel Computing.* Scientific and Engineering Computation. The MIT Press, Cambridge, Massachusetts, 1994.
9. N.N. Ghia, G.A. Osswald, and U. Ghia. A direct method for the solution of unsteady two-dimensional incompressible Navier-Stokes equations. *Aerodynamic Flows*, January.
10. R.W. Hockney. *The Science of Computer Benchmarking.* Software, Environments and Tools. SIAM – Society for Industrial and Applied Mathematics, Philadelphia, USA, 1996.
11. J. Kim and P. Moin. Application of a fractional step method to incompressible Navier-Stokes equations. *Journal of Computational Physics*, 59:308, 1985.
12. M. Kurreck and S. Wittig. A comparative study of pressure correction and block-implicit finite volume algorithms on parallel computers. *International Journal for Numerical Methods in Fluids*, 24:1111–1128, 1997.
13. J.J. McGuirk and J.M.L.M. Palma. The efficiency of alternative pressure-correction formulations for incompressible turbulent flow problems. *Computers & Fluids*, 22(1):77–87, 1993.

14. J.M.L.M. Palma and J. Dongarra, editors. *Vector and Parallel Processing – VECPAR'96. Second International Conference on Vector and Parallel Processing – Systems and Applications, Porto (Portugal). Selected Papers*, volume 1215 of *Lecture Notes in Computer Science*. Springer, Berlin, 1997.
15. S.V. Patankar and D.B. Spalding. A calculation procedure for heat, mass and momentum transfer in three-dimensional parabolic flows. *International Journal of Heat and Mass Transfer*, 15:1787–1806, 1972.
16. C.M. Rhie and W.L. Chow. Numerical study of the turbulent flow past an airfoil with trailing edge separation. *AIAA Journal*, 21(11):1525–1532, 1983.
17. R.L. Simpson. Two-dimensional turbulent separated flows. Technical report, AGARDograph N. 287, 1985.
18. H.L. Stone. Iterative solution of implicit approximations of multidimensional partial differential equations. *SIAM Journal of Numerical Analysis*, 5:530–558, 1968.
19. O.B. Widlund. *Domain decomposition algorithms and the bicentennial of the french revolution*, pages XV–XX. In [5], 1989.
20. Y. Zang, R.L. Street, and J.R. Koseff. A non-staggered grid, fractional step method for time-dependent incompressible Navier-Stokes equations in curvilinear coordinates. *Journal of Computational Physics*, 114:18–33, 1994.

A Low Cost Distributed System for FEM Parallel Structural Analysis

Célio Oda Moretti[1], Túlio Nogueira Bittencourt[1], and Luiz Fernando Martha[2]

[1] Computational Mechanics Laboratory, Department of Structural and Foundation Engineering, Polytechnic School, University of São Paulo,
Av. Prof. Almeida Prado, travessa 2, no. 271,
CEP 05508-900 - São Paulo - Brazil
{moretti, tbitten}@usp.br
http://www.lmc.ep.usp.br

[2] Department of Civil Engineering and Technology Group on Computer Graphics - TeCGraf, Pontifical Catholic University of Rio de Janeiro - PUC-Rio,
Rua Marquês de São Vicente, no. 225,
CEP 22453-900 - Rio de Janeiro - Brazil
lfm@tecgraf.puc-rio.br
http://www.tecgraf.puc-rio.br

Abstract. In this paper, a distributed computational system for finite element structural analysis and some strategies for improving its efficiency are described. The system consists of a set of programs that performs the structural analysis in a distributed computer network environment. This set is composed by a pre-processor, a post-processor, a program responsible for partitioning the model in substructures, and by a structural analysis parallel solver. The domain partitioning is performed interactively by the user through a graphics interface program (PARTDOM). An existing FEM code, based on object oriented programming concepts (FEMOOP), has been adapted to implement the parallel features. Different implementation aspects concerning scalability and performance speed-up are discussed. The computational environment consists of a 100 Mbit Fast-Ethernet network cluster including eight Pentium 200 MHz micro-computers running under LINUX operating system.

1 Introduction

In this work, a distributed computational system for finite element structural analysis will be presented. This system has been developed to perform the parallel analysis using a disposable local area network as the processing environment. This arrangement makes possible the use of low cost parallel capabilities, avoiding the utilisation of expensive supercomputers. A set of integrated components, each one responsible for a specific task in the analysis process, comprises the computational system. The four basic units of this system are presented in Fig. 1.

This set has four components: a pre-processor, a post-processor, a program responsible for partitioning the model in substructures, and a structural analysis

J. Palma, J. Dongarra, and V. Hernández (Eds.): VECPAR'98, LNCS 1573, pp. 250–262, 1999.

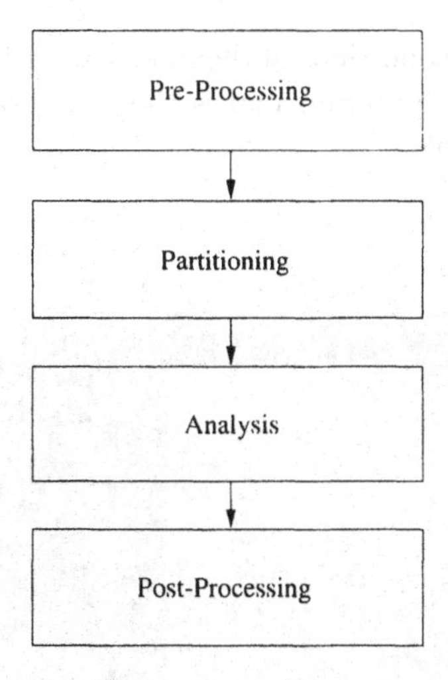

Fig. 1. Basic units of the parallel analysis system

parallel solver. The main focus here is the domain partitioning and the solver, in which the parallel features have been implemented. The domain partitioning is performed interactively by the user through a graphic interface program, called PARTDOM. Three automatic substructuring procedures are available and the user is able to manually edit the resulting sub-domain partitions. An existing FEM code, based on object oriented programming concepts (FEMOOP), has been adapted to implement the parallel features. The main extensions to the code are the introduction of a preconditioned conjugate-gradient parallel algorithm to solve the global equations and the communication procedures among the processors.

In the following sections, the parallel analysis system and its components will be described in detail. Also, different aspects affecting the system performance will be discussed. The performance is measured through the time consumed in each step of the analysis processing for different number of processors. Some strategies for improving the system efficiency have been implemented and their results will be discussed.

2 The Parallel Analysis System

All components of this parallel system are presented in this section. Special attention is placed on the domain partitioning and the analysis programs. The main development and implementations are located in these programs.

2.1 Pre-Processing

The package MTOOL (Bidimensional Mesh Tool) [1] has been used as the pre-processor in this work. MTOOL is an interactive graphics program for bidimensional finite element mesh generation. With this program, the geometry, material properties, and boundary conditions of the model are defined. Through its graphical interface (Fig. 2), the program makes the visualisation and the editing of the generated mesh possible.

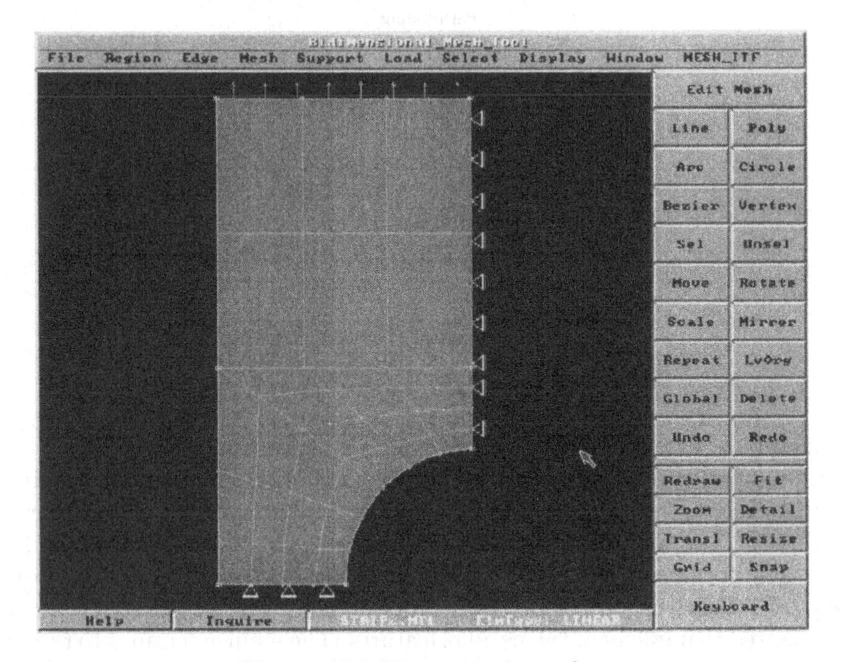

Fig. 2. MTOOL graphical interface

At the end of model creation, a neutral format file is generated. This file contains, in a standard way, all the information about the model, which are necessary in the analysis process. The neutral file is an essential feature for the integration of all components of the parallel system.

2.2 Partitioning

After the model generation, the structure has to be partitioned into a number of substructures to take advantage of the parallel environment. This partitioning is made through the use of automatic domain partitioning algorithms. The main objectives of these algorithms are: (1) to obtain a balanced work distribution among the processors, and (2) to minimise the boundary degrees of freedom (DOF) present between the substructures. These two objectives represent important aspects that affect the overall system performance.

The basic aim of PARTDOM program[2] is to facilitate the user interaction with the automatic domain partitioning algorithms. The program allows the partition of a bidimensional finite element mesh using three different algorithms. Through its graphical interface (Fig. 3), the program permits the visualisation and manipulation of the resulting sub-domain partitions. Thus, PARTDOM has been developed with the following objectives:

- to promote user interaction with the automatic domain partitioning algorithms,
- to facilitate partitioning of complex meshes, through a graphical interface that allows visualisation of resulting partitions,
- to allow interactive user edition of resulting partitions,
- to be portable, i.e., the program code is platform independent.

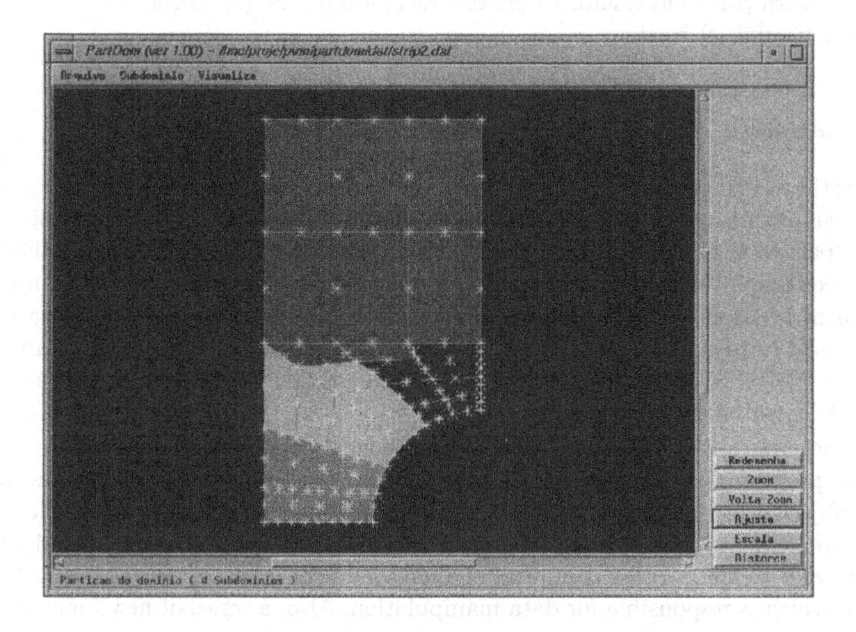

Fig. 3. PARTDOM graphical interface

The three partitioning algorithms used in PARTDOM are: Al-Nasra & Nguyen algorithm [3] and two algorithms from the METIS library [4] called *pmetis* and *kmetis*. Fig. 4 presents the resulting partitions applying the three algorithms over a finite element mesh composed by 587 T6 elements.

The three algorithms present a good balanced work distribution among the processors, with approximately the same number of elements for each substructure. However, the Al-Nasra & Nguyen algorithm generally presents a number of boundary DOF between the substructures greater than the *pmetis* and *kmetis* algorithms. This characteristic increases the requirement of communication between the processors, and, consequently, decreases the parallel analysis performance. The partitions resulting from the use of *pmetis* and *kmetis* algorithms present approximately the same number of boundary DOF.

At the end of this process, the partitioning information is added to the neutral file.

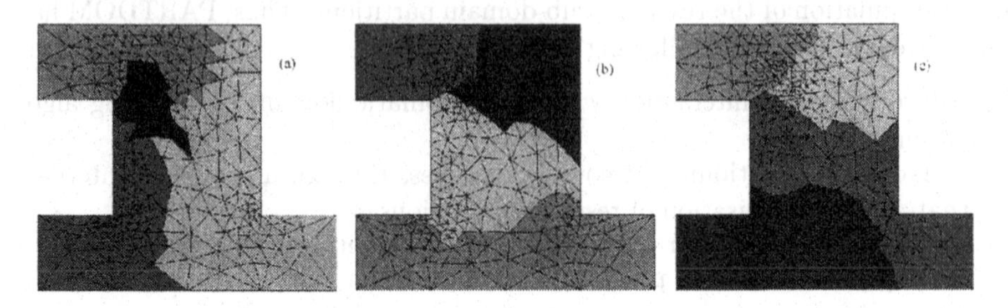

Fig. 4. Mesh partitions resulting from the use of algorithms: (a) Al-Nasra & Nguyen, (b) *pmetis*, and (c) *kmetis*

2.3 Analysis

After the model generation and partitioning, the next step consists of the parallel analysis of this model, with the use of a set of processors, each of one responsible for a part of the computational work. In this work, the Finite Element Method has been employed in the analysis. A analysis program named FEMOOP (Finite Element Method - Object-oriented Programming)[5], developed at the Department of Civil Engineering (PUC-Rio) and at Computational Mechanics Laboratory (Polytechnic School / USP), has been used as the platform for the new parallel capabilities additions.

The program FEMOOP is organized using object-oriented concepts of the C++ programming language[6][7]. One of the most important advantages of the object-oriented programming is the code extensibility. This feature allows new implementations with minimum impact over the existent code. To adapt FEMOOP to the parallel computational environment a new class has been created, which is responsible for data manipulation. Also, a series of new functions has been implemented into existent classes.

The message-passing manager used in this work has been PVM (Parallel Virtual Machine)[8]. PVM is a software system that permits a heterogeneous computer network to be viewed as a single parallel computer. The first step necessary to adapt FEMOOP to the parallel environment has been the implementation of a library responsible for the message passing management. The main objective of this library is to limit the direct access to PVM functions. An eventual change of the message-passing manager is facilitated, which has impact only over the library code. This parallel procedure library contains all functions necessary to perform the message passing in a distributed memory environment. The main functions implemented here are responsible for sending and receiving messages among processors, for parallel process initialisation, and for the identification of program type (if it is either a master or a task program).

The parallel programming paradigm adopted here has been the master-slave model. In this model, the master is a separate program responsible for process spawning, initialisation, reception and display of results, and timing of functions. The task (or slave) programs are executed concurrently and interact through message passing. The actual structural analysis is done by the task programs, each of one responsible for the work corresponding to one substructure. Through interaction between these task programs, the global solution is obtained and then it is sent to the master program.

To obtain the linear system of equilibrium equations a sub-structuring technique[9] has been employed. When this technique is used, the original structure is partitioned into a number of substructures, which are distributed among the processors. The substructure degrees of freedom are classified as internal DOF and boundary DOF, which are shared between neighbour substructures. The substructures interact through these boundary DOF only. Then, the stiffness matrices of each substructure are mounted. In this work, the internal unknowns are eliminated using Crout method. After this step, a condensed system with terms corresponding only to boundary unknowns is obtained. All these procedures can be performed concurrently. To solve the partitioned global system, a parallel iterative solver has been used. A parallel implementation of the preconditioned conjugate gradient (PCG) method[10][9] has been chosen as the solution method adopted in this work. Basically, this parallel implementation of the PCG method consists of parallel operations between matrices and vectors. The sequence of operations is the same both in the parallel and in the sequential versions of the PCG method.

To employ this sub-structuring technique, the stiffness matrix $K^{(i)}$, the force vector $f^{(i)}$, and the nodal unknowns $u^{(i)}$, corresponding to each substructure i, have been mounted with terms partitioned into internal and boundary terms. These partitions are presented in Equation 1, where indices I and S correspond respectively to internal and boundary (or shared) terms.

$$K^{(i)} = \begin{bmatrix} K_I^{(i)} & K_{IS}^{(i)} \\ K_{IS}^{(i)^T} & K_S^{(i)} \end{bmatrix}, f^{(i)} = \begin{bmatrix} f_I^{(i)} \\ f_S^{(i)} \end{bmatrix}, u^{(i)} = \begin{bmatrix} u_I^{(i)} \\ u_S^{(i)} \end{bmatrix}. \tag{1}$$

For each substructure, the linear equation system is written in the form

$$\begin{bmatrix} K_I^{(i)} & K_{IS}^{(i)} \\ K_{IS}^{(i)^T} & K_S^{(i)} \end{bmatrix} \begin{bmatrix} u_I^{(i)} \\ u_S^{(i)} \end{bmatrix} = \begin{bmatrix} f_I^{(i)} \\ f_S^{(i)} \end{bmatrix}. \tag{2}$$

Eliminating the internal unknowns, a condensed equation system is obtained

$$\overline{K}_S^{(i)} u_S^{(i)} = \overline{f}_S^{(i)}, \tag{3}$$

where $\overline{K}_S^{(i)}$ and $\overline{f}_S^{(i)}$ are respectively the condensed stiffness matrix and the condensed force vector, and

$$\overline{K}_S^{(i)} = K_S^{(i)} - K_{IS}^{(i)^T} K_I^{(i)^{-1}} K_{IS}^{(i)}, \tag{4}$$

$$\overline{f}_S^{(i)} = f_S^{(i)} - K_{IS}^{(i)^T} K_I^{(i)^{-1}} f_I^{(i)}. \tag{5}$$

The global linear equation system can be written in the form

$$\left[\sum_{i=1}^{p} L^{(i)^T} \overline{K}_S^{(i)} L^{(i)}\right] = \sum_{i=1}^{p} L^{(i)^T} \overline{f}^{(i)}, \tag{6}$$

where p is the number of substructures and $L^{(i)}$ is a boolean matrix that describes the substructure connectivity in the original structure.

2.4 Post-Processing

A program named MVIEW (Bidimensional Mesh View)[11] has been utilised as the post-processor tool. MVIEW is an interactive graphical program (Fig. 5) for visualisation of structural analysis results. This program provides visualisation of the deformed configuration and contours of scalar results at nodes and Gauss points.

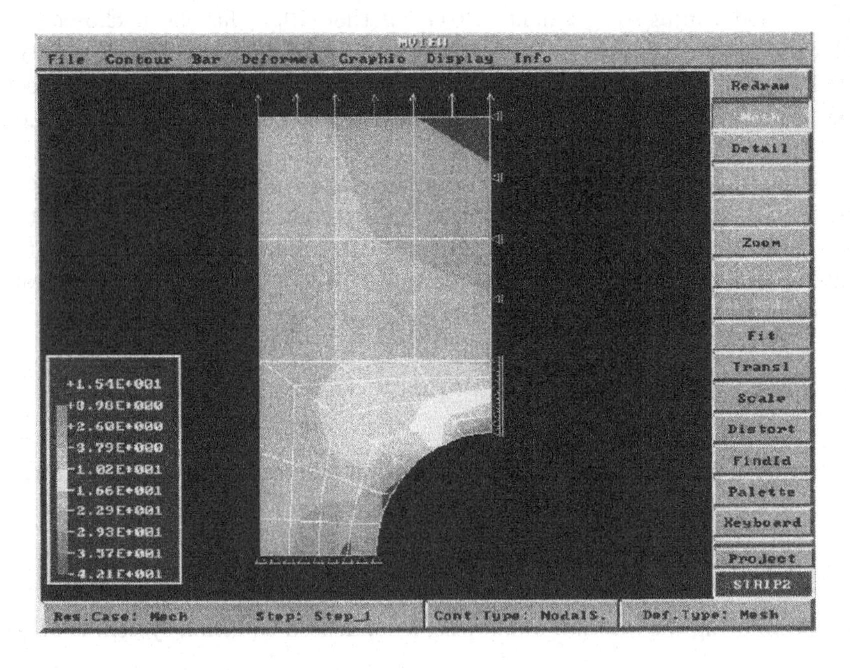

Fig. 5. MVIEW graphical interface

3 Aspects of Parallel Processing

In this section, different aspects that affect the parallel analysis system performance are presented.

The quality of partition obtained from application of automatic domain partitioning algorithms affects the global performance of the parallel system. A good partition generates substructures with approximately the same number of elements, and with minimum number of boundary nodes between neighbour substructures. The equilibrated distribution of elements is important to balance the computational work load among the processors, avoiding occurrence of idle periods. Idle periods occur when a processor interrupts its job to wait for a information from a busy processor. These waiting intervals cause system performance degradation. The number of boundary nodes between substructures is a important aspect that influences the parallel system solver performance. When the number of boundary nodes increases, the number and size of messages exchanged during the solution phase also increase. The three automatic domain partitioning algorithms used in this work present a good work load distribution, but Al-Nasra & Nguyen algorithm usually produces boundaries with a greater number of nodes than *pmetis* and *kmetis* algorithms[12].

The message passing also influences the initialisation process executed by the master program. In this analysis phase, the master program spawns task programs and sends all information needed by tasks to perform the analysis. This information includes geometry data, boundary conditions, material properties, etc. When the model size increases, the message sizes also increase, reducing consequently the system performance.

The way the condensed stiffness matrix (Equation 4) is mounted is the most important aspect that affects the parallel analysis system performance. First, system equation 7 is solved

$$K_I^{(i)} w = k_{IS}^{(i)}, \tag{7}$$

where $k_{IS}^{(i)}$ is a column of $K_{IS}^{(i)}$, and the operation 8 is carried out to obtain a line $\overline{k}_S^{(i)}$ of $\overline{K}_S^{(i)}$

$$\overline{k}_S^{(i)} = k_S^{(i)} - w^T K_{IS}^{(i)}. \tag{8}$$

Therefore, solving equations 7 and 8 to each column of $K_{IS}^{(i)}$, all lines of the condensed stiffness matrix are obtained. This process is highly influenced by the number of internal DOF of the substructure and by the bandwidth of the stiffness matrix $K_I^{(i)}$ since this matrix must be decomposed to solve equation 7.

4 System Performance

The parallel system performance is presented in this section. Two versions of this system have been evaluated and compared for the same model. The second version incorporates implementations to improve the system performance. The hardware setup consists of a 100 Mbit Fast-Ethernet network cluster including eight Pentium 200 MHz micro-computers running under LINUX operating system. This cluster is fully dedicated to parallel processing. The numerical example used to measure system performance is a beam with geometry, boundary conditions and mesh presented in Fig. 6. The model attributes are: E = 7000 kN/cm^2, $\nu = 0.25$ and thickness = 1 cm.

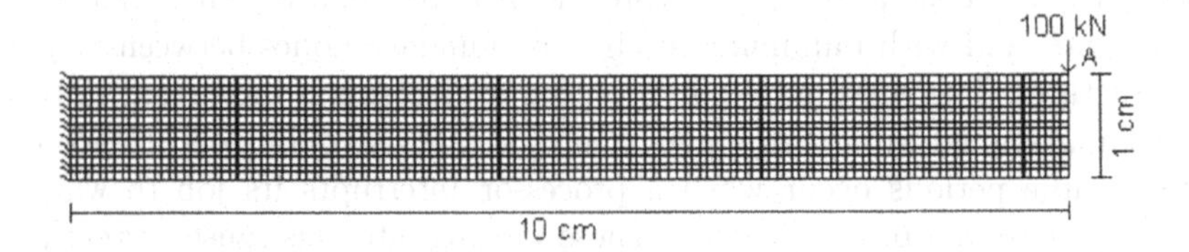

Fig. 6. Model geometry, boundary conditions and mesh

The model is created and discretized using a regular mesh of 13x130 plane stress Q8 elements. The mesh is partitioned into 2 to 8 substructures, using the *kmetis* algorithm available in PARTDOM. Fig. 7 presents the resulting mesh partitioning when 4 substructures are used.

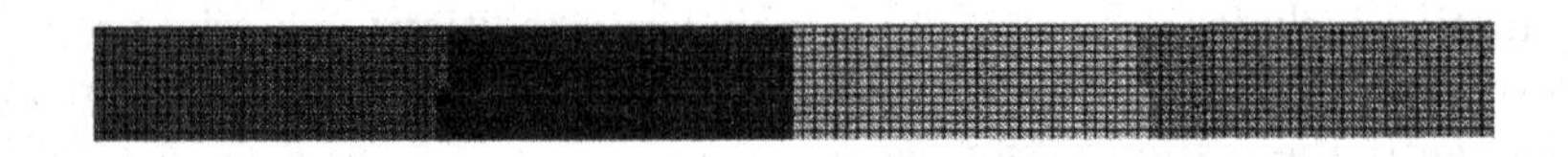

Fig. 7. Mesh partitioning into 4 substructures

Through the analysis of performance of the first version of the parallel system, the main bottleneck aspects have been identified. These aspects have been presented in the last section. The second version of the parallel system incorporates new implementations to improve the performance of critical system routines.

In the first FEMOOP parallel version, the master program is responsible for initialisation, spawning of task programs, and sending all information needed by tasks to perform the actual analysis. This information include all geometry data, topological data, material properties, boundary conditions, and domain partitioning data. With a large scale model, this time step becomes significant. In the second version, the task programs read information directly from data files, avoiding large message exchanging.

Another aspect influencing system performance is the decomposition of $K_I^{(i)}$ matrix, required to mount the condensed stiffness matrix. Usually this step causes an unbalanced computational work distribution among processors because the substructures, generated through automatic domain partitioning algorithms, present a stiffness matrix that is not optimised to numerical analysis. Thus, in the second version, FEMOOP is able to perform nodal reordering for each substructure, reducing the time to mount the condensed stiffness matrix.

In the new FEMOOP version, the pointers to nodal objects are obtained through a vector request, and not through a linked list of pointers. The use of a vector of pointers increases the overall system performance.

The main parallel analysis phases are presented in Fig. 8, 9, 10 and 11. These graphics compare the phase time consuming between first and second FEMOOP versions. Fig. 8 presents time consumed to send substructure information to each processor. Fig. 9 presents time consumed to mount the condensed stiffness matrix. Fig. 10 presents time needed to solve the linear system of equations using the parallel solver. And in Fig. 11 the time consumed to perform complete analysis is depicted.

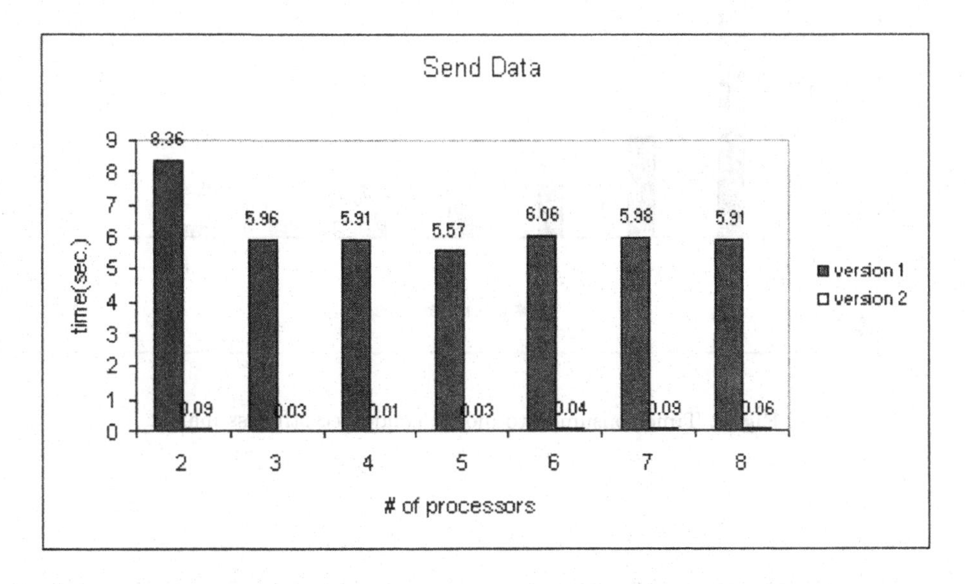

Fig. 8. Time consumed to send substructure data to processors

The effect of direct access to data files is evident (Fig. 8). The time consumed to perform the data transmission is much lower in the second FEMOOP version. The disadvantage of direct access is that an up-to-date data file must be available to each system processor.

The nodal reordering has a crucial effect on the condensed stiffness matrix mounting routine. Fig. 9 shows a great time reduction in the second FEMOOP version, specially for a small number of substructures. To mount the condensed stiffness matrix, decomposition of $K_I^{(i)}$ matrix is needed. The time consumed in this decomposition also decreases greatly with nodal reordering.

The parallel solver is highly affected by synchronization problems (Fig. 10). The first step of the parallel solver is the mounting of pre-conditioning matrix. To accomplish this, message passing is needed. Thus, the time consumed by the solver incorporates idle periods of processors waiting for information from another busy processor. This is a problem that appears in both FEMOOP versions. Nodal reordering has not been sufficient to address this problem.

The new implementations allowed a reduction in the total analysis time (Fig. 11) when 2 to 8 processors are used simultaneously.

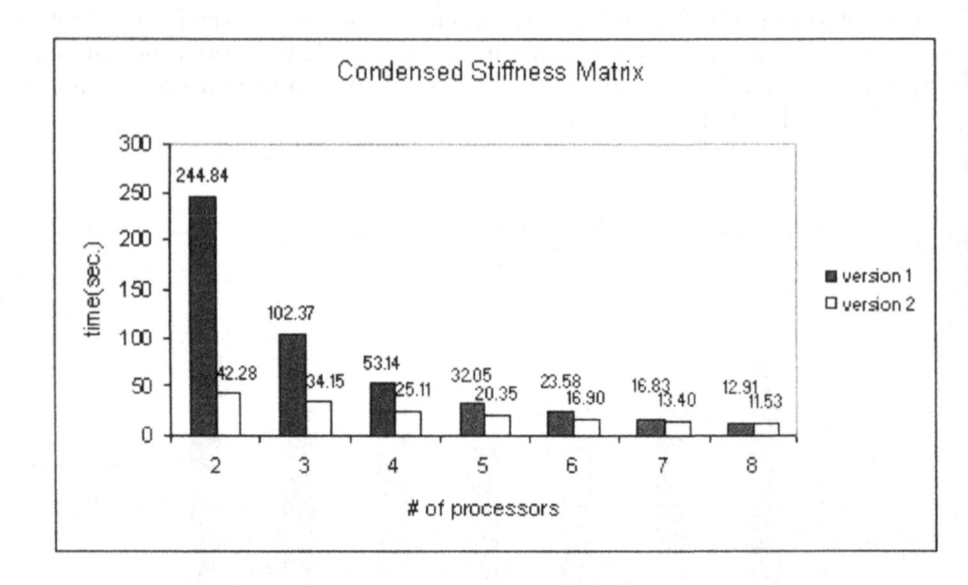

Fig. 9. Time consumed to mount condensed stiffness matrix

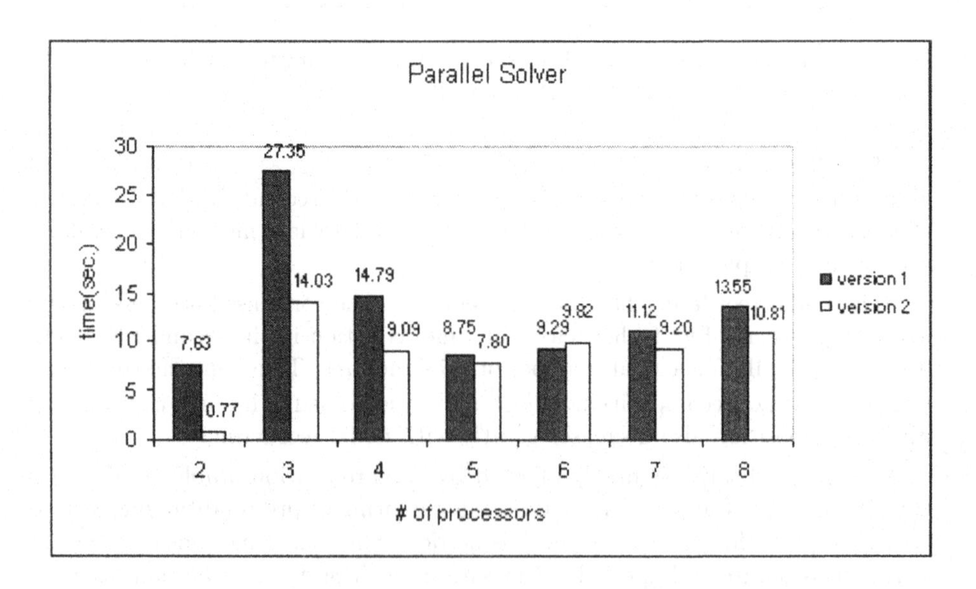

Fig. 10. Time consumed to solve linear equation system using the parallel solver

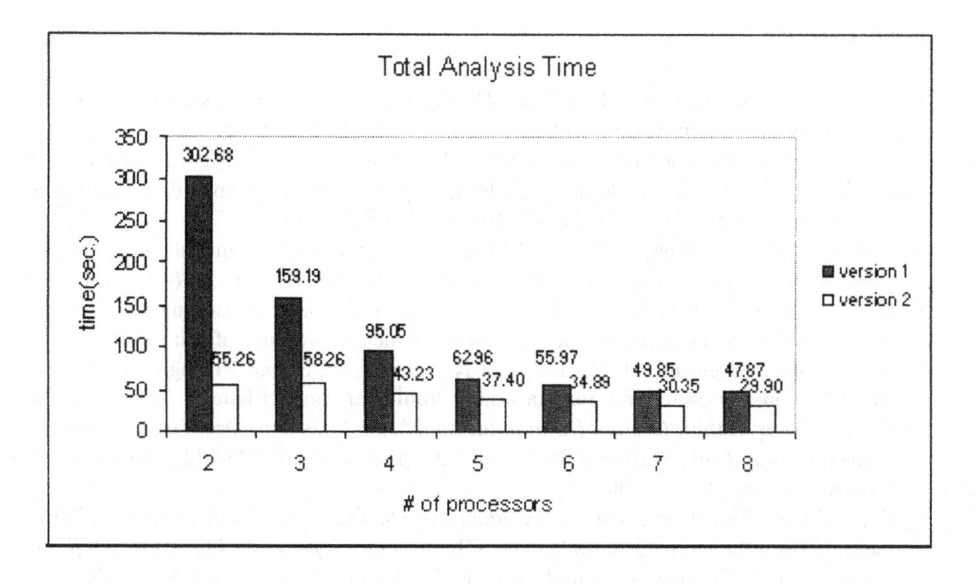

Fig. 11. Total time consumed to complete the analysis

5 Conclusions

A parallel structural analysis system has been presented in this work. This parallel system permits a better utilisation of the resources present in a local area network. This arrangement enables a low cost parallel structural analysis environment, avoiding for certain classes of problems the utilisation of expensive supercomputers.

A first version of the parallel system has been used to identify the main aspects affecting the global system performance. New implementations have been added to the code to improve performance. The development of this system takes advantage of the object-oriented programming concepts, which permits new implementations without great impact over the existing code.

The nodal reordering and the direct access to data files improved greatly the system performance. The reduction of the time consumed to mount the condensed stiffness matrix and to send data over to all processors confirm the adequacy of these new implementations.

Some improvements are still needed to resolve the synchronization problem that appears in the current system. One of these improvements is certainly a better work distribution among processors.

This system will be used in a future work to perform 3D fracture analysis of cracked structures. To accomplish this objective, new features and improvements will be necessary. Automatic 3D domain partitioning and dynamic computational work distribution are some of these improvements.

References

1. MTOOL - Bidimensional Mesh Tool (Versão 1.0) - Manual do Usuário, Grupo de Tecnologia em Computação Gráfica - TeCGraf / PUC-Rio, 1992.
2. Moretti, C.O., Bittencourt, T.N., André, J.C., and Martha, L.F., Algoritmos Automáticos de Partição de Domínio, Boletim Técnico, Departamento de Engenharia de Estruturas e Fundações, Escola Politécnica - USP, 1998.
3. Al-Nasra, M. e Nguyen, D.T., An Algorithm for Domain Decomposition in Finite Element Analysis, *Computers & Structures*, Vol. 39, No. 3/4, pp. 277-289, 1991.
4. Karypis, G. e Kumar, V., METIS: Unstructured Graph Partitioning and Sparse Matrix Ordering, Department of Computer Science, University of Minnesota, 1995.
5. Martha, L.F., Menezes, I.F.M., Lages, E.N., Parente Jr, E. and Pitangueira, R.L.S., An OOP Class Organisation for Materially Nonlinear Finite Element Analysis, *Join Conference of Italian Group of Computational Mechanics and Ibero-Latin American Association of Computational Methods in Engineering*, pp. 229-232, University of Padova, Padova, Italy, 1996.
6. Fujii, G., Análise de Estruturas Tridimensionais: Desenvolvimento de uma Ferramenta Computacional Orientada para Objetos, Dissertação de Mestrado, Dep. de Engenharia de Estruturas e Fundações (PEF), Escola Politécnica, USP, 1997.
7. Guimarães, L.G.S., Menezes, I.F.M. and Martha. L.F., Object Oriented Programming Discipline for Finite Element Analysis Systems (in Portuguese), *Proceedings of XIII CILAMCE*, Porto Alegre, RS, Brasil, Vol. 1, pp. 342-351, 1992.
8. Geist, A., Beguelin, A., Dongarra, J., Jiang, W., Manchek, R., Sunderman, V., PVM: Parallel Virtual Machine - A User's Guide and Tutorial for Networked Parallel Computing, MIT Press, Cambridge, 1994.
9. Nour-Omid, B., Raefsky, A., e Lyzenga, G., Solving Finite Element Equations on Concurrent Computers, in A.K. Noor, Ed., *Parallel Computations and Their Impact on Mechanics*, pp. 209-227, ASME, New York, 1987.
10. Hestenes, M. e Stiefel, E., Methods of Conjugate Gradients for Solving Linear Systems, *Journal of Research of the National Bureau of Standards*, Vol. 49, No. 6, pp. 409-436, Research Paper 2379, December 1952.
11. MVIEW - Bidimensional Mesh View (Versão 1.1) - Manual do Usuário, Grupo de Tecnologia em Computação Gráfica - TeCGraf / PUC-Rio, 1993.
12. Moretti, C.O., Análise de Estruturas Utilizando Técnicas de Processamento Paralelo Distribuído, Dissertação de Mestrado, Dep. de Engenharia de Estruturas e Fundações (PEF), Escola Politécnica, USP, 1997.

Dynamic Load Balancing in Crashworthiness Simulation

Hans Georg Galbas and Otto Kolp

GMD German National Research Center for Information Technology
Institute for Algorithms and Scientific Computing
Schloss Birlinghoven, D-53754 Sankt Augustin, Germany

Abstract. In the numerical simulation of crashworthiness the use of parallel architectures is becoming more and more important. This stems from the desire of engineers in the motorcar industry to get run times which make a dialogue feasible. Parallel computation seems to be the only way to solve these problems in an acceptable time. The computations inherent in crashworthiness simulation can be divided into a contact and a non-contact part. The contact part leads in contrast to the non-contact part to an imbalance due to the uneven (in space and time) distribution of contact. Good scalability becomes a challenge. In this paper we present a dynamic load balancing strategy for crash simulation. It keeps the contact and the non-contact part of the computation separately balanced over the whole simulation time. Results of the dynamic load balancing algorithm are discussed for a contact search algorithm applied to it.

1 Introduction

In the numerical simulation of crashworthiness the use of parallel architectures is becoming more and more important. This stems from the desire of engineers in the motorcar industry to get run times which make a dialogue feasible. Today Finite-Element models for cars, consist of approximately 250000 elements, and more than 100000 time-steps are needed in explicit time-marching schemes. Parallel computation seems to be the only way to solve these problems in an acceptable time. For small parallel systems (less than 16 processors) with shared-memory the obtainable speedups are very satisfactory. But this performance decreases significantly with increasing processor numbers. So for large problems of crash simulation distributed-memory MIMD architectures are the better choice. Standard domain partitioning tools (see [1] and [2]) employing a recursive spectral bisection algorithm and trying to find connected parts of equal size for each processor, generate partitions with a good balanced workload for the finite element part of the crash simulation algorithm. In general however the contact-impact part of the computation is distributed very unevenly (in space and time) by such a partition which leads to an undesired load imbalance. Good scalability becomes a challenge for crash simulation. Some improvements were obtained by giving higher weights (in the partitioning tool) to elements with expected contact and making blocks of the partition with high contact smaller [3].

J. Palma, J. Dongarra, and V. Hernández (Eds.): VECPAR'98, LNCS 1573, pp. 263–270, 1999.

To develop efficient parallel contact search algorithms the KALCRASH Project, funded by the German Ministry for Research and Education (BMBF), was carried out. The research led to a remarkable improvement concerning the contact calculations [4]. This was valid for the sequential as well as the parallel case. A part of the performance improvements was obtained by a static load balancing strategy [5].

2 From Static to Dynamic Load Balancing

In the following discussion we differ between two parts of the program for crash simulation, the contact part (CP) and the non-contact part (NCP). One observation is the fact that the two parts are connected by necessary communications. The synchronising effects of communication between CP and NCP in each simulation step make the strategy to compensate load imbalance in one part by a suitable load imbalance of the other part not very successful. Apart from communication the overall computation time is determined mainly by the sum of the maximal computation times for each part over all processors.
These observations led to the static load balancing strategy to distribute similar workloads to each processor for each part (CP and NCP). Instead of using the partition of the domain partitioning tool directly which optimises the workload for NCP, a refined partition i.e. for a multiple number of processors, was considered (over-partitioning). Then blocks of the refined partition with high workload concerning CP are combined with blocks with low workload concerning CP to form a block of a new partition for the given number of processors. The blocks of the refined partition are now sub-blocks of the new partition. This new partition fulfils the condition of similar workloads for all processors for each part (CP and NCP). For example a block of the refined partition, (of the crashing zone i.e. front of the car) was combined with another block of the refined partition of the rear of the car to build a block of a new partition.
A drawback of this approach is that a block generally consists of different separated sub-blocks which leads to some overhead in the finite element part of crash simulation concerning communication. This effect is investigated in the running EU Project SIM-VR where the discussed contact search algorithm (CSA) was successfully integrated into the crash simulation code PAM-CRASH of ESI. First results show that the gain in the contact computations by far outweighs the loss in the non-contact part.
Another drawback of the static load balancing approach is that some skill and time is needed to put the right sub-blocks together to obtain a good partition. A good partition is a partition which fulfils the condition of similar low workloads for all processors for CP and NCP in the average concerning the whole run time. Since the workload concerning CP is changing a dynamic load balancing strategy based on the static approach is to perform an exchange of sub-blocks from time to time automatically. Once a good partition is found it remains for some time a fairly good partition since the model itself changes in general only slowly (with respect to time-steps). This dynamic load balancing approach lies

with regard to its granularity to repartition every time-step and using connected and compact blocks [6] on the other extreme. This work and the integration of CSA with dynamic load balancing into PAM-CRASH is now to be performed in the BMBF-Project AUTOBENCH.

3 Dynamic Load Balancing by Over-Partitioning

Starting with an arbitrary refined partition (the number of blocks is a multiple of the number of processors) given by a domain partitioning tool each processor is associated with the same number of sub-blocks. This 'guarantees' a good load balance for the non-contact part. Since the time-steps in the simulation are largely synchronised by the communication structure and by construction we get a good load balance for the non-contact part it remains to achieve good load balance in the contact part. This is done by an exchange of sub-blocks.

3.1 When to Make Load Balancing Steps

With a given partition the simulation is performed for a certain number (nstep, see Table 1) of time-steps but in accordance with the multi-level contact search algorithm [5] (i.e. outside the fast loop). At this event each processor measures the time it needs for certain tasks which are representative for the workload of the contact part. These tasks include the creation of lists of neighbouring but non-connected nodes for each slave node which is admissible for contact and the measuring of distances of elements associated with these nodes to the slave nodes. These times are made available in an all-to-all communication to all processors. If these times differ by no more than the special threshold parameter exstop (see Table 1) no load balancing is performed and the simulation is continued for the next nstep time-steps after which the same procedure is repeated. If they differ by more than exstop a load balancing step is initiated.

3.2 How to Make Load Balancing Steps

To combine the right sub-blocks into one block the workload for the contact part of each sub-block has to be determined. Since these sub-blocks are not treated separately in CSA the workload caused by them can not be measured directly. It is measured indirectly through the number of nodes and proximity-pairs (a slave node and a nearby non-connected element) which have been detected by CSA inside the corresponding sub-blocks. A good estimation model for the workload was achieved by a regression of the time spent for certain contact search computations inside a block on the size of these sets of nodes and proximity-pairs. These sets represent the hierarchical strategy of CSA: elimination of nodes and elements from current contact search as early and cheaply as possible in order to minimise computation and communication.

The estimated workload of contact computation for a processor is given by the sum of the estimated workloads of the sub-blocks which form a block. Improvement is assumed if through a recombination of sub-blocks the maximal estimated workload over all processors can be reduced. If improvement is possible the corresponding processors exchange suitable sub-blocks. A typical block for an eight processor run formed by a recursive spectral bisectioning tool is shown by the dark region in Fig. 1. The block is connected and compact. All of it lies in a zone with likely contact. This induces load imbalance in the contact part of the simulation. After an exchange of sub-blocks the block associated with the same

Fig. 1. Block of a BMW model from standard partitioning for 8 processors.

processor has changed into one with two connectivity components (see Fig. 2). The part in the front of the car where the most deformation takes place has intensive contact whereas the part in the rear of the car has almost no contact. The new block has inclusive communication less workload for the contact part of the simulation than the old one.

The new algorithm performs this operation automatically at certain instances (when the observed load imbalance exceeds the given threshold parameter exstop) to minimise the maximal workload over all processors. This operation introduces of cause additional overhead but this is controlled sufficiently by the above-mentioned threshold parameters exstop and nstep.

Fig. 2. Block of a BMW model from over-partitioning for 8 processors.

4 Performance Results

First performance results of the dynamic load balancing strategy have been obtained for the contact search program CSA. CSA is implemented in Fortran77 with the message-passing interface MPI. Since CSA with dynamic load balancing is not yet implemented into PAM-CRASH we use interpolated data as in [5]. They are based on a 40% off-set crash simulation of a BMW benchmark model with around 60000 elements.
Simulations of 40000 time-steps on an IBM SP-2 are considered for 8, 16 and 32 processors. The dynamic load balancing partitions are created from a fourfold over-partitioning, i.e. each processor has got four sub-blocks (see Table 1). There are two 'start' partitions for the dynamic case. The first are the ones given by the partitioning tool directly (the four sub-blocks created through bisection in the last two steps of the partitioning tool are combined to one block) and the second are the static load balancing partitions created from a fourfold over-partitioning which were found to be very good in the static load balancing approach [5]. Comparing the run times for contact search without dynamic load balancing for this two partitions (labelled 'without' and 'static' in Table 1) we see that in case of 8 processors the time is halved and in case of 16 resp. 32 processors an improvement of 36% resp. 40% was obtained. The improvement is measured against run times with no load balancing.

Considering now the dynamic load balancing results we have varied the parameters nstep (number of time-steps until the next workload monitoring) and exstop (maximal difference in measured workload over all processors). The dynamic load balancing starts from the partition without load balancing and from

Table 1. Dynamic load balancing results for 40000 time-steps with fourfold over-partitioning.

number of proc	number of subblocks	load balancing	nstep	exstop (%)	time (sec)	improvement (%)
8	8	without			2615	0
8	32	static			1269	51
8	32	dynamic	500	10	1445	45
8	32	dynamic	1000	5	1449	45
8	32	dynamic	1000	10	1309	50
8	32	dynamic	1000	15	1375	47
8	32	dynamic	2000	10	1386	47
8	32	stat+dyn	1000	5	1347	48
8	32	stat+dyn	1000	10	1295	50
8	32	stat+dyn	1000	15	1321	49
16	16	without			1829	0
16	64	static			1176	36
16	64	dynamic	500	15	958	48
16	64	dynamic	1000	10	949	48
16	64	dynamic	1000	15	941	48
16	64	dynamic	1000	20	933	49
16	64	dynamic	2000	15	945	48
16	64	stat+dyn	1000	10	929	49
16	64	stat+dyn	1000	15	903	51
16	64	stat+dyn	2000	15	888	51
16	64	stat+dyn	2000	15	874	52
32	32	without			1134	0
32	128	static			685	40
32	128	dynamic	500	15	843	26
32	128	dynamic	1000	10	824	28
32	128	dynamic	1000	15	778	31
32	128	dynamic	1000	20	784	31
32	128	dynamic	2000	15	776	32
32	128	stat+dyn	1000	10	750	34
32	128	stat+dyn	1000	15	725	36
32	128	stat+dyn	1000	20	768	32
32	128	stat+dyn	2000	15	721	36

the static load balancing partitions ('stat+dyn', see Table 1). The computing times for the latter case are slightly better because of the better starting partition. For example for 32 processors the improvement rose from 32% to 36%. It can be seen that the times for dynamic load balancing are similar to the times in the static load balancing case. In case of 16 processors the performance was better. The exchanges of sub-blocks become less frequent while the simulation is going on. Altogether between 50 and 100 exchanges within the first 40000 time-steps were observed for the dynamic cases with 32 processors.
The two parameters nstep and exstop have to be adapted to the crash model. They show that one should not try too few or too many times to find a better partition and that the difference in workload between the processors must be worthwhile. In Table 2 the partitions are created from two- or eight-fold over-partitioning. While the twofold over-partitioning leads to big sub-blocks in the front with intensive contact the eight-fold over-partitioning leads to complex blocks with more communication (more neighbours) (see Table 2) but better adjustments. Since the eight-fold over-partitioning implies the fourfold case (the refined partition was created by a bisectioning procedure) a more elaborate designed objective function incorporating communication requirements should lead to some further improvement. It seems that decreasing the granularity further leads to more (untractable) overhead (see Table 2, 8 processor case with 128 sub-blocks).

Table 2. Dynamic load balancing results for 40000 time-steps with two-, eight- and sixteen-fold over-partitioning.

number of proc	number of subblocks	load balancing	nstep	exstop (%)	time (sec)	improvement (%)
8	16	dynamic	1000	15	2004	23
8	32	dynamic	1000	15	1375	47
8	64	dynamic	1000	15	1275	51
8	128	dynamic	1000	15	1428	45
16	32	dynamic	1000	15	1306	29
16	64	dynamic	1000	15	941	48
16	128	dynamic	1000	15	900	51
32	64	dynamic	1000	15	899	21
32	128	dynamic	1000	15	778	31

5 Concluding Remarks

Future work especially the integration of CSA with dynamic load balancing into PAM-CRASH will be done in the project AUTOBENCH. One topic will be the integration of communication into the objective function for determining the sub-blocks to exchange. To find good partitions the number of neighbouring blocks in a partition and the number of boundary elements of these blocks will also be considered. So for instance the rule that neighbours in the front should also be neighbours in the rear seems to be a good strategy since it leads in tendency to less communication.

Acknowledgements

The authors would like to thank their colleagues in GMD-SCAI for their support.

References

1. Floros, N., Reeve, J.S., Clinckemaillie, J., Vlachoutsis, S., Londsdale, G.: Practical aspects and experiences, Comparative efficiencies of domain decompositions. Parallel Computing **21** Elsvier Science B.V., 1995.
2. Persson, P.: Parallel Numerical Procedures for the Solution of Contact-Impact Problems. Linköping Studies in Science and Technology, Thesis No. 584, UniTryck, Linköping 1996.
3. Clinckemaillie, J., Elsner, B., Londsdale, G., Meliciani, S., Vlachoutsis, S., de Bruyne, F., Holzner, M.: Performance Issues of the Parallel PAM-CRASH Code. The International Journal of Supercomputer Applications and High Performance Computing, Vol. 11, No. 1, Sage Publications, Inc., 1997.
4. Elsner, B., Galbas, H.G., Görg, B., Kolp, O., Lonsdale, G.: A Parallel, Multi-level Approach for Contact Problems in Crashworthiness Simulation. Parallel Computing: State-of-the-Art and Perspectives, Proceedings of the Int. Conf. Parco95, Elsvier Science B.V., 1996.
5. Elsner, B., Galbas, H.G., Görg, B., Kolp, O., Lonsdale, G.: A Parallel, Multi-level Contact Search Algorithm in Crashworthiness Simulation. Advances in Computational Structures Technology, Civil-Comp Press, Edinburgh, 1996.
6. Attaway, S., Hendrickson, B., Swegle, J., Vaughan, C., Gardner, D.: Transient Dynamics Simulations: Parallel Algorithms for Contact Detection and Smoothed Particle Hydrodynamics. Supercomputing 96 Technical Papers, Pitsburgh, 1996.

Some Concepts of the Software Package FEAST

Ch. Becker, S. Kilian, S. Turek, and the FEAST Group

Institut für Angewandte Mathematik, Universität Heidelberg, Germany
http://gaia.iwr.uni-heidelberg.de/featflow

Abstract. This paper deals with the basic principles of the new FEM software package FEAST. For the FEAST software, which is mainly designed for high performance simulations, we explain the basic principles of the underlying numerical, algorithmic and implementation concepts. Computational examples illustrate the (expected) numerical and computational efficiency of this new software package, particularly in relation to existing approaches.

1 Introduction

Current trends in the software development for Partial Differential Equations (PDE's), and here in particular for Finite Element (FEM) approaches, go clearly towards object oriented techniques and adaptive methods in any sense. Hereby the employed data and solver structures, and especially the matrix structures, are often in contradiction to modern hardware platforms. As a result, the observed computational efficiency is far away from expected peak rates of almost 1 GFLOP/s nowadays, and the "real life" gap will even further increase. Since high performance calculations may be only reached by explicitly exploiting "caching in" and "pipelining" in combination with sequentially stored arrays (using special machine optimised linear algebra libraries), the corresponding realization seems to be "easier" for simple Finite Difference approaches. So, the question arises how to perform similar techniques for much more sophisticated Finite Element codes?

These discrepancies between complex mathematical approaches and highly structured computational demands often lead to unreasonable calculation times for "real world" problems, e.g. *Computational Fluid Dynamics* (CFD) calculations in 3D, as can be seen from recent benchmarks [6] for commercial as well as research codes. Hence, strategies for efficiency enhancement are necessary, not only from the mathematical (algorithms, discretisations) but also from the software point of view. To realize some of these necessary improvements our new Finite Element package (**FEAST** – **F**inite **E**lement **A**nalysis & **S**olution **T**ools) is under development. This package is based on the following concepts:

- (recursive) "Divide and Conquer" strategies,
- hierarchical data, solver and matrix structures,
- ScaRC as generalisation of multi-grid and domain decomposition techniques,

J. Palma, J. Dongarra, and V. Hernández (Eds.): VECPAR'98, LNCS 1573, pp. 271–284, 1999.

– frequent use of machine optimised linear algebra routines,
– all typical Finite Element facilities included.

The result is going to be a flexible software package with special emphasis on:

– (closer to) peak performance on modern processors,
– typical multi-grid behaviour w.r.t. efficiency and robustness,
– parallelisation tools directly included on low level,
– open for different adaptivity concepts,
– low storage requirements,
– application to many "real life" problems possible.

Figure 1 shows the general structure of the FEAST package:

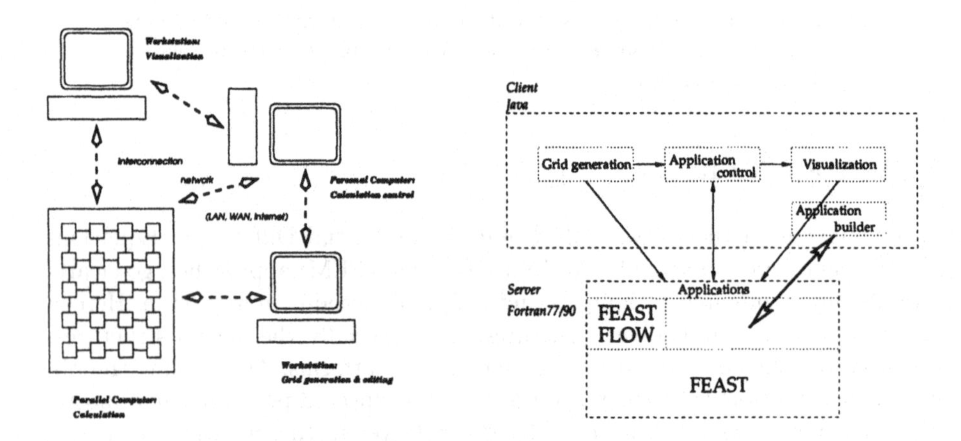

Fig. 1: FEAST structure and configuration

As programming language Fortran (77 and 90) is used. The explicit use of the two Fortran dialects arises from following observations. For Fortran77 very efficient and well tried compilers are available which allow to exploit much of the machine performance. Further it is possible to reuse many reliable parts of the predecessor packages FEAT2D, FEAT3D and FEATFLOW [8]. On the other hand Fortran77 is not more than a better "macro assembler", the very limited language constructs make the project work very hard. Further F77 is no longer the actual standard, so what is about the support in the future? And which developer can be motivated to program in F77?

F90 on the other hand is the new standard and provides new helpful features like records, dynamic memory allocation, etc. But there are several disadvantages. The language is very overloaded and the realization of some features like pointers is not succeeded. The weighty point of criticism is the nowadays very inefficient code generation of many F90 compilers. Programs compiled with F77 and F90 show a difference in runtime up to a factor of 8, depending on algorithm and compiler.

The compromise is to implement the time critical routines from the numerical linear algebra in F77, while the administrative routines are based on F90. If the

F90 compilers achieve the same code quality as their F77 pendants it is no problem to switch completely to F90.

The pre- and post-processing is mainly handled by Java based program parts. Configuring a high performance computer as a FEAST server, the user shall be able to perform the remote calculation by a FEAST client.

In the following we give examples for "real" computational efficiency results of typical numerical tools which help to motivate our hierarchical data, solver and matrix structures. To understand these better, we illustrate shortly the corresponding solution technique ScARC (**Sca**lable **R**ecursive **C**lustering) in combination with the overall "Divide and Conquer" philosophy which is essential for FEAST. We discuss how typical multi-grid rates can be achieved on parallel as well as sequential computers with a very high computational efficiency.

2 Main Principles in FEAST

2.1 Hierarchical Data, Solver and Matrix Structures

One of the most important principles in FEAST is to apply consequently a *(Recursive) Divide and Conquer* strategy. The solution of the complete "global" problem is recursively split into smaller "independent" subproblems on "patches" as part of the complete set of unknowns. Thus the two major aims in this splitting procedure which can be performed by hand or via self-adaptive strategies are:

- *Find locally structured parts.*
- *Find locally anisotropic parts.*

Based on "small" structured sub-domains on the lowest level (in fact, even one single or a small number of elements is allowed), the "higher-level" substructures are generated via clustering of "lower-level" parts such that algebraic or geometric irregularities are hidden inside the new "higher-level" patch. More background for this strategy is given in the following sections which describe the corresponding solvers related to each stage.

Figures 2 and 3 illustrate exemplarily the employed data structure for a (coarse) triangulation of a given domain and its recursive partitioning into several kinds of substructures.

According to this decomposition, a corresponding data tree - the skeleton of the partitioning strategy - describes the hierarchical decomposition process. It consists of a specific collection of elements, macros (Mxxx), matrix blocks (MB), parallel blocks (PB), subdomain blocks (SB), etc.

The *atomic units* in our decomposition are the "macros" which may be of type **structured** (as $n \times n$ collection of quadrilaterals (in 2D) with local Finite Difference data structures) or **unstructured** (any collection of elements, for instance in the case of fully adaptive local grid refinement). These "macros" (one or several) can be clustered to build a "matrix block" which contains the "local matrix parts": only here is the complete matrix information stored. Higher

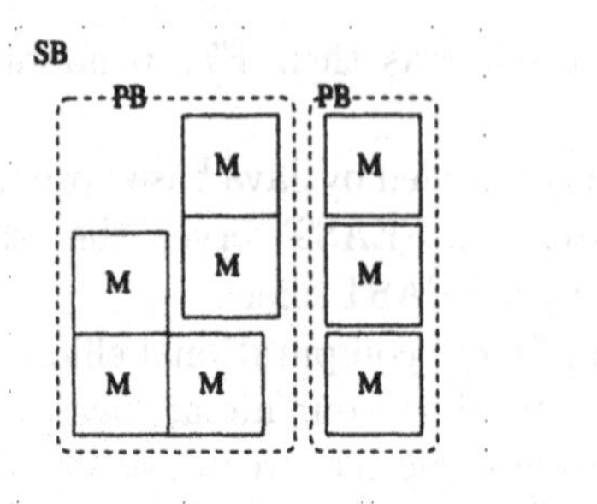

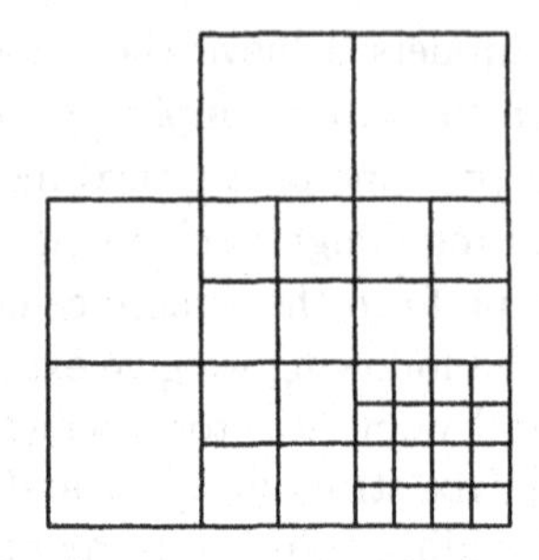

Fig. 2: FEAST domain structure

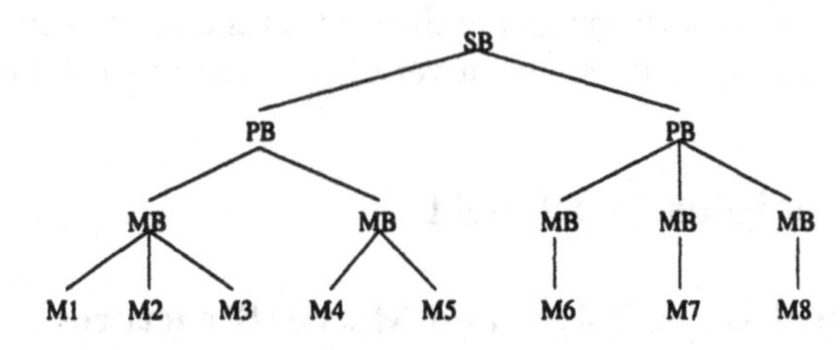

Fig. 3: FEAST data tree

level constructs are "parallel blocks" (for the parallel distribution and the realization of the load balancing) and "subdomain blocks" (with special conformity rules with respect to grid refinement and applied discretisation spaces). They all together build the complete domain, resp. the complete set of unknowns. It is important to realize that each stage in this hierarchical tree can act as independent "father" in relation to its "child" substructures while it is a "child" at the same time in another phase of the solution process (inside of the ScaRC solver, see later).

2.2 Generalized Solver Strategy ScaRC

In short form our long time experience with the numerical and computational runtime behaviour of typical multi-grid (MG) and Domain Decomposition (DD) solvers can be concluded as follows:

Some Observations from Standard Multi-grid Approaches: While in fact the numerical convergence behaviour of (optimised) multi-grid is very satisfying with respect to robustness and efficiency requirements, there still remain some "open" problems: often the parallelisation of powerful recursive smoothers (like SOR or ILU) leads to performance degradations since they can be realized only in a block-wise sense. Thus it is often not clear how the nice numerical behaviour in sequential codes for complicated geometric structures or local anisotropies can be reached in parallel computations. And additionally, the communication overhead especially on coarser grid levels dominates the total CPU time. Even more important is the "computational observation" that the realized performance on modern platforms is often far beyond (sometimes less than 1 %) the expected peak performance. Many codes often reach much less than 10 MFLOP/s, and this on computers which are said (by the vendors) to run with up to 1 GFLOP/s peak. The reason is simply that the single components in multi-grid (smoother, defect calculation, grid transfer) perform too few arithmetic work with respect to each data exchange such that the facilities of modern super-scalar architectures are poorly exploitable. In contrast, we will show that in fact 30 – 70 % can be realistic with appropriate techniques.

Some Observations from Standard Domain Decomposition Approaches: In contrast to standard multi-grid, the parallel efficiency is much higher, at least as long as no large overlap region between processors must be exchanged. While *overlapping* DD methods do not require additional coarse grid problems (however the implementation in 3D for complicated domains or for complex Finite Element spaces is a hard job), *non-overlapping* DD approaches require certain coarse grid problems, as the BPS preconditioner for instance which may lead again to several numerical and computational problems, depending on the geometrical structure or the used discretisation spaces. However the most important difference between Domain Decomposition and multi-grid are the (often) much worse convergence rates of DD although at the same time more arithmetic work is done on each processor.

As a conclusion improvements are enforced by the facts that the **convergence behaviour** is often quite sensitive with respect to (local) geometric/ algebraic **anisotropies** (in "real life" configurations), and that the performed **arithmetic work** (which allows the high performance) is often restricted by (un)necessary **data exchanges**.

An additional observation which is strongly related to the previous data structure in combination with the specific hierarchical ScaRC solver is illustrated in the following figure. We show the resulting "optimal" mesh from a numerical simulation of R.Becker/R.Rannacher for "Flow around the cylinder" which was adaptively refined via rigorous a-posteriori error control mechanisms specified for the required drag coefficient ([5]).

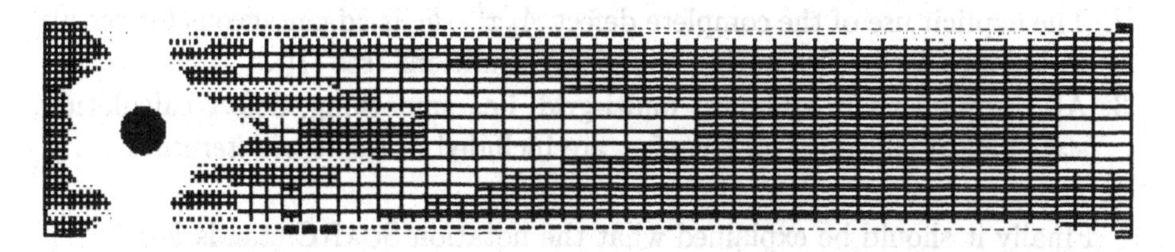

Fig. 4: "Optimal grid" via a–posteriori error estimation

As can be seen the adaptive grid refinement techniques are needed only locally, near the boundaries, while mostly regular substructures (up to 90 %) can be used in the interior of the domain. This is a quite typical result and shows that even for (more or less) complex flow simulations (here as a prototypical example) locally block-wise "Finite Difference" techniques can be applied: these regions can be detected and exploited by the given hierarchical strategies.

The ScaRC approach consists of a separated multi-grid approach for every hierarchical layer, whereby the multi-grid scheme on the outest layer (subdomain layer) gives the final result. The smoothing step of the multi-grid method is based on the following notation:

Smoothing on level h for $A_h x = b_h$:

- **global** outer block Jacobi scheme (with averaging operator 'M')

$$x^{l+1} = x^l - \omega_g \tilde{A}_{h,M}^{-1}(A_h x^l - b_h)$$

  with $\tilde{A}_{h,M}^{-1} := M \circ \tilde{A}_h^{-1}$, $\tilde{A}_h^{-1} := \sum_{i=1}^{N} \tilde{A}_{h,i}^{-1}$, $\tilde{A}_{h,i} :=$ "$A_{h|\Omega_i}$"
- "solve" **local** problems $\tilde{A}_{h,i} y_i = def_i^l := (A_h x^l - b_h)_{|\Omega_i}$ via

$$y_i^{k+1} = y_i^k - \omega_l C_{h,i}^{-1}(\tilde{A}_{h,i} y_i^k - def_i^l)$$

  with $C_{h,i}^{-1}$ preconditioner for $\tilde{A}_{h,i}$, or direct!

The local smoothing operators can be a further multi-grid scheme or any other scheme like Jacobi, Gauß–Seidel, ADI or ILU. The choice of the method depends on the local structure and "hardness" of the given domain. In a first step this decision is taken by the user to choose explicitly the method but in future it is possible to use an "expert system" which makes this decision widely automatically.

There are several reasons why we expliciteIy use **this** *basic iteration*:

1. This general form allows the splitting into matrix–vector multiplication, preconditioning and linear combination. All 3 components can be separately performed with high performance tools if available.
2. The explicit use of the complete defect $A_h x^l - b_h$ is advantageous for certain techniques for implementing boundary conditions (see [7]).
3. All components in standard multi-grid, i.e., smoothing, defect calculation, step–length control, grid transfer, are included in this *basic iteration*.

Finally it should be explained what the notation ScaRC stands for:

- **Scalable**, w.r.t. the number of global ('l') and local solution steps ('k'),
- **Recursive**, since it may be applied to more than 2 global/local levels,
- **Clustering**, since fixed or adaptive blocking of substructures is possible.

For more information about ScaRC see [3], [4].

Numerical Tests: For examining the convergence behaviour of ScARC w.r.t local anisotropies, we defined two types of anisotropic $M \times M$-topologies, $M = 4, 8$, namely $T4(a,b,c)$ and $T8(a,b,c)$. Starting from the equidistant 4×4-topology for the unit square, the "amount of anisotropy" can be parametrised by shifting the inner x- coordinates a, b and c to the left side of the domain as shown in figure 5, whereas the y-coordinates retain their old positions of 0.25, 0.5 and 0.75. The corresponding 8×8-topology is obtained by one regular refinement of the 4×4-topology.

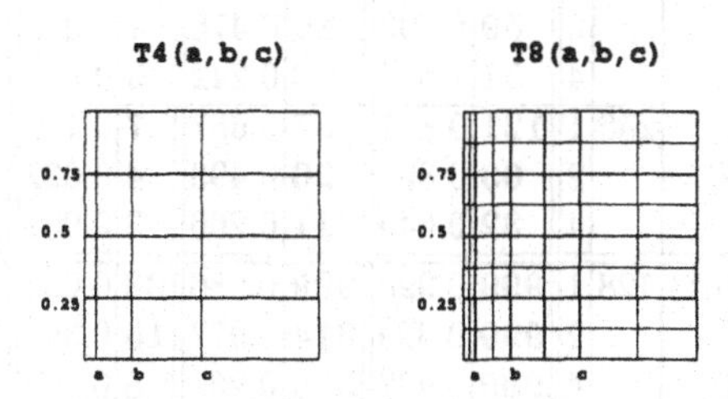

Fig. 5: Parametrisable anisotropic 4×4- and 8×8-topologies

Further, we added a local refinement procedure which provides an additional local anisotropic refinement of the single macros corresponding to some given input parameters (r_1, r_2, r_3), which indicate the amount of local distortion. This procedure is designed in a special way, such that it allows the use of the standard multi-grid transfer operators.

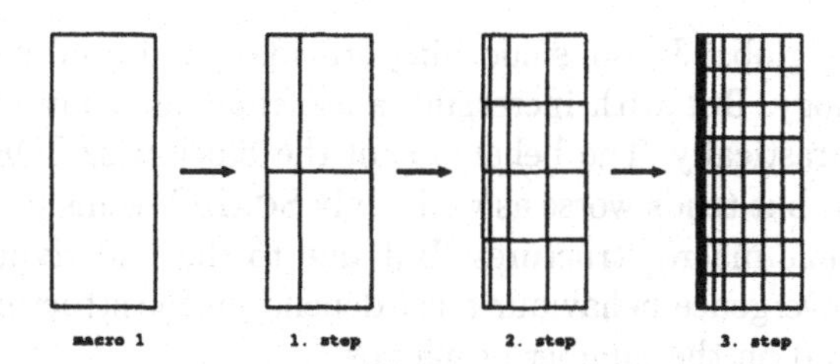

Fig. 6: Anisotropic refinement of a single macro

Figure 6 shows the refinement of the left macros in figure 5, where the single elements are shifted to the left. By means of these both procedures, arbitrarily small step sizes h_{min}, i.e., large aspect ratios AR are obtained.

For different smoothing techniques (global Jacobi, block-wise SOR, ScARC) and various parameter settings of (a,b,c), Table 1 shows the required number of multi-grid iterations # for a relative accuracy of 10^{-6} and corresponding convergence rates ρ. The term N_g denotes the global number of elements in one space direction, distributed equally over the single macros. For all smoothers we performed $l = 1,2,4$ (global) smoothing steps and 999 multi-grid iterations in

(a,b,c)	N_g	l	4×4 Jacobi #	4×4 Jacobi ρ	4×4 SOR #	4×4 SOR ρ	4×4 ScaRC #	4×4 ScaRC ρ	8×8 ScaRC #	8×8 ScaRC ρ
(0.25,0.5,0.75)	128	1	**12**	0.301	**6**	0.079	**6**	0.076	**5**	0.049
		2	**7**	0.126	**5**	0.036	**3**	0.006	**3**	0.003
$AR=1$		4	**5**	0.056	**4**	0.017	**2**	0.001	**2**	0.001
	256	1	**12**	0.304	**6**	0.079	**6**	0.083	**5**	0.060
$h_{min} \approx$ 3.9-3		2	**7**	0.128	**5**	0.036	**3**	0.009	**3**	0.005
		4	**5**	0.057	**4**	0.017	**2**	0.001	**2**	0.001
(0.05,0.2,0.5)	128	1	**114**	0.886	**35**	0.674	**7**	0.109	**9**	0.208
		2	**59**	0.790	**19**	0.475	**4**	0.018	**5**	0.050
$AR=5$		4	**31**	0.635	**10**	0.242	**3**	0.002	**3**	0.004
	256	1	**121**	0.892	**37**	0.687	**7**	0.112	**9**	0.203
$h_{min} \approx$ 7.8-4		2	**63**	0.801	**20**	0.493	**4**	0.022	**5**	0.049
		4	**32**	0.649	**11**	0.266	**3**	0.001	**3**	0.004
(0.001,0.01,0.1)	128	1	**999**	0.990	**999**	0.986	**28**	0.605	**79**	0.839
		2	**999**	0.989	**614**	0.977	**15**	0.391	**43**	0.723
$AR=250$		4	**999**	0.987	**355**	0.961	**8**	0.160	**22**	0.532
	256	1	**999**	0.990	**999**	0.988	**38**	0.692	**113**	0.884
$h_{min} \approx$ 1.6-5		2	**999**	0.990	**999**	0.987	**21**	0.516	**64**	0.804
		4	**999**	0.989	**838**	0.984	**11**	0.278	**34**	0.663

Table 1: Equidistant refinement on different anisotropic $M \times M$ – topologies

the maximum. In case of ScaRC the local problems have been solved "exactly" with pcg-methods.

Obviously the global Jacobi smoothing provides good results for the equidistant 4×4–topology. But with increasing anisotropy the convergence rates deteriorate quite drastically. The behaviour of the block-wise SOR–smoothing is somewhat better, but tends worse as well. Only ScaRC seems to be able to work well for anisotropic macro structures. But due to the underlying block Jacobi character, its convergence behaviour must depend on the structure of the macro decomposition and on the number of macros.

For the settings $(a,b,c) = (0.05, 0.2, 0.5)$ and $(a,b,c) = (0.001, 0.01, 0.1)$, table 2 compares the convergence results for different choices of the **local refinement** parameters (r_1, r_2, r_3). The considered cases $(0.5, 0.5, 0.5)/(0.25, 0.25, 0.5)$ (which have to be compared with the equidistant case $(0.5, 1.0, 1.0)$ from the previous table) involve a moderately to strongly anisotropic refinement of the most left hand side macros of the topology, up to a finest mesh size of $h_{min} \approx 4 \cdot 10^{-9}$.

Here, the global Jacobi smoother leads to divergence in case of strongly anisotropic refinement, and the block-wise SOR smoothing produces relatively bad convergence rates, as well. In contrast, ScaRRC provides the same rates for all kind of local anisotropies, i.e., does not deteriorate with increasing local anisotropy (in contrast to the global macro structure).

(r_1,r_2,r_3)	N_g	l	4×4 Jacobi #	ρ	SOR #	ρ	ScaRC #	ρ	8×8 ScaRC #	ρ
(0.25,0.25,0.5)	128	1	↗	↗	**80**	0.841	**7**	0.108	**7**	0.124
		2	**134**	0.902	**43**	0.725	**4**	0.018	**4**	0.020
$AR \approx 20,559$		4	**70**	0.820	**24**	0.558	**3**	0.001	**3**	0.004
	256	1	↗	↗	**167**	0.921	**7**	0.112	**7**	0.121
$h_{min} \approx$ 1.9-7		2	**284**	0.952	**92**	0.860	**4**	0.021	**4**	0.020
		4	**151**	0.912	**51**	0.762	**3**	0.001	**3**	0.003
(0.25,0.25,0.5)	128	1	↗	↗	**470**	0.971	**24**	0.557	**62**	0.800
		2	↗	↗	**257**	0.948	**13**	0.338	**33**	0.658
$AR \approx 10^6$		4	**430**	0.968	**144**	0.908	**7**	0.112	**18**	0.451
	256	1	↗	↗	**955**	0.986	**32**	0.649	**87**	0.853
$h_{min} \approx$ 3.8-9		2	↗	↗	**539**	0.975	**18**	0.463	**49**	0.752
		4	**947**	0.986	**313**	0.957	**10**	0.235	**26**	0.582

Table 2: Anisotropic refinement for $(a,b,c) = (0.05, 0.2, 0.5)$ and $(a,b,c) = (0.001, 0.01, 0.1)$

2.3 High Performance Linear Algebra

One of the main ideas behind the described *(Recursive) Divide and Conquer* approach in combination with the SCARC solver technology is to detect "locally structured parts". In these "local subdomains" we apply consequently "highly structured tools" as typical for Finite Difference approaches: line– or row-wise numbering of unknowns and storing of matrices as sparse bands (however the matrix entries are calculated via the Finite Element modules). As a result we have "optimal" data structures on each of these patches (which often correspond to the former introduced "matrix blocks") and we can perform very powerful linear algebra tools which explicitely exploit the high performance of specific machine optimised libraries.

We have performed several tests for different tasks and techniques in numerical linear algebra on some selected hardware platforms. In all cases we attempted to use "optimal" compiler options and machine optimised linear algebra libraries.

Gaussian Elimination: While Gaussian Elimination (GE) is presented only to demonstrate the (potentially) available performance of the given processors (often several hundreds of MFLOP/s which are really measured), we are much more interested in the realistic runtime behaviour of several matrix–vector multiplication (MV) techniques. The measured MFLOP for the Gaussian Elimination are for a dense matrix with $N = 1089$ (analogously to the standard *linpack* test).

Matrix–vector Multiplication: We examine more carefully the following variants which all are typical in the context of iterative schemes with sparse matrices. The test matrix is a typical 9–point stencil ("discretized Poisson operator"). We perform tests for two different vector lengths N and give the measured MFLOP rates which are all calculated via $20 \times N/time$ (for MV), resp., $2 \times N/time$ (for DAXPY).

Sparse MV: SMV

The *sparse MV* technique is the standard technique in Finite Element codes (and others), also well known as "compact storage" technique or similar: the matrix plus index arrays or lists are stored as long arrays containing the nonzero elements only. While this approach can be applied for arbitrary meshes and numbering of the unknowns, no explicit advantage of the line-wise numbering can be exploited. We expect a massive loss of performance with respect to the possible peak rates since — at least for larger problems — no "caching in" and "pipelining" can be exploited such that the higher cost of memory access will dominate the resulting MFLOP rates.

Banded MV: BMV

A "natural" way to improve the *sparse MV* is to exploit that the matrix is a banded matrix with 9 bands only. Hence the matrix–vector multiplication is rewritten such that now "band after band" are applied. The obvious advantage of this *banded MV* approach is that these tasks can be performed on the basis of BLAS1–like routines which may exploit the vectorisation facilities of many processors (particularly on vector computers). However for "long" vector lengths the improvements can be absolutely disappointing: For the recent workstation/PC chip technology the processor cache dominates the resulting efficiency!

Banded blocked MV: BBMVA, BBMVL, BBMVC

The final step towards highly efficient components is to rearrange the matrix–vector multiplication in a "block-wise" sense: for a certain set of unknowns, a corresponding part of the matrix is treated such that cache–optimised and fully vectorised operations can be performed. This procedure is called "BLAS 2+"-style since in fact certain techniques for dense matrices which are based on ideas from the BLAS2, resp., BLAS3 library, have now been developed for such sparse banded matrices. The exact procedure has to be carefully developed in dependence of the underlying FEM discretisation, and a more detailed description can be found in [2].

While BBMVA has to be applied in the case of arbitrary matrix coefficients, BBMVL and BBMVC are modified versions which can be used under certain circumstances only (see [2] for technical details). For example PDE's with constant coefficients as the Poisson operator but on a mesh which is adapted in one special direction only, allow the use of BBMVL: This is often the case for the Pressure–Poisson problem in flow simulations (see [7]) on boundary adapted meshes. Additionally version BBMVC may be applied for PDE's with constant coefficients on meshes with equidistant mesh distribution in each (local) direction separately: This is typical for tensor product meshes in the interior domain where the solution is mostly smooth.

Computational Results: The following table illustrates the above discussed linear algebra routines with their performance rates. For further results see [1].

		$\frac{2N}{T}$			$\frac{20N}{T}$			
Computer	N	DAXPY	SMV	BMV	BBMVA	BBMVL	BBMVC	GE
IBM RS6000	4K	420	54	154	154	196	240	365
(166 Mhz)	64K	100	49	71	140	206	234	
'SP2'	256K	100	50	71	143	189	240	
DEC Alpha	4K	390	34	46	81	158	223	309
(433 Mhz)	64K	60	22	42	63	117	189	
'CRAY T3E'	256K	60	22	42	67	110	196	
SUN U450	4K	165	35	102	100	99	127	246
(250 Mhz)	64K	123	15	21	35	85	119	
	256K	30	14	17	36	72	99	

Table 3: Benchmark results

2.4 Several Adaptivity Concepts

As typical for modern FEM packages, we directly incorporate certain tools for grid generation which allow an easy handling of local and global refinement or coarsening strategies: **adaptive mesh moving, macro adaptivity** and **fully local adaptivity**.

Adaptive strategies for moving mesh points, along boundaries or inner structures, allow the same logic structure in each "macro block", and hence the shown performance rates can be preserved. Additionally, we work with adaptivity concepts related to each "macro block". Allowing "blind" or "slave macro nodes" preserves the high performance facilities in each "matrix block", and is a good compromise between fully local adaptivity and optimal efficiency through structured data. Only in that case, that these concepts do not lead to satisfying results, certain macros will loose their "highly structured" features through the (local) use of fully adaptive techniques. On these (hopefully) few patches, the standard "sparse" techniques for unstructured meshes have to be applied.

2.5 Direct Integration of Parallelism

Most software packages are designed for sequential algorithms to solve a given PDE problem, and the subsequent parallelisation of certain methods takes often un-proportionately long. In fact it is easy to say, but hard to realize with most software packages. However the more important step, which makes parallelisation much more easier, is the design of the SCARC solver according to the hierarchical decomposition in different stages. Indeed from an algorithmic point of view, our sequential and parallel versions differ only as analogously Jacobi– and Gauß–Seidel–like schemes work differently. Hence all parallel executions can be identically simulated on single processors which however can additionally improve their numerical behaviour with respect to efficiency and robustness through Gauß–Seidel–like mechanisms.

Hence we only provide in FEAST the "software" tools for including parallelism on low level, while the "numerical parallelism" is incorporated via our ScARC solver and the hierarchical "tree structure". However what will be "non–standard" is our concept of (adaptive) parallel load-balancing which is oriented in "total numerical efficiency" (that means, "how much processor work is spent to achieve a certain accuracy, depending on the local configuration") in contrast to the "classical" criterion of equilibrating the number of local unknowns (see [2] for detailed information and examples in FEAST).

3 Pre– and Post-processing

3.1 General Remarks

As remarked in the introduction the pre– and post-processing should be realized in main tasks by a general framework of Java based programs called DeViSoR. DeViSoR means "Design & Visualisation Software Resource". This framework is intented to perform the main tasks grid generation and editing, control of the calculation and visualisation of the results. These main tasks use the same ground classes (called DFC - DeViSoR Foundation Classes) and the same user interface, so the access to the underlying numerical core parts are performed in the same manner.

As intended in the introduction the various subtasks can be performed on several machines which communicates over a network system. This allows the user to choose the suitable system for the corresponding task, e.g. a Silicon Graphics workstation for the visualisation. The access to a parallel computing system should also be performed by a Java program. This allows not only the developer of a numerical code to use a parallel computer.

DeViSoR is planned to be an "open system" for the developing of pre– and post-processing tools for FEM packages. The DeViSoR foundation classes contain the basic tools to handle and administrate FEM typical structures. Further applications could realize e.g. further visualisation procedures and adaptions to several parallel computer systems.

For this project Java as implementation environment is been chosen. Though Java is a relative "young" programming language the advantages of this system are significant. The "write once, run anywhere" capability reduces the implementation effort widely against combinations like C/C++/OpenGL. It exist only one program which runs without any modification on several different configurations like Unix workstations, Linux PCs, Windows PCs, Macintoshs and many more. A further advantage is the core class library for various subareas like file handling, network functions, visualisation and user interface facilities. These classes are easy to use and produce an pleasing output. The use of additional tools like applications builders is not necessary. The most disadvantage of nowadays Java implementations is the relative low performance because of the fact that Java is an interpreted language. However further developments like more sophisticated interpreter with Just–In–Time compiling facilities and especially the native Java processor will hopefully close this performance leak.

3.2 Preprocessing: DeViSoRGrid

This subprogram should support the generation and editing of 2D domains. The two main parts are the description of the domain boundary and the generation of the grid structure. The program supports several boundary elements like lines, arcs and splines, further it is planned to add a segment which consists of an Fortran subroutine which describes a parametrisation. This allows to use an analytic description. Several triangular and quadrilateral elements are supported. Extensive editing possibilities allow the user to delete, move and adjust the boundaries and elements. For the future it is planned to implement simply automatic grid generators for producing coarse grids. As further tasks this program should be able to read many formats from other tools like CAD systems and professional grid generators and prepare this data for the use in the calculation process.

3.3 Processing: DeViSoRControl

DeViSoRControl enables the user to control the calculation und to follow the calculation progress. Main tasks of this program part are the distribution of the macros to the processing nodes (at the moment manually, in future automatically), the collecting and displaying of the log information of the processing nodes and finally the configuration of the ScARC algorithm with respect to the selection of smoothing/preconditioning methods on a given hierarchical layer, the size of smoothing steps and the stopping criterion. Furthermore this part builds the interface to the other DeViSoR parts for the pre– and post-processing. From the control part the grid program is invoked, a grid is editing. For this grid the user selects the desired solution method and visualise finally the results with the DeViSoRVision program.

3.4 Post-processing: DeViSoRVision

The last part in the current project is the DeViSoRVision program which performs the visualisation task. The program offers several techniques to visualise the results of the calculation like shading techniques, isolines, particle tracing (planned). Furthermore it contains an animation module to create animations for non-stationary problems. The result of the animation can be stored in several formats like MPEG and AnimatedGIF.

4 Conclusions and Outlook

Our numerical examples show the advantages of the proposed method ScARC with respect to efficiency and robustness. This method allows parallelisation and the use of standard numerical methods as basic modules for smoothing and solving. Further it makes it possible to use highly regular data structures which enables high performance facilities.

Our computational examples have shown that there is a large gap between the linpack rates of several hundreds of MFLOP as typical representatives for direct solvers and the more realistic results for iterative schemes. Moreover certain properties of cache management and architecture influence the run time behaviour massively, which leads to further problems for the developer. On the other hand the examples show that appropriate algorithmic and implementation techniques lead to much higher performance rates. However, massive changes in the design of numerical, algorithmic and implementation ingredients are necessary.

The actual status of the FEAST project and further information can always be obtained from our web page: http://gaia.iwr.uni-heidelberg.de/~featflow.

References

1. Altieri, M., Becker, Ch., Turek, S.: On the realistic performance of components in iterative solvers, Proc. FORTWIHR Conference, Munich, March 1998, LNCSE, Springer-Verlag, to appear.
2. Becker, Ch.: The realization of Finite Element software for high-performance applications, Thesis, to appear.
3. Kilian, S.: Efficient parallel iterative solvers of ScARC-type and their application to the incompressible Navier-Stokes equations, Thesis, 1998.
4. Kilian, S., Turek, S.: An example for parallel ScARC and its application to the incompressible Navier-Stokes equations, Proc. ENUMATH-97, Heidelberg, October 1997.
5. Rannacher, R., Becker, R.: A Feed-Back Approach to Error Control in Finite Element Methods: Basic Analysis and Examples, Preprint 96–52, University of Heidelberg, SFB 359, 1996.
6. Schäfer, M., Rannacher, R., Turek, S.: Evaluation of a CFD Benchmark for Laminar Flows, Proc. ENUMATH-97, Heidelberg, October 1997.
7. Turek, S.: Efficient solvers for incompressible flow problems: An algorithmic approach in view of computational aspects, LNCSE, Springer-Verlag, 1998.
8. Turek, S.: FEATFLOW . Finite element software for the incompressible Navier-Stokes equations: User Manual, Release 1.1, 1998

Multilevel Mesh Partitioning for Optimising Aspect Ratio

C. Walshaw[1], M. Cross[1], R. Diekmann[2], and F. Schlimbach[2]

[1] School of Computing and Mathematical Sciences, The University of Greenwich, London, SE18 6PF, UK.
{C.Walshaw,M.Cross}@gre.ac.uk

[2] Department of Computer Science, University of Paderborn, Fürstenallee 11, D-33102 Paderborn, Germany.
{diek,schlimbo}@uni-paderborn.de

Abstract. Multilevel algorithms are a successful class of optimisation techniques which address the mesh partitioning problem. They usually combine a graph contraction algorithm together with a local optimisation method which refines the partition at each graph level. To date these algorithms have been used almost exclusively to minimise the cut-edge weight, however it has been shown that for certain classes of solution algorithm, the convergence of the solver is strongly influenced by the subdomain aspect ratio. In this paper therefore, we modify the multilevel algorithms in order to optimise a cost function based on aspect ratio. Several variants of the algorithms are tested and shown to provide excellent results.

1 Introduction

The need for mesh partitioning arises naturally in many parallel finite element (FE) and finite volume (FV) applications. Meshes composed of elements such as triangles or tetrahedra are often better suited than regularly structured grids for representing completely general geometries and resolving wide variations in behaviour via variable mesh densities. Meanwhile, the modelling of complex behaviour patterns means that the problems are often too large to fit onto serial computers, either because of memory limitations or computational demands, or both. Distributing the mesh across a parallel computer so that the computational load is evenly balanced and the data locality maximised is known as mesh partitioning. It is well known that this problem is NP-complete, so in recent years much attention has been focused on developing suitable heuristics, and some powerful methods, many based on a graph corresponding to the communication requirements of the mesh, have been devised, e.g. [11].

A particularly popular and successful class of algorithms which address this mesh partitioning problem are known as multilevel algorithms. They usually combine a graph contraction algorithm which creates a series of progressively smaller and coarser graphs together with a local optimisation method which, starting with the coarsest graph, refines the partition at each graph level. These

J. Palma, J. Dongarra, and V. Hernández (Eds.): VECPAR'98, LNCS 1573, pp. 285–300, 1999.

algorithms have been used almost exclusively to minimise the cut-edge weight, a cost which approximates the total communications volume in the underlying solver. This is an important goal in any parallel application, to minimise the communications overhead, however, it has been shown, [17], that for certain classes of solution algorithm, the convergence of the solver is actually heavily in fluenced by the shape or aspect ratio (AR) of the subdomains. In this paper therefore, we modify the multilevel algorithms (the matching and local optimisation) in order to optimise a cost function based on AR. We also abstract the process of modification in order to suggest how the multilevel strategy can be modified into a generic technique which can optimise arbitrary cost functions.

1.1 Domain Decomposition Preconditioners and Aspect Ratio

To motivate the need for aspect ratio we consider the requirements of a class of solution techniques. A natural *parallel* solution strategy for the underlying problem is to use an iterative solver such as the conjugate gradient (CG) algorithm together with domain decomposition (DD) preconditioning, e.g. [2]. DD methods take advantage of the partition of the mesh into subdomains by imposing artificial boundary conditions on the subdomain boundaries and solving the original problem on these subdomains, [4]. The subdomain solutions are independent of each other, and thus can be determined in parallel without any communication between processors. In a second step, an 'interface' problem is solved on the inner boundaries which depends on the jump of the subdomain solutions over the boundaries. This interface problem gives new conditions on the inner boundaries for the next step of subdomain solution. Adding the results of the third step to the first gives the new conjugate search direction in the CG algorithm.

The time needed by such a preconditioned CG solver is determined by two factors, the maximum time needed by any of the subdomain solutions and the number of iterations of the global CG. Both are at least partially determined by the shape of the subdomains. Whilst an algorithm such as the multigrid method as the solver on the subdomains is relatively robust against shape, the number of global iterations are heavily influenced by the AR of subdomains, [16]. Essentially, the subdomains can be viewed as elements of the interface problem, [7,8], and just as with the normal finite element method, where the condition of the matrix system is determined by the AR of elements, the co ndition of the preconditioning matrix is here dependent on the AR of subdomains.

1.2 Overview

Below, in Section 2, we introduce the mesh partitioning problem and establish some terminology. We then discuss the mesh partitioning problem as applied to AR optimisation and describe how the graph needs to be modified to carry this out. Next, in Section 3, we describe the multilevel paradigm and present and compare three possible matching algorithms which take account of AR. In Section 4 we then describe a Kernighan-Lin (KL) type iterative local optimisation algorithm and describe two possible modifications which aim to optimise AR. Finally in Section 5 we compare the results with a cut edge partitioner, suggest how the multilevel strategy can be modified into a generic technique and present some ideas for further investigation.

The principal innovations described in this paper are:

- In §2.2 we describe how the graph can be modified to take AR into account.
- In §3.2 we describe three matching algorithms based on AR.
- In §4.3 we describe two ways of using the cost function to optimise for AR.
- In §4.4 we describe how the bucket sort can be modified to take into account non-integer gains.

2 The Mesh Partitioning Problem

To define the mesh partitioning problem, let $G = G(V, E)$ be an undirected graph of vertices V, with edges E which represent the data dependencies in the mesh. We assume that both vertices and edges can be weighted (with positive integer values) and that $|v|$ denotes the weight of a vertex v and similarly for edges and sets of vertices and edges. Given that the mesh needs to be distributed to P processors, define a partition π to be a mapping of V into P disjoint subdomains S_p such that $\bigcup_P S_p = V$. To evenly balance the load, the optimal subdomain weight is given by $\overline{S} := \lceil |V|/P \rceil$ (where the ceiling function $\lceil x \rceil$ returns the smallest integer $\geq x$) and the *imbalance* is then defined as the maximum subdomain weight divided by the optimal (since the computational speed of the underlying application is determined by the most heavily weighted processor).

The definition of the mesh-partitioning problem is to find a partition which evenly balances the load or vertex weight in each subdomain whilst minimising some cost function Γ. Typically this cost function is simply the total weight of cut edges, but in this paper we describe a cost function based on AR. A more precise definition of the mesh-partitioning problem is therefore to find π such that $S_p \leq \overline{S}$ and such that Γ is minimised.

2.1 The Aspect Ratio and Cost Function

We seek to modify the methods by optimising the partition on the basis of AR rather than cut-edge weight. In order to do this it is necessary to define a cost function which we seek to minimise and a logical choice would be $\max_p \mathrm{AR}(S_p)$, where $\mathrm{AR}(S_p)$ is the AR of the subdomain S_p. However maximum functions are notoriously difficult to optimise (indeed it is for this reason that most mesh partitioning algorithms attempt to minimise the total cut-edge weight rather than the maximum between any two subdomains) and so instead we choose to minimise the average AR

$$\Gamma_{AR} = \sum_p \frac{\mathrm{AR}(S_p)}{P}. \tag{1}$$

There are several definitions of AR, however, and for example, for a given polygon S, a typical definition, [14], is the ratio of the largest circle which can be contained entirely within S (inscribed circle) to the smallest circle which ent irely contains S (circumcircle). However these circles are not easy to calculate for arbitrary polygons and in an optimisation code where ARs may need to be calculated very frequently, we do not believe this to be a practical metric. It may also fail to express certain irregularities of shape. A careful discussion of the relative merits of different ways of measuring AR may be found in [15] and for the purposes of this paper we follow the ideas therein and define the AR of a given shape by measuring the ratio of its perimet er length (surface area in 3d) over that of some ideal shape with identical area (volume in 3d).

Suppose then that in 2d the ideal shape is chosen to be a square. Given a polygon S with area ΩS and perimeter length ∂S, the ideal perimeter length (the perimeter length of a square with area ΩS) is $4\sqrt{\Omega S}$ and so the AR is defined as $\partial S/4\sqrt{\Omega S}$. Alternatively, if the ideal shape is chosen to be a circle then the same argument gives the AR of $\partial S/2\sqrt{\pi\Omega S}$. In fact, given the definition of the cost function (1) it can be seen that these two definitions will produce the same optimisation problem (and hence the same results) with the cost just modified by a constant C (where $C = 1/4$ for the squ are and $1/2\sqrt{\pi}$ for circle). These definitions of AR are easily extendible to 3d and given a polyhedron S with volume ΩS and surface area ∂S, the AR can be calculated as $C\partial S/(\Omega S)^{2/3}$, where $C = 1/4$ if the cube is chosen as the optimal shape and $C = 1/(4\pi)^{1/3}3^{2/3}$ for the sphere. Note that henceforth, in order to talk in general terms for both 2d & 3d, given an object S we shall use the terms ∂S or *surface* for the surface area (3d) or perimeter length (2d) of the object and ΩS or *volume* for the vol ume (3d) or area (2d).

Of the above definitions of AR we choose to use the square/cube based formulae for two reasons; firstly because we are attempting to partition a mesh into interlocking subdomains (and circles/spheres are not known for their interlocking qualities) and sec ondly because it gives a convenient formula for the cost function of:

$$\Gamma_{\text{template}} = \frac{1}{C}\sum_{p}\frac{\partial S_p}{(\Omega S_p)^{\frac{d-1}{d}}} \tag{2}$$

where $C = 2dP$ and d ($= 2$ or 3) is the dimension of the mesh. We refer to this cost function as Γ_{template} or Γ_t because of the way it tries to match shapes to chosen templates.

In fact, it will turn out (see for example §3.2) that even this function may be too complex for certain optimisation needs and we can define a simpler one by assuming that all subdomains have approximately the same volume, $\Omega S_p \approx \Omega M/P$, where ΩM is the total volume of the mesh. This assumption may not necessarily be true, but it is likely to be true locally (see §4.5). We can then approximate (2) by

$$\Gamma_{\text{template}} \approx \frac{1}{C'}\sum_{p}\partial S_p \tag{3}$$

where $C' = 2dP^{\frac{1}{d}}(\Omega M)^{\frac{d-1}{d}}$. This can be simplified still further by noting that the surface of each subdomain S_p consists of two components, the *exterior* surface, $\partial^e S_p$, where the surface of the subdomain coincides with the surface of the mesh ∂M, and t he *interior* surface, $\partial^i S_p$, where S_p is adjacent to other subdomains and the surface cuts through the mesh. Thus we can break the $\sum_p \partial S_p$ term in (3) into two parts $\sum_p \partial^i S_p$ and $\sum_p \partial^e S_p$ and simplify (3) further by noting that $\sum_p \partial^e S_p$ is just ∂M, the exterior sur face of the mesh M. This then gives us a second cost function to optimise:

$$\Gamma_{\text{surface}} = \frac{1}{K_1} \sum_p \partial^i S_p + K_2 \tag{4}$$

where $K_1 = 2dP^{\frac{1}{d}}(\Omega M)^{\frac{d-1}{d}}$ and $K_2 = \partial M / K_1$. We refer to this cost function as Γ_{surface} or Γ_s because it is just concerned with optimising surfaces.

2.2 Modifying the Graph

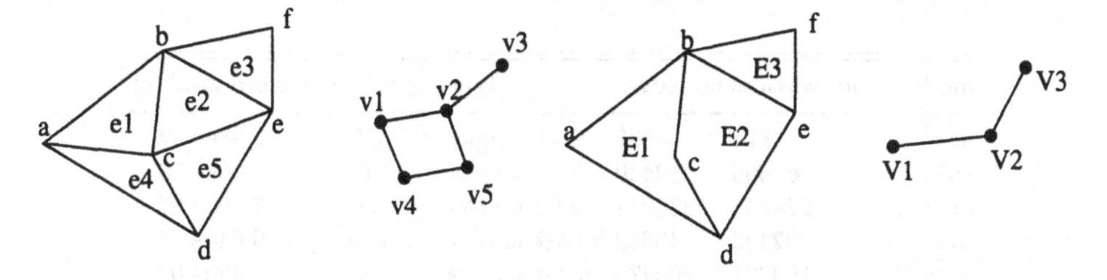

Fig. 1. Left to right: a simple mesh (a), its dual (b), the same mesh with combined elements (c) and its dual (d)

To use these cost functions in a graph-partitioning context, we must add some additional qualities to the graph. Figure 1 shows a very simple mesh (1a) and its dual graph (1b). Each element of the mesh corresponds to a vertex in the graph. The vertices of the graph can be weighted as is usual (to carry out load-balancing) but in addition, vertices store the volume and total surface of their corresponding element (e.g. $\Omega v_1 = \Omega e_1$ and $\partial v_1 = \partial e_1$). We also weight the edges of the graph with the size of the surface they correspond to. Thus, in Figure 1, if $D(b, c)$ refers to the distance between points b and c, then the weight of edge (v_1, v_2) is set to $D(b, c)$. In this way, for vertices v_i corresponding to elements which have no exterior surface, the sum of their edge weights is equivalent to their surface ($\partial v_i = \sum_E |(v_i, v_j)|$). Thus for vertex v_2, $\partial v_2 = \partial e_2 = D(b, c) + D(c, e) + D(e, b) = |(v_2, v_1)| + |(v_2, v_3)| + |(v_2, v_5)|$.

When it comes to combining elements together, either into subdomains, or for the multilevel matching (§3) these properties, volume and surface can be easily combined. Thus in Figure 1c where $E_1 = e_1 + e_4$, $E_2 = e_3 + e_5$ and $E_3 = e_3$ we see that volumes can be directly summed, for example $\Omega V_1 =$

$\Omega E_1 = \Omega e_1 + \Omega e_4 = \Omega v_1 + \Omega v_4$, as can edge weights, e.g. $|(V_1, V_2)| = D(b, c) + D(c, d) = |(v_1, v_2)| + |(v_4, v_5)|$. The surface of a combined object S is the sum of the surfaces of its constituent parts less twice the interior surface, e.g. $\partial V_1 = \partial E_1 = \partial e_1 + \partial e_4 - 2 \times D(a, c) = \partial v_1 + \partial v_1 - 2|(v_1, v_4)|$. These properties are very similar to properties in conventional graph algorithms, where the volume combines in the same way as weight and surfaces combine as the sum of edge weights (although including an additional term which expresses the exterior surfa ce ∂^e). The edge weights function identically.

Note that with these modifications to the graph, it can be seen that if we optimise using the Γ_s cost function (4), the AR mesh partitioning problem is identical to the cut-edge weight mesh partitioning problem with a special edge weighting. However, the inclusion of non integer edge weights does have an effect on the some of the techniques that can be used (e.g. see §4.4).

2.3 Testing the Algorithms

Table 1. Test meshes

mesh	no. vertices	no. edges	type	aspect ratio	mesh grading
uk	4824	6837	2d triangles	3.39	7.98e+02
t60k	60005	89440	2d triangles	1.60	2.00e+00
dime20	224843	336024	2d triangles	1.87	3.70e+03
cs4	22499	43858	3d tetrahedra	1.07	9.64e+01
mesh100	103081	200976	3d tetrahedra	1.63	2.45e+02
cyl3	232362	457853	3d tetrahedra	1.28	8.42e+00

Throughout this paper we compare the effectiveness of different approaches using a set of test meshes. The algorithms have been implemented within the framework of JOSTLE, a mesh partitioning software tool developed at the University of Greenwich and freely available for academic and research purposes under a licensing agreement (available from `http://www.gre.ac.uk/~c.walshaw/-jostle`). The experiments were carried out on a DEC Alpha with a 466 MHz CPU and 1 Gbyte of memory. Due to space considerations we only include 6 test meshes but they have been chosen to be a representative sample of medium to large scale real-life problems and include both 2d and 3d examples. Table 1 gives a list of the meshes and their sizes in terms of the number of vertices and edges. The table also shows the aspect ratio of each entire mesh and the mesh grading, which here we define as the maximum surface of any element over the minimum surface, and these two figures give a guide as to how difficult the optimisation may be. For example, 'uk' is simply a triangulation of the British mainland and hence has a very intricate boundary and therefore a high aspect ratio. Meanwhile, 'dime20' which has a moderate aspect ratio, has been very heavily refined in parts and thus has a high mesh grading – the largest element has a surface around 3,700 times larger than that of the smallest.

Table 2. Final results using template cost matching and surface gain/template cost optimisation

	$P = 16$			$P = 32$			$P = 64$			$P = 128$		
mesh	Γ_t	$\lvert E_c\rvert$	t_s	Γ_t	$\lvert E_c\rvert$	t_s	Γ_t	$\lvert E_c\rvert$	t_s	Γ_t	$\lvert E_c\rvert$	t_s
uk	1.48	206	0.12	1.31	331	0.12	1.23	543	0.22	1.25	917	0.50
t60k	1.16	1003	1.63	1.10	1547	2.07	1.11	2437	2.33	1.11	3647	2.65
dime20	1.22	1623	5.78	1.20	2868	5.17	1.15	4406	5.70	1.12	6620	7.57
cs4	1.22	2727	0.85	1.22	3738	0.90	1.23	5066	1.12	1.23	6747	1.60
mesh100	1.25	5950	3.20	1.24	8752	3.53	1.26	12467	4.13	1.28	17346	5.13
cyl3	1.21	11141	10.05	1.21	15944	10.77	1.23	22378	13.02	1.22	29719	13.18

Table 2 shows the results of the final combination of algorithms – TCM (see §3.2) and SGTC (see §4.3) – which were chosen as a benchmark for the other combinations. For the 4 different values of P (the number of subdomains), the table shows the average aspect ratio as given by Γ_t, the edge cut $|E_c|$ (that is the number of cut edges, not the weight of cut edges weighted by surface size) and the time in sec onds, t_s, to partition the mesh. Notice that with the exception of the 'uk' mesh, all partitions have average aspect ratios of less than 1.30 which is well within the target range suggested in [6]. Indeed for the 'uk' mesh it is no surprise that the results are not optimal because the subdomains inherit some of the poor AR from the original mesh (which has an AR of 3.39) and it is only when the mesh is split into small enough pieces, $P = 64$ or 128, that the optimisation succeeds in ameliorating this effect. Intuitively this also gives a hint as to why DD methods are a very successful technique as a solver.

3 The Multilevel Paradigm

In recent years it has been recognised that an effective way of both speeding up partition refinement and, perhaps more importantly giving it a global perspective is to use multilevel techniques. The idea is to match pairs of vertices to form *clusters*, use the clusters to define a new graph and recursively iterate this procedure until the graph size falls below some threshold. The coarsest graph is then partitioned and the partition is successively optimised on all the graphs starting with the coarsest and ending with the original. This sequence of contraction followed by repeated expansion/optimisation loops is known as the multilevel paradigm and has been successfully developed as a strategy for overcoming the localised nature of the KL (and other) optimisation algorithms. The multilevel idea was first proposed by Barnard & Simon, [1], as a method of speeding up spectral bisection and improved by Hendrickson & Leland, [11], who generalised it to encompass local refinement algorithms. Several algorithms for carrying out the matching have been devised by Karypis & Kumar, [12], while Walshaw & Cross describe a method for utilising imbalance in the coarsest graphs to enhance the final partition quality, [18].

3.1 Implementation

Graph contraction. To create a coarser graph $G_{l+1}(V_{l+1}, E_{l+1})$ from $G_l(V_l, E_l)$ we use a variant of the edge contraction algorithm proposed by Hendrickson & Leland, [11]. The idea is to find a maximal independent subset of graph edges, or a *matching* of vertices, and then collapse them. The set is independent because no two edges in the set are incident on the same vertex (so no two edges in the set are adjacent), and maximal because no more edges can be added to the set without breaking the independence criterion. Having found such a set, each selected edge is collapsed and the vertices, $u_1, u_2 \in V_l$ say, at either end of it are merged to form a new vertex $v \in V_{l+1}$ with weight $|v| = |u_1| + |u_2|$.

The initial partition. Having constructed the series of graphs until the number of vertices in the coarsest graph is smaller than some threshold, the normal practice of the multilevel strategy is to carry out an initial partition. Here, following the idea of Gupta, [10], we contract until the number of vertices in the coarsest graph is the same as the number of subdomains, P, and then simply assign vertex i to subdomain S_i. Unlike Gupta, however, we do not carry out repeated expansion/contraction cycles of the coarsest graphs to find a well balanced initial partition but instead, since our optimisation algorithm incorporates balancing, we commence on the expansion/optimisati on sequence immediately.

Partition expansion. Having optimised the partition on a graph G_l, the partition must be interpolated onto its parent G_{l-1}. The interpolation itself is a trivial matter; if a vertex $v \in V_l$ is in subdomain S_p then the matched pair of vertices that it represents, $v_1, v_2 \in V_{l-1}$, will be in S_p.

3.2 Incorporating Aspect Ratio

The matching part of the multilevel strategy can be easily modified in several ways to take into account AR and in each case the vertices are visited (at most once) using a randomly ordered linked list. Each vertex is then matched with an unmatched neighbour using the chosen matching algorithm and it and its match removed from the list. Vertices with no unmatched neighbours remain unmatched and are also removed. In addition to **Random Matching (RM)**, [11], where vertices are matched with random neighbours, we propose and have tested 3 matching algorithms:

Surface Matching (SM). As we have seen in §2.2, the AR partitioning problem can be approximated by the cut-edge weight problem using (4), the Γ_s cost function, and so the simplest matching is to use the Heavy Edge approach of Karypis & Kum ar, [12], where the vertex matches across the heaviest edge to any of its unmatched neighbours. This is the same as matching across the largest surface (since here edge weights represent surfaces) and we refer to this as *surface matching.*

Template Cost Matching (TCM). A second approach follows the ideas of Bouhmala, [3], and matches with the neighbour which minimises the cost function. In this case, the chosen vertex matches with the unmatched neighbour which gives the resulting element the best aspect ratio. Using the Γ_t cost function, we refer to this as *template cost matching.*

Surface Cost Matching (SCM). This is the same idea as TCM only using the Γ_s cost function, (4), which is faster to calculate.

3.3 Results for Different Matching Functions

Space precludes the full discussion of results for the different matching functions and a full presentation of the results may be found in [19]. Briefly, however, SM, SCM and TCM all produced very similar quality partitions (within 3.5%), with TCM marginally better than the other two. Random matching (RM), on the other hand, produced partitions with ARs about 30% worse on average. This is not altogether surprising since the AR of elements in the coarsest graph is very poor if the matching takes no account of it, and hence the optimisation has to work with badly shaped elements. Overall the matching results suggests that the multilevel strategy is relatively robust to the matching algorithm *provided* the AR is taken into account in some way.

4 The Kernighan-Lin Optimisation Algorithm

In this section we discuss the key features of an optimisation algorithm, fully described in [18] and then in §4.3 describe how it can be modified to optimise for AR. It is a Kernighan-Lin (KL) type algorithm incorporating a hill-climbing mechanism to enable it to escape from local minima. The algorithm uses bucket sorting (§4.4), the linear time complexity improvement of Fiduccia & Mattheyses, [9], and is a partition optimisation formulation; in other words it optimises a partition of P subdomains rather tha n a bisection.

4.1 The Gain Function

A key concept in the method is the idea of *gain.* The gain $g(v, q)$ of a vertex v in subdomain S_p can be calculated for every other subdomain, S_q, $q \neq p$, and expresses how much the cost of a given partition would be improved were v to migrate to S_q. Thus, if π denotes the current partition and π' the partition if v migrates to S_q then for a cost function Γ, the gain $g(v,q) = \Gamma(\pi') - \Gamma(\pi)$. Assuming the migration of v only affects the cost of S_p and S_q (as is true for Γ_t and Γ_s) then we get

$$g(v,q) = \mathrm{AR}(S_q + v) - \mathrm{AR}(S_q) + \mathrm{AR}(S_p - v) - \mathrm{AR}(S_p). \qquad (5)$$

For Γ_t this gives an expression which cannot be further simplified, however, for Γ_s, since

$$\begin{aligned} \mathrm{AR}(S_q + v) - \mathrm{AR}(S_q) &= \frac{1}{K_1}\left\{\partial^i(S_q + v) - \partial^i S_q\right\} \\ &= \frac{1}{K_1}\left\{\partial^i S_q + \partial^i v - 2|(S_q, v)| - \partial^i S_q\right\} \\ &= \frac{1}{K_1}\left\{\partial^i v - 2|(S_q, v)|\right\} \end{aligned}$$

(where $|(S_q, v)|$ denotes the sum of edge weights between S_q and v), we get

$$g_{\mathrm{surface}}(v,q) = \frac{2}{K_1}\left\{|(S_p, v)| - |(S_q, v)|\right\} \qquad (6)$$

Notice in particular that g_{surface} is the same as the cut-edge weight gain function and that it is entirely localised, i.e. the gain of a vertex only depends on the length of its boundaries with a subdomain and not on any intrinsic qual ities of the subdomain which could be changed by non-local migration.

4.2 The Iterative Optimisation Algorithm

The serial optimisation algorithm, as is typical for KL type algorithms, has inner and outer iterative loops with the outer loop terminating when no migration takes place during an inner loop. The optimisation uses two bucket sorting structures or bucket trees (see below, §4.4) and is initialised by calculating the gain for all border vertices and inserting them into one of the bucket trees. These vertices will subsequently be referred to as *candidate* vertices and the tree containing them as the *candidate tree.*

The inner loop proceeds by examining candidate vertices, highest gain first (by always picking vertices from the highest ranked bucket), testing whether the vertex is acceptable for migration and then transferring it to the other bucket tree (the tree of *examined* vertices). This inner loop terminates when the candidate tree is empty although it may terminate early if the partition cost (i.e. the number of cut edges) rises too far above the cost of the best partition found so far. Once the inner loop has terminated any vertices remaining in the candidate tree are transferred to the examined tree and finally pointers to the two trees are swapped ready for the next pass through the inner loop.

The algorithm also uses a KL type hill-climbing strategy; in other words vertex migration from subdomain to subdomain can be *accepted* even if it degrades the partition quality and later, based on the subsequent evolution of the partition, either rej ected or *confirmed.* During each pass through the inner loop, a record of the optimal partition achieved by migration within that loop is maintained together with a list of vertices which have migrated since that value was attained. If subsequent migration finds a 'better' partition then the migration is *confirmed* and the list is reset. Once the inner loop is terminated, any vertices remaining in the list (vertices whose migration has not been confirmed) are migrated back to the subdomains they came from when the optimal cost was attained.

The algorithm, together with conditions for vertex migration acceptance and confirmation is fully described in [18].

4.3 Incorporating Aspect Ratio: Localisation

One of the advantages of using cut-edge weight as a cost function is its localised nature. When a graph vertex migrates from one subdomain to another, only the gains of adjacent vertices are affected. In contrast, when using the graph to optimise AR, if a vertex v migrates from S_p to S_q, the volume and surface of both subdomains will change. This in turn means that, when using the template cost function (2), the gain of all border vertices both within and abutting subdomains S_p and S_q will change. Strictly speaking, all these gains should be adjusted with the huge disadvantage that this may involve thousands of floating point operations and hence be prohibitively expensive. As an alternative, therefore, we propose two localised variants:

Surface Gain/Surface Cost (SGSC).

The simplest way to localise the updating of the gains is to make the assumption in §2.1 that the subdomains all have approximately equal volume and to use the surface cost function Γ_s from (4). As mentioned in §2.2 the problem immediately reduces to the cut-edge weight problem, albeit with non-integer edge weights, and from (6) only the gains of the vertices adjacent to the migrating vertex will need updating.

However, if this assumption is not true, it is not clear how well Γ_s will optimise the AR and below we provide some experimental results.

Surface Gain/Template Cost (SGTC).

The second method we propose for localising the updates of gain relies on the observation that the gain is simply used as a method of rating the elements so that the algorithm always visits those with highest gain first (using the bucket sort). It is not clear how crucial this rating is to the success of the algorithm and indeed Karypis & Kumar demonstrated that (at least when optimising for cut-edge weight) almost as good results can be achieved by simply visiting the vertices in random order, [13]. We therefore propose approximating the gain with the surface cost function Γ_s from (4) to rate the elements and store them in the bucket tree structure, but using the template cost function Γ_t from (2) to a ssess the change in cost when actually migrating an element. This localises the gain function.

4.4 Incorporating Aspect Ratio: Bucket Sorting with Non-integer Gains

The bucket sort is an essential tool for the efficient and rapid sorting and adjustment of vertices by their gain. Space precludes a full description here but the idea is that all vertices of a given gain g are placed together in a 'bucket' which is ranked g. Our implementation of bucket sorting is fully described in [18].

The only difficulty in adapting bucket sorting to AR optimisation is that with non-integer edge weight, the gains are also real non-integer numbers and so in a naive implementation there could be as many buckets as border vertices. This is not a major problem in itself as we can just give buckets an interval of gains rather than a single integer, i.e. the bucket ranked 1 could contain any vertex with gain in the interval $[1.0, 2.0)$. However, if using the surface gain function, the issue of scaling then arises since for a mesh entirely contained within the unit square/cube, all the vertices are likely to end up in one of two buckets (dependent only on whether they have positive or neg ative gains). Fortunately, if using Γ_s as a gain function, as in SGSC and SGTC, we can easily calculate the maximum possible gain. This would occur if the vertex with the largest surface, $v \in S_p$ say, were entirely surrounded by neighbours in S_q. The maximum possible gain is then $2 \max_{v \in V} \partial v$ (strictly speaking $2 \max_{v \in V} \partial^i v$) and similarly the minimum gain is $-2 \max_{v \in V} \partial v$. This means we can easily choose the number of buckets and scale the gain accordingly. A problem still arises for meshes with a high grading because many of the elements will have an insignificant surface area compared to the maximum. However the experiments carried out here all used a scaling which allowed a maximum of 100 buckets and we have tested the algorithm with up to 10,000 buckets without significant penalty in terms either memory or run-time.

4.5 Results for Different Optimisation Functions

Table 3 compares SGSC against the SGTC results in Table 2. Both set of results use template cost matching (TCM). For each value of P, the first column shows the average AR, Γ_t of the partitioning. The second column for each value of P then compares results with those in Table 2 using the metric $\frac{\Gamma(\mathrm{SGSC})-1}{\Gamma(\mathrm{SGTC})-1}$. Thus a figure > 1 means that SGSC has produced worse results than SGTC. These comparisons are then averaged and so it can be seen, e.g. for $P = 64$ that SGSC produces results 4% (1.04) worse on average than SGTC. Indeed we see that there is on average only a tiny difference between the two (SGTC is 0.5% better than SGSC) and again, with the exception of the 'uk' mesh for $P = 16$ & 32, all results have an average AR of less than 1.30. This implication of this table is that the assumption made in §2.1, that all subdomains have approximately the same volume, is reasonably good. However this assumption is not necessarily true, because for example, for $P = 128$, the 'dime20' mesh, with its high grading, has a ratio of $\max \Omega S_p / \min \Omega S_p = 2723$. A possible explanation is that although the assumption is false globally, it is true locally, since the mesh density does not change too gradually (as should be the case with most meshes generated by adaptive refinement) and so the volume of each subdomai n is approximately equal to that of its neighbours.

Table 3. Surface gain/surface cost optimisation compared with surface gain/template cost

	$P = 16$		$P = 32$		$P = 64$		$P = 128$	
mesh	Γ_t	$\frac{\Gamma(SGSC)-1}{\Gamma(SGTC)-1}$	Γ_t	$\frac{\Gamma(SGSC)-1}{\Gamma(SGTC)-1}$	Γ_t	$\frac{\Gamma(SGSC)-1}{\Gamma(SGTC)-1}$	Γ_t	$\frac{\Gamma(SGSC)-1}{\Gamma(SGTC)-1}$
uk	1.49	1.02	1.32	1.05	1.24	1.02	1.23	0.92
t60k	1.15	0.95	1.10	0.96	1.12	1.07	1.12	1.11
dime20	1.23	1.03	1.17	0.86	1.15	0.98	1.11	0.91
cs4	1.20	0.90	1.23	1.05	1.24	1.03	1.22	0.97
mesh100	1.24	0.95	1.26	1.10	1.27	1.06	1.27	0.97
cyl3	1.23	1.10	1.22	1.08	1.24	1.06	1.22	1.00
Average		0.99		1.01		1.04		0.98

We are not not primarily concerned with partitioning times here, but it was surprising to see that SGSC was an average 30% slower than SGTC. A possible explanation is that although the cost function Γ_s is a good approximation, Γ_t is a more global function and so the optimisation converges more quickly.

5 Discussion

5.1 Comparison with Cut-edge Weight Partitioning

In Table 4 we compare AR as produced by the edge cut partitioner (EC) described in [18] with the results in Table 2. On average AR partitioning produces results which are 16% better than those of the edge cut partitioner (as could be expected). However, for the mesh 'cs4' EC partitioning is consistently better and this is a subject for further investigation.

Meanwhile in Table 5 we compare the edge cut produced by the EC partitioner with that of the AR partitioner. Again as expected, EC partitioning produces the best results (about 11% better than AR). These results demonstrate that a good partition for aspect ratio is not necessarily a good partition for edge-cut and vice-versa.

In terms of time, the EC partitioner is about 26% faster than AR on average. Again this is no surprise since the AR partitioning involves floating point operations (assessing cost and combining elements) while EC partitioning only requires integer operations.

Table 4. AR results for the edge cut partitioner compared with the AR partitioner

	$P = 16$		$P = 32$		$P = 64$		$P = 128$	
mesh	Γ_t	$\frac{\Gamma(\mathrm{EC})-1}{\Gamma(\mathrm{AR})-1}$	Γ_t	$\frac{\Gamma(\mathrm{EC})-1}{\Gamma(\mathrm{AR})-1}$	Γ_t	$\frac{\Gamma(\mathrm{EC})-1}{\Gamma(\mathrm{AR})-1}$	Γ_t	$\frac{\Gamma(\mathrm{EC})-1}{\Gamma(\mathrm{AR})-1}$
uk	1.52	1.09	1.33	1.07	1.26	1.09	1.28	1.14
t60k	1.19	1.18	1.18	1.76	1.17	1.47	1.17	1.55
dime20	1.32	1.45	1.26	1.34	1.25	1.65	1.21	1.72
cs4	1.19	0.86	1.21	0.93	1.20	0.87	1.21	0.92
mesh100	1.22	0.89	1.22	0.91	1.26	1.03	1.24	0.86
cyl3	1.22	1.05	1.23	1.09	1.23	1.00	1.23	1.02
Average		1.09		1.18		1.19		1.20

Table 5. $|E_c|$ results for the edge cut partitioner compared with the AR partitioner

	$P = 16$		$P = 32$		$P = 64$		$P = 128$	
mesh	$\lvert E_c \rvert$	$\frac{\lvert E_c \rvert(\mathrm{RM})}{\lvert E_c \rvert(\mathrm{AR})}$	$\lvert E_c \rvert$	$\frac{\lvert E_c \rvert(\mathrm{RM})}{\lvert E_c \rvert(\mathrm{AR})}$	$\lvert E_c \rvert$	$\frac{\lvert E_c \rvert(\mathrm{RM})}{\lvert E_c \rvert(\mathrm{AR})}$	$\lvert E_c \rvert$	$\frac{\lvert E_c \rvert(\mathrm{RM})}{\lvert E_c \rvert(\mathrm{AR})}$
uk	189	0.92	290	0.88	478	0.88	845	0.92
t60k	974	0.97	1588	1.03	2440	1.00	3646	1.00
dime20	1326	0.82	2294	0.80	3637	0.83	5497	0.83
cs4	2343	0.86	3351	0.90	4534	0.89	6101	0.90
mesh100	4577	0.77	7109	0.81	10740	0.86	14313	0.83
cyl3	10458	0.94	14986	0.94	20765	0.93	27869	0.94
Average		0.88		0.89		0.90		0.90

5.2 Conclusion and Future Research

We have shown that the multilevel strategy can be modified to optimise for aspect ratio and that a good partition for aspect ratio is not necessarily a good partition for edge-cut and vice-versa. To fully validate the method, however, we need to demonstrate that the measure of aspect ratio used here does indeed provide the benefits for DD preconditioners that the theoretical results suggest. It is also desirable to measure the correlation between aspect ratio and convergence in the solver.

Also, although parallel implementations of the multilevel strategy do exist, e.g. [20], it is not clear how well AR optimisation, with its more global cost function, will work in parallel and this is another direction for future researc h. Some related work already exists in the context of a parallel dynamic adaptive mesh environment, [5,6,15], but these are not multilevel methods and it was necessary to use a combination of several complex cost functions in order to achieve reasonable results so the question arises whether multilevel techniques can help to overcome this.

Finally in this paper we have adapted a mesh partitioning technique originally designed to solve the edge cut partitioning problem to a different cost function. The question then arises, is the multilevel strategy an appropriate technique for solving partitioning problems (or indeed other optimisation problems) with different cost functions? Clearly this is an impossible question to answer in general but a discussion can be found in [19].

References

1. S. T. Barnard and H. D. Simon. A Fast Multilevel Implementation of Recursive Spectral Bisection for Partitioning Unstructured Problems. *Concurrency: Practice & Experience*, 6(2):101–117, 1994.
2. S. Blazy, W. Borchers, and U. Dralle. Parallelization methods for a characteristic's pressure correction scheme. In E. H. Hirschel, editor, *Flow Simulation with High Performance Computers II, Notes on Numerical Fluid Mechanics*, 1995.
3. N. Bouhmala. *Partitioning of Unstructured Meshes for Parallel Processing*. PhD thesis, Inst. d'Informatique, Univ. Neuchatel, 1998.
4. J. H. Bramble, J. E. Pasciac, and A. H. Schatz. The Construction of Preconditioners for Elliptic Problems by Substructuring I+II. *Math. Comp.*, 47+49, 1986+87.
5. R. Diekmann, B. Meyer, and B. Monien. Parallel Decomposition of Unstructured FEM-Meshes. *Concurrency: Practice & Experience*, 10(1):53–72, 1998.
6. R. Diekmann, F. Schlimbach, and C. Walshaw. Quality Balancing for Parallel Adaptive FEM. In A. Ferreira *et al.*, editor, *Proc. IRREGULAR '98: Solving Irregularly Structured Problems in Parallel*, volume 1457 of *LNCS*, pages 170–181. Springer, 1998.
7. C. Farhat, N. Maman, and G. Brown. Mesh Partitioning for Implicit Computations via Domain Decomposition. *Int. J. Num. Meth. Engng.*, 38:989–1000, 1995.
8. C. Farhat, J. Mandel, and F. X. Roux. Optimal convergence properties of the FETI domain decomposition method. *Comput. Methods Appl. Mech. Engrg.*, 115:365–385, 1994.
9. C. M. Fiduccia and R. M. Mattheyses. A Linear Time Heuristic for Improving Network Partitions. In *Proc. 19th IEEE Design Automation Conf.*, pages 175–181, IEEE, Piscataway, NJ, 1982.
10. A. Gupta. Fast and effective algorithms for graph partitioning and sparse matrix reordering. *IBM Journal of Research and Development*, 41(1/2):171–183, 1996.
11. B. Hendrickson and R. Leland. A Multilevel Algorithm for Partitioning Graphs. In *Proc. Supercomputing '95*, 1995.
12. G. Karypis and V. Kumar. A Fast and High Quality Multilevel Scheme for Partitioning Irregular Graphs. TR 95-035, Dept. Comp. Sci., Univ. Minnesota, Minneapolis, MN 55455, 1995.
13. G. Karypis and V. Kumar. Multilevel k-way partitioning scheme for irregular graphs. TR 95-064, Dept. Comp. Sci., Univ. Minnesota, Minneapolis, MN 55455, 1995.
14. S. A. Mitchell and S. A. Vasavis. Quality Mesh Generation in Three Dimensions. In *Proc. ACM Conf. Comp Geometry*, pages 212–221, 1992.

15. F. Schlimbach. *Load Balancing Heuristics Optimising Subdomain Aspect Ratios for Adaptive Finite Element Simulations.* Diploma Thesis, Dept. Math. Comp. Sci., Univ. Paderborn, 1998.
16. D. Vanderstraeten, C. Farhat, P. S. Chen, R. Keunings, and O. Zone. A Retrofit Based Methodology for the Fast Generation and Optimization of Large-Scale Mesh Partitions: Beyond the Minimum Interface Size Criterion. *Comp. Meth. Appl. Mech. Engrg.*, 133:25–45, 1996.
17. D. Vanderstraeten, R. Keunings, and C. Farhat. Beyond Conventional Mesh Partitioning Algorithms and the Minimum Edge Cut Criterion: Impact on Realistic Applications. In D. Bailey *et al.*, editor, *Parallel Processing for Scientific Computing*, pages 611–614. SIAM, 1995.
18. C. Walshaw and M. Cross. Mesh Partitioning: a Multilevel Balancing and Refinement Algorithm. Tech. Rep. 98/IM/35, Univ. Greenwich, London SE18 6PF, UK, March 1998.
19. C. Walshaw, M. Cross, R. Diekmann, and F. Schlimbach. Multilevel Mesh Partitioning for Optimising Domain Shape. Tech. Rep. 98/IM/38, Univ. Greenwich, London SE18 6PF, UK, July 1998.
20. C. Walshaw, M. Cross, and M. Everett. Parallel Dynamic Graph Partitioning for Adaptive Unstructured Meshes. *J. Par. Dist. Comput.*, 47(2):102–108, 1997.

Chapter 3: Computing in Education

Parallel and Distributed Computing in Education (Invited Talk)

Peter H. Welch

Computing Laboratory,
University of Kent at Canterbury, CT2 7NF, United Kingdom
P.H.Welch@ukc.ac.uk

Abstract. The natural world is certainly not organised through a central thread of control. Things happen as the result of the actions and interactions of unimaginably large numbers of independent agents, operating at all levels of scale from nuclear to astronomic. Computer systems aiming to be of real use in this real world need to model, at the appropriate level of abstraction, that part of it for which it is to be of service. If that modelling can reflect the natural concurrency in the system, it ought to be much simpler

Yet, traditionally, concurrent programming is considered to be an advanced and difficult topic – certainly much harder than serial computing which, therefore, needs to be mastered first. But this tradition is wrong.

This talk presents an intuitive, sound and practical model of parallel computing that can be mastered by undergraduate students in the first year of a computing (major) degree. It is based upon Hoare's mathematical theory of *Communicating Sequential Processes* (CSP), but does not require mathematical maturity from the students – that maturity is pre-engineered in the model. Fluency can be quickly developed in both message-passing and shared-memory concurrency, whilst learning to cope with key issues such as race hazards, deadlock, livelock, process starvation and the efficient use of resources. Practical work can be hosted on commodity PCs or UNIX workstations using either Java or the Occam multiprocessing language. Armed with this maturity, students are well-prepared for coping with real problems on real parallel architectures that have, possibly, less robust mathematical foundations.

1 Introduction

At Kent, we have been teaching parallel computing at the undergraduate level for the past ten years. Originally, this was presented to first-year students *before* they became too set in the ways of serial logic. When this course was expanded into a full unit (about 30 hours of teaching), timetable pressure moved it into

J. Palma, J. Dongarra, and V. Hernández (Eds.): VECPAR'98, LNCS 1573, pp. 301–330, 1999.

the second year. Either way, the material is easy to absorb and, after only a few (around 5) hours of teaching, students have no difficulty in grappling with the interactions of 25 (say) threads of control, appreciating and eliminating race hazards and deadlock.

Parallel computing is still an immature discipline with many conflicting cultures. Our approach to educating people into successful exploitation of parallel mechanisms is based upon focusing on parallelism as a powerful tool for *simplifying* the description of systems, rather than simply as a means for improving their performance. We never start with an existing serial algorithm and say: 'OK, let's parallelise that!'. And we work solely with a model of concurrency that has a semantics that is *compositional* – a fancy word for WYSIWYG – since, without that property, combinatorial explosions of complexity always get us as soon as we step away from simple examples. In our view, this rules out low-level concurrency mechanisms, such as spin-locks, mutexes and semaphores, as well as some of the higher-level ones (like monitors).

Communicating Sequential Processes (CSP)[1–3] is a mathematical theory for specifying and verifying complex patterns of behaviour arising from interactions between concurrent objects. Developed by Tony Hoare in the light of earlier work on monitors, CSP has a compositional semantics that greatly simplifies the design and engineering of such systems – so much so, that parallel design often becomes easier to manage than its serial counterpart. CSP primitives have also proven to be extremely lightweight, with overheads in the order of a few hundred nanoseconds for channel synchronisation (including context-switch) on current microprocessors [4,5].

Recently, the CSP model has been introduced into the Java programming language [6–10]. Implemented as a library of packages [11,12], JavaPP[10] enables multithreaded systems to be designed, implemented and reasoned about entirely in terms of CSP synchronisation primitives (channels, events, etc.) and constructors (parallel, choice, etc.). This allows 20 years of theory, design patterns (with formally proven good properties – such as the absence of race hazards, deadlock, livelock and thread starvation), tools supporting those design patterns, education and experience to be deployed in support of Java-based multithreaded applications.

2 Processes, Channels and Message Passing

This section describes a simple and structured multiprocessing model derived from CSP. It is easy to teach and can describe arbitrarily complex systems. No formal mathematics need be presented – we rely on an intuitive understanding of how the world works.

2.1 Processes

A *process* is a component that encapsulates some data structures and algorithms for manipulating that data. Both its data and algorithms are private. The outside world can neither see that data nor execute those algorithms. Each process is alive, executing its own algorithms on its own data. Because those algorithms are executed by the component in its own thread (or threads) of control, they express the behaviour of the component from its own point of view[1]. This considerably simplifies that expression.

A *sequential process* is simply a process whose algorithms execute in a single thread of control. A *network* is a collection of processes (and is, itself, a process). Note that recursive hierarchies of structure are part of this model: a network is a collection of processes, each of which may be a sub-network or a sequential process.

But how do the processes within a network interact to achieve the behaviour required from the network? They can't see each other's data nor execute each other's algorithms – at least, not if they abide by the rules.

2.2 Synchronising Channels

The simplest form of interaction is synchronised message-passing along channels. The simplest form of channel is zero-buffered and point-to-point. Such channels correspond very closely to our intuitive understanding of a wire connecting two (hardware) components.

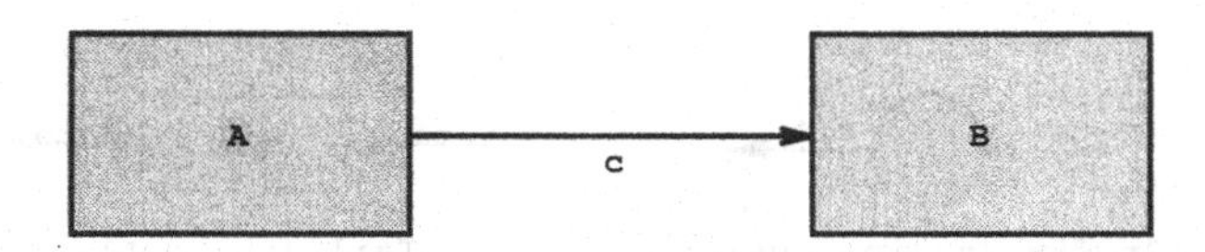

Fig. 1. A simple network

In Figure 1, `A` and `B` are processes and `c` is a channel connecting them. A wire has no capacity to hold data and is only a medium for transmission. To avoid undetected loss of data, channel communication is synchronised. This means that if `A` transmits before `B` is ready to receive, then `A` will block. Similarly, if `B` tries to receive before `A` transmits, `B` will block. When both are ready, a data packet is transferred – directly from the state space of `A` into the state space of `B`. We have a synchronised distributed assignment.

2.3 Legoland

Much can be done, or simplified, just with this basic model – for example the design and simulation of self-timed digital logic, multiprocessor embedded control systems (for which Occam[13–16] was originally designed), GUIs etc.

[1] This is in contrast with simple 'objects' and their 'methods'. A method body normally executes in the thread of control of the invoking object. Consequently, object behaviour is expressed from the point of view of its environment rather than the object itself. This is a slightly confusing property of traditional 'object-oriented' programming.

Here are some simple examples to build up fluency. First we introduce some elementary components from our 'teaching' catalogue – see Figure 2. All processes are cyclic and all transmit and receive just numbers. The `Id` process cycles through waiting for a number to arrive and, then, sending it on. Although inserting an `Id` process in a wire will clearly not affect the data flowing through it, it *does* make a difference. A bare wire has no buffering capacity. A wire containing an `Id` process gives us a one-place FIFO. Connect 20 in series and we get a 20-place FIFO – sophisticated function from a trivial design.

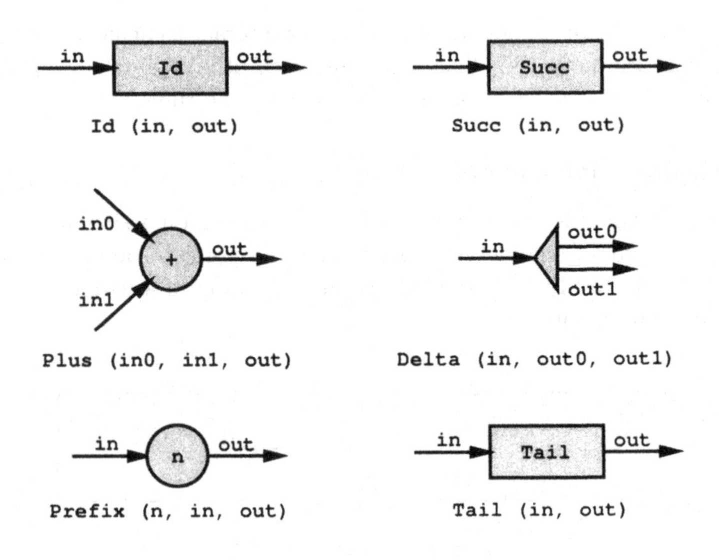

Fig. 2. Extract from a component catalogue

`Succ` is like `Id`, but increments each number as it flows through. The `Plus` component waits until a number arrives on each input line (accepting their arrival in either order) and outputs their sum. `Delta` waits for a number to arrive and, then, broadcasts it in parallel on its two output lines – both those outputs must complete (in either order) before it cycles round to accept further input. `Prefix` first outputs the number stamped on it and then behaves like `Id`. `Tail` swallows its first input without passing it on and then, also, behaves like `Id`. `Prefix` and `Tail` are so named because they perform, respectively, prefixing and tail operations on the streams of data flowing through them.

It's essential to provide a practical environment in which students can develop executable versions of these components and play with them (by plugging them together and seeing what happens). This is easy to do in Occam and now, with the JCSP library [11], in Java. Appendices A and B give some of the details. Here we only give some CSP pseudo-code for our catalogue (because that's shorter than the real code):

```
Id (in, out) = in ? x --> out ! x --> Id (in, out)

Succ (in, out) = in ? x --> out ! (x+1) --> Succ (in, out)

Plus (in0, in1, out)
  = ((in0 ? x0 --> SKIP) || (in1 ? x1 --> SKIP));
    out ! (x0 + x1) --> Plus (in0, in1, out)

Delta (in, out0, out1)
  = in ? x --> ((out0 ! x --> SKIP) || (out1 ! x --> SKIP));
    Delta (in, out0, out1)

Prefix (n, in, out) = out ! n --> Id (in, out)

Tail (in, out) = in ? x --> Id (in, out)
```

[Notes: 'free' variables used in these pseudo-codes are assumed to be locally declared and hidden from outside view. All these components are *sequential* processes. The process `(in ? x --> P (...))` means: "wait until you can engage in the input event `(in ? x)` and, then, become the process `P (...)`". The input operator (`?`) and output operator (`!`) bind more tightly than the `-->`.]

2.4 Plug and Play

Plugging these components together and reasoning about the resulting behaviour is easy. Thanks to the rules on process privacy[2], race hazards leading to unpredictable internal state do not arise. Thanks to the rules on channel synchronisation, data loss or corruption during communication cannot occur[3]. What makes the reasoning simple is that the parallel constructor and channel primitives are deterministic. Non-determinism has to be explicitly designed into a process and coded – it can't sneak in by accident!

Figure 3 shows a simple example of reasoning about network composition. Connect a `Prefix` and a `Tail` and we get two `Id`s:

```
(Prefix (in, c) || Tail (c, out))  =  (Id (in, c) || Id (c, out))
```

Equivalence means that no environment (i.e. eexternal network in which they are placed) can tell them apart. In this case, both circuit fragments implement a 2-place FIFO. The only place where anything different happens is on the internal wire and that's undetectable from outside. The formal proof is a one-liner from the definition of the parallel (`||`), communications (`!`, `?`) and *and-then-becomes* (`-->`) operators in CSP. But the good thing about CSP is that the mathematics

[2] No external access to internal data. No external execution of internal algorithms (methods).

[3] Unreliable communications over a distributed network can be accommodated in this model – the unreliable network being another active process (or set of processes) that happens not to guarantee to pass things through correctly.

engineered into its design and semantics cleanly reflects an intuitive human feel for the model. We can see the equivalence at a glance and this quickly builds confidence both for us and our students.

Figure 4 shows some more interesting circuits with the first two incorporating feedback. What do they do? Ask the students! Here are some CSP pseudo-codes for these circuits:

```
Numbers (out)
  = Prefix (0, c, a) || Delta (a, out, b) || Succ (b, c)

Integrate (in, out)
  = Plus (in, c, a) || Delta (a, out, b) || Prefix (0, b, c)

Pairs (in, out)
  = Delta (in, a, b) || Tail (b, c) || Plus (a, c, out)
```

Again, our rule for these pseudo-codes means that a, b and c are locally declared channels (hidden, in the CSP sense, from the outside world). Appendices A and B list Occam and Java executables – notice how closely they reflect the CSP.

Back to what these circuits do: **Numbers** generates the sequence of natural numbers, **Integrate** computes running sums of its inputs and **Pairs** outputs the sum of its last two inputs. If we wish to be more formal, let c<i> represent the i'th element that passes through channel c – i.e. the first element through is c<1>. Then, for any i >= 1:

```
Numbers:    out<i> = i - 1
Integrate:  out<i> = Sum {in<j> | j = 1..i}
Pairs:      out<i> = in<i> + in<i + 1>
```

Be careful that the above details only *part* of the specification of these circuits: how the values in their output stream(s) relate to the values in their input stream(s). We also have to be aware of how flexible they are in synchronising with their environments, as they generate and consume those streams. The base level components Id, Succ, Plus and Delta each demand one input (or pair of inputs) before generating one output (or pair of outputs). Tail demands two inputs before its first output, but thereafter gives one output for each input. This effect carries over into Pairs. Integrate adds 2-place buffering between its input and output channels (ignoring the transformation in the actual values passed). Numbers will always deliver to anything trying to take input from it.

If necessary, we can make these synchronisation properties mathematically precise. That is, after all, one of the reasons for which CSP was designed.

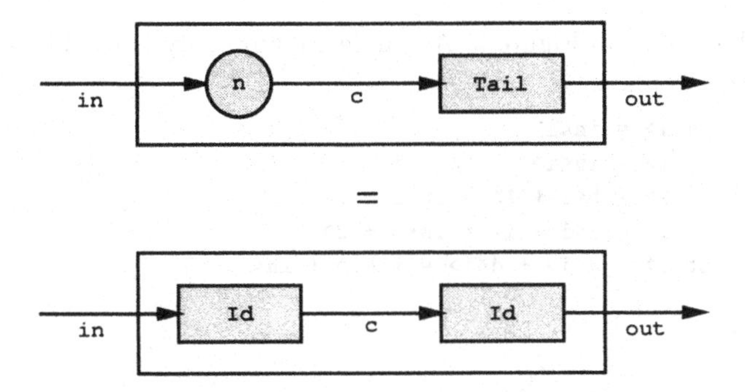

Fig. 3. A simple equivalence

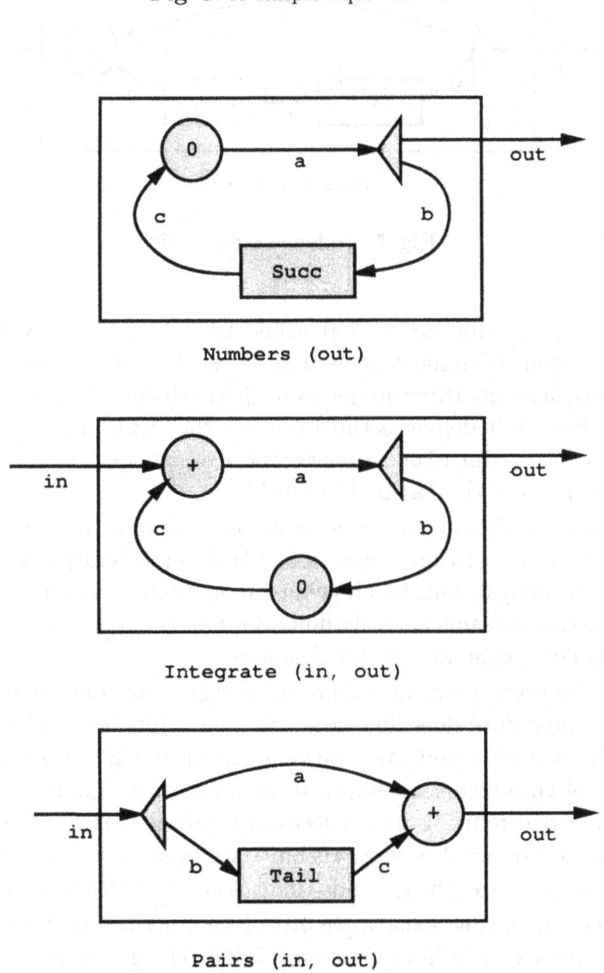

Fig. 4. Some more interesting circuits

2.5 Deadlock – First Contact

Consider the circuit in Figure 5. A simple stream analysis would indicate that:

```
Pairs2:     a<i> = in<i>
Pairs2:     b<i> = in<i>
Pairs2:     c<i> = b<i + 1> = in<i + 1>
Pairs2:     d<i> = c<i + 1> = in<i + 2>
Pairs2:   out<i> = a<i> + d<i> = in<i> + in<i + 2>
```

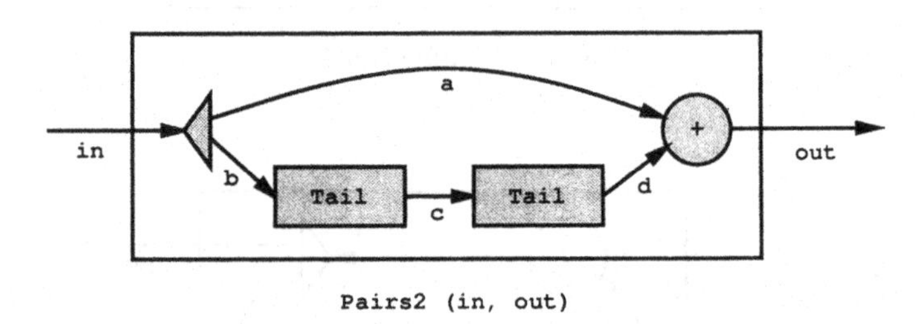

Fig. 5. A dangerous circuit

But this analysis only shows what would be generated *if* anything were generated. In this case, nothing is generated since the system deadlocks. The two `Tail` processes demand three items from `Delta` before delivering anything to `Plus`. But `Delta` can't deliver a third item to the `Tails` until it's got rid of its second item to `Plus`. But `Plus` won't accept a second item from `Delta` until it's had its first item from the `Tails`. Deadlock!

In this case, deadlock can be designed out by inserting an `Id` process on the upper (`a`) channel. `Id` processes (and FIFOs in general) have no impact on stream contents analysis but, by allowing a more decoupled synchronisation, *can* impact on whether streams actually flow. Beware, though, that adding buffering to channels is not a general cure for deadlock.

So, there are always two questions to answer: what data flows through the channels, assuming data does flow, and are the circuits deadlock-free? Deadlock is a monster that must – and can – be vanquished. In CSP, deadlock only occurs from a cycle of committed attempts to communicate (input or output): each process in the cycle refusing its predecessor's call as it tries to contact its successor. Deadlock potential is very visible – we even have a deadlock primitive (`STOP`) to represent it, on the grounds that it is a good idea to know your enemy!

In practice, there now exist a wealth of design rules that provide formally proven guarantees of deadlock freedom[17–22]. Design tools supporting these rules – both constructive and analytical – have been researched[23, 24]. Deadlock, together with related problems such as livelock and starvation, need threaten us no longer – even in the most complex of parallel system.

2.6 Structured Plug and Play

Consider the circuits of Figure 6. They are similar to the previous circuits, but contain components other than those from our base catalogue – they use components we have just constructed. Here is the CSP:

```
Fibonacci (out)
  = Prefix (1, d, a) || Prefix (0, a, b) ||
    Delta (b, out, c) || Pairs (c, d)

Squares (out)
  = Numbers (a) || Integrate (a, b) || Pairs (b, out)

Demo (out)
  = Numbers (a) || Fibonacci (b) || Squares (c) ||
    Tabulate3 (a, b, c, out)
```

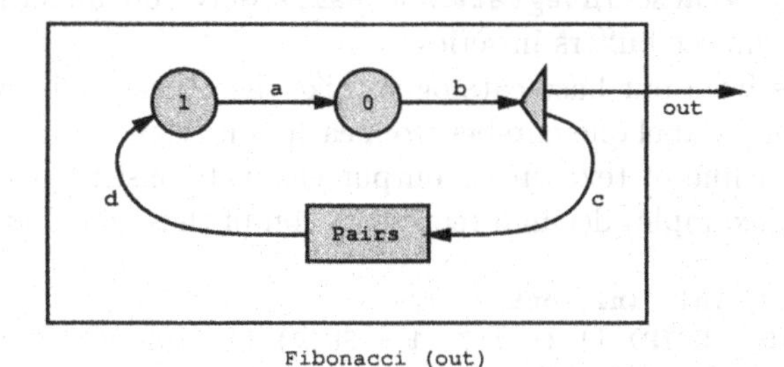

Fibonacci (out)

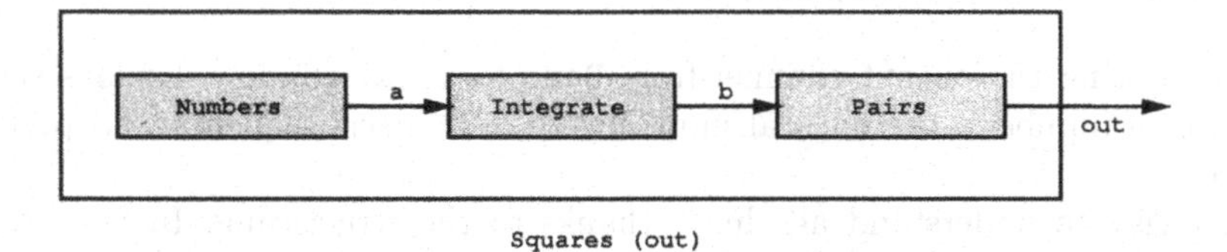

Squares (out)

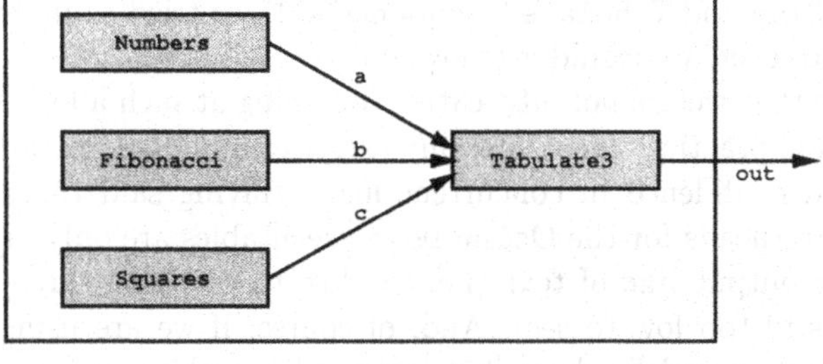

Demo (out)

Fig. 6. Circuits of circuits

One of the powers of CSP is that its semantics obey simple composition rules. To understand the behaviour implemented by a network, we only need to know the behaviour of its nodes – not their implementations.

For example, Fibonacci is a feedback loop of four components. At this level, we can remain happily ignorant of the fact that its Pairs node contains another three. We only need to know that Pairs requires two numbers before it outputs anything and that, thereafter, it outputs once for every input. The two Prefixes initially inject two numbers (0 and 1) into the circuit. Both go into Pairs, but only one (their sum) emerges. After this, the feedback loop just contains a single circulating packet of information (successive elements of the Fibonacci sequence). The Delta process taps this circuit to provide external output.

Squares is a simple pipeline of three components. It's best not to think of the nine processes actually involved. Clearly, for i >= 1:

```
Squares:    a<i> = i - 1
Squares:    b<i> = Sum {j - 1 | j = 1..i} = Sum {j | j = 0..(i - 1)}
Squares: out<i> = Sum {j | j = 0..(i - 1)} + Sum {j | j = 0..i} = i * i
```

So, Squares outputs the increasing sequence of squared natural numbers. It doesn't deadlock because Integrate and Pairs only add buffering properties and it's safe to connect buffers in series.

Tabulate3 is from our base catalogue. Like the others, it is cyclic. In each cycle, it inputs in parallel one number from each of its three input channels and, then, generates a line of text on its output channel consisting of a tabulated (15-wide, in this example) decimal representation of those numbers.

```
Tabulate3 (in0, in1, in2, out)
  = ((in0 ? x0 - SKIP) || (in1 ? x1 - SKIP) || (in2 ? x2 - SKIP));
    print (x0, 15, out); print (x1, 15, out); println (x2, 15, out);
    Tabulate3 (in0, in1, in2, out)
```

Connecting the output channel from Demo to a text window displays three columns of numbers: the natural numbers, the Fibonacci sequence and perfect squares.

It's easy to understand all this – thanks to the structuring. In fact, Demo consists of 27 threads of control, 19 of them permanent with the other 8 being repeatedly created and destroyed by the low-level parallel inputs and outputs in the Delta, Plus and Tabulate3 components. If we tried to understand it on those terms, however, we would get nowhere.

Please note that we are not advocating designing at such a fine level of granularity as normal practice! These are only exercises and demonstrations to build up fluency and confidence in concurrent logic. Having said that, the process management overheads for the Occam Demo executables are only around 30 microseconds per output line of text (i.e. too low to see) and three milliseconds for the Java (still too low to see). And, of course, if we are using these techniques for designing real hardware[25], we will be working at much finer levels of granularity than this.

2.7 Coping with the Real World – Making Choices

The model we have considered so far – parallel processes communicating through dedicated (point-to-point) channels – is *deterministic*. If we input the same data in repeated runs, we will always receive the same results. This is true regardless of how the processes are scheduled or distributed. This provides a very stable base from which to explore the real world, which doesn't always behave like this.

Any machine with externally operatable controls that influence its internal operation, but whose internal operations will continue to run in the absence of that external control, is not deterministic in the above sense. The scheduling of that external control will make a difference. Consider a car and its driver heading for a brick wall. Depending on when the driver applies the brakes, they will end up in very different states!

CSP provides operators for internal and external choice. An external choice is when a process waits for its environment to engage in one of several events – what happens next is something the environment can determine (e.g. a driver can press the accelerator or brake pedal to make the car go faster or slower). An internal choice is when a process changes state for reasons its environment cannot determine (e.g. a self-clocked timeout or the car runs out of petrol). Note that for the combined (parallel) system of car-and-driver, the accelerating and braking become internal choices so far as the rest of the world is concerned.

Occam provides a constructor (`ALT`) that lets a process wait for one of many events. These events are restricted to channel input, timeouts and `SKIP` (a *null* event that has always happened). We can also set pre-conditions – run-time tests on internal state – that mask whether a listed event should be included in any particular execution of the `ALT`. This allows very flexible internal choice within a component as to whether it is prepared to accept an external communication[4]. The JavaPP libraries provide an exact analogue (`Alternative.select`) for these choice mechanisms.

If several events are pending at an `ALT`, an internal choice is normally made between them. However, Occam allows a `PRI ALT` which resolves the choice between pending events in order of their listing. This returns control of the operation to the environment, since the reaction of the `PRI ALT`ing process to multiple communications is now predictable. This control is crucial for the provision of real-time guarantees in multi-process systems and for the design of hardware. Recently, extensions to CSP to provide a formal treatment of these mechanisms have been made[26, 27].

Figure 7 shows two simple components with this kind of control. `Replace` listens for incoming data on its `in` and `inject` lines. Most of the time, data arrives from `in` and is immediately copied to its `out` line. Occasionally, a signal from the `inject` line occurs. When this happens, the signal is copied `out` but, *at the same time*, the next input from `in` is waited for and discarded. In case

[4] This is in contrast to *monitors*, whose methods cannot refuse an external call when they are unlocked and have to *wait* on condition variables should their state prevent them from servicing the call. The close coupling necessary between sibling monitor methods to undo the resulting mess is not WYSIWYG[9].

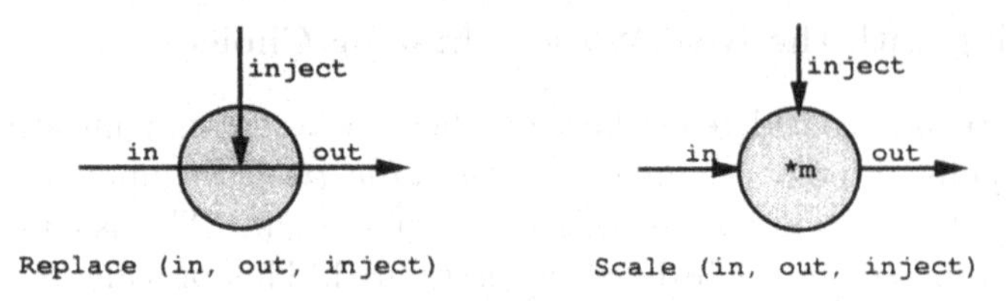

Fig. 7. Two control processes

both `inject` and `in` communications are on offer, priority is given to the (less frequently occurring) `inject`:

```
Replace (in, inject, out)
  = (inject ? signal --> ((in ? x --> SKIP) || (out ! signal --> SKIP))
     [PRI]
     in ? x --> out ! x --> SKIP
    );
    Replace (in, inject, out)
```

`Replace` is something that can be spliced into any channel. If we don't use the `inject` line, all it does is add a one-place buffer to the circuit. If we send something down the `inject` line, it gets injected into the circuit – replacing the next piece of data that would have travelled through that channel.

Figure 8 shows `RNumbers` and `RIntegrate`, which are just `Numbers` and `Integrate` with an added `Replace` component. We now have components that are resettable by their environments. `RNumbers` can be reset at any time to continue its output sequence from any chosen value. `RIntegrate` can have its internal running sum redefined.

Like `Replace`, `Scale` (figure 7) normally copies numbers straight through, but scales them by its factor `m`. An `inject` signal resets the scale factor:

```
Scale (m, in, inject, out)
  = (inject ? m --> SKIP
     [PRI]
     in ? x --> out ! m*x --> SKIP
    );
    Scale (m, in, inject, out)
```

Figure 9 shows `RPairs`, which is `Pairs` with the `Scale` control component added. If we send just `+1` or `-1` down the reset line of `RPairs`, we control whether it's adding or subtracting successive pairs of inputs. When it's subtracting, its behaviour changes to that of a *differentiator* – in the sense that it undoes the effect of `Integrate`.

This allows a nice control demonstration. Figure 10 shows a circuit whose core is a resettable version of the `Squares` pipeline. The `Monitor` process reacts to characters from the `keyboard` channel. Depending on its value, it outputs an appropriate signal down an appropriate reset channel:

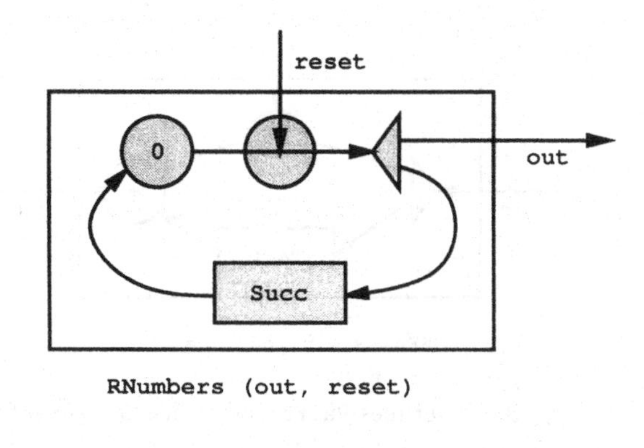

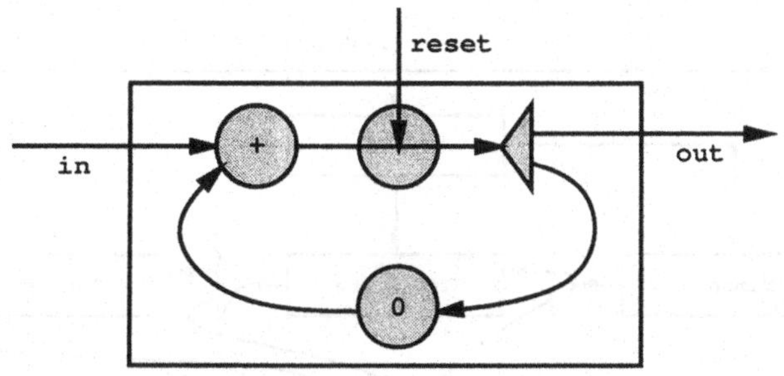

Fig. 8. Two controllable processes

```
Monitor (keyboard, resetN, resetI, resetP)
  = (keyboard ? ch -->
       CASE ch
         'N': resetN !  0 --> SKIP
         'I': resetI !  0 --> SKIP
         '+': resetP ! +1 --> SKIP
         '-': resetP ! -1 --> SKIP
    );
    Monitor (keyboard, resetN, resetI, resetP)
```

When Demo2 runs and we don't type anything, we see the inner workings of the Squares pipeline tabulated in three columns of output. Keying in an 'N', 'I', '+' or '-' character allows the user some control over those workings[5]. Note that after a '-', the output from RPairs should be the same as that taken from RNumbers.

[5] In practice, we need to add another process after Tabulate3 to slow down the rate of output to around 10 lines per second. Otherwise, the user cannot properly appreciate the immediacy of control that has been obtained.

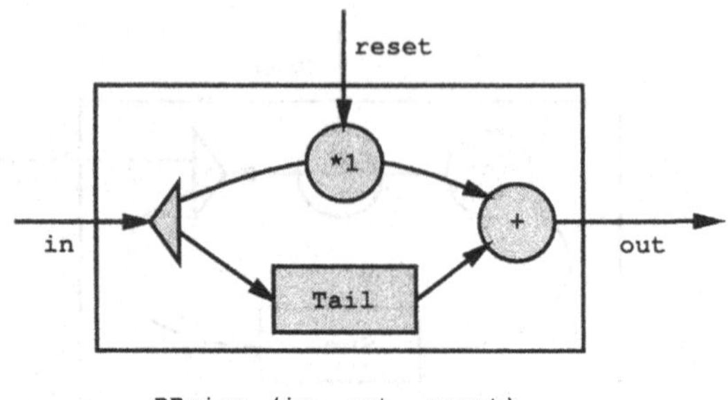

Fig. 9. Sometimes `Pairs`, sometimes `Differentiate`

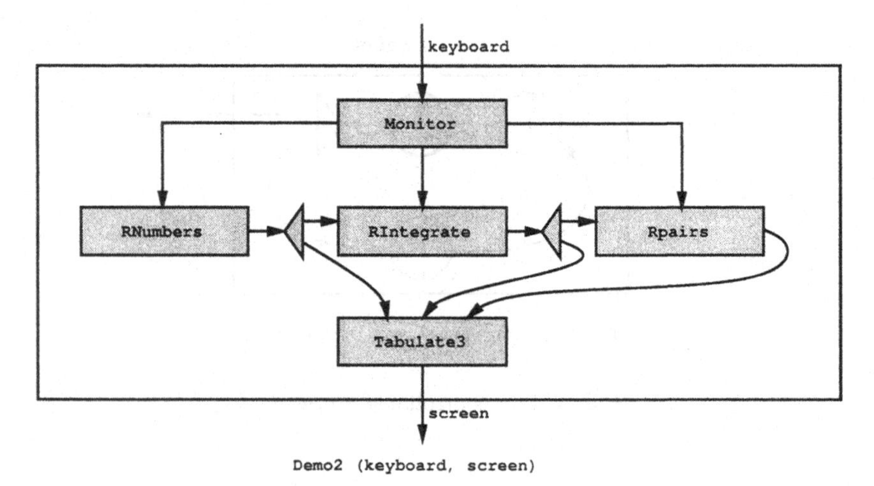

Fig. 10. A user controllable machine

2.8 A Nastier Deadlock

One last exercise should be done. Modify the system so that output freezes if an 'F' is typed and unfreezes following the next character.

Two 'solutions' offer themselves and Figure 11 shows the *wrong* one (`Demo3`). This feeds the output from `Tabulate3` back to a modified `Monitor2` and then on to the `screen`. The `Monitor2` process `PRI ALT`s between the `keyboard` channel and this `feedback`:

```
Monitor2 (keyboard, feedback, resetN, resetI, resetP, screen)
  = (keyboard ? ch -->
       CASE ch
          ...  deal with 'N', 'I', '+', '-' as before
          'F': keyboard ? ch --> SKIP
     [PRI]
     feedback ? x --> screen ! x  --> SKIP
    );
    Monitor2 (keyboard, feedback, resetN, resetI, resetP, screen)
```

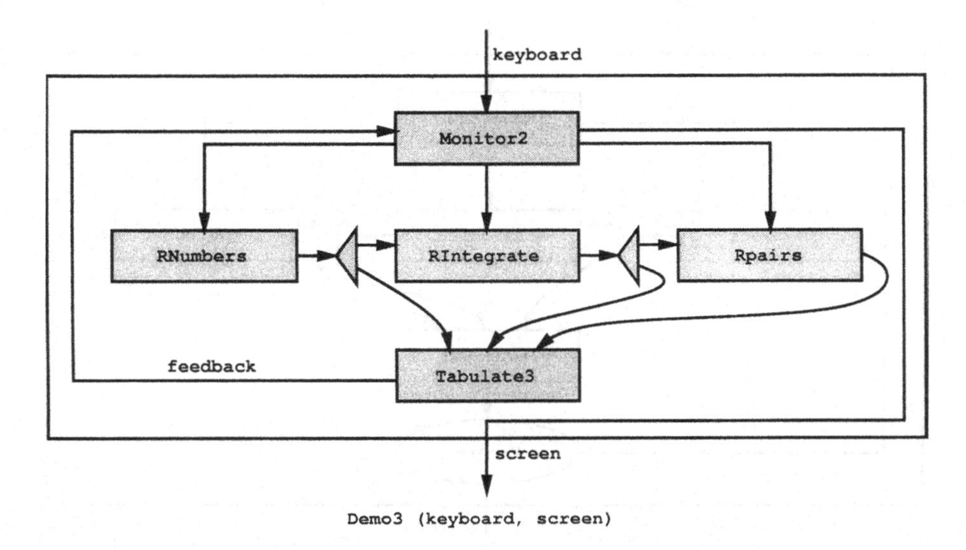

Fig. 11. A machine over which we may lose control

Traffic will normally be flowing along the `feedback-screen` route, interrupted only when `Monitor2` services the `keyboard`. The attraction is that if an 'F' arrives, `Monitor2` simply waits for the next character (and discards it). As a side-effect of this waiting, the `screen` traffic is frozen.

But if we implement this, we get some worrying behaviour. The freeze operation works fine and so, *probably*, do the 'N' and 'I' resets. *Sometimes*, however, a '+' or '-' reset deadlocks the whole system – the `screen` freezes and all further `keyboard` events are refused!

The problem is that one of the rules for deadlock-free design has been broken: *any data-flow circuit must control the number of packets circulating!* If this number rises to the number of sequential (i.e. lowest level) processes in the circuit, deadlock always results. Each node will be trying to output to its successor and refusing input from its predecessor.

The `Numbers`, `RNumbers`, `Integrate`, `RIntegrate` and `Fibonacci` networks all contain data-flow loops, but the number of packets concurrently in flight is kept at one[6].

In `Demo3` however, packets are continually being generated within `RNumbers`, flowing through several paths to `Monitor2` and, then, to the `screen`. Whenever `Monitor2` feeds a reset back into the circuit, deadlock is possible – although not certain. It depends on the scheduling. `RNumbers` is always pressing new packets into the system, so the circuits are likely to be fairly full. If `Monitor2` generates a reset when they are full, the system deadlocks. The shortest feedback loop is from `Monitor2`, `RPairs`, `Tabulate3` and back to `Monitor2` - hence, it is the '+' and '-' inputs from `keyboard` that are most likely to trigger the deadlock.

[6] Initially, `Fibonacci` has two packets, but they combine into one before the end of their first circuit.

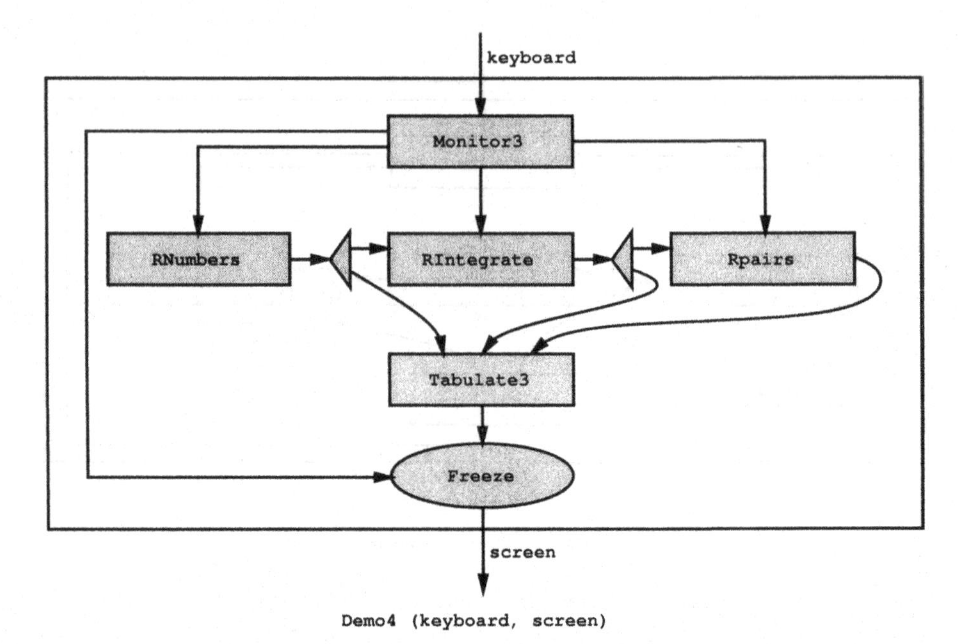

Fig. 12. A machine over which we will not lose control

The design is simply fixed by removing that feedback at this level – see Demo4 in Figure 12. We have abstracted the freezing operation into its own component (and catalogued it). It's never a good idea to try and do too many functions in one sequential process. That needlessly constrains the synchronisation freedom of the network and heightens the risk of deadlock. Note that the idea being pushed here is that, unless there are special circumstances, *parallel design is safer and simpler than its serial counterpart!*

Demo4 obeys another golden rule: *every device should be driven from its own separate process.* The keyboard and screen channels interface to separate devices and should be operated concurrently (in Demo3, both were driven from one sequential process – Monitor2). Here are the driver processes from Demo4:

```
Freeze (in, freeze, out)
  = (freeze ? x --> freeze ? x --> SKIP
      [PRI]
      (in ? x --> out ! x --> SKIP
    );
    Freeze (in, freeze, out)

Monitor3 (keyboard, resetN, resetI, resetP, freeze)
  = (keyboard ? ch -->
       CASE ch
         ... deal with 'N', 'I', '+', '-' as before
         'F': freeze ! ch --> keyboard ? ch --> freeze ! ch --> SKIP
    );
    Monitor3 (keyboard, resetN, resetI, resetP, freeze)
```

2.9 Buffered and Asynchronous Communications

We have seen how fixed capacity FIFO buffers can be added as active processes to CSP channels. For the Occam binding, the overheads for such extra processes are negligible.

With the JavaPP libraries, the same technique *may* be used, but the channel objects can be directly configured to support buffered communications – which saves a couple of context switches. The user may supply objects supporting *any* buffering strategy for channel configuration, including normal blocking buffers, overwrite-when-full buffers, infinite buffers and black-hole buffers (channels that can be written to but not read from – useful for masking off unwanted outputs from components that, otherwise, we wish to reuse intact). However, the user had better stay aware of the semantics of the channels thus created!

Asynchronous communication is commonly found in libraries supporting inter-processor message-passing (such as PVM and MPI). However, the concurrency model usually supported is one for which there is only *one* thread of control on each processor. Asynchronous communication lets that thread of control launch an external communication and continue with its computation. At some point, that computation may need to block until that communication has completed.

These mechanisms are easy to obtain from the concurrency model we are teaching (and which we claim to be general). We don't need anything new. Asynchronous sends are what happen when we output to a buffer (or buffered channel). If we are worried about being blocked when the buffer is full or if we need to block at some later point (should the communication still be unfinished), we can simply spawn off another process[7] to do the send:

```
(out ! packet --> SKIP |PRI| someMoreComputation (...));
continue (...)
```

The `continue` process only starts when both the `packet` has been sent and `someMoreComputation` has finished. `someMoreComputation` and sending the `packet` proceed concurrently. We have used the priority version of the parallel operator (`|PRI|`, which gives priority to its left operand), to ensure that the sending process initiates the transfer before the `someMoreComputation` is scheduled. Asynchronous receives are implemented in the same way:

```
(in ? packet --> SKIP |PRI| someMoreComputation (...));
continue (...)
```

2.10 Shared Channels

CSP channels are strictly point-to-point. occam3[28] introduced the notion of (securely) *shared* channels and channel structures. These are further extended in the KRoC Occam[29] and JavaPP libraries and are included in the teaching model.

[7] The Occam overheads for doing this are less than half a microsecond.

A channel structure is just a record (or object) holding two or more CSP channels. Usually, there would be just two channels – one for each direction of communication. The channel structure is used to conduct a two-way conversation between two processes. To avoid deadlock, of course, they will have to understand protocols for using the channel structure – such as who speaks first and when the conversation finishes. We call the process that opens the conversation a *client* and the process that listens for that call a *server*[8].

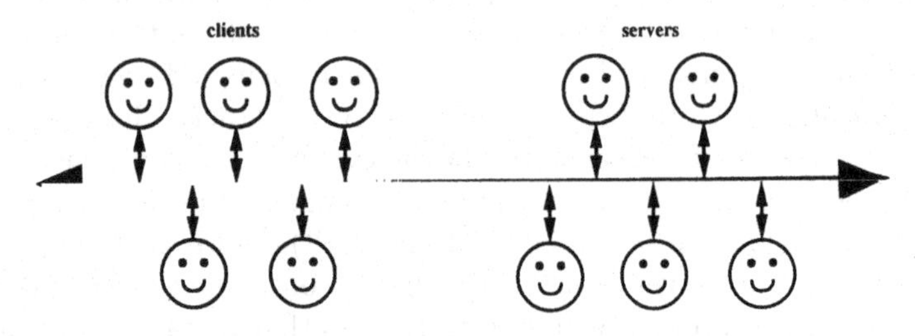

Fig. 13. A many-many shared channel

The CSP model is extended by allowing multiple clients and servers to share the same channel (or channel structure) – see Figure 13. Sanity is preserved by ensuring that only one client and one server use the shared object at any one time. Clients wishing to use the channel queue up first on a client-queue (associated with the shared channel) – servers on a server-queue (also associated with the shared channel). A client only completes its actions on the shared channel when it gets to the front of its queue, finds a server (for which it may have to wait if business is good) and completes its transaction. A server only completes when it reaches the front of its queue, finds a client (for which it may have to wait in times of recession) and completes its transaction.

Note that shared channels – like the choice operator between multiple events – introduce scheduling dependent non-determinism. The order in which processes are granted access to the shared channel depends on the order in which they join the queues.

Shared channels provide a very efficient mechanism for a common form of choice. Any server that offers a non-discriminatory service[9] to multiple clients should use a shared channel, rather than `ALT`ing between individual channels from those clients. The shared channel has a constant time overhead – `ALT`ing is linear on the number of clients. However, if the server needs to discriminate between its clients (e.g. to refuse service to some, depending upon its internal state), `ALT`ing gives us that flexibility. The mechanisms can be efficiently com-

[8] In fact, the client/server relationship is with respect to the channel structure. A process may be both a server on one interface and a client on another.

[9] Examples for such servers include window managers for multiple animation processes, data loggers for recording traces from multiple components from some machine, etc.

bined. Clients can be grouped into equal-treatment partitions, with each group clustered on its own shared channel and the server `ALTing` between them.

For deadlock freedom, each server must guarantee to respond to a client call within some bounded time. During its transaction with the client, it must follow the protocols for communication defined for the channel structure *and* it may engage in separate client transactions with other servers. A client may open a transaction at any time but may not interleave its communications with the server with any other synchronisation (e.g. with another server). These rules have been formalised as CSP specifications[21]. Client-server networks may have plenty of data-flow feedback but, so long as no cycle of client-server relations exist, [21] gives formal proof that the system is deadlock, livelock and starvation free.

Shared channel structures may be stretched across distributed memory (e.g. networked) multiprocessors[15]. Channels may carry all kinds of object – including channels and processes themselves. A shared channel is an excellent means for a client and server to find each other, pass over a private channel and communicate independently of the shared one. Processes will drag pre-attached channels with them as they are moved and can have local channels dynamically (and temporarily) attached when they arrive. See David May's work on *Icarus*[30, 31] for a consistent, simple and practical realisation of this model for distributed and mobile computing.

3 Events and Shared Memory

Shared memory concurrency is often described as being 'easier' than message passing. But great care must be taken to synchronise concurrent access to shared data, else we will be plagued with race hazards and our systems will be useless. CSP primitives provide a sharp set of tools for exercising this control.

3.1 Symmetric Multi-Processing (SMP)

The private memory/algorithm principles of the underlying model – and the security guarantees that go with them – are a powerful way of programming shared memory multiprocessors. Processes can be automatically and dynamically scheduled between available processors (*one object code fits all*). So long as there is an excess of (runnable) processes over processors and the scheduling overheads are sufficiently low, high multiprocessor efficiency can be achieved with guaranteed no race hazards. With the design methods we have been describing, it's very easy to generate *lots* of processes with most of them runnable most of the time.

3.2 Token Passing and Dynamic CREW

Taking advantage of shared memory to communicate between processes is an extension to this model and must be synchronised. The shared data does not belong to any of the sharing processes, but must be globally visible to them – either on the stack (for Occam) or heap (for Java).

The JavaPP channels in previous examples were only used to send data *values* between processes – but they can also be used to send objects. This steps outside the automatic guarantees against race hazard since, unconstrained, it allows parallel access to the same data. One common and useful constraint is only to send *immutable* objects. Another design pattern treats the sent object as a *token* conferring permission to use it – the sending process losing the token as a side-effect of the communication. The trick is to ensure that only one copy of the token ever exists for each sharable object.

Dynamic CREW (Concurrent Read Exclusive Write) operations are also possible with shared memory. Shared channels give us an efficient, elegant and easily provable way to construct an active *guardian* process with which application processes synchronise to effect CREW access to the shared data. Guarantees against starvation of writers by readers – and vice-versa – are made. Details will appear in a later report (available from [32]).

3.3 Structured Barrier Synchronisation and SPMD

Point-to-point channels are just a specialised form of the general CSP multi-process synchronising *event*. The CSP parallel operator binds processes together with events. When *one* process synchronises on an event, *all* processes registered for that event must synchronise on it before that first process may continue. Events give us structured multiway barrier synchronisation[29].

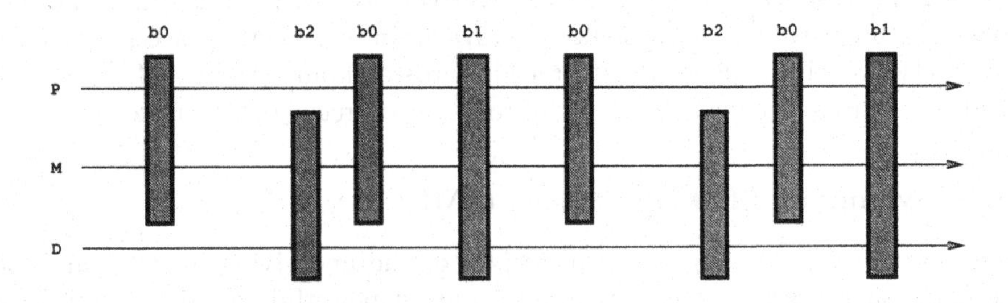

Fig. 14. Multiple barriers to three processes

We can have many event barriers in a system, with different (and not necessarily disjoint) subsets of processes registered for each barrier. Figure 14 shows the execution traces for three processes (P, M and D) with time flowing horizontally. They do not all progress at the same – or even constant – speed. From time to time, tha faster ones will have to wait for their slower partners to reach an agreed barrier before all of them can proceed. We can wrap up the system in typical SPMD form as:

```
|| <i = 0 FOR 3>
  S (i, ..., b0, b1, b2)
```

where `b0`, `b1` and `b2` are events. The replicated parallel operator runs 3 instances of `S` in parallel (with `i` taking the values `0`, `1` and `2` respectively in the different instances). The `S` process simply switches into the required form:

```
S (i, ..., b0, b1, b2)
  = CASE i
      0 : P (..., b0, b1)
      1 : M (..., b0, b1, b2)
      2 : D (..., b1, b2)
```

and where `P`, `M` and `D` are registered only for the events in their parameters. The code for `P` has the form:

```
P (..., b0, b1)
   = someWork (...); b0 --> SKIP;
     moreWork (...); b0 --> SKIP;
     lastBitOfWork (...); b1 --> SKIP;
     P (..., b0, b1)
```

3.4 Non-blocking Barrier Synchronisation

In the same way that asynchronous communications can be expressed (section 2.9), we can also achieve the somewhat contradictory sounding, but potentially useful, *non-blocking* barrier synchronisation.

In terms of serial programming, this is a two-phase commitment to the barrier. The first phase declares that we have done everything we need to do this side of the barrier, but does not block us. We can then continue for a while, doing things that do not disturb what we have set up for our partners in the barrier and do not need whatever it is that they have to set. When we need their work, we enter the second phase of our synchronisation on the barrier. This blocks us only if there is one, or more, of our partners who has not reached the first phase of its synchronisation. With luck, this window on the barrier will enable most processes most of the time to pass through without blocking:

```
doOurWorkNeededByOthers (...);
barrier.firstPhase ();
privateWork (...);
barrier.secondPhase ();
useSharedResourcesProtectedByTheBarrier (...);
```

With our lightweight CSP processes, we do not need these special phases to get the same effect:

```
doOurWorkNeededByOthers (...);
(barrier --> SKIP |PRI| privateWork (...));
useSharedResourcesProtectedByTheBarrier (...);
```

The explanation as to why this works is just the same as for the asynchronous sends and receives.

3.5 Bucket Synchronisation

Although CSP allows choice over general events, the occam and Java bindings do not. The reasons are practical – a concern for run-time overheads[10]. So, synchronising on an event commits a process to wait until everyone registered for the event has synchronised. These multi-way events, therefore, do not introduce non-determinism into a system and provide a stable platform for much scientific and engineering modelling.

Buckets[15] provide a non-deterministic version of events that are useful for when the system being modelled is irregular and dynamic (e.g. motor vehicle traffic[33]). Buckets have just two operations: `jump` and `kick`. There is no limit to the number of processes that can jump into a bucket – where they all block. Usually, there will only be one process with responsibility for kicking over the bucket. This can be done at any time of its own (internal) choosing – hence the non-determinism. The result of kicking over a bucket is the unblocking of all the processes that had jumped into it[11].

4 Conclusions

A simple model for parallel computing has been presented that is easy to learn, teach and use. Based upon the mathematically sound framework of Hoare's CSP, it has a compositional semantics that corresponds well with out intuition about how the world is constructed. The basic model encompasses object-oriented design with active processes (i.e. objects whose methods are exclusively under their own thread of control) communicating via passive, but synchronising, wires. Systems can be composed through natural layers of communicating components so that an understanding of each layer does not depend on an understanding of the inner ones. In this way, systems with arbitrarily complex behaviour can be safely constructed – free from race hazard, deadlock, livelock and process starvation.

A small extension to the model addresses fundamental issues and paradigms for shared memory concurrency (such as token passing, CREW dynamics and bulk synchronisation). We can explore with equal fluency serial, message-passing and shared-memory logic and strike whatever balance between them is appropriate for the problem under study. Applications include hardware design (e.g. FPGAs and ASICs), real-time control systems, animation, GUIs, regular and irregular modelling, distributed and mobile computing.

Occam and Java bindings for the model are available to support practical work on commodity PCs and workstations. Currently, the Occam bindings are

[10] Synchronising on an event in Occam has a unit time overhead, regardless of the number of processes registered. This includes being the last process to synchronise, when all blocked processes are released. These overheads are well below a microsecond for modern microprocessors.

[11] As for events, the `jump` and `kick` operations have constant time overhead, regardless of the number of processes involved. The bucket overheads are slightly lower than those for events.

the fastest (context-switch times under 300 nano-seconds), lightest (in terms of memory demands), most secure (in terms of guaranteed thread safety) and quickest to learn. But Java has the libraries (e.g. for GUIs and graphics) and will get faster. Java thread safety, in this context, depends on following the CSP design patterns – and these are easy to acquire[12].

The JavaPP JCSP library[11] also includes an extension to the Java AWT package that drops channel interfaces on all GUI components[13]. Each item (e.g. a `Button`) is a process with a `configure` and `action` channel interface. These are connected to separate internal handler processes. To change the text or colour of a `Button`, an application process outputs to its `configure` channel. If someone presses the `Button`, it outputs down its `action` channel to an application process (which can accept or refuse the communication as it chooses). Example demonstrations of the use of this package may be found at [11]. Whether GUI programming through the process-channel design pattern is simpler than the listener-callback pattern offered by the underlying AWT, we leave for the interested reader to experiment and decide.

All the primitives described in this paper are available for KRoC occam and Java. Multiprocessor versions of the KRoC kernel targeting NoWs and SMPs will be available later this year. SMP versions of the JCSP[11] and CJT[12] libraries are automatic if your JVM supports SMP threads. Hooks are provided in the channel libraries to allow user-defined network drivers to be installed. Research is continuing on portable/faster kernels and language/tool design for *enforcing* higher level aspects of CSP design patterns (e.g. for shared memory safety and deadlock freedom) that currently rely on self-discipline.

Finally, we stress that this is *undergraduate* material. The concepts are mature and fundamental – *not* advanced – and the earlier they are introduced the better. For developing fluency in concurrent design and implementation, no special hardware is needed. Students can graduate to real parallel systems once they have mastered this fluency. The CSP model is neutral with respect to parallel architecture so that coping with a change in language or paradigm is straightforward. However, even for uni-processor applications, the ability to do safe and lightweight multithreading is becoming crucial *both* to improve response times *and* simplify their design.

The experience at Kent is that students absorb these ideas very quickly and become very creative[14]. Now that they can apply them in the context of Java, they are smiling indeed.

---

[12] Java active objects (processes) do not invoke each other's methods, but communicate only through shared passive objects with carefully designed synchronisation properties (e.g. channels and events). Shared use of *user*-defined passive objects will be automatically thread-safe so long as the usage patterns outlined in Section 3 are kept – their methods should not be `synchronised` (in the sense of Java monitors).

[13] We believe that the new Swing GUI libraries from Sun (that will replace the AWT) can also be extended through a channel interface for secure use in parallel designs – despite the warnings concerning the use of Swing and multithreading[34].

[14] The JCSP libraries used in Appendix B were produced by Paul Austin, an undergraduate student at Kent.

References

1. C.A. Hoare. Communication Sequential Processes. *CACM*, 21(8):666–677, August 1978.
2. C.A. Hoare. *Communication Sequential Processes.* Prentice Hall, 1985.
3. Oxford University Computer Laboratory. *The CSP Archive.* <URL: http:// www.comlab.ox.ac.uk/ archive/ csp.html>, 1997.
4. P.H. Welch and D.C. Wood. KRoC – the Kent Retargetable occam Compiler. In B. O'Neill, editor, *Proceedings of WoTUG 19*, Amsterdam, March 1996. WoTUG, IOS Press. <URL:http:// www.hensa.ac.uk/ parallel/ occam/ projects/ occam-for-all/ kroc/>.
5. Peter H. Welch and Michael D. Poole. occam for Multi-Processor DEC Alphas. In A. Bakkers, editor, *Parallel Programming and Java, Proceedings of WoTUG 20*, volume 50 of *Concurrent Systems Engineering*, pages 189–198, Amsterdam, Netherlands, April 1997. World occam and Transputer User Group (WoTUG), IOS Press.
6. Peter Welch et al. *Java Threads Workshop – Post Workshop Discussion.* <URL:http://www.hensa.ac.uk/parallel/groups/wotug/java/discussion/>, February 1997.
7. Gerald Hilderink, Jan Broenink, Wiek Vervoort, and Andre Bakkers. Communicating Java Threads. In *Parallel Programming and Java, Proceedings of WoTUG 20*, pages 48–76, 1997. (See reference [5]).
8. G.H. Hilderink. Communicating Java Threads Reference Manual. In *Parallel Programming and Java, Proceedings of WoTUG 20*, pages 283–325, 1997. (See reference [5]).
9. Peter Welch. Java Threads in the Light of occam/CSP. In P.H.Welch and A. Bakkers, editors, *Architectures, Languages and Patterns, Proceedings of WoTUG 21*, volume 52 of *Concurrent Systems Engineering*, pages 259–284, Amsterdam, Netherlands, April 1998. World occam and Transputer User Group (WoTUG), IOS Press. ISBN 90-5199-391-9.
10. Alan Chalmers. *JavaPP Page – Bristol.* <URL:http://www.cs.bris.ac.uk/~alan/ javapp.html/>, May 1998.
11. P.D. Austin. *JCSP Home Page.* <URL:http://www.hensa.ac.uk/parallel/languages/ java/jcsp/>, May 1998.
12. Gerald Hilderink. *JavaPP Page – Twente.* <URL:http://www.rt.el.utwente.nl/ javapp/>, May 1998.
13. Ian East. *Parallel Processing with Communication Process Architecture.* UCL press, 1995. ISBN 1-85728-239-6.
14. John Galletly. *occam 2 – including occam 2.1.* UCL Press, 1996. ISBN 1-85728-362-7.
15. occam-for-all Team. *occam-for-all Home Page.* <URL:http://www.hensa.ac.uk/ parallel/occam/occam-for-all/>, February 1997.
16. Mark Debbage, Mark Hill, Sean Wykes, and Denis Nicole. Southampton's Portable occam Compiler (SPoC). In R. Miles and A. Chalmers, editors, *Progress in Transputer and occam Research, Proceedings of WoTUG 17*, Concurrent Systems Engineering, pages 40–55, Amsterdam, Netherlands, April 1994. World occam and Transputer User Group (WoTUG), IOS Press. <URL:http://www.hensa.ac.uk/ parallel/ occam/ compilers/ spoc/>.
17. J.M.R. Martin and S.A. Jassim. How to Design Deadlock-Free Networks Using CSP and Verification Tools – a Tutorial Introduction. In *Parallel Programming and Java, Proceedings of WoTUG 20*, pages 326–338, 1997. (See reference [5]).

18. A.W. Roscoe and N. Dathi. The Pursuit of Deadlock Freedom. Technical Report *Technical Monograph PRG-57*, Oxford University Computing Laboratory, 1986.
19. J. Martin, I. East, and S. Jassim. Design Rules for Deadlock Freedom. *Transputer Communications*, 2(3):121–133, September 1994. John Wiley & Sons, Ltd. ISSN 1070-454X.
20. P.H. Welch, G.R.R. Justo, and C. Willcock. High-Level Paradigms for Deadlock-Free High-Performance Systems. In Grebe et al., editors, *Transputer Applications and Systems '93*, pages 981–1004, Amsterdam, 1993. IOS Press. ISBN 90-5199-140-1.
21. J.M.R. Martin and P.H. Welch. A Design Strategy for Deadlock-Free Concurrent Systems. *Transputer Communications*, 3(4):215–232, October 1996. John Wiley & Sons, Ltd. ISSN 1070-454X.
22. A.W. Roscoe. *Model Checking* CSP, *A Classical Mind*. Prentice Hall, 1994.
23. J.M.R. Martin and S.A. Jassim. A Tool for Proving Deadlock Freedom. In *Parallel Programming and Java, Proceedings of WoTUG 20*, pages 1–16, 1997. (See reference [5]).
24. D.J. Beckett and P.H. Welch. A Strict occam Design Tool. In *Proceedings of UK Parallel '96*, pages 53–69, London, July 1996. BCS PPSIG, Springer-Verlag. ISBN 3-540-76068-7.
25. M. Aubury, I. Page, D. Plunkett, M. Sauer, and J. Saul. Advanced Silicon Prototyping in a Reconfigurable Environment. In *Architectures, Languages and Patterns, Proceedings of WoTUG 21*, pages 81–92, 1998. (See reference [9]).
26. A.E. Lawrence. Extending CSP. In *Architectures, Languages and Patterns, Proceedings of WoTUG 21*, pages 111–132, 1998. (See reference [9]).
27. A.E. Lawrence. HCSP: Extending CSP for Co-design and Shared Memory. In *Architectures, Languages and Patterns, Proceedings of WoTUG 21*, pages 133–156, 1998. (See reference [9]).
28. Geoff Barrett. occam3 reference manual (draft). <URL:http:// www.hensa.ac.uk/ parallel/occam/documents/>, March 1992. (unpublished in paper).
29. Peter H. Welch and David C. Wood. Higher Levels of Process Synchronisation. In *Parallel Programming and Java, Proceedings of WoTUG 20*, pages 104–129, 1997. (See reference [5]).
30. David May and Henk L Muller. Icarus language definition. Technical Report CSTR-97-007, Department of Computer Science, University of Bristol, January 1997.
31. Henk L. Muller and David May. A simple protocol to communicate channels over channels. Technical Report CSTR-98-001, Department of Computer Science, University of Bristol, January 1998.
32. D.J. Beckett. *Java Resources Page.* <URL:http://www.hensa.ac.uk/parallel/ languages/java/>, May 1998.
33. Kang Hsin Lu, Jeff Jones, and Jon Kerridge. Modelling Congested Road Traffic Networks Using a Highly Parallel System. In A. DeGloria, M.R. Jane, and D. Marini, editors, *Transputer Applications and Systems '94*, volume 42 of *Concurrent Systems Engineering*, pages 634–647, Amsterdam, Netherlands, September 1994. The Transputer Consortium, IOS Press. ISBN 90-5199-177-0.
34. Hans Muller and Kathy Walrath. Threads and swing. <URL:http://java.sun.com/ products/jfc/swingdoc-archive/threads.html>, April 1998.

Appendix A: Occam Executables

Space only permits a sample of the examples to be shown here. This first group are from the 'Legoland' catalogue (Section 2.3):

```
PROC Id (CHAN OF INT in, out)
  WHILE TRUE
    INT x:
    SEQ
      in ? x
      out ! x
:
```

```
PROC Succ (CHAN OF INT in, out)
  WHILE TRUE
    INT x:
    SEQ
      in ? x
      out ! x PLUS 1
:
```

```
PROC Plus (CHAN OF INT in0, in1, out)
  WHILE TRUE
    INT x0, x1:
    SEQ
      PAR
        in0 ? x0
        in1 ? x1
      out ! x0 PLUS x1
:
```

```
PROC Prefix (VAL INT n, CHAN OF INT in, out)
  SEQ
    out ! n
    Id (in, out)
:
```

Next come four of the 'Plug and Play' examples from Sections 2.4 and 2.6:

```
PROC Numbers (CHAN OF INT out)
  CHAN OF INT a, b, c:
  PAR
    Prefix (0, c, a)
    Delta (a, out, b)
    Succ (b, c)
:
```

```
PROC Integrate (CHAN OF INT in, out)
  CHAN OF INT a, b, c:
  PAR
    Plus (in, c, a)
    Delta (a, out, b)
    Prefix (0, b, c)
:
```

```
PROC Pairs (CHAN OF INT in, out)
  CHAN OF INT a, b, c:
  PAR
    Delta (in, a, b)
    Tail (b, c)
    Plus (a, c, out)
:
```

```
PROC Squares (CHAN OF INT out)
  CHAN OF INT a, b:
  PAR
    Numbers (a)
    Integrate (a, b)
    Pairs (b, out)
:
```

Here is one of the controllers from Section 2.7:

```
PROC Replace (CHAN OF INT in, inject, out)
  WHILE TRUE
    PRI ALT
      INT x:
      inject ? x
        PAR
          INT discard:
          in ? discard
          out ! x
      INT x:
      in ? x
        out ! x
:
```

Asynchronous receive from Section 2.9:

```
SEQ
  PRI PAR
    in ? packet
    someMoreComputation (...)
  continue (...)
```

Barrier synchronisation from Section 3.3:

```
PROC P (..., EVENT b0, b1)
  ...  local state declarations
  SEQ
    ...  initialise local state
    WHILE TRUE
      SEQ
        someWork (...)
        synchronise.event (b0)
        moreWork (...)
        synchronise.event (b0)
        lastBitOfWork (...)
        synchronise.event (b1)
:
```

Finally, non-blocking barrier synchronisation from Section 3.4:

```
SEQ
  doOurWorkNeededByOthers (...)
  PRI PAR
    synchronise.event (barrier)
    privateWork (...)
  useSharedResourcesProtectedByTheBarrier (...)
```

Appendix B: Java Executables

These examples use the JCSP library for processes and channels[11]. A process is an instance of a class that implements the `CSProcess` interface. This is similar to, but different from, the standard `Runable` interface:

```
package jcsp.lang;

public interface CSProcess {
  public void run ();
}
```

For example, from the 'Legoland' catalogue (Section 2.3):

```
import jcsp.lang.*;         // processes and object carrying channels
import jcsp.lang.ints.*;    // integer versions of channels

class Succ implements CSProcess {

  private ChannelInputInt in;
  private ChannelOutputInt out;

  public Succ (ChannelInputInt in, ChannelOutputInt out) {
    this.in = in;
    this.out = out;
  }

  public void run () {
    while (true) {
      int x = in.read ();
      out.write (x + 1);
    }
  }
}

class Prefix implements CSProcess {

  private int n;
  private ChannelInputInt in;
  private ChannelOutputInt out;

  public Prefix (int n, ChannelInputInt in, ChannelOutputInt out) {
    this.n = n;
    this.in = in;
    this.out = out;
  }
```

```
    public void run () {
      out.write (n);
      new Id (in, out).run ();
    }
  }
```

JCSP provides a `Parallel` class that combines an array of `CSProcesses` into a `CSProcess`. It's execution is the parallel composition of that array. For example, here are two of the 'Plug and Play' examples from Sections 2.4 and 2.6:

```
  class Numbers implements CSProcess {

    private ChannelOutputInt out;

    public Numbers (ChannelOutputInt out) {
      this.out = out;
    }

    public void run () {
      One2OneChannelInt a = new One2OneChannelInt ();
      One2OneChannelInt b = new One2OneChannelInt ();
      One2OneChannelInt c = new One2OneChannelInt ();
      new Parallel (
        new CSProcess[] {
          new Delta (a, out, b),
          new Succ (b, c),
          new Prefix (0, c, a),
        }
      ).run ();
    }
  }

  class Squares implements CSProcess {

    private ChannelOutputInt out;

    public Squares (ChannelOutputInt out) {
      this.out = out;
    }

    public void run () {
      One2OneChannelInt a = new One2OneChannelInt ();
      One2OneChannelInt b = new One2OneChannelInt ();
      new Parallel (
        new CSProcess[] {
          new Numbers (a),
          new Integrate (a, b),
          new Pairs (b, out),
        }
      ).run ();
    }
  }
```

Here is one of the controllers from Section 2.7. The processes ProcessReadInt and ProcessWriteInt just read and write a single integer (into and from a public value field) and, then, terminate:

```
class Replace implements CSProcess {

  private AltingChannelInputInt in;
  private AltingChannelInputInt inject;
  private ChannelOutputInt out;

  public Replace (AltingChannelInputInt in,
                  AltingChannelInputInt inject,
                  ChannelOutputInt out) {
    this.in = in;
    this.inject = inject;
    this.out = out;
  }

  public void run () {

    AlternativeInt alt = new AlternativeInt ();
    AltingChannelInputInt[] alternatives = {inject, in};

    ProcessWriteInt forward = new ProcessWriteInt (out);  // a CSProcess
    ProcessReadInt discard = new ProcessReadInt (in);     // a CSProcess
    CSProcess parIO  = new Parallel (new CSProcess[] {discard, forward});

    while (true) {
      switch (alt.select (alternatives)) {
        case 0:
          forward.value = inject.read ();
          parIO.run ();
        break;
        case 1:
          out.write (in.read ());
        break;
      }
    }
  }
}
```

JCSP also has channels for sending and receiving arbitrary Objects. Here is an asynchronous receive (from Section 2.9) of an expected Packet:

```
// set up processes once (before we start looping ...)

ProcessRead readObj = new ProcessRead (in);                 // a CSProcess
CSProcess someMore = new someMoreComputation (...);
CSProcess async = new PriParallel (new CSProcess[] {readObj, someMore});

while (looping) {
  async.run ();
  Packet packet = (Packet) readObj.value
  continue (...);
}
```

Chapter 4: Computer Organisation, Programming and Benchmarking

Introduction

José Silva Matos

Faculdade de Engenharia da Universidade do Porto
Rua dos Bragas
4099 Porto Codex, Portugal
jsm@fe.up.pt

The invited talk and the 15 articles selected for inclusion in this chapter cover an important set of topics ranging from computer organization to algorithms for parallel execution, including program behaviour analysis, processor and memory interconnecti on and benchmarking. The interdependence relations among these issues are of great relevance in the context of current and future work, and may prove useful to a considerable number of readers. The following is a brief description of the articles in thi s chapter, and an attempt to justify their selection and to place them under a common focus.

The chapter opens with the invited talk by Jean Vuillemin, *Reconfigurable Systems: Past and Next 10 Years*. The author provides an insightful view of the effect of continuous shrinking in feature sizes of the silicon implementation process, in th e performance and computing densities of various chips architectures and technologies. Reconfigurable systems are shown to combine the flexibility of software programming with the high performance level of dedicated hardware, and the architectural featur es that are required in order to be able to take advantage of future reductions of feature sizes are characterised. A key to this goal is the embedding of memory, computation and communication at a deeper level. This may also help providing an answer to the issue of how to bring the level of productivity, in mapping applications into custom hardware, to accompany the pace of current process and technological development.

The article by Seo, Downie, Hearn and Philips, *A Systolic Algorithm for the Factorisation of Matrices Arising in the Field of Hydrodynamics*, simulates a systolic array structure on a non-systolic parallel architecture. The major goal is the produc tion of an efficient systolic matrix factorisation routine for general-purpose distributed parallel systems, including clusters of workstations.

Two articles make use of behaviour modelling to analyse parallel program execution and system configuration. Espinosa, Margalef and Luque, in *Automatic Detection of Parallel Program Performance Problems*, propose an automatic performance analysis tool to find poorly designed structures in an application. This will be of great help as the programmer will not need to under-

J. Palma, J. Dongarra, and V. Hernández (Eds.): VECPAR'98, LNCS 1573, pp. 331–333, 1999.

stand the large amount of information obtained from the execution of a parallel program. The article by Suárez and García, *Behavioural Analysis Methodology Oriented to Configuration of Parallel, Real/Time and Embedded Systems*, describes a methodology to achieve proper system configuration for satisfying a set of specified real-time constraints. It uses metrics obtained from an event trace, to analyse system behaviour and to provide alternatives to improve the system design. Experimental results of applying the methodology to the analysis of a well-known case study are included.

The Hierarchical Processor-And-Memory (HPAM) architecture exploits heterogeneity, computing-in-memory and locality of memory references with respect to degree of parallelism to provide a cost-effective approach to parallel processing. The article by Figueiredo, Fortes and Miled, *Spatial Data Locality With Respect to Degree of Parallelism in Processor-And-Memory Hierarchies*, studies the impact of using multiple transfer sizes across an HPAM hierarchy and proposes an inter-level coherence protocol that supports inclusion and multiple block sizes across the hierarchy. The growing dependence on the efficient use of memory hierarchy in current high performance computers is central to the work presented in the article by Matías, Llorente and Tirado, *Partitioning Regular Domains on Modern Parallel Computers*. Linear rather than higher-dimension partitioning, and avoiding boundaries with poor data locality are two conclusions that were verified experimentally on two microprocessor based computers.

The following articles deal with ways to increase performance in different types of architectures. *New Access Order to Reduce Inter-Vector Conflicts*, by Corral and Llaberia, proposes an access order to the vector elements that avoids or reduces bus and memory module conflicts in vector processors. The article describes a hardware solution for its implementation and reports important performance increases over the classical access case, that are backed-up by simulation results. The article by Villa, Espasa and Valero, *Registers Size Influence on Vector Architectures*, studies the effect of reducing the size of registers on conventional and out-of-order vector architectures. The authors show that the price to pay in performance, by reducing register length in a conventional vector architecture, may be very low, or even lead to a speed-up, in case of an out-of-order execution. Performance improvement of superscalar processors is the subject of the article by González and González, *Limits of Instruction Level Parallelism with Data Value Speculation*. It is shown that data value speculation is a promising technique to be considered for future generation microprocessors, offering interesting potential to boost the limits of instruction level parallelism, its benefits increasing with the instruction wi ndow size. Finally, the article by García, Carretero, Pérez and de Miguel, *High Performance Cache Management for Parallel File Systems*, demonstrates the use of caching at processing nodes in a parallel file system, using a dynamic scheme of cache coherence protocols with different sizes and shapes of granularity.

Task scheduling and load distribution, two central issues in parallel computing, are the subject of the two following articles. *Using Synthetic Workloads for Parallel Task Scheduling Improvement Analysis*, by Kitajima and Porto, presents an experim ental validation of makespan improvements of two scheduling algorithms, a greedy construction algorithm and one based on the tabu search meta-heuristic. Results show low impact of system overhead on makespan improvement estimation, guaranteeing a reliabl e cost function for static scheduling algorithms, and confirming good results in the application tabu search to this type of problem. *Dynamic Routing Balancing in Parallel Computer Interconnection Networks*, by Franco, Garcés and Luque, presents a method for balancing traffic load over all paths in an interconnection network, in order to maintain a low message latency. Results for different traffic patterns show improved throughput and network saturation reached at higher load rates.

The best use of available computing power is central to the two following articles, where the use of widely available networked computers/workstations is considered as a viable alternative to expensive massively parallel processors. Barbosa and Padilha in their article, *Algorithm-Dependent Method to Determine the Optimal Number of Computers in Parallel Virtual Machines*, show how to find, for a given algorithm, the number of processors that must be used in order to get minimum processing time. Their conclusions are supported by experimental results obtained on a network of personal computers. *Low-cost Parallelising: A Way to be Efficient*, by Martin and Chopard, proposes a dynamic load balancing solution, based on a local partitioning scheme, to the problem presented in such a network where a slow node can significantly decrease the overall performance. Experimental results show the benef its of the load balancing strategy with a network of heterogeneous processing elements, unpredictably loaded by other jobs.

The chapter concludes with two articles on computer and architecture benchmarking. *A Performance Analysis of the SGI Origin2000*, by van der Steen and van der Pas, presents a review of this machine and an assessment of its performance using the EuroBen Benchmark. The last article in the chapter, *An ISA Comparison between Superscalar and Vector Processors* is authored by Quint ana, Espasa and Valero, who present a comparison between these two models at the instruction set architecture (ISA) level and an analysis of their behaviour in terms of speculative execution.

Reconfigurable Systems: Past and Next 10 Years

Invited Talk *

Jean Vuillemin

Ecole Normale Supérieure,
45 rue dÚlm, 75230 Paris cedex 05, France.

Abstract. A driving factor in *Digital System DS* architecture is the *feature size* of the silicon implementation process. We present Moore's laws and focus on the shrink laws, which relate chip performance to feature size. The theory is backed with experimental measures from [14], relating performance to feature size, for various memory, processor and FPGA chips from the past decade. Conceptually shrinking back existing chips to a common feature size leads to common architectural measures, which we call *normalised*: area, clock frequency, memory and operations per cycle. We measure and compare the normalised *compute density* of various chips, architectures and silicon technologies.

A *Reconfigurable System RS* is a standard processor tightly coupled to a *Programmable Active Memory PAM*, through a high bandwidth digital link. The PAM is a FPGA and SRAM based coprocessor. Through software configuration, it may emulate any specific custom hardware, within size and speed limits. RS combine the flexibility of software programming to the performance level of application specific integrated circuits ASIC. We analyse the performance achieved by P1, a first generation RS [13]. It still holds some significant absolute speed records: RSA cryptography, applications from high-energy physics, and solving the Heat Equation. We observe how the software versions for these applications have gained performance, through better microprocessors. We compare with the performance gain which can be achieved, through implementation in P2, a second-generation RS [16].

Recent experimental systems, such as the *Dynamically Programmable Arithmetic Array* in [19] and others in [14], present advantages over current FPGA, both in storage and compute density. RS based on such chips are tailored for *video processing*, and similar compute, memory and IO bandwidth intensive. We characterise some of the architectural features that a RS must posses in order to be *fit to shrink*: automatically enjoy the optimal gain in performance through future shrinks. The key to scale, for any general purpose system, is to embed memory, computation and communication at a much deeper level than presently done.

* This research was partly done at *Hewlett Packard Laboratories*, Bristol U.K.

J. Palma, J. Dongarra, and V. Hernández (Eds.): VECPAR'98, LNCS 1573, pp. 334–354, 1999.

1 Moore's Laws

Our modern world relies on an ever increasing number of Digital Systems DS: from home to office, through car, boat, plane and elsewhere. As a point in case, the shear economic magnitude of the Millenium Bug [21], shows how futile it would be t o try and list *all* the functions which DS serve in our brave new digital world.

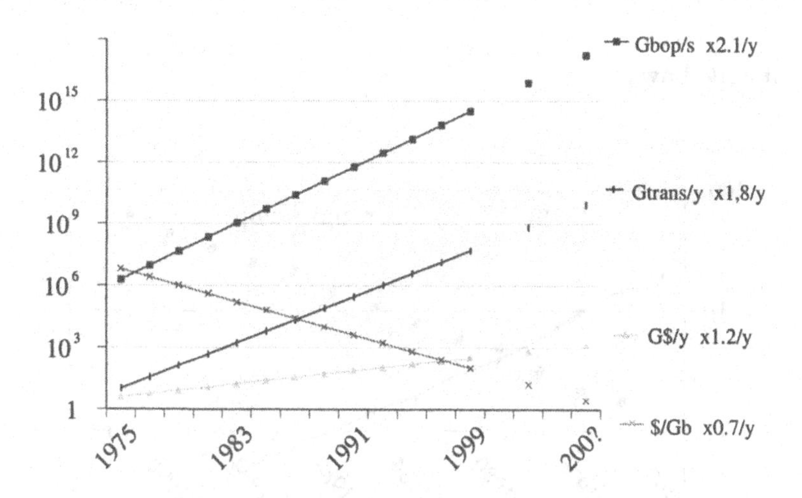

Fig. 1. Estimated number and world wide growth rate: $G = 10^9$ transistors fabricated per year; G bit operations computed each second; Billion $ revenues from silicon sold world wide; $ cost per $G = 2^{30}$ bits of storage.

Through recent decades, earth's combined raw compute power has more than doubled each year. Somehow, the market remains elastic enough to find applications, and people to pay, for having twice as many bits automatically switch state than twelve months ag o. At least, many people did so, each year, for over thirty years - fig. 1.

An ever improving silicon manufacturing technology meets this ever increasing demand for computations: more transistors per unit area, bigger and faster chips. On the average over 30 years, the cost per bit stored in memory goes down by 30% each year. De spite this drop in price, selling 80% more transistors each year increases revenue for the semi-conductor industry by 20% - fig. 1.

The number of transistors per mm^2 grows about 40% each year, and chip size increases by 15%, so:

The number of transistors per chip doubles in about 18 months.

That is how *G. Moore*, one of the founders of *Intel*, famously stated the laws embodied in fig. 1. That was in the late sixties, known since as *Moore's Laws*.

More recently, *G. Moore* [18] points out that we will soon fabricate more transistors per year than there are living ants on earth: an estimated 10^{17}.

Yet, people buy computations, not transistors. How much computation do they buy? Operating all of this year's transistors at 60 MHz amounts to an aggregate *compute power* worth 10^{24} *bop/s* - *bit operation per second.* That would be o n the order of 10 million *bop/s* per ant!

This estimate of the world's compute power could well be off by some order of magnitude. What matters is that *computing power at large has more than doubled each year* for three decades, and it should do so for some years to come.

1.1 Shrink Laws

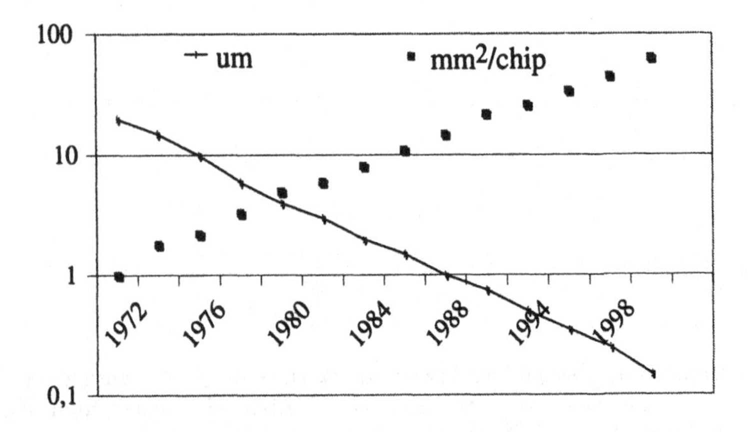

Fig. 2. Shrink of the feature size with time: minimum transistor width, in $\mu m = 10^{-6} m$. Growth of chip area - in mm^2.

The economic factors at work in fig. 1 are separated from their technological consequences in fig 2. The feature size of silicon chips *shrinks*: over the past two decades, the average shrink rate was near 85% per year. Duri ng the same time, chip size has increased: at a yearly rate near 10% for DRAM, and 20% for processors.

The effect on performance of scaling down all dimensions and the voltage of a silicon structure by 1/2: the area reduces by 1/4, the clock delay reduces to 1/2 and the power dissipated per operation by 1/8.

Equivalently, the clock frequency doubles, the transistor density per unit area quadruples, and the number of operations per unit energy is multiplied by 8, see fig. 2. This shrink model was presented by [2] in 1980, and intended to co ver feature sizes down to 0.3 μm - see fig. 3.

Fig. 4 compares the shrink model from fig. 3 with experimental data gathered in [14], for various DRAM chips, published between in the last decade. The last entry - from [15] - accounts for synchronous SDRAM, where access latency is

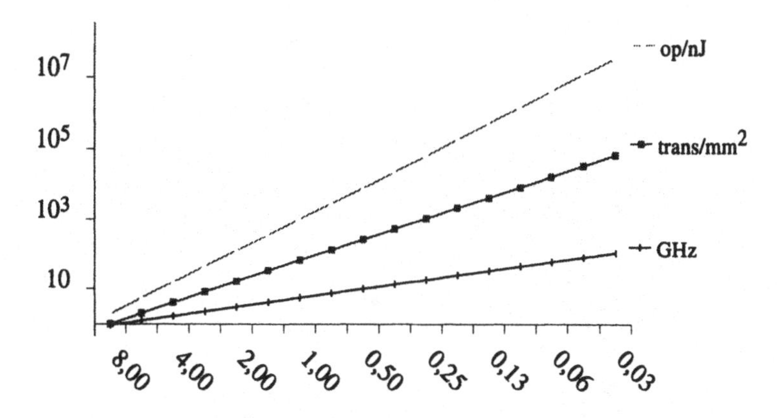

Fig. 3. Theoretical chip performance, as the minimum transistor width (feature size) shrinks from 8 to 0.03 micron μm: transistors per square millimeter; fastest possible chip wide synchronous clock frequency, in giga hertz; number of operations com puted, per nano joule.

traded for throughput. Overall, we find a rather nice fit to the model. In fig. 7, we also find agreement between the theoretical fig. 3 and experimental data for microprocessors and FPGA, although some architectu ral trends appear.

A recent update of the shrink model by Mead [9] covers features down to 0.03 μm. The optimists conclusion, from [9]:

> We can safely count on at least one more order of magnitude of scaling.

The pessimist will observe that it takes 2 pages in [2] to state and justify the *linear* shrink rules; it takes 15 pages in [9], and the rules are *no longer linear*. Indeed, thin oxide is already nearly 20 atoms thick, at cu rrent feature size 0.2 μm. A linear shrink would have it be less than one atom thick, around 0.01 μm. Other fundamental limits (*quantum mechanical effects, thermal noise, light's wavelength, ...*) become dominant as well, near the same lim it. Although *C. Mead* [9] does not *explicitly* cover finer sizes, the *implicit* conclusion is:

> We cannot count on two more orders of magnitude of scaling.

Moore's law will thus eventually either run out of fuel - demands for *bop/s* will some year be under twice that of the previous - or it will be out of an engine - *shrink laws* no longer apply below 0.01 μm. One likely possibility is some comb ination of both: feature size will shrink ever more slowly, from some future time on.

On the other hand, there is no fundamental reason why the size of chips cannot keep on increasing, even if the shrink stops. Likewise, we can expect new architecture to improve the currently understood technology path. No matter what happens, how to *best use the available silicon* will long remain an important question. Another good bet: the amount of storage, computation and communication, available in each system will grow, ever larger.

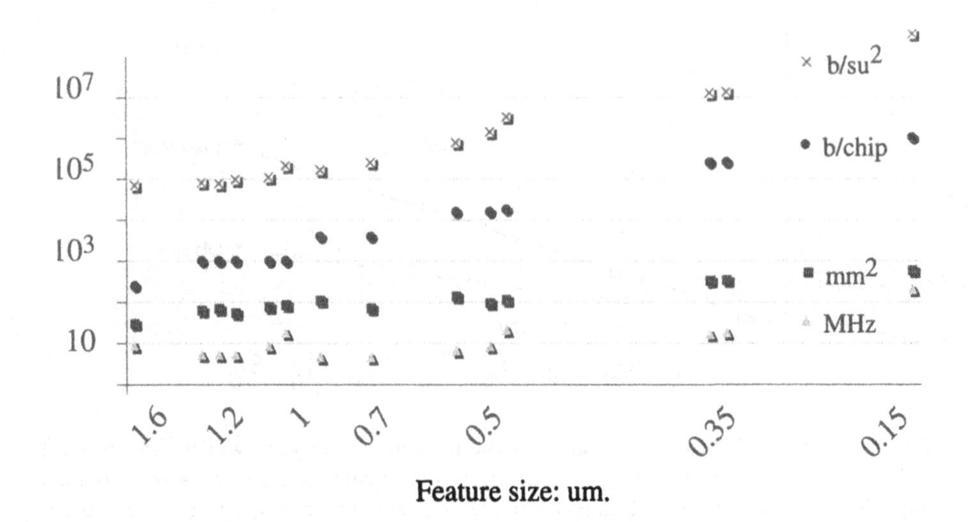

Fig. 4. : Actual DRAM performance as feature size shrinks from 0.8 to 0.075 μm: clock frequency in Mega hertz; square millimeters per chip; bits per chip; power is expressed in bit per second per square micron.

2 Performance Measures for Digital Systems

Communication, processing and storage are the three building blocks of DS. They are intimately combined at all levels. At micron scale, wires, transistors and capacitors implement the required functions. At human scale, the combination of a modem, microprocessor and memory in a PC box does the trick. At planet scale, communication happens through more exotic media - waves in the electromagnetic ether, or optic fiber - at either end of which one finds more memory, and more processing units.

2.1 Theoretical Performance Measures

Shannon's Mathematical Theory of Communication [1] shows that physical measures of information (bits b) and communication (bits per second b/s) are related to the abstract mathematical measure ot statistical entropy H, a positive r eal number $H > 0$. Shannon's theory does not account for the cost of any computation. Indeed, the global function of a communication or storage device is the identity $X = Y$.

On the other hand, source coding for MPEG video is among the most demanding computational tasks. Similarly, *random* channel coding (and decoding), which gets near the optimal for the communication purposes of Shannon as coding blocks become bigger, has a computational complexity which increases exponentially with block size.

The basic question in Complexity Theory is to determine how many operations $C(f)$, are necessary and sufficient for computing a digital function f. All operations in the computation of f are accounted for, down to the bit level, regardless of when, w here, or how the operation is performed. The unit of measure for $C(f)$ is one *Boolean operation bop*. It is applicable to all forms of computations - sequential, parallel, general and special purpose. Some relevant results (see [5] for proofs):

1. The complexity of n bit binary addition is $5n - 3$ bop. The complexity of computing one bit of sum is 1 *add* = 5 *bop* (full adder: 3 in, 2 out).
2. The complexity of n bit binary multiplication can be reduced, from $6n^2$ *bop* for the naive method (and $4n^2$ *bop* through *Booth Encoding*), down to $c(\epsilon)n^{1+\epsilon}$, for any real number $\epsilon > 0$. As $c(\epsilon) \mapsto \infty$ when $\epsilon \mapsto 0$, the practical complexity of binary multiplication is only improved for n large.
3. Most Boolean functions f, with n bits of input and one output, have a *bop* complexity $C(f)$ such that $2n/n < C(f) < 2n/n(2 + \epsilon)$, for all $\epsilon > 0$ and n large enough. To build one, just choose at random! No explicitly described Bool ean function has yet been proved to posses more than linear complexity (including multiplication). An efficient way to compute a *random* Boolean function is through a *Lookup Table LUT*, implemented with a RAM or a ROM.

Computation is free in Shannon's model, while communication and memory are free within Complexity Theory. The *Theory of VLSI Complexity* aims at measuring, for all physical realizations of digital function f, the combined complexity of *communication, memory*, and *computation*. The *VLSI complexity* of function f is defined with respect to all possible chips for computing f. Implementations are all within the same silicon process, defined by some feature size, speed and design rules. Each design computes f within some area A, clock frequency F and T clock periods per IO sample. The silicon area A is used for storage, communication and computation, through transistors and wires. Optimal designs are selected, based on some performance measure. For our purposes: minimize the area A for computing function f, subject to the real time requirement $F/T < F_{io}$. In theory, one has to optimize among all designs for computing f. In practice, the search is reduced to structural decompositions into well known *standard components*: adders, multipliers, shifters, memories, ...

2.2 Trading Size for Speed

VLSI design allows trading area for speed. Consider, for example, the family of adders: their function is to repeatedly compute the binary sum $S = A + B$ of two n bits numbers A, B. Fig. 5 shows four adders, each with a different structure, perfo rmance, and mapping of the operands through time and IO ports. Let us analyze the VLSI performance of these adders, under simplifying assumptions: $a_{fa} = 2a_r$ for the area (based on transistor counts), and $d_{fa} = d_r$ for the combinatorial delays o f *fadd* and *reg* (setup and hold delay).

1. Bit serial (base 2) adder $sA2$. The bits of the binary sum appear through the unique output port as a time sequence $s_0, s_1, ..., s_n, ...$ one bit per clock

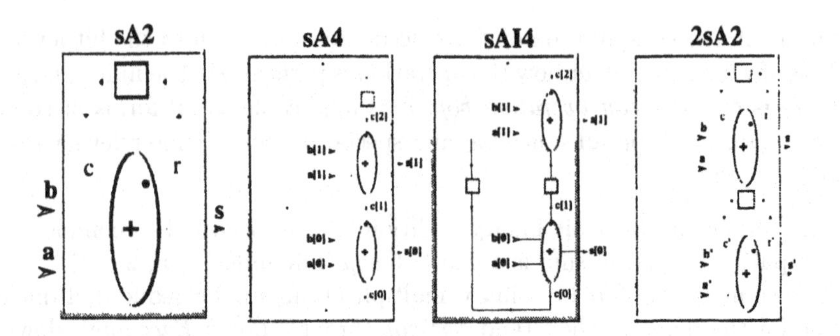

Fig. 5. Four serial adders: $sA2$ - base 2, $sA4$ - base 4, $sAI4$ - base 4 interleaved, and $2sA2$ - two independent $sA2$. An oval represents the full adder *fadd*; a square denotes the register *reg* (one bit synchronous *flip-flop*; the clock is implicit in the schematics).

cycle, from least to most significant. It takes $T = n + 1$ cycles per sum S. The are a is $A = 3a_r$: it is the *smallest of all* adders. The chip operates at clock frequencies up to $F = 1/2d_r$: the *highest possible.*

2. Serial two bits wide (base 4) adder $sA4$. The bits of the binary sum appear as two time sequences s_0, s_2, ..., s_{2n}, ... and s_1, s_3, ... two bits per cycle, through two output ports. Assuming n to be odd, we have $T = (n+1)/2$ cycl es per sum. The area is $A = 5a_r$ and the operating frequency $F = 1/3d_r$.
3. Serial interleaved base 4 adder $sAI4$. The bits of the binary sum S appear as two time sequences s_0, *, s_2, *, ..., s_{2n}, *, ... and *, s_1, *, s_3, ... one bit per clock cycle, even cycles through one output port, odd through the other. The alternate cycles (the *) are used to compute an independent sum S', whose IO bits (and carries) are interleaved with those for sum S. Although it still takes $n+1$ cycles in order to compute each sum S and S', we get *both* sums i n so many cycles, at the rate of $T = (n+1)/2$ cycles per sum. The area is $A = 6a_r$ and the maximum operating frequency $F = 1/2d_r$.
4. Two independent bit serial adders $2sA2$. This circuit achieves the same performance as the previous: $T = (n+1)/2$ cycles per sum, area $A = 6a_r$ and frequency $F = 1/2d_r$.

The transformation that unfolds the base 2 adder $sA2$ into the base 4 adder $sA4$ is a special instance of a general procedure. Consider a circuit C which computes some function f in T cycles, within gate complexity G *bop* and memory M bits. Th e procedure from [11] unfolds C into a circuit C' for computing f: it trades cycles $T' = T/2$ for gates $G' = 2G$, at constant storage $M' = M$.

In the case of serial adders, the area relation is $A' = 5A/3 < 2A$, so that $A'T' < AT$. On the other hand, since $F' = 1/3d$ and $F = 1/2d$, we find that $A'T'/F' > AT/F$. An equivalent way to measure this, is to consider the density of full adders fadd per unit ar ea $a_{fa} = 2a_r$, for both designs C and C': as $2/A = 0.66 < 4/A' = 0.8$, the unfolded design has a better *fadd* density than the original. Yet, since $F' = 1.5F$, the *compute density* - in *fadd* per unit area and time $d_{fa} = d_r$ - is lower for c ircuit C': $F/A = 0.16 > 2/A'F' = 0.13$. When we unfold from base 2 all the way to base $2n$, the carry register may be simplified away: it is always 0. The fadd densities of this n-bit wide carry propagate adder is 1 per unit area, which is optimal ; yet, as clock frequency is $F = 1/n$, the compute density is low: $1/n$.

Circuits $sAI4$ and $2sA2$ present two ways of *optimally* trading time for area, at *constant operator and compute density.* Both are instances of general methods, applicable to any function f, besides binary addition. From any circuit C fo r computing f within area A, time T and frequency F, we can derive circuits C' which optimally trades area $A' = 2A$ for time $T' = T/2$, at constant clock frequency $F' = F$. The trivial unfolding constructs $C' = 2C$ from two independent copies of C, which operate on separate IO. So does the interleaved adder $sAI4$, in a different manner. Generalizing the interleaved unfolding to arbitrary functions does not always lead to an optimal circuit: the extra wiring required may force the area to be mor e than $A' > 2A$. Also note that while these optimal unfolding double the throughput ($T = n/2$ cycles per add), the *latency* for each individual addition is not reduced from the original one ($T = n$ cycles per addition). We may constrain the unfolded ci rcuit to produce the IO samples in the standard order, by adding reformatting circuitry on each side of the IO: a buffer of size n-bit, and a few gates for each input and output suffice. As we account for the extra area (for *corner turning*), we s ee that the unfolded circuit is no longer optimal: $A' > 2A$. For a complex function where a large area is required, the loss in *corner turning* area can be marginal. For simpler functions, it is not.

In the case of addition, area may be optimally traded for time, for all integer data bit width $D = n/T$, as long as $D < \sqrt{n}$. Fast wide $D = n$ parallel adders have area $A = nlog(n)$, and are structured as binary trees. The area is dominated by the wire s connecting the tree nodes, their drivers (the longer the wire, the bigger the driver), and by pipelining registers, whose function is to reduce all combinatorial delays in the circuit below the clock period $1/F$ of the system.

Transitive functions permute their inputs in a rich manner (see [4]): any input bit may be mapped - through an appropriate choice of the external controls - into any output bit position, among N possible per IO sample. It is shown in [4] that computing a transitive function at IO rate $D = NF/T$, requires an area A such that:

$$A > a_m N + a_{io} D + a_w D^2, \qquad (1)$$

where a_m, a_{io} and a_w are proportional to the area per bit respectively required for memory, IO and communication wires. Note that the gate complexity of a transitive function is zero: input bit values are simply permuted on the output. The abov e bound merely accounts for the area - IO ports, wires and registers - which is required to acquire, transport and buffer the data at the required rate. Bound (1) applies to shifters, and thus also to multipliers. Consider a multiplier that compu tes $2n$-bit products on each cycle, at frequency F. The wire area of any such multiplier is proportional to n^2, as $T = 1$ in (1). For high bandwidth multipliers, the area required for wires and pipelining registers is bigger than that for ar ithmetic operations.

The bit serial multiplier (see [11]) has a minimal area $A = n$, high operating frequency F, and it requires $T = 2n$ cycles per product. A parallel nave multiplier has area $A' = n^2$ and $T' = 1$ cycle per product. In order to maintain high frequen cy $F' = F$, one has to introduce on the order of n^2 *pipelining registers*, so (perhaps) $A' = 2n^2$ for the *fully pipelined* multiplier. These are two extreme points in a range of optimal multipliers: according to bound (1), and within a constant factor. Both are based on nave multiplication, and compute n^2 mul per product. High frequency is achieved through deep pipelining, and the latency per multiplication remains proportional to n. In theory, latency can be reduced to T, b y using reduced complexity $n^{1+\epsilon}$ shallow multipliers (see [3]); yet, shallow multipliers have so far proved bigger than nave ones, for practical values such as $n < 256$.

2.3 Experimental Performance Measures

Consider a VLSI design with area A and clock frequency F, which computes function f in T cycles per N-bit sample. In theory, there is another design for f which optimally trades area $A' = 2A$ for cycles $T' = T/2$, at constant frequency $F' = F$. T he frequency F and the AT product remain invariant in such an optimal tradeoff. Also invariant:

- The gate density (in *bop*/mm^2), given by $D_{op} = c(f)/A = C(f)/AT$. Here $c(f)$ is the *bop* complexity of f per cycle, while $C(f)$ is the bop complexity per sample.
- The compute density (in *bop*/smm^2) is $c(f)F/A = FD_{op}$.

Note that trading area for time at constant gate and compute density is equivalent to keeping F and AT invariant.

Let us examine how various architectures trade size for performance, in practice. The data from [14] tabulates the area, frequency, and feature size, for a representative collection of chips from the previous decade: sRAM, DRAM, mPROC, FPGA, MUL.

The normalised area A/λ^2 provides a performance measure that is independent of the specific feature size λ. It leads [14] to a quantitative assessment of the gate density for the various chips, fig. 6 and 7.

Unlike [14], we also normalise clock frequency: the product by the operation density is the normalised compute power. To define the normalised the system clock frequency ϕ, we follow [9] and use $\phi = 1/100\tau(\lambda)$, where $\tau(\lambda)$ is the minimal inverter delay corresponding to feature size λ.

- The non linear formula used for $\tau((l) = cl^e$ is taken from [9]: the exponent $e = 1 - \epsilon(l)$ decreases from 1 to 0.9 as l shrinks from 0.3 to 0.03 μm. The non linear effect is not yet apparent in the reported data. It wil l become more significant with finer feature sizes, and clock frequency will cease to increase some time before the shrink itself stops.
- The factor 100 leads to normalised clock frequencies whose average value is 0.2 for DRAM, 0.9 for SRAM, 2 for processors and 2 for FPGA.

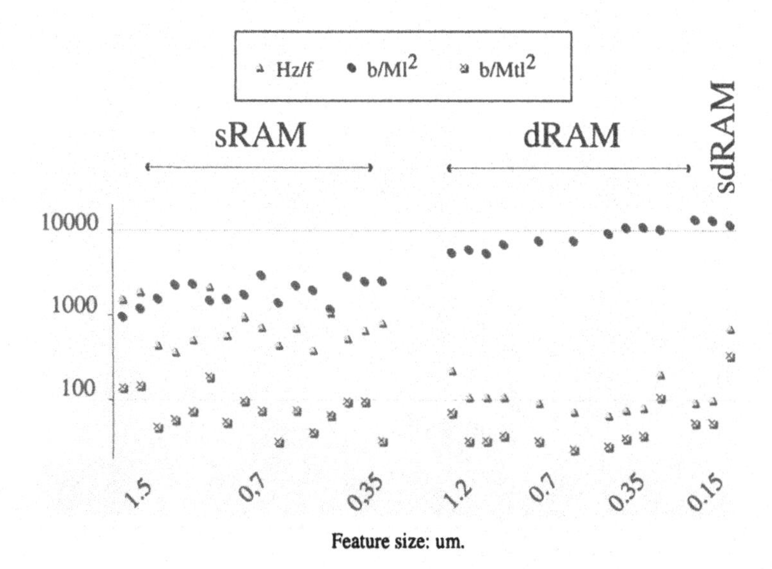

Fig. 6. Performance of various SRAM and DRAM chips, within to a common feature size technology: normalised clock frequency Hz/ϕ; bit density per normalised area $10^6\lambda^2$; binary gate operations per normalised area per normalised clock period $1/\phi$.

In the absence of architectural improvement, the normalised gate and compute density of the same function on two different feature size silicon implementations should be the same, and this indicates an optimal shrink.

- The normalised performance figures for SRAM chips in fig. 6 are all within range: from one half to twice the average value.
- The normalised bit density for DRAM chips in the data set is 4.5 times that of SRAM. Observe in fig. 6 that it has increased over the past decade, as the result of improvements in the architecture of the memory cell (*trench* capacitor s). The average normalised speed of DRAM is 4.5 times slower than SRAM. As a consequence the average normalised compute density of SRAM equals that of DRAM. The situation is different with SDRAM (last entry in fig. 6): with the storage density of DRAM and nearly the speed of SRAM, the normalised compute density of SDRAM is 4 times that of either: a genuine improvement in memory architecture.

A *Field Programmable Gate Array FPGA* is a mesh made of programmable gates and interconnect [17]. The specific function - Boolean or register - of each gate in the mesh, and the interconnection between the gates, is coded in some binary bit stream, specific to function f, which must first be downloaded into the *configuration memory* of the device. At the end of configuration, the FPGA switches to user mode: it then computes function f, by operating just as any regular ASIC would.

The comparative normalised performance figures for various recent microprocessors and FPGA is found in fig. 7.

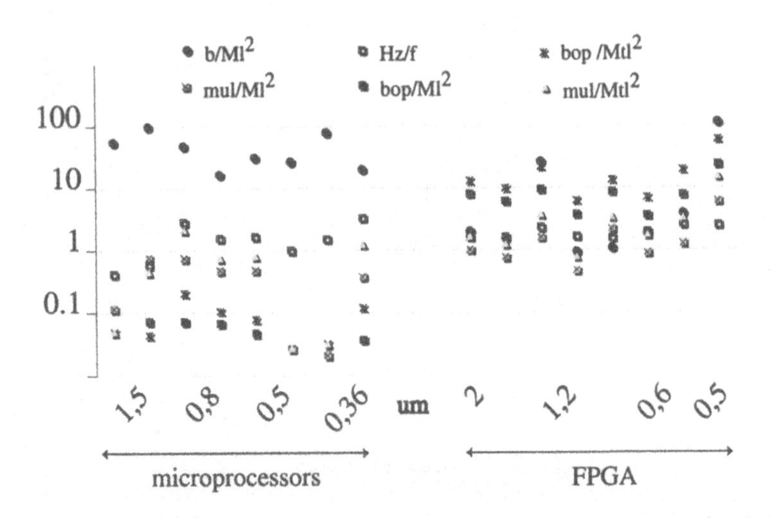

Fig. 7. Performance of various microprocessor and FPGA chips from [14], within a common feature size technology: normalised clock frequency Hz/ϕ; normalised bit density; normalised gate and compute density: for Boolean operations, additions and multiplication's.

- Microprocessors in the survey appear to have maintained their normalised compute density, by trading *lower* normalised operation density, for a *higher* normalised clock frequency, as feature size has shrunk. Only the microprocessors wit h a built-in multiplier have kept the normalised compute density constant. If we exclude multipliers, the normalised compute density of microprocessors has actually decreased through the sample data.
- FPGA have stayed much closer to the model, and normalised performances do not appear to have changed significantly over the survey (rightmost entry excluded).

3 Reconfigurable Systems

A *Reconfigurable System RS* is a standard sequential processor (the host) tightly coupled to a *Programmable Active Memory PAM*, through a high bandwidth link. The PAM is a reconfigurable processor, based on FPGA and SRAM. Through software conf iguration, the PAM emulate any specific custom hardware, within size and speed limits. The host can write into, and read data from the PAM, as with any memory. Unlike conventional RAM, the PAM processes data between write and read cycles: it an active mem ory. The specific processing is determined by the contents of its configuration memory. The content of configuration memory can be updated by the host, in a matter of milliseconds: it is programmable.

RS combine the flexibility of software programming to the performance level of application specific integrated circuits ASIC. As a point in case, consider the system P1 described in [13]. From the abstract of that paper:

> We exhibit a dozen applications where PAM technology proves superior, both in performance and cost, to every other existing technology, including supercomputers, massively parallel machines, and conventional custom hardware.
>
> The fields covered include computer arithmetics, cryptography, error correction, image analysis, stereo vision, video compression, sound synthesis, neural networks, high-energy physics, thermodynamics, biology and astronomy.
>
> At comparable cost, the computing power virtually available in a PAM exceeds that of conventional processors by a factor 10 to 1000, depending on the specific application, in 1992.

RS P1 is built from chips available in 92 - SRAM, FPGA and processor. Six long technology years later, it still holds at least 4 significant absolute speed records. In theory, it is a straightforward matter to port these applications on a state of the art RS, and enjoy the performance gain from the shrink. In the practical state of our CAD tools, porting the highly optimised P1 designs on oher systems would require time and skills. On the other hand, it is straightforward to estimate the performance witho ut doing the actual hardware implementation. We use the Reconfigurable System P2 [16] - built in 97 - to conceptually implement the same applications as P1, and compare. The P2 system has 1/4 the physical size and chip count of P1. Both have rou ghly the same logical size (4k CLB), so the applications can be transferred without any redesign. The clock frequency is 66 MHz on P2, and 25MHz on P1 (and 33MHz for RSA). So, the applications will run at least twice faster on P2 than on P1. Of course, i f we compare equal size and cost systems, we have to match P1 against 4P2, and the compute power has been multiplied by at least 8. This is expected by the theory, as the feature size of chips in P1 is twice that of chips in P2.

What has been done [20] is to port and run on recent fast processors, the software version for some of the original P1 applications. That provides us with a technology update on the respective compute power of RS and processors.

3.1 3D Heat Equation

The fastest reported software for to solving the Heat Equation on a supercomputer, is presented in [6]. It is based on the finite differences method. The Heat Equation can be solved more efficiently on specific hardware structures [7]:

- Start from an initial state - at time $t\Delta t$ - of the discrete temperatures in a discrete 3D domain, all stored in RAM.
- Move to the next state - at time $(t+1)\Delta t$ - by traversing the RAM three times, along the x, y and z axis.

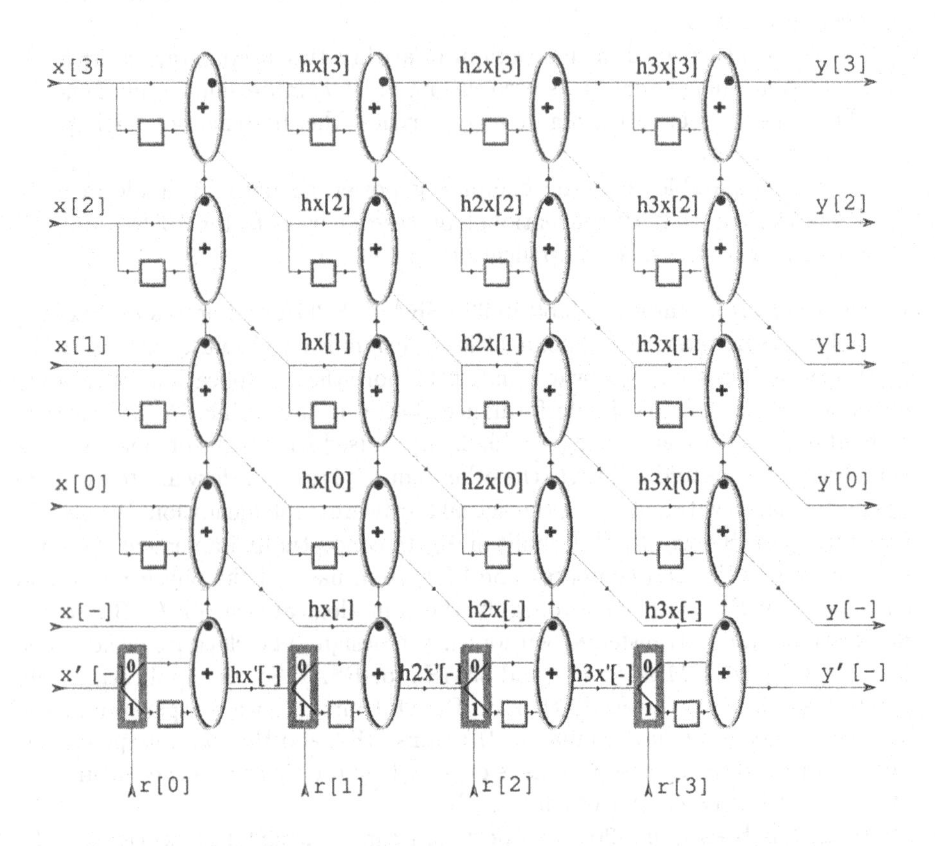

Fig. 8. Schematics of a hardware pipeline for solving the Heat equation. It is drawn with a pipeline depth of 4, and bit width of 4, plus 2 bits for randomised round off. The actual 1 pipeline is 256 deep, and 16+2 wide. Pipelining registers, which allow the network to operate at maximum clock frequency, are not indicated here. Neither is the random bit generator.

– On each traversal, the data from the RAM feeds a pipeline of *averaging operators*, and the output of the pipeline is stored back in RAM.

Each *averaging operator* computes the average value $(a_t + a_{t+1})/2$ of two consecutive samples a_t and a_{t+1}. In order to be able to reduce the precision of internal temperatures down to 16 bits, it is necessary, when division by two is odd, to distribute that low-order bit randomly between the sample and its neighbour. All deterministic round-off schemes lead to parasitic effects that can significantly perturb the result. The pseudo-randomness is generated by a 64-bit *linear feedback shift-register LFSR*. The resulting pipeline is shown in fig. 8. Instead of being shifted, the least significant sum bit is either delayed or not, based on a random choice in the LFSR.

P1 standing design can accurately simulate the evolution of temperature over time in a 3D volume, mapped on 512^3 discrete points, with arbitrary power source distributions on the boundaries. In order to reproduce that computation in real time, it take s a 40,000 MIPS equivalent processing power: 40 G instructions per second, on $32b$ data. This is out of the reach of microprocessors, at least until 2001.

3.2 High Energy Physics

The *Transition Radiation Tracker TRT* is part in a suite of benchmarks proposed by CERN [12]. The goal is to measure the performance of various computer architectures in order to build the electronics required for the Large Hadron Collider L HC, soon after the turn of the millennium. Both benchmarks are challenging, and well documented for a wide variety of processing technologies, including some of the fastest current computers, DSP-based multiprocessors, systolic arrays, massively parallel arrays, Reconfigurable Systems, and full custom ASIC based solutions.

The TRT problem is to find straight lines (particle trajectories) in a noisy digital black and white image. The rate of images is at 100 kHz; the implied IO rate close to 200 MB/s, and the low latency requirement (2 images) preclude any implementation so lution other specialised hardware, as shown by [12].

The P1 implementation of the TRT is based on the *Fast Hough Transform* [10], an algorithm whose hardware implementation trades computation for wiring complexity. To reproduce the P1 performance reported in [12], a 64-bit sequential proc essor needs to run at over 1.2 GHz. That is about the amount of computation one gets, in 1998, with a dual processor, 64-bit machine, at 600 MHz. The required external bandwidth (up to 300 MB/s) is what still keeps such application out of current microp rocessor reach.

3.3 RSA Cryptography

The P1 design for RSA cryptography combines a number of algorithm techniques, presented in [8]. For 512-bit keys, it delivers a decryption rate in excess of 300 kb/s, although it uses only half the logical resources available in P1.

The implementation takes advantage of hardware reconfiguration in many ways: a rather different design is used for RSA encryption and decryption; a different hardware modular multiplier is generated for each different prime modulus: the coefficients of th e binary representation of each modulus is hardwired into the logical equations of the design. None of these techniques is readily applicable to ASIC implementations, where the same chip must do both encryption and decryption, for all keys.

As of printing time, this design still holds the acknowledged shortest time per block of RSA, all digital species included. It is surprising that it has held five years against other RSA hardware. According to [20], the record will go to a (soon to be announced) Alpha processor (one 64b multiply per cycle, at 750MHz) running (a modified version of) the original software version in [8]. We expect the record to be claimed back in the future by a P2 RSA design; yet, the speedup between P1 wa s 10x reported in 92, and we estimate that it should be only be 6x on 2P2, in 97. The reason: the fully pipelined multiplier, found in recent processors, is fully utilised by RSA software. A normalised measure of the impact of multiplier on theoretical performance can be observed in fig. 7.

For the Heat Equation, the actual performance ratio between P1 and the fastest processor (64b, 250MHz) was 100x in 92; with 4P2 against the 64b, 750MHz processor, the ratio should be over 200x in 98. Indeed, the computation in fig. 8 combines 16 b add and shift, with Boolean operations on three low order bits: software is not efficient, and the multiplier is not used.

4 What Will Digital Systems Shrink to?

Consider a DS whose function and real time frequency remain fixed, once and for all. Examples: digital watch, $56kb/s$ modem and GPS.

How does such DS shrink with feature size?

To answer, start from the first chip (feature size l) which computes function f: area A, time T, and clock frequency F. Move in time, and shrink feature size to l/2. The design now has area $A' = A/4$, and the clock frequency doubles $F' = 2F$ ($F' = (2-\epsilon)F$ with non-linear shrink). The number of cycles per sample remains the same: $T' = T$. The new design has twice (or $2-\epsilon$) the required real time bandwidth: we can (in theory) further fold space in time: produce a design C'' for computi ng f within area $A'' = A'/2 = A/8$ and $T'' = 2T$ cycles, still at frequency $F'' = F' = 2F$. The size of any fixed real time DS shrinks very fast with technology, indeed. At the end of that road, after so many hardware shrinks, the DS gets implemented in softw are.

On the other hand, microprocessors, memories and FPGA actually grow in area, as feature size shrinks. So far, such commodity products have each aimed at delivering ever more compute power, on one single chip. Indeed, if you look inside some recent digital device, chances are that you will see mostly three types of chips: RAM, processor and FPGA. While a specific DS shrinks with feature size, a general purpose DS gains performance through the shrink, ideally at constant normalised density.

4.1 System on a Chip

There are compelling reasons for wanting a Digital System to fit on a single chip. Cost per system is one. Performance is another:

- Off-chip communication is expensive, in area, latency and power. The bandwidth available across some on-chip boundary is orders of magnitude that across the corresponding off-chip boundary.
- If one quadruples the area of a square, the perimeter just doubles. As a consequence, when feature size shrinks by $1/x$, the internal communication bandwidth grows faster than the external IO bandwidth: $x^{3-\epsilon}$ against $x^{2-\epsilon}$. T his is true as long as silicon technology remains planar: transistors within a chip, and chips within a printed circuit board, must all layed out side by side (not on top of each other).

4.2 Ready to Shrink Architecture

So far, normalised performance density has been maintained, through the successive generations of chip architecture.

Can this be sustained in future shrinks?

A dominant consideration is to keep up the system clock frequency F. The formula for the normalised clock frequency $1/\phi = 100\tau(\lambda)$ implies that each combinatorial sub-circuit within the chip must have delay less than 100x that of a minimal size inverter. The depth of combinatorial gates that may be traversed along any path between two registers is limited. The length of combinatorial paths is limited by wire delays. It follows that only finitely many combinatorial structures can operate at normalised clock frequency ϕ. There is a limit to the number N of IO bits to any combinatorial structure which can operate at such a high frequency. In particular, this applies to combinatorial adders (say $N < 256$), multipliers (say $N < 64$) and m emories.

4.3 Reconfigurable Memory

The use of fast SRAM with small block size is common in microprocessors: for registers, data and instruction caches. Large and fast current memories are made of many small monolithic blocks. A recent SDRAM is described in [15]: $1Gb$ stored as 32 combinatorial blocks of $32Mb$ each. A 1.6 GB/s bandwidth is obtained: data is $64b$ wide at 200MHz.

By the argument from the preceding section, a large N bit memory must be broken into N/B combinatorial blocks of size B, in order to operate at normalised clock frequency $F = \phi$. A N bit memory with minimum latency may be constructed, through r ecursive decomposition into 4 quad memories, each of size $N/4$ - layed out within one quarter of the chip. The decomposition stops for $N = B$, when a block of combinatorial RAM is used. The access latency is proportional to the depth $log(N/B)$ of the hie rarchical decomposition.

A *Reconfigurable Memory RM* is an array of high speed dense combinatorial memory blocks. The blocks are connected through a reconfigurable pipelined wiring structure. As with FPGA, the RM has a configuration mode, during which the configuration par t of the RM is loaded. In user mode, the RM is some group of memories, whose specific interconnect and block decomposition is coded by the configuration. One can trade data width for address depth, from $1 \times N$ to $N/B \times B$ in the extreme case s.

A natural way to design a RM is to imbed blocks of SRAM within a FPGA structure. In CHESS [19], the atomic SRAM block has size 8×256. The SRAM blocks form a regular pitch matrix within the logic, and it occupies about 30% of the area. As a consequence, the storage density of CHESS is over 1/3 that of a monolithic SRAM. This is comparable to the storage density of current microprocessors; it is much higher than the storage density of FPGA, which rely (so far) on off-chip memories.

After configuration, the FPGA is a large array of small SRAM: each is used as LUT - typically LUT4. Yet, most of the configuration memory itself is not accessible as a computational resource by the application. In most current FPGA, the process of downl oading the configuration is serial, and it writes the entire configuration memory. In a 0.5x shrink, the download time doubles: 4x bits at (2-e)x the frequency. As a consequence, the download takes about 20 ms on P1, and 40 ms on P2.

A more efficient alternative is found in the X6k [17] and CHESS: in configuration mode, configuration memory is viewed as a single SRAM by the host system. This allows for faster complete download. An important feature is the ability to randomly access the elements of the configuration memory. For the RSA design, this allows for very fast partial reconfigurations: as we change the value of the 512*b* key which is hardwired into the logical equations, only few of the configuration bits have to up dated. Configuration memory can also be used as a general-purpose communication channel between the host and the application.

4.4 Reconfigurable Arithmetic Array

The normalised gate density of current FPGA is over 10x that of processors, both for Boolean operations and additions - fig. 7. This is no longer true for the multiply density, where common FPGA barely meets the multiply density of processors which recent ly integrate one (or more) pipelined floating point multiplier.

The arithmetical density of RS can be raised: MATRIX [DeHon], which is an array of 8b ALU, with Reconfigurable Interconnect, does better than FPGA. CHESS is based on 4*b* ALU, which are packed as the *white* squares in a chessboard. It follows that CHESS has an arithmetic density which is near 1/3 that of custom multipliers. The synchronous registers in CHESS are 4*b* wide, and they are found both within ALU and routing network, to as to facilitate high speed systematic pipelining.

Another feature of CHESS [19], is that each *black* square in the chessboard may be used either as a switch-box, or as a memory, based on a local configuration bit. As a switch-box, it operates on 4*b* *nibbles*, which are all routed tog ether. In memory mode, it may implement various specialised memories, such as a depth 8 shift register, in place of eight 4b wide synchronous registers. In memory mode, it can also be used as a 4b in, 4b out $4LUT4$. This feature provides CHESS with a LUT 4 density which is as high as for any FPGA.

4.5 Hardware or Software?

In order to implement digital function $Y = f(X)$, start from a specification by a program in some high level language. Some work is usually required to have the code match the digital specification, bit per bit - high level languages provide little support for funny bit formats and operations beneath the word size.

Once done, compile and unwind this code so as to obtain the run-code C_f. It is the sequence of machine instructions, which a sequential processor executes, in order to compute output sample Y_t from input sample X_t. This computation is to be repe ated indefinitely, for consecutive samples: t=0, 1, For the sake of simplicity, assume the run-code to be straight-line: each instruction is executed once in sequence, regardless of individual data values; there is no conditional branch. In theory, th e run-code should be one of minimal length, among all possible for function f, within some given instruction set. Operations are performed in sequence through the Arithmetic and Logic Unit ALU of the processor. Internal memory is used to feed the ALU, a nd provide (memory-mapped) external IO. For W the data width of the processor, the complexity of so computing f is $W|C_f|$ bop per sample. It is greater than the gate complexity $G(f)$. Equality $|C_f| = G(f)/W$ only happens in ideal cases. In pract ice, the ratio between the two can be kept close to one, at least for straight-line code.

The execution of run-code Cf on a processor chip at frequency F computes function f at the rate of F/C samples per second, with $C = |C_f|$. The feasibility of a software implementation of the DS on that processor depends on the real time requirement F_{io} - in samples per second.

1. If $F/C > F_{io}$, the DS can be implemented on the sequential processor at hand, through straightforward software.
2. If $F/C < F_{io}$, one needs a more parallel implementation of the digital system.

In case 1, the full computing power - WF in *bop/s* - of the processor is only used when $F/C = F_{io}$. When that is not the case, say $F/C > 2F_{io}$, one can attempt to trade time for area, by reducing the data width to $W/2$, while increasing the c ode length to $2C$: each operation on W bits is replaced by two operations on $W/2$ bits, performed in sequence. The invariant is the product CW, which gives the complexity of f in *bop* per sample. One can thus find the smallest processor on which some sequential code for f can be executed within the real time specification. The end of that road is reached for $W = 1$: a *single bit wide sequential processor*, whose run-code has length proportional to $G(f)$.

In case 2, and when one is not far away from meeting the real time requirement - say $F/C < 8F_{io}$ - it is advised to check if code C could be further reduced, or moved to a wider and faster processor (either existing or soon to come when the featur e size shrinks again). Failing that software solution, one has to find a hardware one. A common case mandating a hardware implementation, is when $F \approx F_{io}$: the real time external IO frequency F_{io} is near the internal clock frequency F of the chip.

4.6 Dynamic Reconfiguration

We have seen how to fold time in space: from a small design into a larger one, with more performance. The inverse operation, which folds space in time, is not always possible: how to fold any bit serial circuit (such as the adder from fig 5) into a half-size and half-rate structure is not obvious. Known solutions involve dynamic reconfiguration.

Suppose that function f may be computed on some RS of size $2A$, at twice the real-time frequency $F = 2F_{io}$. We need to compute f on a RS of size A at frequency F_{io} per sample. One technique, which is commonly used in [13], works when $Y = f(X) = g(h(X))$, and both g and h fit within size A.

1. Change the RS configuration to design h.
2. Process N input samples X; store each output sample $Z = h(X)$ in an external buffer.
3. Change the RS configuration to design g.
4. Process the N samples Z from the buffer, and produce the final output $Y = g(Z)$.
5. Go to 1, and process the next batch of N samples.

Reconfiguration takes time R/F, and the time to process N samples is $2(N + R)/F = (N + R)/F_{io}$. The frequency per sample $F_{io}/(1 + R/N)$ gets close to real-time F_{io}, as N gets large. Buffer size and latency are also proportional to N, and th is form of dynamic reconfiguration may only happen at a low frequency.

The opposite situation is found in the ALU of a sequential processor: the operation may change on every cycle. The same holds in *dynamically programmable* systems, such as arrays of processors and DPGA [14]. With such a system, one can reduce by half the number of processors for computing f, by having each execute twice more code. Note that this is a more efficient way to fold space in time than previously: no external memory is required, and the latency is not significantly affected.

The ALU in CHESS is also dynamically programmable. Although no specialised memory is provided for storing instructions (unlike DPGA), it is possible to build specialised *dynamically programmed* sequential processors, within the otherwise *stati cally configured* CHESS array. Through this feature, one can modulate the amount of parallelism in the implementation of a function f, in the range between serial hardware and sequential software, which is not accessible without dynamic reconfiguration.

5 Conclusion

We expect it to be possible to build *Reconfigurable Systems* of arbitrary size, which are fit to shrink: they can exploit all the available silicon, with a high normalised density for storage, arithmetic and Boolean operations, and operate at high n ormalised clock frequency.

For how long will the market demands for operations keep-up with the supply which such RS promise?

Can the productivity in mapping applications to massively parallel custom hardware be raised at the pace set by technology?

Let us take the conclusion from Carver Mead [9]:

> There is far more potential in a square centimetre of silicon than we have developed the paradigms to use.

References

1. C. E. Shannon, W. Weaver, *The Mathematical Theory of Communication*, University of Illinois Press, Urbana, 1949.
2. C. Mead, L. Conway *Introduction to VLSI systems*, Addison Wesley, 1980.
3. F.P. Preparata and J. Vuillemin. *Area-time optimal VLSI networks for computing integer multiplication and Discrete Fourier Transform*, Proceedings of I.C.A.L.P (Springer-Verlag), Haifa, Israel, Jul. 1981.
4. J. Vuillemin, *A combinatorial limit to the computing power of VLSI circuits*, IEEE Transactions on Computers, C-32:3:294-300, 1983.
5. I. Wegener *The Complexity of Boolean Functions*, John Wiley & sons, 1987.
6. O. A. McBryan, P. O. Frederickson, J. Linden, A. Schüller, K. Solchenbach, K. Stüben, C-A. Thole and U. Trottenberg, "Multigrid methods on parallel computers—a survey of recent developments", *Impact of Computing in Science and Engineering*, vol. 3(1), pp. 1–75, Academic Press, 1991.
7. J. Vuillemin. *Contribution à la résolution numérique des équations de Laplace et de la chaleur*, Mathematical Modelling and Numerical Analysis, AFCET, Gauthier-Villars, RAIRO, 27:5:591–611, 1993.
8. M. Shand and J. Vuillemin. *Fast implementation of RSA cryptography*, 11-th IEEE Symposium on Computer Arithmetic, Windsor, Ontario, Canada, 1993.
9. C. Mead, *Scaling of MOS Technology to Submicrometre Feature Sizes*, Journal of VLSI Signal Processing, V 8, N 1, pp. 9-26, 1994.
10. J.Vuillemin. *Fast linear Hough transform*, The International Conference on Application-Specific Array Processors, IEEE press, 1–9, 1994.
11. J. Vuillemin. *On circuits and numbers*, IEEE Trans. on Computers, 43:8:868–79, 1994.
12. L. Moll, J. Vuillemin, P. Boucard and L. Lundheim, *Real-time High-Energy Physics Applications on DECPeRLe-1 Programmable Active Memory,* Journal of VLSI Signal Processing, Vol 12, pp. 21-33, 1996.
13. J. Vuillemin, P. Bertin , D. Roncin, M. Shand, H. Touati, P. Boucard *Programmable Active Memories: the Coming of Age*, IEEE Trans. on VLSI, Vol. 4, NO. 1, pp. 56-69, 1996.
14. A. DeHon *Reconfigurable Architectures for General-Purpose Computing*, MIT, Artificial Intelligence Laboratory, AI Technical Report No. 1586, 1996.
15. N. Sakashita & al., *A 1.6-GB/s Data-Rate 1-Gb Synchronous DRAM with Hierarchical Square-Shaped Memory Block and Distributed Bank Architecture*, IEEE Journal of Solid-state Circuits, vol. 31, No 11, pp 1645-54, 1996.

16. M. Shand, *Pamette*, a Reconfigurable System for the PCI Bus, 1998. http://www.research.digital.com/SRC/pamette/
17. Xilinx, Inc., *The Programmable Gate Array Data Book*, Xilinx, 2100 Logic Drive, San Jose, CA 95124 USA, 1998.
18. G. Moore. *An Update on Moore's Law*, 1998. http://www.intel.com/pressroom/archive/speeches/gem93097.htm
19. Alan Marshall, Tony Stansfield, Jean Vuillemin *CHESS: a Dynamically Programmable Arithmetic Array for Multimedia Processing*, Hewlett Packard Laboratories, Bristol, 1998.
20. M. Shand. *An Update on RSA software performance*, private communication, 1998.
21. *The millenium bug: how much did it really cost?*, your newspaper, 2001.

A Systolic Algorithm for the Factorisation of Matrices Arising in the Field of Hydrodynamics

S.-G. Seo[1], M. J. Downie[1], G. E. Hearn[1] and C. Phillips[2]

[1]Department of Marine Technology, University of Newcastle upon Tyne, Newcastle upon Tyne, NE1 7RU, UK

[2]Department of Computing Science, University of Newcastle upon Tyne, Newcastle upon Tyne, NE1 7RU, UK

Abstract. Systolic algorithms often present an attractive parallel programming paradigm. However, the unavailability of specialised hardware for efficient implementation means that such algorithms are often dismissed as being of theoretical interest only. In this paper we report on experience with implementing a systolic algorithm for matrix factorisation and present a modified version that is expected to lead to acceptable performance on a distributed memory multicomputer. The origin of the problem that generates the full complex matrix in the first place is in the field of hydrodynamics.

1. Forming the Linear System of Equations

The efficient, safe and economic design of large floating offshore structures and vessels requires a knowledge of how they respond in an often hostile wave environment [1]. Prediction of the hydrodynamic forces experienced by them and their resulting responses, which occur with six rigid degrees of freedom, involves the use of complex mathematical models leading to the implementation of computationally demanding software. The solution to such problems can be formulated in terms of a velocity potential involving an integral expression that can be thought of as representing a distribution of sources over the wetted surface of the body. In most applications there is no closed solution to the problem and it has to be solved numerically using a discretisation procedure in which the surface of the body is represented by a number of panels, or facets. The accuracy of the solution depends on a number of factors, one of which is the resolution of the discretisation. The solution converges as resolution becomes finer and complicated geometries can require very large numbers of facets to attain an acceptable solution.

In the simplest approach a source is associated with each panel and the interaction of the sources is modelled by a Green function which automatically satisfies relevant 'wave boundary conditions'. The strength of the sources is determined by satisfying a velocity continuity condition over the mean wetted surface of the body. This is

J. Palma, J. Dongarra, and V. Hernández (Eds.): VECPAR'98, LNCS 1573, pp. 355-364, 1999.

achieved by setting up an influence matrix, A, for the sources based on the Green functions and solving a linear set of algebraic equations in which the unknowns, $\mathbf{x}$, are either the source strengths or the velocity potential values and the right-hand side, $\mathbf{b}$, are related to the appropriate normal velocities at a representative point on each facet. By construction, the diagonal elements of A represent the influence of the sources on themselves. Since the off-diagonal elements are inversely proportional to the distance between any two sources, they tend to be of increasingly smaller magnitude as we move away from the diagonal. Hence if geometric symmetry is not used and the full matrix A is considered, it is guaranteed to be diagonally dominant. For a given wave frequency, each facet has a separate source contributing to the wave potential representing the incoming and diffracted waves, ϕ_0 and ϕ_7, and one radiation velocity potential for each degree of freedom of the motion, ϕ_i : $i=1,2,\ldots,6$. When the velocity potentials have been determined, once the source strengths are known, the pressure can be computed at every facet and the resultant forces and moments on the body computed by integrating them over the wetted surface. The forces can then be introduced into the equations of motion and the response of the vessel at the given wave frequency calculated.

The complexity of the mathematical form of the Green functions and the requirement to refine the discretisation of the wetted surfaces within practical offshore analyses, significantly increases the memory and computational load associated with the formulation of the required fluid-structure interactions. Similarly the solution of the very large dense square matrix equations formulated in terms of complex variables requires considerable effort to provide the complex variable solution. In some problems 5,000 panels might be required using constant plane elements or 5,000 nodes using higher order boundary elements, leading to a matrix with 25,000,000 locations. Since double precision arithmetic is required, and the numbers are complex, this will require memory of the order of 4 gigabytes. The number of operations for direct inversion or solution by iteration is large and of the order of n^3, e.g. 5,000 elements requires 125,000 x 10^6 operations. Furthermore, the sea-state for a particular wave environment has a continuous frequency spectrum which can be modelled as the sum of a number of regular waves of different frequencies with random phase angles and amplitudes determined by the nature of the spectrum. Determination of vessel responses in a realistic sea-state requires the solution of the boundary integral problem described above over a range of discrete frequencies sufficiently large to encompass the region in which the wave energy of the irregular seas is concentrated. In view of the size and complexity of such problems, and the importance of being able to treat them, it is essential to develop methods to speed up their formulation and solution times. One possible means of achieving this is through the use of parallel computers [2] [3].

Numerical techniques for solving linear systems of equations neatly divide into direct methods that involve some transformation of A, and iterative methods that do not change A but typically involve a matrix-vector multiplication involving A at each iteration. (Iterative techniques, such as those based on the conjugate gradient method, often employ preconditioning, which requires the formulation of a pseudo-transformation of A in order to improve the convergence characteristics.) Iterative techniques are normally restricted to the case that A is sparse, when the reduction in operations count and space requirements caused by fill-in merit their use. For our application A is a full, square matrix, and the use of a direct method of solution based on elimination techniques is therefore the most attractive proposition. The method of LU-decomposition has been chosen because in this scheme only one factorisation is required for multiple unknown right-hand side vectors **b**. We recall that A is diagonally dominant and hence pivoting, which often limits the performance of parallel algorithms, is not an issue here. It is well known that this factorisation is computationally intensive, involving order n^3 arithmetic operations (multiplications and additions). In contrast the forward- and backward-substitution required to solve the original system once the factorisation has been performed involves an order of magnitude less computation, namely order n^2, which becomes insignificant as the matrix size increases. Consequently, we limit consideration to the factorisation process only.

Formally, we have that the elements u_{ij} of U are given by

$$u_{rj} = a_{rj} - \sum_{k=1}^{r-1} l_{rk} u_{kj} \qquad j = r, \ldots, n \tag{1}$$

and l_{ij} of L are given by

$$l_{ir} = \left(a_{ir} - \sum_{k=1}^{r-1} l_{ik} u_{kr} \right) / u_{rr} \qquad i = r+1, \ldots, n \tag{2}$$

(Doolittle factorisation) leading to an upper-triangular U and a unit lower-triangular L.

2. A Naive Systolic Algorithm Solution

A systolic system can be envisaged as an array of synchronised processing elements (PEs), or cells, which process data in parallel by passing them from cell to cell in a regular rhythmic pattern. Systolic arrays have gained popularity because of their ability to exploit massive parallelism and pipelining to produce high performance computing [4] [5]. Although systolic algorithms support a high degree of concurrency, they are often regarded as being appropriate only for those machines specially built for the particular algorithm in mind. This is because of the inherent high communication/computation ratio.

In a soft-systolic algorithm, the emphasis is on retaining systolic computation as a design principle and mapping the algorithm onto an available (non-systolic) parallel architecture, with inevitable trade-offs in speed and efficiency due to communication and processor overheads incurred in simulating the systolic array structure.

Initially a systolic matrix factorisation routine was written in Encore Parallel Fortran (epf) and tested on an Encore Multimax 520. This machine has seven dual processor cards (a maximum of ten can be accommodated), each of which contains two independent 10 MHz National Semiconductor 32532 32-bit PEs with LSI

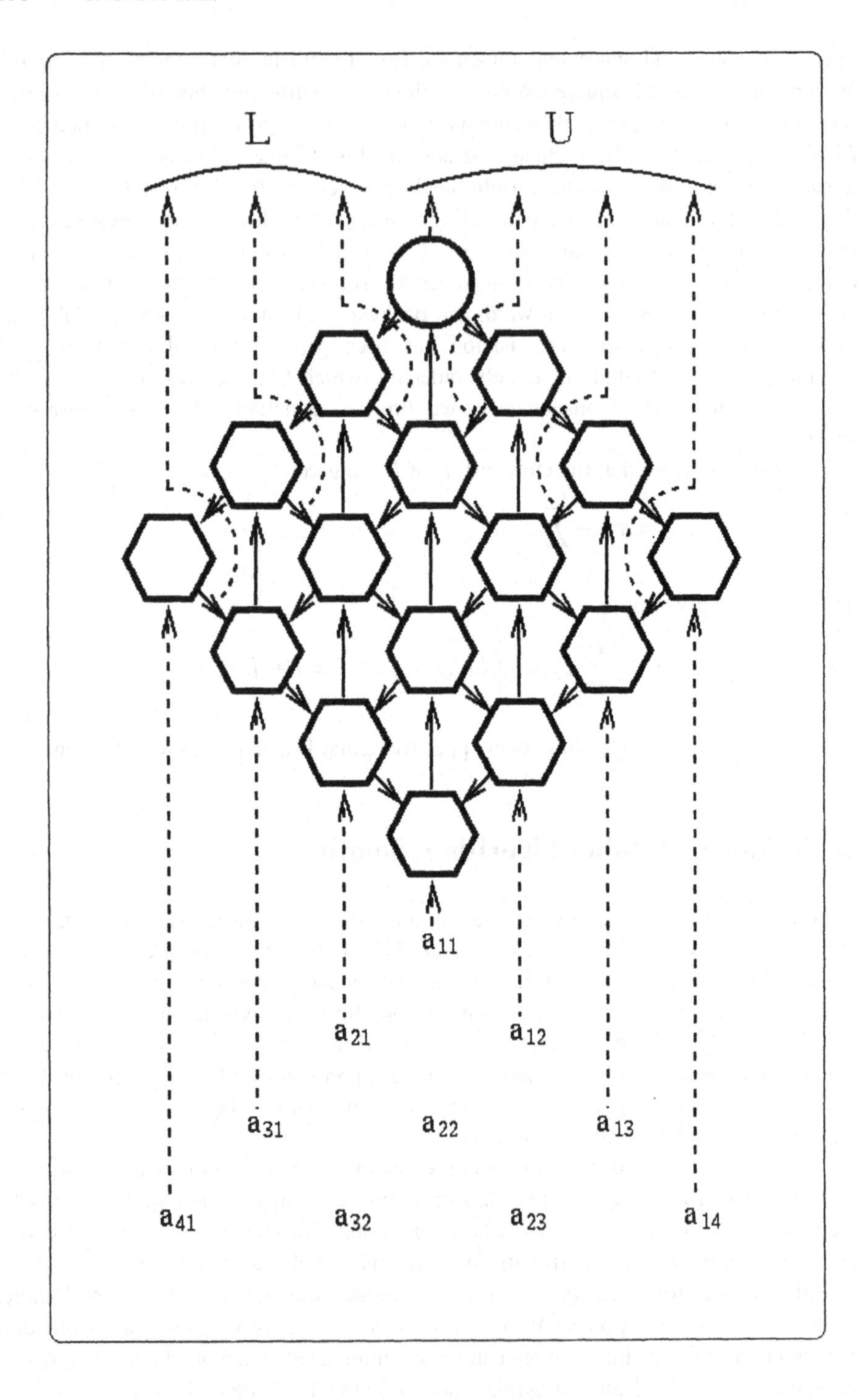

Fig. 1. Data flow for LU factorisation

memory management units and floating-point co-processors. This work was undertaken on a well-established machine platform that was able to provide a mechanism for validating the model of computation, and the software developed from that model, with a view to refining the model for subsequent development on a state-of-the-art distributed memory parallel machine. epf's parallel program paradigm is much simpler to work with than that for message passing on distributed memory architectures and produces a convenient means of providing validation data for subsequent developments. As long as the underlying algorithm is retained this implementation can be used to check the correctness of the equations developed and the validity of the algorithm with a view to an eventual port.

The task of systolic array design may be defined more precisely by investigating the cell types required, the way they are to be connected, and how data moves through the array in order to achieve the desired computational effect. The hexagonal shaped cells employed here are due to Kung and Leiserson [6].

The manner in which the elements of L and U are computed in the array and passed on to the cells that require them is demonstrated in Fig. 1. An example of a 4x4 cell array is shown for the purposes of illustrating the paradigm employed. Of the PEs on the upper-right boundary, the top-most has three roles to perform:

1. Produce the diagonal components of L (which, for a Doolittle-based factorisation, are all unit).
2. Produce the diagonal components of U using elements of A which have filtered through the cells along the diagonal (and been modified, as appropriate).
3. Pass the reciprocal of the diagonal components of U down and left.

The cells on the upper left boundary are responsible for computing the multipliers (the elements of L), having received the appropriate reciprocals of the diagonal elements of U.

The flow of data through the systolic array is shown in Fig. 1. Elements of A flow in an upward direction; elements of L (computed from (**1**)) flow in a right-and-down direction; and elements of U (computed from (**2**)) flow in a left-and-down direction. Each cell computes a value every 3 clock ticks, although they start at different times. Note that at each time step each cell responsible for forming an element of L or U calculates one multiplication only in forming a partial summation. Data flowing out correspond to the required elements of the L and U factors of A. Formally, the elements of A are fed into the cells as indicated, although for efficiency the cells are directly assigned the appropriate elements of A.

Table 1 shows the execution times (T_p) of the parallel code with a fixed-size (100*100 double complex) matrix and various numbers of PEs (p). The gradual algorithmic speed-up (S_p), defined as the ratio of the time to execute the program on p processors to the time to execute the same parallel program on a single processor, is clearly seen all the way up to twelve PEs. The (generally) decreasing efficiency (E_p), defined as the ratio of speed-up to the number of PEs times 100, is a consequence of the von Neumann bottleneck. The results show some minor anomalies, but this is not atypical when attempting to obtain accurate timings on a shared resource, with other processes - system or those of other users - interfering with program behaviour, even at times of low activity. At this level, the results are encouraging.

Table 1. Shared memory implementation

p	**1**	**2**	**3**	**4**	**5**	**6**	**7**	**8**	**9**	**10**	**11**	**12**
T_p	48.9	28.3	20.2	14.9	12.3	10.6	8.9	8.0	7.3	6.6	6.2	6.1
S_p	1	1.7	2.4	3.3	4.0	4.6	5.5	6.1	6.7	7.4	7.9	8.0
E_p	100	87	81	82	80	77	79	77	74	74	72	68

The algorithm was compared with an existing parallel matrix factorisation routine [7], which uses more conventional techniques, available in the Department of Marine Technology (DMT). The results are summarised in Fig 2, where $S_{p(systolic)}$ denotes the speedup for the systolic algorithm (from Table 1) and $S_{p(DMT)}$ the speedup for the DMT algorithm. The extra cost of the systolic algorithm due to overheads

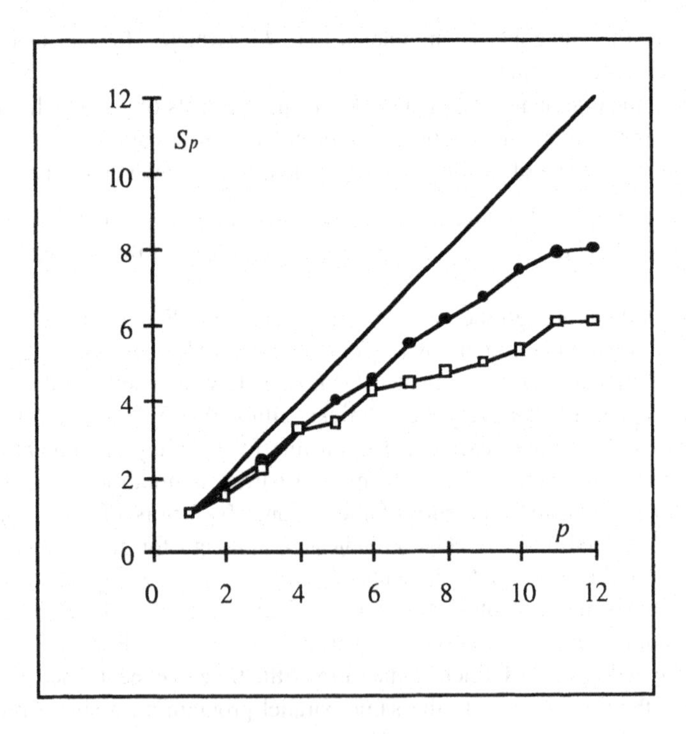

Fig. 2. Comparison of speedup for systolic and DMT algorithms.

associated with index calculation and array accesses causes significant delays in the computation resulting in much poorer performance of the systolic algorithm in comparison to the DMT routine in terms of elapsed time. Nevertheless, the systolic algorithm shows better speedup characteristics than the DMT algorithm, as illustrated by Fig. 2.

If A is an n by n dense matrix then the systolic algorithm implemented on n^2 PEs can compute L and U in $4n$ clock ticks, giving a cell efficiency of 33%. Assuming a hardware system in which the number of PEs is much less than the number of cells, and using an appropriate mapping of cells to PEs, we can improve this position considerably, and we now address this issue.

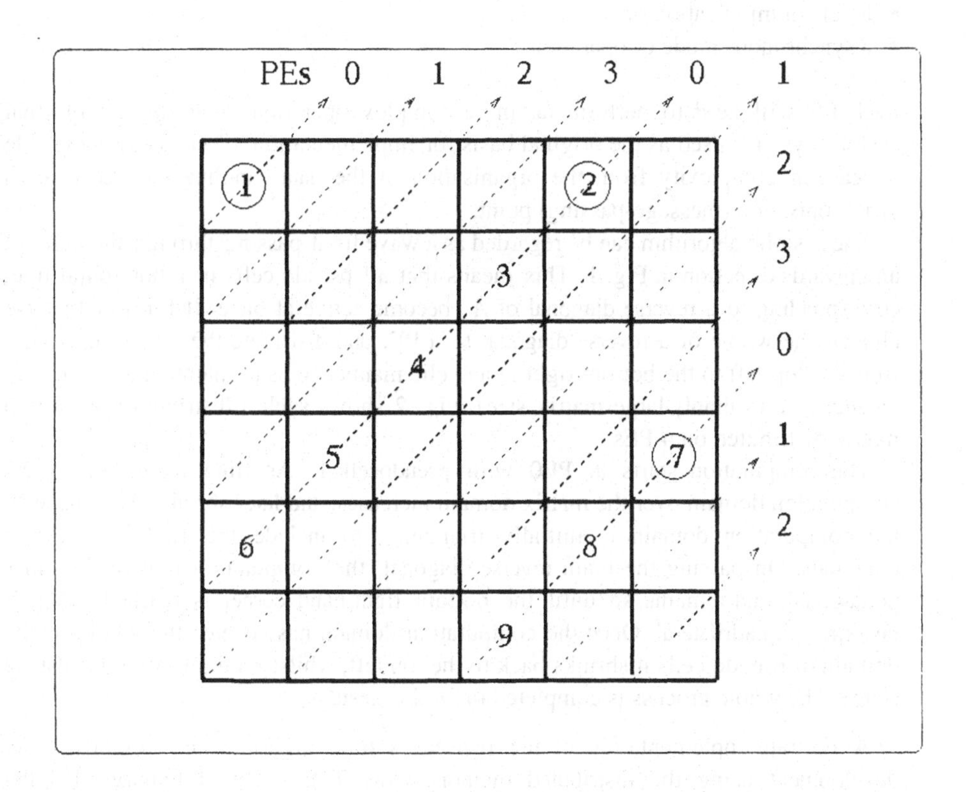

Fig. 3. Allocation of pseudo cells to PEs for a 6x6 matrix.

3. An Improved Systolic Algorithm

As already indicated, the ultimate goal is to produce an efficient systolic matrix factorisation routine for general-purpose distributed parallel systems including clusters of workstations. This is to be achieved by increasing the granularity of the computation within the algorithm and thus reducing the communication/ computation ratio, while balancing the load on each PE to minimise the adverse effect due to enforced synchronisation. Nevertheless, the characteristics peculiar to systolic algorithms should be retained. Thus we aim at

- massive, distributed parallelism
- local communication only
- a synchronous mode of operation

Each PE will need to perform far more complex operations than in the original systolic systems used as the original basis for implementation. There is an inevitable increase in complexity from the organisation of the data handling required at each synchronisation (message-passing) point.

The systolic algorithm can be regarded as a wave front passing through the cells in an upwards direction in Fig. 1. This means that all pseudo cells in a horizontal line, corresponding to a reverse diagonal of A, become active at once. It follows that we allocate the whole of a reverse diagonal to a PE, and distribute the reverse diagonals from the top left to the bottom right in a cyclic manner so as to maintain an even load balance (for a suitably large matrix size). Fig. 2 shows such a distribution for a 6x6 matrix distributed on 4 PEs.

The computation starts at PE0 with pseudocell 1. As time increases, so the computation domain over the matrix domain increases, and later shrinks. The shape of the computation domain is initially triangular, to include the first few reverse diagonals. On passing the main reverse diagonal, the computation domain becomes pentagonal, and remains so until the bottom right-hand corner is reached, when it becomes a quadrilateral. Once the computation domain has covered the whole of the domain of pseudo cells it shrinks back to the top left, whilst retaining its quadrilateral shape. The whole process is completed in $3n-2$ timesteps.

A Fortran implementation of the revised systolic algorithm is currently under development using the distributed memory Cray T3D at the Edinburgh Parallel Computer Centre (EPCC), and the Fujitsu AP1000 at Imperial College, London, and MPI [8] [9] for message passing.

In this implementation the elements in each reverse (top-right to bottom-left) diagonal of the matrix are bundled together so that they are dealt with by a single PE and all of the reverse diagonals which are active at a given time step are again grouped together to be passed around as a single message. Preliminary experience with the revised algorithm indicates that it scales well, as shown by the speed up figures for a 400 by 400 array computed on the Fujitsu machine and illustrated in Fig. 4.

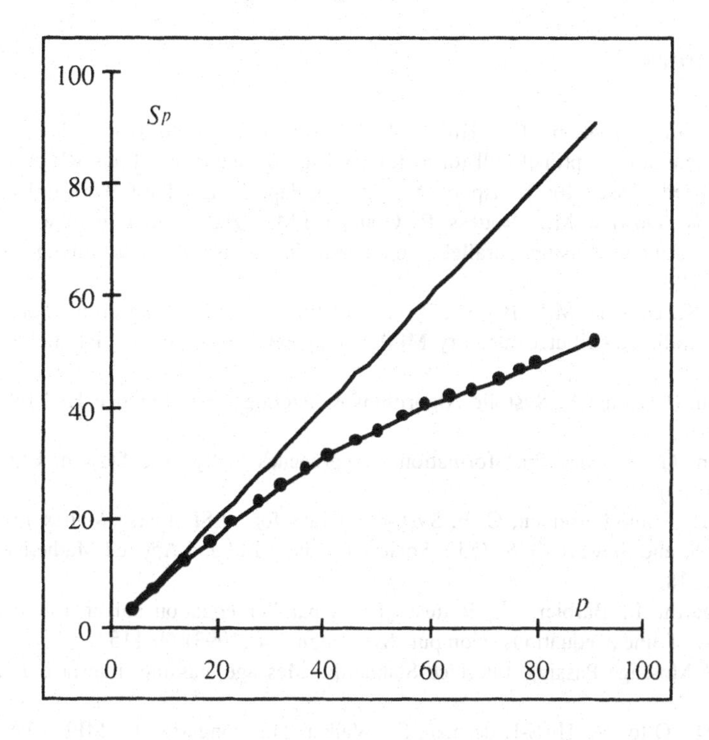

Fig. 4. Speedup on distributed memory machine for 400 by 400 array

4. Conclusions

The algorithm described initially was a precursor to a more generalized one designed with the intention of increasing the granularity of computation and with the solution of large systems of equations in mind. As expected, it performed poorly in terms of elapsed time due to the penalties imposed by an unfavourable balance between processor communication and computation. However, the speedup characteristics compared favourably with those of a more conventional approach and pointed to the potential for the generalised algorithm.

Accordingly, a generalised version of the algorithm has been developed and is currently being tested. The speedup obtained on the distributed memory machine suggest that the approach taken is valid. It remains to benchmark the algorithm against a conventional solver, such as the one available within the public domain software package, ScaLAPACK.

References

1. Hearn, G.E., Yaakob, O., Hills, W.: Seakeeping for design: identification of hydrodynamically optimal hull forms for high speed catamarans. Proc. RINA Int. Symp. High Speed Vessels for Transport and Defence, Paper 4, London (1995), pp15.
2. Hardy, N., Downie, M.J., Bettess, P., Graham, J.M.: The calculation of wave forces on offshore structures using parallel computers., Int. J. Num. Meth. Engng. **36** (1993) 1765-1783
3. Hardy, N., Downie, M.J., Bettess, P.: Calculation of fluid loading on offshore structures using parallel distributed memory MIMD computers. Proceedings, Parallel CFD, Paris (1993).
4. Quinton, P., Craig, I.: Systolic Algorithms & Architectures. Prentice Hall International (1991)
5. Megson, G. M. (ed.): Transformational Approaches to Systolic Design. Chapman and Hall (1994)
6. Kung, H. T. and Leiserson, C. E. Systolic Arrays for VLSI, Sparse Matrix Proceedings, Duff, I.S. and Stewart, G.W. (ed.). Society for Industrial and Applied Mathematics (1979) PP.256-282.
7. Applegarth, I., Barbier, C., Bettess, P.: A parallel equation solver for unsymmetric systems of linear equations. Comput. Sys. Engng. **4** (1993) 99-115
8. MPI: A Message-Passing Interface Standard. Message Passing Interface Forum (May 1994)
9. Snir, M., Otto, S., Huss-Lederman, S., Walker, D., Dongarra, J.: MPI: The Complete Reference. MIT Press (1996)

Automatic Detection of Parallel Program Performance Problems [1]

A. Espinosa, T. Margalef, E. Luque.

Computer Science Department
Universitat Autònoma de Barcelona.
08193 Bellaterra, Barcelona, SPAIN.
e-mail: iinfd@cc.uab.es

Abstract. *Actual behaviour of parallel programs is of capital importance for the development of an application. Programs will be considered matured applications when their performance is under acceptable limits. Traditional parallel programming forces the programmer to understand the enormous amount of performance information obtained from the execution of a program. In this paper, we propose an automatic analysis tool that lets the programmers of applications avoid this difficult task. This automatic performance analysis tool main objective is to find poor designed structures in the application. It considers the trace file obtained from the execution of the application in order to locate the most important behaviour problems of the application. Then, the tool relates them with the corresponding application code and scans the code looking for any design decision which could be changed to improve the behaviour*

1. Introduction

The performance of a parallel program is one of the main reasons for designing and building a parallel program [1]. When facing the problem of analysing the performance of a parallel program, programmers, designers or occasional parallel systems users must acquire the necessary knowledge to become performance analysis experts.

Traditional parallel program performance analysis has been based on the visualization of several execution graphical views [2, 3, 4, 5]. These high level graphical views represent an abstract description of the execution data obtained from many possible sources and even different executions of the same program [6].

[1] This work has been supported by the CICYT under contract TIC 95-0868

J. Palma, J. Dongarra, and V. Hernández (Eds.): VECPAR'98, LNCS 1573, pp. 365-377, 1999.

The amount of data to be visualized and analyzed, together with the huge number of sources of information (parallel processors and interconnecting network states, messages between processes, etc.) make this task of becoming a performance expert difficult. Programmers need a high level of experience to be able to derive any conclusions about the program behaviour using these visualisation tools. Moreover, they also need to have a deep knowledge of the parallel system because the analysis of many performance features must consider architectural aspects like the topology of the system and the interconnection network.

In this paper we describe a Knowledge-based Automatic Parallel Program Analyser for Performance Improvement (KAPPA-PI tool) that eases the performance analysis of a parallel program. Analysis experts look for special configurations of the graphical representations of the execution which refer to problems at the execution of the application. Our purpose is to substitute the expert with an automatic analysis tool which, based on a certain knowledge of what the most important performance problems of the parallel applications are, detects the critical execution problems of the application and shows them to the application programmer, together with source code references of the problem found, and indications on how to overcome the problem.

We can find other automatic performance analysis tools:

-Paradyn [7] focuses on minimising the monitoring overhead. The Paradyn tool performs the analysis "on the fly", not having to generate a trace file to analyse the behaviour of the application. It also has a list of hypotheses of execution problems that drive the dynamic monitoring.

- AIMS tool [8], is a similar approach to the problem of performance analysis. The tool builds a hierarchical account of program execution time spent on different operations, analyzing in detail the communications performed between the processes.

-Another approach to addressing the problem of analysing parallel program performance is carried out by [9] and [10]. The solution proposed is to build an abstract representation of the program with the help of an assumed programming model of the parallel system. This abstract representation of the program is analysed to predict some future aspects of the program behaviour. The main problem of this approach is that, if the program is modelled from a high level view, some important aspects of its performance may not be considered, as they will be hidden under the abstract representation.

- Performance of a program can also be measured by a pre-compiler, like Fortran approaches (P3T [11], this approach is not applicable to all parallel programs, especially those where the programmer expresses dynamic unstructured behaviour.

Our KAPPA-PI tool is currently implemented (in Perl language [12]) to analyse applications programmed under the PVM [13] programming model. The KAPPA-PI tool bases the search for performance problems on its knowledge of their causes. The analysis tool makes a "pattern matching" between those execution intervals which degrade performance and the "knowledge base" of causes of the problems. This is a process of identification of problems and creation of recommendations for their solution. This working model allows the "performance problem data base" to adapt to new possibilities of analysis with the incorporation of new problems (new knowledge data) derived from the experimentation with programs and new types of programming models.

In section 2, we describe the analysis methodology briefly, explaining the basis of its operations and the processing steps to detect a performance problem. Section 3 presents the actual analysis of a performance problem detected in an example application. Finally, section 4 exposes the conclusions and future work on the tool development.

2.- Automatic analysis overview

The objective of the automatic performance analysis of parallel programs is to provide information regarding the behaviour of the user's application code.

This information may be obtained analysing statically the code of the parallel program. However, due to the dynamic behaviour of the processes that form the program and the parallel system features, this static analysis may not be sufficient.

Then, execution information is needed to effectively draw any conclusion about the behaviour of the program. This execution information can be collected in a trace file that includes all the events related to the execution of the parallel program. However, the information included in the trace file is not significant to the user who is only concerned with the code of the application.

The automatic performance analysis tool concentrates on analysing the behaviour of the parallel application expressed in the trace file in order to detect the most important performance problems. Nonetheless, the analysis process can not stop there and must relate the problems found with the actual code of the application. In this way, user receives meaningful information about the application behaviour.

In figure 1, we represent the basic analysis cycle followed by the tool to analyse the behaviour of a parallel application.

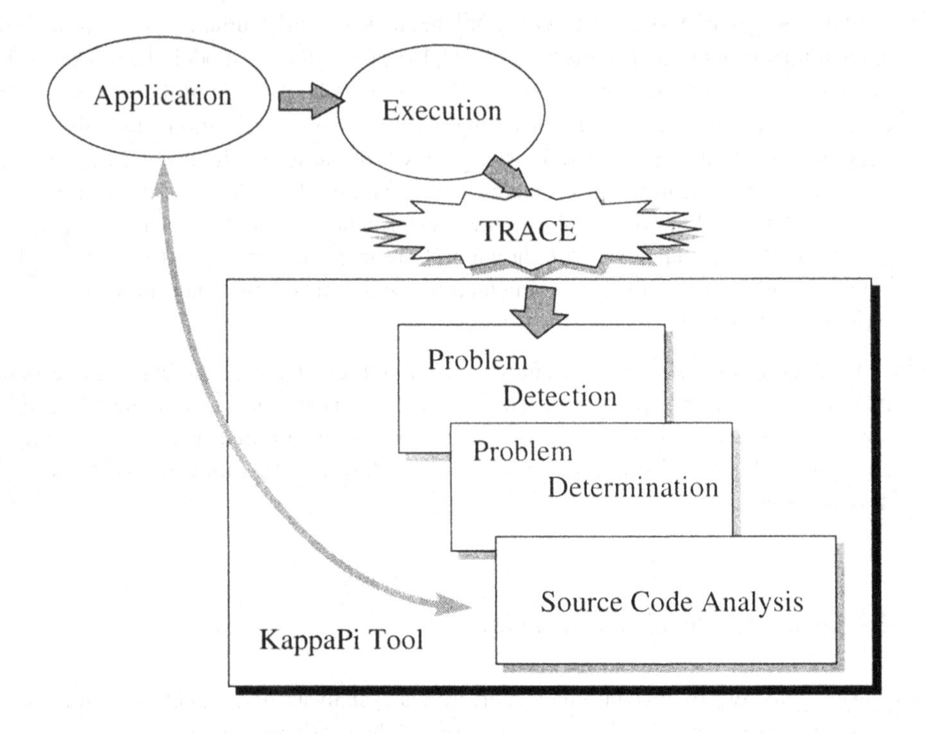

Fig. 1. Schema of the analysis of a parallel application

The analysis first considers the study of the trace file in order to locate the most important performance problems occurring at the execution. Once those problematic execution intervals have been found, they are studied individually to determinate the type of performance problem for each execution interval.

When the problem is classified under a specific category, the analysis tool scans the segment of application source code related to the execution data previously studied. This analysis of the code brings out any design problem that may have produced the performance problem. Finally, the analysis tool produces an explanation of the problems found at this application design level and recommends what should be changed in the application code to improve its execution behaviour.

In the following points, the operations performed by the analysis tool are explained in detail.

2.1. Problem Detection

The first part of the analysis is the study of the trace file obtained from the execution of the application. In this phase, the analysis tool scans the trace file, obtained with the use of TapePVM [14], with the purpose of following the evolution of the efficiency of the application. The application efficiency is basically found by measuring the number of processors that are executing the application during a certain time.

The analysis tool collects those execution time intervals when the efficiency is minimum. These intervals represent those situations where the application is not using all the capabilities of the parallel machine. They could be evidence of an application design fault. In order to analyse these intervals further, the analysis tool selects the most important inefficiencies found at the trace file. More importance is given to those inefficiency intervals that affect the most number of processors for the longest time.

2.2. Problem Determination

Once the most important inefficiencies are found, the analysis tool proceeds to classify the performance with the help of a "knowledge base" of performance problems. This classification is implemented in the form of a problem tree, as seen in figure 2.

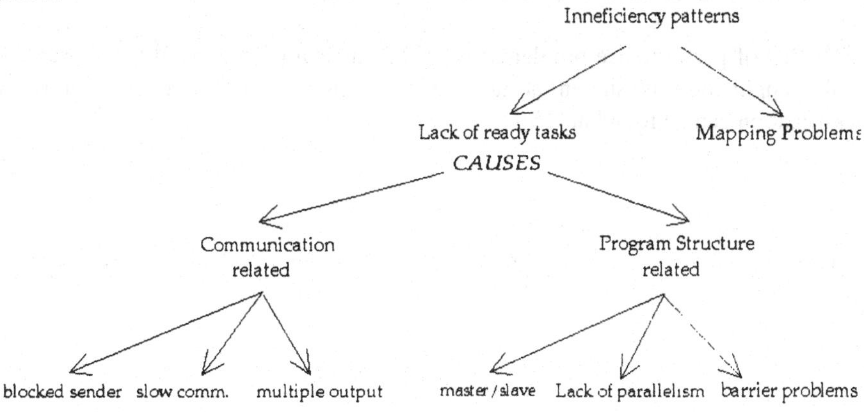

Fig. 2. *Classification of the performance problems of an application*

Each inefficiency interval at the trace is exhaustively studied in order to find which branches in the tree describe the problem in a more accurate way. When the classification of the problem arrives at the lowest level of the tree, the tool can proceed to the next stage, the source code analysis

2.3. Application of the source code analysis

At this stage of the program evaluation, the analysis tool has found a performance problem in the execution trace file and has classified it under one category.

The aim of the analysis tool at this point is to point out any relationship between the application structure and the performance problem found. This detailed analysis differ from one performance problem to another, but basically consists of the application of several techniques of pattern recognition to the code of the application.

First of all, the analysis tool must select those portions of source code of the application that generated the performance problem when executed. In order to establish a relationship between the executed processes and the program code, the analysis tool builds up a table of process identificators and their corresponding code modules names.

With the help of the trace file, the tool is able to relate the execution events of certain operations, like sending or receiving a message, to a certain line number in the program code. Therefore, the analysis tool is able to find which instructions in the source code generated a certain behaviour at execution time. Each pattern-matching technique tries to test a certain condition of the source code related to the problem found. For each of the matches obtained in this phase, the analysis tool will generate some explanations of the problem found, the bounds of the problem and what possible alternatives there are to alleviate the problem.

The list of performance problems, as well as their implications of the source code of the application is shown at table 1. A more exhaustive description of the classification can be found at [15].

NAME	DESCRIPTION	TRACE INFORMATION	SOURCE CODE IMPLICATIONS
Mapping Problems			
Mapping problem	There are idle processors and ready-to-execute processes in busy processors	Processes assignments to busy processors, number of ready processors	Solutions affect the process-processor mapping
Lack of Ready Tasks Problems			
Communication Related			
Blocked Sender	A blocked process is waiting for a message from another process that is already blocked for reception.	Waiting receive times of the blocked processes. Process identifiers of the sender partner of each receive.	Study of the dependencies between the processes to eliminate waiting.
Multiple Output	Serialization of the output messages of a process.	Identification of the sender process and the messages sent by this process.	Study of the dependencies between the messages sent to all receiving processes.
Long Communication	Long communications block the execution of parts of the program.	Time spent waiting. Operations performed by the sender at that time.	Study of the size of data transmitted and delays of the interconnection network.
Program Structure Related			
Master/ Slave problems	The number of masters and collaborating slaves is not optimum.	Synchronization times of the slaves and master processes.	Modications of the number of slaves/masters.
Barrier problems	Barrier primitive blocks the execution for too much time.	Identification of barrier processes and time spent waiting for barrier end.	Study of the latest processes to arrive at the barrier.
Lack of parallelism	Application design does not produce enough processes to fill all processors	Analysis of the dependences of the next processes to execute.	Possibilities of increasing parallelism by dividing processes

Table 1. *Performance problems detected by the analysis tool.*

In the next section, we illustrate the process of analysing a parallel application with the use of an example.

3. Example: analysis of an application

In this example we analyse a tree-like application with important amount of communications between processes. The application is executed mapping each process to a different processor. From the execution of the application we obtain a trace file, which is shown as a time-space diagram, together with the application structure, in figure 3.

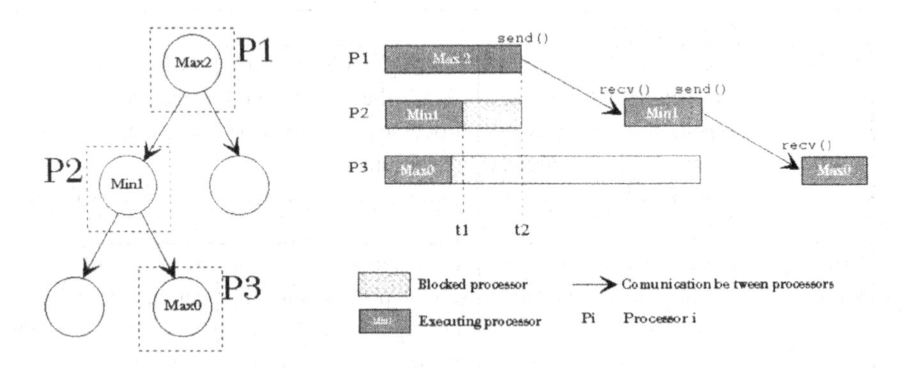

Fig. 3. *Application trace file space-time diagram*

In the next points we follow the operations carried out by the tool when analysing the behaviour of the parallel application.

3.1. Problem Detection

First of all, the trace is scanned to look for low efficiency intervals. The analysis tool finds an interval of low efficiency when processors P2 and P3 are idle due to the blocking of the processes "Min1" and "Max0". Then, the execution interval (t1,t2) is considered for further study.

3.2. Problem Determination

The analysis tool tries to classify this problem found under one of the categories. To do so, it studies the number of ready-to-execute processes in the interval. As there are no such kind of processes, it classifies the problem as "lack of ready processes". The analysis tool also finds that the processors are not just idle, but waiting for a message to arrive, so the problem is classified as a communication related.

Then, the analysis tool must find out what the appropriate communication problem is. It starts analyzing the last process (Max0) which is waiting for a message from Min1 process. When the tool tries to study what the Min1 process was doing at that time, it finds that Min1 was already waiting for a message from Max2, so the analysis tool classifies this problem as a blocked sender problem, sorting the process sequence: Max2 sends a message to Min1 and Min1 sends a message to Max0.

3.3. Analysis of the source code

In this phase of the analysis, the analysis tool wants to analyse the data dependencies between the messages sent by processes Max2, Min1 and Max0 (see figure 3).

First of all, the analysis tool builds up a table of the process identifiers and each source C program name of the processes.

When the program names are known, the analysis tool opens the source code file of process Min1 and scans it looking for the send and the receive operations performed. From there, it collects the name of the variables which are actually used to send and receive the messages. This part of the code is expressed on figure 4.

```
 1  pvm_recv(-1,-1);
 2
 3  pvm_upkfl(&calc,1,1);
 4
 5  calc1 = min(calc,1);
 6
 7  for(i=0;i<sons;i++)
 8  {
 9      pvm_initsend(PvmDataDefault);
10
11      pvm_pkfl(&calc1,1,1);
12
13      pvm_send(tid_son[i],1);
14  }
```

Fig. 4. *"Min1.c" relevant portion of source code*

When the variables are found ("calc" and "calc1" at the example) , the analysis tool starts searching the source code of process "Min1" to find all possible relationships between both variables. As these variables define the communication dependence of the processes, the results of these tests will describe the designed relationship between the processes.

In this example, the dependency test is found true due to the instruction found at line 5, which relates "calc1" with the value of "calc". This dependency means that the message sent to process "Max0" depends on the message received from process "Max2".

The recommendation produced to the user explains this situation of dependency found. The analysis tool suggests the modification of the design of the parallel application in order to distribute part of the code of process "Min1" (the instructions that modify the variable to send) to process "Max0", and then send the same message to "Min1" and to "Max0". This message shown to the user is expressed in figure 5.

```
Analysing MaxMin....

A Blocked Sender situation has been found in the
execution.

Processes involved are:
Max0, Min1, Max2
Recommendation: A dependency between Max2 and Max0 has
been found.
The design of the application should be revised.
Line 25 of Min1 process should be distributed to Max0.
```

Fig. 5. *Output of the analysis tool*

The line referred in the recommendations of the tool (Line 5 of Min1 Process) should be executed in the process Max0, so variable "calc" must be sent to Max0 to solve the expression. Then, the codes of the processes may be changed as follows in figure 6.

```
...
pvm_recv(-1,-1);
pvm_upkfl(&calc,1,1);
calc1 = min(calc,1);.
...
```

Process Max0

```
...
pvm_recv(-1,-1);
pvm_upkfl(&calc,1,1);
calc1 = min(calc,1);
...
```

Process Min1

```
...
calc = min(old,myvalue);
pvm_initsend(PvmDataDefault);
pvm_pkfl(&calc1,1,1);
pvm_send(tid_Min1,1);
pvm_send(tid_Max2,1);
...
```

Process Max2

Fig. 6. *New process code*

In the new processes code, the dependencies between Min1 and Max2 processes have been eliminated. From the execution of these processes we obtain a new trace file, shown in figure 7. In the figure, the process Max0 does not have to wait so long until the message arrives. As a consequence, the execution time of this part of the application has been reduced.

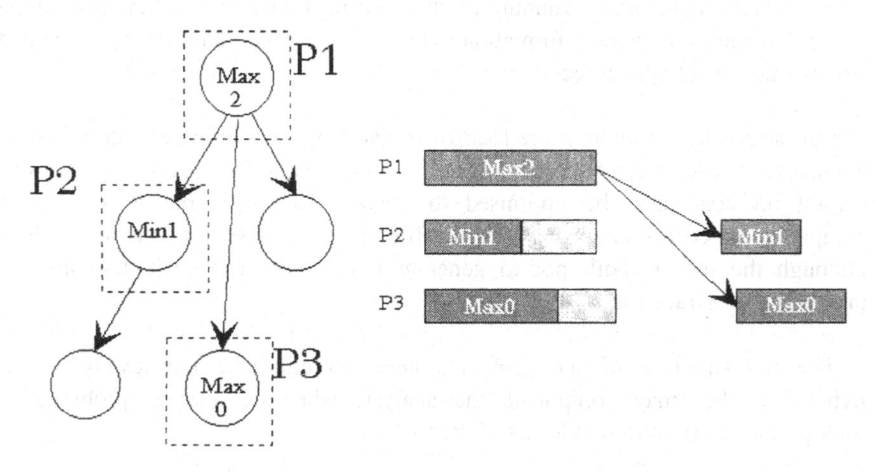

Fig. 7. *Space-state diagram of the new execution of the application*

4. Conclusions

This automatic analysis tool is designed for programmers of parallel applications that want to improve the behaviour of their applications. The application programmers' view of the tool is quite simple: the application is brought to the analysis tool as input and, after the analysis, the programmer receives a list of suggestions to improve the performance of the program. Those suggestions explain, at programmer level, which problems have been found in the execution of the application and how to solve them changing the program code.

Nonetheless, when applying the suggested changes to the application code, other new performance problems could appear. Programmers must be aware of the behaviour side-effects of introducing changes in the applications. Hence, once the application code is rebuilt, new analysis should be considered. This new analysis must be tested to find a set of representative input data in order to analyse the execution of the application comprehensively with a trace file.

Moreover, some problems may be produced by more than one cause. Sometimes it is difficult to separate the different causes of the problems and propose the most adequate solution. This process of progressive analysis of problems with multiple causes is one of the future fields of tool development.

Future work on the tool will consider the increment and refinement of the causes of performance problems, the "knowledge base". The programming model of the analysed applications must also be extended from the currently used (PVM) to other parallel programming paradigms.

Due to the general use of a few parallel execution trace formats [16, 4] and programming libraries, it is possible to have similar kind of performance data of many different applications running on different parallel systems. Although we have found that additional trace information (which is not easily obtained) can alleviate the analysis task to a high degree.

But far greater efforts must be focused on the optimisation of the search phases of the program. The search for problems in the trace file and the analysis of causes for a certain problem must be optimised to operate on very large trace files. The computational cost of analysing the trace file to derive these results is not irrelevant, although the tool is built not to generate much more overhead than the visual processing of a trace file.

The tree-structure of the problems helps to eliminate the testing of some hypotheses, but may complicate the analysis when considering problems with multiple causes (at different levels of the tree).

References

[1] Pancake, C. M., Simmons, M. L., Yan J. C.: Performance Evaluation Tools for Parallel and Distributed Systems. IEEE Computer, November 1995, vol. 28, p. 16-19.

[2] Heath, M. T., Etheridge, J. A.: Visualizing the performance of parallel programs. IEEE Computer, November 1995, vol. 28, p. 21-28 .

[3] Kohl, J.A. and Geist, G.A.: "*XPVM Users Guide*". Tech. Report. Oak Ridge National Laboratory, 1995.

[4] Reed, D. A., Aydt , R. A., Noe , R. J., Roth, P. C., Shields, K. A., Schwartz , B. W. and Tavera, L. F .: Scalable Performance Analysis: The Pablo Performance Analysis Environment. Proceedings of Scalable Parallel Libraries Conference. IEEE Computer Society, 1993.

[5] Reed, D. A., Giles, R. C., Catlett, C. E.. Distributed Data and Immersive Collaboration. Communications of the ACM. November 1997. Vol. 40, No 11. p. 39-48.

[6] Karavanic, K. L., Miller, B. P.. Experiment Management Support for Performance Tuning. In Proceedings of SC'97 (San Jose, CA, USA, November 1997).

[7] Hollingsworth, J. K., Miller, B, P. Dynamic Control of Performance Monitoring on Large Scale Parallel Systems. International Conference on Supercomputing (Tokyo, July 19-23, 1993).

[8] Yan, Y. C., Sarukhai, S. R.: Analyzing parallel program performance using normalized performance indices and trace transformation techniques. Parallel Computing 22 (1996) 1215-1237.

[9] Crovella, M.E. and LeBlanc, T. J. . The search for Lost Cycles: A New approach to parallel performance evaluation. TR479. The University of Rochester, Computer Science Department, Rochester, New York, December 1994.

[10] Meira W. Jr. Modelling performance of parallel programs. TR859. Computer Science Department, University of Rochester, June 1995.

[11] Fahringer T. , Automatic Performance Prediction of Parallel Programs. Kluwer Academic Publishers. 1996.

[12] Wall, L. , Christiansen, T. ., Schwartz, R. L. , : Programming Perl. O'Reilly and Associates, 2nd Edition, Nov 96.

[13] Geist, A. , Beguelin, A. , Dongarra, J. , Jiang, W. , Manchek, R. and Sunderam, V. , PVM: Parallel Virtual Machine, A User's Guide and Tutorial for Network Parallel Computing. MIT Press, Cambridge, MA, 1994.

[14] Maillet, E. : *TAPE/PVM a*n efficient performance monitor for PVM applications-user guide, LMC-IMAG Grenoble, France. June 1995.

[15] Espinosa, A. , Margalef, T. and Luque, E. . Automatic Performance Evaluation of Parallel Programs. Proc. of the 6th EUROMICRO Workshop on Parallel and Distributed Processing, pp. 43-49. IEEE CS. 1998.

[16] Geist, G. A., Heath, M.T. , Peyton, B. W. and Worley, P. H. . PICL. A portable instrumented communication library. Tech. Report ORNL/TM-11130, Oak Ridge National Laboratory, July 1990.

Behavioural Analysis Methodology Oriented to Configuration of Parallel, Real-Time and Embedded Systems

F.J. Suárez and D.F. García

Universidad de Oviedo
Area de Arquitectura y Tecnología de Computadores
Campus de Viesques s/n, 33204 Gijón, Spain
suarez@etsiig.uniovi.es

Abstract. This paper [1] describes a methodology suitable for behavioural analysis of parallel real-time and embedded systems. The main goal of the is to achieve a proper configuration of the system in order to fulfill the real-time constraints specified for it. The analysis is based on the measurement of a prototype of the system and is supported by a behavioural model. The main components of this model are known as "macro-activities", that is, the sequences of activities which are carried out in response to input events, causing the corresponding output events. This supposes a behavioural view in the analysis that complements the more usual structural and resource views. The methodology incorporates steps of diagnosis (evaluation of the causes of system behaviour) and configuration (planning of alternatives for design improvement after diagnosis). The experimental results of applying the methodology to the analysis of a well-known case study are also an important part of this paper.

1 Introduction

The motivation of this work comes from the lack of research works addressing jointly the three following aspects related to the analysis of systems: 1) Development of a methodology of system behavioural analysis; 2) Addressing the particular problems of real-time and embedded systems; and 3) Use of analysis metrics obtained from a event trace after system execution. Effectively, none of the research works among the more relevant ones in the area addresses together the three aspects. In [5] and [3] only the metric and real-time aspects are considered respectively. In [10], [2] and [9] aspects in metric based analysis are shown. In [1] a set of metrics for real-time systems is used. Besides, regarding the analysis metrics, this work defines metrics corresponding to the three possible system views [8], that is, behavioural, structural and resource views. Structural view is static view of the system and provides information about its design. Resource view provides dynamic information about resource use an is,

[1] This research work has been supported by the ESPRIT HPC 8169 project ESCORT.

J. Palma, J. Dongarra, and V. Hernández (Eds.): VECPAR'98, LNCS 1573, pp. 378–395, 1999.

together with the structural view, the more usual view in system behavioural analysis. The last view, behavioural view, provides dynamic information about the temporal behaviour of the system in terms of sequences of activities along system execution.

The organisation of the rest of the paper is as follows: in point 2 the model used in the behavioural view is presented; in point 3 the main steps and aspects of the analysis methodology are shown, including the metrics that support the methodology; in point 4 a well known case study is analysed using the methodology; finally in point 5 the conclusions and future work are presented. Although the methodology can be applied to real-time systems in general, this work deals with parallel real-time embedded systems, hereafter referee to simply as *real-time systems* (RTS).

2 Behavioural Model

To describe the temporal behaviour of RTS in terms of events, delays and actions, several approaches or behavioural models can be employed. A model based on an event-ordering graph has been selected as the behavioural model for this research work. The selection was made as a result of its simplicity, wide applicability and suitability for the proposed methodology. The model is composed of two main elements (see [8]): 1) Activities, which are represented by a sequence of three events: *Ready*, when the activity is ready to start; *Begin*, when it starts; and *End*, when it finishes; and 2) A set of precedence and synchronization relationships defined on the *ready* and *end* events, which establish partial ordering for a group of activities. These relationships give rise to the following kinds of activities: sequential activity (SEQ); activity of synchronization with other macro-activities (SYN); alternative or conditional execution activity (ALT); replicated activity (REP); and activity executed in parallel with others (PAR). These kinds of activities are represented in Figure 1.

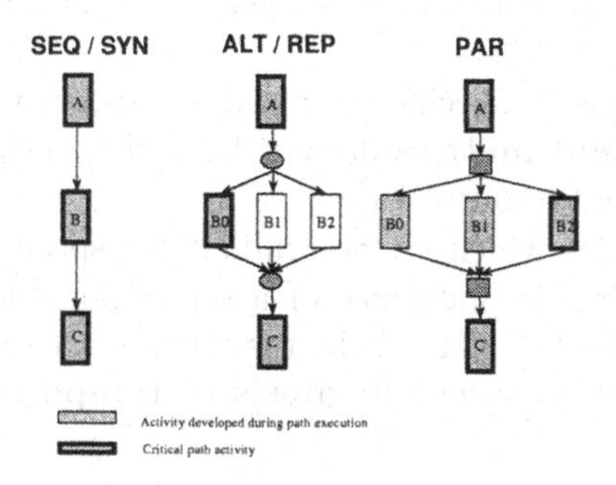

Fig. 1. *Kinds of activities*

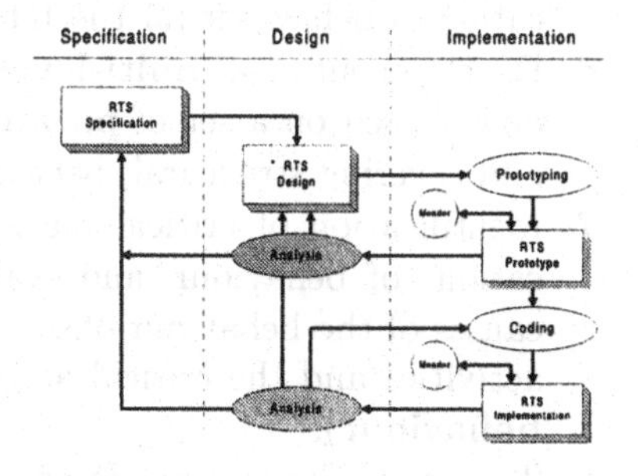

Fig. 2. *RTS development cycle*

In the model, three principal times are associated to each one of these activities to explain its temporal behaviour: Waiting Time (the time between the time-stamps of its *ready* and *begin* events), Service Time (the time between the time-stamps of its *begin* and *end* events) and Response Time (the sum of Waiting and Service Times). The model represents sequences of activities executed in response to the main input events which the RTS must deal with. These sequences, called *macro-activities* in the methodology, are equivalent to the conventional *end-to-end tasks* found in related literature. Construction of the model implies an understanding of the functional and structural models provided by the RTS design methodology.

3 Analysis Methodology

The goal of the methodology is the analysis of RTS behaviour from the early design phase to its final implementation. To achieve this, the methodology involves the construction of a synthetic prototype [4] of the RTS design, which is analysed and refined until the validation of the timing requirements is achieved. It also serves as a skeleton for the implementation phase. In Figure 2 the integration of the methodology in the RTS development cycle is shown. Two refining cycles can be derived from the methodology; one working with early designs through prototypes of the RTS, and the other with the implementation of the RTS. The methodology approach can be resumed in the following 10 steps:

1. Prototyping of the initial RTS design under analysis.
2. Understanding of the sequences of activities to be executed as RTS responses to events (macro-activities), selecting the ones to be considered in the analysis, and establishing a **specification of behaviour** for them.
3. Instrumentation of the RTS software prototype to enable the **monitoring system** to obtain information about the behaviour of macro-activities during the RTS execution.
4. Execution of the instrumented RTS under specific operational conditions (scenario) over a period of time long enough to obtain a representative event trace.
5. Checking of the fulfilment of the real-time constraints defined in the specification of behaviour for the RTS.
6. Development of a **multi-level analysis** in specific temporal analysis windows based on a set of **parameters and metrics** derived from the trace, and covering structural, behavioural and resource views.
7. Identification of critical macro-activities which do not fulfil their specifications of behaviour, and evaluation of the incidence of a set of possible causes of the behaviour observed in the RTS as a whole, the critical macro-activities and the critical activities within them (**diagnosis of temporal behaviour**).
8. Tuning of the system design according to the incidence of each cause of behaviour, establishing a suitable **configuration** of the RTS.

9. Repetition of the analysis cycle until a final prototype which permits timing validation is obtained.
10. Implementation of RTS design and repetition of the analysis cycle until its final implementation.

The main aspects of the methodology, highlighted above in bold face, are briefly described in the following points.

3.1 Specification of Behaviour

The specification of behaviour of the RTS consists of specifications of the behaviour of each macro-activity. These specifications consider load characteristics and real-time constraints in macro-activities. The most typical macro-activities in a RTS respond to periodic events and, consequently, are characterised by a periodic execution. These are called *periodic macro-activities.* Macro-activities responding to aperiodic events are called *aperiodic macro-activities.* The period of activation is the main load characteristic for periodic macro-activities, while the mean activation period and the typical deviation characterise aperiodic macro-activities. The real-time constraints considered for both periodic and aperiodic macro-activities are: the *absorption of productivity* and the *deadline* (*end-to-end deadline*). The fulfilment of the first constraint implies capacity of the RTS to respond to all the input events produced during execution.

3.2 Monitoring System

The monitoring system, or simply the monitor, is highly dependent on the target system for which it is developed. In the context of this research work, a full software monitor for a multiprocessor based on T9000 transputers was developed [7]. The function of the monitor is to trace the occurrence of the most relevant software events during an application execution, and to store information related to them in a set of trace files. So, the functionality of the monitoring system consists of *run-time events* (communications, synchronization operations, I/O operations, etc.), *macro-activity events* (start) and *activity events* (ready, begin and end). The monitor is structured in three main components: a set of *distributed monitoring processes*, a collection of *instrumentation probes* spread over the application processes, and one *instrumentation data structure* per application process. In Figure 3 the instrumentation of one activity is shown. Finally, two steps are taken in order to improve the quality of measurement: precise synchronization of the system clocks (error < 10 microsec. with clock resolution $= 1$ microsec.) and reduction of the monitor intrusiveness (26-37 microsec./probe) to a minimum.

3.3 Multi-level Analysis

This methodology allows the analysis of the system at three possible levels of abstraction. The first level considers the analysis of the RTS as a whole. The

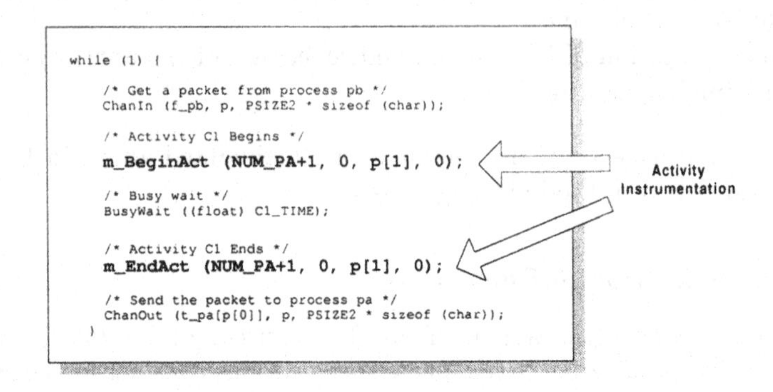

while (1) {
/* Get a packet from process pb */
ChanIn (f_pb, p, PSIZE2 * sizeof (char));
/* Activity C1 Begins */
m_BeginAct (NUM_PA+1, 0, p[1], 0);
/* Busy wait */
BusyWait ((float) C1_TIME);
/* Activity C1 Ends */
m_EndAct (NUM_PA+1, 0, p[1], 0);
/* Send the packet to process pa */
ChanOut (t_pa[p[0]], p, PSIZE2 * sizeof (char));
}
Activity
Instrumentation

Fig. 3. *Software instrumentation*

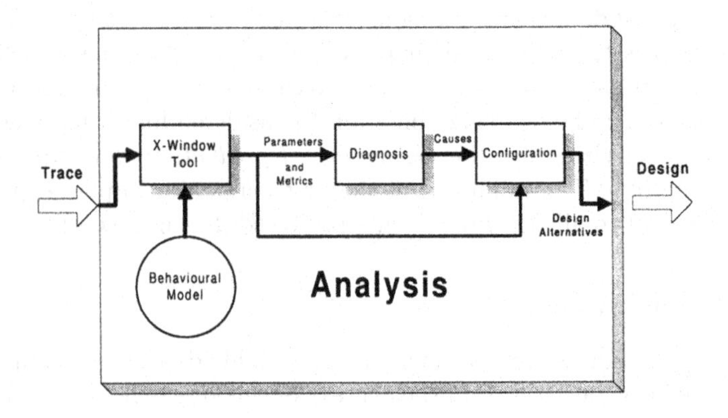

Fig. 4. *Analysis approach*

second level considers the analysis of each macro-activity of the RTS. Finally, the third level considers the analysis of each of the activities composing the macro-activities. This multi-level character of the analysis permits a top-down approach, which is very useful in explaining the behaviour observed in the RTS. Figure 4 resumes the steps carried out during the analysis process. Starting from the event trace obtained by the monitor after RTS execution, an X-window tool permits the validation of the trace according to the behavioural model and generates the parameter and metric values. From these values, the diagnosis process is carried out, checking the fulfilment of the real-time constraints and obtaining the causes of behaviour of the RTS. Finally, with the causes of behaviour, parameters and metrics, the configuration process suggests alternatives for design improvement.

According to the multi-level analysis character described above, the methodology considers three possible analysis windows: RTS Window (a temporal window long enough to represent all the system behaviour characteristics for the scenario under analysis), Macro-activity Window (which corresponds to the longest response interval of a macro-activity within the RTS window) and Activity Window (which corresponds to the response interval of an activity within the macro-activity window). The RTS window is obtained from the *basic RTS period*, which corresponds to the Least Common Multiple (LCM) of all the periods specified for the periodic macro-activities, as seen in figure 5. The number of basic RTS periods in the RTS window is fixed considering factors such as the transient effect caused by pipelining, the statistic characteristics of the aperiodic macro-activities, and the variability of the response times in the macro-activities.

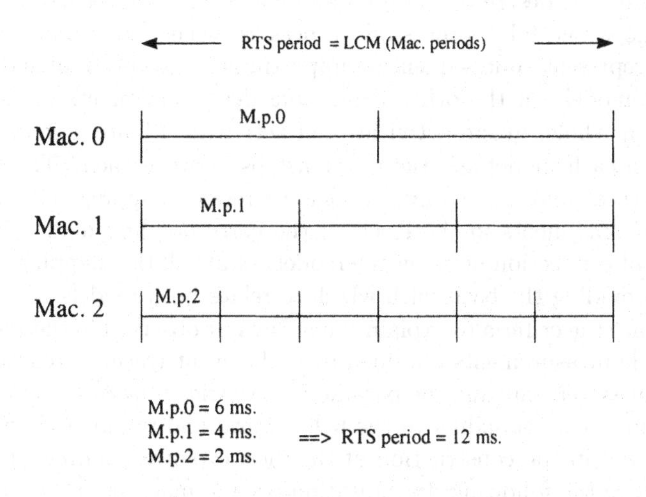

Fig. 5. *Basic RTS period*

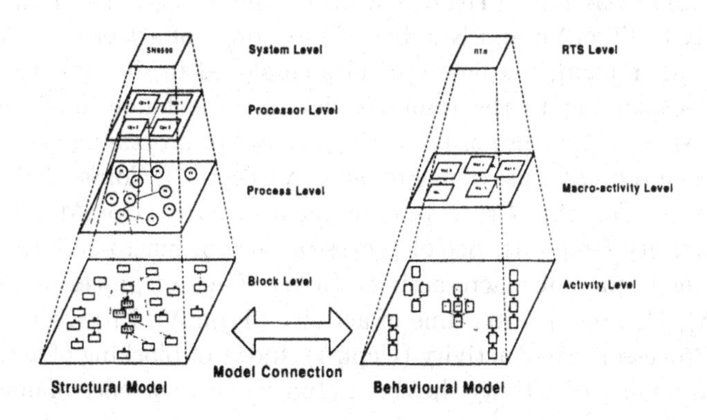

Fig. 6. *Parameters in methodology*

3.4 Parameters and Metrics

Parameters resume all the known information about the RTS before execution. So, while they give only a static view of the system, they are necessary to determine the influence of the design components on the behaviour of the system. Four kinds of parameters are considered in the methodology: load parameters, parameters of the structural model, parameters of the behavioural model, and parameters of connection between models. Load parameters define the demands of service on the system from the environment, and reflect the load characteristics of macro-activities in the RTS behavioural model. The structural model defines the current design of the RTS and considers components on four levels or layers: the whole RTS, processors, processes (schedulable units) and *blocks*. The blocks represent code sequences supporting the execution of activities. The behavioural model, on the other hand, considers components on three levels: the RTS as a whole, macro-activities and activities. Figure 6 shows the levels considered in each model and the relationships between them. The parameters of the structural and behavioural models represent mapping relationships between their components in the levels of the corresponding model. Finally, the parameters of connection between the models establish the mapping of activities in blocks, providing the basis with which to relate both models.

Metrics are the criteria to explain the behaviour observed in the system. They can be simple measurements obtained from the event trace, or relationships between the measurements and the parameters. Metrics provide information which feed the models corresponding to the behavioural and resource views. The variables employed in the construction of the metrics corresponding to the behavioural view are the following: Initialisation time of macro-activity (TiniM); Response time of macro-activity (TresM); Waiting time of macro-activity (TwaiM); Service time of macro-activity (TserM); Number of executions of macro-activity (NeM); Theoretic number of executions of macro-activity (NeTM); Deadline of macro-activity (DM); Number of failures of deadline in macro-activity (NfDM); Response time of activity (TresA); Waiting time of activity (TwaiA); Service time of activity (TserA); and Number of executions of activity (NeA).

On the other hand, the new variables employed in the construction of the metrics corresponding to the resource view are the following: Process use of RTS (UproS); Processor use of RTS (UcpuS); Set of processors use of RTS (UcpusS); Time of processor use of macro-activity (TcpuM); Communication time of macro-activity (TcomM); Process use of macro-activity (UproM); Processor use of macro-activity (UcpuM); Set of processors use of macro-activity (UcpusM); Index of concurrence of macro-activity (IcM); Time of processor use of activity (TcpuA); Communication time of activity (TcomA); Process use of activity (UproA); Processor use of activity (UcpuA); Index of blocking of activity (IbA); Index of parallelism of activity (IpA); and Index of concurrence of macro-activity (IcA).

The metrics used in this methodology can be classified according to the level of analysis in which they are applied. So, three different levels can be distinguished: RTS level metrics, macro-activity level metrics and activity level metrics. For a specific level of analysis and a specific view, the metrics can be calculated using all three analysis windows, that is the RTS window (Wrts), the macro-activity window (Wmac) and the activity window (Wact). Tables 1 and 2 show all the metrics at activity level corresponding to the behavioural and resource views, respectively. M and D prefixes in metrics refer to the mean and deviation values respectively.

Activity Level		
Wrts	Wmac	Wact
MTwaiA/MTresA		TwaiA/TresA
MTresA/MTresM		TresA/TresM
DTresA/MTresA		
NeA/NeM		

Table 1. *Behavioural view*

Activity Level		
Wrts	Wmac	Wact
MTcpuA/MTserA		TcpuA/TserA
DTcpuA/MTcpuA		
MTcomA/MTwaiA		TcomA/TwaiA
MIbA		IbA, IpA, IcA
UproA	UproA	UproA
UcpuA	UcpuA	UcpuA

Table 2. *Resource view*

The Index of blocking (Ib) helps to identify the cause of blocking time in an activity. This index compares the activity response time with the macro-activity period, in order to establish if the blocking time is caused by overlapping of macro-activity executions (when the sum of the service and communication times is greater than the macro-activity activation period). Therefore, index values over 1 indicate overlapping. The Index of parallelism (Ip) provides information about the level of concurrence of parallel activities (PAR activities) of the behavioural model, in a given execution. The Index of concurrence (Ic) provides information about the level of concurrence of all the activities composing the macro-activity in a given execution.

3.5 Diagnosis of Temporal Behaviour

The stages to follow in the diagnosis of temporal behaviour are the following:

- *Stage 1: Diagnosis of the RTS.* In this first stage, global causes of behaviour of the set of macro-activities in the behavioural model are evaluated.
- *Stage 2: Identification of critical macro-activities.* The objective of this stage is the identification of macro-activities which do not fulfill one or more of the real-time constraints defined in the specification of behaviour.
- *Stage 3: Diagnosis of each critical macro-activity.* In this stage, the causes of behaviour which explain the response time of each critical macro-activity are found.
- *Stage 4: Identification of critical activities.*The objective of this stage is the identification of the most significant or critical activities (bottlenecks) in each critical macro-activity.

- *Stage 5: Diagnosis of each critical activity.* In this last stage, the causes of behaviour which explain the response time of each critical activity are found.

The support provided by parameters and metrics in the diagnosis is shown in figure 7. The figure resumes the parameters and metrics (including the analysis window considered for them) useful at each diagnosis stage.

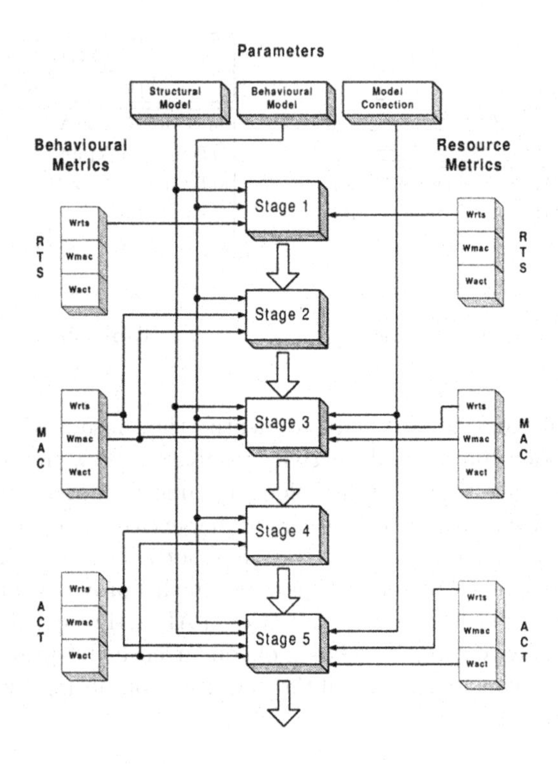

Fig. 7. *Parameter and metric support in diagnosis*

When considering the causes of temporal behaviour in the diagnosis, the differences between the causes at activity level and the causes at macro-activity and RTS levels must be clearly established. At activity level, the response time of the critical activities must be explained, and three levels of diagnosis are considered. The first level of diagnosis evaluates the waiting and service times of each activity. Evaluation of communication and blocking times during waiting time, processing during service time and resource contention during both waiting and service times, correspond to the second level of diagnosis. Finally, the third level of diagnosis evaluates the specific causes of communication, processing, blocking or contention. Based on these three diagnosis levels, a set of causes of temporal behaviour of the critical activities can be established. Figure 8 represents the three levels of diagnosis. In table 3 all the causes considered, with a brief

Cause	Description
WCM	Communication with other activities of the same Macro-activity, during the Waiting time of the activity
WCS	Communication for Synchronization with other macro-activities during the Waiting time
WTB	ConTention of CPU due to competition with other activities supported by the same Block during the Waiting time
WTP	ConTention of CPU due to competition with activities supported by other blocks of the same Process
WTH	ConTention of CPU due to competition with Higher priority activities during the Waiting time
WTE	ConTention of CPU due to competition with activities of Equal priority during the Waiting time
WBO	Blocking during the Waiting time due to execution Overlapping in the macro-activity
WBS	Blocking during the Waiting time due to synchronization with other macro-activities
WBP	Blocking during the Waiting time of a PAR activity caused by the previous synchronization of other PAR activities
WBC	Blocking during the Waiting time due to other overheads associated with Communication
STH	ConTention of CPU due to competition with Higher priority activities during the Service time of the activity
STE	ConTention of CPU due to competition with activities of Equal priority during the Service time
SPP	Load imbalance of a PAR activity during the Processing part of the Service time
SPE	Execution of code during the Processing part of the Service time

Table 3. *Causes of behaviour. Activity level*

description of each, are detailed. The incidence of each cause of behaviour is evaluated with a value in the range of 0–1. This value of incidence is the product of the partial incidence values obtained at each level of diagnosis.

At RTS level, the causes of behaviour of the RTS as a whole must be found, whereas at macro-activity level, the response time of critical macro-activities must be explained. To achieve those objectives, the aggregation of macro-activity and activity metrics are considered respectively. Therefore it is not possible to distinguish all the causes of behaviour considered at activity level. Only the first two levels of diagnosis are considered, and so the causes of behaviour are reduced to five. These causes are the following: INI (initialisation time of macro-activities); WAI_COM (communication during waiting time); WAI_BLO (blocking during waiting time, contention included); SER_CON (contention during service time); and SER_PRO (processing during service time);

3.6 Configuration

Once the incidence of the causes of the behaviour of each critical activity composing the critical macro-activities has been established, those with high incidence will be considered, in order to tune the RTS design and establish its proper configuration. The goal of configuration can be either the fulfilment of the real-time constraints using the available resources, or the reduction of resources in the RTS design, while maintaining the fulfilment of the real-time constraints. The proper design alternatives for the causes of behaviour of critical activities within the critical macro-activities are the following: Re-mapping of blocks on processes (RBL); Re-mapping of processes on processors (RPR); Change of process priority (CPR); Segmentation of an activity (SEG); Replication of an activity (REP); Parallelisation of an activity (PAR); Balance of load in PAR activities (BAL); and Optimisation of block code (OPT).

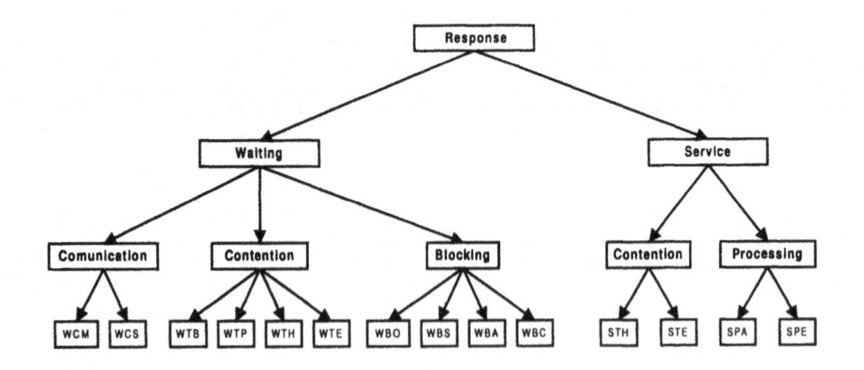

Fig. 8. *Levels of diagnosis*

4 Case Study

The case study considered here to show the use of the analysis methodology and described below has been widely studied in other papers, such as [6].

A Remote Speed Sensor (RSS) measures the speed of a number of motors, and reports them to a remote host computer. The speed of each motor is obtained by periodically reading a corresponding digital tachometer. The interval between speed readings (10-1000 ms.) for each motor is specified by the host computer. An Analogic to Digital Converter (ADC) with a set of multiplexed channels is used to measure the speed signal provided by tachometers coupled to the motors. The ADC accepts reading-requests in the form of motor numbers (integers in the range of 0–15). After a request has been received, the converter reads the speed of the motor, stores it in a hardware buffer, and generates an interruption. The converter can only read the speed of one motor at a time. The interval between readings for a given motor is specified in a control packet which is sent from the host computer to the RSS. The speed of a motor is reported to the host via a data packet. When a control packet is received from the host, it is checked for validity. If the message is valid, an acknowledgement (ACK) is sent to the host. If it is not, a negative acknowledgement (NAK) is sent. When a data packet is sent, the RSS waits to receive either an ACK or a NAK from the host. If a NAK is received, or neither an ACK nor a NAK is received within half the reading interval, the message emission is marked as a failure.

4.1 Design Structure

The design structure of the case study has four principal parts: the tachometer, the motors, the host interface and the host. Each of these parts is composed of one or more software processes, as seen in Figure 9. The tachometer process can access the speeds of all the motors and is constantly waiting to read requests. When a request is received, it reads the corresponding speed and sends the data to the motor process. Motor processes, one per motor, periodically send the reading-requests to the tachometer process. Once data is received, it is filtered and sent to the host. These processes have associated processes which inform them about new reading intervals requested from the host. The Host interface is implemented with four processes: *inport*, *outport*, *inmsg* and *outmsg*. Inport receives packets from the host. If the packet is a control packet, it is sent to inmsg. If it is an ACK or a NAK, it is sent to outmsg. Outport receives packets from outmsg and inmsg and sends them to the host. Inmsg receives control packets from inport. If the packet is not valid, it sends NAK to the host through outport. If it is valid, it sends ACK to the host through outport and the new interval to the motor process. Outmsg receives speeds from the motor processes and ACKs and NAKs from inport. All of them are sent to the host. The host has three processes. *Phost* is the main process and *phostin* and *phostout* are input and output processes for communication with their corresponding processes in the RSS host interface. The embedded system was implemented in a PARSYS SN9500 machine, a distributed memory parallel machine based on Transputers with 8

CPUs. In this machine the process communications are established through a virtual channel network. RSS processes were implemented over 4 CPUs. For simplicity in the case study, host processes were also implemented in the same machine using an extra CPU.

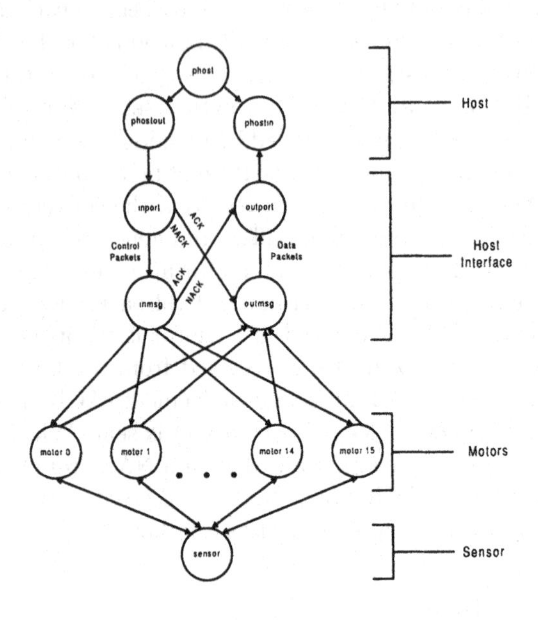

Fig. 9. *Process Structure*

4.2 Behavioural Model and Real-Time Constraints

According to the methodology, the relevant action-reaction event couples which demand system response must first be identified. The action events are the periodic requests for speed-reading from the motor processes to the tachometer process. The reaction events are the arrivals at outmsg process of ACKs coming from the host in response to the emission of packets with speed data. A total of 16 kinds of events, one per motor, are considered. The macro-activities include all the activities executed from the speed-reading request to the ACK reception from the host. All 16 macro-activities considered in the analysis have the same structure, as shown in Figure 10, corresponding to the RTS behavioural model.

The real-time constraints of each macro-activity in the case study depend on the speed of the corresponding motor. The deadline of each macro-activity has the same value as its period of execution, as shown in table 4, where *Mac.* refers to the macro-activity number, and times are given in milliseconds.

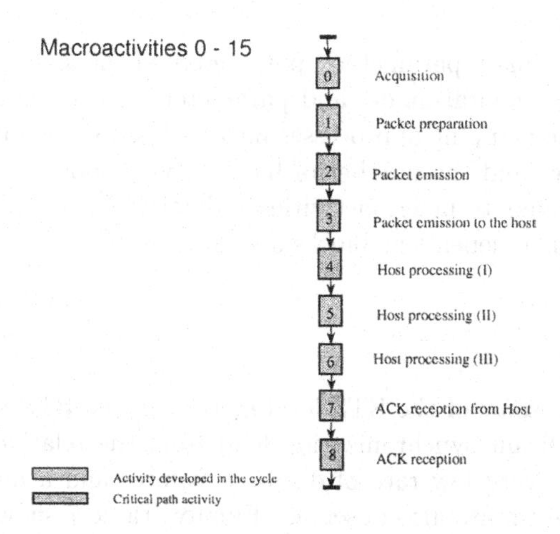

Fig. 10. *Behavioural model*

Mac.	Period	DM	Mac.	Period	DM
0	12	12	8	80	80
1	20	20	9	100	100
2	25	25	10	120	120
3	30	30	11	150	150
4	40	40	12	200	200
5	50	50	13	300	300
6	60	60	14	400	400
7	75	75	15	600	600

Table 4. *Real-time constraints*

CPU	Processes	Activities	Nblocks
T9000[0]	phost	[i,5] i=0-15	1
	phostin	[i,4] i=0-15	1
	phostout	[i,6] i=0-15	1
T9000[1]	inport	[i,7] i=0-15	1
	outport	[i,3] i=0-15	1
	inmsg	-	-
	outmsg	[i,2] [i,8] i=0-15	1
T9000[2]	motor[i]	[i,1] i=0,2...(even)	8
T9000[3]	motor[i]	[i,1] i=1,3...(odd)	8
T9000[4]	tacometer	[i,0] i=0-15	1

Table 5. *Parameters*

4.3 Analysis

Table 5 resumes the main parameters: parameters of the structural model, parameters of the behavioural model and parameters of connection between the models. It shows the mapping of processes into processors, the mapping of activities in the processes and the number of blocks giving support to the activities (*Nblocks*). Index i refers to macro-activities and activities [i,1] are the only ones supported in various independent blocks and processes.

Diagnosis

Stage 1. In Table 6 some of the RTS level metrics in the RTS analysis window are shown. A significant synchronization load from the relative value of Twai can be observed. A very low rate of deadline failures and a medium level use of the set of processors are also observed. Finally, Table 7 shows the incidence of the causes of behaviour for the RTS as a whole. Here, blocking is the most important cause of behaviour.

Metric	Value
M(MTiniM/MTresM)	0.025
M(MTwaiM/MTresM)	0.606
M(MTresM/DM)	0.213
M(NeM/NeTM)	1.000
M(NfDM/NeM)	0.001
M(MTcomM/MTwaiM)	0.183
M(MTcpuM/MTserM)	0.778
UcpusS	0.506

Table 6. *RTS metrics*

Cause	Incidence
INI	0.03
WAI_COM	0.11
WAI_BLO	0.50
SER_CON	0.08
SER_PRO	0.29

Table 7. *RTS diagnosis*

Stage 2. In Figure 11 the relative value of macro-activity response times with regard to their deadlines is shown. Only macro-activity 0 in its macro-activity window exceeds its deadline. The rest of metrics establish that the constraint of productivity absorption is fulfilled in all macro-activities and the deadline constraint is not fulfilled in macro-activity 0. So, macro-activity 0 is selected as a critical macro-activity.

Stage 3. Processing (0.29 incidence), blocking (0.50 incidence) and communication (0.11 incidence) result to be the principal causes of the behaviour in the macro-activity window for critical macro-activity 0.

Stage 4. Activity level metrics considered here correspond to activities composing the critical macro-activity 0. In figure 12 the relative value of each activity

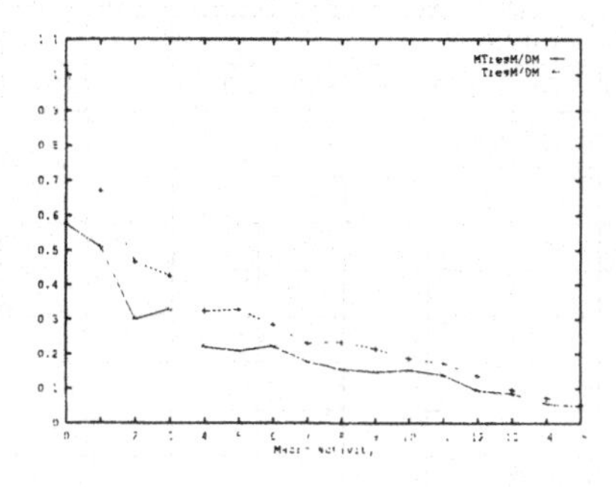

Fig. 11. *Macro-activity level metrics*

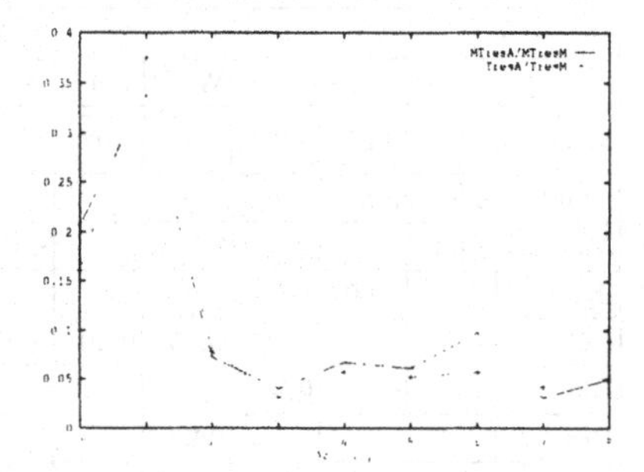

Fig. 12. *Activity level metrics*

response time with respect to the macro-activity response time in the activity and RTS windows is shown. Activity 1 is seen to be the longest activity with nearly 40% incidence in the macro-activity response time. So, activity 1 is selected as a critical activity.

Stage 5. Table 8 shows the incidence of the causes of the behaviour in the activity window of critical activity 1. It also shows the partial incidence of each diagnosis level. The main causes of behaviour are: contention with other activities of equal priority (the corresponding activities [i,1] in the other macro-activities) supported by different processes; and execution of code. Both causes correspond to the service time.

Configuration. The parameters show that critical activity 1 of critical macro-activity 0 is placed in CPU2. Checking the use of this CPU by each macro-activity in the critical activity window, macro-activities 2 and 10 can be observed in strong competition with macro-activity 0. A new mapping with motor processes motor[2] and motor[10] in CPU2 eliminates deadline failures in all macro-activities.

5 Conclusions and Future Work

An analysis methodology applicable to configuration of parallel, real-time and embedded systems has been presented. The methodology is based on temporal behaviour analysis of the system and considers a behavioural model of the RTS composed of real-time *macro-activities* giving response to the input events. The methodology involves the construction of a synthetic prototype for the initial design of the RTS, which is refined until the final implementation. Configuration of the RTS is achieved by previous diagnosis of its temporal behaviour, based on a set of parameters and metrics covering three complementary views of the system: behavioural view, structural view and resource view. A well known case study has been analyzed with the methodology and demonstrated its possibilities in configuration of RTS.

	Causes of behaviour													
Comp	WCM	WCS	WTB	WTP	WTH	WTE	WBO	WBS	WBP	WBC	STH	STE	SPP	SPE
WAI	0.07	0.07	0.07	0.07	0.07	0.07	0.07	0.07	0.07	0.07	-	-	-	-
COM	0.80	0.80	-	-	-	-	-	-	-	-	-	-	-	-
Mac	1.00	-	-	-	-	-	-	-	-	-	-	-	-	-
Syn	-	0.00	-	-	-	-	-	-	-	-	-	-	-	-
CON	-	-	0.20	0.20	0.20	0.20	-	-	-	-	-	-	-	-
Blo	-	-	0.00	-	-	-	-	-	-	-	-	-	-	-
Pro	-	-	-	0.00	-	-	-	-	-	-	-	-	-	-
Hpr	-	-	-	-	0.00	-	-	-	-	-	-	-	-	-
Epr	-	-	-	-	-	1.00	-	-	-	-	-	-	-	-
BLO	-	-	-	-	-	-	0.00	0.00	0.00	0.00	-	-	-	-
Ovr	-	-	-	-	-	-	0.00	-	-	-	-	-	-	-
Syn	-	-	-	-	-	-	-	0.00	-	-	-	-	-	-
Par	-	-	-	-	-	-	-	-	0.00	-	-	-	-	-
Com	-	-	-	-	-	-	-	-	-	0.00	-	-	-	-
SER	-	-	-	-	-	-	-	-	-	-	0.93	0.93	0.93	0.93
CON	-	-	-	-	-	-	-	-	-	-	0.51	0.51	-	-
Hpr	-	-	-	-	-	-	-	-	-	-	0.00	-	-	-
Epr	-	-	-	-	-	-	-	-	-	-	-	1.00	-	-
PRO	-	-	-	-	-	-	-	-	-	-	-	-	0.49	0.49
Par	-	-	-	-	-	-	-	-	-	-	-	-	0.00	-
Exe	-	-	-	-	-	-	-	-	-	-	-	-	-	1.00
TOTAL	0.06	0.00	0.00	0.00	0.00	0.01	0.00	0.00	0.00	0.00	0.00	0.47	0.00	0.46

Table 8. *Diagnosis of behaviour for activity 1*

Future work has two main objectives. Firstly, to derive automatic rules for proper configuration of systems from the expertise gained with the use of the methodology. Secondly, to apply the methodology to real-time POSIX applications implemented with either parallel or distributed architectures.

References

1. Rolf Borgeest, Bernward Dimke, and Olav Hansen. A trace based performance evaluation tool for parallel real-time systems. *Parallel Computing*, 21(4):551–564, April 1995.
2. J.K. Hollingsworth. *Finding Bottlenecks in Large Scale Parallel Programs.* PhD thesis, University of Wisconsin-Madison, 1994.
3. O. Pasquier, J.P. Calvez, and V. Henault. A complete toolset for prototyping and validating multi-transputer applications. In Monique Becker, Luc Litzler, and Michel Trhel, editors, *Transputers'94. Advanced Research and Industrial Applications*, pages 71–86. IOS Press, 1994.
4. David A. Poplawsky. Synthetic models od distributed-memory parallel programs. *Journal of Parallel and Distributed Computing*, 12:423–426, 1991.

5. L. Schäfers, C. Scheidler, and O. Krämer-Fuhrmann. Trapper: A graphical programming environment for embedded mimd computers. In R Grebe et Al., editor, *Transputer Applications and Systems'93*, pages 1023–1034. IOS Press, 1993.
6. C.U. Smith. Software performance engineering: A case study including performance comparison with design alternatives. *IEEE Transactions on Software Engineering*, 19(7):120–141, July 1993.
7. F.J. Suárez, J. García, S. Grana, D. García, and P. de Miguel. A toolset for visualization and behavioural analysis of parallel real-time systems based on fast prototyping techniques. In *6th Euromicro Workshop on Parallel and Distributed Processing*. IEEE Computer Society, January 1998.
8. C.M. Woodside. A three-view model of performance engineering of concurrent software. *IEEE Transactions on Software Engineering*, 21(9):754–767, September 1995.
9. J.C. Yan and S.R. Sarukkai. Analyzing parallel program performance using normalized performance indices and trace transformation techniques. *Parallel Computing*, 22:1215–1237, 1996.
10. C. Yang and B.P. Miller. Performance measurement for parallel and distributed programs: a structured and automatic approach. *IEEE Transactions on Software Engineering*, 15(12):1615–1629, December 1989.

Spatial Data Locality with Respect to Degree of Parallelism in Processor-and-Memory Hierarchies

Renato J. O. Figueiredo[1], José A. B. Fortes[1], and Zina Ben Miled[2]

[1] School of ECE - Purdue University, West Lafayette, IN 47907
{figueire,fortes}@ecn.purdue.edu
[2] Department of EE - Purdue University, Indianapolis, IN 46202
miled@engr.iupui.edu

Abstract. A system organised as a Hierarchy of Processor-And-Memory (HPAM) extends the familiar notion of memory hierarchy by including processors with different performance in different levels of the hierarchy. Tasks are assigned to different hierarchy levels according to their degree of parallelism. This paper studies the spatial locality (with respect to degree of parallelism) behaviour of simulated parallelised benchmarks in multi-level HPAM systems, and presents an inter-level cache coherence protocol that supports inclusion and multiple block sizes on an HPAM architecture. Inter-level miss rates and traffic simulation results show that the use of multiple data transfer sizes (as opposed to a unique size) across the HPAM hierarchy allows the reduction of data traffic between the uppermost levels in the hierarchy while not degrading the miss rate in the lowest level.

1 Introduction

The Hierarchical Processor-And-Memory (HPAM) architecture [15] has been proposed as a cost/effective approach to parallel processing. The HPAM concept is based on a heterogeneous, hierarchical organisation of resources that is similar to conventional memory hierarchies. However, each level of the hierarchy has not only storage but also processing capabilities. Assuming that the top (i.e. first) level of the hierarchy is the fastest, any given memory level is extended with processors that are slower, less expensive and in larger number than those in the preceding level. Figure 1 depicts a generic 3-level HPAM machine.

The mapping of an application to an HPAM system is based on the degrees of parallelism that the application exhibits during its execution. Each level of an HPAM hierarchy handles portions of code whose parallelism degree is within a certain range. Levels with large number of slow processors and large memory capacity (bottom levels) are responsible for the highly parallel fractions of an application, whereas levels with small number of fast processors and memories are responsible for the execution of sequential and moderately parallel code.

J. Palma, J. Dongarra, and V. Hernández (Eds.): VECPAR'98, LNCS 1573, pp. 396–410, 1999.

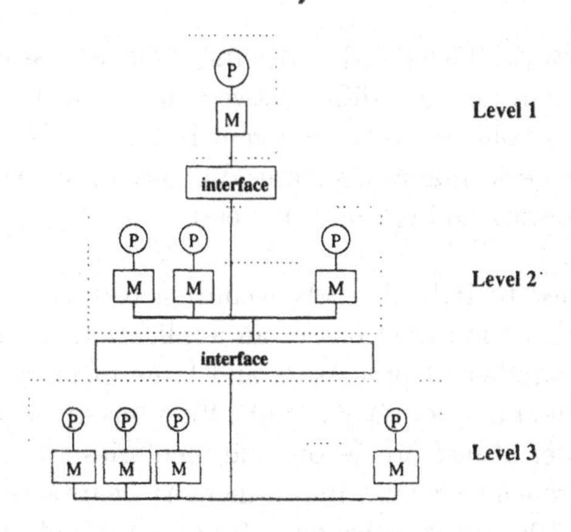

Fig. 1. Processor and memory organisation of a 3-level HPAM

An HPAM machine exploits heterogeneity, computing-in-memory and locality of memory references with respect to degree of parallelism to provide superior cost/performance over conventional homogeneous multiprocessors. Previous studies [14, 15] have empirically established that applications exhibit temporal locality with respect to degree of parallelism. This paper extends these studies by empirically establishing that applications also exhibit *spatial* data locality. Furthermore, this pap er studies the impact of using multiple transfer sizes across an HPAM hierarchy with more that two levels in inter-level miss rates and traffic. To this end, this paper proposes an inter-level coherence protocol that supports inclusion and multiple block sizes across the hierarchy.

The data locality studies have been performed through execution-driven simulation of parallelised benchmarks. Several benchmarks from three different suites (CMU [9], Perfect Club [7] and Spec95 [11]) have been instrumented to detect do-loop parallelism with the Polaris [12] parallel compiler. The stream of memory references generated by a benchmark is simulated by a multi-level memory hierarchy simulator that implements the proposed inter-level coherence protocol to obtain measurements of data locality, data traffic and invalidation traffic at each HPAM level.

The rest of this paper is organised as follows. Section 2 introduces the HPAM machine model and discusses coherence protocols for an HPAM architecture. Section 3 presents the methodology used to perform the data locality studies, and Section 4 presents simulation results and analysis of the locality behaviour of applications and of the proposed inter-level coherence protocol. Section 5 concludes the paper.

2 HPAM Architecture

A hierarchical processor-and-memory (HPAM) machine is a heterogeneous, multilevel parallel computer. Each HPAM level contains processors, memory and interconnection network. The speed and number of processors, latency and capacity of memories and network differ between levels in the hierarchy. The following characteristics hold across the different HPAM levels from top to bottom: individual processor performance decreases, number of processors increases, and memory/network latency and capacity increase.

Tasks are assigned to HPAM levels according to their *degree of parallelism* (DoP). Highly parallel code fractions of an application are assigned to bottom levels, where large number of processors and large memory capacity are available, while sequent ial and modestly parallel fractions are assigned to top levels, where a small number of fast processors and memories are available.

The HPAM approach to computing-in-memory bears similarities with IRAM and PIM efforts [2, 10], but it relies on heterogeneity and locality with respect to degree of parallelism to build a hierarchy of processor-and-memory subsystems w here each subsystem is designed to be cost-efficient in its parallelism range. A massively parallel system implemented with dense, relatively inexpensive and slow memory technology, can be very efficient for highly parallel code, while a tightly-coupled s ymmetric multiprocessor containing a smaller amount of fast, expensive memory, can be very efficient for code mostly sequential or with moderate parallelism. The merging of these systems under the HPAM concept provides a cost-efficient solution for applic ations with different levels of parallelism.

2.1 Inter-Level Coherence Protocols

A distributed shared-memory (DSM) implementation of HPAM is assumed in this paper. Each level of such shared-memory HPAM machine relies on caching of data from remote levels to reduce inter-level bandwidth requirements and improve remote access latency. T herefore, cache coherence has to be enforced both inside an HPAM level and among different levels. Cache-coherence solutions that use a combination of different protocols (snoopy and directory-based) have been proposed and implemented [6] for ho mogeneous DSMs, and can be reused in an HPAM context. However, an HPAM machine can take advantage of coherence solutions that exploit its heterogeneous nature.

In this paper, the potential advantages of having multiple line sizes across the hierarchy are studied. Similar to conventional uniprocessor memory hierarchies, multiple line sizes in an HPAM context can provide low miss rates in the bottom levels of the hierarchy while not sacrificing traffic and miss penalty in the upper hierarchy levels. In this section, a coherence protocol that allows requests to be generated by any level of the hierarchy (as opposed to a conventional memory hierarchy, where all acce sses are generated in the topmost level), and supports both inclusion and multiple line sizes, is described. Similar to the MESI coherence protocol [8], the protocol assigns one of four states to each memory block and relies on invalidatio ns to maintain coherence.

Let the hierarchical organisation have h levels, where level 1 is the top level and h is the bottom hierarchy level. Let ls_i be the line size (also referred to as block size) in level i. All data transfers between adjacent levels i and $i+1$ ha ve size ls_i. Let $B_j(i)$ be block i in level j, where $i \geq 0$ and $1 \leq j \leq h$. Assume that block sizes across the hierarchy satisfy the relation:

$$\frac{ls_j}{ls_k} = 2^x, j \geq k, x \in N \tag{1}$$

Given this alignment, let the $\frac{ls_j}{ls_k}$ *sub-blocks* in level k of a block $B_j(i)$ be defined as:

$$B_k\left(\frac{ls_j}{ls_k} * i\right), B_k\left(\frac{ls_j}{ls_k} * i + 1\right), ..., B_k\left(\frac{ls_j}{ls_k} * i + \frac{ls_j}{ls_k} - 1\right), j \geq k \tag{2}$$

and the unique *superblock* in level l of $B_j(i)$, $l \geq j$, be defined as:

$$B_l\left(\lfloor \frac{ls_j}{ls_l} * i \rfloor\right) \tag{3}$$

For the proposed inter-level coherence protocol, blocks can be in any of the following four states:

- **Invalid (I)**: data in the block is entirely non-valid
- **Accessible (A)**: data in the block is valid and may be shared (read-only) by one or more processors
- **Reserved (R)**: data in the block is the only valid copy in the hierarchy
- **Partially Invalid (P)**: at least one sub-block of the memory block is outdated (due to a write in an upper-level memory)

Memory access operations consist of read and write commands that can be issued from any level j of the hierarchy. These operations are considered atomic. The read/write commands are defined in terms of four primitive coherence operations, as follows: (t he algorithms used to implement these basic inter-level coherence operations are defined in Appendix A)

- **ULI($B_j(i)$)(Upper-Level Invalidate)**: Invalidates all sub-blocks of $B_j(i)$
- **ULW($B_j(i)$)(Upper-Level Writeback)**: Writes back dirty data to $B_j(i)$ from upper-level sub-blocks; sets sub-blocks to *Accessible* (read-only)
- **LLP($B_j(i)$)(Lower-Level Partial-Invalidate)**: Sets all superblocks of $B_j(i)$ *Partially Invalid*
- **LLR($B_j(i)$)(Lower-Level Read)**: Fetches block $B_j(i)$ from lower-level superblocks; sets super-blocks to *Accessible* (read-only)

The coherence protocol implements read/write operations as combinations of these four primitives. In order to allow for multi-level inclusion, i.e., (definition here), the coherence protocol enforces the following properties:

1. If a block $B_j(i)$ is *Partially Invalid*, then all of its superblocks must also be *Partially Invalid*
2. If a block $B_j(i)$ is *Invalid*, then all of its sub-blocks must also be *Invalid*

3. If a block $B_j(i)$ is *Reserved*, then all of its sub-blocks must be *Invalid* and all of its superblocks must be *Partially Invalid*
4. If a block $B_j(i)$ is *Accessible*, then all of its sub-blocks are either *Invalid* or *Accessible*, and all of its superblocks are either *Partially Invalid* or *Accessible*

Algorithms for the coherent write and read operations of a block $B_j(i)$ in level j are presented in Appendix A. Property **1** allows the coherence controller to fetch the most recent copy of a block $B_j(i)$ if any sub-block of it has been modified by an upper-level processor before completing a read or write request. Property **3** ensures that a processor can complete a write to block $B_j(i)$ when the state of $B_j(i)$ is *Reserved* without involving other processors of the write, since it is the only processor that has a valid copy of the block.

An example of the inter-level coherence protocol operation on a 3-level configuration is shown in Figure 2. Each memory block is represented in this figure by both its state (gray-shaded boxes) and contents of each of its sub-blocks. Note th at the block sizes differ among the levels; a block in level 2 is twice larger than a block in level 1 and four times larger than a block in level 0.

The example begins with the configuration of Figure 2(a): the bottom level has valid data in the *Reserved* state, and the other levels have invalid data. Level 0 then issues a memory *read* of block $B_0(1)$. The protocol issues a *lower-level read* primitive, bringing a sub-block of level 2 to level 1, containing values **x** and **y**, then a subblock of level 1 to level 0, containing the desired data (**y**). All blocks involved in this transaction become *Accessibl e* (Figure 2(b)).

Suppose level 1 issues a *write* to block $B_1(0)$ and let **t** and **u** denote the new contents of the respective sub-blocks. The protocol handles this request by invalidating upper level sub-blocks ($B_0(0)$ and $B_0(1)$), setting the lower le vel superblock $B_2(0)$ to *Partially Invalid* and setting the state of $B_1(0)$ to *Reserved* (Figure 2(c)).

The next memory reference in this example is a *read* of block $B_0(3)$ (Figure 2(c)). Similarly to the first read, data is brought from level 2 to level 1, then to level 0. The states of the blocks in levels 0 and 1 become *Accessib le.* However, the state of the block $B_2(0)$ in level 2 remains *Partially Invalid* to flag that at least one of its sub-blocks ($B_1(0)$ in this case) contains data that needs to be written back, as Figure 2(d) shows.

The last memory reference of this example is a *read* of $B_2(0)$. This reference generates a write-back request to the *Reserved* sub-block $B_1(0)$ in the upper level. The *Reserved* block in level 1 becomes *Accessible* (Figure 2(e)). This assumes the existence of state bits associated with each sub-block of the adjacent upper level.

The implementation of read and write fences for relaxed consistency models requires that the coherence protocol handles acknowledgements of all messages exchanged between adjacent levels (not shown in the primitives of Appendix A). The completion of an op eration (read or write) issued by processor P with respect to all other processors is signalled by the arrival of positive acknowledgement from upper and lower adjacent levels.

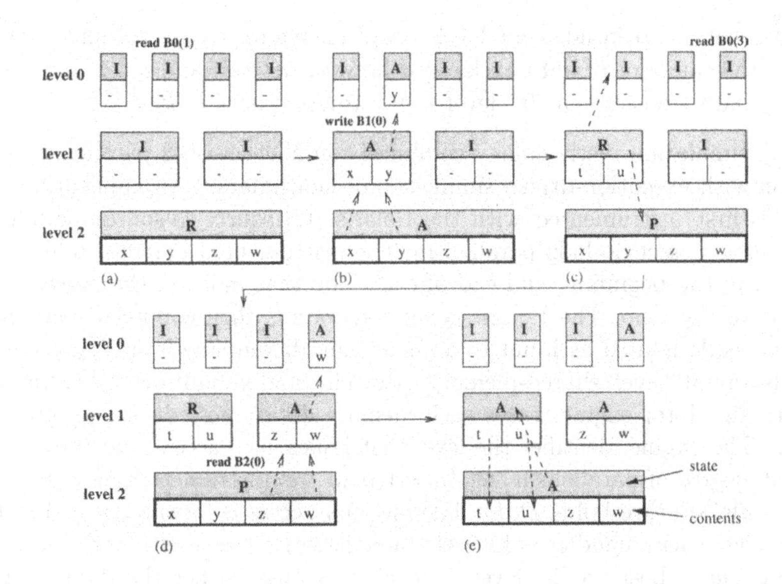

Fig. 2. Example of coherence protocol **I** for 3-level configuration. Line states are shown in the top of each box: P (Partially Invalid), I (Invalid), A (Accessible) and R (Reserved) and contents of blocks are shown below the state.

3 Simulation Models and Methodology

The locality studies presented in this paper are based on a generic four-level HPAM architecture. Each level is labelled from *0* to *3*, where *0* represents the topmost level (smallest degree of parallelism). Level i in the hierarchy is re sponsible for executing fractions of code which have degree of parallelism greater or equal to DoP_i and less than DoP_{i+1} (or infinity for the bottom level). The degrees of parallelism in this study were fixed as powers of ten, $DoP_0 = 1$, $DoP_1 = 10$, $DoP_2 = 100$ and $DoP_3 = 1000$. The variables used in the locality experiments in this paper are defined as follows:

- Line size of level i (ls_i): represented in log_2 notation in this paper (i.e. $ls_i = 8$ means a line size of $2^8 = 256$ bytes). This parameter determines the amount of data that is transferred to level i when a data miss occurs in this level. It is assumed that each level has an inter-level cache that is shared by all level processors. The shared caches at each level are assumed to be ideal (fully associative and infinite) in the simulations performed.
- Data miss rate at level i (mr_i): percentage of memory references (loads and stores) in level i that miss. Misses in level i can be serviced either by disk (cold misses) or by another level j, $i \neq j$. The latter case will be referred t o as *inter-level misses* throughout this paper. Intra-level misses, which are serviced by processors in the same level, are not modeled.
- Data traffic between levels i and $i+1$ ($tr_{i,i+1}$): aggregate amount of data transferred between levels i and $i+1$. Inter-level communication is assumed

to occur between adjacent levels only. Therefore, $tr_{i,i+1}$ accounts not only for th e amount of data exchanged between levels i and $i+1$, but also for any data transfer from/to level j, $j \leq i$ to/from level k, $k \geq i+1$.

The simulation methodology combines compiler-assisted parallelism identification with execution-driven simulation of benchmarks. An application under study is first instrumented with the Polaris [12] source-to-source parallelising compiler to detect do-loop parallelism. The instrumented Fortran code has tags inserted in the beginning and end of each loop that indicate the degree of parallelism of the loop. The Polaris-generated code is then compiled, and the executable code is used as input to an execution-driven simulator. The simulator models a multi-level, shared-memory hierarchy, and is built on top of Shade [1].

The simulator engine traces each memory data access during program execution. The engine identifies the level that issues each access by comparing the current degree-of-parallelism tag (inserted in the instrumentation phase) with the parallelism thresholds DoP_i. The memory access is forwarded to the appropriate level cache handler, which characterises the access either as a hit (data has previously been in the level's cache) or a miss (either the data is present in another level's cache or needs to be fetched from disk). Coherence messages are sent by the cache handler to other level caches on misses, according to the cache coherence protocol under use. Hence, the inter-level coherence protocol behaviour is modelled. However, the intra-level coherence protocol is not modelled. The miss rate and traffic results obtained with such model are therefore optimistic, since ideal caches and intra-level sharing are assumed. Nonetheless, this simplified model is able to capture the inherent spatial locality with respect to degree of parallelism of the applications under study. Miss rates degrade when finite caches and intra-level sharing are considered, but locality with respect to degree of parallelism is still evident for non-ideal memory systems [3].

The following benchmarks from the CMU, Perfect Club and Spec95 suites have been used in the spatial locality studies: Radar, Stereo, FFT2 and Airshed (CMU Parallel Suite [9]); TRFD, FLO52, ARC2D, OCEAN and MDG (Perfect Club Suite [7]); Hydro2d and Swim (Spec95 Suite).

4 Simulation Results and Analysis

For each benchmark, simulations have been performed for various line sizes (ls_i), and measurements of miss rates (mr_i) and data traffic ($tr_{i,i+1}$) have been collected. Two inter-level coherence protocols have been considered in this study. Initi ally, a homogeneous solution analogous to a cache-only (COMA) protocol [4] is used to observe the inherent locality behaviour of the set of benchmarks under study. In this scenario, a block can migrate to any level of the hierarchy, i.e., there i s no fixed home node associated with a given memory block. Such scenario is referred to as "migration protocol" in the rest of this paper. The other coherence solution considered is the heterogeneous protocol introduced in Section 2.1. Such scenario is referred to as "inclusion protocol" in the rest of this paper.

Subsection 4.1 presents data obtained when a migration protocol with unique line sizes across the hierarchy is assumed. Such scenario is used as a basis for the analysis of (level) data locality. Subsection 4.2 compares the resu lts obtained from the migration scenario to results obtained when the coherence protocol enforcing inclusion presented in Section 2.1 is used and line sizes are allowed to be different across levels.

4.1 Migration Protocol with Unique Line Size Across Levels

The results obtained for this protocol configuration confirm empirically that applications have good spatial locality with respect to degree of parallelism, in addition to previously observed [14] temporal locality with respect to degree of paral lelism. Inter-level miss rates are low for small line sizes: a highest miss rate of 17.3% occurs in *Swim* for 16-Byte line sizes, but typical values are around 1%. Furthermore, inter-level miss rates tend to decrease as the line size gets larger.

Figure 3 shows this trend for the benchmark *FLO52*, assuming a 4-level HPAM. The figure is divided into four sub-plots, each corresponding to an HPAM level, labelled *lvl0* through *lvl3* in the x-axis. Each sub-plot is fu rther divided into six line sizes, ranging from $ls = 2^4$ to $ls = 2^{14}$ bytes. For each level and line size, the *absolute number* of misses is plotted in the y-axis in log scale, and the corresponding *miss rate* is indicated in the x-axis betwee n parentheses. The different shades of the bars in the y-axis correspond to the *percentage* of inter-level misses that are either cold misses or serviced by another level. To illustrate this notation, consider the case where FLO52 runs on a 4-level H PAM with line size of $2^6 = 64$ bytes in all levels. The inter-level miss rates for levels 0 through 3 are 3.40%, 0.35%, 0.20% and 0.63%. For level *lvl2* and line size of 64 bytes, approximately half of the inter-level misses are serviced by level 3, 20% are serviced by level 1, 30% by level 0, and a negligible fraction is due to cold misses.

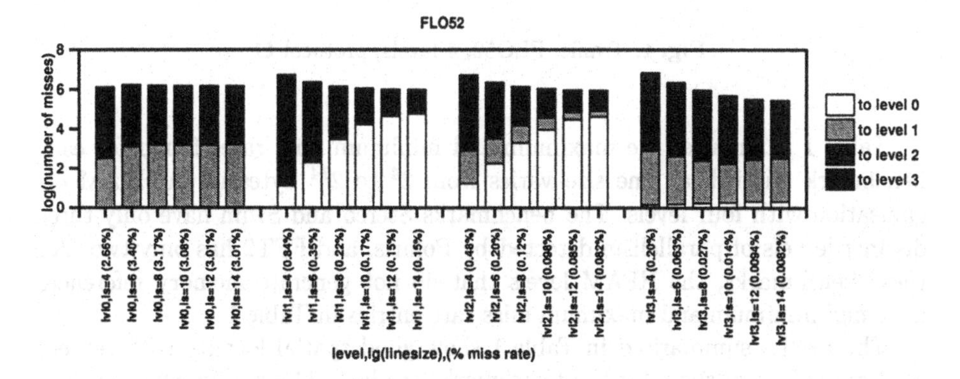

Fig. 3. Misses: FLO52, 4 levels, protocol **C**

Figure 3 shows that the spatial locality behaviour for *FLO52* varies across hierarchy levels. For level 0, the inter-level miss rate remains approximately constant, slightly degrading as the line size increases. In contrast, the m iss rate decreases about two orders of magnitude as line size increases from 2^4 to 2^{14} in level 3. Such behaviour suggests that the parallel fraction of FLO52 that executes in the lowest hierarchy level operates on large, regular data structures t hat benefit from fetching large data lines on a miss.

While a larger line size tends to improve inter-level miss rates, it also tends to increase inter-level data traffic. Figure 4 shows how data traffic between adjacent levels varies as a function of line size for the benchmark FLO52. Notice that the traffic between levels 0 and 1 in this case increases by about three orders of magnitude for the range of line sizes considered, while the traffic between levels 2 and 3 increases only by about two orders of magnitude across the same line size range. Such behaviour can be explained with the aid of the inter-level miss rate profile for FLO52 (Figure 3). The larger line sizes brought to levels 2 and 3 often contain data that is likely to be used in future references, whi le larger line sizes in levels 0 and 1 most often bring data that remains unused. Since traffic is proportional to the product of number of misses and line size, if the number of misses do not decrease as line size increases, the traffic increases.

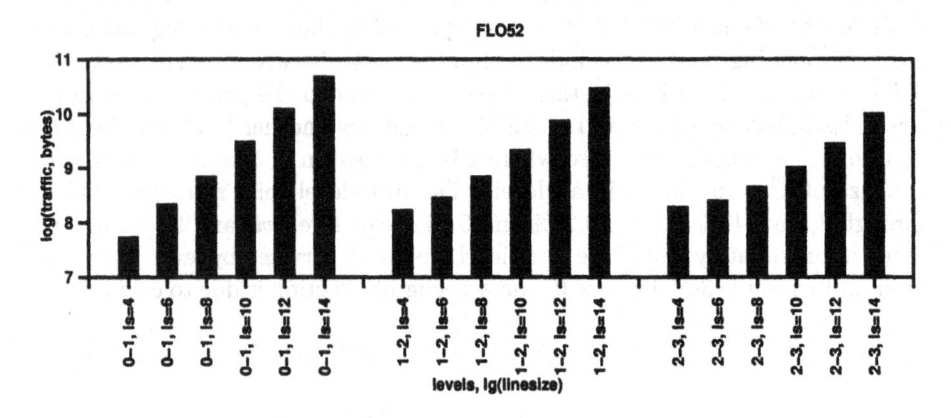

Fig. 4. Traffic: FLO52, 4 levels, protocol C

Table 1 summaries the maximum and minimum miss rates found for each benchmark studied, as line size varies from 2^4 to 2^{14} bytes in an HPAM organisation with four levels. The benchmarks Stereo and Swim have only three distinct le vels of parallelism detected by Polaris, and FFT2 has only two. For these benchmarks, the HPAM levels that do not generate memory references have null minimum and maximum miss rate entries in Table 1.

The results summarised in Table 1 show good spatial locality with respect to degree of parallelism for the benchmarks studied. The maximum inter-level miss rate observed for the lowest level of the hierarchy is 0.9%; for the topmost

Table 1. Min/Max inter-level miss rates for a 4-level HPAM configuration

Benchmark	Inter-level Miss rate							
	Level 0		Level 1		Level 2		Level 3	
	min	max	min	max	min	max	min	max
FLO52	2.66%	3.40%	0.15%	0.84%	0.082%	0.48%	0.0082%	0.20%
TRFD	0.12%	1.30%	5.28%	6.97%	0.87%	1.08%	0.046%	0.086%
OCEAN	0.0052%	0.17%	0.096%	1.63%	0.27%	1.70%	0.023%	0.45%
MDG	0.39%	1.50%	0.77%	3.16%	0.049%	3.76%	0.0012%	0.097%
ARC2D	1.03%	2.17%	0.0043%	1.45%	0.0071%	1.41%	0.0004%	0.062%
Airshed	2.92%	4.01%	0.48%	0.72%	0.049%	0.049%	0.018%	0.037%
Stereo	0.021%	1.48%	-	-	0.049%	1.51%	0.011%	0.12%
Radar	0.00039%	0.066%	0.27%	1.63%	0.035%	3.79%	0.0015%	0.90%
FFT2	0.00008%	0.065%	-	-	-	-	0.0050%	0.65%
Hydro2d	0.0053%	2.88%	1.78%	10.60%	0.29%	6.55%	0.00038%	0.0074%
Swim	12.81%	17.30%	0.80%	9.48%	0.0010%	0.011%	-	-

le vel, the maximum inter-level miss rate observed is 17.3% for *Swim*, and less than or equal to 4.01% for all other benchmarks. In general, the best miss rates are found in the lowest level of the architecture, where loops with high degree of paralle lism are executed.

4.2 Inclusion Protocol

Identical Line Sizes: The inclusion protocol was initially studied assuming that a unique line size is used across all HPAM levels, similar to the migration protocol discussed in Subsection 4.1. Since the level caches are assumed to be ideal, and intra-level shar ing is not modelled, the results for the inclusion coherence protocol with unique line size do not differ considerably from the results obtained from the migration protocol simulations, in general. Assuming such idealised memory model, both protocols yiel d measurements that characterise the *inherent* sharing behaviour of the benchmarks. The inter-level miss rates obtained for this scenario differ by at most 17% from the migration scenario for *TRFD*, with an average difference of 1.4% across al l benchmarks. scenarios, most of those misses are serviced by level 1 in the inclusion scenario for small line sizes, while they are serviced mostly by level 3 in the COMA scenario. The inclusion protocol has succeeded in this case to bring data "closer" to level 0 (i.e., to level 1 instead of level 3). If communication from level 0 to level 3 is slower than communication between adjacent levels 0 and 1, inter-level misses at level 0 can be serviced faster in the inclusion protocol scenario.

Distinct Line Sizes: Conventional uniprocessor cache hierarchies typically use distinct line sizes across the cache levels; large lines are desirable in large caches to improve miss rates, while small cache lines are desirable in small caches to avoid excessive bandwidth requ irements and increases in miss penalties and conflict misses [5,13]. An HPAM machine can also benefit from distinct line sizes across levels by reducing inter-level traffic while not sacrificing inter-level miss rates.

The inter-level miss rate and traffic profiles for the benchmark *FLO52* (Figures 3 and 4) illustrate a scenario commonly observed in the simulations performed, where a large line size effectively reduces the inter-level miss rate in level 3, but unnecessarily increases the traffic between levels 0 and 1 without improving the inter-level miss rates in these levels. The inter-level inclusion coherence protocol described in Section 2.1 supports a c onfiguration with distinct line sizes across HPAM levels; the effects of distinct line sizes on inter-level miss rates and traffic have been captured quantitatively through simulation and are discussed in this subsection.

Line sizes have been set up such that the relationship $ls_{i+1} > ls_i$ holds for any adjacent levels $i, i+1$. Hence, levels executing highly parallel code are assigned line sizes strictly larger than levels executing moderately parallel or sequential code . Table 2 shows how inter-level miss rates and traffic for a configuration with multiple line sizes compare to configurations with unique line sizes, for the benchmark *MDG*. The first row of Table 2 shows inter-lev el miss rates for the multiple line size configuration. Rows 2 through 5 of the table show miss rates obtained in four different simulations with unique line sizes, each corresponding to a line size chosen for the multiple line size scenario. The remainin g rows of Table 2 show the total traffic in the level boundaries for three scenarios: multiple line sizes, unique line of smallest size (2^6 Bytes), and unique line of largest size (2^{14} Bytes).

Table 2. Traffic and inter-level miss rates for MDG with multiple line sizes: 2^6, 2^8, 2^{10} and 2^{14}

	Level 0 (ls=2^6)	Level 1 (ls=2^8)	Level 2 (ls=2^{10})	Level 3 (ls=2^{14})
Miss rate (multiple)	**1.28%**	**1.38%**	**0.10%**	**0.0012%**
Miss rate (unique, 2^6)	**0.82%**	2.17%	1.00%	0.025%
Miss rate (unique, 2^8)	0.48%	**1.44%**	0.29%	0.0073%
Miss rate (unique, 2^{10})	0.56%	0.98%	**0.10%**	0.0026%
Miss rate (unique, 2^{14})	0.39%	0.77%	0.049%	**0.0012%**
	levels 0-1	levels 1-2	levels 2-3	
Traffic (multiple)	26.6GB	12.9MB	17.7MB	-
Traffic (single, ls=2^6)	15.6GB	10.8MB	11.3MB	-
Traffic (single, ls=2^{14})	1.6TB	71.8MB	127.7MB	-

Table 2 shows that the miss rate observed in the multiple line size scenario is at most 56% larger than the rate observed for the corresponding unique line size rate (values in bold face) for each level. For levels 2 and 3, in particula r, the inter-level miss rates are equal for both scenarios. When inter-level traffics are compared, the multiple line size scenario demands about an order of magnitude less traffic than the unique line size scenario with the largest line size, while deman ding no more than 70% more traffic than the unique line size scenario with the smallest line size. In this example, the multiple line size configuration is therefore capable of providing very low miss rates at the lowest hierarchy level without generatin g excessive traffic in the upper level boundaries. When a unique line size is used, either the miss rate in the lowest level or traffic in the topmost level degrades. The same motivations for using multiple line sizes across uniprocessor memory hierarchie s thus apply to a hierarchy of processor-and-memories: maintaining good miss rates across the hierarchy while avoiding the generation of unnecessary traffic in the upper levels.

Table 3 shows the average increase in the inter-level miss rate of the multiple line size configuration compared to the miss rate of a corresponding unique line size configuration for simulations performed in six of the studied benchma rks. The inter-level miss rates of the lowest levels in the hierarchy remain unchanged, with respect to the unique line size scenario, when multiple line sizes are used. Miss rates at the topmost level increase by 31% in average.

Table 3. Average ratio: miss_rate(multi)/miss_rate(single)

	Level 0	Level 1	Level 2	Level 3
Average mr_{multi}/mr_{single}	1.31	1.05	1.01	1.00

5 Conclusions

The conclusions reached in this paper provide guidelines to the design of the memory and network subsystems of an HPAM machine. The implementation of a coherence controller that supports multiple line sizes across the hierarchy is an ongoing research subj ect. The inclusion coherence protocol presented in this paper has been used as a proof of concept to study the advantages of fetching larger blocks of data to lower levels of the hierarchy as a means of increasing spatial locality without sacrificing traf fic in the upper levels of the hierarchy. One solution under investigation that may require minimal modifications to the existing directory controllers relies on hardware-assisted pre-fetching. In this scheme, the coherence unit size is kept constant acro ss the hierarchy. However, lower hierarchy levels pre-fetch larger number of coherence units on a miss than upper levels. Such scheme allows reusing of cache coherence implementations found in homogeneous machines.

The experimental results obtained in this study for inter-level miss rates among different parallelism levels confirm that there is spatial locality with respect to degree of parallelism in parallel applications, in addition to previously observed temporal locality. The differences in degrees of parallelism and memory capacity across HPAM levels motivate the use of multiple line sizes across the hierarchy as a means of reducing inter-level traffic associated with large line sizes while keeping miss rates comparable to the case where a unique line size is used across all levels An invalidation-based inter-level coherence protocol that supports such multiple line size configuration across processor-and-memory levels has been proposed, and the experimental results obtained with simulations using such protocol have confirmed that more balanced inter-level miss rate and traffic characteristics can be achieved with line sizes that increase from the top to the bottom of the hierarchy. A distribution with larger data blocks at the lowest levels of the hierarchy is consistent with the proposed HPAM organization, where lowest levels have larger amounts of memory.

Idealised caches and line sizes ranging from very small to very large have been used in the experiments in order to observe the inherent locality behaviour of the studied benchmarks. The authors believe that the overall inter-level locality behaviour in s ystems with non-ideal caches can be derived from the results obtained.

An HPAM machine combines heterogeneity, data locality with respect to degree of parallelism and computing near memory to provide a cost-effective solution to high-performance parallel computing. The data locality studies presented in this study confirm that HPAM machines have the potential to competitively exploit the trend towards merging processor and memory technologies and the increasingly more powerful but also extremely expensive fabrication processes needed for billion-transistor chips.

Acknowledgements

This research was supported in part by NSF grants ASC-9612133, ASC-9612023 and CDA-9617372. Renato Figueiredo is also supported by a CAPES grant.

References

1. B. Cmelik and D. Keppel. Shade: A fast instruction-set simulator for execution profiling. In *Proceedings of the 1994 SIGMETRICS Conf. on Measurement and Modeling of Computer Systems*, 1994.
2. D. Patterson et al. A Case for Intelligent RAM: IRAM. *IEEE Micro*, Apr 1997.
3. Figueiredo, R. J. O. and Fortes, J. A. B. and Ben-Miled, Z. and Taylor, V. and Eigenmann, R. Impact of Computing-in-Memory on the Performance of Processor-and-Memory Hierarchies. Technical Report TR-ECE-98-1, Electrical and Computer Engineering Department, Purdue University, 1998.
4. Hagersten, E. and Landin, A. and Haridi, S. DDM - A Cache-Only Memory Architecture. *IEEE Computer*, Sep. 1992.
5. J.L. Hennessy and D.A. Patterson. *Computer Architecture: A Quantitative Approach.* Morgan Kaufmann, 1996.
6. Lenosky, D. and Laudon, J. and Gharacharloo, K. and Gupta, A. and Hennessy, J. The Directory-Based Cache Coherence Protocol for the DASH Multiprocessor. In *Proc. of the 17th Annual Int. Symp. on Computer Architecture*, May 1990.

7. M. Berry et al. The Perfect Club Benchmarks: Effective Performance Evaluation on Supercomputers. Technical Report UIUC-CSRD-827, Center for Supercomputing Research and Development, University of Illinois at Urbana-Champaign, July 1994.
8. M. Papamarcos and J. Patel. A Low Overhead Coherence Solution for Multiprocessors with Private Cache Memories. In *Proc. of 11th Annual Int. Symp. on Computer Architecture*, 1984.
9. P. Dinda et al. The CMU Task parallel Program Suite. Technical Report CMU-CS-94-131, School of Computer Science, Carnegie Mellon University, Pittsburgh, Pennsylvania, Jan 1994.
10. P.M. Kogge and T. Sunaga and H. Miyataka and K. Kitamura and E. Retter. Combined DRAM and Logic Chip for Massively Parallel Systems. *16th Conference on Advanced Research in VLSI*, 1995.
11. Standard Performance Evaluation Corporation. Spec newsletter, Sep 1995.
12. W. Blume, R. Doallo, R. Eigenmann, J. Grout, J. Hoeflinger, T. Lawrence, J. Lee, D. Padua, Y. Paek, B. Pottenger, L. Rauchwerger and P. Tu. Parallel Programming with Polaris. *IEEE Computer*, Dec 1996.
13. Y.-S. Chen and M. Dubois. Cache Protocols with Partial Block Invalidations. In *Proc. 7th Int. Parallel Processing Symp.*, 1993.
14. Z. Ben-Miled and J.A.B. Fortes. A Heterogeneous Hierarchical Solution to Cost-efficient High Performance Computing. *Par. and Dist. Processing Symp.*, Oct 1996.
15. Z. Ben-Miled, R. Eigenmann, J.A.B. Fortes, and V. Taylor. Hierarchical Processors-and-Memory Architecture for High Performance Computing. *Frontiers of Massively Parallel Computation Symp.*, Oct 1996.

Appendix A: Inter-Level Coherence Protocol Messages

```
ULW(Bj(i))  // UPPER-LVL WRITEBACK
 for all level-(j-1) sub-blocks
  if (sub-block is PART-INV) then
   ULW(sub-block)
  if (sub-block is not INV) then
   write-back sub-block from
     level j-1 to level j
   state (sub-block) = ACC
 state (Bj(i)) = ACC
```

```
LLR(Bj(i))    // LOWER-LEVELS READ
 temp = Bj(i)
 L = j
 while (temp is INV)
  increment L
  temp = level-L superblock of temp
 decrement L
 while (L >= j)
  read level-L superblock
       from level L+1
  state(level-L superblock) = ACC
  decrement L
```

```
ULI(Bj(i)) // UPPER-LVL INVALIDATE
 if (j is not the first level)
  for all level-(j-1) sub-blocks
   if (sub-block is not INV)
    state(sub-block) = INV
    ULI(sub-block)
```

```
LLP(Bj(i))  // LOWER-LEVELS
            // PARTIAL-INVALIDATE
 temp = superblock of Bj(i)
 L = j+2
 while (temp is not PART-INV)
  state(temp) = PART-INV
  increment L
  temp=level-L superblock of temp
```

```
READ(Bj(i))    // MEMORY READ
 if (Bj(i) is INV)
  LLR(Bj(i))
  state(Bj(i)) = ACC
 if (Bj(i) is PART-INV) then
  ULW(Bj(i))
 read Bj(i) from level j
```

```
WRITE(Bj(i))    // MEMORY WRITE
 if (Bj(i) is RES or ACC)
  ULI(Bj(i)); LLP(Bj(i));
 if (Bj(i) is PART-INV)
  ULW(Bj(i)); ULI(Bj(i));
 if (Bj(i) is INV)
  LLR(Bj(i)); LLP(Bj(i));
 state(Bj(i)) = RES
 write Bj(i) to level j
```

Partitioning Regular Domains on Modern Parallel Computers

Manuel Prieto-Matías, Ignacio Martín-Llorente and Francisco Tirado

Departamento de Arquitectura de Computadores y Automatica
Facultad de Ciencias Fisicas
Universidad Complutense
28040 Madrid, Spain
{mpmatias,llorente,ptirado}@dacya.ucm.es

Abstract. It has become apparent in recent years that the performance of current high performance computers, from powerful workstations to massively parallel processors, is strongly dependent on the behaviour of the memory hierarchy. In fact, it does not only affect the computation time but the time consumed in performing communications. In this research, the impact of the memory hierarchy usage on the partitioning of multidimensional regular domain problems is studied. We use as an example the numerical solution of a three-dimensional partial differential equation in a regular mesh, by means of a multigrid-like iterative method. Experimental results contradict the traditional regular partitioning techniques on some present parallel computers like the Cray T3E or the SGI Origin 2000: a linear decomposition is more efficient than a three dimensional one due to the better exploitation of the spatial data locality. For similar reasons, computation-communication overlapping increases also execution time.

1. Introduction

The performance of current parallel computers, composed of up to hundreds of superscalar commodity microprocessors, presents an increasing dependence on the efficient use of their hierarchical memory structures. Indeed, the maximum performance that can be obtained in current microprocessors is limited by the memory access. The peak performance of the microprocessors has increased by a factor of 4-5 every 3 years by exploiting the increasing integration density, reducing the clock cycle, and by implementing architectural techniques to take advantage of the multiple levels of parallelism. However, the memory access time has been reduced by a factor of just 1.5-2 over the same period. Thus, the latency of memory access in terms of processor performance grows by a factor of 2-3 every three years. This situation seems likely to continue over the next few years and it has been suggested that such

J. Palma, J. Dongarra, and V. Hernández (Eds.): VECPAR'98, LNCS 1573, pp. 411-424, 1999.

trends may result in a "memory wall" in which application performance is entirely dominated by memory access time [1][2].

The common technique to bridge this gap and hide the problem is by using a hierarchical memory structure with large and fast cache memories close to the processor. As a result, the memory structure has a strong impact on the design and development of a code, and the programs must exhibit spatial and temporal locality to make efficient use of the cache memory and so keep the processor busy. The effectiveness of data locality has been well demonstrated in the LAPACK project, and major research has just begun to develop cache-friendly iterative methods [3] [4]. However, to the best of the authors' knowledge, the impact of the memory hierarchy usage on the partitioning has not previously been studied.

In this research, we have studied applications where the main computational portion of the program belongs to a class of kernels known as stencils. A stencil is a matrix computation in which groups of neighbouring data elements are combined to calculate a new value. This type of computation is common in image processing, geometric modelling and solving partial differential equations by means of finite difference or finite volume. The simplest approach to parallelizing these kinds of regular applications distributes the data among the processes, and each process runs essentially the same program on its share of the data. For three-dimensional applications, decompositions in the x, y, and/or z dimensions are possible.

During the last decade, a d-dimensional mesh of processors has been considered as the best partitioning to split a d-dimensional regular domain because in this way the interconnection network is more efficiently exploited [5][6]. Furthermore, communication-computation overlapping techniques are performed to keep the processor busy and so improve the parallel efficiency. However, our results show that in modern parallel computers it is more important to make effective use of the local memory hierarchy than to reduce the overheads due to network delay cost. The interconnection systems have also taken advantage of the increasing integration density offered by the integrated circuit processing technology and the effective bandwidth and latency are now hundreds of times faster than ten years ago.

This paper is organised as follows. In Section 2 we describe the sample code that has been used in our research. The effect of spatial locality on message sending is described in Section 3. Based on this analysis, the choice of an optimal partition is presented in Section 4. The influence of overlapping computations with communications is presented in Section 5. The paper ends with some conclusions to guide the partitioning of regular applications in current parallel computers.

2. Sample code

In this research, we are only interested in a qualitative description of the most important aspects that affect the performance, and that should be considered in further developments. As a sample problem, we have studied the numerical solution of a time-dependent partial differential equation, the three-dimensional Bose-Einstein equation [7], in a regular mesh subject to Dirichlet boundary conditions. The problem is to describe the evolution of a physical field (a complex function) given an initial condition. An implicit finite difference method has been used to carry-out the simulation, and the systems of equations are solved by means of a multigrid-like iterative method [8]. The execution times that we present in this paper are the result of a single time step simulation using only one multigrid iteration.

Like in other regular applications, the parallel program execution is a sequence of computation and communication steps. Values relative to different subdomains are independently computed on the different processors of the network, and then, a communication step between neighbouring logical processors updates the boundary values in these subdomains.

The code used in this study parallelizes well because the discretization is regular, and the same operations are applied at each grid point, even though the evolution of the system is non-linear. Thus, the problem can be statically load-balanced at the beginning of the code.

3. Spatial locality impact on message sending

Message sending between two tasks located on different processors can be divided into three phases: two of them are where the processors interface with the communication system (the send and receive overhead phases), and a network delay phase, where the data is transmitted between the physical processors. Details of what the system does during these phases varies. Typically, however, during the send overhead phase the message is copied into a system buffer area, and control information is appended to the message. In the same way, on the receiving process, the message is copied from a system-controlled buffering area into user-controlled memory (receive overhead is usually larger than send overhead).

In several out-of-date parallel computers, like the Thinking Machines CM5, the Parsys Supernode 1000 or the Meiko CS-2, the most important component was the network delay [9]. However, in current machines like the Cray T3E or the SGI Origin 2000, as the interconnection networks increase their bandwidth, the send and receive overheads are becoming important. The factors determining these overheads are different in each system, but they are mainly due to uncached operation, misses and synchronisation instructions, generally considered to be infrequent events and therefore a low priority for architectural optimisations of current microprocessors. These components improve very rapidly in terms of performance, and also present very low price to performance ratios, but microprocessors are well designed mostly for workstations and for parallel servers with modest number of processors. Large scale multiprocessors integrate very specific environments requiring appropriate developments for efficiency. For example, the memory interfaces are cache line based, making references to single words (corresponding to strided or scatter/gather references in a vector machine) inefficient [10]. Therefore, the cost of communication depends not only on the amount of communication but also on how the data are structured in the local memories (mainly the spatial data locality).

3.1 The Cray T3E message passing performance

The T3E used in this study had 32 DEC Alpha 21164 running at 300 MHz at the beginning of our research, and has recently been upgraded with 450 MHz processors. Like the T3D, The T3E contains no board-level cache, but the Alpha 21164 has two levels of caching on-chip: 8 KB first-level instructions and data caches, and a unified, 3-way associative, 96-Kbyte write-back second-level cache. The local memory is distributed across eight banks, and its bandwidth is enhanced by a set of hardware stream buffers. These buffers, which exploit spatial locality alone, can take the place of a large board-level cache, which is designed to exploit both spatial and temporal locality. Each node augments the memory interface of the processor with 640 (512 user and 128 system) external registers (E-registers). They serve as the interface for message sending; packets are transmitted by first assembling them in an aligned block of 8 E-registers.

The processors are connected via a 3D torus with an inter-processor communication bandwidth of 480 Mbytes/sec. Using MPI, however, the effective bandwidth is smaller due to overhead associated with buffering and with deadlock detection. The library message passing mechanism uses the E-registers to implement transfers, directly from memory to memory. Data does not cross the processor bus; it flows from memory into E-registers and out to memory again in the receiving processor. E-registers enhance performance when no locality is available by allowing the on-chip caches to be bypassed. However, if the data to be loaded were in the data cache, then accessing that data via E-registers would be sub-optimal because the cache-backmap would first have to flush the data from data cache to memory [9][10][11].

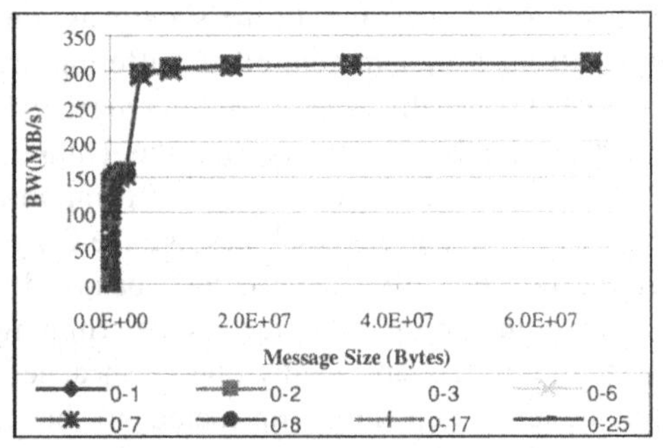

Fig. 1. CRAY T3E point-to-point bandwidth between different processor pairs using contiguous data

Figure 1 shows the measured one-way communication bandwidth for different message sizes using MPI. The test program uses all of the 28 processors available in the system. There is always the same sender processor and one receiver processor that varies. The sender initiates an immediate send followed by an immediate receive, then it waits until both the send and the receive have been completed. The receiver begins by starting an immediate receive operation, then waits until it is finished. It replies with another message using a send/wait combination. Because this operation is repeated many times, if all the data fits into the cache then, except for the first echo, the required data will be found in the cache. But, on the CRAY T3E, the `suppress` directive [12] can be used to invalidate the entire cache and so, it forces all entities in the cache to be read from memory. The measures demonstrate that there is no difference between close and distant processors in the CRAY T3E if there is no contention on the communication network. For clarity we present the results of only a few pairs of processors.

Figure 2 shows the impact of the spatial data locality. We use also the simple echo test, but we modify the data locality by means of different strides between successive elements of the message. The stride is the number of double precision data between successive elements of the message, so stride-1 represents contiguous data. We use MPI datatypes (`MPI_Type_vector`) instead of the `MPI_Pack` / `MPI_Unpack` routines, because they may allow certain performance optimisations. However, we must be careful because the use of certain MPI datatypes can dramatically slow down communication performance, e.g., the `MPI_Type_hvector` type in the T3E implementation. We send buffers that are 8-byte aligned because the T3E copies non-aligned data slowly. This is automatic for the usual case of sending double precision data. Due to memory constraints the larger message is limited to 32Kbytes, although it is not big enough to obtain the asymptotic bandwidth for the stride-1 case, these sizes are similar to the messages used in our application program.

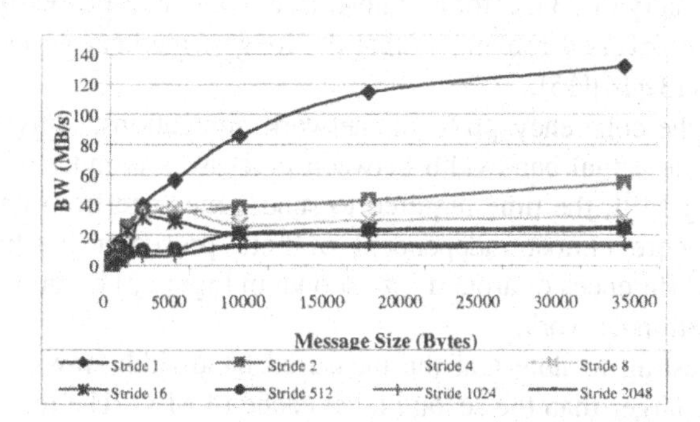

Fig. 2. CRAY T3E message passing performance using non-contiguous data

It is interesting to note that almost the same effective bandwidth is obtained for strides between 16 and 512 double precision data. For 32 KB messages, stride-1 bandwidth is around 5 times larger than stride-16. Beyond Stride-1024 this difference grows, with stride-1 bandwidth being 10 times larger than that of stride-2048 for example.

3.2 SGI Origin 2000 message passing performance

We repeated these tests on a SGI Origin 2000, a cache-coherent distributed shared memory system. Each node of the Origin contains two processors connected by a system bus, a portion of the shared main memory on the machine, a directory for cache coherence, the Hub (which is the combined communication/coherence controller and network interface) and an I/O interface called Xbow.

The system used in this study has the MIPS R10000 running at 195 MHz. Each processor has a 32 Kbyte two-way set-associative primary data cache and a 4-Mbyte two-way set-associative secondary data cache.

The peak bandwidths of the bus that connects the two processors and the Hub's connection to memory are 780 MB/s. However the local Memory peak bandwidth is only about 670 MB/s . One important difference between this system and the T3E is that it caches remote data, while the T3E does not. All cache misses, whether to local or remote memory, go through the Hub, which implements the coherence protocol

The Hub's connections to the off-board network router chip and the I/O interface are 1.56 GB/s each. The SGI Origin network is based on a flexible switch, called SPIDER, that supports six pairs of unidirectional links, each pair providing over 1.56 GB/s of total bandwidth in the two directions. Two nodes (four processors) are connected to each switch so there are four pairs of links to connect to other switches. This building block is configured in a family of topologies related to hypercubes.

Latencies to the memory modules of the Origin 2000 system depend on the network distance from the issuing processor to the destination memory node. Accesses to local memory take 80 clock cycles (CC) (400 ns), while latencies to remote nodes are the local memory time plus 22 CC (110 ns) for each network router, plus a one-time penalty of 33 CC for a remote access. On a 32-processor machine, the maximum distance covers 4 routers, so that the longest memory access is about 201 CC (1005 ns) [9] [13][14][15].

Due to the cache-coherency protocol, network contentions, software layers and other overheads, the actual bandwidth between processors is much lower. As in the CRAY T3E, using MPI, the time required to send a message from one processor to another is once more almost independent of both processor locations. We have measured erratic differences of around 7% (shown in Figure 3) because of contentions on the communication network.

It is also interesting to note that the measured bandwidth slows down when the message sizes are larger than the second level cache (4 MB). The reason for this fact could be the internal message buffers that are used in the SGI MPI implementation. For longer message sizes, they do no fit in the secondary cache and extra main memory references are needed.

Figure 4 shows the impact of the spatial data locality; the legend at the botton is the number of double precision data between successive elements. For non-contiguous data, the reduction in the effective bandwidth is even greater than in the T3E case. For 256 KB messages, the stride-1 bandwidth is around 6.3 times longer than stride-2. This difference grows with the stride, being 23 times for stride-256. The memory interface of the Origin is cache line based, making references to single data less efficient than in the Cray T3E. Moreover, the current MPI implementation on the Origin 2000 requires one extra buffer copy.

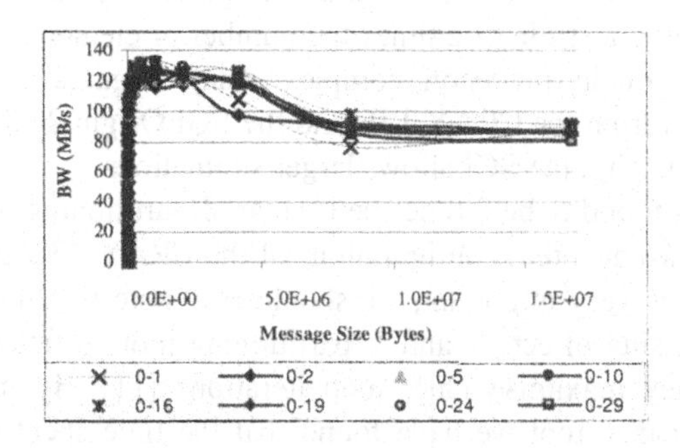

Fig. 3. SGI Origin 2000 point-to-point bandwidth between different processor pairs using contiguous data

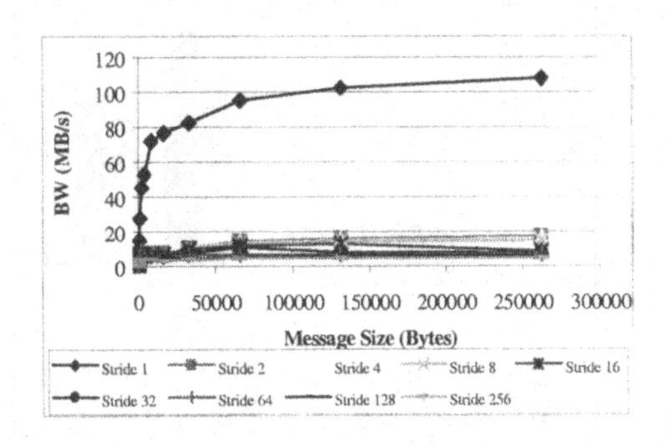

Fig. 4. SGI Origin 2000 message passing performance using non-contiguous data

3.3 Experimental results in our sample code

Although the communication pattern that we found in our application program is not a one-way transfer, but a message exchange between neighbouring logical processors, we notice the impact of the spatial locality as well. In this data exchange, advantage can be taken of bi-directional links, and a greater bandwidth can be reached, compared to what has been observed in the echo test. The code was written in C, so a three dimensional domain is stored in a row-ordered (x,y,z)-array. It can be distributed across a 1D mesh of processors following three possible partitionings: x-direction, y-direction and z-direction. The x and y-direction partitioning were found to be more efficient, because the message data exhibits a better spatial locality. X and Y boundaries are stride-1 data, except for the strides between different Z-columns (two complex data, i.e. four doubles, for X-partitioning and this quantity plus two times the number of elements in a x-plane for Y-partitioning). A message using Z-partitioning references data having a stride of 2 times the number of elements in dimension z (all the elements are double precision complex data). Figures 5 and 6 show the experimental results from the CRAY T3E and the SGI Origin 2000 respectively. Due to main memory capacity, the SGI allows larger simulations.

X-partitioning is found to be 2 times better than Z-partitioning for the 128-element simulation on the two different configurations of the CRAY T3E. Although message-passing bandwidth is very important, we should also note that this difference is not only a message passing effect. X and Y-partitioning more efficiency exploit stream buffers because they maximise inner loop iterations [11]. By means of the MPP Apprentice performance tool we have found that the time spent in the initiation of message sending is 5 times larger in the Z-partitioning simulations. This fact fits in with what we measure in the echo test.

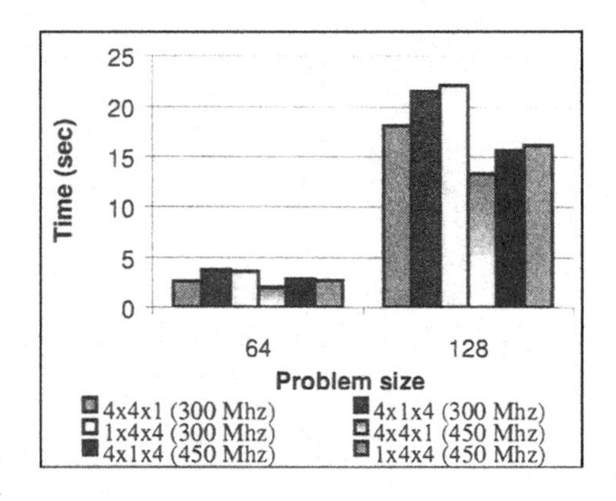

Fig. 5. Different linear partitioning of our sample application using sixteen processor in the CRAY T3E. The problem size is the number of cells in each dimension for the finest grid in the multigrid hierarchy.

Equivalent differences in the Origin 2000 (Fig 6) can be observed, but are less important than in the T3E case. For the 128-element problem, X partitioning is only 1.2 times better, while for the 256-element one, it grows to 1.4. The large second-level cache of this system, which allows the best exploitation of the temporal locality, influences these results (it is less important the spatial locality) [16].

Using 2D and 3D decompositions, we notice the same effects. Z-plane boundaries slow down the performance of the application because they are discontinuous in memory. Therefore, as figure 7 shows, a 2D decomposition using a 4x4x1 topology of processors (4 subdomains in the x and y dimensions and no decomposition in the z direction) is better than 4x1x4 and 1x4x4 topologies (the differences are around 15 %

in the Cray T3E). In the same way, a 3D decomposition using a 4x2x2 array is better than a 2x2x4 one.

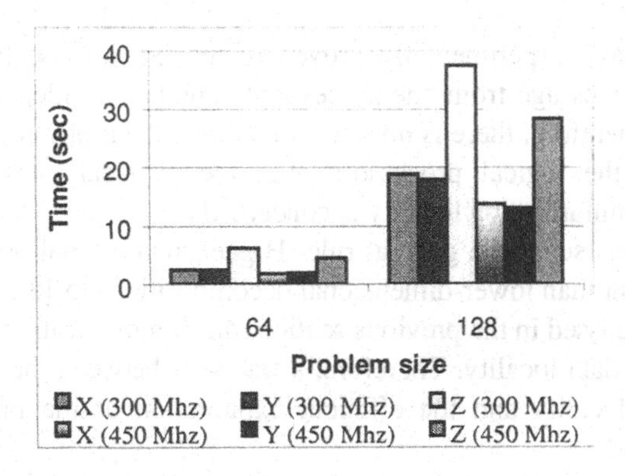

Fig. 6. Different linear partitioning of our sample application using 32 processors in the SGI Origin 2000. The problem size is the number of cells in each dimension for the finest grid in the multigrid hierarchy.

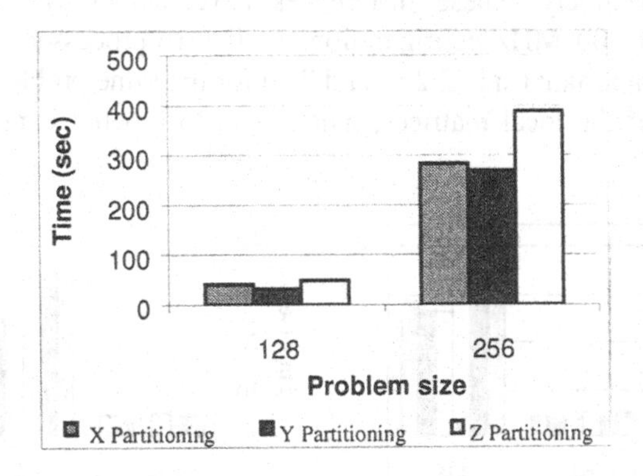

Fig. 7. Different 2D decompositions of our sample application using 16 processors in the CRAY T3E. The problem size is the number of cells in each dimension for the finest grid in the multigrid hierarchy.

4. Partitioning for Performance

Over the last decade partitioning strategies were mainly oriented towards the reduction of communications that are inherent to the parallel program. As it is well known, for a d-dimensional problem, the communication requirements for a process grow proportionally to the size of the boundaries, while computations grow proportionally to the size of its entire partition. The communication to computation ratio is thus a perimeter-to-surface area ratio in a two-dimensional problem, and similarly, a surface area to volume ratio in three-dimensions. So, the three dimensional decomposition leads to a lower inherent communication-to-computation ratio.

Moreover, as we have experimentally proved in the previous section, the time required for sending a message from one processor to another is independent of both processor locations. Therefore, there is no sense in talking about physical neighbours, and the mapping of the logical processors over the physical ones is not very important, as far as communication locality is concerned.

Therefore, these ideas suggest a general rule: Higher-dimensional decompositions tend to be more efficient than lower-dimensional decompositions [5][8].

However, as we discussed in the previous section, the communication cost is also a function of the spatial data locality. Therefore, a trade-off between the improvement of the message data locality and the efficient exploitation of the interconnection network has to be found.

The following figures compare timing obtained on the Cray T3E for different decompositions of our sample application. In the case of largest problem size, and with an 8-450 MHz processors configuration (see Fig 8, right chart) the best 1D-decomposition achieves improvements of 6.5% and 14,5% over the best 2D and 3D-decompositions respectively. These differences have grown by 2% and 10 % compared to the old 300 MHz configuration. In the 16-processor simulation, the differences are less important (only 2.2 % and 7%) for the same problem size because of the smaller size of the local matrices, which lead to a better exploitation of the temporal locality.

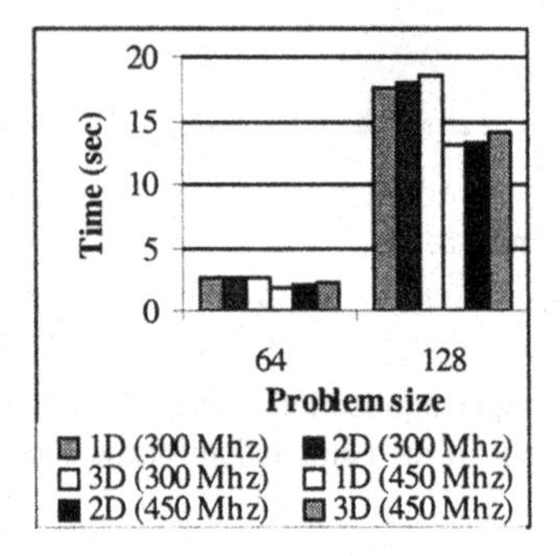

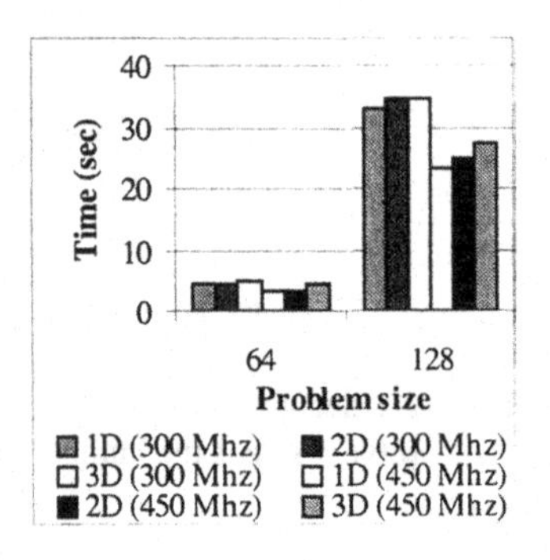

Fig. 8. Different decompositions for our sample program in the CRAY T3E using 16 (on the left) and 8 processors (on the right). The problem size is the number of cells in each dimension for the finest grid in the multigrid hierarchy.

In the SGI Origin 2000 (Fig 9), we have observed lower differences. Using 8 processors, the best choice is also a linear decomposition, but it is only 5% and 7% better than the 2D and 3D decompositions. However, for the 16-processor simulation, the 2D decomposition is 15 % and 1% better than the 1D and 3D decompositions respectively. The large second-level cache of this system is again the reason of these results. Cray T3E is more sensitive to spatial data locality than the SGI because its performance depends significantly on the effective use of the stream buffers system.

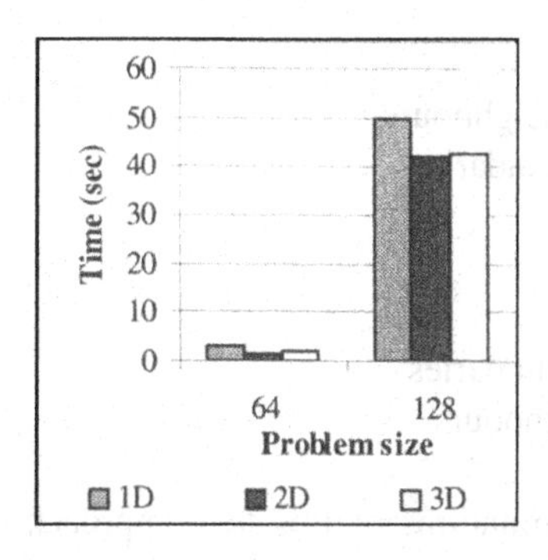

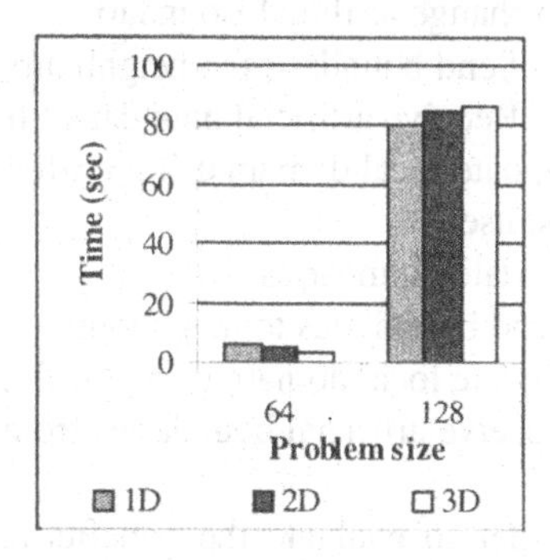

Fig. 9. Different decompositions for our sample program in the SGI Origin 2000 using 16 (on the left) and 8 processors (on the right). The problem size is the number of cells in each dimension for the finest grid in the multigrid hierarchy.

Therefore, in both multiprocessors, it is more important to efficiently exploit the local memory hierarchy than to reduce the overheads due to network delay cost. Indeed, the best performance is usually obtained by means of a simple linear decomposition.

We should also note that, although we have considered execution time as the performance metric, there are many aspects to the evaluation of a parallel program. A lower-dimensional partitioning program is easier to code, so if we consider implementation cost, a one-dimensional partitioning is also the best choice. Besides, it allows the implementation of fast sequential algorithms in the non-partitioned directions. In a workstation cluster a linear data distribution is also the best because the fewer the number of neighbours, the fewer the number of messages to be sent. Therefore, a one-dimensional decomposition reduces TCP/IP overheads as well. So, if we consider portability, a one-dimensional partitioning is also the best choice.

5. Computation – Communication overlapping

A typical approach for dealing with the communication cost due to the transit latency, the bandwidth-related cost, and contention, is to hide it by overlapping this part of the communication with other useful work. The results in the previous sections have been obtained without overlapping, but these types of algorithms can be structured so that every process request for remote data is interleaved explicitly with local computation. For this purpose, it is necessary to deal with the boundaries before than with the inner domain. In this way, it is possible to initiate an immediate send operation before the point where it would naturally appear in the program and the message may reach the receiver before it is actually needed. Thus, the receive operation does not stall waiting for the message to arrive; it will copy the data straight away from an incoming buffer into the application address space. Therefore, instead of using the simple pattern:

1- Exchange artificial Boundaries:
    Send boundaries to neighbours
    Receive artificial boundaries from neighbours
2- Update local domain using artificial boundaries

we must use:

1- Update boundaries
2- Send boundaries to neighbours
3- Update local domain using artificial boundaries
4- Receive artificial boundaries from neighbours

In order to evaluate the benefits and limitations of this new approach, we will assume that message initiation and reception costs are the same in the two structures, so the execution time can be estimated as:

$$\text{Twithout\_overlapping} = \text{Tlocal} + \text{Tcom\_overhead} + \text{Tcom}\ . \tag{1}$$

$$\text{Toverlapping} = \text{Tboundaries} + \text{Tcom\_overhead} + \max(\text{Tinner}, \text{Tcom})\ . \tag{2}$$

Tlocal is the time spent in the local domain update, Tinner is the cost of inner domain actualisation, Tboundaries is the time required for updating the boundaries, Tcom_overhead is the send and receive overheads (it is important to recall that these overheads incurred on the processors cannot be hidden) and Tcom is the network delay. For a real problem, Tcom is lower than Tinner. Therefore, the overlapping pattern is better than the simple approach while:

$$\text{Tboundaries} + \text{Tinner} < \text{Tlocal} + \text{Tcom}\ . \tag{3}$$

Tlocal can be divided into a Tinner and a Tboundaries_2, and the last inequality can then be reduced to:

$$\text{Tboundaries} - \text{Tboundaries\_2} < \text{Tcom}\ . \tag{4}$$

This latter boundary actualisation time is different from the previous one. Usually, the cost of updating the boundaries in the non-overlapping approach (which are updated together with the inner local domain) is lower than in the overlapping pattern due to the better exploitation of the memory hierarchy.

The overlapping approach has been successfully used in old parallel computers like the Parys Supernode SN 1000, where the network bandwidth-related cost is very important. In workstations clusters, the benefits are even greater because the network is usually a non-private resource. However, as we have discussed in the previous sections, in the current generation of parallel computers Tcom is not so important. Therefore, the increase due to the boundary actualisation may be greater than the reduction obtained by way of the overlapping.

We have verified these ideas with our test program. Figure 10 illustrates both patterns using a linear decomposition. In the CRAY T3E the non-overlapping approach performance is 7.3% higher than the overlap pattern for the 16-processor simulation (for the larger problem size with the 450 MHz processor) and 5% higher using 8 processors. These differences have grown compared to the old configuration where the differences are 6.4% and 4% respectively. Using 2D and 3D decompositions we observed the similar differences [16].

In the SGI, the differences are about the same. In the 32 processor-simulation, using a linear decomposition, the difference for the larger problem is 7.5 % [16].

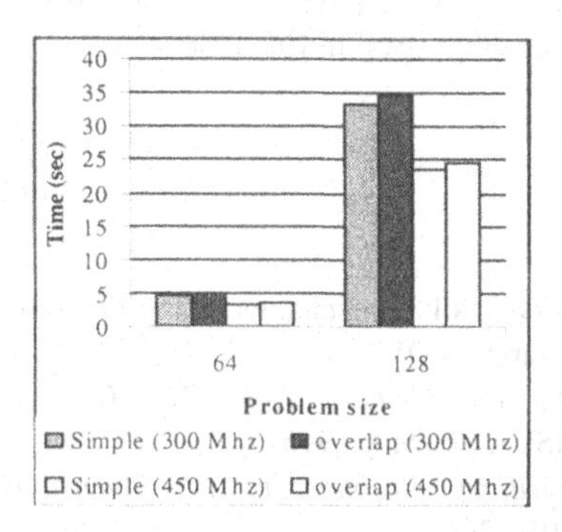

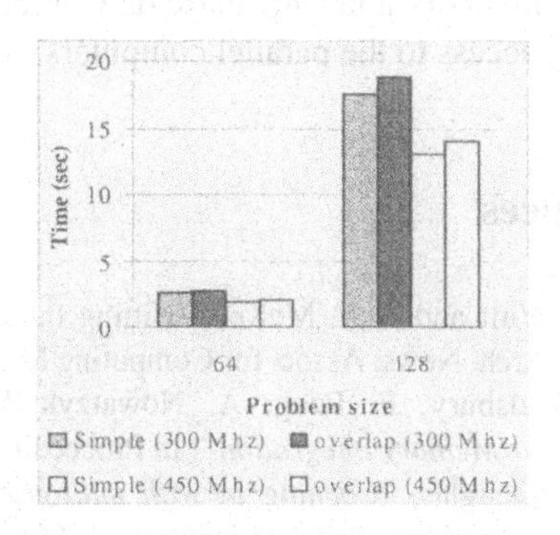

Fig. 10. Overlapping versus non-overlapping approach on the Cray T3E using 8 (on the left) and 16 processors (on the right). The problem size is the number of cells on each dimension for the finest grid in the multigrid hierarchy.

6. Conclusions

We have shown how the optimal data partitioning of regular domains is a trade off between the improvement of the message data locality and the computation/communication ratio. In older parallel computers the performance depends mainly on the efficient exploitation of the interconnection network. However, the performance obtained on current parallel computers, based on the replication of commodity microprocessors, presents a growing dependence on the efficient use of the memory hierarchy.

The main conclusions of the paper can be summarized in the following points, that contradict to a certain extent the traditional view of data partitioning: (1) the partitioning of the domain must avoid boundaries with poor data locality which may reduce the effective bandwidth, (2) 1D partitioning is becoming more efficient than higher dimension partitioning (futhermore, it is easier to code, it allows the inclusion of fast sequential algorithms in non-partitioned directions and it is more portable), and (3) communication/computation overlapping does not reduce in general the execution time. These conclusions have been verified by experimental results on two microprocessor based computers: the Cray T3E and the SGI Origin 2000.

Acknowledgements

This work has been supported by the Spanish research grants TIC 96-1071 and TIC IN96-0510, the Human Mobility Network CHRX-CT94-0459 and the Access to Large-Scale Facilities (LSF) Activity of the European Community's Training and Mobility of Researchers (TMR) Programme.

We would like to thank Ciemat, the Department of Computer Architecture at Malaga University and C4(Centre de Computatió i Comunicacions de Catalunya) for providing access to the parallel computers that have been used in this research.

References

[1] W. A Wulf and S. A. McKee, "Hitting the Memory Wall: Implications of the Obvious," Comp. Arch. News, Assoc. for Computing Mach., March, 1995.

[2] A. Saulsbury, F. Pong, A. Nowatzyk.*"Missing the Memory Wall: The Case for Processor/Memory Integration"*. In Proceeding of ISCA'96. May 1996.

[3] C. C. Douglas, *"Caching in with multigrid algorithms: Problems in two dimensions"* Parallel Algorithms and Applications, (1996), pp. 195 - 204.

[4] L. Stals and U. Rude. *"Techniques for improving the data locality of iterative methods"*. Tech. Report MRR97-038, School of Math. Sc. of the Australian National University, 1997.

[5] Ian T. Foster. "Designing and building parallel programs. Concepts and tools for parallel software engineering", Addison-Wesley Publishing Company 1995.

[6] I. M. Llorente, F. Tirado y L. Vázquez, *"Some Aspects about the Scalability of Scientific Applications on Parallel Computers"*, Parallel Computing, Vol. 22, pp. 1169-1195, 1997

[7] V. M. Pérez-García, et al. *"Low Energy Excitations of a Bose-Einstein Condensate"*, Physical Review Letters, Vol 77, pp. 5320-5323, 1996

[8] I. M. Llorente y F. Tirado, *"Relationships between Efficiency an Execution Time of Full Multigrid Methods on Parallel Computers"*, IEEE Trans. on Parallel and Distributed Systems, Vol. 8, Nº 6, 1997

[9] David Culler, Jaswinder Pal Singh, Annop Gupta. Parallel Computer Architecture. A hardware /software approach. Morgan-Kaufmann Publishers 1998.

[10] S. L. Scott. *"Synchronization and Communication in the T3E Multiprocessor"*, Proceeding of the ASPLOS VII, October 1996.

[11] E. Anderson, J. Brooks, C. Grassl, S. Scott. "Performance of the CRAY T3E Multiprocessor". In Proceeding of SC97, November 1997.

[12] Cray C/C++ Reference Manual, SR-2179 3.0.

[13] J. Laudon and D. Lenoski. *"The SGI Origin: A ccNUMA Highly Scalable Server"*. In Proceeding of ISCA'97.May 1997.

[14] H. J. Wassermann, O. M. Lubeck, F. Bassetti. "Performance Evaluation of the SGI Origin 2000: A Memory-Centric Characterization of LANL ASCI Applications". In Proceeding of the SC97, November 1997.

[15] Silicon Graphics Inc., Origin Servers, Technical Report, April 1997.

[16] M. P. Matías, D. Espadas, I. M. Llorente, F. Tirado, *"Experimental results of different partitionings of a regular domain on the Cray T3E, the SGI Origin 2000 and the IBM SP2"*, Tech. Report 98-001, Dept. of Computer Architecture at Complutense University, Madrid, Spain, 1998.

New Access Order to Reduce Inter-Vector-Conflicts

A. M. del Corral & J. M. Llaberia
Department d'Arquitectura de Computadors.
Universitat Politecnica de Catalunya. Barcelona (Spain)
anna@ac.upc.es, llaberia@ac.upc.es

Abstract. In vector processors, when several vector streams concurrently access the memory system, references of different vectors can interfere in the access to the memory modules, causing module conflicts. Besides, in a memory system where several modules are mapped in every bus, delays due to bus conflicts are added to module conflict delays. This paper proposes an access order to the vector elements that avoids conflicts when the concurrent access corresponds to vectors of a subfamily, and the request rate to the memory modules is less than or equal to the service rate. For other cases of concurrent access, the proposal dramatically reduces conflicts.

1 Introduction

In vector processors, the ideal execution of a memory vector instruction would permit to obtain a datum at every cycle after an initial latency. As, in general as the memory module reservation time is much longer than the processor cycle time, the memory system usually consists of multiple memory modules with independent access paths.

Usually, vector processors have more than one port to the memory subsystem to allow several memory vector instructions to proceed concurrently. Under these conditions, conflicts appear in the access to the memory modules when two or more references are simultaneously issued to the same module. Besides, a reference to a busy module also causes a memory module conflict.

In vector processors with several paths to the memory system, or in multi-vector processors, another factor that affects the performance of the memory system is the interconnection network between processors and memory modules. In the design of some memory systems, the decision of reducing the number of independent access paths to the memory modules (several modules are mapped on every bus) [2][6], implies a reduction in its economic cost. However, this solution implies assuming the presence of conflicts in the access to the interconnection network, as well as the memory module conflicts mentioned above. Both type of conflicts appear even in the specially common case of several one-strided vector streams concurrent access. The main effect of the conflicts is the starvation of the functional units, with the subsequent loss of performance.

Memory vector instructions with regular access patterns generate periodical conflicts as these kind of instructions generate periodical streams of references (vector streams with a constant stride). In the context of this paper, our interest is the reduction, and the elimination when possible, of the memory conflicts (interconnection and memory module conflicts) caused by concurrent constant-strided vector streams.

Several kind of methods have been proposed to reduce the number of cycles lost due to memory conflicts. Some authors propose to accurately place in memory the vectors to be concurrently accessed [10][14][17]. This technique implies that patterns must be known in compilation time, and, the access to a vector stream in different context of a program could decrease its effectiveness. Other authors propose the use of buffers in the memory modules [17] or in the interconnection network [19]. Buffers allow the requesting processor to keep sending requests without waiting, but this technique requires labelling the memory references to allow their reordering before being used by the processor; the cost of the interconnection network increases as the tag must be sent along with the request [17]. In addition, buffers do not directly solve the problem of the convergence to a single port of the requests in the return network [21].

Our proposal consists of a new access order to the vector stream elements. In parallel with our work, other authors have studied this kind of solution [15]. This new order working with a new arbitration algorithm will help concurrent vector streams perform their memory request with no conflicts or less number of conflicts than the *classical access* implies.

One of the cases for which our proposal completely avoids conflicts is the very common case of the concurrent access of several one-strided vector streams. J. Fu and J.H. Patel in [7] show that between 7% and 54% of the vector streams in four programs of the Perfect Club benchmark set [1] (ADM, ARC2D, BDNA and DYESM) access the memory with stride 1.

Section 2 outlines the architecture model, on which the present study is based, and the characterization of the interleaving mapping and vector access functions. The interaction between vector streams in a complex memory system is studied in Section 3. Section 4 presents the proposed access order to the memory modules and presents its hardware support. Finally, Section 5 deals with the comparison between the proposal and the method used in a classical system, like the CRAY X-MP.

2 Architecture

The memory architecture presented in Fig. 1 is an example of a complex memory system, similar to the one used in the CRAY X-MP [2].

The memory subsystem consists of $M = 2^m$ memory modules (memory cycle, $n_c = 2^c$ clock cycles), connected to $P = \lfloor M/n_c \rfloor$ memory ports through an interconnection network. To reduce the number of access paths to the memory subsystem the memory modules are distributed into SC sections. A memory module request occupies the section path where the module is located during one cycle. It is supposed that $SC = 2^{sc}$, and the number of memory modules is a multiple of SC.

In each cycle, every port requests an element of a vector stream except when a conflict appears in the interconnection network or in a memory module. In case of conflict, only one vector stream obtains the access and the other requests must wait; a priority rule must determine which port will be able to proceed. In the present paper, we use the arbitration implemented in the CRAY X-MP [2], to measure the performance of the *classical access* (Definition 5) and in the examples of concurrent access when another algorithm is not specified. This arbitration gives priority to the vector stream with the lower 2^s stride factor; for ports with the same stride, the priority is fixed.

The memory is organized as an interleaved address mapping model (*section* = A_i *mod SC, memory module* = A_i *mod M, offset* = $\lfloor A_i/M \rfloor$). The interleaving function which maps the address into memory modules has a period of P=M.

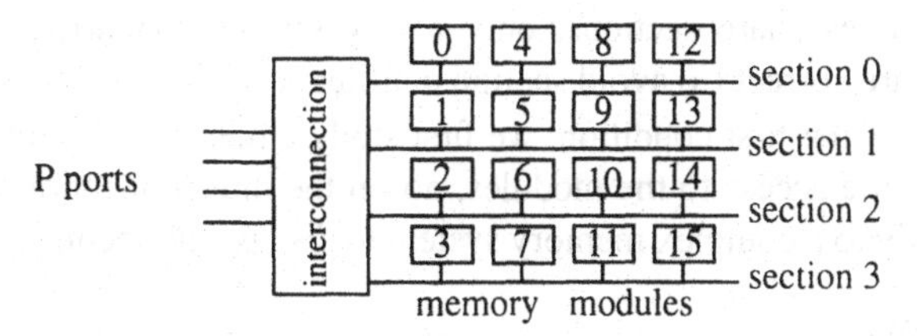

Fig. 1. Complex Memory System.

The following definitions will help the reader to follow the method.

Definition 1: A *vector stream* $A = (A_0, S, VL)$ is the set of references to memory modules $\{A_i | A_i = A_0 + i \times S, 0 \leq i < VL\}$, where A_0 is the address of the first reference, S (*stride*) is the distance between two consecutive references and VL is the vector length, or number of references. If the length is not relevant a stream is specified as $A = (A_0, S)$.

Vector streams can be classified into different families according to their stride.

Definition 2: A *stride family* (F_s) is the set of vector streams with strides $S = \sigma \times 2^s$, where σ is an odd factor [9].

A vector stream with a stride $S = \sigma \times 2^s$ references $P_s = M / gcd(M, 2^s)$ memory modules periodically, and the period is P_s.

Definition 3: The memory module set (MMS) of the vector stream $A = (A_0, S)$ is the set of all the memory modules accessed by the vector stream $A=(A_0,S,P_s)$. $MMS = \{m_i | m_i=(A_0+i \times S) mod M, 0 \leq i < P_s\}$.

Definition 4: A *stride subfamily* ($SF_s^{m_0}$) is the set of vector streams of a family that reference the same set of memory modules.

To give some examples, the family F_0 (odd-strided vector streams) only has one subfamily SF_0^0, and the family F_1 (even-strided vector streams) has two subfamilies, SF_1^0 references the even memory modules, and SF_1^1 references the odd modules.

Definition 5: *Classical access* is the access order that uses the recurrence $A_{i+1} = A_i + S$ (S=*Stride*) to compute vector stream addresses.

Since the vector length is usually greater than the vector register length, the compiler is required to transform the code using strip-mining. Under this condition, a great proportion of memory accesses from vector streams are issued by vector instructions *load* and *store*, which are of a fixed length equal to the vector register length. Let us assume that, in order to simplify the explanation of the proposed method, the vector stream length ($VL = 2^{vl}$) is a multiple of the vector register length $MVL = 2^{mvl}$ which is assumed to be a multiple of the number of memory modules $M = 2^m$.

3 Characterization of the Conflicts

Only in the case that the memory request rate imposed by concurrent vector streams is equal to or less than the memory module response rate, the concurrent access can be conflict-free. When the request rate is equal to the response rate, it is said that the memory system (or similarly, the memory modules) works tight, and when the request rate is less than the response rate, the memory system works loose.

To obtain a conflict-free access, not only the system must work loose or tight, in addition, the concurrent access of the vector streams must fulfil two conditions:

- consecutive references to a memory module must be distanced at least n_c cycles (to avoid memory module conflicts).

- since memory modules share sections, only a few sets of concurrent memory module references are correct (to avoid section conflicts).

To analyse the effect of the first condition, we first study a memory system that can only present conflicts in the access to the modules, not in the interconnection network. Then, we extend the study to a complex memory system to discuss the second condition.

Simple Memory System

A simple memory system has an independent access path from every port to every memory module, thereby its interconnection network does not present conflicts. In a system like that, the concurrent *classical access* of vector streams that have the same stride has a conflict-free steady state when the request rate they imply is less than or equal to memory modules response rate (the system works loose or tight) [16][17].

Fig. 2 presents the concurrent *classical access* of four one-strided vector streams in a memory system with 16 memory modules and an n_c of 4 cycles. Vector streams start their concurrent access in different memory modules. In the figure, it is possible to observe for every cycle the memory module that begins to be occupied by every vector stream (the module remains occupied during latency cycles). A delay due to a memory module conflict is depicted in black.

	Cycles	**0**	**1**	**2**	**3**	**4**	**5**	**6**	**7**	**8**	**9**	**10**	**11**	**12**	**13**	**14**	**15**	**16**
Modules	***A***	0				1	2	3		4	5	6	7	8	9	10	11	12
	B	1	2	3		4	5	6	7	8	9	10	11	12	13	14	15	0
	C	4	5	6	7	8	9	10	11	12	13	14	15	0	1	2	3	4
	D	8	9	10	11	12	13	14	15	0	1	2	3	4	5	6	7	8

Fig. 2. 16-way interleaved memory with $n_c = 4$. Conflicts with the *classical access*.

This concurrent access presents conflicts at the very beginning, but the steady state, that starts at cycle 8, is conflict-free. At the steady state, four sets of concurrent memory module references ({0, 4, 8, 12}, {1, 5, 9, 13}, {2, 6, 10, 14} and {3, 7, 11, 15}) are periodically repeated every n_c cycles, thereby, consecutive references to the same memory module are distanced n_c cycles. The periodicity of these four sets (called *CMR -Concurrent memory Module References-* from now on) can be guaranteed because vector streams reference the memory modules with the same order.

R. Raghavan and J.P. Hayes stated with theorem 6 of [17] the conditions the concurrent vector streams must fulfil to obtain a conflict-free *classical access* in a simple memory system. These conditions can be fulfilled only by vector streams that belong to the same subfamily. All the combinations of vector streams that have the same stride have a conflict-free access whenever the system works loose or tight. The concurrent *classical access* of vector streams of different subfamilies is always conflictive (corollary 3 of [4]).

Complex Memory System

Combinations of vector streams that obtain a conflict-free access in a simple memory system, may not have a good behaviour in a complex memory system. The sets *CMR* that are suitable in a simple memory system may not be appropriated in a system where several memory modules are mapped in the same section. As an example, none of the *CMR* of the concurrent access of Fig. 2 are appropriated in a complex memory system where the 16 memory modules are interleavedly mapped in 4 sections (Fig. 1): all the memory modules of every *CMR* are mapped in the same section, then, they can not be concurrently accessed.

Fig. 3 shows the conflictive *classical access* of four one-strided vector streams in the system of Fig. 1. The delay due to a section conflict is represented in light grey, and a memory module conflict is depicted in black; a section is locked during one cycle in the access to a memory module. In this concurrent access, conflicts are linked and periodically repeated: a section conflict causes a memory module conflict which also causes a section conflict, and so on.

Cycles		0	1	2	3	4	5	6	7	8	9	10	11	12	13	14	15	16
Sections	*A*	0	1	2	3	0	0	1	2	3	0	0	1	2	3	0	0	1
	B		0	1	2	3		0	1	2	3		0	1	2	3		0
	C			0	1	2	3		0	1	2	3		0	1	2	3	
	D				0	1	2	3		0	1	2	3		0	1	2	3
Modules	*A*	**0**	**1**	**2**	**3**		**4**	**5**	**6**	**7**		**8**	**9**	**10**	**11**		**12**	**13**
	B		**4**	**5**	**6**	**7**		**8**	**9**	**10**	**11**		**12**	**13**	**14**	**15**		**0**
	C			**8**	**9**	**10**	**11**		**12**	**13**	**14**	**15**		**0**	**1**	**2**	**3**	
	D				**12**	**13**	**14**	**15**		**0**	**1**	**2**	**3**		**4**	**5**	**6**	**7**

Fig. 3. 16-way interleaved memory system with $n_c = 4$ and SC=4. Conflicts with the *classical access*.

T. Cheung and J.E. Smith characterize in [2] the linked conflicts that appear in the concurrent *classical access* of two one-strided vector streams and use the term complex linked conflict (complex conflict) when three or more vector streams interfere with each other in a less precise way. Authors prove that the steady-state linked conflicts and complex conflicts reduce the effective bandwidth.

Authors of [2] show that in the concurrent *classical access* of three one-strided vector streams (the system works loose), in 34% of the cases (combinations of initial memory modules) linked conflicts appear, in 7% of the cases complex conflicts are generated, and performance can be degraded by 20%.

To solve these conflicts, W. Oed and O. Lange conclude in [16] that n_c and SC must be coprime (theorem 9). A solution with a prime SC is proposed in [15]. In [2], authors give some alternatives to avoid linked conflicts, i.e. a solution with odd values of n_c. For all the proposals, if vector streams have different strides conflicts persist and, in any case, complex conflicts do not disappear.

Tab. 1 shows the asymptotic number of operations per cycle[1] (R_∞) the *classical access* obtains in average for four types of combinations of vector streams, in a simple memory system (M=16 and n_c=4) and in the corresponding complex system (M=16, n_c=4 and SC=4). The concurrent accesses simulated are all the combinations of four, three and two odd strided vector streams, two odd strided with one even strided vector streams, and two even strided vector streams. For the simple memory system, the average R_∞ for the *classical access* is far away from the ideal, even for combinations for

1. $R_\infty = ops \times r_\infty \times t_c$, where t_c is the processor cycle time, ops is the number of concurrent vector streams, and r_∞ is the asymptotic performance [12].

which the system works very loose and vector streams belong to the same subfamily. Comparing the results for both memory systems, it can be easily concluded that in a complex memory system, the results are worst because of interferences in the interconnection network.

Tab. 1. R_∞ for the *classical access* and Ideal.

Combinations of Strides		*Complex Mem. Syst.* $M=16$ $n_c=4$ $SC=4$		*Simple Mem. Syst.* $M=16$ $n_c=4$	
Odd	*Even*	R_∞ *Classical*	R_∞ *Ideal*	R_∞ *Classical*	R_∞ *Ideal*
4	0	1.57	4	1.86	4
3	0	1.51	3	1.66	3
2	0	1.35	2	1.38	2
2	1	1.39	3	1.60	3
0	2	1.05	2	1.27	2

The next section presents an access method that completely avoids conflicts in the concurrent access of vector streams of the same subfamily when the system works loose or tight. This method also dramatically reduces conflicts for other cases of concurrent access. The name of the proposal is *Skewed Sequence of memory Modules* (*SSM*).

4 Proposal *SSM*

To reduce the number of memory module conflicts, we propose that concurrent vector streams reference the memory modules with the same order. All the vector streams of a subfamily reference the same set of P_s memory modules ($P_s = M/\gcd(M,2^s)$), but with the *classical access*, the order every vector uses to access them depends on the σ-factor of the stride. We propose to construct a σ-independent access order, then all the vector streams of a subfamily will reference the P_s modules with the same order.

To avoid section conflicts, this σ-independent access order must be constructed considering that the resulting *CMR* sets must comprise memory modules mapped in different sections.

This new sequence of memory modules will be called *SSM* (*Skewed Sequence of memory Modules*). Fig. 4 shows the *SSM* proposed for different subfamilies in a memory system that has $M=16$, $n_c=4$ and $SC=4$ (Fig. 1). For every sequence *SSM* it is also shown the sequence of sections referenced and the corresponding *CMR*.

Subfamily SF^0_0 ***(odd strides, all modules)*** - *CMR = {{0,7,10,13}, {1,4,11,14}, {2,5,8,15}, {3,6,9,12}}*

SSM	0	1	2	3	7	4	5	6	10	11	8	9	13	14	15	12
sections	0	1	2	3	3	0	1	2	2	3	0	1	1	2	3	0
	subperiod 0				***subperiod 1***				***subperiod 2***				***subperiod 3***			

Subfamily SF^0_1 ***(even strides, even modules)***

CMR = {{0,10}, {2,8}, {4,14}, {6,12}}

SSM	0	2	4	6	10	8	14	12
sections	0	2	0	2	2	0	2	0
	subperiod 0				***subperiod 1***			

Subfamily SF^0_1 ***(even strides, odd modules)***

CMR = {{1,11}, {5,15}, {7,13}, {3,9}}

SSM	1	3	7	5	11	9	13	15
sections	1	3	3	1	3	1	1	
	subperiod 0				***subperiod 1***			

Fig. 4. 16-way interleaved memory with $n_c = 4$ and SC=4. *SSM* for several subfamilies.

Each one of the *SSM* we propose has n_c *CMR* sets of P_s/n_c memory modules. In consequence, P_s/n_c concurrent vector streams of a subfamily can concurrently reference memory modules of different sections, avoiding section conflicts. Besides, module conflicts are also avoided as consecutive references to a *CMR* are distanced n_c cycles.

Fig. 5 shows the conflict-free access of four odd-strided vector streams in the system of Fig. 1, when the corresponding *SSM* is used. This *SSM* has n_c=4 *CMR* sets with P_s/n_c =16/4 modules, so four odd-strided vector streams could have a conflict-free access.

Cycles		0	1	2	3	4	5	6	7	8	9	10	11	12	13	14	15	16
Sections	*A*	1	2	3	0	0	1	2	3	3	0	1	2	2	3	0	1	1
	B	2	3	0	1	1	2	3	0	0	1	2	3	3	0	1	2	2
	C	3	0	1	2	2	3	0	1	1	2	3	0	0	1	2	3	3
	D	0	1	2	3	3	0	1	2	2	3	0	1	1	2	3	0	0
Modules	*A*	**13**	**14**	**15**	**12**	**0**	**1**	**2**	**3**	**7**	**4**	**5**	**6**	**10**	**11**	**8**	**9**	**13**
	B	**10**	**11**	**8**	**9**	**13**	**14**	**15**	**12**	**0**	**1**	**2**	**3**	**7**	**4**	**5**	**6**	**10**
	C	**7**	**4**	**5**	**6**	**10**	**11**	**8**	**9**	**13**	**14**	**15**	**12**	**0**	**1**	**2**	**3**	**7**
	D	**0**	**1**	**2**	**3**	**7**	**4**	**5**	**6**	**10**	**11**	**8**	**9**	**13**	**14**	**15**	**12**	**0**

Fig. 5. 16-way interleaved memory system with n_c = 4 and *SC*=4. Conflict-free concurrent access of four odd-strided vector streams using *SSM*.

Vector streams of Fig. 5 start their concurrent access in correct memory modules (same *CMR*), so the concurrent access synchronizes from the beginning. When the start addresses do not correspond to a *CMR*, an arbitration algorithm is necessary. Section 4.2 presents a dynamic arbitration that forces vector streams to concurrently access memory modules of the appropriate *CMR* [3].

4.1 *Skewed Sequence of memory Modules - SSM*

The new sequence of memory modules is called "*Skewed*" as the *SSM* we define for every subfamily is the result of applying a skew function to the subfamily *MMS* lexicographically ordered.

Definition 6: For a vector stream $A = (A0, S, P_s)$, of the subfamily $SF_s^{M_0}$, ($M_0 = A_0 \bmod \gcd(M,2^s)$), we call *Skewed Sequence of memory Modules* (*SSM*) to the sequence determined by the expression:

$$k = f(m_i) = ((m_i+\lfloor m_i/n_c \rfloor) \bmod n_c + \lfloor m_i/n_c \rfloor \times n_c)/\gcd(M,2^s),$$

where k is the position that the memory module m_i ($0 \le m_i < M$) occupies in the sequence and m_i belongs to the vector stream *MMS*.

The function $f'(m_i)$, that gives the memory module from a position in the sequence (reverse function of $f(m_i)$), will permit to generate the *SSM* sequence. We express $f'(m_i)$ as an algorithm, but before presenting it, we will make some considerations (Fig. 6 helps to follow the explanation):

- The first module a vector references with the *SSM* is $M_0 = A_0 \bmod \gcd(M,2^s)$.
- Every set of $\lceil n_c/\gcd(M,2^s) \rceil$ consecutive memory modules of the *MMS* lexicographically ordered suffers a skew. We call *GS* to every one of these sets, and in a *SSM* there are $(M/n_c)/\lceil \gcd(M,2^s)/n_c \rceil$ *GS* sets.

- The same skew is applied to gcd$(M,2^s)$ consecutive GS sets, but the first skew is applied to at most gcd$(M,2^s)$ consecutive GS. If M_0 is not the memory module 0, only the gcd$(M,2^s)$-M_0 first GS sets suffer the same skew.

To give an example of the former considerations, ina system with M=16, $n_c = 4$ and SC=4, the SSM of the subfamily SF_1^1 has $(M/n_c)/\lceil \text{gcd}(M,2^s)/n_c \rceil = 4$ GS sets. The first skew is applied only to gcd$(M,2^s)$-M_0=1 GS as M_0 is the memory module 1, but the second skew is applied to gcd$(M,2^s)$=2 consecutive GS.

Subfamily SF_0 **(odd strides, all modules)**

	skew 0				skew 1				skew 2				skew 3			
MMS	0	1	2	3	4	5	6	7	8	9	10	11	12	13	14	15
SSM	0	1	2	3	7	4	5	6	10	11	8	9	13	14	15	12
sections	0	1	2	3	3	0	1	2	2	3	0	1	1	2	3	0

Subfamily SF_1^0 **(even strides, even modules)**

	skew 0				skew 1			
SSM	0	2	4	6	10	8	14	12
sections	0	2	0	2	2	0	2	0

Subfamily SF_1^1 **(even strides, odd modules)**

	skew 0		skew 1				skew 2	
SSM	1	3	7	5	11	9	13	15
sections	1	3	3	1	3	1	1	3

Fig. 6. 16-way interleaved memory system with $n_c = 4$ and SC=4. SSM construction for several subfamilies.

The algorithm used to generate the SSM sequence for any subfamily is:

```
M0 = A0 mod gcd(M,2^s)
control = M0
skew = 0
for NGS = M0 / nc to M/nc -1 step ⌈gcd(M,2^s)/nc⌉
        for I = M0 mod nc to nc-1 step gcd(M,2^s)
                module = ((I - skew × gcd(M,2^s)) mod nc + NGS × nc) mod M
        endfor
        control = (control + ⌈gcd(M,2^s)/nc⌉) mod gcd(M,2^s)
        if ( control = 0 )  then skew = skew + 1
endfor
```

In the algorithm, NGS controls the generation of the memory module references for every GS set. The variable I controls the generation of the memory module references within a GS set. Control controls the skew changes after the generation of gcd$(M,2^s)$ consecutive GS. If M_0 is not the memory module 0, only the gcd$(M,2^s)$-M_0 first sets GS suffer the same skew.

4.2 Arbitration algorithm

An arbitration algorithm is needed in order to synchronize vector streams to reach a conflict-free steady-state phase, or to dramatically reduce inter-conflicts, for any combination of initial memory modules.

The SSM sequences can be divided in P_s/n_c *subperiods* of n_c memory modules. In Fig. 4, we can observe that each *subperiod* of a SSM references the sections following a predetermined order which is different for every *subperiod*. Thus, in the concurrent access of P_s/n_c vector streams of a subfamily, we obtain a conflict-free access if we overlap different *subperiods* (different sections are simultaneously referenced as in the example of Fig. 5 with family F_0). The main idea is that, in every cycle concurrent vector streams reference memory modules of a different *subperiod*, and these different *subperiods* must be aligned.

The arbitration algorithm controls the *subperiod* changes between vector streams; when all *subperiod* changes have been detected for all the vector streams, *subperiods* are assigned using a fixed priority. The *subperiod* change is detected by computing the expression $subperiod = \lfloor m_i / (n_c \times \gcd(M, 2^s)) \rfloor mod SC$ ($m_i = A_i^{SSR} mod M$) for two consecutive memory module requests of a vector stream (the current and the previous).

4.3 *Skew Sequence of memory References - SSR*

The *SSM* is the order in which memory modules must be referenced periodically, then vector stream memory references must be generated to periodically access the modules with this new order.

Definition 7: For a vector stream $A = (A0, S)$ of the subfamily $SF_s^{M_0}$ ($M_0 = A_0$ mod gcd$(M,2^s)$), the *Skewed Sequence of References* (*SSR*) is the sequence of memory references that permits to reference the memory modules following the *SSM* periodically.

The algorithm that generates *SSR* is a modification of the algorithm that generates *SSM*. The following definition will help designing the algorithm.

Definition 8: The *order number* (*ON*) of a vector stream element, is the position on which its address is generated using the *classical access*, $0 \leq ON < VL$.

The address of an element of a vector stream $A = (A_0, S)$, can be computed using its order number as $Addr = A_0 + ON \times S$. With the *classical access*, addresses of elements with consecutive *ON* are consecutively generated ($ON_{i+1} = ON_i + 1$). This is not the case with the *SSR*, but, if we know how to generate the sequence of order numbers that fulfil *SSM*, we will be able to generate *SSR*.

First, we suppose *SSR* is P_s references long, then we extend the study to any length.

P_s references long (one Period)

Vector elements placed in memory modules adjacent in the *MMS* lexicographically ordered have *order numbers* separated by a constant distance, C_s [3]. Then, we can compute the *ON* of a vector element from the *ON* of any other vector element if we know the distance between the memory modules[1] where they both are placed: $ON_j = ON_i + K \times C_s$, where *K* is the distance.

To compute the sequence of *order numbers* the *SSM* implies, the *K* we can use can be the distance between the memory module to be referenced and the first memory module referenced using the *SSM* that is $M_0 = A_0$ mod gcd$(M,2^s)$. Then, we must use the *order number* of the first vector stream element referenced using the *SSR*, *NO0*, that can be easily computed. In this case, the order number of a vector element placed in the memory module m_j is:

$$ON_j = NO0 + ((m_j - M_0) mod\ M / gcd(M,2^s)) \times C_s.$$

Any Length (any number of Periods)

As the distance between memory modules can be computed within a period, the former recurrence actually gives the *order number* relative to a period (*ONR*). To extend the

---

1. distance is the number of memory modules between them in the *MMS* lexicographically ordered.

computation of the *order number* to any number of periods, we can consider that every period has a *base order number* (*BN*), to be added to the *ONR* to obtain the *ON*. From period to period this *BN* must be increased in P_s units.

The next algorithm is based in the algorithm proposed in Section 4.1, adding the computation of the *order number* and the loop that controls the period. The bold lines are the ones added.

```
M0 = A0 mod gcd(M,2^s)
BN = 0
for Q = 0 to ⌈VL/Ps⌉-1
        control = M0
        skew = 0
        for NGS = M0/ nc to M/nc -1 step ⌈gcd(M,2^s)/nc⌉
                for I = M0 mod nc to nc-1 step gcd(M,2^s)
                        module=((I-skew×gcd(M,2^s))mod nc+NGS×nc) mod M
                        K = ((module - M0) mod M)/gcd(M,2^s)
                        ONR = (ON0 + K × Cs) mod Ps
                        Addr^SSR= A0 + (BN + ONR) × S
                endfor
                control = (control +⌈gcd(M,2^s)/nc⌉) mod gcd(M,2^s)
                if ( control = 0 ) then skew = skew + 1
        endfor
        BN = BN + Ps
endfor
```

As a synopsis, the recurrences that compute the vector memory references are:

$$A_i^{SSR} = A_0 + \text{Base\_Addr} + A_i^r \text{ and } A_i^r = (A_j^r + K \times C_s \times S) \bmod (P_s \times S)$$

where A_i^r is the vector element address relative to a period, A_i^{SSR} is its absolute address, K is the distance between memory modules where A_i^r and A_j^r are placed, and Base_Addr is the base address of a period *($BN \times S$)*.

4.4 Hardware Support to Reduce Conflicts

To design the hardware that computes the *SSR*, we must rewrite the algorithm to make it easier to implement. There are two issues that must be solved: the presence of a multiplier and a modulo operation in the critical path of the address computation (every iteration).

To avoid the use of a multiplier, the relative addresses A_i^r are computed using the relative addresses A_j^r, so only two precomputed products $K \times C_s \times S \bmod (P_s \times S)$ must be used (K=1 and K=n_c). This implies using three registers to store different previous values A_j^r.

The modulo operation (mod ($P_s \times S$)) can be performed by subtracting $P_s \times S$ if necessary, as demonstrated in [5]. In fact, the two values, $A_j^r + K \times C_s \times S$ and $A_j^r + K \times C_s \times S - P_s \times S$, are computed in parallel, and the selection between them is performed by a signal that comes from the vector register index computation [5]. This signal indicates if $ON_j^r + K \times C_s \geq P_s$, easier to compute as P_s is a power of two number.

Fig. 7 shows a hardware design of the data-path. The hardware cost is moderate, two adders in the critical path and a CSA, and it is not more complex than that needed by other solutions [8][18] proposed to reduce the average memory latency time in vector processors.

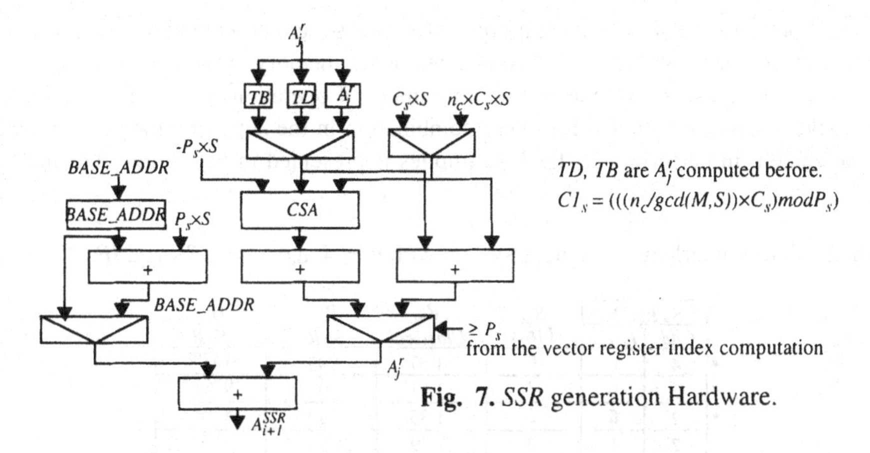

Fig. 7. *SSR* generation Hardware.

The rate at which a memory request can be issued is limited by the rate at which additions can be performed. The design can be pipelined to obtain a reduction of the cycle time (this would be also needed in the *classical access*). The additional hardware introduces a initial delay of a few cycles in the memory path. The number of clock cycles needed to access the memory is of the order of *14 + MVL* for the CRAY X-MP, *17 + MVL* for the CRAY Y-MP and *23 + MVL* for the C90 [20]. However, as the processor speed continues to increase faster than the memory speed, an extra initial delay of some cycles introduced by the hardware proposed is acceptable.

The number of parameters to be calculated is comparable to the number needed for other proposals [8][18][22], and most of them can be determined by the compiler.

The hardware needed to access the vector registers is similar to the hardware shown at Fig. 7 but simpler.

The cost of the hardware components can be considered a minor part of the cost of the memory subsystem. Additionally, in contrast with other solutions, which include a significant number of buffers to eliminate the effect of unsuitable temporal distributions [8][18], this proposal does not need buffers.

5 New method performance

In this section we present the advantages of the method proposed in this paper. Tab. 2 shows the comparison between the *SSM* and the *classical access* in a memory system with $M=16$ memory modules, interleavedly mapped in $SC=4$ sections, with an $n_c=4$.

Some considerations about the simulations:

a) We obtain the value R_∞ for the concurrent access, using the *classical access* and the proposal, of all the possible combinations of four, three or two vector streams of the families F_0 and SF_1^0.

b) All the combinations of vector streams whose concurrent access has been simulated have a non void intersection of *MMS* sets.

c) The parameter we use to perform the comparison is the increment in performance (IR_∞) implied by the proposal, and it is computed as $IR_\infty = ((R_{\infty SSM} - R_{\infty classical})/R_{\infty classical}) \times 100$[11].

d) The results presented under the name R_∞ are harmonic means of the asymptotic number of operations per cycle that the *classical access* and the *SSM* obtain for combinations of vector streams we group in types.

Tab. 2 presents the R_∞ for several types of vector stream combinations the *classical access* and the *SSM* obtain in a 16-way interleaved memory system with $n_c = 4$ and *SC*=4. The table also shows the maximum number of operations per cycle (IR_∞ *Ideal*) that could be ideally obtained for every combination in the supposed memory system. The increment in performance the *SSM* implies is presented in the column labelled as IR_∞.

Tab. 2. 16-way interleaved memory system with $n_c = 4$ and *SC*=4. R_∞ and IR_∞ for *SSM*.

	STRIDE		R_∞	R_∞	R_∞	IR_∞
	Odd	Even	Ideal	Classical	SSR	SSR
*	4	0	4	1.57	3.95	152%
*	3	0	3	1.51	2.98	97%
*	2	0	2	1.35	1.99	47%
	2	1	3	1.39	1.99	43%
	1	1	2	1.18	1.33	13%
	1	2	2.4	1.19	1.99	67%
	2	2	2.67	1.34	2.65	98%
*	0	2	2	1.05	1.99	90%
	0	3	2	1.03	1.50	46%
	0	4	2	1.04	1.99	91%

In the table, the types with an asterisk ('*') correspond to combinations of vector streams of the same subfamily that make the system work loose or tight. For these types the R_∞ the *SSM* obtains is almost R_∞ *Ideal*, and the IR_∞ is very important, between 47% and 152%. For the other types, the IR_∞ is also important, between 13% and 98%.

Fig. 8.a presents the IR_∞ the use of the *SSM* implies in function of the σ–factor of the stride, in the concurrent access of: four vector streams of the family F_0 (dark bars), four vector streams of the subfamily SF_1^0 (medium grey bars), and two vector streams of F_0 with two vector streams of SF_1^0 (light bars). For every case we grouped combinations that have four (bars labelled as "four"), three, two o zero (bars labelled as "zero") vector stream with the same σ-factor.

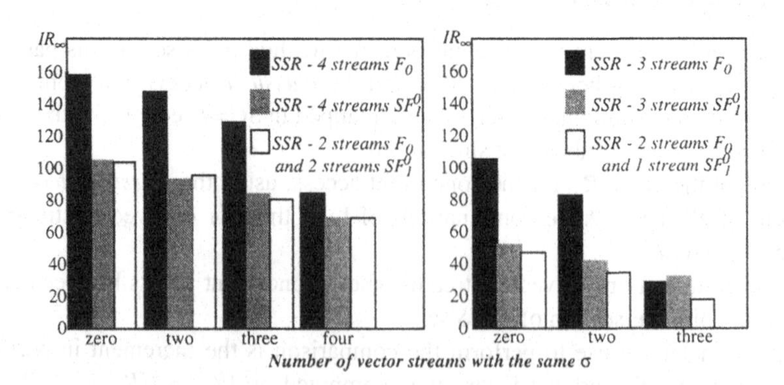

Fig. 8. 16-way interleaved memory system with $n_c = 4$ and *SC*=4. IR_∞ for *SSM*, in the concurrent access of four (a) or three vector streams (b).

When four F_0 vector streams (odd-strides, dark bars) access the memory system, the memory works tight, but the concurrent access with the *SSM* is conflict-free and IR_∞ is substantial, between 85% and 159%. Even when all the concurrent vector streams have the same σ-factor (same stride), *SSM* overworks the *classical access*, as this access does not avoids section conflicts.

For combinations of four SF_1^0 vector streams (even-strides, medium grey bars) the concurrent access with the *SSM* is not conflict-free as there are more than P_s/n_c (=8/4=2) concurrent vector streams, but the IR_∞ is important, between 69% and 105%.

When in the concurrent access there are two F_0 vector streams and two SF_1^0 vector streams the concurrent access with the *SSM* is not conflict-free as there are vector streams of different subfamilies but the IR_∞ is important, it ranges from 69% and 104%

Fig. 8.b presents the IR_∞ the *SSM* represents in function of the σ–factor, in the concurrent access of: three vector streams of the family F_0 (dark bars), three vector streams of the subfamily SF_1^0 (medium grey bars), and two vector streams of F_0 with one vector stream of SF_1^0 (light bars). For every case we grouped combinations that have three (bars labelled as "three"), two or zero (bars labelled as "zero") vector stream with the same σ-factor in the stride. For these cases, the IR_∞ the *SSM* obtains is lower than in the case of four vector streams, as the *classical access* finds the system working looser and, in consequence, there are less conflicts or they have less effect.

Vectors and matrices are the most common data structures in vector processors. In Fortran, the most frequent accesses to matrices are made by columns, rows and diagonals, that correspond to the strides 1, n and $n+1$ respectively, where n is the column length, which is dependent on the problem size that varies widely. Present compilation technology detects if n is even, then the matrix size can be increased in one row (odd stride), and the number of referenced memory modules is M. Thus, in row-major and column-major accesses the use of *SSM* performs equally well, and there are no conflicts. When n is even and there is no possibility of increasing the number of rows, the *SSM* reduces the number of conflicts.

6 Conclusions

The interferences between concurrent vector streams accessing the memory system of a vector or multivector processor cause conflicts in the memory that reduce the processor efficiency.

The present paper has proposed a σ-independent access order to the vector stream elements (*SSM*), for which all the vector streams of a subfamily reference the memory modules with the same order. The use of the *SSM* associated with the proposed arbitration algorithm, avoids conflicts when the concurrent access correspond to vector streams of the same subfamily and the system works loose or tight. The proposal significantly reduces conflicts for other types of concurrent accesses.

The hardware solution that generates the *SSM* and the hardware used to access the vector registers have a moderate cost.

The simulations confirmed that the proposal can achieve the maximum number of operations per cycle, and the results showed that the *SSM* always outperforms the *classical access*, with performance increments between 13% and 152% for combinations of even and odd strided vector streams. In the interesting case of the concurrent access of 4 one-strided vector streams the increment in performance is 85%.

Acknowledgments
Work supported by the Ministry of Education of Spain, contract TIC-95-429.

References

1. M. Berry et al. Perfect Club Benchmarks: Effective Performance Evaluation of Supercomputers. Int. Journal for Supercomputer Applications. 1989.
2. T. Cheung and J.E. Smith.. A Simulation Study of the CRAY X-MP Memory System. IEEE Transactions on Computers, Vol. C-35, no 7, october 1980.
3. A.M. del Corral and J.M. Llaberia. Reduce Conflicts between Vector Streams in Complex Memory Systems. CEPBA Report. DAC-UPC Report. June 1994.
4. A.M. del Corral and J.M. Llaberia. Eliminating Conflicts between Vector Streams in Interleaved Memory Systems. CEPBA Report. DAC-UPC Report. August 1995.
5. A. M. del Corral and J. M. Llaberia. Avoiding Inter-Vector-Conflicts in Complex Memory Systems. CEPBA Report, DAC-UPC Report. March 1996.
6. U. Detert and G. Hofemann. CRAY X-MP and Y-MP memory performance. Parallel Computing, North-Holland, n0 17, 1991.
7. J.W.C. FU and J.H.Patel. Memory Reference Behavior of Compiler Optimized Programs on High Speed Architectures. International Conference on Parallel Processing, Vol II. 1993.
8. D.T. Harper III and J.R. Jump. Vector Access Performance in Parallel Memories Using a Skewed Storage Scheme.IEEE Transactions on Computers, Vol. C-36, no 12, december 1987.
9. D.T. Harper III and D.A. Linebarger. Conflict-free Vector Access Using a Dynamic Storage Scheme. IEEE Transactions on Computers, Vol. C-40, no 3, march 1991.
10. D.T. Harper III and J.R. Jump. Vector Access Performance in Parallel Memories Using a Skewed Storage Scheme. IEEE Transactions on Computers, Vol. C-36, no 12, december 1987.
11. J.L. Hennessy and D.A. Patterson. Computer Architecture. A Quantitative Approach. Morgan Kaufmann Publishers, inc. 1990.
12. R.W. Hockney and C.R. Jesshope. *Parallel Computers* 2, Adam Hilger. 1988.
13. K. Kitai, T. Isobre, T. Sakakibara, S. Yazawa, Y. Tamaki, T. Tanaka and K. Ishii. Distributed Storage Control Unit for the Hitachi S-3800 Multivector Supercomputer. Int. Conference on Supercomputing, 1994.
14. L. Kurian, B. Choi, P.T. Hulina and L.D. Coroor. Module Partitioning and Interleaved Data Placement Schemes to Reduce Conflicts in Interleaved Memories. International Conference on Parallel Processing, Vol. 1. 1994.
15. D.L. Lee. Prime-way Interleaved Memory. International Conference on Parallel Processing, Vol. I. 1993.
16. W. Oed and O. Lange. On the Effective Bandwidth of Interleaved Memories in Vector Processor Systems. IEEE Transactions on Computers, Vol. C-34, *no 10*. October 1985.
17. R. Raghavan and J. Hayes. Reducing Interference Among Vector Accesses in Interleaved Memories. IEEE Transactions on Computers, Vol. 42, n.4. April 1993.
18. R.Raghavan and J.P.Hayes. On Randomly Interleaved Memories. Proceedings of the Supercomputing'90. November 1990.
19. J.E. Smith y W.R. Taylor. Accurate Modelling of Interconnection Networks. Int. Conference on Supercomputing, pp. 264 - 273. 1991.
20. J.E. Smith, W.C. Hsu and C. Hsiung. Future General Purpose Supercomputer Architectures. Proc. Supercomputing'90. 1990.
21. J. Torrellas and Z. Zhang. The Performance of Cedar Multistage Switching Network. Proceeding of the Supercomputing'94. November 1994.
22. M. Valero, T. Lang, J.M. Llaberia, M. Peiron, E. Ayguade y J.J. Navarro. Increasing the Number of Strides for Conflict-free Vector Access. Int. Symp. on Comp. Architecture. 1992.

Registers Size Influence on Vector Architectures

Luis Villa*, Roger Espasa**, and Mateo Valero

Departament d'Arquitectura de Computadors,
Universitat Politècnica de Catalunya–Barcelona, Spain
{luisv,roger,mateo}@ac.upc.es
http://www.ac.upc.es/hpc

Abstract. In this work we studied the influence of the vector register size over two different concepts of vector architectures. Long vector registers play an important role in a conventional vector architecture, however, even using highly vectorisable codes, only a small fraction of that large vector registers is used. Reducing vector register size on a conventional vector architecture results in a severe performance degradation, providing slowdowns in the range of 1.8 to 3.8. When we included an out-of-order execution on a vector architecture, the need for long vector registers was reduced. We used a trace driven approach to simulate a selection of the Perfect Club and Specfp92 programs. The results of the simulations showed that the reduction of the register size on an out-of-order vector architecture led to slowdowns in the range of 1.04 to 1.9. These compare favourably with the values found for a conventional vector machine. Even when reducing the registers size to 1/4 of the original size on an out-of-order machine, the slowdown was between 1.04 and 1.5, and was better still than on a conventional vector machine. Finally, when comparing both architectures, using the same register file size (8kb) we found that the gains in performance using out-of-order execution were between 1.13 and 1.40.

1 Introduction

Numerical applications has been the area where vector architectures have proved their efficiency. These vector architectures have used in-order execution, limited form of ILP techniques and large latencies memory systems. To achieve good performance and t olerate the large latencies, this kind of processors have exploited the data level parallelism embedded in each vector instruction and have allowed the overlapping of vector and scalar instructions, when possible. Conventional vector architectures have us ed large vector registers as one of the principal resources to hide latency. When a vector instruction is started, it pays

* On leave from the Centro de Investigación en Cómputo, Instituto Politécnico Nacional – México D.F. This work was supported by the Instituto de Cooperación Iberoamericana (ICI), Consejo Nacional de Ciencia y Tecnologia (CONACYT).

** This work was supported by the Ministry of Education of Spain, under contract 0429/95, and by the CEPBA.

J. Palma, J. Dongarra, and V. Hernández (Eds.): VECPAR'98, LNCS 1573, pp. 439–451, 1999.

for some initial (potentially long) latency, but then it works on a long stream of elements and effectively amortises this latency a cross all elements.

Taking into account this point of view, we can understand why vector machines have been designed with vector registers as large as possible. Unfortunately, large registers have several disadvantages:

- When the application cannot make full use of the vector register size, a precious hardware resource is being wasted [1].
- Large registers mean a large number of transistors and expensive cost; this implies that on ly a few of them can be implemented on the design.
- If the number of registers that the compiler sees is small, then the amount of spill code introduced to support all live variables is considerable [2].

Reducing the vector registers length is certainly a solution to the problems just outlined. If most applications cannot fully use all elements in each vector register then, reducing the vector register length will reduce cost and increase the fraction of usage of registers.

Some related research works, have been developed on vector register design. Lee and J. Smith [3] studies in a form of partitioned vector register in order to improve the use of this resource. The objective was to provide more vector registers w ithout increasing the size of the crossbar that connect the vector registers to the rest of the data path. We can see similar ideas in architectures like Arden Titan [4] and the Fujitsu [5] architectures lineage, which have imp lemented a partitioned vector register file. However, partitioned register files increase the control logic of the architecture and require special algorithms for accessing vector registers.

The drawback of register length reduction is the associated performance penalty. Each time a vector instruction is executed, its associated latencies are amortised over a smaller number of elements. This can degrade the performance, especially for memory accesses. Besides, more instructions have to be executed with a shorter effective length, increasing the number of times that latencies must be paid.

Unless some extra latency tolerance mechanism is introduced in a vector architecture, vector length cannot be reduced without a severe performance penalty. Many techniques have been developed to tolerate memory latency in superscalar processors, only a fe w studies have considered the same problem in the context of vector architectures [6,7,2].

We study in this paper the influence of the vector register size over two different concepts of vector architectures: a conventional vector architecture and an out-of-order vector machine. We show that the vector register size on conventional vector arch itectures cannot be reduced without a severe performance penalty. Combining an out-of-order execution and short registers, the performance degradation was lower than the observed on a conventional vector machine. This combination allows a vector register reduction with a good performance and a higher performance of the new out-of-order vector machine.

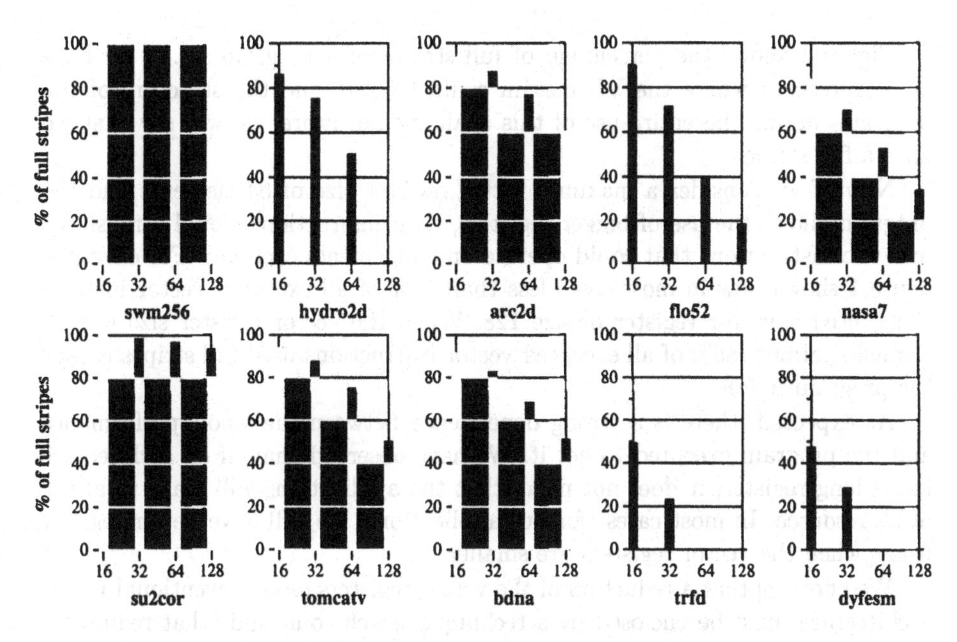

Fig. 1. Percentage of full stripes for different vector register sizes

2 Vector Registers Usage

In this section, we investigate the relationship between the Vector Register Size (VRZ) and the benchmark programs.

High memory latencies are common in vector architectures. In order to hide that latency, large vector registers have been a norm in the design of this kind of architectures. Unfortunately, with large vector registers there are also a few drawbacks:

- Because large registers mean large hardware space and more cost, the number of large registers is usually reduced to, for instance 8 or 16 with 128 element each.
- A restricted number of registers is a drawback for the compiler, because the quality of the code that it can generate is quite poor.

We have seen [1] that when a machine has large registers, programs do not make use of their hardware. Many researchers in areas as physics, chemistry or mathematics where this kind of architectures still excel, are developing new algorithms t o execute their calculus as fast as possible. The characteristic algorithms are quite varied and the different architectures are trying to apply all their capacity, however some times the data structures from the applications are the major difficulty.

To know how some applications use the register file on a vector architecture, after having a set of registers, we executed our set of programs, using vector register size with 16, 32, 64 and 128 elements.

Figure 1 shows the percentage of full stripes of a program set. If we have an architecture where the VL maxim could be 128, and the structure of the programs allows the entire use of this available hardware, we will say that we have a full stripe.

Now, if we consider a maximal vector register size of 64 elements and the program allows the use of bigger registers, then instructions would "translate" into two instructions that could operate on 64 elements each one. For example, figure 1 shows how in most cases less than 50% of all executed vector instructions, used a vector register of size 128. When the vector register size was 16 elements, almost 85% of all executed vector instructions used full stripes except the program *dyfesm.*

As expected, there is a strong dependence between the whole performance and the program executed to get it. We have observed that, if an architecture has a long register, it does not mean that the applications will make total use of its resource. In most cases (for our applications) we will have better register usage when the vector registers are smaller.

We know [8] that a reduction of the vector registers on a conventional vector architecture must be enclosed by a technique which could hide that reduction, in order to keep or in the best of the cases improve the performance.

3 Reducing Vector Registers Length

The architecture and compiler are reflected on the characteristics of the code that they generate from an application. If these are an intelligent pair, it could be easy to obtain programs that use different vector register sizes; sections of a register, where each section could be considerate a independent register. The Fujitsu VPP500 [5] is an example of that kind of architectures. The VPP500 has a vector register file organised on 256 registers, each with 64 elements (8 bytes each). Differ ent combinations of register and elements are possible, from 256 registers of 64 elements each until 8 registers of 2048 elements each. For our purpose, the lower limit size (64 elements) is not enough, because we want to study shorter vector register, in order to have better register usage (see section 2). Unfortunately, most of vector architectures do not have the VPP500 vector register reorganisation. Our reference architecture falls into this category.

The procedure that we followed, in order to obtain a set of binaries (from benchmarks) assuming different vector register lengths, was the following:

- For each program and with the help of the compiler information, we searched all the highly vectorised loops.
- First, we manually modified the benchmark sources and then, we manually added strip-mined loop (see figure 2) performing steps of desired length VLZ (vector length size).
- In this way, we constructed four different configurations for each source program using VLS=16, 32, 64 and 128 elements by register.

```
DO 40 J=2,JL
  DO 40 I=2,IL
    DW(I,J,1) = DW(I,J,1) +FW(I,J,1)
    DW(I,J,2) = DW(I,J,2) +FW(I,J,1)
    DW(I,J,3) = DW(I,J,3) +FW(I,J,3)
    DW(I,J,4) = DW(I,J,4) +FW(I,J,4)
40 CONTINUE

(a)
```

```
DO 40 J=2,JL
   DO 40 STRIPV=2,IL,VLZ
C$DIR MAX_TRIPS(32)
    DO 40 I=STRIPV,MIN(IL,STRIPV+VLZ)
        DW(I,J,1) = DW(I,J,1) +FW(I,J,1)
        DW(I,J,2) = DW(I,J,2) +FW(I,J,2)
        DW(I,J,3) = DW(I,J,3) +FW(I,J,3)
        DW(I,J,4) = DW(I,J,4) +FW(I,J,4)
40 CONTINUE

(b)
```

Fig. 2. (a) Flo52 loop without Strip-Mining, (b) Adding Strip-mining.

After applying this technique, we may notice that the architecture sees more scalar and vector instructions. The vectorisable loop will need more iterations to complete the same number of vector operations and because the scalar operations are inside the loop, these are executed more often.

4 Vector Architectures and Simulation Tools

In this section, we describe the main characteristics of the architectures used in this work. First, we show the reference, baseline, vector architecture and introduce the *out-of-order* vector architecture used. Finally, we describe the tools used t o generate traces and simulate each architecture.

4.1 The Baseline Architecture

As a baseline vector architecture, we used a machine loosely based on a Convex C3400 [9]. Although this is a multiprocessor architecture, our work assumes a uniprocessor vector machine. Figure 3 shows a basic description of a C3400.

- Scalar Unit
  - The scalar unit (with a maximum of 1 instruction per cycle) executes all instructions that involve scalar registers (A and S registers), adds, subtracts, compares, shifts, logical operations and integer converts.
  - The scalar unit has eight 32 bits address registers and eight 64 bit scalar registers.
  - This unit has a 16kb data cache, with 32 bytes line size.
- Vector Unit
  - The vector unit consists of two computation units (FU1 and FU2) and one memory accessing unit. FU2 is a general-purpose arithmetic unit capable of executing all vector instructions. FU1 is a restricted functional unit that executes all vector instructions apart from multiplication, division and square root.
  - The vector unit has 8 vector registers, grouped in pairs. Each register holds up to 128 elements of 64 bits each. Each group shares two read ports and one write port that links them to the functional units.

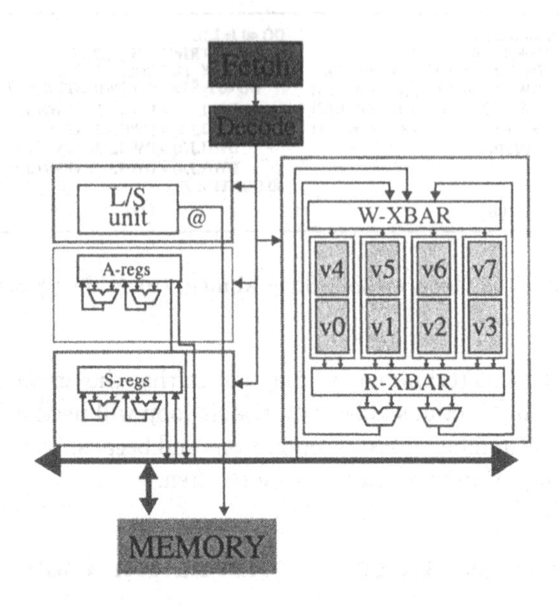

Fig. 3. The reference vector architecture.

- The request of memory is done through only one data bus (Loads and Stores).
- The reference machine implements vector chaining, from functional units to other functional units and to store unit. Memory load does not chain with any functional unit.

4.2 The Out-of-order Vector Architecture

For our simulations we used the out-of-order vector architecture introduced in [2]. The out-of-order and renaming version of the reference architecture is shown in figure 4. We will refer to this architecture as OOO. It has the same computing capacity as the reference machine but it is extended to use a renaming technique very similar to that found in the R10000 [10].Instructions flow in-order through the Fetch and Decode/Rename stages and then go to one of the four queues present in the architecture based on instruction type. At the rename stage, a mapping table translates each virtual register into a physical register. There are 4 independent mapping tables, one for each type of register: A, S, V and mask register s. Each mapping table has its own associated list of free registers. When instructions are accepted into the decode stage, a slot in the reorder buffer is allocated also. Instructions enter and exit the reorder buffer in strict program order. When an inst ruction defines a new logical register, a physical register is taken from the free list, the mapping table entry is updated with the new physical register number and the old mapping is stored in the reorder buffer slot allocated to the instruction. When t he instruction commits the old physical register, it is returned to the free list.

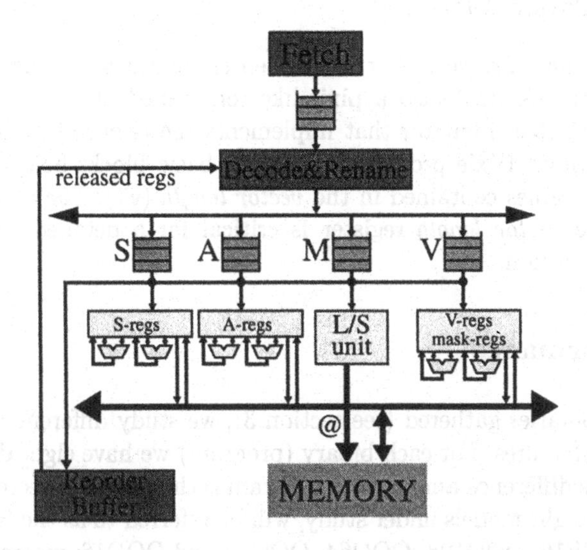

Fig. 4. The Out-of-Order vector architecture studied in this paper.

The A, S and V queues monitor the ready status of all instructions held in the queue and as soon as one instruction is ready, it is sent to the appropriate functional unit for execution. All instruction queues can hold up to 16 instructions. The machine h as a 64 entry BTB, where each entry has a 2-bit saturating counter for predicting the outcome of branches. Both scalar register files (A and S) have 64 physicals registers each. The mask register file has 8 physical registers. The fetch stage, the decode stage and all four queues only process a maximum of 1 instruction per cycle. Committing instructions proceeds at a faster rate, and up to 4 instructions may commit per cycle. The functional unit latencies of the architecture are very similar to the R10000 ones. See [2] for further details of the architecture.

The most important aspect of the architecture when considering final performance is the number of physical vector registers available for renaming vector instructions. In [2] it is shown that 16 physical vector registers is the optimum poin t that maximises performance at a reasonable cost. Unless otherwise stated, we will use 16 physical vector registers for our simulations. In section 5, we will vary the number of physical vector registers from 16 to 32 and to 64 to study how the number of physical registers interacts with the length of each register.

As we did for the traditional machine, we define four different versions of the OOO architecture, each having a different vector register length. The four versions will be referred to as the OOO128, OOO64, OOO32 and OOO16 architectures and will have a vec tor length of 128, 64, 32 and 16 elements, respectively.

4.3 Simulation Tools

For our simulations, we used trace-driven simulations to generate all the data being shown. We used also a pixie-like tool called Dixie [11] and Jinks [12], a parameterisable simulator that implements the reference architecture m odel described above. Dixie produces a trace of basic blocks being executed and a trace of the values contained in the *vector length* (`vl`). The ability to trace the value of the *vector length* register is critical for a detailed simulation of the program execution.

5 Performance

Using the binaries gathered (see section 3), we study different variations of our vector architectures. For each binary (program) we have eight different configurations. The difference among each program is the maximal vector register size al lowed. The eight models under study, will be referred to as the REF128, REF64, REF32, REF16, OOO128, OO064, OO032 and OO016, where 128, 64, 32 and 16 are the vector register size used by each model.

Both architectures have the same number of logical registers, which means that the same code was introduced in both architectures. However, because the OOO architecture implements renaming, it uses 16 physical registers, which are invisible both to the c ompiler and to the user.

In this section, for latencies of 1, 50 and 100 cycles, firstly we show how the architectures tolerate the vector register reduction plus memory latencies effect, and secondly we show the performance (Speed-Up) of each architecture, using different vector register sizes.

5.1 Reference Architecture

Figure 5 shows the effect of reducing vector register sizes on the reference vector architecture. REF128 was selected as the baseline in order to study the register reduction effect. Using 1 cycle latency and register sizes of 128 and 64, t he behaviour seems to be constant, an ideal vector architecture behaviour. When we reduce the register size and use more realistic memory latency, of 50 and 100 cycles, the effect is clearly negative. It is most remarkable when the memory latency is large r and the vector registers are shorter.

Although, the architecture uses large registers (REF128), the performance degradation is quite important. The slowdown degradation varies between 1.1 and 1.7. This is an important point to emphasise because large registers on vector architectures have bee n one of the best tools used to attack memory latency, but we can see that it is not enough.

If we compare REF128 with configurations REF64, REF32 and REF16, the slowdown can reach up to 3.5. This is not a surprise, because it was expected that reducing the vector register size on a conventional vector architecture would degrade the performance.

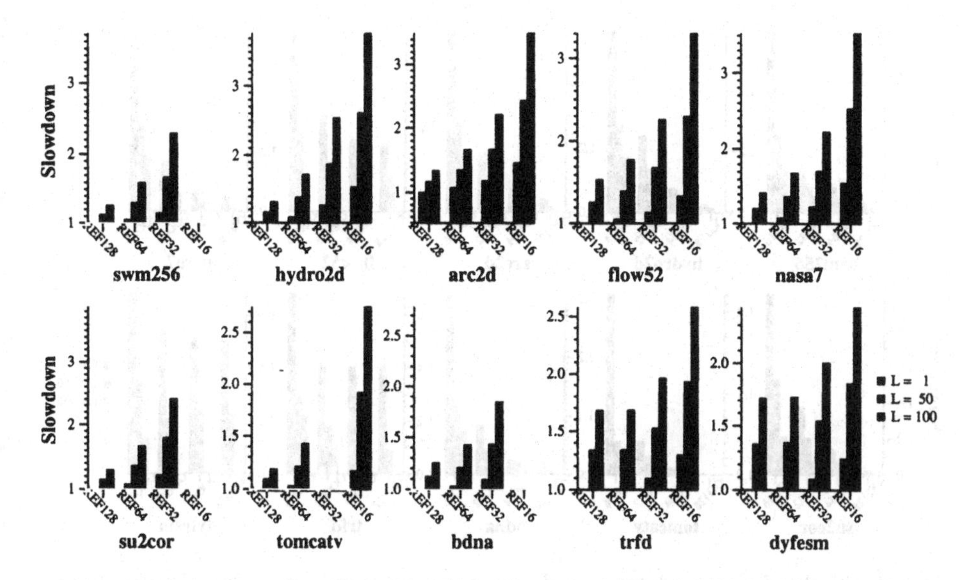

Fig. 5. Effects of memory latency and vector register size on a Conventional Vector Architecture. X-axis is memory latency

5.2 OOO Vector Architecture

Figure 6 shows the vector register reduction effect but now on the *out-of-order* vector architecture. Again, the baseline (OOO128 in this case) was the best configuration. Clearly, we can observe that this architecture has better vect or reduction tolerance. Reducing the vector register size up to 1/4 (from 128 to 32), line OOO32, the execution time is degraded by a factor of 1.0 to 1.5.

When we evaluated the memory latency effect, we saw that in most cases OOO128, OOO64 and OOO32 (programs *swm256, hydro2d, arc2d, nasa7, tomcatv, bdna*) have a high memory latency tolerance, with slowdown between 1.0 and 1.3. Other programs, such as *flow52, trfd, dyfesm and su2cor*, do not have good behaviour using short registers, but it is still better than the tolerance showed by the reference architecture, with slowdowns between 1.22 and 1.98.

Until this point we can conclude that if an architecture uses advanced ILP techniques like an *out-of-order*, it will be able to tolerate the vector register reduction better, even across large latency range.

5.3 Performance Comparison

In this section we compare the performance of both architectures, using identical or lower register file size. We compare REF128 against OOO16, OOO32 and OOO64.

Figure 7 shows the simulated performance using three different memory latencies. For each program, configuration and value of memory latency, we compute the speedup relative to the performance of the REF128 configuration at latency 1. Using the same register file size, 8kb, REF128 and OOO64 lines, the

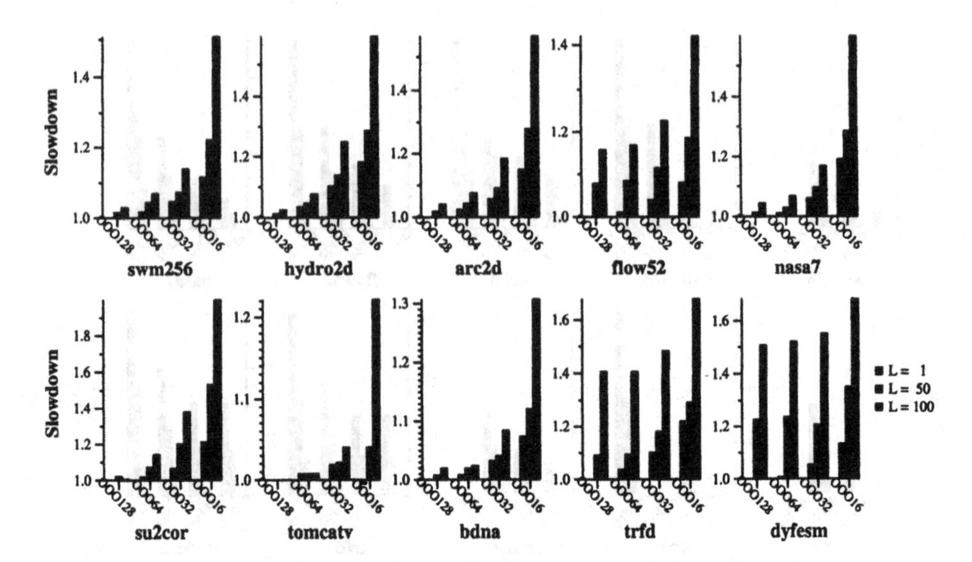

Fig. 6. Effects of memory latency and vector register length on a *Out-of-order* Vector Architecture. X-axis is memory latency.

performance with respect to the REF128 is much better for all the programs and all the memory latencies, with speedups between 1.09 and 1.4. Reducing the register file size (on OOO64) up to 1/2, line OOO32, it is still better than the reference machine with large registers, for all programs and all memory latencies, with speedups between 1.04 and 1.34. Nevertheless, when reducing the size up to 1/4 on OOO64, OOO16 line, the performance of the OOO machine is not always better than the REF128. The programs *hydro2d, flo52, tomcatv, bdna* and *trfd*, show better performance than the reference machine with speedups between 1.03 and 1.1. The programs *swm256, arc2d and nasa7*, have performance that is slightly better or slightly worse than the REF128, but the difference is typically around the 8%. Finally, the worst case was the performance of the pr ogram *su2cor*, with a slowdown around 40%.

One way of looking at the advantages of out-of-order execution over the reference architecture is that it allows long memory latencies to be hidden. The out-of-order issue feature allows memory access instructions to slip ahead of computation instructions , resulting in a compression of memory access operations and allows for much higher usage of the system resources. The register renaming used in the OOO architecture cannot wholly eliminate the spill code generated by the compiler due to the lack of physical registers, but the negative effect has less influence over the performance than on the reference machine.

In Figure 7 we note that programs have different behaviours when the vector registers are shorter or when the memory latency is higher. True dependencies and false dependencies between instructions like gather and scatter instructions, and programs with large basic blocks are two possible explanations for this behaviour. A gather instruction cannot start execution because the hardware

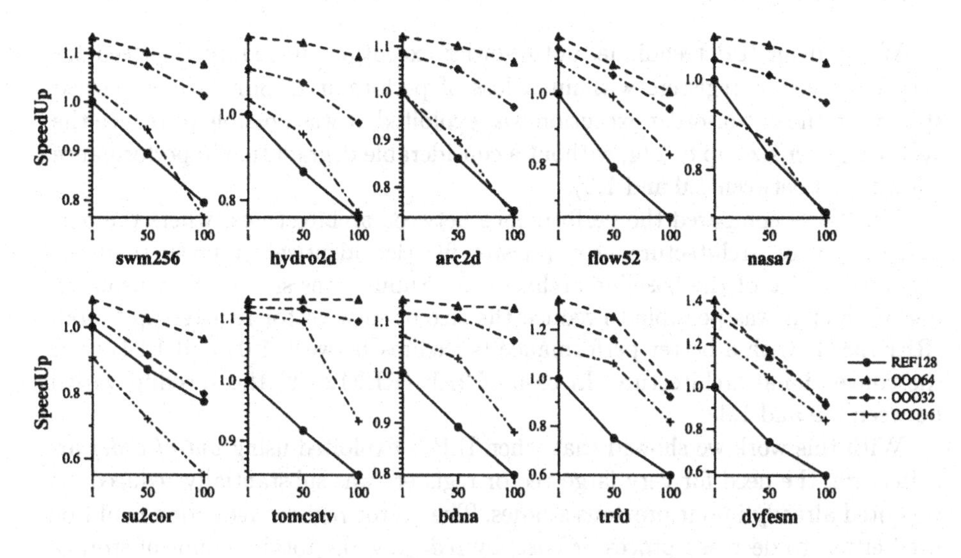

Fig. 7. Performance comparison of the OOO architecture and the Reference Architecture using the same or less, register file size. X-axis is memory latency in cycles and Y-axis represents Sp eedUp.

is unable to determine whether the scatter instruction has a conflict or not, whereupon the execution is completely serialised, exposing the entire memory latency in each memory access. Rem ember that one access in the OOO128 model (with a VL=128) represents eight accesses on the OOO16 model, whereupon the memory latency could be eight times higher. Furthermore, when the vector registers are shorter this latency will be amortised over a redu ced number of elements. In case of programs with large basic blocks (*bdna, tomcatv*), the spill code may induce a lot of memory conflicts in the memory queue.

6 Summary

In this paper we studied the influence of reducing the vector register size, on two different concepts of vector architectures. The in order execution, traditionally used on vector architectures, and the long latencies paid on a memory request have been always used with long vector registers in order to hide and amortise this latency and this strict program order. Nevertheless, we have showed that long registers were rarely fully used for a set of highly vectorisable programs. Less than 40% of all the registers being used were completely filled up with 128 elements of data.

As expected, reducing the vector register length on a traditional vector machine results in a loss of performance. The cost saving is out-weighted by the execution time degradation. Halving the vector length yields slowdowns between 1.1 and 3.5. Unless some latency tolerance technique is added to a traditional vector machine, vector register length should be kept as long as possible.

We used an ILP technique, out-of-order execution, to reduce the need for very large vector registers without a loss of performance. Simulations showed that when the out-of-order execution was exploited, it was possible to reduce the vector register size up to 1/4, without a considerable degradation in performance (slowdowns between 1.0 and 1.5).

Finally we compared the performance between architectures, where the *out-of-order* vector architecture used register file size identical or lower than the register file size of the baseline architecture. Simulations showed that using an *out-of-order* it was possible to reduce the size of each vector register up to 4kb (REF128/4) with a better performance (speedups between 1.04 and 1.34) than the conventional architecture. In case of 2Kb (REF128/8) the speedup varied between 0.9 and 1.3.

With this work we showed that when ILP is exploited using *out-of-order* architecture, the need for very large vector registers was substantially reduced, as we noted already in our previous studies. The vector register reduction could be used either to decrease processor cost, by reducing the total amount of storage devoted to register values, or to improve performance by more effective use of the available storage. Using *out-of-order* execution and short register, the vector architecture concept like a big and expensive supercomputers could change, because designers could use the current technology and ideas (ca ches, memory systems, no blocking loads, clustering, etc.) in order to improve the performance.

References

1. Luis Villa, Roger Espasa, and Mateo Valero. Effective usage of vector registers in advanced vector architectures. In *International Conference on Parallel Architectures and Compilation Techniques (PACT97)*, San Francisco, California, 1997.
2. Roger Espasa, Mateo Valero, and James E. Smith. Out-of-order Vector Architectures. In *MICRO-30*, pages 160–170, IEEE Press, December 1997.
3. Corinna G. Lee and James E. Smith. A study of partitioned vector register files. *Supercomputing*, pages 94–103, 1992.
4. Tom Diede, Carl F. Hagenmaier, Glen S. Miranker, Jonathan J. Rubinstein, and Jr. William S. Worley. The Titan graphics supercomputer architecture. *IEEE Computer*, 21(9):13–30, September 1988.
5. T. Utsumi, M. Ikeda, and M. Takamura. Architecture of the VPP500 Parallel Supercomputer. In *Proceedings of Supercomputing'94*. Washington D.C., IEEE Computer Society Press, November 1997.
6. Roger Espasa and Mateo Valero. Decoupled vector architectures. In *HPCA-2*, pages 281–290, Computer Society Press, February 1996.
7. Roger Espasa and Mateo Valero. Multithread vector architectures. In *HPCA-3*, pages 237–249, Computer Society Press, February 1997.
8. Luis Villa, Roger Espasa, and Mateo Valero. Effective usage of vector registers in advanced vector architectures. In *Parallel and Distributed Computing (PDP98)*. Madrid, Spain, 1998.
9. Convex Press, Richardson, Texas, USA. *CONVEX Architecture Reference Manual (C series)*, 6 edition, April 1992.
10. Keneth C. Yager. The Mips R1000 Superscalar Microprocessor. *IEEE Micro*, pages 28–40, April 1996.
11. Roger Espasa and Xavier Martorell. Dixie: a trace generation system for the C3480. Technical report: CEPBA-RR-94-08, Universitat Politècnica de Catalunya, 1994.
12. Roger Espasa. JINKS: A parametrizable simulator for vector architectures. Technical report: UPC-CEPBA-1995-31, Universitat Politècnica de Catalunya, 1995.

Limits of Instruction Level Parallelism with Data Value Speculation

José González and Antonio González
Departament d'Arquitectura de Computadors
Universitat Politècnica de Catalunya
c/Jordi Girona 1-3, Modul C6 - E208
08034 - Barcelona, Spain
{joseg,antonio}@ac.upc.es

Abstract. Increasing the instruction level parallelism (ILP) is one of the key issues to boost the performance of future generation processors. Current processor organizations include different mechanisms to overcome the limitations imposed by name and control dependences but no mechanisms targeting to data dependences. Thus, these dependences will become one of the main bottlenecks in the future. Data value speculation is gaining popularity as a mechanism to overcome the limitations imposed by data dependences by predicting the values that flow through them. In this work, we present a study of the potential of data value speculation to boost the limits of instruction level parallelism using both perfect and realistic predictors. Speedups obtained by data value speculation are very huge for an infinite window and still significant for a limited window. Different prediction schemes oriented to single thread and multiple threads (from a single program) architectures have been studied. The latter shows a significant improvement respect to the former for FP benchmarks although the difference is much smaller for integer programs.

1 Introduction

The performance of superscalar processors is limited by the necessity to obey the dependences existing among the program instructions. These dependences can be classified into three types[5]: name dependences, control dependences and data dependences.

Name dependences appear when the values generated by two instructions are to be written in the same storage location, either a register or memory. They can be eliminated by renaming the storage location that causes the dependence (i.e. changing the name of the locations where the values are to be written). Register renaming is a well known technique that deals with this kind of dependences. It is implemented dynamically by many current microprocessors such as DEC Alpha 21264 [4] or MIPS R10000 [23].

Control dependences are caused by branch instructions. They slow down the processor since it has to stall the fetch of instructions until the branch is solved, i.e. the destination address is computed and the condition is evaluated. Branch prediction is the mechanism that current microprocessors implement in order to overcome control dependences. It is based on the prediction of the outcome of branches which allows instructions that depend on a branch to be executed before the result of such branch is known.

J. Palma, J. Dongarra, and V. Hernández (Eds.): VECPAR'98, LNCS 1573, pp. 452-465, 1999.

Data dependences or true dependences appear when an instruction consumes the value produced by another previous instruction. These dependences are enforced in current microprocessors by executing the consumer after the producer. Thus, data dependences limit the amount of instruction level parallelism (ILP) by imposing a serialization on the execution of some instructions.

In the same way as control dependences are managed predicting the behavior of branches, it may be feasible to predict the result of some instructions in order to avoid the ordering imposed by data dependences, allowing the consumer instruction to be issued before the execution of the producer. The term *data value speculation* is used to refer to those mechanisms that predict the operands of an instruction, either source or destination, and execute speculatively the instructions dependent on it before the actual value is computed, allowing the processor to avoid the ordering imposed by data dependences.

In this work, we present a study of the ILP improvement that data value speculation techniques can provide. We present an evaluation of the limits of ILP that can be exploited by dynamically scheduled processors with infinite resources and data value speculation, and compare it with that of the same processor without data value speculation. We evaluate the benefits of predicting individual types of instructions (loads, stores, simple arithmetic, and multiplications) and the improvement achieved by predicting all of them. We consider both ideal prediction schemes and realistic ones. Finally, the impact of data value speculation for a limited instruction window is also evaluated. The results shows that data value speculation can significantly increase the ILP that dynamically scheduled processors can exploit, and therefore, it is a promising technique to be considered for future generation microprocessors.

The rest of this paper is organized as follows. Section 2 reviews the related work. The methodology to evaluate the ILP that can be exploited by an ideal processor, either with or without data value speculation, is described in section 3. The value predictors considered in this work are presented in section 4. The results of this study are detailed in section 5. Finally, section 6 summarizes the main conclusions of this work.

2 Related work

There have been a plethora of works dealing with the limits of the ILP [1][2][6][10][16][20][21]. Each work studies the ILP that could be exploited under some constraints such as fetch width, instruction window size, branch prediction, register renaming, memory aliasing, etc. A conclusion that can be extracted from all these works is that one of the main features that limit the parallelism are data dependences. For instance, in [5] it is shown that the maximum ILP that a processor could achieve with infinite resources and perfect branch prediction is not much higher than a few hundred instructions per cycle (IPC) and for some applications it is about a few tens of IPC.

Data value speculation has been the focus of several recent works. It is performed in [14] by predicting the address of load instructions whereas in [9] the address of stores is also predicted. In both cases the prediction is carried out using a history table of memory instructions and a stride based predictor. In [12], data value speculation is based on predicting the value that load instructions read from memory. The proposed mechanism exploits the feature that the authors call *value locality*, which refers to the fact that many load instructions repeatedly bring the same value from memory. Value locality is extended for all type of instructions in [11]. In [8] data value speculation is performed by predicting the value read by load instructions. Unlike the mechanism proposed in [12], the load values are predicted by predicting their effective address and prefetching the data from memory into the history table. In [15] Sazeides and Smith show that the results that an instruction generates may follow a repetitive pattern that stride predictors cannot predict and propose a context-based predictor. In [22] Wang and Franklin present a hybrid predictor. The implementation of this predictor is similar to that of a 2-level branch predictor. In [7] the impact of different value predictors on the performance of a processor is studied using a limited instruction window.

The main contributions of this work are the following: This is the first work to our knowledge that evaluates the limits of ILP in an ideal dynamically scheduled superscalar processor that exploits data value speculation and compares it with that of the same processor without data value speculation. In [11], value prediction is evaluated for a perfect machine, as it is called by the authors. However, that machine is limited by a finite instruction window (4096 entries), branch prediction and fetch bandwidth. Besides, in this paper we study the benefits of predicting individual types of instructions for both ideal and realistic predictors.

3 Methodology

This section describes the methodology that we have used to obtain the ILP under different scenarios regarding prediction schemes and hardware resources.

3.1 Experimental framework

The evaluation methodology is trace-driven. The trace of each program has been generated using the ATOM tool [19]. For each instruction, the instrumentation routine obtains: its operation code, the source and target registers, the effective address (if the instruction is either a load or a store), and the value generated in the case of arithmetic and load instructions. These data are fed into the analysis program, which computes the performance achieved by the particular architectural model. Performance is reported as Instructions per Cycle (IPC).

The whole SPEC95 benchmark suite has been used for the different experiments. All the benchmarks have been compiled for a DEC AlphaStation600 5/266 with '-O4' optimization flag, and executed with their largest input set. Each program has been run for 5 billion of instructions, except gcc and ijpeg, which have been run until completion (1,569,885,184 and 684,497,921 instructions respectively). Figure 1 details the percentage of different types of instructions executed for the whole SPEC95 benchmark suite.

3.2 Architectural model

The first study of the limits of ILP is achieved assuming an ideal microprocessor with infinite resources, perfect branch prediction, infinite instruction fetch bandwidth, an infinite cache memory with infinite number of ports, perfect memory disambiguation, dynamic renaming with an infinite number of registers and memory renaming with infi-

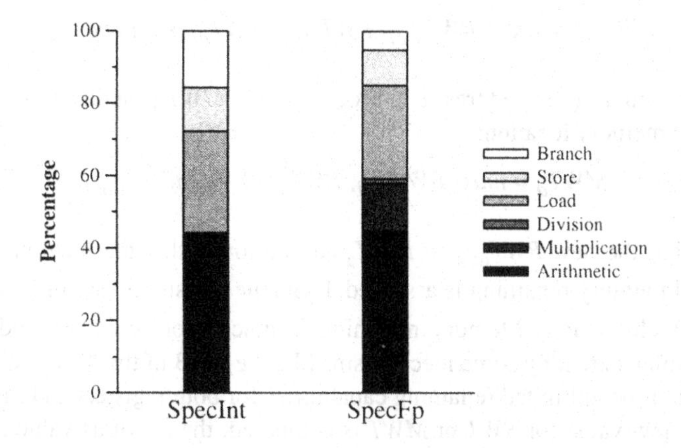

Figure 1. Dynamic percentage of each type of instructions

nite storage locations for renaming. Both an infinite and a limited instruction window are considered. In all the cases, precise exceptions [17] and an infinite retirement (commit) bandwidth are assumed.

3.3 IPC computation for an ideal architecture without data value speculation

The IPC of a given program for a particular architectural model is obtained by determining the time (measured in number of cycles) when the latest result of any instruction of the program is computed, and then, dividing the number of executed instructions by such number of cycles.

We will refer to the cycle when the result of an instruction i is available as the *completion time* of i, or CT_i for short. CT_i is computed as the maximum CT_j for any j such that j produces a result that is a source operand of i plus the latency of the operation i. This approach is similar to the one used in [1].

Each instruction of the trace produced by the execution of the instrumented program is analyzed in order to know the time when its operands are available. For each storage location the analysis program keeps track of the CT of the last instruction that wrote to it. This is implemented by means of two tables that are called the *register write table (RWT)* and the *memory write table (MWT)*. RWT_r stores the CT of the last instruction so far that its destination operand was the logical register r. MWT_a stores the CT of the last store that wrote into address a.

Therefore, when an arithmetic instruction is processed, the RWT is accessed in order to obtain the cycle that the source operands are available. Then, its CT is computed and the RWT entry associated to its destination register is updated with the new computed CT. That is:

$$RWT_{dest} = max\ (RWT_{src1}, RWT_{src2}) + Latency_{operation} \tag{1}$$

In a similar way, when a load from address a is processed, the MWT is accessed to obtain the cycle that a previous store wrote into that memory position. Then, the RWT is updated as follows:

$$RWT_{dest} = max\,(RWT_{src1}, RWT_{src2}, MWT_a) + Latency_{load} \tag{2}$$

Finally, when a store to address *a* is processed, the *MWT* is updated to reflect the new write to this memory location:

$$MWT_a = max\,(RWT_{src1}, RWT_{src2}) + Latency_{store} \tag{3}$$

Notice that the new RWT_{dest} or MWT_a can be lower that the previous one because register and memory renaming is assumed. Dynamic register renaming is very common in current architectures. Memory renaming is much more complex and it is implemented to some extent by some mechanisms like the ARB of the Multiscalar [3]. In this paper, we assume unlimited renaming capabilities for both registers and memory.

When a new value for *RWT* or *MWT* is computed, the previous value is overwritten because any further instruction in the trace will always refer to the last value stored into a register or a memory location. However, in order to compute the IPC, we have to determine the maximum *CT* for any instruction of the program. To obtain such value, the analysis program keeps a variable that stores the maximum *CT* up to the current execution point (*Max_CT*).

3.4 IPC computation for a limited instruction window

A limited instruction window with W entries and in-order retirement implies that an instruction cannot start execution until the instruction W locations above in the trace and all previous instructions have completed and retired. Thus, the restriction of having a limited instruction window can be modeled by keeping track of the *CT* of the last W instructions. This is accomplished by means of a table, which is called *window retirement time* (*WRT*), that has W entries and stores the retirement time of the last W instructions processed so far.

Thus, when computing the *CT* of an instruction, in addition to consider the *CT* of its source operands, the *WRT* of the instruction W locations above has also to be considered. For instance, for each arithmetic instruction processed by the analysis program, the corresponding entry in the *RWT* is updated as follows:

$$RWT_{dest} = max\,(RWT_{src1}, RWT_{src2}, WRT_{n\_inst\%W}) + Latency_{operation} \tag{4}$$

where *n_inst* refers to the ordinal number of the current instruction in the trace. Expressions (2) and (3) are modified in a similar way to account for the effect of the limited instruction window.

For each new instruction, the *WRT* is updated to reflect the retirement (commit) time of the current instruction. This time is the maximum *CT* of any previous instruction, including the current one, and it is stored in the same entry of the *WRT* that was occupied by the instruction W locations above since it is not useful any more:

$$WRT_{n\_inst\%W} = Max\_CT \tag{5}$$

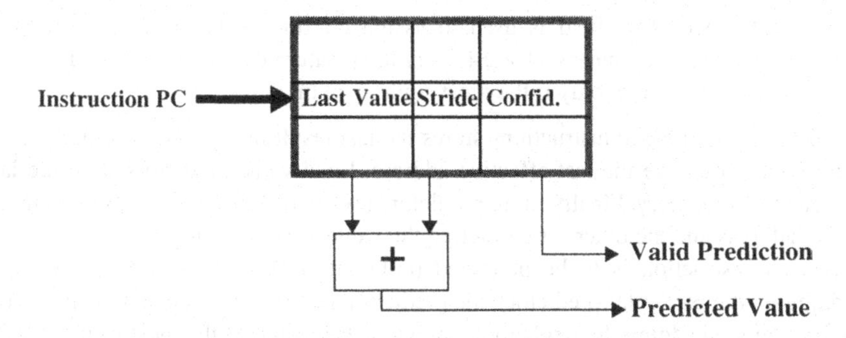

Figure 2. A stride-based predictor.

3.5 IPC computation for data value speculation

Data value speculation is based on predicting the source and/or the destination operands of some instructions. In this section, we present a methodology to compute the IPC when data value speculation is incorporated into a superscalar processor, independently of the particular predictor being used. In this way, we consider a predictor as a system that given an instruction (usually its program counter), provides its source and/or destination operands. In addition, each individual prediction is characterized by the time when the prediction is available (*PT*) and the correctness of the prediction.

In this paper, we consider data value speculation for the following type of instructions: Loads, Stores, Integer Arithmetic, Integer Multiplication, Float Arithmetic and Float Multiplication.

In all the cases, if a prediction is not correct, the *RWT* and *MWT* are updated as if prediction were not used. If the prediction is correct, the *RWT* and *MWT* are updated with the minimum between the completion time, given by expressions (1), (2) and (3), and the prediction time, which is a characteristic of the particular predictor being used. Section 4 discusses the predictors considered in this work and in particular, the time when predictions are available.

4 Data predictors

In this work we consider stride-based predictors, although the presented methodology could be applied for any other data predictor. A stride predictor has the structure shown in Figure 2. It is implemented by means of a table of 4096 entries that is direct-mapped, non-tagged and it is indexed with the least significant bits of the instruction address (PC) whose source or destination operands are to be predicted. Each entry stores the following information:

- Last value: This is the last value seen by that instruction. This value corresponds to the destination operand for all predictors except for the load and store address predictors. In these cases, it corresponds to the last effective address.
- Stride: This field contains the stride observed for the values of the corresponding instruction.

- Confidence: This field is used to assign confidence to the prediction. It is implemented by means of a 2-bit up/down saturated counter. A prediction is considered correct only if the most significant bit is set.

Predictor for arithmetic instructions stores the last result in the last value field. Load address predictors store the last effective address. Load value predictors store the last value read from memory. Finally, store predictors uses two tables: one for predicting the effective address and the other for predicting the value to be written.

When an instruction is to be predicted (either its result or its effective address, depending on the particular predictor), the prediction table is accessed and the predicted value is computed adding the stride to the previous last value. If the most significant bit of the confidence field is set (i.e., the prediction is considered to be correct) and the prediction is correct, the predicted value can be used instead of the actual value if the former is available earlier. The stride field is only updated if the confidence counter is below 10_2 after being updated.

In addition, we consider a perfect predictor that is assumed to produce always correct predictions. This is used to determine the upper bound of the performance that data value speculation can achieve.

4.1 Prediction time

An important feature of a predictor is the time when the predicted value is available. This time is used to update the *RWT* and *MWT* as explained in section 3.5.

Regarding the prediction time, two different types of predictors have been considered:

- Serialized: Every time the prediction table is accessed, only one prediction per static instruction can be performed at most. That is, an instruction is not predicted until the last execution of the same static instruction has been predicted.
- Non-serialized: Every time the prediction table is accessed, multiple predictions for each static instruction can be performed. In particular, all the subsequent executions of the same static instruction are predicted until the first one that is incorrect. That is, once the corresponding entry of the table has the correct stride, successive executions of the same static instructions can be predicted all at once.

The serialized predictors may be suitable for superscalar processors. In fact, most of the studies on value prediction assume this type of predictors [8][9][11][12][14]. A non-serialized predictor could be useful for architectures supporting multiple threads of control obtained from a single program, such as multiscalar processors [18] and the speculative multithreaded processors [13].

To determine the time when a prediction is available we consider a parameter that reflects the time required to perform a prediction operation (either of a single value for the serialized approach or multiple values for the non-serialized one). This parameter is called the *prediction latency (PL)*. This is the time required for a table look-up plus its update.

The prediction time of each instruction is determined by means of an additional field that is added to each entry of the prediction table for evaluation purposes. This field stores the cycle in which the entry has been used/updated for the last time. This field will be called *last update time* (*LUT*).

The prediction time for an instruction is just the sum of the last update time plus the prediction latency. That is:

$$PT = LUT + PL \qquad (6)$$

The *LUT* is updated in a different way for serialized and non-serialized predictors. For the former, for each new instruction of the trace, the corresponding *LUT* is updated with the time when its operand is available (either computed or predicted, whichever occurs first):

$$LUT = RWT_{dest} \text{ for load and arithmetic instructions with destination register dest}$$

$$LUT = MWT_a \text{ for stores to address a} \qquad (7)$$

For non-serialized predictors, the *LUT* field is updated in the same way as the serialized case but only for those instructions that are mispredicted or are considered not predictable as stated by the confidence field.

5 Results

The results of this section assume a one-cycle latency for all instructions and one-cycle prediction latency.

Table 1 shows the IPC achieved by the ideal processor described in section 3.2 with an infinite instruction window and without data value speculation

This results will be used as a baseline to compare the performance of data value speculation techniques. They represent the maximum parallelism that is possible to achieve in an ideal processor that is only constrained by data dependences whereas data value speculation removes this constraint. Notice that even for this ideal machine, the average IPC is only 37.39 for integer programs and 790.29 for floating point applications. When we add the constraint of a limited instruction window of 128 instructions, the IPC goes down to 9.64 and 17.51 respectively. This may suggest that relieving the restrictions imposed by data dependences through data value speculation can be an interesting mechanism to boost performance. In the following results, only the average result for integer and FP programs will be shown.

Figure 3 shows the speedup (in logarithmic scale) achieved by data value speculation with perfect prediction in relation to the infinite machine without data value speculation. In this figure and the following ones the speedup is computed as follows:

$$\text{Speedup} = \frac{\text{IPC with data value speculation}}{\text{IPC without data value speculation}}$$

In each bar, only a single type of instructions is predicted individually. With perfect prediction, when an instruction is predicted its result is considered to be available at cycle 0. Looking at the graphs, one can see that the potential performance of predicting

Table 1. IPC achieved with infinite resources and no data value speculation

SpecInt	**IPC**	**SpecFP**	**IPC**
go	89.45	tomcatv	397.79
m88ksim	17.14	swim	1403.82
gcc	47.02	su2cor	56.64
compress	35.71	hydro2d	181.09
li	27.62	applu	578.31
ijpeg	34.12	mgrid	4735.11
perl	18.72	turb3d	140.19
vortex	29.34	apsi	231.21
		fpppp	105.71
		wave5	73.02
Average	37.39	**Average**	790.29

memory instructions, both loads and stores, is less than the speedup achieved by predicting arithmetic instructions. This suggests that for the analyzed programs, there are much more arithmetic than memory instructions on critical paths. The speedup achieved by predicting multiplications is almost negligible. In addition to not being on critical paths, this may be due to the small percentage of multiplication operations, as shown in Figure 1.

Figure 4 shows the speedup obtained for a realistic prediction scheme based on a stride predictor, as it was described in previous sections. The instruction window is considered to be infinite and the prediction is non-serialized. The speedup achieved by predicting arithmetic instruction is very huge and it suggests that arithmetic prediction may be the most effective approach to remove the serialization imposed by data dependences. The IPC of data value speculation just for arithmetic instructions is 531 times higher than the IPC achieved without data value speculation, for an infinite machine and the FP benchmarks. When data value speculation is implemented for all the instructions, the speedup goes up to 2368. The speedup for integer programs is not so high (42 when predicting all the instructions). On the other hand, the speedup achieved by predicting memory instructions is much more limited (1.4 and 4.8 for integer and FP benchmarks respectively when predicting stores and load values). Predicting multiplications is not considered any more due to the poor results observed for the perfect predictor.

The speedup obtained with a serialized predictor is depicted in Figure 5. Notice that, as pointed out before, this scheme would correspond to the implementation of data value speculation on a superscalar processor since in such processors there is only one

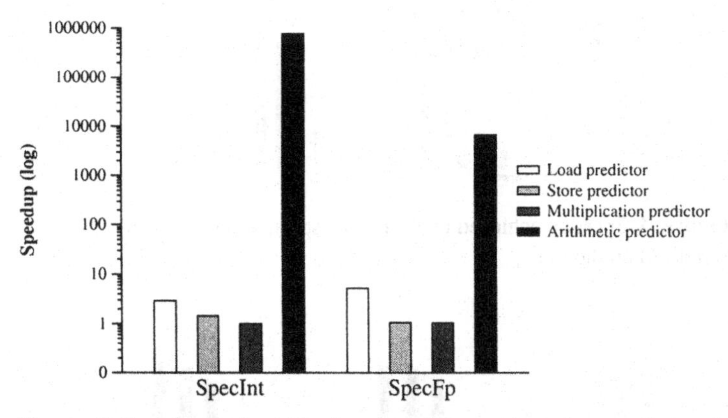

Figure 3. Speedup achieved by data speculation with perfect prediction, for different types of predictors.

flow of control and a given execution of a static instruction can be predicted only if its previous execution has updated the prediction table. On the other hand, a non-serialized predictor can be exploited by an architecture supporting multiple threads of control.

The speedup achieved by serialized prediction is still quite significant. The IPC achieved by these schemes is 30 and 35 times higher than the IPC achieved without data value speculation for integer and FP programs respectively. These results also show that the potential gain that load prediction may achieve is slightly higher for value prediction than for address prediction, but this gain is insignificant when compared to arithmetic prediction.

If we compare the speedup achieved by non-serialized prediction (Figure 4) against the speedup achieved by serialized prediction (Figure 5) we can observe that for integer benchmarks there is not much difference (e.g. it goes from 42 to 30 when predicting all the instructions) whereas for FP benchmarks the difference is huge (e.g. it goes from 2368 to 35 when predicting all the instructions). The main reason for this different behavior in the two types of benchmarks can be explained through the figures in Table 2. This table shows the percentage of correctly predicted arithmetic instructions for which the completion time (*CT*) is lower than prediction time (*PT*). For these instructions, the prediction does not provide any improvement in spite of being correct. As expected, this percentage is greater when the predictions are serialized than when they are not since the prediction time of the serialized scheme is in general higher. Besides, the difference between serialized and non-serialized schemes for FP benchmarks is much higher than for integer benchmarks, which explains the higher impact of serialized prediction for FP benchmarks, as observed in Figure 4 and Figure 5.

The speedup achieved by predicting instructions relies on the amount of strided values existing among the applications. Figure 6 shows the percentage of strided values for the different instruction types for the whole Spec95 benchmark suite. It can be seen that load addresses have the greatest percentage of strided references and therefore one may expect a speedup for load address speculation higher than it actually is (see Figure 4 and Figure 5). However, even when the address of a load is predicted, it has to wait for previous stores to the same address to finish. On the other hand, predicting the value of a

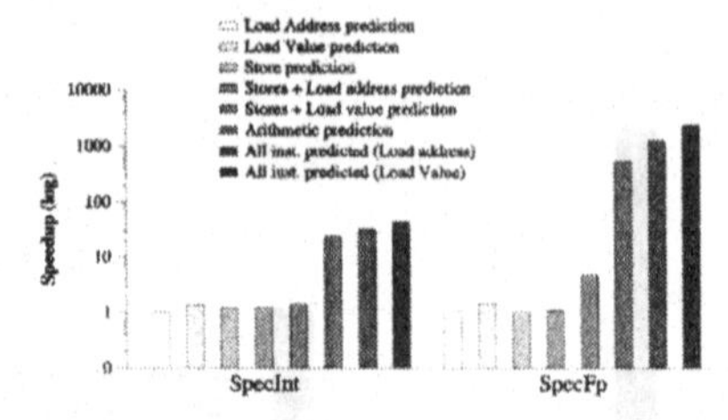

Figure 4. Speedup achieved by data value speculation with non-serialized prediction

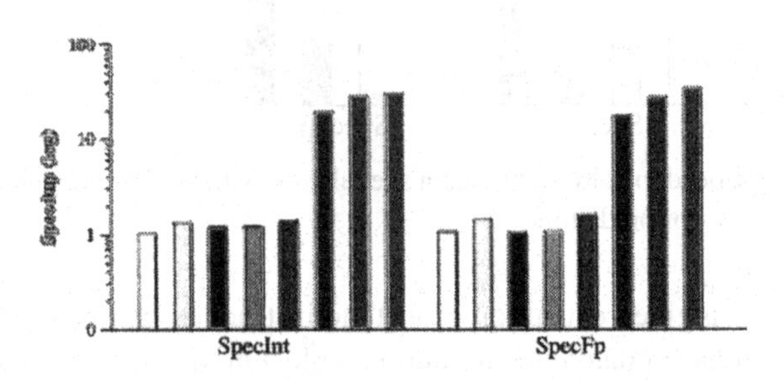

Figure 5. Speedup achieved by data value speculation with serialized predictions

Table 2. Percentage of correctly predicted instructions whose *CT* is lower than its *PT*.

	Non-serialized	**Serialized**
SpecInt	59.65	70.85
SpecFp	48.31	90.64

load or the result of any other instruction avoids completely the order imposed by data dependences. Simple arithmetic instructions (mainly integer arithmetic) has a high percentage of strided values. This fact, along with the significant weight of arithmetic instructions on the critical path (as confirmed in the evaluation of the prefect prediction scheme), makes arithmetic prediction to be the most effective type of speculation among the ones evaluated in this work.

Finally, we consider the impact of data value speculation with a limited instruction window. Figure 7 shows the speedup of data value speculation (IPC achieved by data value speculation divided by IPC achieved without data value speculation) when all types of instructions are predicted using separate history tables for each class, and predicting the value of loads. A non-serialized predictor is considered since it outperforms a serialized predictor for an infinite window (notice that the speedup is not depicted in logarithmic scale but in linear scale). It can be seen in this figure that the impact of the size of the instruction window its very significant since, for instance, the speedup is decreased from 2368 to only 1.75 for a window of 512 instructions in the SpecFp pro-

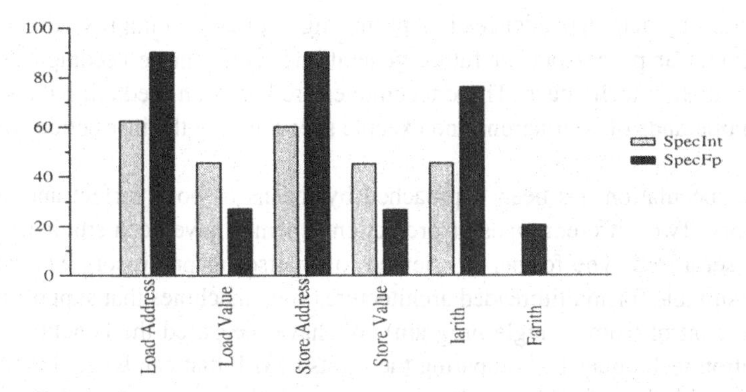

Figure 6. Percentage of strided values for each type of instruction

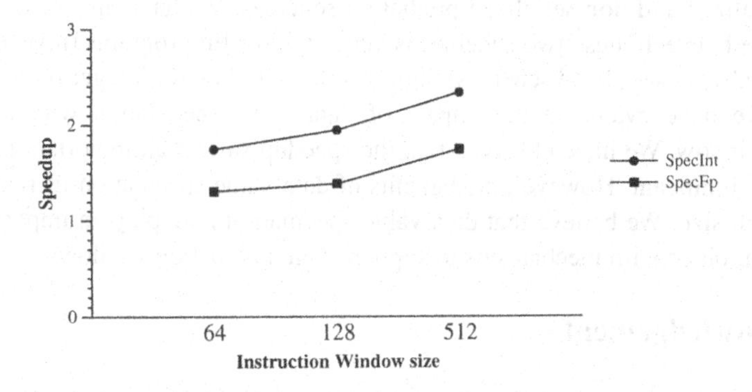

Figure 7. Speedup achieved with a finite instruction

grams. Furthermore, the gain due to data value speculation for the SpecInt outperforms the gain for SpecFp, which is the opposite to what happened with an infinite instruction window.

A main conclusion of the study of the effect of data value speculation on a limited instruction window is that it is an effective technique that could be considered for future generation microprocessors. A speedup around 2 can be achieved with simple stride-based predictors. However, the potential benefits of data value speculation are much higher for very large instructions windows. In this scenario, conventional superscalar microprocessors have been shown to be rather limited in the amount ILP that they can exploit due mainly to data dependences. This limitation can be significantly relieved by data value speculation techniques. Thus, novel organizations to support large instructions windows, like the multiscalar architecture [18] and speculative multithreaded processor[13] can be benefitted from data value speculation to a larger extent than superscalar processors.

6 Conclusions

In this work we have presented a study of the limits of instruction level parallelism (ILP) that can be exploited by a machine with infinite resources, infinite instruction window, perfect branch prediction and ideal memory. We have shown that avoiding the ordering imposed by data dependences is a promising approach to improve the performance of superscalar processors for future generations. This can be accomplished by data value speculation techniques. These techniques are based on predicting the source or destination operands of instructions and execute speculatively the instructions dependent on them.

Data value speculation has been approached by means of both perfect and stride-based predictors. Two different types of prediction schemes have been studied: serialized and non-serialized. The former is oriented to superscalar processors whereas the latter is more suitable for multithreaded architectures (i.e., machines that support multiple threads of control from a single program). We have measured the benefits of data value speculation techniques by comparing the limits of ILP that can be exploited with such technique with that of a superscalar processor with the same features but without data value speculation. Results show an important speedup for arithmetic instructions both for serialized and non-serialized prediction schemes. We have also observed that the difference between these two schemes is very high for FP programs (non-serialized outperforms always serialized schemes) but it is relatively low for integer programs.

Finally, we have evaluated the impact of data value speculation with a limited instruction window. We have observed that the speedup suffers an important reduction but it is still significant. However, the benefits of data value speculation increases with the instruction size. We believe that data value speculation may play an important role when it is combined with mechanisms to support large instruction windows.

7 Acknowledgements

This work has been supported by the Spanish Ministry of Education under grant CYCIT TIC 429/95, the ESPRIT project MHAOTEU (24942) and the Direcció General de Recerca of the Generalitat de Catalunya under grant 1996FI-03039-APDT.

The research described in this paper has been developed using the resources of the Center of Parallelism of Barcelona (CEPBA).

References

1. T.M. Austin and G. S. Sohi. "Dynamic Dependency Analysis of Ordinary Programs". In *Proc. of Int. Symp. on Computer Architecture*, pp 342-351, 1992.
2. M. Butle T.Y. Yeh, Y. Patt, M. Alsup, H. Scales and M. Shebanowr. "Single Instruction Stream Parallelism is Greater than Two". In *Proc. of Int. Symp. on Computer Architecture*, pp. 276-286, 1991.
3. M. Franklin and G. S. Sohi. "ARB: A Hardware Mechanism for Dynamic Reordering of Memory References". *IEEE Transactions on Computer*,45(6), pp. 552-571, May 1996.
4. L. Gweunnap. "Digital 21264 Sets New Standard". *Microprocessor Report*, 9(3), March 1995.
5. J.L Hennessy and D.A. Patterson. *Computer Architecture. A Quantitative Approach*. Second Edition. Morgan Kaufmann Publishers, San Francisco 1996.
6. N.P. Jouppi and D.W. Wall. "Available Instruction-Level Parallelism for Superscalar and Superpipelined Machines". In *Proc. of the ACM Conf. on Architectural Support for Programming Languages and Operating Systems*, 1989.

7. F. Gabbay and A. Mendelson. "Speculative Execution Based on Value Prediction". Technical Report, Technion, 1997
8. J. Gonzalez and A.Gonzalez. "Speculative Execution via Address Prediction and Data Prefetching". In *Proc. of the International Symposium on Supercomputing (ICS)*, pp 196,203, 1997.
9. J. Gonzalez and A. Gonzalez. "Memory Address Prediction for Data Speculation". *In proceedings of the Europar Conference*, 1997.
10. M.S. Lam and R.P. Wilson. "Limits on Control Flow on Parallelism". In *Proc. of Int. Symp. on Computer Architecture*, pp 46-57, 1992
11. M.H. Lipasti and J.P. Shen. "Exceeding the Dataflow Limit via Value Prediction". In *Proc. of Int. Symp. on Microarchitecture*, 1996.
12. M.H. Lipasti, C.B. Wilkerson and J.P. Shen. "Value Locality and Load Value Prediction". In *Proc. of the ACM Conf. on Architectural Support for Programming Languages and Operating Systems*, 1996.
13 P. Marcuello, A. Gonzalez and J. Tubella. "Speculative Multithreaded Processors". In *Proc. of the International Symposium on Supercomputing (ICS)*, 1998.
14. Y. Sazeides, S Vassiliadis and J.E. Smith. "The Performance Potential of Data Dependence Speculation & Collapsing". In *Proc. of Int. Symp. on Microarchitecture*, 1996.
15. Y. Sazeides and J.E. Smith. "The Predictability of Data Values". In *Proc. of Int. Symp. on Microarchitecture*, pp 248-258. 1997.
16. M.D. Smith, M. Johnson and M.A. Horowitz. "Limits on Multiple Instruction Issue". In *Proc. of the ACM Conf. on Architectural Support for Programming Languages and Operating Systems*, 1989.
17. J.E. Smith and A.R. Pleszkun. "Implementing Precise Interrupts in Pipelined Processors". *IEEE Transaction on Computers*, 37(5), pp. 562-573, May 1988
18. G.Sohi, S.Breach and T. Vijaykumar."Multiscalar Processors". In *Proc. of Int. Symp. on Computer Architecture*, pp 414-425, 1995
19. A. Srivastava and A. Eustace. "ATOM: A system for building customized program analysis tools". In *Proc of the 1994 Conf. on Programming Languages Design and Implementation*, 1994.
20. K.B. Theobald, G.R. Gao and L.J. Hendren. "On the Limits of Program Parallelism and its Smoothability" In *Proc. of Int. Symp. on Microarchitecture*, pp 10-19, 1992.
21. D.W. Wall. "Limits of Instruction-Level Parallelism". Technical Report WRL 93/6 Digital Western Research Laboratory, 1993.
22 K. Wang and M. Franklin. "Highly Accurate Data Value Prediction using Hybrid Predictors". In *Proc. of Int. Symp. on Microarchitecture*, pp 281-290, 1997.
23. K.C. Yeager. "The MIPS R10000 Superscalar Microprocessor" *IEEE Micro*, 16(2), pp. 28-40, April 1996.

High Performance Cache Management for Parallel File Systems

F. García, J. Carretero, F. Pérez, and P. de Miguel

Facultad de Informática, Universidad Politécnica de Madrid (UPM)
Campus de Montegancedo, Boadilla del Monte, 28660 Madrid, Spain
fgarcia@fi.upm.es

Abstract. Caching has been intensively used in memory and traditional file systems to improve system performance. However, the use of caching in parallel file systems has been limited to I/O nodes to avoid cache coherence problems. In this paper we present the cache mechanisms implemented in *ParFiSys*, a parallel file system developed at the UPM. *ParFiSys* exploits the use of cache, both at processing and I/O nodes, with aggressive pre-fetching and delayed-write techniques. The cache coherence problem is solved by using a dynamic scheme of cache coherence protocols with different sizes and shapes of granularity. Performance results, obtained on an IBM SP2, are presented to demonstrate the advantages offered by the cache management methods used in *ParFiSys*.

1 Introduction

There is a general trend to use *parallelism* in the I/O systems to alleviate the growing disparity in computational and I/O capability of the parallel and distributed architectures. Parallelism in the I/O subsystem is obtained using several independent I/O nodes supporting one or more secondary storage devices. Data are *declustered* among these nodes and devices to allow parallel access to different files, and parallel access to the same file. Parallelism has been used in some parallel file systems and I/O libraries described in the bibliography (CFS [20], Vesta [5], *ParFiSys* [2], PASSION [4], Galley [19], Scotch [11], PIOUS [17]).

Caching has been a technique frequently used in memory and traditional file systems to improve system performance. Caching [14,2] can be used in parallel file systems, by allocating a buffer cache at the processing nodes (PN) and I/O nodes (ION). This approach improves I/O performance by avoiding unnecessary disk traffic, network traffic and servers load, and also by allowing pre-fetching and delayed-write techniques [14,2]. However, the use of caching in parallel file systems has been limited to I/O nodes because any attempt to maintain caching at the processing nodes would require a cache coherence protocol.

In this paper we demonstrate that the use of caching at processing nodes is feasible in parallel file systems. With this aim we show the cache management policies and mechanisms implemented in *ParFiSys*[1], a parallel file system de-

[1] http://laurel.datsi.fi.upm.es/~gp/parfisys.html

J. Palma, J. Dongarra, and V. Hernández (Eds.): VECPAR'98, LNCS 1573, pp. 466–479, 1999.

veloped at the UPM[2]. *ParFiSys* exploits the use of caching both at processing and I/O nodes. It avoids the cache coherence and false sharing problems in an efficient manner by using a dynamic scheme of cache coherence protocols with different sizes and shapes of granularity.

The rest of the paper is organised as follows. Section 2 describes *ParFiSys* main architectural features. Section 3 presents the cache management implemented in *ParFiSys*. The cache coherence problem and how *ParFiSys* solved this problem is explained in section 4. Performance measurements are shown in section 5. Finally, section 6 summarises our conclusions.

2 *ParFiSys* Architecture

ParFiSys [2, 3] is a parallel and distributed file system developed at the UPM to provide parallel I/O services for parallel and distributed systems. To fully exploit all the parallel and distributed features of the I/O hardware, the architecture of *ParFiSys* is clearly divided in two levels: file services and block services (see figure 1.) The first level is comprised into a component named *ParClient*. It provides file services that can be obtained using two mechanisms: linked library or message passing library. The first is preferred for parallel machines [4, 13], where a single user is usually expected per processing node (PN). The second is aimed to be used in distributed systems, where several users may be requesting I/O services to the *ParClient*. Both modalities provide the users with a *ParClient* library that includes the POSIX interface and some high-performance extensions [2] . The main architectural difference between the former models, is that with the linked library approach the *ParClient* has to be present on every PN requesting I/O, while the message passing option allows the existence of remote users for a *ParClient*. This option is specifically though for distributed file systems or big scale parallel machines, and it can be used to define *groups* of users related with a single *ParClient* to increase scalability [2]. The *ParClient* translates user addresses to logical blocks establishing the connections with the *ParServer*. All communication is handled through a high performance I/O library, named *ParServer* library. This library optimises the I/O requests and sends them to the I/O servers via message passing. It also controls the flow of data from the I/O servers to the application's address space and vice-versa. The *ParServer*, located at the input/output nodes (IONs), deal with logical block requests issued by the *ParClient*, translating them to the local secondary storage devices. Both levels intensively use caching to optimise I/O operations: the *ParClient* to avoid remote operations, and the *ParServer* to reduce accesses to devices.

ParFiSys uses a very generic *distributed partition* which allows to create several types of file systems on any kind of parallel I/O system. A distributed partition has a unique identifier, physical layout, list of sub-partitions, etc. The physical layout, represented as the tuple $(\{NODE_n\}, \{CTLR_c\}_n, \{DEV_d\}_{nc})$, describes the set of I/O nodes, controllers per node, and devices per controller.

[2] *ParFiSys* has been developed under EU's ESPRIT Project GPMIMD (P-5404)

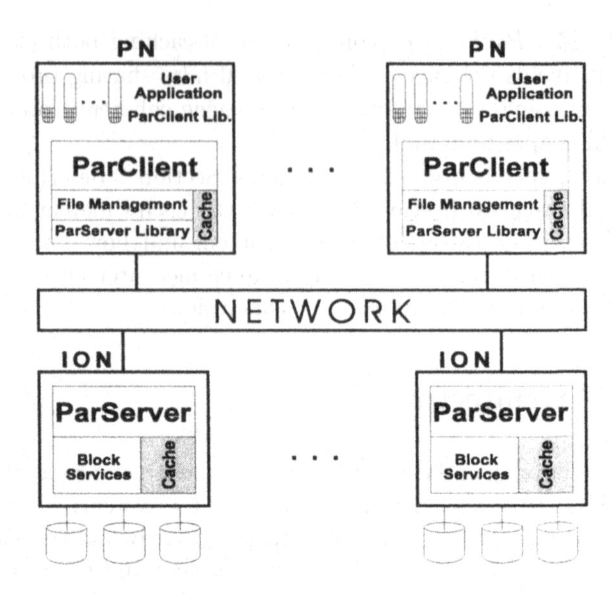

Fig. 1. *ParFiSys* Architecture

ParFiSys partitions can be modified by the administrator freely, adding or removing devices dynamically. The only restriction to be considered is that devices being used by the existing applications should not be affected. The current implementation of *ParFiSys* supports three kinds of predefined file systems on the partition structure (see figure 2):

- UNIX-like non-distributed file systems, where $|NODE| = 1$, $|CTLR| = 1$, $|DEV| = 1$.
- Extended distributed file systems with sequential layout, where $|NODE| = k$, $|CTLR| = n$, $|DEV| = m$. They can be seen as a concatenation of UNIX-like partitions.
- Distributed file systems striped with cyclic layout, where $|NODE| = k$, $|CTLR| = n$, $|DEV| = m$. Blocks are distributed through the partition devices following a *round-robin* pattern.

3 Cache Management in *ParFiSys*

ParFiSys exploits the use of caching both in *ParClient* and *ParServer* (see figure 1), allowing multiple readers and writers concurrently. Specially important is the use of cache at *ParClient*. Each *ParClient* has a buffer cache that is maintained by a *ParClient cache manager*. When a user request is received in a *ParClient*, the whole buffer is analysed to obtain the list of file blocks associated to the buffer.

Once the user buffer has been mapped to file blocks, the whole block list is searched in the *ParClient* cache with a single search operation. After this step,

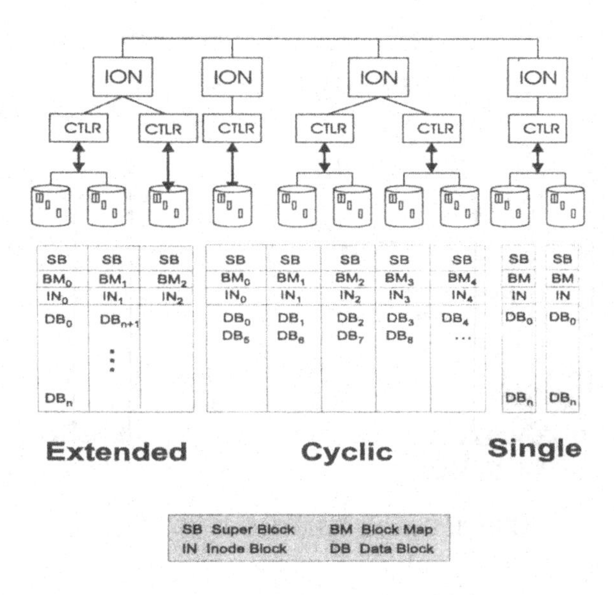

Fig. 2. File Systems Available on *ParFiSys*

two new lists of blocks are obtained: *present* and *absent* blocks. The *present* blocks, those found on the cache, are immediately copied to the user space. The *absent* blocks, those not found in the cache, have to be requested to the I/O devices through the *ParServer* library. This library is conceived as a high-level device driver that concurrently manages the ION related operations. Requests not serviced are enqueued and consumed later by a thread that explores the list of blocks from the queue and generates an independent list of blocks per each ION. These blocks are requested concurrently to each *ParServer*, overlapping I/O and computation (see figure 3.) The result of this stage will be different depending on the mapping function associated to each logical device. Anyway, once the set of blocks stored into an I/O node has been determined, a thread is awaken to take care of the *ParClient-ParServer* operations related to this set of blocks. All the threads execute concurrently, notifying the end of their operations by synchronising themselves with a barrier previously established by the *ParServer* library. An I/O operation is finished at this level only when all the threads have reached the barrier. At this moment, the result is notified to the cache management procedures and the involved I/O operations are finished. The number of client-server interactions needed depends on the maximum number of blocks, named *grouping factor*, that can be requested from each file system on a single operation.

This cache scheme exploits the parallelism in the accesses, having two main advantages: there is not synchronization among the *ParServer* involved in an I/O operation, and there is not sequentialization in the *ParClient-ParServer* communication.

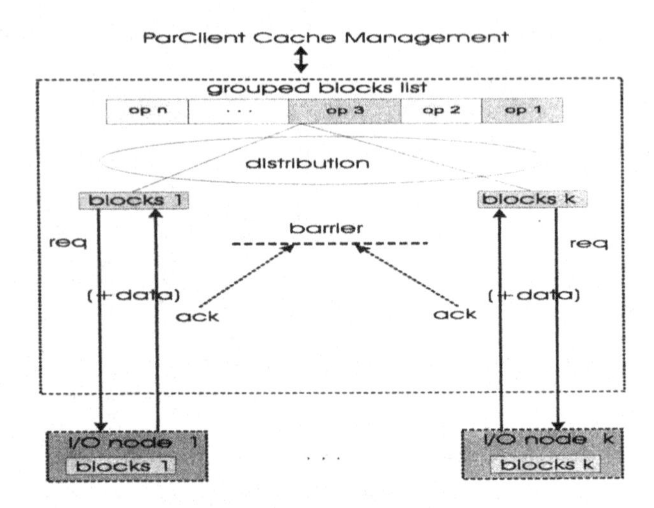

Fig. 3. Parallel Operations in *ParFiSys*

Two additional mechanisms have been used to enhance the behaviour of the cache in *ParFiSys*:

Read-ahead Each *ParServer* reads ahead data using an *Infinite Block Lookahead* (IBL) predictor, that computes the number of blocks (n) to be read in advance. Pre-fetch is executed in the *ParClient* using an *adaptive* predictor, valid for sequential and interleaved patterns, whose behaviour depends on the I/O patterns exhibited by the local processes. Pre-fetch is executed asynchronously to the user requests to enhance the answer time.

Write-before-full This is a *delayed-write* policy that flush dirty blocks from the *ParClient* to the *ParServer*, and from the *ParServer* to the I/O devices, before free blocks may be needed in the cache. *Pre-flush* is activated when a low threshold, calculated by the write-before-full daemon, is reached. When a write request is executed, the number of dirty blocks in the cache is computed. If it is larger than a fixed threshold, a massive flush is executed for the dirty blocks belonging to the file system storing the file. All the operations are executed asynchronously to the user requests to avoid delaying the answer time.

4 Avoiding the Cache Coherence Problem

The main problem of using caching at the client nodes is the possibility of having shared writing of a file from different clients [1,18,10], which might lead to an incoherent view of data. Nelson [18] describes two forms of write-sharing: *sequential write-sharing* (SWS), that occurs when a client reads or writes a file that was previously written by another client, and *concurrent write-sharing* (CWS), that occurs when a file is simultaneously open for reading and writing on more than one client. Concurrent write-sharing is not usual in distributed file systems [1, 18], but it is very frequent in parallel file systems [16] and *meta-computing* [22].

A cache coherence protocol is required to avoid this problem. The use of cache coherence protocols has been unpopular in parallel file systems because of its overhead [16]. Thus, most parallel file systems usually have caching schemes such as the ones implemented in CFS [20], where only the IONs maintain a buffer cache for files. This solution avoids cache coherence problems because there only is a single copy of the data in the whole system. Distributed file systems, where write sharing is infrequent [18], use cache coherence protocols mostly based on weak [12] or coarse grain models [18]. However, most existing file systems with cache coherence protocols fail to provide efficient solutions to the problem of cache coherence for parallel applications that concurrently write a file. NFS [23], very popular in commercial environments, is unable to maintain a consistent view of the file system for parallel applications. AFS [12] does not support concurrent write-sharing due to the session semantic implemented, which makes it not suitable for parallel applications. Sprite [18] ensures concurrent write-sharing coherence by disabling client caches, thus limiting the potential benefits of caching for many parallel applications. There are very few cache coherence solutions in parallel file systems. ENWRICH [21] provides a cache coherence solution for parallel file systems, but it is not a general one because client caches are only used for writing.

Recently other approaches to avoid cache coherence problem in parallel and distributed file systems have been proposed. An example is the cooperative cache scheme proposed in [7], that eliminates the cache coherence problem by avoiding data replication. This solution, however, reduces the potential parallelism that the use of data replication may offer.

4.1 *ParFiSys* Coherence Model

In order to solve the write-sharing problem, *ParFiSys* uses a dynamic cache coherence scheme [9, 10] based on the following protocols:

- *Sequential coherence protocol* (SCP), that solves the SWS problem and detects the CWS on a file.
- *Concurrent coherence protocol* (CCP), that solves the CWS problem after being activated when the SCP detects a CWS situation on a file.

Sequential Coherence Protocol This protocol ensures coherence in SWS situations and detects CWS situations on a file. It has a behaviour similar to the Sprite protocol [18]: the servers track *open* and *close* operations to know not only which clients are currently using a file, but whether any of them are potential writers. When a client opens a file, the event is notified to the server storing the file descriptor. If there is no CWS for the file, the server looks whether another client has updated the file data in its local cache, because of the delayed-write policy, and requests it to flush the data. When the server has the most up to date copy of the file, a message is sent to the client to enable local caching for the file. No more interactions with the server are needed to maintain coherence, thus alleviating overhead. When a CWS situation is detected by the server, a message is sent to all the clients with the file open to activate the CCP. When the CWS situation disappears, a message is sent to all the clients with the file open to deactivate the CCP.

SCP has been optimised to reduce client-server interactions by sending coherence messages to the servers only when a change of the client local state of the file occurs. The local state of a file changes when it is open the first time, when it was open for read and it is open for write, when it is closed for write and it remains open for read, and when it is closed by the last user in the client. SCP has also been optimised to reduce servers load by distributing the protocol overhead among all servers. Each server executes SCP only for the files whose descriptors are stored on it, which alleviates the bottleneck of a centralised service and improves scalability.

Concurrent Coherence Protocol This protocol solves the CWS problem. It is activated when a CWS is detected on a file, being executed on each access to the file while the CWS situation remains. CCP is based on invalidations, directories and the existence of a exclusive write-shared copy of data.

The main problem in cache coherence protocols is the false sharing situation generated when multiple processes, belonging to the same parallel application, access the same file for writing. To alleviate false sharing problems, it would be desirable to allow the parallel applications to adjust the granularity of the protocol to their I/O patterns. *ParFiSys* ensures this by allowing the users to define *coherence regions*. A coherence region is a disjoint subset of the file used as the coherence unit. It has two main features: *size* and *shape*. The size of a region may range from the whole file to byte. The shape of a region can be defined according to the most frequent parallel access patterns: sequential and interleaved (see figure 4).

The applications can define the mapping of the regions on a file using four parameters (see figure 4):

- *Register size*, minimum unit for coherence.
- *Register stride*, width of the register groups into each segment.
- *Segment size*, number of register groups in a segment.
- *Segment stride*, distance between two segments with the same pattern.

This model of region is suitable to map very different parallel I/O patterns. Moreover, it allows to define *optimal regions*, the best suited to the I/O access pattern, to minimise coherency overhead. Optimal regions are defined on a file when:

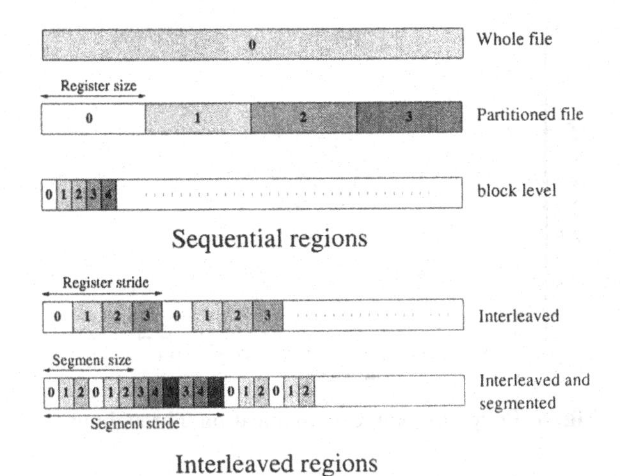

Fig. 4. Some Coherence Region Patterns

1. The number of regions is equal to the number of process in the parallel application.
2. Each process only accesses the data of a single region.

The use of optimal regions allows to adjust perfectly the protocol granularity and the I/O pattern of the applications, offering coherence with a minimal cost. This coherence regions model can be applied to the High Performance Fortran distributions [8], Vesta interface [5] and MPI-IO interface [6].

CWP compels the clients to check the coherence state of the region on every access to it by acquiring the appropriate read or write *tokens.* When a client does not have the appropriate token, a message to the region's manager must be sent to request the desired rights on the region. The server stores a *callback* for each region in a coherence directory to trace the coherence state of the region. If a conflict is detected, the call-backs are revoked. The region's manager guarantees that at any given time there is a single read-write token or any number of read-only tokens. When a write token is revoked in a client, the client must flush any dirty data of the region. If the new token is for read the token in the previous writer is changed from writing to read-only. When the appropriated rights on a region are acquired, they remain until explicit revocation, thus eliminating the overhead for future accesses. Two main design mechanisms were used in *ParFiSys* to define the coherence protocols: callback and directory location and management, and callback revocation policy. Two policies can be used for the first issue: *centralised* (C), where coherence for all the regions of a file is maintained by the server storing the file descriptor (region manager), and *distributed* (D), where the information of the coherence region is distributed among several servers, each one being responsible of maintaining the coherence of the regions allocated to it. Two approaches can be used to revoke call-backs when a conflict appears: *server driven* (SD) and *client driven* (CD). In SD, the server

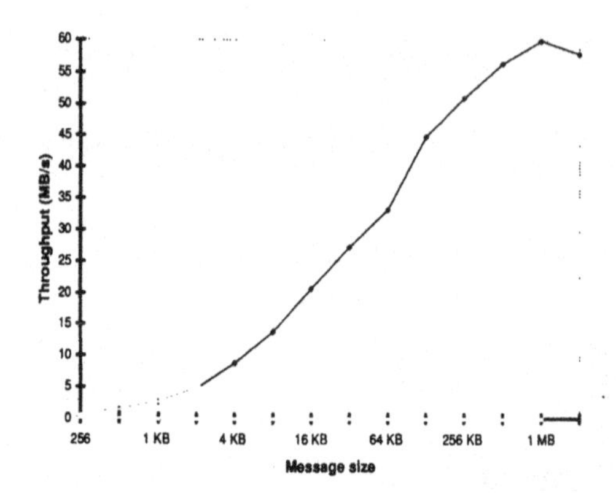

Fig. 5. Point-to-point Communication Throughput

sends revocation messages to all the clients caching data from the conflictive region. In CD, the client generating the conflict sends revocation messages, on behalf of the servers, to other clients caching data from the conflictive region. Several CCP have implemented in *ParFiSys* by combining different management and revocation policies: C-SD, C-CD, D-SD, and D-CD.

5 Performance Evaluation

This section describes the performance experiments designed to test cache management in *ParFiSys* and the results obtained by running them on an IBM SP2 machine.

The IBM SP2 used is a distributed-memory MIMD machine with 14 nodes available to execute parallel applications. Each node has a 66 MHz POWER2 RISC System/6000 processor with 256 MB of memory, being connected to both an Ethernet and IBM's high performance switch. Because of the IBM's message-passing libraries (PVM, MPL or MPI) cannot operate in a multi-threaded environment, we have implemented a multi-threaded subset of MPI using TCP/IP on top of the high performance switch. To characterise the message-passing performance of our communication library, we executed two simple benchmarks on the SP2. The first evaluates point-to-point communication throughput by engaging two processes in a sort of ping-pong. One process reads the value of a wall-time clock before invoking a send operation and it then blocks in a receive operation. Once the latter operation finishes, the clock is read again. The throughput achieved is computed with a half of the communication time and the message size. Figure 5 shows the results obtained for this benchmark. The maximum throughput between two nodes is approximately 60 MB/s.

The second benchmark emulates *ParFiSys* reading and writing activity. We used 4 servers and varied the number of clients from 1 to 8. Clients send (write to servers) or receive (read from servers) 100 MB declustered across the servers. Each client sends, or receives, the same amount of data to, or from, the servers

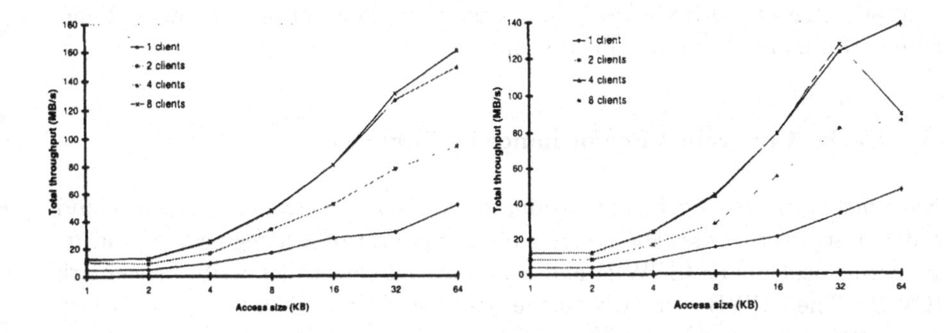

Fig. 6. Total Message Passing Throughput in *ParFiSys*

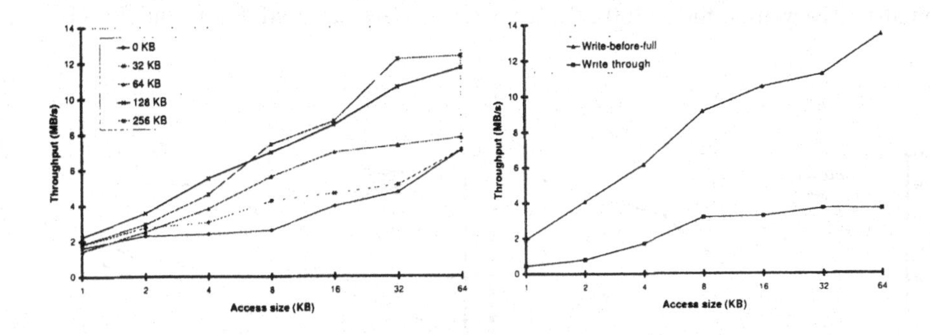

Fig. 7. Pre-fetching and Write-before-full Performance

using a fixed record size. Figure 6 shows the total throughput obtained for reading and writing operations. As shown in the figure, the maximum throughput achieved increases with the number of clients and the record size.

All experiments described below were executed using 4 ION, with a single simulated disk with a bandwidth of 5 MB/s on each. In all tests a 4 MB per-client cache and a 16 MB per-ION cache were used. The file size used in all experiments was 100 MB. The stripe-width was 4 and the stripe-unit size was 8 KB.

5.1 Pre-fetch and Write-before-full Evaluation

Figure 7, left, shows the throughput obtained when a client sequentially reads a file of 100 MB. This test varied the read-ahead value, from 0 KB to 256 KB, and the access size, from 1 KB to 64 KB.

Figure 7, right, shows the throughput obtained when a client sequentially writes a file of 100 MB using either write-through or write-before-full. The threshold used in write-before-full was a 95 % of the *ParClient* cache size. The results obtained demonstrate that having a cache at processing nodes, managed using pre-fetch and write-before-full, is a useful mechanism to increase read and write performance in parallel file systems.

5.2 Cache Coherence Performance in *ParFiSys*

Two benchmarks were defined to evaluate the *ParFiSys* coherence protocol and to demonstrate their feasibility for parallel applications: a segmented concurrent write benchmark (SCWB) and a interleaved concurrent write benchmark (ICWB). The SCWB, similarly to the one described in [17], is a parallel program with partitioned access that divides a file into contiguous segments, one per process, with each segment accessed sequentially by a different process. In the ICWB, each process concurrently writes the file in a interleaved fashion. The parallel program for each benchmark consists of 1, 2, 4 and 8 processes that concurrently write a file of 100 MB. Two access sizes are used: 8 KB and 32 KB.

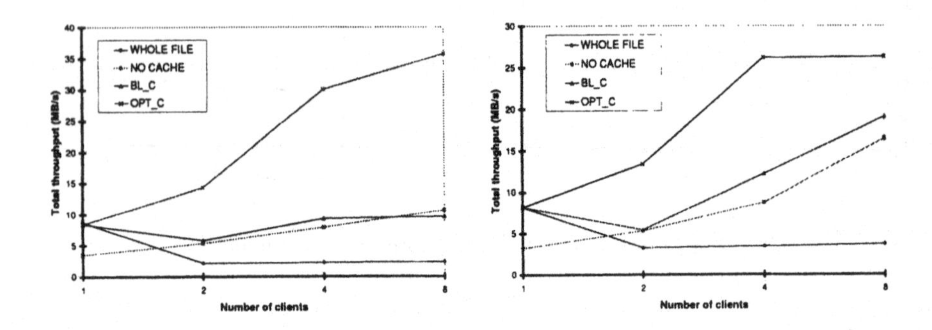

Fig. 8. Concurrent Segmented and Interleaved Benchmark for 8 KB Access Size

Figures 8 and 9 show the aggregated bandwidth for concurrent segmented and interleaved I/O patterns, respectively. Several cache coherence protocols using SCWB and ICWB are evaluated: client cache deactivated (NO CACHE), file granularity (WHOLE F.), block granularity for centralised protocol (BL_C), and optimal regions for centralised protocol (OPT_C). The first relevant result obtained shows that maintaining coherency with file granularity is the worst method, mainly due to false sharing. Deactivating cache is very similar to the block centralised, because the number of client-server interactions is almost the same. The small difference observed between deactivating cache and block distributed is mainly due to the lower contention generated on the servers by the coherence protocols. This feature also makes the block distributed protocol more scalable. The best results are obtained using optimal regions, both centralised and distributed. This behaviour is mainly due to the false sharing elimination,

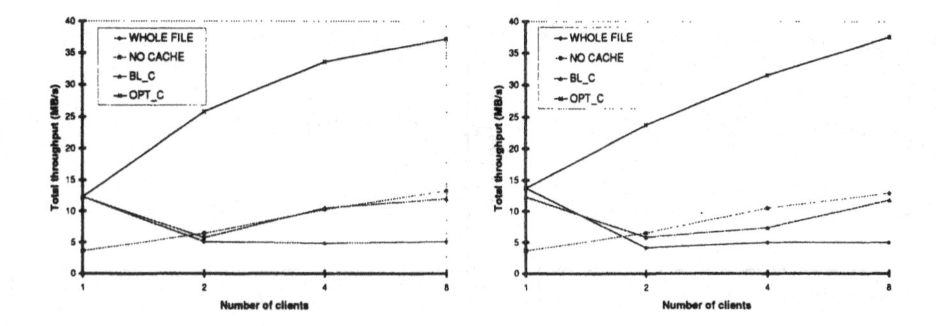

Fig. 9. Concurrent Segmented and Interleaved Benchmark for 32 KB Access Size

the main problem with whole file granularity, the minimisation of the coherence load, the main problem in block granularity protocols, and the local cache utilisation, the main problem in cache deactivation.

Concurrent write benchmarks for segmented and interleaved I/O patterns show that optimal regions provide a performance very close to the ideal one. Moreover, they show a good scalability compared with the other protocols, whose performance decreases very quickly as the number of clients is increased.

Figure 10 compares the bandwidth obtained in the SCWB using normal files with optimal regions for centralised and distributed protocols, versus the use of multi-files. A multi-file [15, 2] is a collection of subfiles, each of which is a separate sequence of bytes. A multi-file is created by a parallel program with a certain number of subfiles, usually equal to the number of processes in the program, with each process accessing its own subfile. A multi-file combines the advantages of a single file, single name for a single data set, with those of multi-files, that are independently addressable. Multi-files are an alternative mechanism to the segmented patterns used in SCWB. The results are very similar in all cases, being a little better for multi-files because of the lighter writing operation on each subfile than on a normal file with concurrent access. The proposed protocols with normal files offer an efficient alternative to the use of multi-files for segmented I/O patterns, because present a performance very similar with the advantage of using less specialised and more portable interfaces.

6 Conclusions

This paper has presented the design of the cache scheme implemented in *ParFiSys*. *ParFiSys* allocates buffer caches at the processing and I/O nodes, improving I/O performance by avoiding unnecessary disk traffic, network traffic and servers load, and by allowing pre-fetching and delayed-write techniques. The cache coherence problem is solved, without loss of scalability, by using a dynamic scheme of cache coherence protocols where data coherence is maintained on user-defined *coherence regions* for the conflictive file. The utilisation

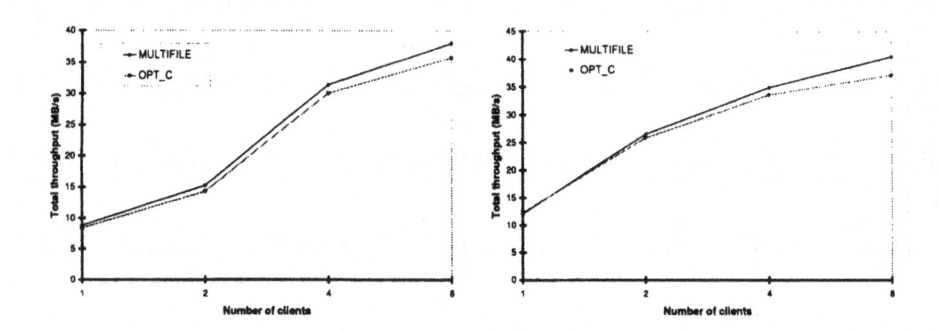

Fig. 10. Optimal Regions in SCWB versus Multi-files for 8 KB and 32 KB Access Size

of two protocols, SCP and CCP, allows to afford all the conflictive situations for SWS and CWS patterns, as demonstrated with the evaluation results obtained by running *ParFiSys* on an IBM SP2. The benchmarks used to test the model show considerably better results for our model than for other existing models. The aggregated bandwidth obtained is higher when using our model, mainly because false sharing is reduced, coherence load is minimised, and local caches at processing nodes are heavily used. The proposed protocols also offer an efficient alternative to the use of multi-files for segmented I/O patterns, because present a very similar performance with the advantage of using less specialised and more portable interfaces.

Acknowledgments

We want to express our grateful acknowledgment to the CESCA institution for giving us access to their IBM SP2 machine.

References

1. Burrows, M.: Efficient Data Sharing. PhD thesis, Computer Laboratory, University of Cambridge, December 1988.
2. Carretero, J.: Un Sistema de Ficheros Paralelo con Coherencia de Cache para Multiprocesadores de Proposito General. PhD thesis, Universidad Politécnica de Madrid, May 1995.
3. Carretero, J., Pérez, F., De Miguel, P., García, F., Alonso, L.: Performance Increase Mechanisms for Parallel and Distributed File Systems. *Parallel Computing*, (23):525–542, August 1997.
4. Choudhary, A., Bordawekar, R., More, S., Sivaram, K., Thakur, R.: PASSION Runtime Library for the Intel Paragon. Proceedings of the Intel Supercomputer User's Group Conference. June 1995.
5. Corbett, P., Feitelson, D.: The Vesta Parallel File System. ACM Transactions on Computer Systems, 14(3):225–264, August 1996.

6. Corbett, P., et al.: MPI-IO: A Parallel File I/O Interface for MPI. Technical Report NAS-95-002, NASA Ames Research Center, June 1995.
7. Cortes, T., Girona, S., Labarta, J.: Design Issues of a Cooperative Cache with no Coherence Problems. 5th Workshop on I/O in Parallel and Distributed Systems (IOPADS'97), San Jose, CA, November, 1997.
8. High Performance Fortran Forum.: High Performance Fortran Language Specification. May 1993.
9. García, F.: Coherencia de Cache en Sistemas de Ficheros para Entornos Distribuidos y Paralelos. PhD thesis, Universidad Politécnica de Madrid, España, September 1996.
10. García, F., Carretero, J., Pérez, F., De Miguel, P., Alonso, L.: Cache Coherence in Parallel and Distributed File Systems. 5th EUROMICRO Workshop on Parallel and Distributed Processing. IEEE. London, pages 60–65, January 1997.
11. Gibson, G.: The Scotch Paralell Storage Systems. Technical Report CMU-CS-95-107, Scholl of Computer Science, Carnegie Mellon University, Pittsburbh, Pennsylvania, 1995.
12. Howard, J., et al.: Scale and Performance in a Distributed File System. ACM Transactions on Computer Systems, 6(1):51–81, February 1988.
13. Huber, J., et al.: PPFS: A High Performance Portable Parallel File System. Proceedings of the 9th ACM International Conference on Supercomputing, pages 385-394, July 1995.
14. Kotz, D.: Prefetching and Caching Techniques in File Systems for MIMD Multiprocessors. PhD thesis, Duke University, USA, April 1991.
15. Kotz, D.: Multiprocessor file system interfaces. Proceedings of the 2nd. International Conference on Parallel and Distributed Information Systems, pages 194–201, May 1993.
16. Kotz, D., Nieuwejaar, N.: File System Workload on a Scientific Multiprocessor. IEEE Parallel and Distributed Technology. Systems and Applications, pages 134–154, Spring 1995.
17. Moyer, S. A., Sunderam, V. S.: Scalable Concurrency Control for Parallel File Systems. Technical Report CSTR-950202, Department of Math and Computer Science, Emory University, Atlanta, GA 30322, USA, 1995.
18. Nelson, M., Welch, B., Ousterhout, J.: Caching in the Sprite Network File System. ACM Transactions on Computer Systems, 6(1):134–154, February 1988.
19. Nieuwejaar, N., Kotz, D.: The Galley Parallel File System. Technical Report PCS-TR96-286, Darmouth College Computer Science, 1996.
20. Pieper, J.: Parallel I/O Systems for Multicomputers. Technical Report CMU-CS-89-143, Carnegie Mellon University, Computer Science Department, Pittsburgh, USA, 1989.
21. Purakayastha, A., Ellis, C. S., Kotz, D.: ENWRICH: A Computer-Processor Write Caching Scheme for Parallel File Systems. Technical Report TRCS-1995-22, Department of Computer Science, Duke University, Durhan North Carolina, October 1995.
22. Del Rosario, J., Choudhary, A.: High-Performance I/O for Massively Parallel Computers. IEEE Computer, pages 59–68, March 1994.
23. Sandberg, R., et al.: Design and Implementation of the SUN Network Filesystem. Proceedings of the 1985 USENIX Conference. USENIX, 1985.

Using Synthetic Workloads for Parallel Task Scheduling Improvement Analysis

João Paulo Kitajima[1] and Stella Porto[2]

[1] Departamento de Ciência da Computação
Universidade Federal de Minas Gerais
Caixa Postal 702 30123-970 Belo Horizonte - MG Brazil
kita@dcc.ufmg.br

[2] Computação Aplicada e Automação
Universidade Federal Fluminense
Rua Passo da Pátria 156 24210-240 Niterói - RJ Brazil
stella@caa.uff.br

Abstract. This paper presents an experimental validation of makespan improvements of two scheduling algorithms: a greedy construction algorithm and a tabu search based algorithm. Synthetic parallel executions were performed using the scheduled graph costs. These synthetic executions were performed on a real parallel machine (IBM SP). The estimated and observed response times improvements are very similar, representing the low impact of system overhead on makespan improvement estimation. This guarantees a reliable cost function for static scheduling algorithms and confirms the actual better results of the tabu search meta-heuristic applied to scheduling problems.

1 Introduction

Parallel applications with regular and well-known behaviour, where task execution time estimates are fairly reliable, are suited to static task scheduling (in opposition to dynamic scheduling, performed during the execution of the application). This is the case of a great majority of scientific applications. For these applications, the static scheduling algorithm is executed once, before the execution of the parallel program, which is then actually run several times according to the previously obtained schedule. Consequently, even if the scheduling algorithm is a costly procedure, this cost will be amortised throughout the numerous executions of the parallel application, i.e. the obtained schedule is re-applied repeatedly.

Static task scheduling is, thus, performed based on estimated data about the parallel application and the system architecture. Therefore, realistic performance evaluation of a task scheduling algorithm can only be fully accomplished if practical results are also considered. In this sense, the present work analyses the quality of greedy and tabu search task scheduling algorithms comparing estimated deterministic results with the actual observed makespan of several parallel

J. Palma, J. Dongarra, and V. Hernández (Eds.): VECPAR'98, LNCS 1573, pp. 480–493, 1999.

synthetic applications executing on real heterogeneous parallel machines following the static schedule previously determined. The following section presents the schedule system model. In Section 3, both the greedy and tabu search algorithms are described. In Section 4, we report the overall experimentation, including: (i) a description of the testing platform and problem instances considered during the testing phase; (ii) the most significant numerical results, and (iii) the comparative solution quality analysis according to different parameters. Section 5 presents some brief concluding remarks.

2 The Scheduling Model

A parallel application Π with a set of n tasks $T = \{t_1, \cdots, t_n\}$ and a heterogeneous multiprocessor system composed by a set of m interconnected processors $P = \{p_1, \cdots, p_m\}$ can be represented by a task precedence graph $G(\Pi)$ and an $n \times m$ matrix μ, where $\mu_{kj} = \mu(t_k, p_j)$ is the estimated execution time of a task $t_k \in T$ at processor $p_j \in P$. Each processor can run one task at a time, all tasks can be executed by any processor, and processors are said to be uniform in the sense that $\mu_{kj}/\mu_{ki} = \mu_{lj}/\mu_{li}, \forall t_k, t_l \in T, \forall p_i, p_j \in P$. This implies that processors may be ranked according to their processing speeds. In a framework with one single faster (heterogeneous) processor, the heterogeneity may be expressed by a unique parameter called *processor power ratio*, PPR, which is the ratio between the processing speed of the fastest processor and that of the remaining ones (those in the subset of homogeneous processors). Thus, an instance of our scheduling problem is characterised by the workload and parallel system models.

Given a solution s for the scheduling problem, a processor assignment function is designed as the mapping $\mathcal{A}_s : T \to P$. A task t_k is said to be assigned to processor $p_j \in P$ in solution s if $\mathcal{A}_s(t_k) = p_j$. The task scheduling problem can then be formulated as the search for an optimal assignment of the set of tasks onto that of the processors, in terms of the *makespan* $c(s)$ of the parallel application (cost of the solution s), i.e. the completion time of the last task being executed. At the end of the scheduling process, each processor ends up with an ordered list of tasks that will run on it as soon as they become executable.

3 Heuristic Task Scheduling Algorithms

We consider two algorithms in this work, namely: a greedy algorithm called DES+ MFT and a parallel tabu search algorithm, here referred as TSpar. Although both of them are heuristic, they present different fundamental characteristics. The former is a construction algorithm, which iteratively assigns tasks to processors based on heuristic criteria, taking into account the static information of the system model. On the other hand, the TSpar is a synchronous parallel implementation of a tabu search meta-heuristic algorithm, which guides an aggressive local search procedure over the task scheduling solution space.

3.1 The DES+MFT Greedy Algorithm

DES+MFT stands for *Deterministic Execution Simulation with Minimum Finish Time* [4]. This algorithm iteratively schedules tasks in a partial order according to the simulated execution of the parallel application (DES), based on the estimated task execution times, while scheduling decisions are made according to the minimum finishing time (MFT) for each "schedulable" task. Figure 1 describes the DES+MFT in a procedural scheme. In this scheme, the *clock* variable measures the evolution of the execution. At the end of this procedure, $c(s) = clock$ is the cost of the obtained solution, i.e., the makespan of the parallel application when submitted to the DES+MFT processor assignment. At each iteration, certain tasks are scheduled to processors, building an ordered list of tasks associated to each processor. This is the actual execution order if tasks were to be executed in an ideal system with estimated execution times. During this deterministic execution simulation, each task $t_k \in T$ assumes one of the following states at each time instant: *non-executable*, *executable*, *executing*, *executed*. At the same time, each processor $p_j \in P$ alternates between two different states: *free* and *busy*. A processor p_j is said to be *busy* if it has a task in the *executing* state allocated to it.

It should be noticed that DES+MFT, like most greedy algorithms, does not come back to re-evaluate the scheduling decisions taken in previous iterations. This means that besides the "look-ahead" feature, it is not capable of making changes in scheduling decisions made in previous iterations, which were based on snapshots of the simulated execution. Consequently, these scheduling decisions depend on how strongly tasks are tied through precedence relations, because they determine the order in which tasks may possibly be scheduled. Differently, the TSpar algorithm, departing from the initial solution obtained by the DES+MFT algorithm, evaluates many other possible assignments, which eventually improve the makespan of the parallel application, as we can see in the following section.

3.2 The Parallel Tabu Search Algorithm

To describe the TSpar algorithm, we first consider a general combinatorial optimisation problem (P) formulated as to

$$\begin{array}{l} \text{minimize } \; c(s) \\ \text{subject to } s \in S, \end{array}$$

where S is a discrete set of feasible solutions. Local search approaches for solving problem (P) are based on search procedures in the solution space S starting from an initial solution $s_0 \in S$. At each iteration, a heuristic is used to obtain a new solution s' in the neighbourhood $N(s)$ of the current solution s, through slight changes in s. A move is an atomic change which transforms the current solution, s, into one of its neighbours, say $\bar{s}$. Thus, $movevalue = c(\bar{s}) - c(s)$ is the difference between the value of the cost function after the move, $c(\bar{s})$, and the value of the cost function before the move, $c(s)$. Every feasible solution $\bar{s} \in N(s)$ is evaluated

```
DES+MFT algorithm
begin
    clock ← 0
    state(p_j) ← free ∀p_j ∈ P
    start(t_k), finish(t_k) ← 0 ∀t_k ∈ T
    while (∃t_k ∈ T | state(t_k) ≠ executed) do
    begin
        for (each t_k ∈ T | state(t_k) = executable and p_j ∈ P) do
            obtain the pair (t_l, p_i) with the minimum finish time
            if (state(p_i) = free) then
            begin
                state(t_l) ← executing
                A_s(t_l) = p_i
                state(p_i) ← busy
                start(t_l) ← clock
                finish(t_l) ← start(t_l) + μ(t_l, p_i)
            end
        Let i be such that finish(t_i) = min_{t_k ∈ T | state(t_k)=executing} {finish(t_k)}
        clock ← finish(t_i)
        for (each t_k ∈ T | state(t_k) = executing and finish(t_k) = clock) do
        begin
            state(t_k) ← executed
            state(A_s(t_k)) ← free
        end
    end
    c(s) ← clock
end
```

Fig. 1. DES+MFT algorithm description.

according to the cost function $c(.)$, which is eventually optimised. The current solution moves smoothly towards better neighbour solutions, enhancing the best obtained solution s^*.

Tabu search [1, 2] may be described as a higher level heuristic for solving minimisation problems, designed to guide other hill-descending heuristics in order to escape from local optima. Thus, tabu search is an adaptive search technique that aims to intelligently exploring the solution space in search of good, hopefully optimal, solutions. The learning capability determines that tabu search supplies richer knowledge about the instance of the problem to be solved than that generated in other iterative algorithms. In the case of the task scheduling problem considered in this paper, the cost of a solution is given by its makespan, i.e., the overall execution time of the parallel application. The neighbourhood $N(s)$ of the current solution s is the set of all solutions differing from it by only a single assignment. If $\bar{s} \in N(s)$, then there is only one task $t_i \in T$ for which $\mathcal{A}_s(t_i) \neq \mathcal{A}_{\bar{s}}(t_i)$. Each move may be characterised by a simple representation given by $(\mathcal{A}_s(t_i), t_i, p_l)$, as far as the position task t_i will occupy in the task list of processor p_l is uniquely defined. If the best move takes the current solution s to a best neighbour solution s' degenerating its cost function, i.e. $c(s') \geq c(s)$, then the reverse move must be prohibited during a certain number of iterations (tabu tenure) in order to avoid cycling. However, there are situations in which a recently prohibited move, if applied after some iterations, will provide a better solution than the best one found by the algorithm so far, despite its prohibited status. In these cases, an *aspiration criterion* is used to override this prohibition, enabling the move to be executed. In [6] and [7] the reader will find more detailed description of the tabu search algorithm.

The promising results obtained through parallelisation led to the possibility of more effectively evaluating solution quality of the proposed tabu search task scheduling algorithm using a parallel implementation. Considering both sequential and parallel implementation, solution quality was analysed according to different parameters and strategies, which needed to fully specify the tabu search algorithm with a certain variety of application model parameters (such as task graph structures, number of tasks, serial fraction and task service demands) and system configurations (such as number of processors and architecture heterogeneity measured by the processor power ratio). It was shown that the tabu search algorithm obtained better results, i.e. shorter completion times for parallel applications, improving up to 40% the makes pan obtained by the DES+MFT algorithm, which in fact is the most appropriate greedy algorithm previously published in the literature [6,8]. We have used the MS-MP parallel version to carry out the experimentation reported here, because it has demonstrated the best speedup results in most of the studied cases [7].

4 Experimental Results

In this section, we depict some experimental results obtained from the execution of synthetic parallel programs scheduled with both the greedy and tabu search algorithms. We first present some results derived from the estimated improvement analysis of tabu search schedules over those generated by the DES+MFT, which is the initial solution for the tabu search algorithm. The performance criterion is the makespan (solution cost) estimated by both algorithms. In the following, we describe *ANDES* [3], a framework for performance evaluation using parallel program models and synthetic programs. Finally, using this framework, we compare execution times of synthetic parallel programs scheduled by DES+MFT and TSpar algorithms.

4.1 Estimated Performance Analysis

DES+MFT and TSpar scheduling algorithms were implemented using ANSI C and PVM (*Parallel Virtual Machine*) [9]. The schedule quality is estimated based on the computed makespan. In other words, the makespan represents the schedule cost, $c(.)$, which is to be minimised.

One of the main goals is to achieve makespan reduction when changing from the schedule produced by DES+MFT to the one produced by TSpar. Thus, solution quality is measured by relative cost reduction, $\mathcal{R}$, computed as

$$\mathcal{R} = \frac{c(s_0) - c(s^*)}{c(s_0)}$$

where s_0 is the initial solution obtained by the greedy algorithm DES+MFT and s^* is the best solution found by the TSpar algorithm.

In [6], relative cost reduction values of up to 30% were obtained considering applications modelled by diamond-shaped precedence graphs. In [8], new results were presented considering other structures for the parallel applications. Part of the *ANDES* benchmark was then used: other types of diamond-shaped graphs (`Diamond3` and `Diamond4`), iterative graphs (`FFT` and `PDE2`), divide-and-conquer strategies (`Divconq`), typical matrix computation structures (`Gauss`), and master-slave models (`MS3`). We can summarise the following results of these above experiments:

- A parallel application is said to be *serialised* by a certain processor assignment algorithm when all of its tasks are scheduled to one unique processor. When the serial fraction (F_s) and/or the processor power ratio (PPR) are very high, the best solution is usually obtained through the serialisation of the application over the heterogeneous processor, which has greater processing capacity. This seems to be clear if we imagine two extreme cases: $F_s = 1$ or $PPR \longrightarrow \infty$. In the first case, we face a totally serial application, which must be executed on the heterogeneous processor (F_s corresponds to the serial fraction defined as the fraction of the total parallel execution time when just one task is executing even if infinite processors were available). In the latter case, the heterogeneous processor is able to execute any task in infinitesimal time, consequently serialisation determines again the best performance.
  In certain circumstances, serialisation will be performed by the DES+MFT algorithm, when there is still available parallelism to be explored in the parallel application. In these cases, the tabu search algorithm will start from a serialised initial schedule, and more easily will be capable of finding different assignments which greatly reduce the overall makespan of the application, augmenting the relative cost reduction.
  For very low and very high PPR values low or null makespan improvements are obtained. In the case of very low values of PPR, which determine an almost completely homogeneous processor set, the greedy algorithm exploits well the potential application parallelism, thus the tabu search algorithm has a very narrow range of possible choices to improve the task allocation achieved during the construction of the initial solution. On the other end of the heterogeneity range, very high PPR values mean that *serialisation* on the very fast processor is the best solution. In these cases both the DES+MFT and TSpar algorithms are able of serialising the application, so makespan improvements are not observed;

- Between the two extremes of the PPR value range, we find a mountain-like peak of improvements, culminating with a PPR that gives the best relative performance achieved by the TSpar algorithm. This point is referred as the PPR_{peak} point. The PPR_{peak} point is highly dependent on the shape of the input task graph. Groups of similar task graphs have a similar behaviour. For example, diamond-shaped graphs present a low PPR_{peak} (around 5). On the other hand, iterative graphs produce a more smooth improvement curve, with higher PPR_{peak} (around 20 or 30), depending on the size of the task graph;
- Not only the structure of the task graph is critical in the relative quality improvement analysis. The number of processors available for scheduling assignments influence the results. The relationship between solution quality improvements and the number of processors is variable depending on the structure of the task graph. On one hand, the greater the number of processors we have, the less heterogeneous the system becomes and thus lower relative cost reduction is achieved. However, a greater number of processors also represents more available parallelism and therefore a greater number of different scheduling possibilities arise.

The above results can be verified in Figures 3 and 2. Figure 3 presents some estimated relative cost reduction values computed between DES+MFT and TSpar algorithms. In [8], Porto *et al.* measured improvements for discrete values of PPR (2, 5, 10, 20, ..., depending on the input). The sizes of the graphs, in number of tasks, are presented below:

Graph	*size 1*	*size 2*
`Diamond3`	66	218
`Divconq`	46	190
`FFT`	83	194
`Gauss`	47	192
`MS3`	52	294

Figure 2 presents a more detailed experiment, with a fine variation of PPR values and number of processors, considering the `Diamond3` benchmark with 66 tasks.

4.2 The Experimental Framework

The *ANDES* Environment - *ANDES* [8] is a PVM-based parallel tool that supports performance evaluation of parallel programs at the prediction level. *ANDES* considers the existing complex overheads of parallel computers. This is achieved through the use of *synthetic parallel executions* directly on the parallel machine. In a synthetic parallel execution, the resources of the parallel computer are used in a controlled way, but no code is generated. All the steps from the interpretation of the parallel program graph-based and of the parallel machine models to the synthetic execution on the target parallel machine

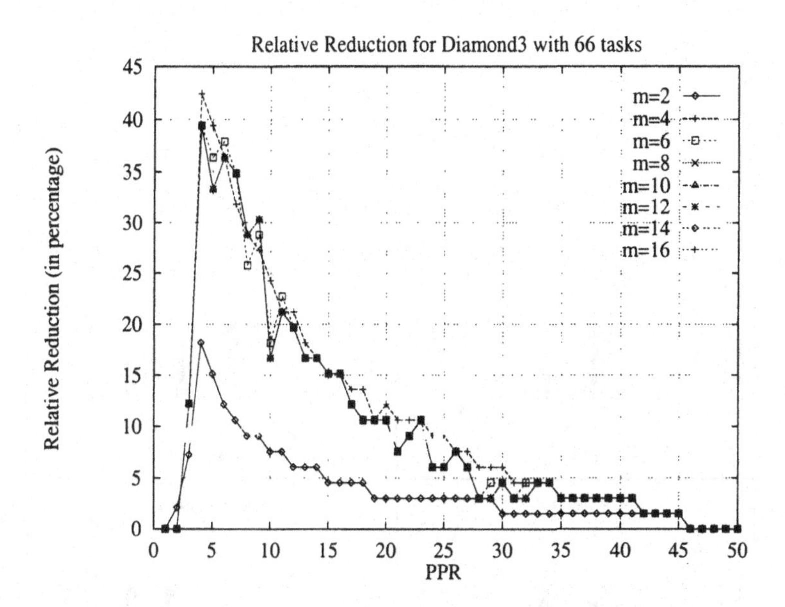

Fig. 2. Detailed relative cost reduction $\mathcal{R}$ versus PPR for `Diamond3` graph (m corresponds to the number of processors to which the tasks are scheduled). Black squares correspond to the intersection of different curves.

are automatically managed by *ANDES*. ANDES finally computes performance metrics along the execution of that workload implemented according to mapping and/or scheduling strategies. Synthetic execution was chosen as the performance technique due to the easy control of parameters as well as the possibility of using a real environment. The idea is to conjugate the best of model-based approaches with the best of realistic parallel executions. *ANDES* has been used to refine analytical and simulation analysis. With the current high availability of parallel systems, the results of *ANDES* have been proved to be precise and useful.

The Parallel System – *ANDES* along with the synthetic parallel programs were executed on an IBM SP multicomputer composed of 32 RS6000 RISC microprocessors with 64 megabytes of RAM. The processors are interconnected by a high-speed switch (bidirectional with nominal speed of 80 megabytes per second).

The Benchmarks – In order to compare estimated and observed improvements of the overall execution times of real parallel synthetic programs, we have used the following benchmark (part of the *ANDES* package): (i) `Diamond3` with 66 tasks; (ii) `FFT` with 194 tasks; (iii)`Gauss` with 192 tasks; and (iv) `Divconq` with 46 tasks.

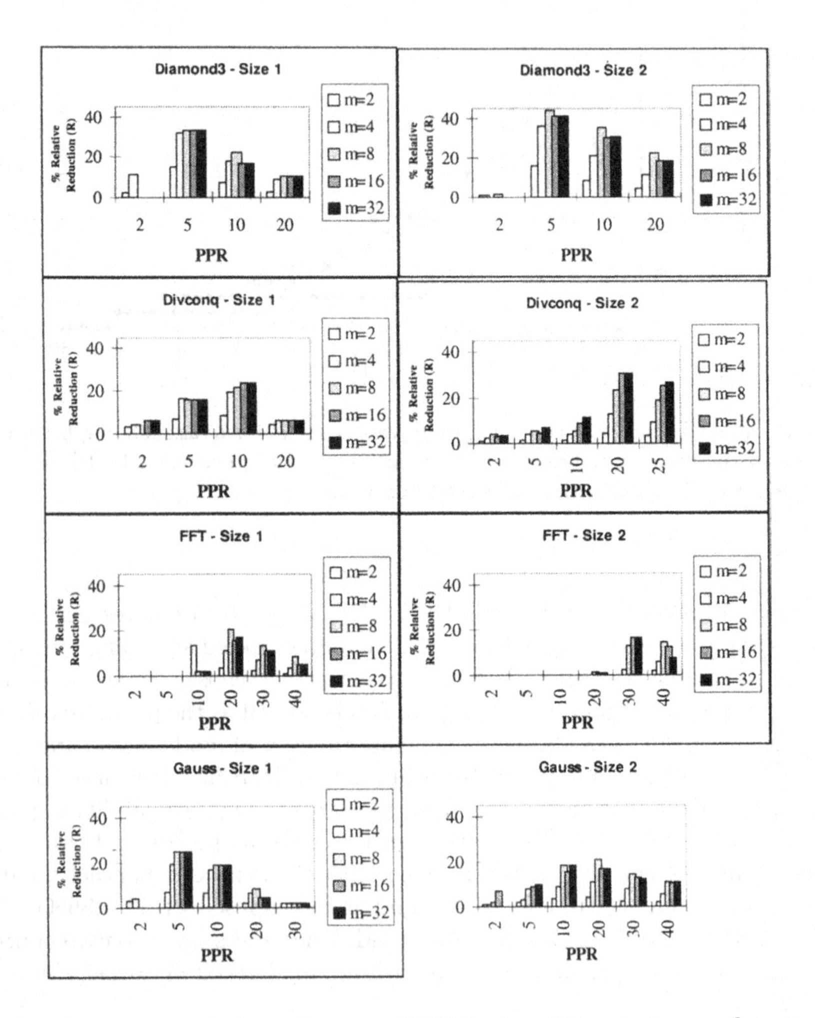

Fig. 3. Relative cost reduction $\mathcal{R}$ versus PPR for two different sizes – cf. text – of `Diamond3`, `Divconq`, `FFT`, `Gauss`, and `MS3` graphs (m corresponds to the number of processors to which the tasks are scheduled).

This benchmark picks representative task graphs from the ones studied in [8]. Small and larger task graphs are used. The TSpar was executed using 4 processors of the IBM SP. The estimated quality of both TSpar and DES+MFT algorithms is computed using a conventional C procedure for computing the makespan of the task graphs, detailed in Figure 4 (very similar to the DES+ MFT description). The final value of *clock* is the actual makespan. Each graph of the benchmark is scheduled to 2, 4, 8, and 16 processors.

---

makespan computation algorithm
begin
  Let $s = (\mathcal{A}_s(t_1), \ldots, \mathcal{A}_s(t_n))$ be a feasible solution for the scheduling problem, i.e.,
    for every $k = 1, \ldots, n$, $\mathcal{A}_s(t_k) = p_j$ for some $p_j \in P$
  $clock \leftarrow 0$
  $state(p_j) \leftarrow$ free $\forall p_j \in P$
  $start(t_k), finish(t_k) \leftarrow 0\ \forall t_k \in T$
  while ($\exists t_k \in T \mid state(t_k) \neq$ executed) **do**
  begin
    for (each $t_k \in T \mid state(t_k) =$ executable) **do**
      if ($state(\mathcal{A}_s(t_k)) =$ free) **then**
      begin
        $state(t_k) \leftarrow$ executing
        $state(\mathcal{A}_s(t_k)) \leftarrow$ busy
        $start(t_k) \leftarrow clock$
        $finish(t_k) \leftarrow start(t_k) + \mu(t_k, \mathcal{A}_s(t_k))$
      end
    Let i be such that $finish(t_i) = \min_{t_k \in T \mid state(t_k) = \text{executing}} \{finish(t_k)\}$
    $clock \leftarrow finish(t_i)$
    for (each $t_k \in T \mid state(t_k) =$ executing and $finish(t_k) = clock$) **do**
    begin
      $state(t_k) \leftarrow$ executed
      $state(\mathcal{A}_s(t_k)) \leftarrow$ free
    end
  end
  $c(s) \leftarrow clock$
end

Fig. 4. Computation of the makespan of a given schedule.

The generated schedules are read by *ANDES* which generates the synthetic load to be interpreted by *ANDES-Synth*, the synthetic execution kernel. Synthetic loads are then executed according to the given schedules.

In order to simulate heterogeneity, the size of synthetic loops corresponding to tasks allocated to the faster processor are reduced by a factor corresponding to the PPR itself. Thus, a PPR of 2 means that loops to be executed on the heterogeneous processor are reduced by half. The scheduling algorithms consider communications with zero overhead. This corresponds in *ANDES* to communications of a single byte (in the IBM SP machine, such message transmitted through the switch determines a latency of around 47.03 microseconds [5]).

Preliminary experiments were performed on an idle machine. The standard deviation was always under 1% for 10 consecutive executions. Considering this low degree of variability, we have performed measures using a sample of size 5.

4.3 Results and Analysis

Figures 5, 6, 7, and 8 present, in the same graphic, estimated and measured relative cost reduction values. The chosen PPR value range includes, for all graphics, the higher relative cost reduction values achieved by TSpar. Differences between estimated and observed improvements are under 5% for all experiments.

Our results demonstrated by the similarity between estimated and observed relative cost reduction values that the makespan computation used in both scheduling algorithms is in fact reliable. This computation is completely deterministic. On the other hand, the observed execution times are definitely non-deterministic due to the overhead from the operating system and the communication subsystem. However, the execution times presented very low variability. Therefore, this overhead does not influence significantly the experimental execution times, i.e. the makespan algorithm shows itself to be very useful to the static scheduling decisions based on estimated data. Although intuitive, this conclusion is not obvious and experiments were necessary to validate it.

Taking into account a precise makespan computation, one important consequence is that tabu search improvements are real and significant. This was foreseen from previous work, based on the estimated relative cost reduction values between DES+MFT and TSpar algorithms. In this paper, we demonstrate that these improvements also occur in more realistic execution environments.

Another interesting result is that the PPR_{peak} is not always the same. As a matter of fact, there is a range of PPR values where the best relative cost reduction varies. This irregular behaviour occurs due to the irregular search through the solution space performed by the tabu search algorithm, which depends on different heuristic parameters such as tabu list size, number of iterations without improvements, and aspiration criteria. Metaheuristics, such as tabu search, frequently depend on a fine tuning stage, where parameters are tested and calibrated. After this step, they remain unchanged, and in some test cases they are not always set to achieve the best results.

Finally, *ANDES* has been proven to be a useful tool in the validation of scheduling algorithms. The direct combination of both scheduling algorithms and the synthetic execution runtime system provided an environment where response time measurements could be quickly obtained.

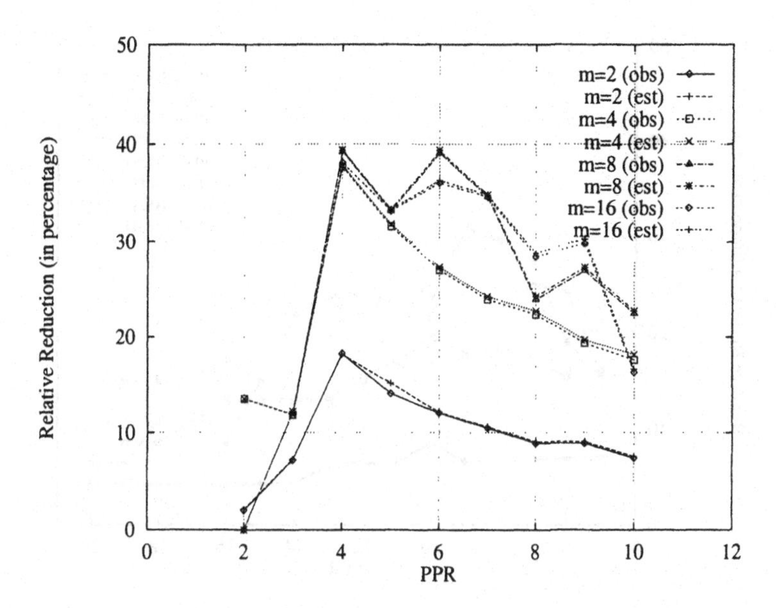

Fig. 5. Estimated (est) and observed (obs) relative cost reduction $\mathcal{R}$ versus PPR for `Diamond3` graph (m corresponds to the number of processors on which the tasks are scheduled).

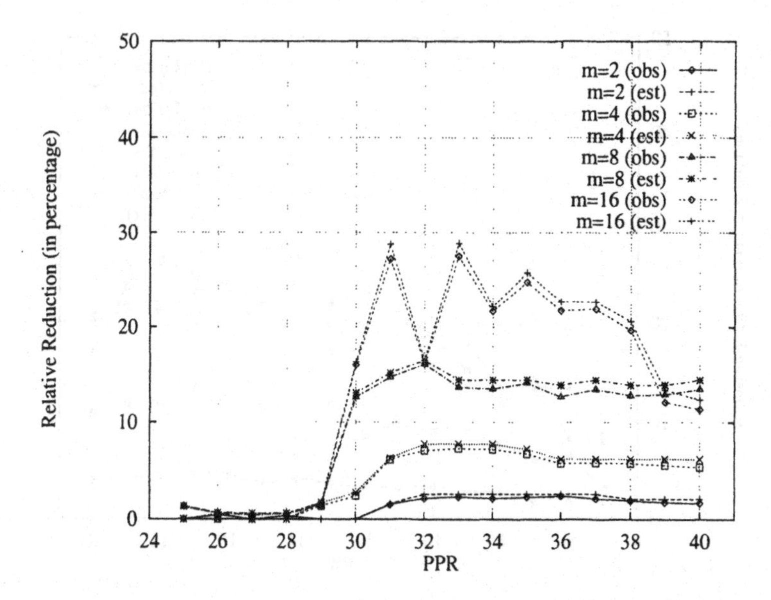

Fig. 6. Estimated (est) and observed (obs) relative cost reduction $\mathcal{R}$ versus PPR for `FFT` graph (m corresponds to the number of processors on which the tasks are scheduled).

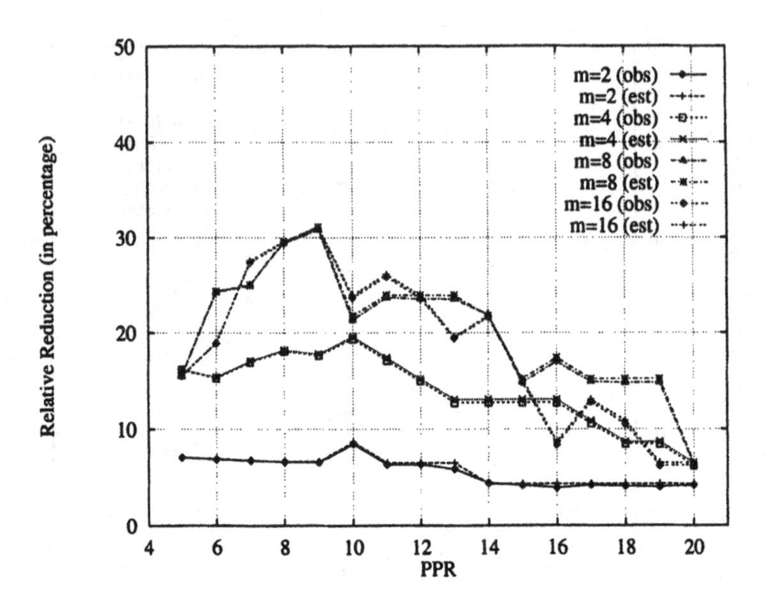

Fig. 7. Estimated (est) and observed (obs) relative cost reduction $\mathcal{R}$ versus PPR for `Divconq` graph (m corresponds to the number of processors on which the tasks are scheduled).

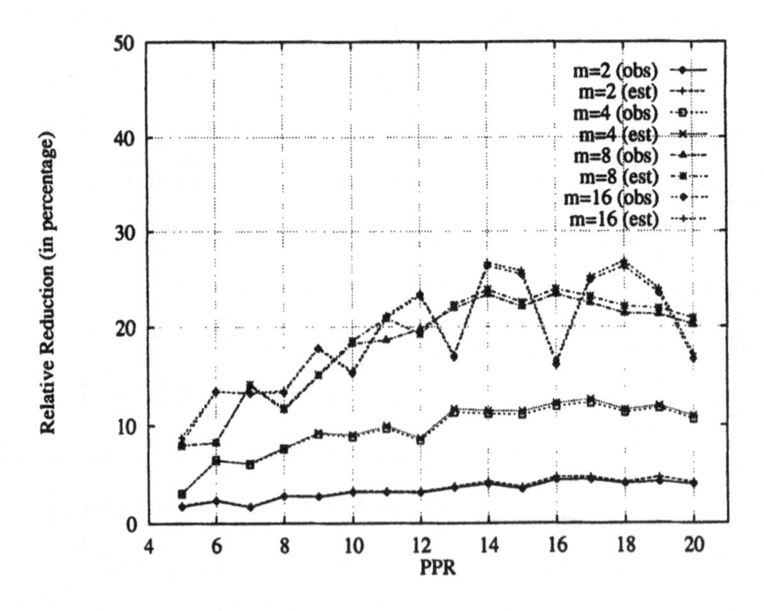

Fig. 8. Estimated (est) and observed (obs) relative cost reduction $\mathcal{R}$ versus PPR for `Gauss` graph (m corresponds to the number of processors on which the tasks are scheduled).

5 Final Remarks

This paper presents an experimental validation of makespan improvements of two scheduling algorithms: a greedy construction algorithm and a tabu search based algorithm. Synthetic parallel executions were performed given data on task execution times, task precedence relations, and task scheduling. These synthetic executions were performed on a real parallel machine (IBM SP). The estimated and observed response times improvements are very similar, representing the low impact of system overhead on makespan improvement estimation. This guarantees a reliable cost function for static scheduling algorithms and confirms the actual better results of the tabu search meta-heuristic applied to scheduling problems.

References

1. F. GLOVER and M. LAGUNA, "Tabu Search", Chapter 3 in *Modern Heuristic Techniques for Combinatorial Problems* (C.R. Reeves, ed.), 70–150, Blackwell Scientific Publications, Oxford, 1992.
2. F. GLOVER, E. TAILLARD, and D. DE WERRA, "A User's Guide to Tabu Search", *Annals of Operations Research* 41 (1993), 3–28.
3. J.P. KITAJIMA, B. PLATEAU, P.BOUVRY, and D. TRYSTRAM, "A method and a tool for performance evaluation. A case study: Evaluating mapping strategies", *Proceedings of the 1994 Cray Users Group Meeting*, Tours, 1994.
4. D.A. MENASCÉ and S.C.S. PORTO, "Processor Assignment in Heterogeneous Parallel Architectures", *Proceedings of the IEEE International Parallel Processing Symposium*, 186–191, Beverly Hills, 1992.
5. J. MIGUEL, A. ARRUABARRENA, R. BEIVIDE, and J. A. GREGORIO, "Assessing the performance of the new IBM SP2 communication subsystem", *IEEE Parallel & Distributed Technology* 4(1996), 12–22.
6. S.C.S. PORTO and C.C. RIBEIRO, "A Tabu Search Approach to Task Scheduling on Heterogeneous Processors under Precedence Constraints", *International Journal of High-Speed Computing* 7 (1995), 45–71.
7. S.C.S. PORTO and C.C. RIBEIRO, "Parallel Tabu Search Message-Passing Synchronous Strategies for Task Scheduling under Precedence Constraints", *Journal of Heuristics* 1 (1996), 207–233.
8. S.C.S. PORTO, J.P.W. KITAJIMA, and C.C. RIBEIRO, "Performance Evaluation of a Parallel Tabu Search Scheduling Algorithm", *Solving Combinatorial Problems in Parallel* (joint workshop with the *International Parallel Processing Symposium'97*), April 1-5 1997, Geneva.
9. V. S. SUNDERAM, G. A. GEIST, J. DONGARRA, and R. MANCHEK, "The PVM concurrent computing system: evolution, experiences, and trends", *Parallel Computing* 20(1994), 531–546.

Dynamic Routing Balancing in Parallel Computer Interconnection Networks[1]

D.Franco, I.Garcés, E.Luque

Unitat d'Arquitectura d'Ordinadors i Sistemes Operatius-Departament d'Informàtica
Universitat Autònoma de Barcelona - 08193-Bellaterra, Barcelona, Spain
Ph. +34-3-581.19.90 Fx.+34-3-581.24.78
E-mail: {iarq23,d.franco,iinfd}@cc.uab.es; Web: http://aows1.uab.es

Abstract. In creating interconnection networks, an efficient design is crucial because of its impact on the parallel computer performance. A routing scheme that minimises contention and avoids the formation of hot-spots should be included in the design. Static schemes are not able to adapt to traffic conditions. We have developed a new method to uniformly distribute traffic over the network, called Distributed Routing Balancing (DRB), that is based on limited and load-controlled path expansion in order to maintain a low message latency. The method uniformly balances the communication load between all links of the interconnection network and maintains a controlled latency, provided that total bandwidth requirements do not exceed the total link bandwidth available in the interconnection network. DRB defines how to create alternative paths to expand single paths (expanded path definition) and when to use them depending on traffic load (expanded path selection policies). We explain the DRB principles and show the performance evaluation of the method carried out by simulation.

1 Introduction

In the evolution of multi-computers, communication performance becomes more and more important. One of the most crucial problems that affect performance in communications is message contention. A sustained contention can produce hotspots [Pfi85]. A hotspot is a saturated region of the network, i.e. there exists more bandwidth demand than the network can offer and then, messages that enter this region suffer a very high latency, while other regions of the network can be less loaded, or even far away from saturation. The problem here is that there exists a poor communication load distribution and that, although the total communication bandwidth requirements do not surpass the total offered bandwidth of the interconnection network, this uneven distribution causes saturated points as if the whole interconnection network were collapsed. In addition, the hot-spot propagates rapidly to contiguous areas in a domino effect, which is even worse in the case of

---

[1] *This work has been supported by the Spanish Comisión Interministerial de Ciencia y Tecnología (CICYT) under contract number TIC 95/0868.*

J. Palma, J. Dongarra, and V. Hernández (Eds.): VECPAR'98, LNCS 1573, pp. 494-507, 1999.

wormhole routing because a blocked packet occupies a large number of links spread in the network.

Latency must be avoided in order to make communications faster, but some amount of latency can be tolerated and it is much more important to avoid big latency variations. This is because latency can be hidden by the mapping task assigning an excess of parallelism, i.e. having enough processes per processor and scheduling any ready process while other processes wait for their messages. However, in order to be able to assign processes to processors correctly, the mapping task must know the process computation and communication volumes and, to some extent, the latency that messages will suffer. But if latency undergoes big unpredictable variations from the expected values, due to hotspots, for example, the mapping will fail and idle processors will appear, increasing the total execution time of the application. This is the reason for the importance of a low and uniform contention latency. In addition, in Distributed Shared Memory Multicomputers latency uniformity is a key issue for scheduling and mapping in such systems.

In order to avoid hotspot generation, static or oblivious routing can not provide any help because the message route is completely determined by the source-destination pair, independently of traffic conditions. Therefore, other mechanisms have been developed to avoid hotspot generation in interconnection networks, like adaptive routing algorithms that try to adapt to traffic conditions such as Planar Adaptive Routing [CK92], the Turn Model [NG92], Duato's Algorithm [Dua93], Comprensionless Routing [KLC94], Chaos Routing [Ksn91], Random Routing [Val81] [May93] and other methods presented in [DYN97]. The main disadvantage of adaptive routing is the high overhead it introduces for information monitoring, path changing and the necessity to guarantee deadlock, livelock and starvation freedom. These drawbacks have limited the implementation of these techniques in commercial machines.

The work presented in this paper focuses on developing new methods to distribute paths in the interconnection network using network-load controlled path expansion. The method is called Distributed Routing Balancing (DRB) and its objective is to uniformly balance traffic load over all paths in the whole interconnection network by creating alternative paths between each source and the destination nodes in order to maintain a low message latency. DRB defines how to create alternative paths to expand single paths (multi-lane path definition) and when to use them depending on traffic load (multi-lane path selection policy).

The next section explains the Distributed Routing Balancing technique. DRB has two components: first, a systematic methodology to generate the multi-lane paths and second, policies to monitor traffic load and select multi-lane paths to get the message distribution according to traffic load. Both are explained in Section 3 and Section 4, respectively. Sections 5 presents the evaluation of the first DRB component and Section 6 the validation of the selection policies. Section 7 presents the conclusions.

2 Distributed Routing Balancing

Distributed Routing Balancing is a method to create alternative source-destination paths in the interconnection network using a load-controlled path expansion. DRB distributes every source-destination message load over a multi-lane path made of several paths. The objective of DRB is a uniform distribution of the traffic load over the whole interconnection network in order to maintain a low message latency and avoid the generation of hotspots. When a single source-destination path is becoming saturated, the method looks for low loaded paths to form a multi-lane path. This distribution will maintain a uniform and low latency on the whole interconnection network provided that total communication bandwidth demand does not exceed the interconnection network capacity.

The DRB method fulfills the following objectives:

1 Reduction of the message latency under a certain threshold value by varying the number of alternative paths used by the source-destination pair, while maintaining a uniform latency for all messages.

2 Minimisation of path-lengthening. This is important for Store&Forward networks because Transmission Delay depends directly on the message path lengths. For Wormhole and Cut-Through flow controls, it is important because the more nodes used by the message, the more collisions with other messages, causing latency increments and more bandwidth use.

3 Maximisation of the use of the source and destination node links (node grade), distributing messages fairly over all processor links.

In order to show how DRB works to create and use alternative paths, we make the following definitions:

Definition 0:

An *interconnection network I* is defined as a directed graph *I=(N,E)*, where N is a set of nodes $N=\bigcup_{i=0}^{MaxN} N_i$ and E a set of links. Every node is composed of a router and is connected to other nodes by means of links. In regular networks, a *regular topology* is defined, with a *dimension* and *size*. For example, for *k*-ary *n*-cubes [Dal90], *n* is the dimension and *k* the size. For irregular networks, an *irregular topology* is defined.

•If two nodes N_i and N_j are directly connected by a link, then, N_i and N_j are *adjacent nodes*.

•*Distance*(N_i , N_j) is the minimum number of links that must be traversed to go from N_i to N_j according to the graph *I*.

•A *path* $P(N_i, N_j)$ between two nodes *Ni* and *Nj* is the set of nodes selected between N_i and N_j according to the minimal static routing defined for the interconnection network. *Ni* is the *source* node and *Nj* the *destination* node. ***Length*** *of a path P, Length(P),* is the number of links traversed between N_i *and* N_j.

Definition 1:

A *Supernode* $S(type, size, N_0^S, V(S)) = \bigcup_{l=0}^{l} N_i^S$ is defined as a structured region of the interconnection network consisting of adjacent nodes N_i^S around a "central" node N_0^S provided that N_i^S comply with a given property specified in *type* and that *distance(* N_i^S , N_0^S *)<= size*. Each node has an associated weight w_i^S stored in the array *V(S)*.

V(S)= $\{0 \le w_i^S \le 1 \in R; i = 0..l\}$ is a linear array of weights associated with each N_i^S where w_i^S is the weight of N_i^S and $\sum_{i=0}^{l} w_i^S = 1$. This weights are used by the DRB selection policy (Sect. 4). As particular cases, any single node and the whole interconnection network are Supernodes. A node can belong to more than one Supernode. Section 3 presents different Supernode types.

Definition 2:

A *Multi-step Path* P_s*(SOrigin,* $N_i^{SOrigin}$, N_j^{SDest} *, SDest)* is the path generated between two Supernodes *SOrigin* and *SDest* as $P_s = \prod (N_0^{SOrigin}, N_i^{SOrigin}, N_j^{SDest}, N_0^{SDest})$=
$P(N_0^{SOrigin}, N_i^{SOrigin})*P(N_i^{SOrigin}, N_j^{SDest})*P(N_j^{SDest}, N_0^{SDest})$, where * means path concatenation, composed of the following steps:

Step 1: From the central node of the Supernode *Supernode_Origin*, $N_0^{SOrigin}$, to a node belonging to *Supernode_Origin*, $N_i^{SOrigin}$.

Step 2: From the $N_i^{SOrigin}$ to a node belonging to *Supernode_Destination*, N_j^{SDest}

Step 3: From the N_j^{SDest} to the central node of the Supernode *Supernode_Destination*, N_0^{SDest} .

In the most general case, there are three steps. However, if one of the Supernodes is reduced to a single node, i.e. *Supernode_Origin={* $N_0^{SOrigin}$ *}* or *Supernode_Destination* ={ N_0^{SDest} }, the number of steps is two; and there is one step, the one which follows static routing, if *Supernode_Origin={* $N_0^{SOrigin}$ *}* and *Supernode_Destination={* N_0^{SDest} *}*.

***Length** of a multi-step path Length(Ps)* is defined as the sum of each individual step length following static routing. From this definition, it can be seen that Multi-step Paths between $N_0^{SOrigin}$ and N_0^{SDest} can be of non minimal length.

Definition 3:

A *Metapath P*(Supernode_Origin,Supernode_Destination)* is the set of all multi-step paths P_i generated between the Supernodes *Supernode_Origin* and *Supernode_Destination:* $P^* = \bigcup_{\forall i,j} P_S(N_0^{SO}, N_i^{SO}, N_j^{SD}, N_0^{SD})$. Suppose l is the number of nodes of *Supernode_Origin* and k the number of nodes of *Supernode_Destination*. The number of Multi-step Paths which compose the Metapath is $s=l*k$. **Metapath Length** (ML) is the average of all the individual multi-step path lengths that compose it, $Length(P^*) = (1/s)\sum_{\forall s} length(P_s)$ and **Metapath Relative Bandwidth** (MRB) is the number of multi-step paths s.

Now, we explain how communication is managed under DRB to get path distribution. Suppose there is a parallel program described as a collection of processes and channels and a process mapping which assigns each process to a processor. Processes are executed concurrently and communicate by channels.

The routing run-time support configures a Metapath P^* for each channel by assigning a *Source Supernode* to the source node and a *Destination Supernode* to the destination node. A Metapath selection policy called Metapath Issuing through Latency Evaluation (MILE) which is described in Section 4 carries out this function. The Source Supernode is a **Message Scattering Area (MeSA)** from the Source node. The Destination Supernode is a **Message Gathering Area (MeGA)** to the Destination node.

Then, for each message that the source process wants to send, the channel manager, in the routing run-time support, selects two nodes, one $N_i^{SOrigin}$ from the *Source Supernode,* and the other N_j^{SDest} from the *Destination Supernode* to form a *Multi-step Path* P_s*(SOrigin,* $N_i^{SOrigin}$, N_j^{SDest}, *SDest)* belonging to the Metapath P^*. These node selections are made based on the weight arrays that are used as probability distributions for each node $N_i^{SuperNode}$. Then, the message travels along the selected *Multi-step Path.*

Under this scheme, the communication between source and destination can be seen as if it were using a wider multi-lane "Metapath" of potentially higher bandwidth than the original path from a source "Supernode" and to a destination "Supernode". This multi-lane path can be likened to a highway and the MeSA and MeGA, the highway access and exit areas, respectively.

It is important to remark that, in order to achieve an effective uniform load distribution, a global action is needed. For this reason, all source-destination nodes are able to expand their paths depending on the message traffic load between them during program execution.

The mentioned DRB Metapath Selection Policy defines the Metapath *type* and *size* to determine a Metapath of specific length and bandwidth depending on traffic conditions (Section 4). The policy needs to know the length and bandwidth of a Metapath given the type and size. Therefore, a Metapath Characterisation is needed to determine the Length and Relative Bandwidth of a Metapath given its type and size. This characterisation has been carried out by experimentation and is explained in Section 5.

Comparison with existing methods

Many adaptive methods try to modify the current path when a message arrives to a congested node. This is the case, for example, of Chaos routing [KSn91], which uses randomisation to misroute messages when the message is blocked. DRB does not act at the individual message level, but tries to adapt communication flow between source and destination nodes to non-congested paths.

Random routing algorithms [Val81] [May93] uniformly distribute bandwidth requirements over the whole machine, independent of the traffic pattern generated by the application but at the expense of doubling the path length. A closer view shows that paths of maximum length are not lengthened but paths of length one are lengthened, in average, up to the average distance for regular networks. So, the shortest paths are extremely affected. This is due to the blindness of the method that does not take into account current traffic but distributes all messages at "brute force" over the entire machine. Although DRB shares some objectives with random routing, the difference is that DRB does not only try to maintain throughput but also maintains limited individual message latency, because path lengthening can be controlled. Random routing, however, doubles in average the lengthening with the negative effect over the latency we mentioned above. It can be seen that static routing is an extreme case of DRB in which both Supernodes, source and destination, contain only the source or destination node, respectively; and that random routing is the other extreme in which the source Supernode contains all nodes of the interconnection network.

A similar but restricted, less flexible and non-adaptive solution is offered by the IBM SP2 routing algorithm, RTG, that statically selects four paths for each source-destination node which are used in a "round-robin" fashion to more uniformly utilise the network [Sni95]. The Meiko CS-2 machine also pre-establishes all source-destination paths and selects four alternative paths to balance the network traffic [Bok96].

Like other adaptive methods, DRB can introduce the possibility of deadlock in which case one of the existing techniques for deadlock avoidance should be used (such as structured buffer-pool [Mer80], virtual channels [Dal87] or virtual networks [Yan89]) depending on network characteristics (topology, flow control, etc.).

It can be seen that, by definition DRB is livelock free, because it never produces infinite path lengths, and also starvation free, because no node is prevented from injecting their messages. Also, message ordering must be preserved and it is the system's responsibility to deliver messages belonging to the same logical channel. Message prefetching can be used to hide message disordering.

3 Supernode Types

We have defined two *Supernode types* suitable for any topology. The first one is called *Gravity Area* and the second *Subtopology*. These two types define a broad range of Supernodes and allow a choice according to the desired trade-off between Metapath length (ML) and Relative Bandwidth (MRB).

The *type* and *size* parameters of the Supernode determine which nodes are included in the Supernode. A given Supernode has the following characteristics: Topological shape, Number of nodes *l*, Grade (number of *Ni* node links not connected to other *Ni* node links, i.e. links connected outside the Supernode) and Grade usage of the central node N_0 in the first step.

Gravity Area Supernode

The first Supernode type, *S("Gravity Area",size,* N_0^S *,V(S)),* is called *Gravity Area* and defines, for a *n*-grade network, a *n*-ary tree with the root at the central node N_0^S and a depth *size*. This type maps a tree of maximum grade over the topology. This tree expands at maximum and includes all nodes that are at distance size or less from the root. It is suitable for regular or irregular networks. A *Gravity Area* Supernode is the set of nodes at a distance smaller or equal than the *size* of the node N_0^S. Metapaths configured using Gravity Area Supernodes fulfill the above mentioned objectives of maximising the number of paths while minimising path-lengthening and maximising node link usage since they make use of all node links.

Subtopology Supernode:

The second Supernode type is called *Subtopology* and defines the topological shape of the Supernode. It can be applied to regular networks with a structured topology, dimension and size. A Supernode *S("Subtopology",size,* N_0^S *,V(S))* has the same full/partial topological shape as the interconnection network but its *dimension* and/or *size* is reduced. Therefore, the *Subtopology* Supernode should be considered as a kind of topological "projection" of the network topology. For example, in a k-ary n-cube a *Subtopology* Supernode is any j-ary m-cube with j<k and/or m<n. For Midimiew networks, which are a special case of wraparound torus (k-ary 2-cubes),

Subtopology Supernodes are k-ary 1-cubes, i.e. linear structures which follow specific wraparound links.

4 Metapath Selection Policy

This section describes the Metapath selection policy to select a Metapath for source destination pairs according to the current message load. We present here a dynamic policy for DRB called "Metapath Issuing through Latency Evaluation (MILE)", which has been designed to minimise overhead and to be scalable. To this end, there is no periodic information exchange and MILE is fully distributed. It has the characteristic that, under low traffic load, the monitoring activity is minimal and the paths follow minimal static routing. The monitoring activity objective is to identify the current traffic pattern.

Policy objectives are to select the supernode *size* and *type* and to distribute the load among the Multi-step Paths of the Metapath. The policy consists of three phases: Traffic Load Monitoring, Dynamic Supernode Configuration and Multi-Step Path Selection. Traffic load monitoring is carried out by the messages. Latency suffered is recorded and carried by the message itself. The message records information of the contention it suffers at each node it traverses when it is blocked by contention with other messages. When messages arrive at their destination carrying latency information, the destination node takes a decision about the Supernode configuration depending on the contention that the messages encountered on the Metapath. This Metapath Configuration is sent to the Source node by means of an acknowledgement message that distributes messages following DRB specification.

The policy algorithm pseudo-code for each source-destination pair is presented as follows:

```
Traffic Load Monitoring()/*Actions performed by the messages*/
Begin
   For each hop,
      Accumulate latency
   EndFor
   Deliver latency to the Destination algorithm.
End
```

```
Dynamic Supernode Configuration(threshold latency ThL)
/*Algorithm executed at destination nodes*/
/*Threshold latency is the change latency from the flat region
to the rise region defined in Sec. 2*/
Begin
   For each Multi-step Path of the Metapath
     Receive Multi-step Latency recorded by the message itself.
   EndFor
   Order Multi-step Paths according to the Latency they suffered
   Classify Multi-step Paths as saturated or non-saturated
depending on whether their latencies exceed the ThL or not.
   Calculate Total Latency (TL) adding each Multi-step Latency.
```

```
  Calculate Total Threshold Latency (TThL) multiplying
TL*Metapath Relative Bandwidth(MRB)
  /*Load Balancing: Distribute traffic load among the Multi-
step Metapaths*/
  If (TL does not exceed TThL and there exist saturated Multi-
step Paths)
      Redistribute Supernode node w_i^S weights to move load from
saturated Multi-step Paths to non-saturated Multi-step Paths
  ElseIf (TL does not exceed TThL and there do not exist
saturated Multi-step Paths)
      Reduce Multi-step Paths (reducing Supernode
configurations)
      Redistribute Supernode node w_i^S weights to move load from
disappearing Multi-step Paths to the other Multi-step Paths
  ElseIf (TL exceed TThL)
      Add new Multi-step Paths (expanding Supernode
configurations) as non-saturated Multi-step Paths
      Redistribute Supernode node w_i^S weights to move load from
saturated Multi-step Paths to non-saturated Multi-step Paths
  EndIf
  Send new Supernode Configurations and weights to the Source
node by means of an acknowledge message.
End Destination
```

```
Multi-Step Path Selection() /*Actions executed by sender nodes*/
Begin
  Receive new Supernode configurations
  Distribute subsequent messages among the Supernode nodes
N_i^S, according to their corresponding weights w_i^S.
End
```

This DRB MILE takes advantage of the spatial and temporal locality of parallel program communications, like cache memory systems do with memory references. The algorithm adapts the Metapath configurations to the current traffic pattern. While this pattern is constant, latencies will be low and the MILE is not activated. If the application changes to a new traffic pattern and message latencies change, the MILE will adapt Metapaths to the new situation. DRB is useful for persistent communication patterns, which are the ones that can cause the worst hotspot situations. This Metapath adaptability is specific and can be different for each source-destination pair depending on their static distance or latency conditions.

Memory space and the execution time overhead of the policy are very low because the implied actions are very simple. In addition, these activities are executed a number of times which is linearly dependent of the number of logical channels of the application and the number of messages. Regarding the time overhead, it can be seen that monitoring is just latency record by the message itself, i.e. storing a few integers, and that the decision algorithm is a local and simple computation applied only each time a change on the latency is detected. Regarding the space overhead, the latency record is one or a few integers that the message carries itself in its header, and the new Supernode configuration information is a short message of integers.

5 Metapath characterisation

This section explains the metapath characterisation that has been carried out to determine the Metapath Length (ML) and Metapath Relative Bandwidth (MRB) of a Metapath given its *type* and *size*. A series of experiments have been carried out for all Metapath types and sizes and for k-ary n-cubes and Midimews from 8 to 64K nodes.

For a given *topology* and metapath *type* and *size*, the experiment consisted in calculating the average ML and average MRB by averaging ML and MRB for all generated Metapaths when changing the source and destination nodes.

This average ML for a specific Metapath is considered as the average network distance of the interconnection network under the routing defined by the Metapath. This average network distance has been compared to the average network distance for static and random routings.

In addition, the Standard deviation of the lengths of the Multi-step Paths that compose the Metapath was calculated. The standard deviation shows the uniform fairness of the method in relation to path lengthening.

As an example of the results obtained, fig. 1 shows a chart for a 1024-node (32x32) 2D Torus and a 1024 10D HyperCube. The chart shows the Relative Metapath Average Length, i.e. the percentage relation between the average ML and average network distance for static routing, and the Metapath Relative Bandwidth. The X axis represents different Metapaths *sizes* for the Metapaths $M(S_O, S_D)$ where S_O(*"Gravity Area",size,* N_i^{So}*,V(S))* and S_D(*"Gravity Area",size,* N_j^{Sd}*,V(S))*. Each Xi axis point is the average of the MLs for the Metapaths of *size*=Xi generated for all source and destination nodes.

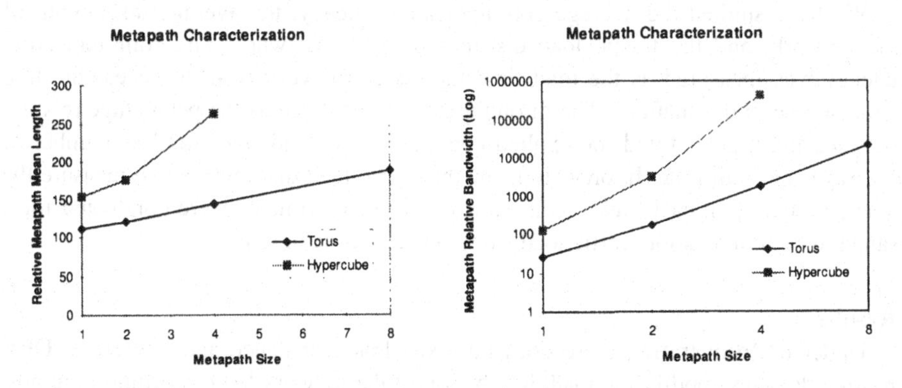

Fig. 1. *.Metapath characterisation*

We can make the following observations from the above charts for Gravity Area methods. Looking at Relative Metapath Average Length values, it can be seen that Metapath lengthening growth is proportional to the supernode size for Torus and HyperCubes. Regarding Metapath Relative Bandwidth, it can be seen that it has a much higher growth with respect to metapath size than Metapath Length has.

In addition, the results show a similar behaviour for equivalent Supernodes in different topologies, which demonstrates method uniformity. Besides the broad range of alternatives offered by the methods, gravity area methods offer a higher extra bandwidth/ latency rate than Subtopology methods. Similar results have been found for all topology sizes from 9 to 64K processors which proves the good scalability of the methods. A complete presentation of these results is found in [Gar97].

6 DRB MILE Policy Evaluation

This section shows the results for different traffic patterns and network loads with a fixed message length and compares the performance with that of static routing.

The simulations consisted of sending packets through the network links according to a specific traffic pattern. The simulations were conducted for an 8x8 torus with bi-directional links. We have assumed wormhole flow control and 10 flits per packet. Each link was designed to have only one flit buffer associated with it. The packet generation rate followed an exponential distribution whose average was the message interarrival time. The results were run many times with different seeds and were observed to be consistent. The simulation was carried out for 100,000 packets. The effects of the first 20,000 delivered packets are not included in the results in order to lessen the transient effects in the simulations.

We have chosen some of the communication patterns commonly used to evaluate interconnection networks [DYN97]: Uniform, hot-spot, Matrix transposition and butterfly.

We have studied the average communication latency, the average throughput of the network, and the traffic load distribution in the network. The communication latency was measured as the total time the packets have to wait to access the link from source to destination. The throughput was calculated as the percentage relation between the accepted and the applied communication loads measured as number of messages per unit time. In order to show the traffic load distribution, we measure the average latency in each link of the network. The experiments were conducted for a range of communication traffic load from low load to saturation.

Results

Under uniform traffic, there does not exist load unbalance and, therefore, DRB routing does not modify the load distribution of the network load, resulting in almost the same average latency and average throughput of the network for all ranges of load. This is the behaviour expected according to DRB's definition, which can not improve this situation. We do not show the figures for this case. For the other three traffic patterns the behaviour changes in a very different way.

Fig.2 shows the latency results for the hot-spot traffic, the bit-reversal and butterfly traffic patterns. DRB routing demonstrates better performance. It can be seen that at low load rates, DRB behaves nearly equal to static routing. This means that the DRB method does not overload the network when it is not necessary. While load is increasing, latency improvements are increasing too, resulting in latency reductions bigger than 50% at the highest load.

At the same time as these latency improvements are achieved, the throughput is increased as can be seen in Fig. 2 for all traffic patterns. The throughput is improved up to a 50%. The conclusions are that more messages are sent and with less latency and that the network saturation point is reached at higher load rates because DRB routing maintains uniform load distribution getting a better use of network resources for all tested traffic patterns.

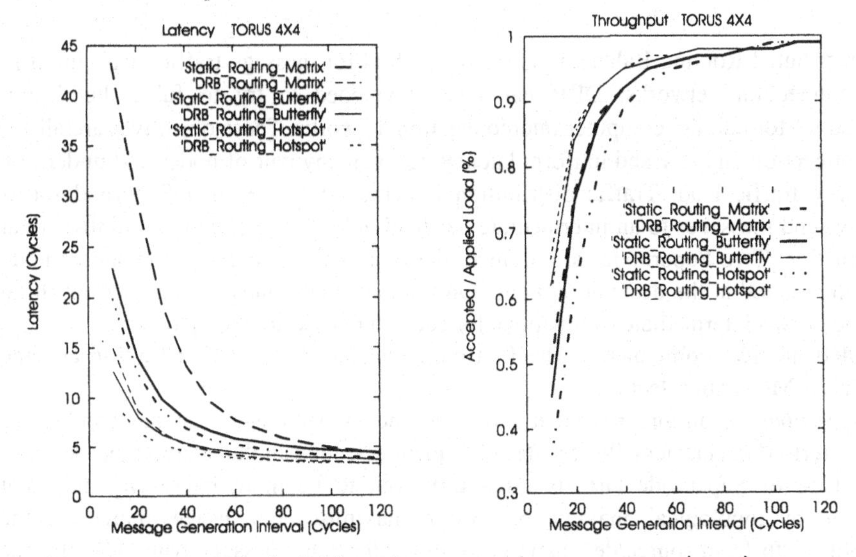

Fig. 2. *Performance results for DRB routing: average latency, average throughput.*

In order to show how DRB Routing distributes load and eliminates hot-spots, Fig. 3 shows the latency surface for network links for the hot-spot traffic pattern at a load rate of 30 cycles as message generation interval. Each grid point represents the average latency of the node links of the torus. It can be seen that, using Static Routing, big hot-spots appear in the network while other regions of the network are only slightly used. The maximum average latency in the hot-spots is around 15 cycles. When using DRB Routing, these hot-spots are effectively eliminated because the overload of the hot-spot nodes is distributed among other links. The maximum average latency in this case is about 3.5 cycles.

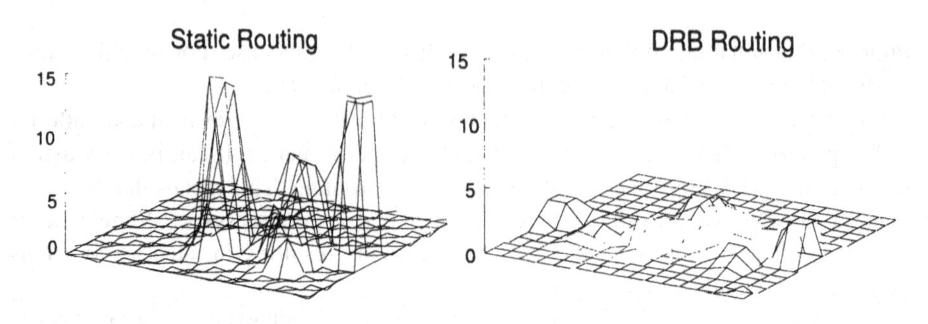

Fig. 3. *Latency distribution for the hot-spot pattern*

7 Conclusions

Distributed Routing Balancing is a new method for message traffic distribution in interconnection networks. DRB has been developed to try to fulfil the design objectives for parallel computer interconnection networks. These objectives are all-to-all connection and low and uniform latency between any pair of nodes and under any message traffic load. Traffic distribution is achieved by defining Supernodes that firstly send messages to an intermediate destination before sending them to their final destination. Two Supernodes are defined, the first one is centered at the source node and the second at the destination node. Only one or both kinds can be used resulting in one or two intermediate destinations for each source-destination pair.

DRB has two components. The first component is Supernode definition and the second is Metapath selection.

Supernode definition has been explained and its parameters (latency/bandwidth) characterised experimentally for its subsequent selection in the adaptive phase. The new type of Supernode Gravity Area turns out to be more interesting than that defined by topological analogy, because it maximises link usage, increasing the output width from source/destination, not only along the message path. Therefore, a methodology for Supernode definition has been created for each topology. DRB offers a set of alternative paths to choose from, depending on the trade-off between throughput and latency.

The second component of DRB are the policies to select a specific Supernode. The dynamic policy we present monitors traffic load and dynamically configures Supernode parameters depending on the temporary requirements of message load in the network. The policy does not waste significant computation or communication resources because it is fully distributed, and the monitoring and decision overhead is linearly dependent of the number of messages in the network. DRB is useful for persistent communication patterns, which are the ones that can cause the worst hotspot situations. The evaluation done to validate DRB shows us very good improvements in latency, effectively eliminating hotspots from the network.

References

[Bei92] R. Beivide et al. "Optimal Distance Networks of Low Degree for Parallel Computers". IEEE Trans. on Computers. Vol. 40. N. 10. pp. 1109-1124. Oct 1992.

[Bok96] Bokhari S. "Multiphase Complete Exchange on Paragon, SP2 and CS2". IEEE Parallel and Distributed Technology, Vol.4, N.3, Fall 1996, pp. 45-49

[CK92] Chien AA, Kim JH, "Planar Adaptive Routing: Low-Cost Adaptive Networks for Multiprocessors". Proc. 19th Symp. on Computer Architecture. May 1992, pp. 268-277

[Dal87] Dally WJ. Seitz CL. "Deadlock-Free Message Routing in Multiprocessor Interconnection Networks" IEEE Trans. On Computers. V. C-36. N.5, May 1987. 547-553.

[Dal90] Dally WJ. "Performance analysis of k-ary n-cube interconnection networks". IEEE Trans. On Comput. Vol. 39. Jun. 1990.

[Dua93] Duato J. "A new theory of Dead-lock free adaptive routing in wormhole networks" IEEE Transactions on Parallel and Distributed Systems, 4(12), Dec 1993, pp.1320-1331

[DYN97] Duato J, Yalamanchili S, Ni L. "Interconnection Networks, an Engineering Approach". IEEE Computer Society Press. 1997

[Gar97] Garces I, Franco D, Luque E "Improving Parallel Computer Communication: Dynamic Routing Balancing". Proc. 6th Euromicro Workshop on Parallel and Distributed Processing. (IEEE-Euromicro) PDP98. Madrid. Spain. January 21-23, 1998 p. 111-119

[KLC94] Kim J, Liu Z, Chien A. "Comprensionless Routing: A Framework for Adaptive and Fault-Tolerant Routing". Proc. of the 21st Int. Symp. On Comp. Arch. Apr 1994, p.289-300

[Ksn91] Konstanyinidou S, Snyder L. "Chaos Router: Architecure and Performance". Proc. of the 18th International Symposium on Computer Architecture, May 1991, pp.212-221

[May93] May MD, Thompson PW, PH Welch Eds. "Networks, Routers and Transputers: Function, Performance and application". IOS Press 1993

[Mer80] Merlin PM, Schweitzer PJ "Deadlock Avoidance in Store-and-Forward Networks-I: Store-and-Forward Deadlock" IEEE Trans. On Comm. V. Com-28, N.3, Mar 1980,345-354

[NG92] Ni L, Glass C. "The Turn model for Adaptive Routing". Proc. of the 19th Intl. Symp. on Computer Architecture, IEEE Computer Society, May 1992, pp. 278-287

[Pfi85] Pfister GF. Norton A. "Hot-Spot Contention and Combining in Multistage Interconnection Networks". IEEE Trans. On Computers. Vol. 34. N.10 Oct 1985.

[PG+94] Pifarre GD et al. "Fully adaptive Minimal Deadlock Free Packet Routing in Hypercubes, Meshes and Other Networks: Algorithms and Simulations" IEEE Transactions on Parallel and Distributed Systems, V. 5, N.3, Mar 1994

[Sni95] Snir M, Hochschild P, Frye DD, Gildea KJ "The communication software and parallel environment of the IBM SP2". IBM Systems Journal. Vol.34, N.2, pp. 205-221.

[Val81] Valiant LG. Brebner GJ. "Universal Schemes for Parallel Communication". ACM STOC. Milwaukee 1981. pp. 263-277

[Yan89] Yantchev JT, Jesshope CR. "Adaptive, Low Latency, Deadlock Free Packet Routing for Networks of Processors". IEE Proceedings. Vol 136, Pt.E, N.3; May 1989

Algorithm-Dependant Method to Determine the Optimal Number of Computers in Parallel Virtual Machines

Jorge Barbosa* and Armando Padilha

FEUP-INEB, Praça Coronel Pacheco, 1, 4050 Porto, Portugal
jbarbosa@tom.fe.up.pt

Abstract. Presently computer networks are becoming common in every place, connecting from only a few to hundreds of personal computers or workstations. The idea of getting the most out of the computing power installed is not new and several studies showed that for long periods of the day most of the computers are in the idle state. The work herein refers to a study aiming to find, for a given algorithm, the number of processors that should be used in order to get the minimum processing time. To support the parallel execution the WPVM software was used, under a Windows NT network of personal computers.

1 Introduction

A parallel computer is composed by a set of processors connected by one network according to one of several possible topologies (mesh, hypercube, crossbar, central bus, etc [9]). If the processors are connected in order to maximise their communication performance and together operate exclusively for the solution of one problem, then it is called a Supercomputer. If the processors are of the general purpose type, each one being a workstation connected by a general purpose network (e.g. Ethernet), then when they operate together for the solution of a given problem, it is called a Parallel Virtual Computer.

There are significant differences between a Supercomputer and a Parallel Virtual Computer, such as the interconnection network. The general purpose network allows only the communication between two processors simultaneously, and it could be also shared by other computers not belonging to the Parallel Virtual Computer, resulting in low communication rates. The fine grain parallelisation, common in Supercomputers, becomes impractical in Parallel Virtual Computers, where medium or coarse grain parallelisations are used, at the program or procedure level.

The aim in the utilisation of a Parallel Virtual Computer, as for a Supercomputer, is to reduce the processing time of a given program. This is achieved by utilising execution cycles of several computers that would not be used in another way. The more conclusive measure of the parallelisation performance is the reduction of the processing time obtained, or equivalently the Speedup obtained.

* PhD grant BD/2850/94 from PRAXIS XXI programme

J. Palma, J. Dongarra, and V. Hernández (Eds.): VECPAR'98, LNCS 1573, pp. 508–521, 1999.

This report presents a method to calculate the number of processors that should participate in the execution of a given algorithm according to its characteristics.

2 Interconnection Networks

A parallel computer is built of processing elements and memories, called nodes, and an interconnection network, composed by switches, to route messages between those nodes. Interconnection network topology is the pattern by which the switches are connected to each other and to nodes. The network topology can be classified as direct or indirect. Direct topologies connect each switch directly to a node, resulting a static network that will not change during program execution. Examples of static networks are the mesh, hypercube and ring.

Indirect topologies connect at least some of the switches to other switches, resulting dynamic networks that can be configured in order to match the communication demand in user program. Examples of dynamic networks are the multistage interconnection network, the fat-tree, crossbar switches and buses [8].

The interconnection network of a parallel virtual computer is similar to a bus as shown in Figure 1.

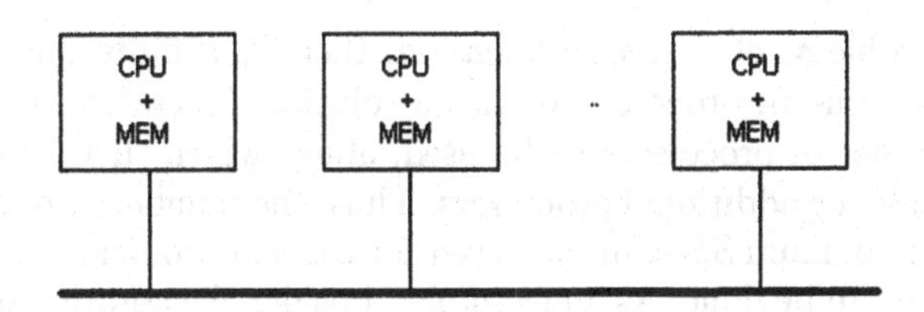

Fig. 1. Interconnection Network (Ethernet) of a Virtual Parallel Computer

The logical topology of an Ethernet provides a single channel, or bus, that carries Ethernet signals to all stations allowing broadcast communication. The physical topology may include bus cables or a star cable layout; however, no matter how computers are connected together, there is only one signal channel delivering packets over those cables to all stations on a given Ethernet LAN [11].

Each message is divided in packets of 46 to 1500 bytes of data, to be sent sequentially and individually onto the shared channel. For each packet the computer has to gain access to the channel. This division of a message into packets leads to a latency time for each message that is proportional to the number of packets into which it is split,

$$T_c = kT_L + \frac{nbytes}{w} \tag{1}$$

T_c being the total communication time for the given message, k the number of packets into which the message is split, T_L the latency time for one packet, *nbytes* the message length in bytes and w the network bandwidth. For a particular system, equation 1 is used to estimate the value of T_L. For the network used in the Virtual Computers, M1 and M2, referred to in the results section, the mean value for T_L was estimated as being 500 microseconds and the packet size to be 1Kbyte.

3 Performance Measures

A Parallel Computer can be evaluated according to several characteristics, such as the processing capacity (Mflop/s), the network bandwidth (Mbytes/s), the processing capacity of each processor individually, the memory access method and time, etc. However, its performance is always referred to a given algorithm.

The ratio between the serial processing time and the parallel processing time is referred to as Speedup and reflects the gain obtained with the parallelisation:

$$Speedup = \frac{T_{Serial}}{T_{Parallel}} \tag{2}$$

3.1 Speedup Limits

For a given problem, of a given dimension, there is a finite quantity of work required to be done in order to obtain its solution. Therefore, there will be a maximum number of processors to be used, above which there will not be any work to schedule for additional processors. Thus, the number of processors to be used and the maximum Speedup achieved for a given problem is limited by the quantity of work to be done. As an example, consider the addition of two vectors of dimension n, which involves n additions. If one uses more than n processors, the remaining processors will not have any work to do, resulting a maximum relative Speedup of n.

Amdahl [6] has defined a rule to demonstrate that the Speedup value is limited by the inherently sequential part of the program: let s represent the sequential part of the program, non parallelizable, and p the part of the program susceptible of being parallelised, that can execute with Speedup P in a computer with P processors, then the observed Speedup will be:

$$Speedup = \frac{s+p}{s+\frac{p}{P}} \tag{3}$$

$s+p$ being the processing time of the sequential program, and P the number of processors used. As an example, if the serial program runs in 93s in which 90s is susceptible of being parallelised, then $p = 90s/93s = 0.9677$ or 96.77% and s, the inherently sequential part, assumes the value of 0.0323 which is 3.23% of the code [7]. The sequential part is composed mainly by input/output operations.

From the speedup definition one can obtain its limit: as p approaches infinite, Speedup equals $1/s$. For the example presented above the Maximum Speedup

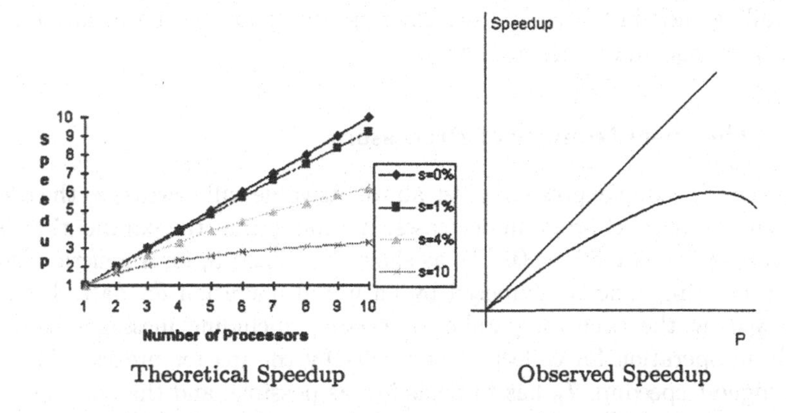

Theoretical Speedup Observed Speedup

Fig. 2. Comparison between Theoretical and Observed Speedup

is $1/0.032 = 31.25$ whatever the number of processors used, being useless to use more than 31 processors.

Figure 2 (left) shows the Theoretical Speedup for several values of s, which is assumed constant for a given algorithm, whichever the value assumed by P. The Amdahl law introduced two important factors in parallelism. First, it allows to have a more realistic expectation of the results that can be achieved, and second, it shows that for the Speedup to achieve high values, one must reduce or eliminate the sequential parts of a given algorithm [7].

Amdahl made s constant; however, in most algorithms, the increase in the number of processors leads to an increase in the communication overheads. If those are considered as bottlenecks and added to s, the Speedup behaviour will be like the one shown in Figure 2 (right). The observed speedup presents a shape that increases until a given value of P is reached, after which it decreases. In conclusion, the ideal number of processors to be used for the solution of a given algorithm will be below of the number obtained by the limit of Amdahl's law.

Several authors contend against the above scenario, noting that in many practical cases the problem size is not kept constant as the number of processors increases. On the contrary, they argue that, realistically, the run time is roughly maintained and the problem size scales with the number of processors. In this context, alternative definitions of the speedup should be used, e.g. the Gustafson law [5] for scaled speedup.

However, the problem domain we are addressing is best governed by the original Amdahl law, as we are mostly concerned with automatically optimising every single run of a given, fixed problem in relation with the uncontrolled and varying availability of computing resources in a distributed environment. For example, assume that a physician uses a specific software tool for analysing and measuring his/her ultrasound images; parts of the software may be computationally demanding and amenable to parallelisation; our goal is to design such software using a scheme where, in a user transparent manner, the response time obtained is, in each run, the best that is achievable for the specific conditions prevailing then in the computing environment (number, speed, memory of the non-busy machines in the network).

3.2 The Ideal Number of Processors

From the Speedup expression given above, its value will increase as the execution time of the parallel program decreases. Assuming that the parallel time is given by $T_p(n,P) = s(n,P) + p(n,P)$, as shown in Figure 3, for a generic algorithm, the processing time is composed by an initial operation for data distribution, followed by the time for parallel processing, including messages, and ending with an operation for collection of results by the master process. Then to get the highest speedup, T_p has to be as low as possible, and the optimal value for P that satisfies this condition is such that, for a given algorithm, the increase of the serial component due to the addition of one more processor will balance the gain obtained in the processing time of the parallel component.

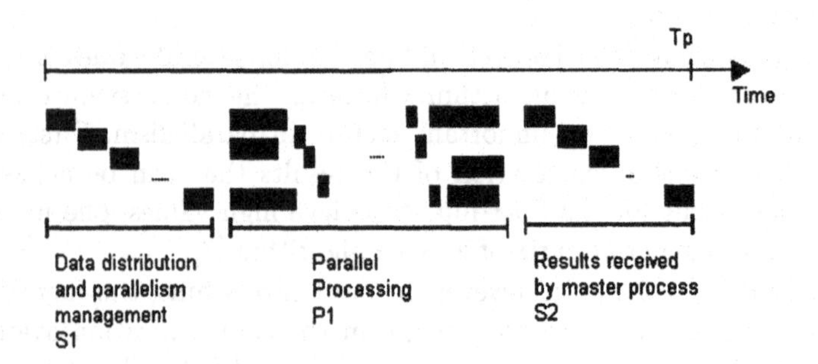

Fig. 3. Timing diagram of a parallel algorithm

The function representing the sequential component of the algorithm, $s(n,P)$, depends on the problem dimension, n, and on the number of processors used, P. It represents the input/output operations, the sequential code required to manage the parallelism and the communication overheads. Assuming an homogeneous computer network, where each processor has an individual processing capacity of $S M flop/s$, a network bandwidth of $W Mbytes/s$, and n input data elements to be distributed, one gets the following expression:

$$s(n,P) = \frac{C_1 P}{S} + \frac{n}{P}\left(\frac{1}{W} + T_E\right)P + k_1 T_L + \left(\frac{1}{W} + T_E\right)b\,(P-1)$$
$$+k_2 T_L\,(P-1) + \frac{k_3}{P}\left(\frac{1}{W} + T_E\right)P + k_4 T_L \qquad (4)$$

In the expression above it was assumed that each process communicates only with the neighbour processes which is certainly not true for all algorithms. In practice the communication components have to be modelled for every algorithm. The first factor represents the time spent for the parallelism management, where C_1 is a constant dependant on the number of instructions being used. The second factor represents the time required to distribute the n elements of the input data among P processors, where T_E is the packing time per byte. The third factor represents the latency time of P initial messages required to distribute the input data, where k_1 is the number of packets required to send the data. The fourth factor represents the time spent by each processor in communications with the next processor for the parallel algorithm, transmission and packing for sending and receiving. The fifth factor represents the latency time of those messages per processor, k_2 being the number of packets required. The sixth factor represents the time required by the master process to receive the results from all processes, k_3 being the number of bytes to receive. The last factor is the latency time for the results messages, where k_4 is the number of packets required.

The function representing the algorithm's parallel component, $p(n, P)$, depends also on the problem dimension and on the number of processors used. It can be calculated as:

$$p(n, P) = \frac{n\beta^\alpha}{PS} \tag{5}$$

β being the number of instructions computed α times for each element of n. From the addition of both expressions, $s + p$, one can obtain the total parallel processing time:

$$\begin{aligned} T_p(n, P) = \frac{C_1 P}{S} + \frac{n}{P}\left(\frac{1}{W} + T_E\right) P + k_1 T_L + \left(\frac{1}{W} + T_E\right) b\,(P-1) \\ + k_2 T_L\,(P-1) + \frac{k_3}{P}\left(\frac{1}{W} + T_E\right) P + k_4 T_L + \frac{n\beta^\alpha}{PS} \end{aligned} \tag{6}$$

For a given problem of dimension n and assuming the constants α and β are known, the minimum value for $T_p(n, P)$ is given by $\frac{\partial T_p}{\partial P} = 0$:

$$\frac{C_1}{S} + \left(\frac{1}{W} + T_E\right) b + k_2 T_L - \frac{S n \beta^\alpha}{(PS)^2} = 0 \tag{7}$$

resulting for P the value

$$P = \sqrt{\frac{n\beta^\alpha / S}{C_1/S + k_2 T_L + (1/W + T_E)\, b}} \tag{8}$$

This expression shows that P is obtained by the square root of the ratio between the useful processing time by the time spent in communications and parallelism management. The value of P is then used to compute the expected processing time for the parallel program, T_P, which is then compared to the estimated serial processing time, T_S. If it happens that $T_P > T_S$ then the serial version of the algorithm is used.

3.3 Application to the Parallel Virtual Computer

The nodes of the Parallel Virtual Computer are composed by processors of different characteristics, mainly with respect to the processing capacity and memory available, forming an heterogeneous system. Therefore, the value of S, made constant in the computation of P, needs to be replaced by a value that represents the Heterogeneous Parallel Virtual Computer. Thus, S can be replaced by a weighted mean such as:

$$\overline{S} = \sum_{i=1}^{M} S_i w_i / \sum_{i=1}^{M} w_i \qquad (9)$$

w_i being the weight given to processor i, defined as the ratio between the processing capacity of processor i and the capacity of the fastest processor in the Virtual Computer: $w_i = S_i/S_{MAX}$. With this weight the fastest processors have more influence, leading to lower values of P, and probably to the utilisation of only the fastest processors. The distribution of computational load (l_i) is made proportional to the relative processing capacity of the i processor:

$$l_i = S_i / \sum_{k=1}^{P} S_k \qquad (10)$$

4 Implementation

The parallel implementation of the image processing algorithms referred to in the results section is done under the WPVM software, which is an implementation of PVM for the MS Windows operating system, developed at University of Coimbra, Portugal [1, 2]. The software can be downloaded from http://dsg.dei.uc.pt /wpvm. WPVM offers the same set of functions as standard PVM [4] and allows the interaction between WPVM and PVM hosts.

5 Results

To validate the presented methodology, two image processing algorithms were implemented: a step edge detection algorithm [10] and an algorithm for histogram computation [3]. Also, two parallel virtual computers were used, M1 and M2, composed by the following processor capacities, in Mflops: M1={80, 80, 80, 80, 45, 45, 40, 40, 35, 35, 35} and M2={161, 161, 105, 91, 80}, with equivalent processing capacities, S1 and S2, of 68.8 and 129.7 Mflops, respectively.

For both algorithms the computational load is evenly distributed on the domain space, since the same operation is carried out for every input data element (an image pixel).

5.1 Step Edge Detection Algorithm

Edge detection is an important subject in image processing because the edges correspond in general to objects that one wants to segment. The edge detector operator described in [10] is an optimal linear operator of a infinite window size. The operator is an infinite impulse response filter realized by a recursive algorithm. To find the filter response for each pixel, the components along x and y are first computed. Each of them requires four basic operations to be executed, as shown in Figure 4.

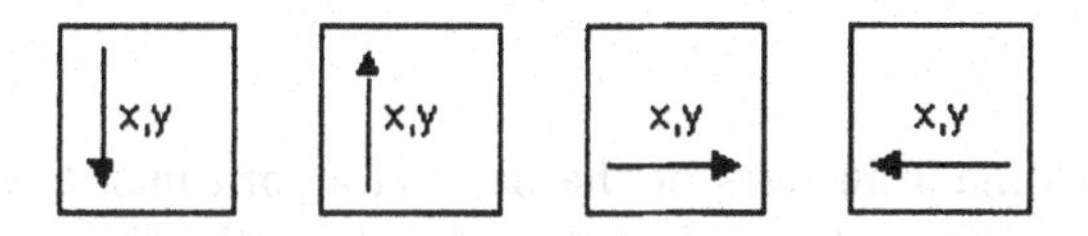

Fig. 4. Edge detection algorithm

The operations are independent from each other; however, in each basic operation, the result for the previous pixel has to be known. This dependency is important in the context of data distribution, which should minimise the number of accesses to non local data. For this algorithm, a square blocked distribution (see Figure 5), requires more messages than the column or row blocked strategies. The data distribution selected was row blocked, because the image is stored row by row in the computer memory, requiring only one packing instruction to pack several contiguous rows. Row and column blocked distributions require the same amount of data to be transferred among processors.

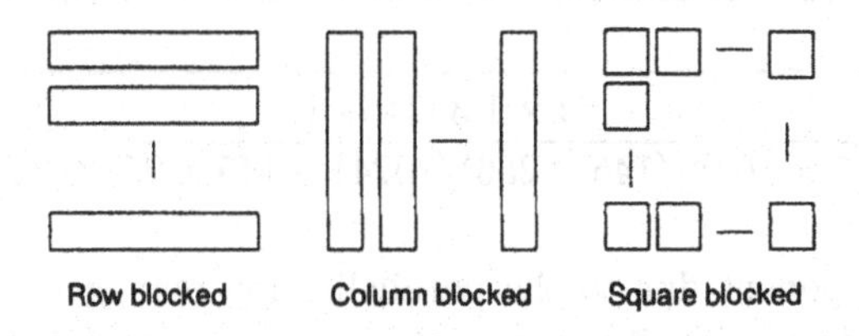

Fig. 5. Data distribution strategies

Due to the row blocked distribution, the result along columns has to be sent to the neighbouring processors. The parallel implementation of the algorithm should allow each processor to start any of the four basic operations as soon as they have the data to start, avoiding the idle state. Figure 6 shows an optimised timing diagram for 3 processors, where processing starts as soon as the data is available and processors give priority to the operation results that others may be waiting for; in this case the priority is given to operations along columns.

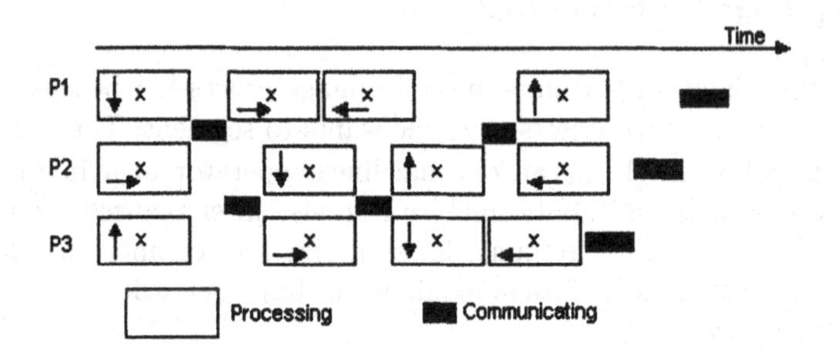

Fig. 6. Timing diagram for 3 processors

By measures made in tests of the parallel algorithms, the parameters that characterise them were obtained as shown in Table 1. The interconnection network, a Fast Ethernet at 100Mbit/s, has a transmission time of $0.08\mu s/byte$. The packing time T_E, of value $0.07\mu s$, was measured indirectly.

Algorithm	β^α(Mflop/kb)	k_3	b (bytes)	C_1 (Mflops)
Edge Det.	1.333	$\lceil b/1024 \rceil$	145×ncolumns	0.8
Histogram	0.150	$\lceil b/1024 \rceil$	$N(P-1)/P$	0.8

Table 1. Algorithm parameters

The edge detection algorithm was run in machine M1 with 68.8 Mflops, the master process being in a 80 Mflops computer. By replacing all the parameters in the expression of P, for an image of 64 kb (256 × 256), one gets:

$$P = \sqrt{\frac{64 \times 1.333/68.8}{0.8/80 + 0.5 \times 10^{-3}\lceil (145 \times 256)/1024 \rceil + 145 \times 256(0.15 \times 10^{-6})}} = 6.03$$

The serial component due to the parallelisation management is run in the master process and therefore it is the master speed that divides the constant C_1.

Figure 7 represents the processing time of the parallel algorithm when executed in the virtual computer M1 and for P varying from 1 to 10 processors, the fastest ones being chosen in each case. There is a decrease in the processing time until P reaches 6; above that number it increases. From these results one concludes that the ideal number of processors to be used for the 64kb image is 6 processors, as obtained theoretically.

The second example is for an image of 144 × 144 pixels, 20.25 kb, where the value obtained for P, for the M1 virtual computer, was 4.07 processors. Experimentally, as shown in Figure 7, the best performance is obtained for 4 processors. This result confirms again the validity of the theoretical model used.

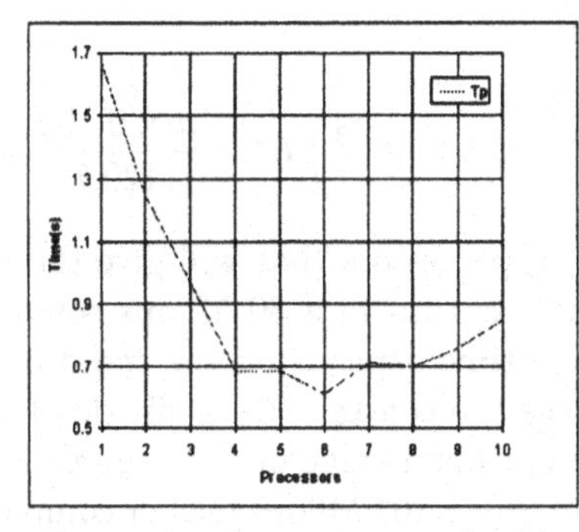

Proc. time for an image of 64kb

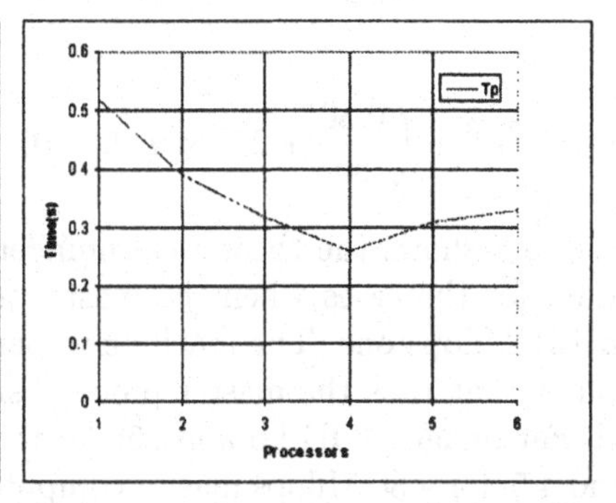

Proc. time for an image of 20.25 kb

Fig. 7. Processing time of the parallel algorithm in a Parallel Virtual Computer

5.2 Histogram Algorithm

Histogram computation is also an extensively used algorithm in image processing, often as a preprocessing stage of more elaborated algorithms. Basically, the algorithm consists in counting the occurrences of each pixel level. The input is an image and the output is a vector, of integer values, with length equal to the number of values that a pixel can assume. For an 8 bit representation, the length is $2^8 = 256$.

Each processor computes a segment of the histogram requiring for that to collect that segment from the other processors. Therefore, the amount of data that is required to exchange is independent of the distribution used, although dependant on the number of processors used, since each processor has to receive from the other $P-1$ processors the histogram segment that it is assigned to compute. As shown in Table 1, the amount of data to exchange, b, is given as a function of P. This changes the expression to compute P, the ideal number of processors, which now becomes:

$$T(n,P) = \frac{C_1 P}{S} + \frac{n}{P}\left(\frac{1}{W} + T_E\right)P + k_1 T_L + \left(\frac{1}{W} + T_E\right)\frac{N}{P}(P-1)P$$
$$+k_2 T_L (P-1)P + \frac{k_3}{P}\left(\frac{1}{W} + T_E\right)P + k_4 T_L + \frac{n\beta^\alpha}{PS} \qquad (11)$$

$$\frac{\partial T}{\partial P} = \frac{C_1}{S} + \left(\frac{1}{W} + T_E\right)N + 2k_2 T_L P - k_2 T_L - \frac{Sn\beta^\alpha}{P^2 S^2} = 0 \qquad (12)$$
$$= 2T_L k_2 P^3 + \left(\frac{C_1}{S} + N\left(\frac{1}{W} + T_E\right) - T_L k_2\right)P^2 - \frac{n\beta^\alpha}{S} = 0$$

The equation is a polynomial in P, of degree 3. Assuming an 8 bit representation for image pixels, N assumes the value of 256. The value of n is the image size in kb. The network parameters assume the same values as before. Since the histogram vector can be sent in a single 1024 packet, k_2 equals 1. Replacing these values in the above expression, one gets:

$$10 \times 10^{-3} P^3 + \left(\frac{0.80}{80} + 256 \times 0.15 \times 10^{-6} - 0.5 \times 10^{-3} \right) P^2 - \frac{n \times 0.150}{129.7} = 0$$

Figure 8 shows the theoretical number of processors that will give the best speedup, for the cases when the master process runs in a 80 Mflops computer or an 161 Mflops one. The machine response time depends on the speed of the computer that runs the master process, since the constant C_1 is divided by its speed. For images of 64 kb and 256 kb the optimum value of P is, respectively, 2.4 and 4.5, for a 80 Mflops master computer. For a 161 Mflops master computer, the values are 3.1 and 5.4 processors.

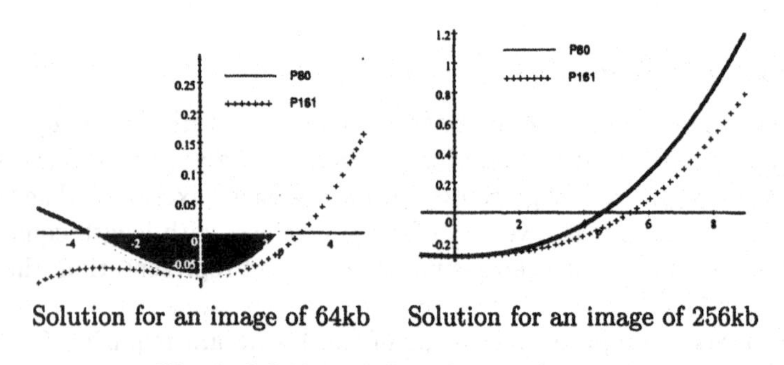

Solution for an image of 64kb   Solution for an image of 256kb

Fig. 8. Solution of the polynomial equation

Figure 9 shows the measures of the processing time made with the virtual machine M2 for images of 64 kb and 256 kb. As obtained theoretically the optimum value practically found for P was, respectively, 2 and 4 when the master process runs in the 80 Mflops computer. When the master process is executed on the 161 Mflops computer, the optimum measured value is 2 and 5 processors. The value of $P = 5$ corresponds to the theoretical value for the 256 kb image; however, for the 64kb image, the theoretical value is 3 as opposed to the value 2 found practically. This discrepancy is in fact not very significant as the practical processing time found for 2 and 3 processors differs only slightly, as shown in Figure 9.

Additionally, the discrepancy can be justified by the fact that the processor running the master process also runs an instance of the slave process; as a consequence one can consider that the available capacity of this processor, as used by the slave process, is decreased; in fact for a lower capacity, the optimum theoretical number of processors would decrease.

Figure 9 also shows an important feature of the parallel virtual computer, namely the advantage of using the parallel version of the algorithm even when the user opts to launch a single slave process; in fact, when the user is logged on a slow machine, the serial version of the algorithm uses the same machine for the whole workload; if the parallel version is selected, then the master process is run on the slow machine, but the slave process is assigned to the fastest available computer, thus reducing the global processing time. This is confirmed by inspecting the measurements displayed in Figure 9 for the 80 Mflops curves: $P = 0$ corresponds to the serial algorithm and $P = 1$ corresponds to the parallel one. Notice also that when the user is logged on the fastest machine available (161 Mflops curves), the serial version is naturally faster, as the parallel one has communication overheads that are unnecessary in this situation of a single slave process.

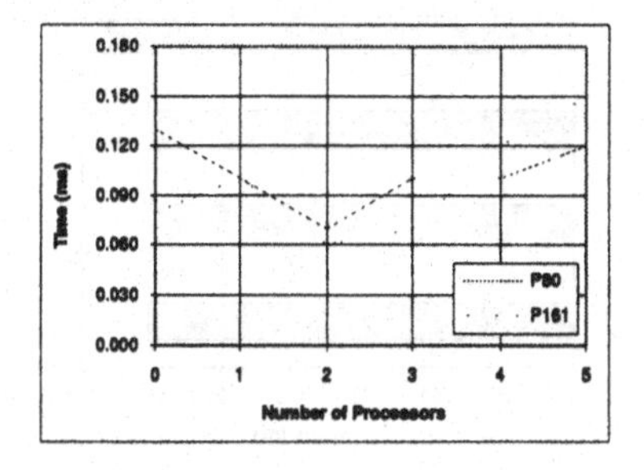

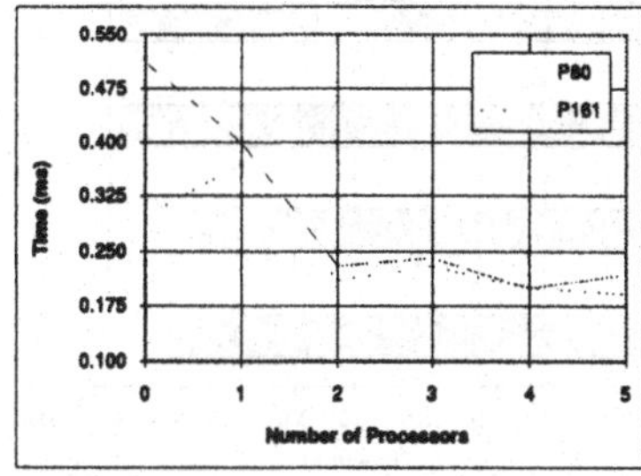

Proc. time for an image of 64kb Proc. time for an image of 256kb

Fig. 9. Processing time of the histogram algorithm

5.3 Load Balancing

In a virtual parallel computer there are frequently machines of many different processing capacities, therefore the load distribution should be managed in order to assign more work to faster processors, so that all processors finish at about the same time. A test with machine M2 and the edge detection algorithm was carried out to verify the validity of the strategy suggested above. Since the computational load is evenly distributed over the domain it is straightforward to make the size of each row block proportional to the processor capacity. Figure 10 shows the response time and the time to process the edge detection algorithm kernel for each processor, for two situations: 3 and 4 processors working. Processors P1, P2, P3 and P4 have processing capacities of 161, 161, 105 and 91 Mflops respectively. The master process runs on processor P1.

From Figure 10 it turns out that the load is well distributed, that is, a balanced workload was obtained even for largely different processing speeds of the Virtual Computer nodes. For algorithms where the load is not so evenly distributed over the domain, one has to have the additional work of blocking the domain in blocks of similar workload, for a static implementation of load balancing.

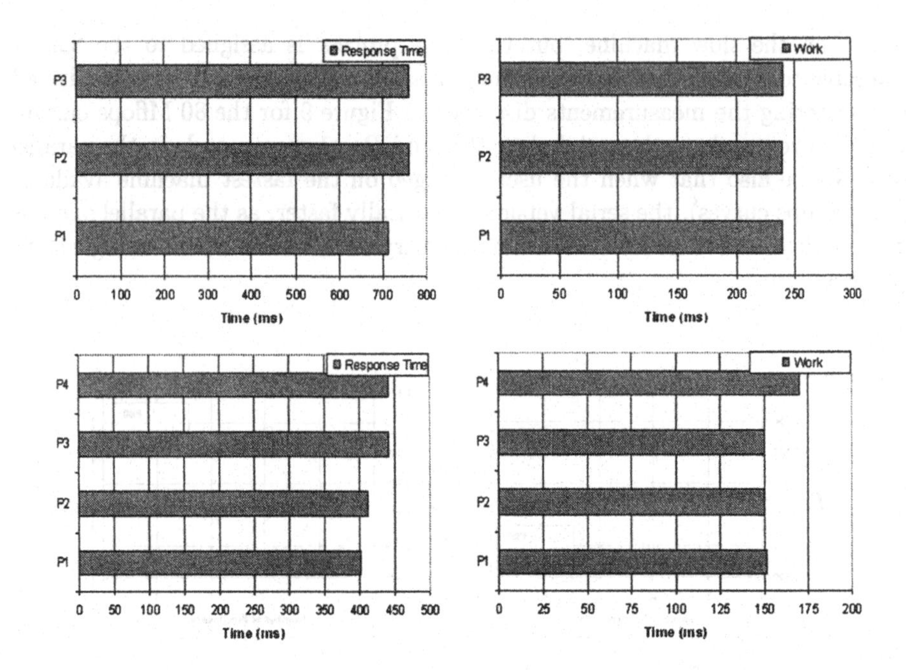

Fig. 10. Response time and work done by each processor

6 Conclusions

It was proved that for obtaining a good performance of the Parallel Virtual Computer, it is required to know the algorithm parameters, in order to compute the correct number of processors to use for its execution in the virtual computer. A methodology to obtain this number was presented, as well as some test results that proved its satisfactory accuracy.

The results suggest that a slow processor can be used to log on the user (and to run the master process), providing also some computation according to its capacity. This type of network allows a staged upgrade, since adding a fast computer to the network has a direct and positive impact on the global performance, no matter which processor is used to launch the algorithm.

References

1. Alves, A., Silva, L., Carreira, J., Silva, J.: WPVM: Parallel Computing for the People. In: Springer Verlag Lecture Notes in Computer Science: Proceedings of HPCN'95, High Performance Computing and Network Conference, Milan, Italy (1995) 582-587
2. Alves, A., Silva, J.: Evaluating the Performance of WPVM. Byte (May 1996)
3. Bader, D., JáJá, J.: Parallel Algorithms for Image Histogramming and Connected Components with an Experimental Study. CS-TR-3384, University of Maryland (December 94)

4. Al Geist et al: PVM 3 User's Guide and Reference Manual. Oak Ridge National Laboratory (1994)
5. Gustafson, J.L.: Reevaluating Amdahl's Law. CACM, 31(5), 1988, pp. 532-533
6. Hwang, K.: Advanced Computer Architecture: Parallelism, Scalability, Programmability. McGraw-Hill (1993)
7. Pancake, C. M.: Is Parallelism for you? IEEE Computational Science & Engineering (Summer 1996) 18-37
8. Siegel, H. J.: Inside Parallel Computers: Trends in Interconnection Networks. IEEE Computational Science and Engineering (Fall 1996) 69-71
9. Steen, A. J.: Overview of Recent Supercomputers. Publication of the NCF, Nationale Computer Faciliteiten, The Netherlands (January 1997). *http://www.sara.nl/now/ncf*
10. Shen, J. and Castan, S.: An Optimal Linear Operator for Step Edge Detection. CVGIP: Graphical Models and Image Processing, Vol. 54. Number 2 (1992) 112-133
11. Spurgeon, C.: Ethernet Configuration Guidelines. Peer-to-Peer Communications, Inc. (Jan 96)

Low Cost Parallelizing: A Way to be Efficient

Marc Martin and Bastien Chopard

CUI, University of Geneva
24 rue General-Dufour,
CH-1211 Geneva 4, Switzerland

Abstract. In order to minimise the heavy cost of running parallel applications on expensive MPPs, we consider the use of workstation clusters. The main difficulty to obtain efficiency on such architectures is the fact that a slow node may dramatically decrease the overall performance of the machine. We propose a dynamic load balancing solution, based on a local partitioning scheme, consisting of moving sub-domain boundaries so as to adjust the processor load according to its available CPU capability.

1 Introduction

Traditionally, heavy scientific applications are run on large and expensive mainframes such as massively parallel computers (MPPs) which are efficient but rather expensive. Price is not the single problem: in order to regulate the available CPU resources, system managers configure the computer for accepting large jobs only in batch mode. Then, even if the mainframe is very powerful, the user may wait for a long time before the job has completed.

On the other hand, commodity computers like workstations or personnel computers become more and more powerful and cheaper. With an appropriate message passing library such as PVM, they represent an interesting alternative to MPPs in order to run heavy parallel applications in a fully interactive way and at better cost.

However, using such a solution is not as simple as porting the code from the parallel mainframe to the workstation cluster. A crucial problem of load balancing may appear because, usually, the processing nodes enrolled in a cluster may not have all the same amount of CPU resources. This may be due to a difference in the chip power, or because they are not running in a single-user mode. Another user may need the resources making the CPU balance very uneven.

Such unbalanced situations are very crucial problems because, in most cases, processing nodes must synchronised from time to time in a parallel application. Since the global time process is lined up to the slowest node, one slow node is enough to reduce considerably the benefit of parallelisation. Unfortunately, the more nodes we use, the more critical this situation will be. Thus, it is essential to implement algorithms handling load-unbalanced problems and, in particular, dynamical algorithms which may re-distribute the load at run time.

J. Palma, J. Dongarra, and V. Hernández (Eds.): VECPAR'98, LNCS 1573, pp. 522–533, 1999.

2 Load Balancing Algorithms

Load balancing algorithms have been studied by several authors (see for instance [4]). Here we are interested in situations where the computational domain is a Cartesian grid. The applications we have in mind are (i) wave propagation in urban areas [6] and (ii) pollution transport in the atmosphere [7]. These application are characterised by regular communications with neighbouring grid points.

In the context of air quality models, several load balancing approach have been proposed: reference [5] uses a one-dimensional static decomposition in which a master processing element (PE) distributes as fairly as possible sets of grid points to other PEs. This methods leads to non-rectangular sub-domains. An advantage of this decomposition is the easiness of the sequential to parallel program conversion. However, the efficiency is low: a typical speedup of 30 is reported on 256 nodes.

Reference [8] uses arbitrary sub-domains, not necessarily rectangular. A dynamical load-balancing is obtained by modifying locally sub-domains boundaries, with a fine grain resolution. This results in complicated inter-processor communications. The calculation of a new data distribution may thus cost a lot of time. Such a technique yields the following scalability: 176 Mflop/s on 4 PE and 1592 Mflop/s on 64 PE.

Schattler [9] suggests to use the time of the previous step to compute a new work distribution. Such a solution is model independent. He also notes that non-rectangular decompositions require more complex communications.

Our dynamic load balancing algorithm is designed for a 2D regular grid. The domain decomposition is performed along the north/south and east/west directions (see fig. 1). This partitioning leads to a set of rectangular sub-domains, each assigned to one PE. Rectangular shapes have the significant advantage that they make it possible fast regular communications between adjacent processors [10], as opposed, for instance, to a cyclic partitioning.

Our algorithm determines each sub-domain size according to the work load of the PE. The sub-domain owned by an over-loaded PE will shrink while that of an unloaded PE will grow. A variation of size is obtained by moving sub-domain boundaries in a coherent way: when a sub-domain attempts to grow, the neighbouring sub-domains are forced to shrink, as illustrated in fig. 1. The problem is then to compute, in a minimal amount of time, the boundary motions which yield a fair work distribution.

The measure of load imbalance we consider here is based on the time each PE has spent during the last computation step. This measure, which has the advantage of being problem-independent, is appropriate in all applications where the same computation is iterated many times.

The load balancing problem can be addressed in two ways [3] [2]: on the one hand we can use a global policy which is very stable and accurate (because it accounts for every processor load), but may not be scalable. On the other hand, we can consider a local policy which may have a slow convergence and lack of stability, but which only requires local communications.

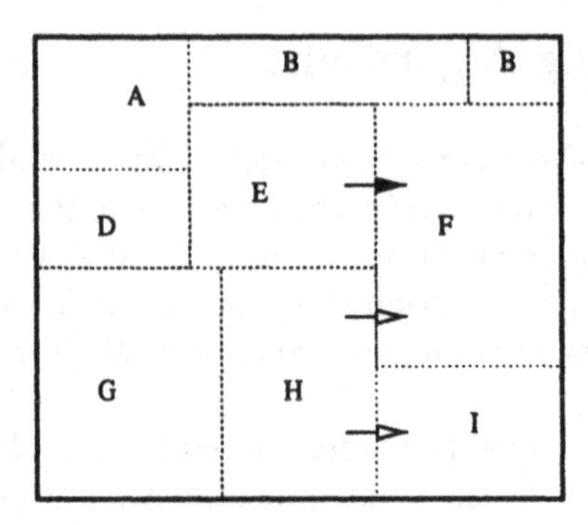

Fig. 1. Moving east boundary of node E will act on node F west boundary, and also on nodes H and I boundaries.

2.1 The Global Scheme

We consider n PEs involved in a parallel computation. At the end of step t, PE number i knows $T_{i,t}$, the time necessary to complete this step on its rectangular sub-domain of sizes $S_{i,t}$. Using an all-to-all communication, the values $T_{i,t}$ and $S_{i,t}$ are broadcast to all other PEs. From these values, each PE computes the ideal time Ta_t

$$Ta_t = \frac{1}{n} \sum_{i=1,n} T_{i,t} \tag{1}$$

and $Tc_{i,t}$ the time used by PE i for computing one single cells of the domain $S_{i,t}$.

$$Tc_{i,t} = \frac{T_{i,t}}{S_{i,t}} \tag{2}$$

Thus, optimising domain distribution means to find $S_{i,t+1}$ minimising the expression

$$\sum_{i=1,n} |Ta_t - Tc_{i,t}S_{i,t+1}| \tag{3}$$

A solution to this optimisation is obtained as follows: we first define $S_{i,t+1} = S_{i,t}$ and sort all PEs according to the value $|Ta_t - Tc_{i,t}S_{i,t+1}|$ (see fig. 2). In this way, we find the worst sized domain $S_{\gamma(1),t+1}$ (either oversized or undersized). Then we try to adjust the size of $S_{\gamma(1),t+1}$ by virtually exchanging segments of cells with the neighbouring sub-domains so as to make $Tc_{i,t}S_{i,t+1} \approx Ta_t$. The new sub-domains resulting from the motion of the boundary of rectangle $\gamma(1)$ can be computed by a recursive procedure. The same algorithm is run by each processor since all of them have a complete list of the coordinates of the all rectangles. This new domain decomposition is performed four times, for an east, west, north and south boundary motion and the best of these solutions is selected to give $S_{i,t+1}$. However, this optimisation may not be possible because: (i) the new partition is more unbalanced than the previous one; (ii) a domain may shrunk beyond an acceptable size (e.g. in finite difference schemes, a specific number of neighbour cells are required); in such a case the optimisation is considered for $S_{\gamma(next)}$;

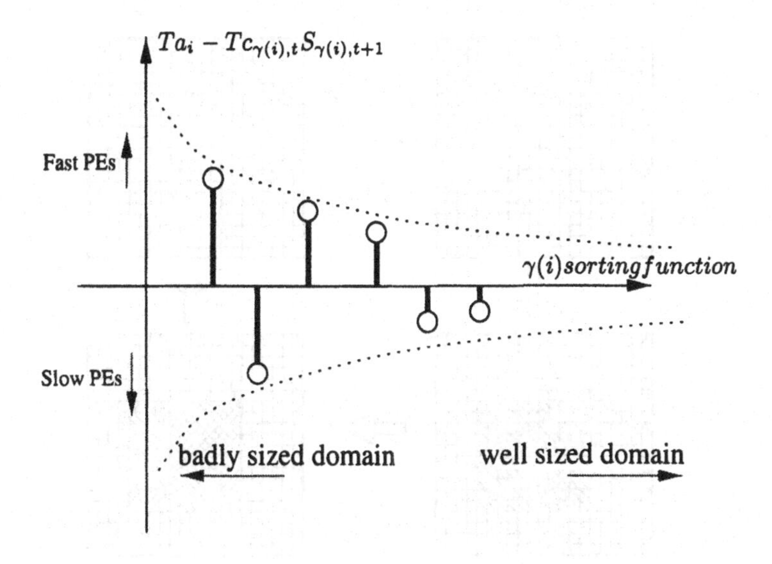

Fig. 2. PE are sorted according to their size.

When the optimisation is profitable, the whole process is repeated (sorting and optimising S) until a pre-assigned maximum number of steps is performed or until the difference of time between the fastest and slowest PE is less than a given threshold.

Once the new partition is determined, the PEs whose domain has shrunk must send the unused data to the PEs whose domain has increased. Obviously, the regular communications between adjacent PEs are affected by the new partition and a new communication pattern has to be generated. Figure 3 shows two configurations, before and after load balancing (upper-left and upper-right partitions, respectively). The data motion problem is solved by superimposing the shape of the new configuration, including ghost cells used for communication, on top of the old one. From the intersections, one can determine which block should be sent and which should be received.

This global dynamic load balancing algorithm is very efficient only when used with a restricted amount of PEs (up to 10-16), because its complexity grows very rapidly as the number of PEs increases. It is then well adapted to a coarse grain parallelism.

2.2 The Local Scheme

For a fine grain parallelism, a local load balancing policy is more suitable. The main problem is for a PE to know whether it can move its boundary since, usually, such a modification affects more than the immediate neighbours. In our local strategy, each PE negotiates several time with its immediate neighbours, in order to let information propagate.

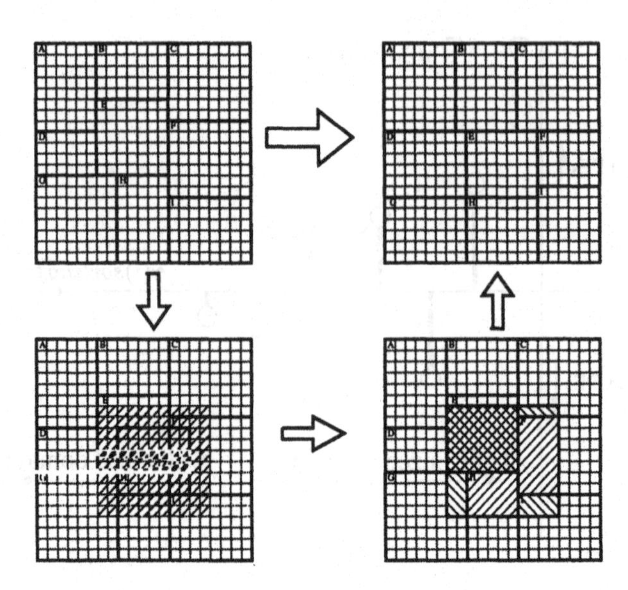

Fig. 3. Use of masks to determine data migration between two configurations (the upper-left to upper-right configuration). We apply to PE E the new partition shape (including ghost borders). One block need not migrate, whereas five blocks must be imported from neighbour PEs.

First, each PE compares the CPU time of the previous iteration with those of its adjacent neighbours. The locally slower PEs try to shrink their domains, while the faster ones try to obtain a larger domain. This decision is taken jointly by propagating a request among the PEs concerned by the change (see fig. 4). The algorithm proceeds as follows: let us assume that n PEs have a north/south boundary along the same vertical line. In the worst case, a full column of PEs could be concerned and n would be of the order $\sqrt{p}$, where p is the total number of processor. When PE number i wants to move its east boundary, it exchanges with all its adjacent east neighbours a message of size $2*n-1$ with, in position i, a request containing the amplitude of the desired motion. The message size $2*n-1$ ensures that all PEs along the same vertical boundary can insert in the message the requested motion. Once all slots of the message are filled in (i.e. after $2*n-1$ iterations), the processors may take a common decision, which is, in the current implementation, to move the boundary by the average displacement. The same procedure can then be repeated across the east/west boundaries.

When each PE knows in which way it will have to move its boundaries, it must determine if its new neighbours lists resulting from the modified partition has changed (see fig. 5). To obtain this information, it must know the new coordinates of the domains owned by the surrounding PEs. This can be obtained by propagating each PE coordinates a number $\max(m, n)$ of times along the connected sub-domains, where m is the number of PEs sharing the same horizontal

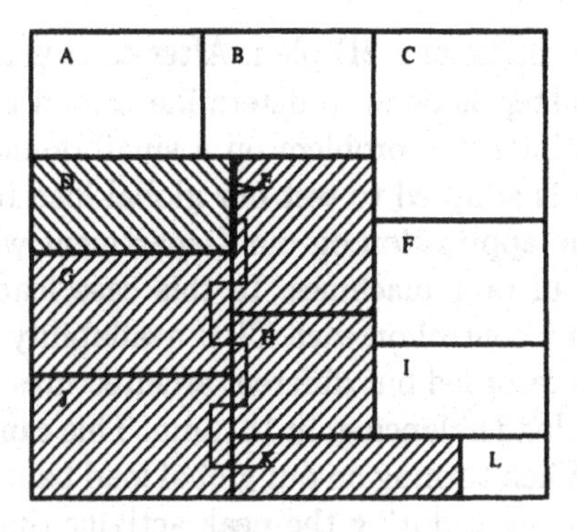

Fig. 4. Taking a coherent decision about moving a boundary is difficult because this affects non-local neighbours: if PE D wants to move its eastern boundary, it will move PE K western one. The decision of how much each boundary will move is made by propagating request messages across the affected region.

boundary. Then, it is possible to reorganise the data using the same technique as discussed in section 2.1, but with a local computation.

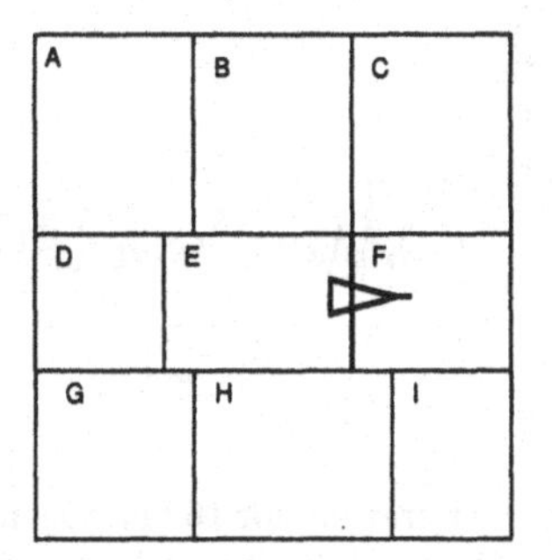

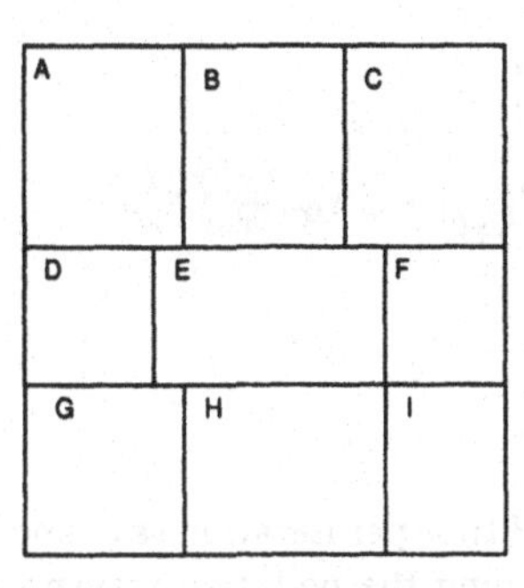

Fig. 5. Due to the east boundary modification, PE E will have to add PE C in its neighbours list.

3 Applications

3.1 Modelling Pollution Transport

We have parallelised an atmospheric pollution transport application involving 35 chemical species [7], distributed on a 3D regular grid and implementing more than one hundred chemical equations depending on various atmospheric parameters like wind, temperature and humidity. For design reasons, the 3D spatial domain is divided in vertical columns along the horizontal plane. Therefore, the domain partitioning is made on a 2D plan. After each useful calculation step, a global communication step is done to determine if load balancing is necessary. This kind of compute intensive problem on a small domain (grid dimension is about 50x40 columns) is adapted to test our global load balancing approach.

We have ported the application on our Department workstations which are Sun sparc 4 & 5 and ultra 1 machines. In this case load balancing is critical because there is not any control on each CPU availability. Early in the morning (where little activity is recorded but different architectures are used) our problem can be solved, without load balancing, in 1h13mn. This can easily turn to 2 hours in middle of the day. When turning on our global load balancing algorithm, a run takes only about 42mn, even during the peak activity time. Figure 6 shows the time per computation step as a function of the iteration. Without load balancing, we can see a significant variation of performance among the machines. With load balancing, the time spent by each PE is much more homogeneous.

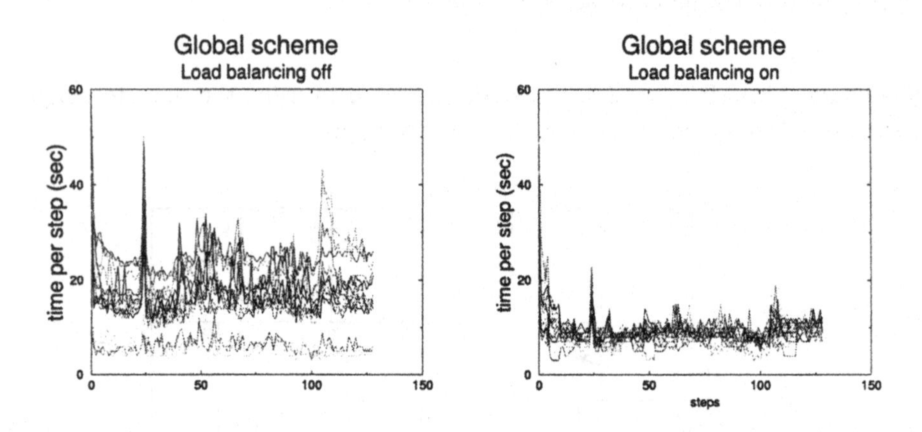

Fig. 6. CPU time per iteration as a function of the iteration, for 16 heterogeneous workstations running the pollution transport code. On the left, the global load balancing algorithm is off and, on the right, it is on.

3.2 Wave Propagation

A second application we have parallelised on a cluster of workstations is the ParFlow wave propagation model for urban environment. This application is based on the lattice Boltzmann method [6] and uses a numerical scheme closed to the so-called TLM method (Transmission Line Matrix). Wave propagation is a important problem when designing a mobile communication network. Figure 7 illustrates the wave intensity pattern predicted by our application in the case where an antenna is located in the middle of an urban area.

This wave propagation model requires 15 floating point operations per grid site and four communications steps (with the east, north, west and south neighbours). Since the computation step is rather light, synchronization of the PE is frequent and this problem may be difficult to parallelise in a efficient way on a workstation cluster with low communication bandwidth and high latency. With such conditions it seems very difficult to obtain a good load balancing without a significant overhead.

This application corresponds to a real-life problem. The simplicity of the numerical scheme makes it a good candidate for an interesting benchmark of a fine grain parallel application. Table 1 shows the performance obtained for a sequential implementation of this code.

CPU	frequency MHz	memory cache	200x200x800 grid (Mflop/s)	600x600x1600 grid (Mflop/s)
DEC alpha	150	96KB	6.0	n/a
Sun Sparc4	110	256KB	7.2	5.4
Intel Ppro	200	512KB	17.1	16.2
Sun Ultra1	166	512KB	18.5	11.8
IBM RS6000	66	64KB	24.9	10.7
Intel PII	266	1MB	25.9	22.0
SG RS10000	175	1MB	33.1	22.4

Table 1. Some benchmark results with the best optimised sequential version of the ParFlow wave propagation model. Note that all these figures are obtained for simulations on square domains, without buildings. Two tests were done: the first one corresponds to running the application on a grid of size 200×200 for 800 iterations; the second test is for a 800×800 grid during 1600 iterations. Due to cache miss augmentation, performances decrease when domain sizes increases. The RS10000 results may be pessimistic because of a possible overload of the machine when the benchmarks were done.

The parallelised version of the ParFlow simulations has been run on a 32-PC cluster running Linux and PVM [11], each node being a Pentium II 266 MHz with 64MB of memory . All PCs are interconnected using two 18-entry switches, with fast-Internet links. A Sun Sparc 5 with NFS is used as file server. The advantage of a such solution is a very low price (about 60'000 dollars) with pretty good performances, a full control on each node since the machine can be dedicated to only parallel applications.

Load imbalances are due to the fact that, during the first iterations, only the processor containing the antenna site must perform useful computations. As the wave front propagates, all processors become active. The presence of buildings may be another reason to produce an uneven load distribution among the processors because no computation is needed on building sites.

Due to some specific optimisation techniques, only valid in the sequential version of the code (memory moves replaced by alias pointers), the parallel version is slower by a factor 2, even if run on a single processor: on a 200×200 domain, 800 iterations take 16.2 seconds to complete for the sequential code and

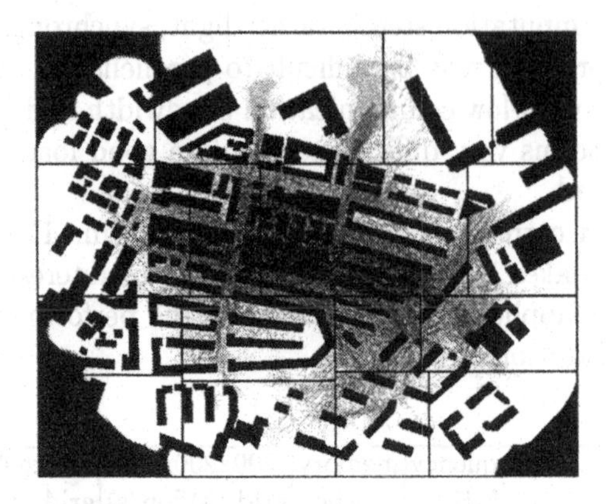

Fig. 7. Processing wave propagation in urban environment with 16 workstations, using our load balancing scheme. The rectangles show the block of data assigned to each processor.

32.8 for the parallel one. Therefore, we do no expect an efficiency larger than 50% for the parallel implementation. However, as the problem size increases (e.g. 1600×1600 during 4800 iterations), the code no longer runs on a single PE without swapping. This makes the parallel implementation very effective.

For this problem size, the parallel code runs in 1026 sec, without load balancing and using 16 free PEs. In fig. 8 we show the evolution of the time needed to perform a chunk of computation corresponding to 16 propagation steps. After an initial stage (where some PEs have not yet been reached by the wave), the computation time becomes uniform and each chunk of 16 consecutive iterations takes about 4.5 seconds.

In order to test our local load-balancing algorithm we launch on some nodes a program aimed at perturbing the computation. This program, given in the appendix, uses a significant amount of CPU power. As a consequence of this extra load, the ParFlow execution time increases to 4976 sec and the infected node performs the 16 steps in about 25 sec, thus slowing down the whole system.

In order to mask the effect of the disturbance program, the local load balancing algorithm modifies the domain partition and the total CPU time goes down to 2243 sec. This situation is illustrated in fig 8 where the effect of the perturbation is erased. However the execution is still slower than the version without perturbation. This factor is due to stability problems with the present method which modifies the domain sizes in a too simplistic way. Stability will improved in our future implementation.

Figure 9 shows the speedup obtained with and without load balancing. Comparison is made with the most optimised sequential version [1].

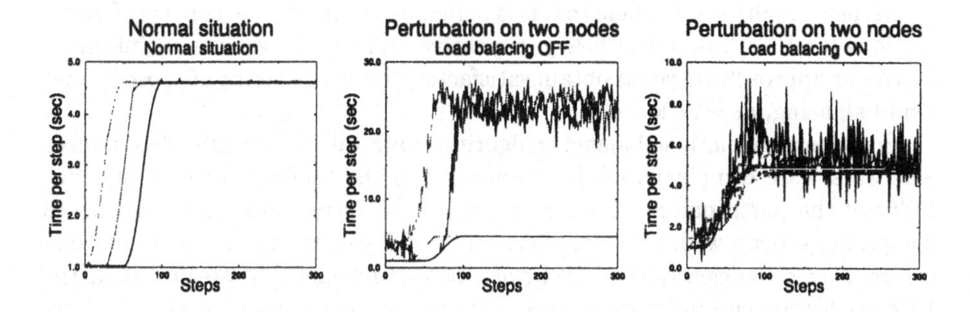

Fig. 8. Evolution of the CPU time needed to perform 16 iterations of the ParFlow problem, with 16 Pentium II-266. On the left, all PEs are dedicated to the computation and a balanced situation is observed. The middle plot corresponds to a situation where 2 nodes are disturbed with a intensive CPU consumer program. A clear load imbalance is observed, unless our local algorithm is turned on, as shown on the left.

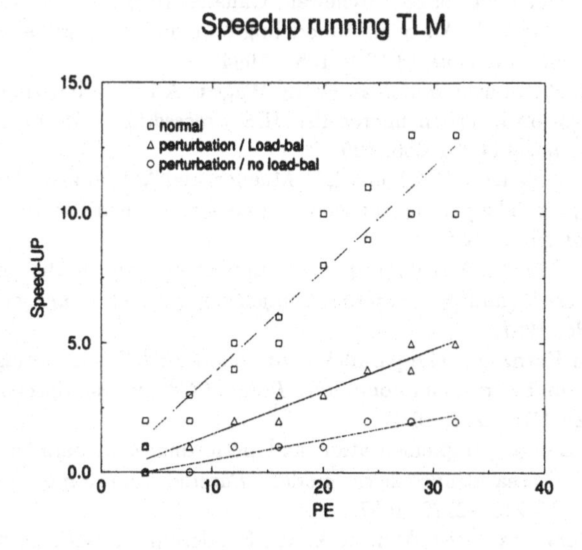

Fig. 9. Speedup computed for problems of sizes 1600×1600 and 800×800, both run for 4800 iterations. The best sequential time, estimated to 118 minutes, is interpolated from the timing of the 800 problem size (since the large grid size would cause swapping on one single PE). Performances of the load-balancing algorithm might improved by decreasing the instability seen in fig 8.

4 Conclusion

From these results we conclude that it is perfectly possible to use clusters of workstations (or PCs) instead of heavy, expensive and less convenient mainframes. With our approach we could obtain satisfactory results in terms of speedup and load balancing, at very low cost.

Some parts of our load balancing algorithms are still heuristically determined. The worst case complexity of the overhead, and the convergence time and stability of the partitioning procedure must still be investigated. Although more comparative tests with other applications could still be done, the results discussed here show clearly the benefit of our load balancing strategies when the PEs are heterogeneous or experience a change of load in an unpredictable way by other jobs.

References

1. Luthi. *lattice wave automata.* PhD thesis, University of Geneva, 1998.
2. Ian Foster. *Designing and building Parallel programs.* Addison-Wesley, 1995.
3. Kuchen and Wagener. Comparaison of dynamic load balancing strategie. http://fiachra.ucd.ie/ david/loadbalancingpapers.html.
4. Dash. Load balancing data parallel computation on workstation network. Master dissertation, University of Saskatchewan, Canada, 1998.
5. Dabdub and Seinfield. Air quality modeling on massively parallel computers. *Atmospheric environnement*, 18:1679–1687, 1994.
6. B. Chopard, P.O. Luthi, and Jean-Frdric Wagen. A lattice boltzmann method for wave propagation in urban microcells. *IEE Proceedings - Microwaves, Antennas and Propagation*, 144:251–255, 1997.
7. M. Martin, O. Oberson, B. Chopard, F. Mueller, and A. Clappier. Atmospheric pollution transport: The parallelization of a transport & chemistry code. *Atmospheric Environment*, submitted.
8. Michalakes. Mm90: A scalable parallel implementation of the penn state/ncar mesocale model (mm5). *Parallel Computing, Special Issues on Applications*, 23:2173–2186, 1997.
9. Schttler and Krenzien. The parallel 'deutschand-modell' - a message passing version for distributed memory computer. *Parallel Computing, Special Issues on Applications*, 23:2215–2226, 1997.
10. Skalin and Bjorge. Implementation and performance of parallel version of the hirlam limited area atmpospheric model. *Parallel Computing, Special Issues on Applications*, 23:2161–2172, 1997.
11. Beguelin, Dongarra, Geist, Mancheck, and Sunderam. A user's guide to pvm 'parallel virtual machine'. *ORNTNLTM-11826*, 1991.

This research is supported by the Swiss National Science Foundation.

Appendix: Slowing Down a Node

This program is used to slow down a Unix system: it starts to fork and produces n copies of itself. Each copy allocates a big block of memory and accesses it randomly. When the parent process is killed with the `kill -USR1 pid` command, all child processes are killed too. Note that the CPU overload created is not constant because swapping events may randomly appear.

```
/*
 * little program used to slow down a UNIX station
 */
#include <stdlib.h>
#include <stdio.h>
#include <signal.h>

/* memory used per process (1Mo) */
#define  BLOCK 0x100000

/* some global datas */
int* pids;
int nb_childs;

/* to kill all processes */
void kill_childs()
{ int i;
  for (i=0;i<nb_childs;i++) kill(pids[i],SIGUSR2);
  printf("\nOk, I'm dead.\n");
  exit(0);
}

void main(char argc,char** argv)
{ double* datas;
  int  i;
  int  nb_el = BLOCK/sizeof(double);
  int  pid   = 1;
  signal(SIGUSR1, kill_childs); /* to kill all processes */

  if (argc!=2) {printf ("Usage : %s nb_processes\n",argv[0]);exit(0);}
  nb_childs = atoi(argv[1])-1;  /* get nb processes from command line*/

  printf  ("Starting with %d processes\n",nb_childs+1);
  printf  (" to kill me use the \"kill -USR1 %d\" command\n",getpid());

  pids  = (int*) malloc (nb_childs*sizeof(int)); /* child ids reminder */
  datas = (double*) calloc(nb_el,sizeof(double)); /* memory allocation */

  for (i=0;i<nb_childs;i++) /* creating child */
     if (pid) {pid=fork();pids[i]=pid;}

  srand (getpid()); /* pseudo-random-number generator init */
  while (1){datas[rand()%nb_el]++;} /* random accesses in memory */
}
```

A Performance Analysis of the SGI Origin2000

Aad J. van der Steen[1] and Ruud van der Pas[2]

[1] Computational Physics, Utrecht University
P.O. Box 80195, 3508 TD Utrecht
The Netherlands
steen@phys.uu.nl

[2] Ruud van der Pas, European HPC Team
Silicon Graphics
Veldzigt 2a, 3454 PW De Meern
De Meern, The Netherlands
ruud@demeern.sgi.com

Abstract. In this paper we present the results of benchmark experiments carried out on a Silicon Graphics Origin2000. We used the three modules of the EuroBen Benchmark ([1]) to assess the performance of a single node, as a shared memory system, and as a distributed memory system. Where the situation calls for it, we compare the results with those obtained on a Cray T3E and an IBM SP2. The results obtained from this benchmark give a good impression of what performances can be attained on the Origin2000 under what circumstances and expose the weak and strong points of the system.

1 Introduction

The Silicon Graphics Origin2000 has been introduced in the last quarter of 1996. Since then a considerable amount of these systems have been installed, ranging from 4-128 processors per system. The Origin2000 machine has a rather complicated architecture and, like most high-performance computers, shows a wide range of performance levels depending on memory access patterns, loop content, fitness for and grain size of parallelism, etc. It was our intention to make a *performance profile* of the Origin2000 which will allow to obtain a fair estimate of the performance under a variety of realistic operating circumstances. At the same time, architectural bottlenecks can be identified. This may be valuable for future system development and will in the end be of benefit for end users.

To assess the performance of the Origin2000 we used the EuroBen Benchmark, version 3.2 ([1]). This benchmark was initially designed for testing shared-memory MIMD systems. However, for a limited number of important cases also message-passing codes have been developed.

This paper has the following structure: first the Origin2000 and the EuroBen Benchmark are briefly described, next we present the most relevant results of our benchmark study and we conclude with a summary and issues that might be addressed in further research.

J. Palma, J. Dongarra, and V. Hernández (Eds.): VECPAR'98, LNCS 1573, pp. 534–547, 1999.

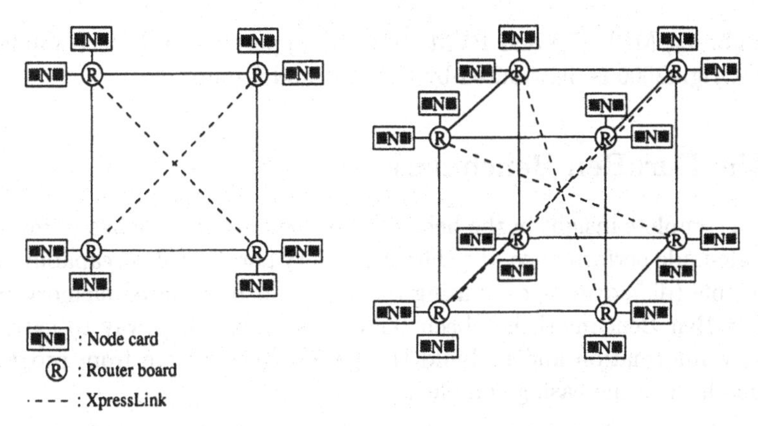

Fig. 1. *Configurations of Origin2000 systems with 16 and 32 processors.*

2 The Origin2000 System

The Origin2000 is a cache coherent, logically shared, physically distributed memory system with 4–128 MIPS R10000 RISC processors. The features of the processors are extensively described in [2,3]. These include out-of-order execution of instructions and pre-fetching of operands in order to hide data-access latency.

The system as we have benchmarked contained 195 MHz processors with a theoretical peak performance of 390 Mflop/s. The processors have 32 KB, two-way set-associative primary instruction and data caches and a combined secondary instruction and data cache of 4 MB. In parallel processing the caches of the processors involved are kept coherent via a directory memory, see [2]. The memory of the total system was in our case 16 GB.

Two processors are mounted on a node card together with a local part of the memory and a *HUB chip*, an ASIC which connects all components on the node card with each other. In addition, the HUB chip also connects the node card to the other node cards and the I/O facilities of the system. The raw bandwidth of the connections on the node card and between node cards is 780 MB/s, see [4]. However, the two processors have to share this bandwidth when accessing data from memory. For the actual point-to-point bandwidth between processors on the user level Silicon Graphics quotes a bandwidth of 150 MB/s. This is due to various overheads and the cache-coherency that is enforced by the system.

Node cards are, via their HUB chip, connected by routers to the rest of the system. The interconnection of the routers has a hypercube topology. However, for up to 32 processors, so-called XpressLinks can be added to reduce the system diameter Ω to 3. Figure 1 shows some system configurations.

Silicon Graphics provides auto-parallelising compilers that attempt to spread the content of loops evenly over the processors. In addition, the user may add parallelisation directives in various styles. Next to SGI-proprietary, also ANSI X3H5 recommended ([5]) and OpenMP ([6]) directives are accepted. Also distributed memory message passing libraries are available. Apart from the SGI/Cray-style `shmem` library, MPI ([7]) and PVM ([8]) are supported. An HPF compiler ([9]) for the Origin2000 is distributed by the Portland Group.

3 The EuroBen Benchmark

To get a complete insight in the behaviour of the machine one has to investigate the single-node performance, the shared-memory parallelisation capabilities, and the possible (dis)advantages of using the system as distributed memory system. The EuroBen Benchmark has been build in a hierarchical way to extract the necessary information and to build the performance profile from programs in three modules of increasing complexity:

- The first module contains programs that identify the machine parameters that govern upper and lower bounds of the performance.
- The second module contains simple but basic algorithms: full and sparse linear systems solvers, FFTs, random number generation, etc.
- The third module places the algorithms in a compact application setting and applies them in various PDE and ODE problem implementations. In addition, linear and non-linear least-squares problems and some I/O-bound problems are considered.

For a full description of the benchmark one is referred to [1].

3.1 Testing Circumstances

The full benchmark applied on single nodes, together with the parallel execution of relevant programs from the benchmark both with a shared-memory and a distributed-memory message-passing programming model gives a sufficient insight in the machine behaviour to enable reasonable performance estimates in many circumstances. For the shared-memory programming model we used both the SGI-proprietary as well as the ANSI X3H5 directives, for the message-passing programs MPI was used. Moreover, features like Inter Procedural Analysis and the quality of the numerical libraries provided by Silicon Graphics have been assessed to complete the profile of the machine. Where relevant, to compare and contrast the distributed memory results we also have done similar tests on two other widely available DM-MIMD systems, a Cray T3E Classic and a IBM SP. In addition some results from a Hitachi SR2201 were used.

We had the following testing circumstances for the systems quoted in this paper:

- **Origin2000** The FORTRAN 77 MIPSPro compiler, version 7.20, compiler options `-O3 -64 -OPT:IEEE:arithmetic=3:roundoff=3`, Operating System IRIX 6.4 02121744. For the hardware specifications seen section 2.
- **IBM RS6000/SP** We used IBM RS6000/SP Thinnodes with 160MHz P2SC processors and 512 MB memory per node. The Fortran 90 compiler was xlf, version 4.1, compiler options were `-O3 -qarch=pwr2`, Operating System AIX, version 2.4 002006959400.

- **Cray T3E Classic** We used 300 MHz DEC Alpha 21164 processors with 128 MB memory per node. The Fortran 90 compiler was CF90, version 3.0.1.3, compiler options were `-O3 -dp`, Operating System UNICOS/mk, version 2.0.2.19.
- **Hitachi SR2201** We used 200 MHz PA-RISC 720 processors with 256 MB of memory per node. The Fortran 90 compiler was OFORT90, version V02-05-/A, compiler option was `-O3`, Operating System HI-UX/MPP, version SR220001 02-02 0.

In all cases we used the system clock with resolutions ranging from 0.5–15 μs. We took care to use timing measurement intervals of at least a few hundred ms to exclude measuring artefacts, repeating measurements where necessary.

4 Benchmark Results

From each of the three benchmark modules we present some representative results as the complete discussion of all results is far to extensive for this paper. One is referred to the report [3] for a comprehensive presentation. The report is downloadable from: `http://www.phys.uu.nl/~steen/euroben/reports/` as a compressed PostScript file.

4.1 Module 1 Results

Program `mod1ac` measures the speed of a number of important basic operations as a function of the array length. With the bandwidth to the CPU known we should be able to assess whether the code generated by the compiler is optimal. In Table 1 we list the single-node speeds for these operations with stride 1 access to the operands as found for operation from the level 1 and level 2 cache.

Program `mod1ac` obtains which the speeds of the operations with stride 1, 3, and 4 memory access. Moreover, also the speeds of the same operations is measured when accessing the operands via an index vector. Non-unit stride access turns out to have quite little influence on the performance. Indirect indexed operations incur a loss of roughly 30% in speed due to address operations. So, we present only the stride-1 values. The first and fourth column show the maximum observed performance, r_{max}, when accessed from the primary and secondary cache, respectively. As the secondary cache is quite large (4 MB), a relatively small proportion of data references will have to be done to the main memory.

The dependency of the execution time of the array length can be modelled with considerable precision by a linear model $t(n) = a + bn$ where a is the latency and b is the time per operation per element. These parameters are given as the third and second column entries of Table 1. It enables us to draw definite conclusions about the optimality of the generated code for the operations considered.

The dyadic operations addition, subtraction, and multiplication operate at $1/6^{\text{th}}$ of the Theoretical Peak Performance, 390 Mflop/s, when accessed from the primary cache as the total operation takes 3 cycles. With an ideal bandwidth situation, transferring two operands to the relevant functional unit and shipping

	Operation	L-1 cache r_{max} Mflop/s	L-1 cache Cycles per op/element	L-1 cache Latency cycles	L-2 cache r_{max} Mflop/s
1	Broadcast	195.60	1	23	61.92
2	Copy	95.29	2	15	43.65
3	Addition	64.66	3	21	34.36
4	Subtraction	64.48	3	18	34.57
5	Multiplication	64.45	3	18	34.52
6	Division	9.23	21	0	9.22
7	Dotproduct	194.46	2	14	137.61
8	$x = x + \alpha y$	128.92	3	19	69.13
9	$z = x + \alpha y$	128.62	3	17	66.99
10	$y = x_1x_2 + x_3x_4$	107.39	6	23	56.97
11	1st order recurs.	96.39	2	23	46.04
12	2nd order recurs.	96.69	4	22	80.31
13	2nd difference	242.31	2.5	36	132.54
14	9th Degr. Polynomial	376.92	9	31	351.17

Table 1. r_{max}, *the number of cycles per operation per element, and the latency values for the primary cache operations on a single processor of the Origin2000. Only results of the first 14 of kernels are shown. The operations all have unit stride access. The operation latency from secondary cache is completely hidden by the data access.*

one result back per clock cycle, the performance should approximately be half the Theoretical Peak Performance. One can conclude that only one 8-byte data item can be transferred per cycle. This is in agreement with the bandwidth quoted by the vendor. The dot-product and the **daxpy** operation (kernel 7 and 8) also show speeds that closely agree with this bandwidth with computational intensities of 1 and 2/3, respectively ([10]). It shows that, at least for these simple operations, the compiler is able to generate optimal code given the limited bandwidth of one operand/cycle. With a high reuse of operands, like the evaluation of a 9th-degree polynomial and a computational intensity of 9, a large fraction of the Theoretical Peak Performance can be obtained: kernel 14 shows a performance of 96% of the Theoretical Peak Performance.

Shared-memory Parallel Performance of Program `modlac` Ideally, the simple, vector-oriented operations in program `modlac` should speed up almost linearly with the number of processors when executed in parallel. There are two effects that will decrease the potential speedup: the parallelisation overhead inherent in the distribution of the data and the synchronisation of the multiple processes and, secondly, the slowdown per processor when the array length per processor decreases because of the latency of the operation. In Figure 2 the speeds on 1, 8, and 32 processors is displayed for the first 14 kernels of program `modlac`.

The FORTRAN compiler uses heuristics to determine whether the computational content of a loop is sufficient to warrant parallel processing. If not, the loop is executed sequentially. When recurrences are detected, the loop is also executed sequentially. This is the case with kernels 11 and 12 representing first and second

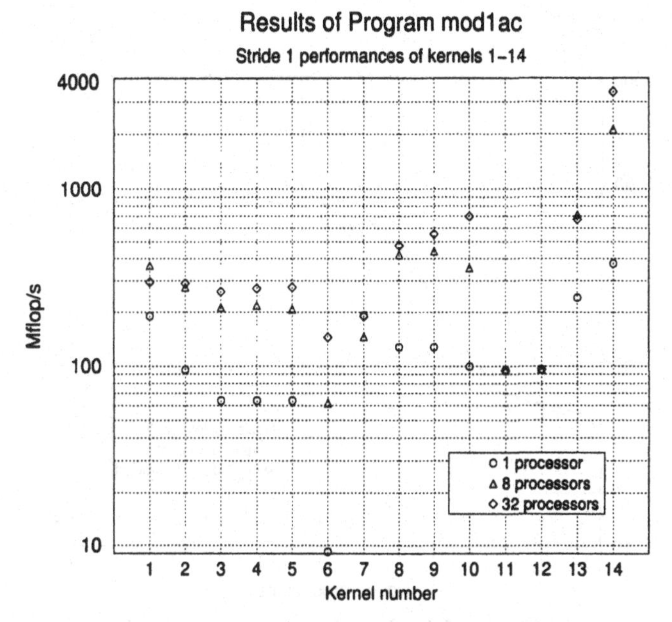

Fig. 2. *Speeds in Mflop/s of the first 14 kernels of program* `mod1ac` *on 1, 8, and 32 processors.*

order recurrences, respectively. All other kernels but one are executed in parallel. For all these kernel there turns out at least to be some benefit in parallel execution. The exception is the dot-product that shows a lower performance on 8 processors in parallel and is executed sequentially on 32 processors. It shows that the heuristics used to determine a sufficient amount of parallelism basically are correct in that the parallel execution is not slower than the sequential one.

In many cases, however, the speedup is not very high. The inherently slow division (kernel 6) and kernel 14, the evaluation of a 9th-degree polynomial, which have both a large computational content benefit the most while a kernel like the `daxpy` operation (kernel 8) show a speedup of only 12% from 8 to 32 processors. Here also the latency of the operation plays a rôle: the array length on 32 processors is only 31 elements. With this array length the speed per processor is already 15% lower than $r_{\max}$.

In summary one can conclude that the computational content of a loop should preferably not be below 10 flops to attain a sizable speedup at 32 processors.

Distributed-memory Parallel Dot-product From Figure 2 it was clear that the use of the shared-memory programming model is not suited for parallel execution of the dot-product. We also executed the dot-product with a distributed-memory programming model using MPI. Three implementations were considered: a "naive" implementation, in which all partial sums are sent to a root processor which also distributes the global sum back directly to all other processors, a FORTRAN-implemented tree algorithm for gathering the partial sums and broadcasting the global sum, and an implementation based on `MPI_Reduce` and `MPI_Broadcast`. The last implementation contains MPI functions that should be optimised by the vendor and perform at least as good as the FORTRAN-

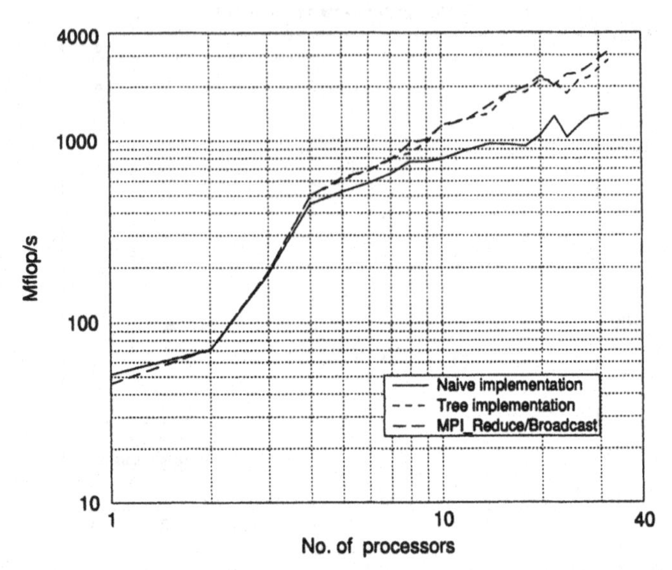

Fig. 3. *Performance in Mflop/s of the three distributed-memory dot-product implementation on 1–32 processors.*

implemented tree algorithm. Figure 3 shows the result of this distributed-memory dot-product.

The first observation that can be made is that the FORTRAN-based tree implementation and the `MPI_Reduce/Broadcast` implementation indeed are quite close in performance. So, `MPI_Reduce` and `MPI_Broadcast` are optimised communication functions. Both perform considerably better than the naive implementation, especially for a larger number of processors. The second observation is that the distributed-memory version of the dot-product scales well with the number of processors: at 32 processors a speed of 3167 Mflop/s is attained: about 100 Mflop/s, including the time lost in communication. So, the distributed-memory version is preferable by far over the shared-memory version from a performance point of view.

Point-to-point Communication The program `mod1h` measures bandwidth and latency between two processors using the MPI library functions `MPI_Send` and `MPI_Receive` with message lengths varying from 40–10,000,000 bytes. This covers the full range of possibilities: communication from the primary cache, from the secondary cache, and from the main memory. The interprocessor communication speed with point-to-point communication is not negligible in comparison with the speed between the local memory and the CPUs. Therefore, it is useful to consider this full range as it may affect the communication patterns one wants to use.

The same program has also been run on a Cray T3E Classic, an IBM SP2 and a Hitachi SR2201. As the cache sizes of these systems are different, one might expect to see different behaviour for these systems as indeed is the case. This is,

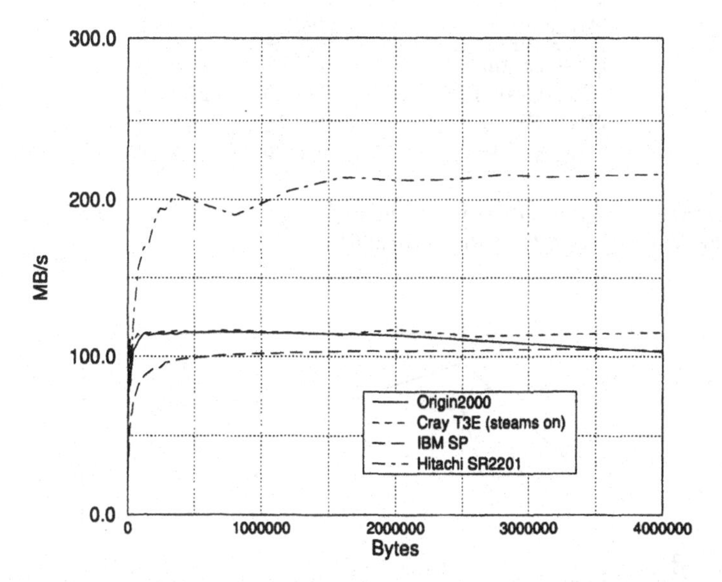

Fig. 4. *Graph of bandwidths in point-to-point message passing using* `MPI_Send` *and* `MPI_Recieve`. *Results for the Origin2000, the SGI/Cray T3E-Classic, and the IBM SP are shown. On the T3E the stream buffers were on.*

however, not only due to the different access speed in the memory hierarchies. In MPI the strategy in `MPI_Send` of buffering messages, or not, is left to the implementator. As it may be assumed that different implementation decisions have been made for different machines, observed differences in bandwidth may originate from differences in local access times, another message buffer strategy or both. Therefore, the best decision seems to be to give the bandwidth as a function of the message length and the latency as derived from very short messages (e.g., up to 400 bytes). For these short messages one may assume that no auxiliary buffering is required and one may obtain a fair idea of the latency as experienced through the software. In addition, this information is important because of the frequency that messages of only one data item are exchanged which enables an estimate for the slow-down caused by such messages. The bandwidth versus the message length is shown in Figure 4.

Note that the bandwidth of the Origin2000 is decreasing from about 115 MB/s for sufficiently long messages up to 2 MB to 102 MB/s at 4 MB. As already mentioned in section 2, the bandwidth available at the application level is 150 MB/s, so the bandwidth found reasonably matches this figure. For messages longer than 4 MB the bandwidth even drops to about 78 MB/s. We do not observe this behaviour on the other three systems. We ascribe the decreasing bandwidth on the Origin to the fact that buffer copies above 4 MB do not fit in the secondary cache anymore and therefore the memory must be accessed. The less than ideal MPI implementation might be at the base of this effect. In table 2 we summarise the maximal bandwidths and latencies for the four systems.

System	Bandwidth Mbyte/s	Latency μs
SGI Origin2000	115.75	14.6
SGI/Cray T3E-Classic	117.30	22.3
IBM SP	104.85	34.7
Hitachi SR2201	216.69	29.7

Table 2. *Maximum bandwidths and latencies for the Origin2000, the SGI/Cray T3E-Classic, the IBM SP, and the Hitachi SR2201.*

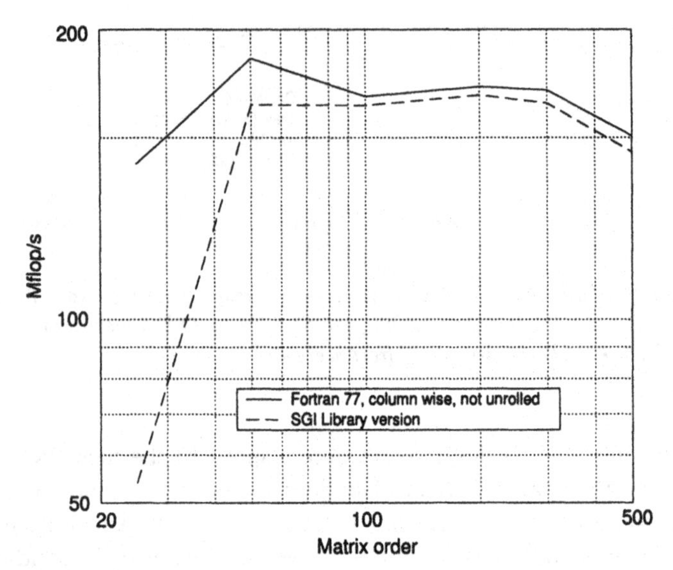

Fig. 5. *Performance for $y = Ax$. Only the fastest* FORTRAN 77 *and the SGI library routine are shown.*

4.2 Module 2 Results

Of module 2 we present two programs. Program `mod2a`, which measures the speed of a matrix-vector multiplication and `mod2e` which solves a large sparse eigen value problem system. For the discussion of all programs of module 2 one is referred to [3].

`mod2a`, single-node In In the single-node version problem sizes of n = 25, 50, 100, 200, 300, and 500 are considered for each of five implementations. For the sake of clearness, we show only the fastest of the FORTRAN 77 implementations together with the result of the library version of the BLAS 2 routine `dgemv` in Figure 5. The implementations actually used are a dot-product, or row-wise implementation, a `daxpy` or column-wise implementation and the four times unrolled versions of these two methods. On many systems the unrolled versions perform better than their not unrolled equivalents. This is, however, not the case on the Origin. The reason is that the FORTRAN 77 compiler itself already unrolls

Order	Not unrolled Row-wise Mflop/s	4×unrolled Row-wise Mflop/s	Not unrolled Column-wise Mflop/s	4×unrolled Column-wise Mflop/s	Library version Mflop/s
25	135.8	78.2	181.3	130.4	53.9
50	173.3	102.3	269.0	166.5	226.2
100	167.7	123.7	233.4	169.5	225.6
200	184.2	138.4	242.3	184.4	234.5
300	186.9	138.1	239.0	187.6	227.6
500	187.1	77.3	201.4	181.2	189.5

Table 3. *Performances on the Origin2000 for $y = Ax$. Four different* FORTRAN 77 *implementations and the SGI libary version are shown.*

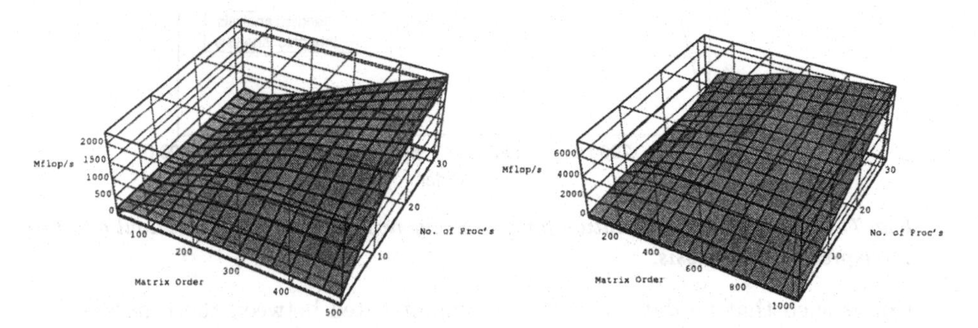

Fig. 6. *Performance surface of a parallel shared-memory implementation (left) and a distributed memory implementation (right) of a matrix-vector product.*

loops where possible and this is certainly so for the simple inner loops used in the various not unrolled implementations. For the implementations where a hand unrolling is done the compiler is not able to generate code of comparable quality and the performance of the unrolled versions lag behind as shown in Table 3. So, a fairly obvious hand optimisation does not work out very well here. The lesson could be *not* to do these kind of optimisations on the Origin to give the compiler a better chance for automatic optimisation. One of the objectives of program `mod2a` is to make users aware of such facts.

Note that in the column-wise version, using **daxpy** operations a speed is attained that is twice as high as found with program `mod1ac` for kernel 8 (see Table 1). Within the context of a matrix-vector multiplication with the **daxpy** as an inner loop, the compiler is able to overlap two successive iterations of the inner loop, thus winning a factor of 2 in speed.

mod2a, parallel versions Of `mod2a` also a shared-memory and a distributed-memory version were executed to assess the potential benefit of the parallelisation in both programming models. In Figure 6 the results for the two implementations is shown. It is clear from the Figure that the distributed-memory version is much faster than its shared-memory counterpart: 7.3 vs. 2.7 Gflop/s on 32 processors. In the distributed-memory implementation the data distribu-

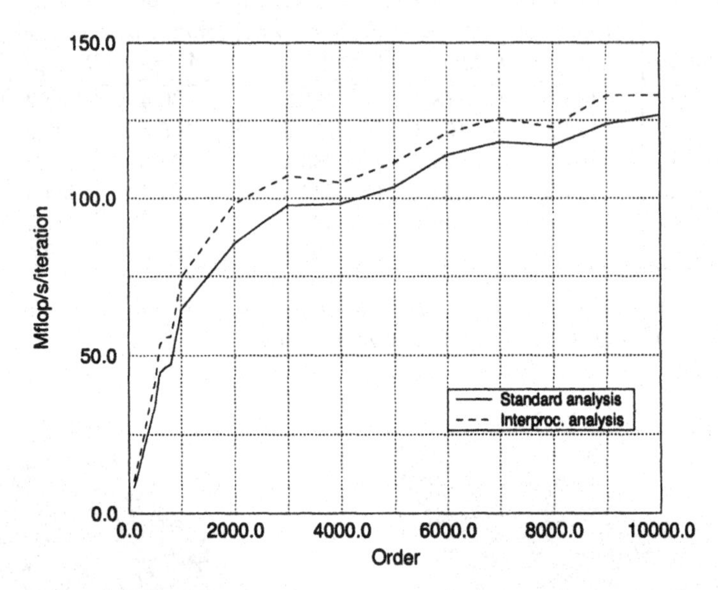

Fig. 7. *Performance per iteration for sparse eigenvalue computation without and with inter-procedural analysis.*

tion is such that no data have to be communicated between the processors. In this situation the distributed-memory is preferable. However, when the transposed matrix-vector product is performed, all-to-all communication is required. The overhead in sending messages turns out to be so high in this case that the shared-memory version is now faster then the distributed-memory version: 2.5 vs. 0.15 Gflop/s on 32 processors.

Program `mod2e` In program `mod2e` the 10 smallest eigenvalues of penta-diagonal, symmetric systems with matrix orders n =100,...,10000 are computed by a generalised Lanczos iteration scheme. In Figure 7 we show the speed per iteration for the range of system orders both without and with inter-procedural analysis.

Figure 7 shows that the inter-procedural analysis results in a small but consistently better performance over the whole problem range. The difference becomes slightly larger for larger problem size because in this case the floating-point operations in the generalised Lanczos routine more strongly dominate the computation.

The floating-point operations on the diagonals of the matrices are typical vector operations as were measured in program `mod1ac` and therefore the kernels from `mod1ac` should predict the speed of the Lanczos routine to a reasonable extent. The mix of floating-point operations as measured in `mod1ac` was as follows:

Dotproduct	34.3%
Kernel 10	25.7%
axpy	22.9%
Dyadic mult.	17.1%

System	mod3c seconds	mod3g seconds	mod3h seconds
Cray T3E-Classic	2.424	0.114	10.083
IBM RS/6000 SP	0.970	0.083	3.670
SGI Origin2000	1.486	0.065	2.366
SGI Origin2000 Interproc. analysis	—	0.062	—

Table 4. *Execution times for three PDE solvers on the Cray T3E-Classic, the IBM RS/6000 SP, and the Origin2000.*

The weighted average of the peak speeds of these operations in the primary cache is 134.5 Mflop/s. From Figure 7 we see that with inter-procedural analysis the speed for the largest problem is 133 Mflop/s and without inter-procedural analysis we find 127 Mflop/s. This is in excellent agreement with the speeds found for the kernels of `mod1ac`. This consistency shows that in the right context the prediction of the performance from kernel speeds might help to understand the observed performance. The right context is important though, as was demonstrated with program `mod2a`.

In the present form program `mod2e` is badly suited for parallelisation. Therefore no parallel results are presented.

4.3 Module 3 Results

In module 3 various programs are considered that represent important classes of applications. The programs have been tailored in the sense that only the essential floating-point parts have been retained as this is our main concern. However, the first two programs in this module are designed to test important I/O patterns to obtain an idea of the I/O capabilities of the systems considered. Again, we do not discuss the full range of programs in this module. See [3] for the complete results.

Most of the programs in this module have a complexity that makes it difficult to estimate their Mflop-rate. So, mainly execution times are reported. In addition, only one of the programs was amenable for parallelisation (program `mod3h`). On the other hand, many module 3 programs have a complexity that made it worthwhile to subject them to inter-procedural analysis.

To place the results in context, we added timings of two other systems: the T3E-Classic and the IBM RS/6000 SP.

PDE Programs In module 3 three implementations of Elliptic/Parabolic PDE solvers are included, programs `mod3c` a Multigrid solver, `mod3g` a Fast Elliptic solver, and `mod3h` a Block Relaxation solver, respectively. They all solve the same model problem: a Laplace equation on the unit square. They differ vastly in their solution speed for this particular problem but each method has its own virtues that make them more or less complementary. The execution times are given in Table 4. As can be seen from the Table, a single node of the the T3E is consistently slower than those of the IBM SP and the Origin2000. Note that

System	Execution time seconds
Cray T3E-600	16.003
IBM RS/6000 SP	8.5646
SGI Origin2000	8.7060
SGI Origin2000 Interproc. analysis	7.6141

Table 5. *Performances in seconds in program* **mod3f** *for various systems (single-node performance).*

only in program `mod3c` the IBM SP is significantly faster than the Origin2000, although the theoretical peak performance is much higher: 640 vs. 390 Mflop/s. Furthermore, it turns out that inter-procedural analysis gives a very slight advantage over the normal analysis. In general, for the programs of this module the effects of inter-procedural analysis were not large.

ODE Program In program `mod3f` the problem of gas diffusion into a porous medium is considered. In this program two gases with different diffusion coefficients are modelled. The implementation is such that a time sequence of stiff two-point boundary-value ODEs is solved. The timing results for the program are displayed in Table 5. Table 5 shows the same general pattern as was found for the PDEs: the T3E is notably slower than the other two machines while the IBM SP is only marginally faster than the Origin2000 with standard code analysis, notwithstanding its higher Theoretical Peak Performance. With inter-procedural analysis, the Origin2000 is about 15% faster than with standard code analysis.

5 Summary and Future Work

The amount of information from our experiments has been vast and, although we have discussed them to a fair extent, we are sure that a more extensive analysis would still bring up new points in the interpretation. It would almost certainly also would give grounds for new experiments. In this study we also have refrained from hand-optimisation: we just let the compiler do the work with the appropriate complier options. Other subjects not considered but probably important are: the explicit placement of data on the Origin2000 system and the migration of data by the operating system to the processor that most uses them. On the other hand, a number of useful conclusions can be drawn from this study of which we list the main ones below:

- In many cases a large proportion of the Theoretical Peak Performance can be attained when operating from the primary cache. The performance with access from the secondary is generally 2–3 times slower, except for the division operation.
- The experiments in program `mod1ac` showed that one 8-byte operand can be loaded or stored from/to the primary cache. From the secondary cache this is about one operand per two cycles.

- When automatic parallelisation is applied, the default choices whether or not to parallelise a certain loop seem to be adequate in most cases we observed.
- The point-to-point bandwidth measured with MPI is about 110 MB/s, about 70% of the bandwidth of 150 MB/s quoted by SGI.
- The automatic shared-memory parallelisation of codes generates a non-negligible parallelisation overhead as shown by program `mod2a`, a matrix-vector multiplication. Compared with the distributed-memory version it gives a large performance loss. On the other hand, as soon as also messages must be exchanged, the shared-memory implementation is clearly faster than the MPI version. The similar phenomenon was observed in the FFT program `mod2f`. Communication timings suggest that MPI implementation we used in the present tests is not optimal.
- In the rather small programs of module 3 inter-procedural analysis generally had a quite modest influence on the execution time (5–15% decrease).

Acknowledgements

We would like to thank Silicon Graphics Inc. for making their Origin2000 system at the Advanced Technology Center in Cortaillod, Switzerland, available to us to conduct the experiments described in this report.

References

1. A.J. van der Steen, *The benchmark of the EuroBen Group*, Parallel Computing, **17**, (1991) 1211-1221.
2. Silicon Graphics Inc., *Origin Servers*, Technical Report, April 1997.
3. A.J. van der Steen, R. van der Pas, *Benchmarking the Silicon Graphics Origin2000 system*, Technical Report WFI-98-2, Utrecht University, May 1998.
4. M. Galles, *Spider: A High-Speed Network Interconnect*, IEEE Micro, **17**, 1, (1997).
5. ANSI Standard Commitee X3H5, *Fortran language binding*, X3H5 Document Number X3H5/91-0023 Revision B, 1992.
6. `http://www.openmp.org/`
7. M. Snir, S. Otto, S. Huss-Lederman, D. Walker, J. Dongarra, *MPI: The Complete Reference*, MIT Press, Boston, 1996.
8. A. Geist, A. Beguelin, J. Dongarra, R. Manchek, W. Jaing, and V. Sunderam, *PVM: A Users' Guide and Tutorial for Networked Parallel Computing*, MIT Press, Boston, 1994.
9. High Performance Fortran Forum, *High Performance Fortran Language Specification*, Scientific Programming, **2**, 13, (1993) 1–170.
10. R.W. Hockney, *$f_{1/2}$: A parameter to characterize memory and communication bottlenecks*, Parallel Computing, **10** (1989) 277-286.

An ISA Comparison Between Superscalar and Vector Processors

Francisca Quintana[1], Roger Espasa[2], and Mateo Valero[2]

[1] University of Las Palmas de Gran Canaria, Edificio de Informática
35017 Las Palmas de Gran Canaria, Canary Islands, Spain
fquintan@dis.ulpgc.es

[2] U. Politècnica Catalunya - Barcelona, Computer Architecture Department,
Campus Nord
{roger,mateo}@ac.upc.es

Abstract. This paper presents a comparison between superscalar and vector processors. First, we start with a detailed ISA analysis of the vector machine, including data related to masked execution, vector length and vector first facilities. Then we present a comparison of the two models at the instruction set architecture (ISA) level that shows that the vector model has several advantages: executes fewer instructions, fewer overall operations, and generally executes fewer memory accesses. We then analyse both models in terms of speculative execution, each one in its context. Results show that superscalar processors make an extensive use of speculation and that there is a large amount of misspeculated instructions. In the vector model, speculation is achieved using vector masks and, in general, fewer operations are misspeculated.

1 Introduction

Traditionally, there have been different approaches aimed at improving microprocessor performance. One of them has been the exploitation of data level parallelism (DLP). The DLP paradigm uses vectorisation techniques to discover data level parallelism in a sequentially specified program and expresses this parallelism using vector instructions [1] [2] [3]. A single vector instruction specifies a series of operations to be performed on a stream of data. Each operation performed on each individual element is independent of all others and, therefore, a vector instruction is easily pipelineable and highly parallel [4] [5] [6]. Another approach aimed at reaching high performance in a program's execution is the exploitation of instruction level parallelism (ILP). Current state-of-the-art microprocessors all include 4-wide fetch engines coupled with sophisticated branch predictors, large reorder buffers to dynamically schedule instructions and non-blocking caches to allow multiple outstanding misses. All these techniques focus on a single goal: executing several instructions that are known to be independent, in parallel [7]. The larger the number of instructions that can be launched on each cycle, the better the performance achieved.

J. Palma, J. Dongarra, and V. Hernández (Eds.): VECPAR'98, LNCS 1573, pp. 548–560, 1999.

There are two very important advantages in using vector instructions to express data-level parallelism. First, the total number of instructions that have to be executed to complete a program is reduced because each vector instruction has more semantic content that the corresponding scalar instructions. Second, the fact that the individual operations in a single vector instruction are independent allows a more efficient execution: once a vector instruction is issued to a functional unit, it will use it with useful work for many cycles. During those cycles, the processor can look for other vector instructions to be launched to the same or other functional units. It is very likely that, by the time a vector instruction completes all its work, there is already another vector instruction ready to occupy the functional unit. Meanwhile, in a scalar processor, when an instruction is launched to a functional unit, another instruction is required at the very next cycle to keep the functional unit busy. Unfortunately, many hazards can get in the way of this requirement: true data dependencies, cache misses, branch misspeculation, etc.

The combination of these two effects has many related advantages. First, the pressure on the fetch unit is greatly reduced. By specifying many operations with a single instruction, the total number of different instructions that have to be fetched is reduced. Many branches disappear embedded in the semantics of vector instructions. A second advantage is the simplicity of the control unit. With relatively few control effort, a vector architecture can control the execution of many different functional units, since most of them work in parallel in a fully synchronous way. A third advantage is related to the way the memory system is accessed: a single vector instruction can exactly specify a long sequence of memory addresses. Consequently, the hardware has considerable advance knowledge regarding memory references, can schedule these accesses in an efficient way [8], and needs to access no more data than is actually needed. In addition, a vector memory operation is able to amortise startup latencies over a potentially long stream of vector elements.

In this paper we make a comparison between vector and superscalar processors by analysing the behaviour of a Mips R10000 [9] superscalar processor and a Convex C4 [10] vector processor. This study is carried out from different points of view. First of all we introduce an initial analysis of the Convex C4 vector processor. This includes an overview of several intrinsic characteristics of vector processing: we will analyse the effect of execution under mask and execution using the vector first facility. Then we will compare the superscalar and vector approaches from the ISA point of view. We will present data about the number of instructions and operations executed in both processors. Finally, we will present a comparison about speculative execution in the two approaches.

2 Convex C4 Analysis

We will start by analysing the vector length and vector mask facilities of vector processors. We will also present the vector first facility which is specific of the

Convex C4 machine. Then we will compare the number of instructions, operations and memory traffic of vector processors and superscalars.

This study will be carried out using the six more vectorisable programs from Specfp92. We have measured the vectorisation percentage using the Dixie tool [12]. We have generated the execution traces of the Specfp92 programs when running on a Convex C4 machine, and then we have used the Jinks simulator to measure the amount of vector and scalar operations carried out by the programs. The vectorisation percentage has been calculated as the ratio between vector operations and the addition of vector and scalar operations.

2.1 Operation Distribution

Table 1 presents the basic operation distribution for the five more vectorisable programs of the Specfp92. First column shows the total number of basic blocks (in millions) executed for each program. Next two columns present the total number of instructions broken down into scalar and vector instructions. We will distinguish between instructions and operations. A scalar instruction performs only one operation, while a vector instruction performs several operations, depending on the value of the vector length (VL) register. Fifth column is the percentage of vectorisation for each program, defined as the ratio between the number of vector operations and the total number of operations performed. Finally column sixth presents the average vector length used in vector instructions. An interesting point from this table is the average vector length observed in the programs, which is not heavily related to the percentage of vectorisation.

Table 1. Operations Distribution

Program	# basic blocks	# instructions Scalar	# instructions Vector	# vector operations	% Vect	Avg. VL
Swm256	2.57	27.46	74.82	8127.98	99.7	93
Hydro2d	4.74	38.85	35.43	3684.89	99.0	101
Nasa7	16.79	139.80	55.98	3885.02	96.5	62
Su2cor	22.53	143.95	24.08	3066.07	95.5	125
Tomcatv	19.95	126.66	6.37	644.41	83.6	99
Wave5	48.99	579.77	35.88	1615.04	73.6	43

2.2 Vector Length Distributions

Vector execution is based on executing a certain operation specified in one instruction over a large amount of independent data. The amount of data specified in each instruction is dynamically specified with the value of the Vector Length register (VL).The latency of the operation being carried out is then amortised across all VL elements. Therefore, the larger the VL, the better the performance. Fig. 1 presents the VL distribution for the six more vectorisable Specfp92 benchmarks. As we can see, the vector length distributions follow several patterns. Swim256, Tomcatv and Su2cor have the majority of their vector lengths clustered around 128. Hydro2d has a single dominant vector length which is the number of grid points used in the z-direction of the problem. Nasa7 and Wave5 have a distribution that follows a staircase, having several dominant vector lengths. All this data suggest that even among vectorisable programs the utilisation of the vector registers varies a lot.

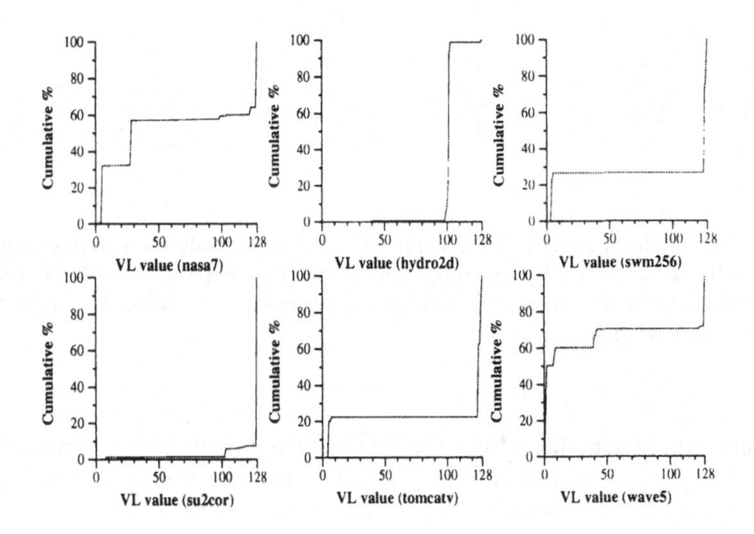

Fig. 1. VL Distribution for the Specfp92 programs

2.3 Vector First Capability

A new capability in the Convex C4 processor is the Vector First facility which allows specifying the first element in the vector register on which the instruction will be executed. That is, an instruction executes VL operations starting at element VF. This facility avoids having to reload data in the cases of recurrences as those presented in Fig. 2(a). In these cases, instead of executing two load instructions for matrix B (for position I and I+1, as presented in Fig. 2(b)), only one load instruction is executed. Fig. 2(b) shows the assembly code without vector first. Every add instruction involves two vector load instructions, which is redundant. In Fig. 2(c), using vector first, the same data can be reused in the loop body just using the appropriate vector first value, so just one vector load is

```
DO  J = 1, N
  DO  I = 1, N
     A(...,I,...) = B(...,I+1,...) + B(...,I,...)
  ENDDO
ENDDO
```

(a)

```
     mov N -> vl
L1: ...
    load (B), v0
    load (B+4), vl
    add vl, v0 -> v2
```

(b)

```
     mov N -> v4
     add #1,a4 -> a5
     mov #1 -> vf
L1: ...
     mov a5 -> vl
     load (B)
     mov a4 -> vl
     add ^v0, v0 -> vl
```

(c)

Fig. 2. Typical vector loop at hydro2d benchmark. (a) Source code for a vector loop with recurrence of distance 1. (b) Assembly code without using vector first facility, with add involving two load instructions. (c) Assembly code using vector first so that every data must be loaded just once

needed for each add instruction. [Note that the notation '^v0' means execution under vector first]. Table 2 presents the distribution of the vector first values for the same Specfp92 benchmarks as Fig. 1. This table shows the total number of operations carried out under vector first and the respective percentages of operations that have been executed with vector first equal to 1, 2 or other values. The compiler is not able to use the vector first neither in benchmark Nasa7 nor in Su2cor. Moreover, these programs only present low order recurrences (with distance 1or 2).

Table 2. Vector First distribution for Specfp92 programs

Program	# Ops under VF(x10^6)	# VF Value (in %) 1	2	Other
Swm256	2.841	76	24	0
Hydro2d	11.060	100	0	0
Nasa7	–	–	–	–
Su2cor	–	–	–	–
Tomcatv	1.124	50	50	0
Wave5	1.449	97	3	0

2.4 Vector Mask Execution

The Convex C4 vector processor allows the execution of instructions under a calculated mask stored in the Vector Mask (VM) register. The VL operations will be carried out, but only those that have the correct value stored in the ith position of the mask will be finally stored in the destination register of the instruction. We have made an analysis of the masks used during the execution of the benchmarks so to test the effectiveness of masked execution. Table 3 shows the total amount of instructions executed under mask and the percentage of instructions with respect to the total amount of instructions. This data shows a relatively small use of the execution under mask in the C4 vector processor. However, taking into account that each vector instruction implies the execution of VL operations, table 3 also shows the total amount of operations executed under mask and the percentage of operations referred to the total amount of operations. From this table we can see that the most intensive use of the masked execution is made by the Hydro2d benchmark with more than 15% of their operations executed under mask. Programs Su2cor and Wave5 execute 3.95% and 3.64% of their operations under mask, respectively. The remaining programs execute either very few operations under mask (Swm256 and Nasa7) or none at all (Tomcatv).

The execution of operations under mask can be considered as speculative execution, as all VL operations are carried out but only those that correspond to the right value in the mask are used. We can think of the extra operations as misspeculative execution. The analysis of the masks, as we will show, has allowed us to measure the amount of speculative work carried out by the vector processor.

Table 3. Instructions and Operations executed under vector mask

Program	Instructions executed under vector mask		Operations executed under vector mask	
	Total Number ($x10^6$)	% over total instructions	Total Number ($x10^6$)	% over total operations
Swm256	0.01	0.015	0.13	0.016
Hydro2d	5.75	7.750	582.91	15.650
Nasa7	0.07	0.036	8.02	0.20
Su2cor	1.06	0.630	130.75	3.95
Tomcatv	0.00	0.000	0.00	0.00
Wave5	5.17	0.840	80.00	3.64

3 Scalar and Vector ISA's Comparison

In this section we present a comparison between superscalar and vector processors at the instruction set architecture level. We will look at three different issues that are determined by the instruction set being used and by the compiler: number of instructions executed, number of operations executed and memory traffic generated. the distinction between instructions and operations is necessary because in the vector architecture, a vector instruction executes several operations (between 1 and 128 in our case).

3.1 Instructions Executed

As already mentioned, vector instructions contain a high semantic content in terms of operations specified. The result is that, to perform a given task, a vector program executes many fewer instructions than a scalar program, since the scalar program has to specify more address calculations, loop counter increments and branch computations that are typically implicit in vector instructions. The net effect of vector instructions is that, in order to specify all the computations required for a certain program, much less instructions are needed. Fig. 3(a) presents the total number of instructions executed in the Mips R10000 (using Mips IV Instruction Set [11]) and the Convex C4 machines for the six benchmark programs. In the Mips R10000 case, we use the values of graduated instructions gathered using the hardware performance counters. In the Convex C4 case we use the traces provided by Dixie [12]. As it can be seen, the differences are huge. Obviously, as vectorisation degree decreases, this gap is diminished. Although several compiler optimisations (loop unrolling, for example) can be used to lower the overhead of typical loop control instructions in superscalar code, vector instructions are inherently more expressive. Having vector instructions allows a loop to do a task in fewer iterations. This implies fewer computations for address calculations and loop control, as well as less instructions dispatched to execute the loop body itself. As a direct consequence of executing less instructions, the instruction fetch bandwidth required, the pressure on the fetch engine and the negative impact of branches are all three reduced in comparison to a superscalar processor. Also, relatively simple control unit is enough to dispatch a large number of operations in a single go, whereas the superscalar processor devotes an always increasing part of its area to manage out-of-order execution and multiple issue. This simple control, in turn, can potentially yield a faster clocking of the whole datapath. It is interesting to note that the ratio of number of instructions can be larger than 128. Consider, for example, Swm256. In vector mode, it requires 102.28 million instructions while in superscalar mode requires 11466 million instructions. If, on average, each vector instruction performs 93 iterations then all these vector instructions would be roughly equivalent to 102.28*93 = 9512 million superscalar instructions. The difference between 9512 and 11466 is the extra overhead that the supercalar machine has to pay due to the larger number of loop iterations it performs.

3.2 Operations Executed

Although the comparison in terms of instructions is important from the point of view of the pressure on the fetch engine, a more accurate comparison between the superscalar and vector model comes from looking at the total number of operations performed. As already mentioned in the previous section, the reduction of overhead due to the semantic content of vector instructions should translate into an smaller number of operations executed in the vector model. Fig. 4(b) plots the total number of operations executed on each platform for each program. These data has been gathered from the internal performance counters of the Mips R1000 processor, and from the traces obtained with Dixie. As expected, the total number of operations in the superscalar platform is greater than in the vector machine, for all programs. The ratio of superscalar operations to vector operations can be favourable to the vector model by factors that go from 1.24 up to 1.88.

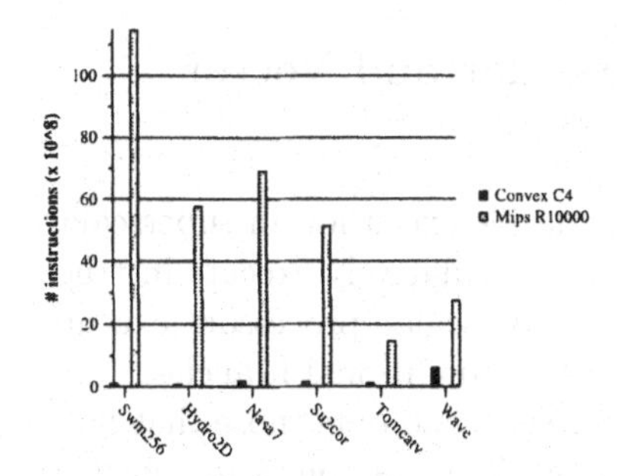

Fig. 3. Vector - Superscalar instructions executed

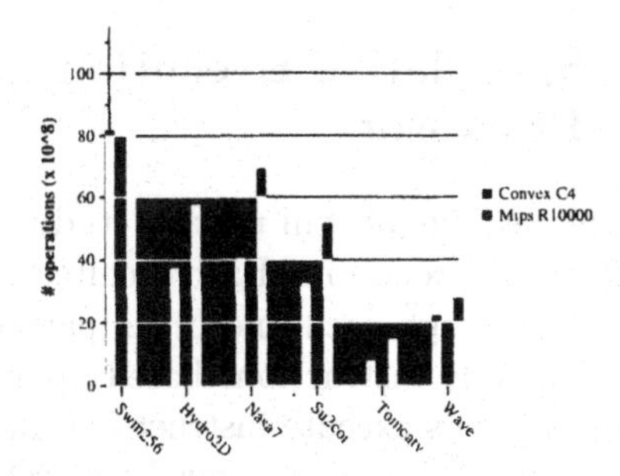

Fig. 4. Vector - Superscalar operations executed

3.3 Memory Traffic

Another analysis that we have carried out is the study of memory traffic both in vector and superscalar processors. Superescalar processors have a memory hierarchy in which data is moved up and down in terms of cache lines. Some of this data is thrown away from the cache before it is used so there is an amount of traffic that is not strictly useful. In vector processors, every data item that is brought from main memory is used, so there is no useless traffic in vector processors. Moreover, depending on the data size of the program there will be different behaviours in superscalar processors. If data fits in L1, there will be almost no traffic between the L1 and the L2 caches. However, if data doesn't fit in L1 but fits in L2, there will be a lot of traffic between the L1 and L2 caches because of conflicts. If data doesn't fit in the L2 cache, traffic will increase a lot between the two memory hierarchy levels. These behaviours can be seen if Fig. 5.

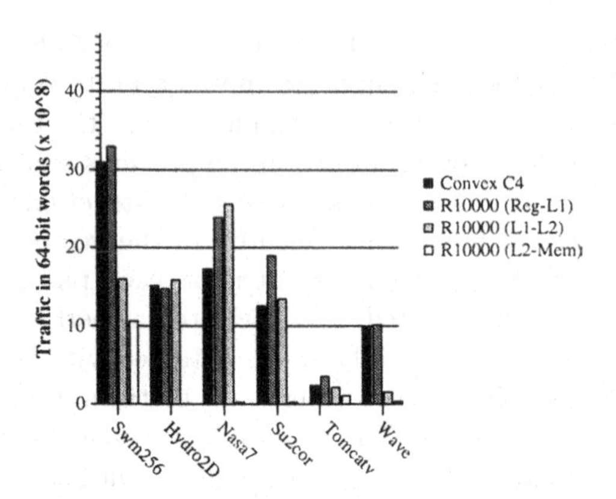

Fig. 5. Vector - Superscalar Memory traffic comparison

4 Speculative Execution in Superscalar and Vector Processors

In this section we will make a study about speculative execution in superscalar and vector processors. Each architecture is able to speculatively execute instructions, although each one in its particular way. Superscalar processors execute speculatively instructions based upon predictions of conditional branches. Vector processors execute instructions under vector masks and only those that have the correct value in the mask are definitely stored. This section is intended to study the effectiveness of the speculative execution in both architectures.

4.1 Speculation in Superscalar Processors

The increase in SS processors aggressiveness regarding issue width and out of order execution has made branch prediction and speculative execution essential techniques in taking advantage of processor capabilities. When a branch is reached, and the result of the condition evaluation is not known, a speculation of the final result of the branch is made, so that the execution continues along the speculated direction. When the actual result of the branch condition is obtained, the executed instructions are validated if the prediction was correct, and rejected if not.

The amount of misspeculative instructions in the SS processor is presented in Fig. 6 and Fig. 7. This data has been gathered using the Mips R10000 performance internal counters. This speculative work includes all types of instructions. As we can see in Fig. 6 the misspeculated execution of instructions (referred to the total number of issued instructions) for the six programs goes from 14% to 25%.

Among the misspeculative work, the load/store misspeculation is specially important because it wastes nonblocking cache resources, bandwidth, and can pollute the cache (and memory hierarchy in general) by making data movements between different levels that won't be used in the future. Fig. 7 shows the load/store misspeculation degree for the benchmarks with respect to the total number of load/store instructions. In some of them, the misspeculation percentage is as large as 40%, although the mean value is about 15%.

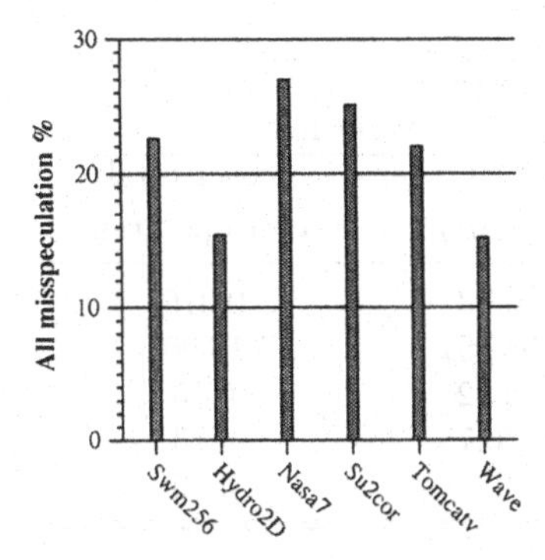

Fig. 6. Misspeculative execution in superscalar processors

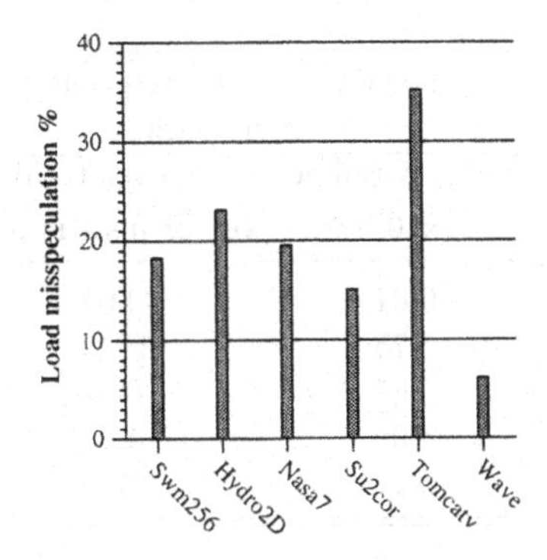

Fig. 7. Load Misspeculation in superscalar processors

4.2 Speculation in Vector Processors

Vector processors are also able to speculatively execute instructions, but in a different way than superscalar processors. It is based on the execution under vector mask. When an instruction is executed under vector mask, all the operations are carried out, but only those having the correct value in the ith position of the vector mask is definitely stored in the destination register. We have previously presented the values of masked executions referred to the total number of instructions and operations carried out by the programs. However, as masked execution is only carried out in vector mode, a more precise measure about the use of masked execution is presented in table 4. Measures in table 4 show that the behaviour differs from one program to another. Program Hydro2d executes a considerable amount of operations under mask (16%). Swm256 and Nasa7 make almost no use of the execution under mask and finally, Su2cor and Wave5 execute 4.23% and 4.95% of their operations under mask.

An interesting analysis, independent from the use of masked execution, is the effectiveness of masked execution. All these instructions executed under mask, are properly speculated or not? An operation is speculated "right" if after the operation has been carried out the result is effectively stored in its destination. All those operation that were carried out but not stored are misspeculated work. Fig. 8 shows the distribution of right and wrong speculated operations in the five programs (recall that program Tomcatv does not execute instructions speculatively). Three of the programs (Nasa7, Su2cor and Wave5) have good values of right prediction: Nasa7 and Wave5 are above 63% of right speculation and Su2cor is more than 56%. The other two programs (Swm256 and Hydro2d) have low values of right speculation, with Swim256 being the program with the worst behaviour (only 2.58% of right speculation). Another interesting consideration

Table 4. Instructions and Operations executed under vector mask

Program	Instructions executed under vector mask		Operations executed under vector mask	
	Total Number ($x10^6$)	% over total vector instructions	Total Number ($x10^6$)	% over total vector operations
Swm256	0.01	0.02	0.13	0.016
Hydro2d	5.75	16.25	582.91	16.00
Nasa7	0.07	0.12	8.02	0.20
Su2cor	1.06	4.41	130.75	4.23
Wave5	5.17	14.40	80.00	4.95

that we have studied regards the distribution of operations executed under mask among the different instruction types. This study has allowed us to establish the amount of instructions executed under mask for each type of instructions. We have considered six types of instructions: add-like, mul-like, div, diadic, load and store. The first consideration comes from the fact that none of the programs

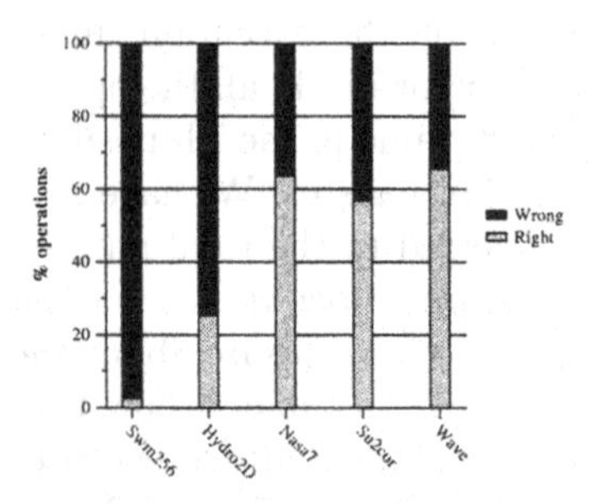

Fig. 8. Distribution of Right and Wrong speculative operations in vector processors

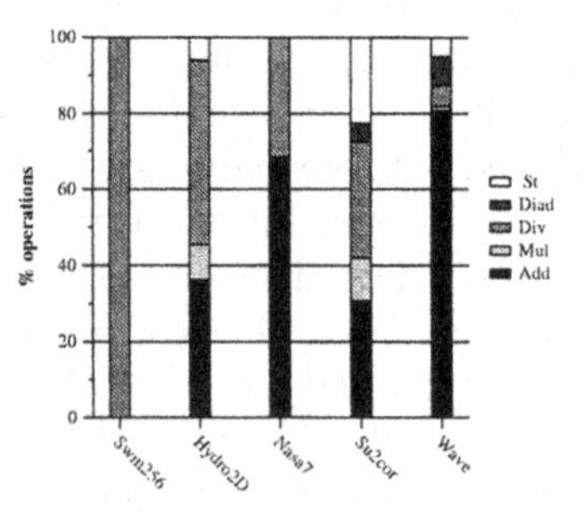

Fig. 9. Distribution of instructions executed under vector mask among the different instruction types

execute load instructions under mask, which may be explained because of the possibility of gather instructions. Fig. 9 shows the breakdown of instructions executed under mask among the different instruction types. Division and add-like instructions are the most used instructions for execution under mask.

Finally, we have also studied the effectiveness of execution under mask among the different types of instructions. Results in Fig. 10 show that, in general, there is not a clear correlation between the instruction type and the misspeculation rate. Division instructions are an exception. For divisions the misspeculation rates are higher than for the rest of instruction in all cases. This result is not unreasonable since division instructions are typically executed in statements such as the following,

$$if A(i) <> 0 then B(i) = B(i)/A(i) \tag{1}$$

In such a case, misspeculation is determined by the value stored in A(i). In our programs, the A(i) vector is sparsely populated and causes large numbers of misspeculations.

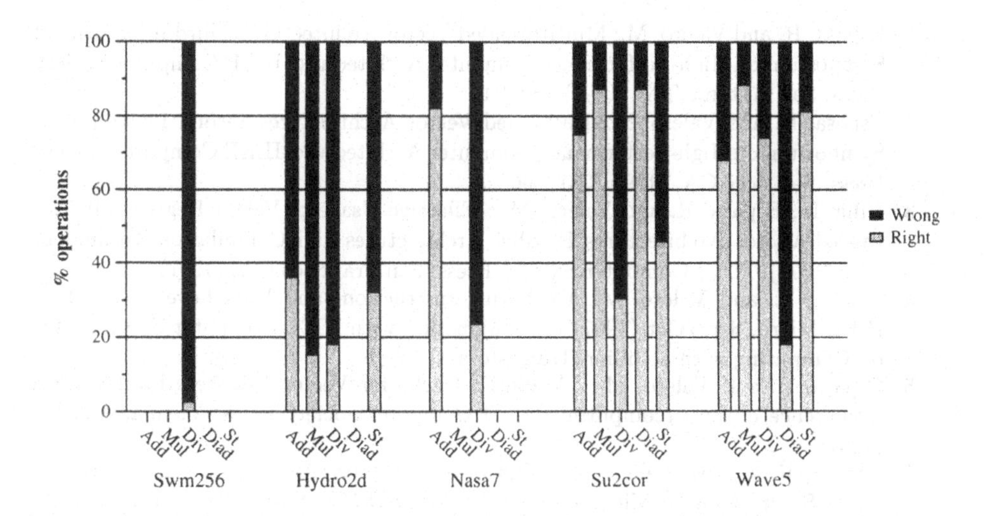

Fig. 10. Breakdown of Right - Wrong speculated vector operations

5 Conclusions

We have outlined a comparison between superscalar and vector processors from several points of view. Vector processors have different possibilities that allow them to decrease the memory traffic and branch impact in a program's execution. Their SIMD model is especially interesting because the initial latency of the operations is amortised across the VL operations that each instruction executes.

We have studied the behaviour at the ISA level of the superscalar and vector processors. We looked at total number of instructions executed, number of operations executed and memory traffic. The vector processor executes much less instructions than the superscalar machine due to the higher semantic content of its instructions. This translates into a lower pressure on the fetch engine and the branch unit. Moreover, the vector model executes less operations than the superscalar machine. The analysis of memory traffic reveals that, in general, and ignoring spill code effects, the vector machine performs less data movements than the superscalar machine.

We have also studied the speculative execution behaviour of superscalar and vector processors. Superscalar processors make an extensive use of speculative execution and the misspeculation rates are important. On the other hand, vector processors execute speculatively by using the vector mask. Vector processors make a lower use of execution under mask and the misspeculation rates are also important, although many of them are produced because of prediction in div instructions.

References

1. Espasa, R. and Valero, M.: Multithreaded Vector Architectures Third International Symposium on High-Performance Computer Architecture, IEEE Computer Society Press, San Antonio,TX, (1997) 237–249
2. Espasa, R. and Valero, M.: Decoupled Vector Architectures Second International Symposium on High-Performance Computer Architecture, IEEE Computer Society Press, San Jose,CA, (1996) 281–290
3. Villa, L., Espasa, R. and Valero, M. : Effective Usage of Vector Registers in Advanced Vector Architectures Parallel Architectures and Compilation Techniques (PACT'97), IEEE Computer Society Press, San Francisco,CA, (1997)
4. Espasa, R. and Valero, M.: Exploiting Instruction- and Data-Level Parallelism IEEE Micro, Sept/Oct (1997) (In conjunction with IEEE Computer Special Issue on Computing with a Billion Transistors)
5. Espasa, R. and Valero, M. : A Victim Cache for Vector Registers International Conference on Supercomputing, ACM Computer Society Press, Vienna, Austria, (1997)
6. Espasa, R. and M. Valero, M. : Out-of-order Vector Architectures 30th International Symposium on Microarchitecture (MICRO-30), North Carolina, (1997)
7. Jouppi, N. and Wall, D.: Available Instruction level parallelism for superscalar and superpipelined machines, ASPLOS, (1989) 272–282.
8. Peiron, M., Valero, Ayguade, E. and Lang, T.: Vector Multiprocessors with Arbitrated Memory Access. 22nd International Symposium on Computer Architecture. Santa Margherita Liguria, Italy (1995) 243–252
9. Yeager, K. et al.: The MIPS R10000 Superscalar Microprocessor. IEEE Micro, vol 16, No 2, April (1996) 28–40
10. Convex Assembly Language Reference Manual (C Series) Convex Press (1991).
11. Price, C.: MIPS IV Instruction Set, revision 3.1 MIPS Technologies, Inc., Mountain View, California (1995)
12. Espasa, R. and Martorell, X.: Dixie: a trace generation system for the C3480. Technical Report CEPBA-TR-94-08, Universitat Politecnica de Catalunya (1994)
13. Burger, D., Austin, T. and Bennett, S.: Evaluating Future Microprocessors: The SimpleScalar ToolSet University of Wisconsin-Madison. Computer Sciences Department. Technical Report CS-TR-1308 (1996)
14. Quintana, F., Espasa, R. and Valero, M.: A Case for Merging the ILP and DLP Paradigms. 6th Euromicro Workshop on Parallel and Distributed Processing, Madrid, Spain (1998) 217–224

Chapter 5: Image, Analysis and Synthesis

Introduction

Thierry Priol

IRISA/INRIA
Campus de Beaulieu, 35042 Rennes Cedex, France
Thierry.Priol@irisa.fr

This chapter includes one regular paper and one invited paper. These two contributions deal with the parallelisation of image processing or image synthesis algorithms. Applications based on the processing or the synthesis of images have always been shown as good examples for high performance parallel computers. But some of them are quite challenging...

The first paper from T. Priol, that corresponds to an invited talk, entitled *High Performance Computing for Image Synthesis* is a survey of parallelisation techniques for image synthesis algorithms such as ray-tracing and radiosity. These two techniques offer a large degree of parallelism. For ray-tracing, the degree of parallelism is equal to the number of pixels since each pixel can be computed independently. However, ray-tracing needs a geometric model of the virtual scene that is quite large (several hundred megabytes). Several solutions are presented to tackle this problem using either a data domain decomposition or a virtual shared memory. Radiosity is even more difficult to parallelise. Indeed, radiosity does not generate directly an image. It is a pre-processing step that is used to compute the indirect ambient illumination provided by light reflected among diffuse surfaces in a virtual scene. However, the number of interaction between objects are so large that they can be computed in parallel. Contrary to the ray-tracing algorithms, radiosity algorithms do not have a data locality property that make their parallelisation not straightforward. This paper gives a solution to this difficult problem based on a data domain decomposition technique. Results are quite impressive since the proposed solution allows the radiosity computation of complex scenes(more than one million of polygons) with a quite good speedup.

The second paper from C. Laurent and J. Roman, entitled *Parallel Implementation of Morphological Connected Operators Based on Irregular Data Structures*, deals with image processing. It gives a solution to the parallelisation of connected operators that can be used for image segmentation, pattern recognition, and more generally for all applications where contour information is essential. The major contribution of this paper is a set of parallel algorithms that require advanced communication schemes and irregular data structures. Despite the irregular nature of these algorithms, that make the parallelisation a non

J. Palma, J. Dongarra, and V. Hernández (Eds.): VECPAR'98, LNCS 1573, pp. 561–562, 1999.

trivial job, the authors show that their solution is scalable when the image size, to be processed, increases. Efficiency varies from 63% to 91% when using 16 processors.

High Performance Computing for Image Synthesis
Invited Talk

Thierry Priol

IRISA/INRIA, Campus de Beaulieu, 35042 Rennes Cedex, France
Thierry.Priol@irisa.fr

Abstract. For more than 20 years, several research works have been carried out to design algorithms for image synthesis able to produce photorealistic images. To reach this level of perfection, it is necessary to use both a geometrical model which accurately represents an existing scene to be rendered and a light propagation model which simulates the light propagation into the environment. These two requirements imply the use of high performance computers which provide both a huge amount of memory to store the geometrical model and fast processing elements for the computation of the light propagation model. Moreover, parallel computing is the only available technology which satisfies these two requirements. Since 1985, several technology tracks have been investigated to design efficient parallel computers. This variety forced designers of parallel algorithms for image synthesis to study several strategies. This paper present these different parallelisation strategies for the two well known computer graphics techniques: ray-tracing and radiosity.

1 Introduction

Since the beginning of the last decade, a lot of research works were made to design fast and efficient rendering algorithms for the producing of photorealistic images. Such efforts were aimed at both having more realistic light propagation models and at reducing the algorithm complexity. Such objectives were driven by the need to produce high quality images having several millions of polygons. Despite these efforts and the increasing performance of new microprocessors, computation times remains at unacceptable level. Using both a realistic light propagation model and an accurate geometrical model require huge computing resources both in term of computing power and memory. Only parallel computers can provide such resources to produce images in a reasonable time frame. For more than 10 years, the design of parallel computers was in constant evolution due to the availability of new technologies. During the last decade, most of the parallel computers were based on the distribution of memories, each processor having its own local memory with its own address space. Such Distributed Memory Parallel Computers (DMPC) have to programmed using a communication model based on the exchange of messages. Such machine were either SIMD (Single Instruction Multiple Data) or MIMD (Multiple Instruction Multiple Data).

J. Palma, J. Dongarra, and V. Hernández (Eds.): VECPAR'98, LNCS 1573, pp. 563–578, 1999.

Among those machines, one can cite the latest machines CM-5, Paragon XP/S, Meiko CS-2, IBM SP-2. More recently, a new kind of parallel systems were available. Scalable Shared Memory Parallel Computers (SMPCs) are still based on distributed memories but provide a single address space. So that, parallel programming can be performed using shared variables instead of message passing. The latest SMPCs are HP/Convex Exemplar or SGI Origin 2000. Designing parallel rendering algorithms for such various machines is not a simple task since, to be efficient, algorithms have to take benefits of the specificities of each of these parallel machines. For instance, the availability of a single address space may simplify greatly the design of a parallel rendering algorithm. This paper aims at presenting different parallelisation strategies for two well known rendering techniques: ray-tracing and radiosity. These two techniques address two different problems when realistic images have to be generated. Ray-tracing is able to take into account direct light sources, transparency and specular effects. However, such technique does not take into account indirect lights coming from the objects belonging to the scene to be rendered. This problem is addressed by the radiosity technique which is able to simulate one of the most important form of illumination, the indirect ambient illumination provided by light reflected among the many diffuse surfaces that typically make up an environment. The paper is organised as follows. The next section gives some insights on parallelisation techniques. Section 3 gives an overview of techniques for the parallelisation of both the ray-tracing algorithms and the radiosity algorithms. Section 4 describes briefly a technique we designed to enhance data locality for a progressive radiosity algorithm. Conclusions are presented in section 5.

2 Parallelisation Techniques

Using parallel computers require the parallelisation of the algorithm prior to their execution. Due to the different technology tracks followed by parallel computer designers, such parallel algorithms have to deal with different parallel programming paradigms (message-passing or shared variables). However, although the paradigms are different, designing efficient parallel algorithms require to pay attention to data locality and load-balancing issues. Exploiting data locality aims at reducing communication between processors. Such communication can be either message-passing when there are several disjoint address spaces or remote memory access when a global address space is provided. Data locality can be exploited using different ways.

A first approach consist in distributing data among processors followed by the distribution of computations. This later distribution has to be performed in such a way that computations assigned to a processor will access as much as possible local data which have been previously distributed. It consists in partitioning the data domain of the algorithm into sub-domains. Each of them is associated with a processor. Computations are assigned to the processor which owns the data involved by these computations. If these latter generate new works that might require some other data not located in the processor, they are sent to relevant processors by means of messages or remote memory accesses. This approach is called *data oriented* parallelisation.

The second approach consist in distributing computations to processors in such a way that computations assigned to a processor will reused as much as possible the same data. Such approach requires a global address space since data have to be fetched and cached into a local processor. Such approach can be applied to the parallelisation of loops. Loops are analysed in order to discover dependencies. A set of tasks is then created, representing a subset of iterations. This approach is called *control oriented* parallelisation.

The last parallel approach, also called *systolic parallelisation*, breaks down the algorithm into a set of tasks, each one being associated with one processor. The data are then passed from processor to processor. A simplified form of systolic parallelisation is the well know pipelining technique.

For DMPCs, the physical distribution of processing elements makes *data oriented* parallelisation the natural way. However, to be efficient, such technique has to be applied to algorithms where the relationship between computation and data accesses is known. If such relationship is unknown, achieving a load balancing will be a tough problem to solve. For SMPCs, the availability of a global shared address space makes this task easier. When a processor has not enough computations to perform, it can synchronise with other processors to get more computations. However, such approach suffer by an increasing number of communications, since the idle processor will have to get data used by these new computations. A tradeoff has often to be found to get the maximum performance of the machine.

3 Parallel Rendering

3.1 Ray-tracing

Principle The ray tracing algorithm is used in computer graphics for rendering high quality images. It is based on simple optical laws which take effects such as shading, reflection and refraction into account. It acts as a light probe, following light rays in the reverse direction. The basic operation consists in tracing a ray from an origin point towards a direction in order to evaluate a light contribution. Computing realistic images requires the evaluation of several million light contributions to a scene described by several hundred thousand objects. This large number of ray/object intersections makes ray tracing a very expensive method. Several attempts have been proposed to reduce this number. They are based on an object access data structure which allows a fast search for objects along a ray path. These data structures are based either on a tree of bounding boxes or on space subdivision.

Parallelisation Strategies Ray tracing is intrinsically parallel since the evaluation of one pixel is independent of the others. The difficulty in exploiting this parallelism is to simultaneously ensure that the load be balanced and that the database be distributed evenly among the memory of the processors. The parallelisation of such an algorithm raises a classical problem when using distributed parallel computers: how to ensure both data distribution and load balancing when no obvious relation between computation and data can be found? This problem can be illustrated by the following schematic ray tracing algorithm:

for $i = 1, xpix$ **do**
  for $j = 1, ypix$ **do**
    $pixel[i, j] = \Sigma(contrib(\ldots, space[f_x(\ldots), f_y(\ldots), f_z(\ldots)], \ldots))$
  done
done

The computation of one *pixel* is the accumulation of various light contributions *contrib*() depending on the lighting model. Their evaluations require the access to a database which models the scene to be rendered. In the ray tracing algorithm, the database *space* is both an object access data structure (space subdivision) and objects. The data accesses entail the evaluation of functions f_x, f_y and f_z. These functions are known only during the execution of the ray tracing algorithm and depend on the ray paths. Therefore, relationships between computation and data are unknown.

Parallelisation strategies explained in section 2 can be applied to the ray-tracing in the following manner. A *data oriented* parallelisation approach consists in distributed geometrical objects and their associated data structures (tree of extents or space subdivision) among the local memories of a DMPC. Each processor is assigned one part of the whole database. There are mainly two techniques for distributing the database, depending on the objects access data structure which is chosen. The first one partitions the scene according to a tree of extents while the second subdivides the scene extent into 3D regions (or voxels). Rays are communicated as soon as they leave out the region associated with one processor. From now on, this technique will be named **processing with ray dataflow**.

The *control oriented* parallelisation consist in distributing the two nested loops of the ray-tracing algorithms as shown previously. Such technique may apply to DMPCs but requires the duplication of the entire object data structure in the local memory of each processor. In that case, there is no dataflow between processors since each of them has the whole database. Pixels are distributed to processors using a master/slave approach. However, the limited size of the local memory associated with each processor of a DMPC prohibits its use for rendering complex scenes. A more realistic approach consists in emulating a shared memory when using a DMPC or to choose a SMPC which provides a single address space. The whole database is stored in the shared memory and accessed whenever it is needed. As said in section 2, such technique relies mainly of the exploitation of data locality. Ray-tracing has such property. Indeed, two rays shot from the observer through two adjacent pixels have a high probability of intersecting the same objects. This property is also true for all the rays spawned from the two primary rays. Such property can be exploited to limit the number of remote accesses when computing pixels. Those algorithms use a scheme of **processing with object dataflow**. Both data oriented and control oriented approaches can be used simultaneously. Such hybrid approach has been investigated recently to provide a better load balance for the design of a scalable and efficient parallel implementation of ray-tracing. Concerning *systolic oriented* parallelisation, several studies have been carried out but the irregular nature of ray-tracing algorithms make such approach ineffective. Table 1 gives a list of references dealing with the parallelisation of ray-tracing depending on their parallelisation strategies.

Type of parallelism	Communication	Data Structure	References
Control	No dataflow	Tree of extents	Nishimura et al. [29]
			Bouville et al. [7]
			Naruse et al. [27]
	Object dataflow	Space subdivision	Green et al. [21, 20]
			Badouel et al. [4, 3]
			Keates et al. [24]
		Bounding volumes	Potmesil et al. [30]
Data	Ray dataflow	Space subdivision	Dippé et al. [16]
			Cleary et al. [12]
			Isler et al. [23]
			Nemoto et al. [28]
			Kobayashi et al. [25]
			Priol et al. [31]
			Caubet et al. [9]
		Tree of extents	Salmon et al. [36]
			Caspary et al. [8]
Hybrid	Object+Ray dataflow	Space subdivision	Reinhard et al. [34]

Table 1. Parallel ray tracing algorithms.

3.2 Radiosity

Principle Contrary to the ray-tracing technique, the radiosity method does not produce an image. It is a technique aiming at computing the indirect ambient illumination provided by inter-reflections of lights between diffuse objects. Such computations are independent of the view direction. Once such computations have been performed, the image of the scene is then computed by applying Gouraud shading or ray-tracing. The radiosity method assumes that all surfaces are perfectly diffuse, i.e. they reflect light with equal radiance in all directions. The surfaces are subdivided into planar patches for which the radiosity at each point is assumed to be constant. The radiosity computation can be represented by this equation:

$$B_i = E_i + \rho_i \sum_{j=1}^{j=N} F_{ji} B_j$$

where,

- B_i : Exitance of patch i (Radiosity) ;
- E_i: self-emitted radiosity of patch i;
- ρ_i : reflectivity of patch i;
- F_{ji} : form-factor giving the fraction of the energy leaving patch j that arrives at patch i;
- N : number of patches.

The solution of this system is the patch radiosities which provide a discrete representation of the diffuse shading of the scene. In this equation, the most important component is the calculation of the form factor F_{ji} which gives the fraction of the energy leaving patch j that arrives at patch i. The computation of each form factor corresponds to the evaluation of an integral which represent the major computation bottleneck of the radiosity method. Form factor computations are carried out using several projection techniques such as the hemi-cube or the hemisphere [14]. Form-factors must be computed from every patch to every other patch resulting in memory and time complexities of $O(n^2)$. The very large memory required for the storage of these form-factors limits the radiosity algorithm practically. This difficulty was addressed by the progressive radiosity approach [13]. In the conventional radiosity approach, the system of radiosity equations is solved using Gauss-Siedel method. At each step the radiosity of a single patch is updated based on the current radiosities of all the patches. At each step, illumination from all other patches is gathered into a single receiving patch. Progressive radiosity method can be represented by the following schematic algorithm (N is the number of patches):

```
real Fi[N];                              /* column of form-factors */
real ΔRad;
real B[N];                               /* array of radiosities */
real ΔB[N];                              /* array of delta radiosities */

for all patches i do                     /* Initialisation:
                                         /* delta radiosity =self-emittance */
   ΔB[i] = E_i;
/* iterative resolution process */
while no convergence() do {              /* emission loop */
   i = patch-of-max-flux() ;
   compute form-factors Fi[j];           /* form-factor loop */
   for all patches j do {                /* update loop */
      ΔRad = ρ_j ΔB[i] × Fi[j]Ai/Aj;
      ΔB[j] = ΔB[j] + ΔRad ;
      B[j] = B[j] + ΔRad ;
   }                                     /* end of update loop */
   ΔB[i] = 0.0;
}                                        /* end of emission loop */
```

At each step, the illumination due to a single patch is distributed to all other patches within the scene. Form factors can be computed either using a hemicube with classical projective rendering techniques or using a hemisphere with a ray-tracing technique. In the first steps, the light source patches are chosen to shoot their energy since the other patches will have received very little energy. The subsequent steps will select secondary sources, starting with those surfaces that receive the most light directly from the light sources, and so on. Each step increases the accuracy of the result that can be displayed. Useful images can thus be produced very early in the shooting process. Note that, at each step, only a column of the system matrix is calculated, avoiding thus the memory storage problem.

Parallelisation Strategies The parallel radiosity algorithms proposed in the literature are difficult to classify, since numerous criteria can be considered: target architectures (SMPC, DMPC), type of parallelism (control oriented, data oriented, systolic), level at which the parallelism is exploited, form-factor calculation method, etc... The most important problem which arises when parallelising radiosity is the data access. Indeed, the form-factor calculation and the radiosity update require the access to the whole database. Another important point to be accounted for is the level at which the parallelism must be exploited in the algorithm. Indeed, a coarse grain reduces the cost entailed by managing the parallelism, but generates less parallelism than a fine grain. Moreover, data locality has to be exploited to keep as low as possible the amount of communications between processors. In the radiosity algorithm, choosing a coarse grain parallelism would not allow the exploitation of the data locality. Therefore, a trade-off between selecting a grain size and exploiting data locality has to be found. Several parallelisation strategies have been studied for both SMPCs and DMPCs. Table 2 gives references for some of these studies. Most of them focuses on the parallelisation of the progressive radiosity approach. In such approach, several levels of parallelism can be exploited. These levels of parallelism correspond to the three nested loops of the progressive radiosity algorithm as given above.

The first level consists in letting several patches to emit (or to shoot) their energy in parallel. Each processor is in charge of an emitter patch, and computes the form-factors between this patch and the others. Three cases can thus be considered. The first case is when all the processors are able to access all the patches' radiosities. In this case, each processor shoots the energy of its emitter patch and selects the next emitter patch. The second case occurs when each processor manages only a subset of radiosities. The form-factors computed by a processor are then sent to the other processors which update, in parallel, their own radiosities. The last case is when only one processor manages all the radiosities. The other ones take care of the calculation of a vector of form-factors (column of the linear system matrix) and send this vector to the master processor which updates all the radiosities and selects the next emitter patches. Note that such parallelisation changes the semantic of the sequential algorithm since the emission order of patches varies with the number of processors available in the parallel machine. In addition, the selection of the emitter patches requires the access to all the patches' radiosities (global vision of the radiosities). Indeed, each processor selects the emitter patch among those it is responsible for, which slows down the convergence of the progressive radiosity algorithm.

The second level of parallelism consists in the computation of the form-factors between a given patch and the other ones by several processors.

The third level of parallelism corresponds to the computation of the delta form-factors in parallel. This level is strongly related to the classical projective techniques and z-buffering used for form-factor calculation.

Architecture	scene	Parallelism	References
SMPC	shared	-	Baum et al. [5]
		-	Renambot et al. [35]
DMPC	shared	control	Bouatouch et al. [6]
	distributed	data	Arnaldi et al. [1]
			Varshney et al.[38]
			Drucker et al. [17]
			Guitton et al. [22]
		control	Chalmers et al.[10]
	duplicated	control	Chen [11]
			Recker et al. [33]
			Lepretre et al. [26]
	passed	systolic	Purgathofer et al. [32]

Table 2. Classification of parallel radiosity algorithms

3.3 Discussion

As shown in the last section, parallelisation of the ray-tracing algorithm has been widely investigated and efficient strategies are now identified. Concerning radiosity computation, several parallelisation strategies have been studied. However, none of them really dealt with data locality. Therefore, solving the radiosity equation for complex scenes[1] in parallel cannot achieve good performance. In [39, 37, 2], several ideas have been proposed to deal with complex environments. They are mostly based on Divide and Conquer strategies: the complex environment is subdivided into several local environments where the radiosity computation is applied. In the following subsection, we present a new parallelisation strategy able to render complex scenes using a progressive radiosity approach as explained in section 3.2. It extends the work published in [2].

[1] scenes that have more than one million of polygons

4 Enhancing Data Locality in a Progressive Radiosity Algorithm

The main goal of our work was to design a data domain decomposition well suited for the radiosity computation. Data domain decomposition aims at dividing data structure into sub-domains, which are associated to processors. Each processor performs its computation on its own sub-domain and send computations to other processors when it does not have the required data. Applied to the radiosity computation, our solution focuses on the ability to compute the radiosity on local environments instead of solving the problem for the whole environment. By splitting the problem into subproblems, using Virtual Interface and Visibility Masks, our technique is able to achieve better data locality than other standard solutions. This property is capital when using either a modern sequential computer to reduce data movement in the memory hierarchy or a multiprocessors to keep as low as possible communication between processors whatever the communication paradigm is: either message passing or shared memory.

4.1 Data Domain Decomposition

This section summarizes briefly the virtual interface that is the basic concept to perform a data domain decomposition. A more detailed description can be found in [1]. A virtual interface is a technique to split the environment (the scene bounding box) into local environments where the radiosity computation can be applied independently from other local environments (figure 1). It also addresses the energy transfer between local environments. When a source, located in a local environment, has to distribute its energy to other neighbouring local environments, its geometry and its emissivity has to be sent. We introduced a new structure called the visibility mask, to the source (figure 2).

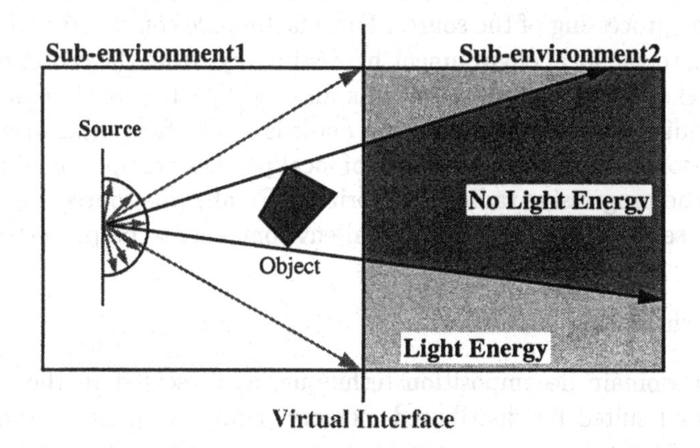

Fig. 1. Virtual Interface.

The visibility mask stores in the source structure all the occlusions encountered during the processing in each local environment. With our virtual interface concept, the energy of each selected patch, called a source, is first distributed in its local environment. Then, its energy is propagated to other local environments. However, to propagate efficiently the energy of a given patch to another local environment, it is necessary to determine the visibility of the patch according to the current local environment: an object along the given direction may hide the source (figure 2). We introduced the visibility mask that is a sub-sampled hemisphere identical to the one involved in the computation of the form factors. To each pixel of the hemisphere, used for form factor computation, corresponds a boolean value in the visibility mask.

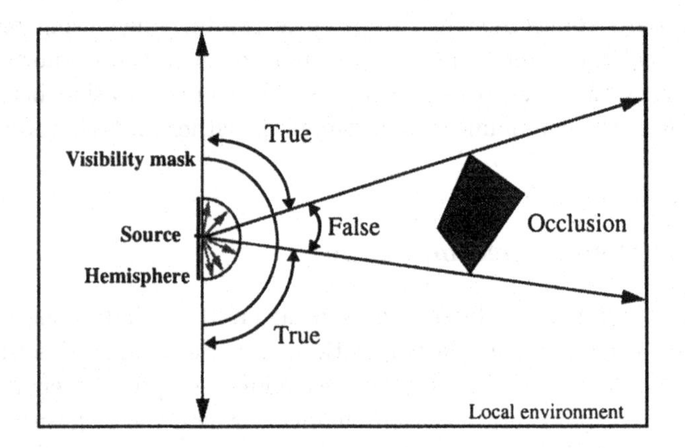

Fig. 2. Initialisation of the visibility mask for a local source.

The visibility mask allows the distribution of energy to local environments in a step by step basis. If the source belongs to the local environment, a visibility mask is created, otherwise the visibility mask already exists and will be updated during the processing of the source. Form factors are computed with the patches belonging to the local environment by casting rays from the center of the source through the hemisphere. If a ray hits an object in the local environment, the corresponding value in the visibility mask is set to false otherwise there is no modification. Afterwards, radiosities of local patches are updated during an iteration of the progressive radiosity algorithm. Finally, the source and its visibility mask are sent to the neighboring local environments to be processed later.

4.2 Parallel Algorithm

The data domain decomposition technique, as presented in the previous section, is well suited for distributed memory parallel computers. Indeed, a local environment can be associated with a processor performing the local radiosity computation. Energy transfer between local environments is performed using message passing. The parallel algorithm running on each processor consists in three successive steps: an initialisation, a radiosity computation and a termination detection as shown in figure 3. These three steps are described in the following paragraphs.

```
void ComputeNode()
{
    Initialisation()
    do {
        do { /* Computing */
                Choose source among region and network queue;
                Shooting the source;
                Read all the sources arrived from network, put them in queue;
        } while(! local_convergence);
        TerminationDetection(done)
    } while ( not done);
}
```

Fig. 3. Algorithm running on a processor.

Each processor performs an initialisation step which consists in reading the local environment geometries that have been assigned to it. Once each processor has the description of its local environment geometries, it reads the scene database to extract polygons that belongs to its local environments. The last processing step consists in subdividing the local environment into cells using a regular spatial subdivision [18] in order to speedup the ray-tracing process when form factors are computed.

Once the initialisation has been performed, each processor selects an emitter patch which has the greater delta-radiosity from either its local environment or from a receive queue which contains patches and visibility masks that have been sent by other processors since the last patch selection. Each selected patch is associated with a visibility mask that represents the distribution of energy. Initially, when a processor selects a patch that belongs to its local environment, the visibility mask is set to true. As the form-factor calculation progress, using a ray-tracing technique, part of the visibility masks is set to false if the rays hit an object in the local environment. Once the energy has been shot in the local environment, it is necessary to determine if there is still some energy to be sent to the neighbouring local environments. A copy of the visibility mask is performed for each neighbouring local environment. A ray-tracing is then performed for all pixels which have a true value. Intersection with the considered plane is performed to determine if there is energy entering in the neighbouring local environment. If a ray hits the plane that separates the two local environments, the corresponding pixel of the visibility mask is left unchanged otherwise it is set to false. At the end of this process, if the copies of the visibility mask have still some pixel values to true, they are sent to processors which own the neighbouring local environments together with some information related to the geometry and the photometry of the emitter patch.

Each processor performs its computation independently from the other processors. Therefore, it has no knowledge about the termination of the other compute nodes. Termination detection is carried out using two steps. The first step consist in deciding to stop the selection of a local patch if its energy is under a given threshold. The service node is in charge of collecting, on a regular basis, the sum of the delta radiosities of the local environments as well as the energy that is under transfer between compute nodes. Depending of this information, the service node can inform compute nodes to stop the selection of new local patches. Once the first step is reached, buffers that contain sources coming from other compute nodes have to be processed. To detect that all the buffers are empty, a termination detection is carried out using a distributed algorithm based on the circulation of a token [15]. Compute nodes are organized as a ring. A token, whose value is initially true, is sent through the ring. If a compute node has sent a patch to another node in the ring since the last visit, the value of the token is changed to false. It is then communicated to the next compute node. If the compute node, that initiates the sending of the token, received it later with a true value, it broadcasts a message to inform all the compute nodes that the radiosity computation has ended.

4.3 Results

Experiments were performed using an 56 processors Intel Paragon XP/S using three different scenes. The *Office* scene represents an office with tables, chairs and shelves. The scene contains two lights on the ceiling. It's an open scene with few occlusions. It is made of roughly six hundred polygons. After meshing, 7440 patches were obtained. The second scene is a set of 32 similar rooms. Four tables, four doors open onto next rooms and one light source compose a room. This is a symmetrical scene with many occlusions. This file comes from benchmarks scenes presented at the 5th *Eurographics workshop on rendering*. After meshing 17280 patches were obtained. The last scene is the biggest one. It represents five floors of a building without any furniture and with one thousand light sources. This scene represents the *Soda Hall Berkeley Computer Science* building [19]. After meshing, 71545 patches were generated.

Since the computing time of the sequential version depends on the number of local environments, we took a decomposition of the scene into 56 local environments in order to avoid super-linear speedup. As said previously, decomposition is a straightforward automatic process without optimisation to balance the load among the processors. Despite that no effort was spent to solve the load balancing problem, speedups were quite good comparing to other parallelisation strategies previously published. We got a speedup of 11 for the *Office* scene, 40 for the *Rooms* scene and 24 for the *Building* scene when using 56 processors.

Unfortunately, the Paragon XP/S we used for our first experiments did not have enough memory for handling complex scenes. Moreover, due to the lack of hardware monitoring available within the processor, it was impossible to study the impact of our technique on the memory hierarchy. We performed several experiments using a 32 processors Silicon Graphics Origin 2000 having 4 Gbytes of physical memory. A complete description of this work has been published in [35]. Although this machine provides a global address space, we did not use it to access data. It has been used to emulate a message passing mechanism to exchange data between processors. Therefore, the parallel algorithm is quite similar to the one described in the previous section. We used two new scenes that have a larger number of polygons. The first, named *Csb*, represents the Soda Hall Building. The five floors are made of many furnished rooms, resulting in a scene of over 400.000 polygons. It's an occluded scene. The second scene, named *Parking*, represents an underground car park with accurate cars models. The scene is over 1.000.000 polygons. It is a regular and open scene. We use a straightforward decomposition algorithm that places virtual interfaces evenly along each axis.

The first experiment we did concerns the study of the impact of our technique to exploit efficiently the memory hierarchy when using one processor. We ran our algorithm using one processor. For that purpose, we designed a sequential algorithm able to process several local environments instead of only one [35]. For the two considered scenes, we subdivided the initial environment into 100 local environments for the *Parking* scene and 125 local environments for the *Csb* scene. A gain factor of 4.2 on the execution times can be achieved for the *Parking* scene with 100 sub-environments, and 5.5 for the *Csb* scene. The main gain is given by a reduction of memory overhead due to a dramatic reduction of secondary data cache access time up to a factor of 30 for the *Parking* scene and a factor of 11 for the *Csb* scene. With the reduction of the working set, we enhance data locality and make a better use of the L2 cache. Data locality reduces memory latency and allows the processor to issue more instructions per cycle, which is a great challenge on a superscalar processor. The overall performance goes from 10 *Mflops* to 28 *Mflops*.

The last experiment was performed using different number of processors. For this experiment, we subdivided the environment into 32 and 96 local environments. Since the number of local environments has an impact of the sequential time, as shown in the previous paragraph, we decided to compute the speedup using the sequential times obtained with the same number of local environments. This protocol aims at avoiding super-linear speedup due to the memory hierarchy. Using 32 processors, we obtained a speedup of 12 for the *Parking* and the *Csb* scenes when using 32 local environments. By increasing the number of local environments to 96, we increased the speedup to 21 for the *Parking* scene and 14 for the *Scene*. This increasing performance is mainly due to a better load balance between processors. Since there are several local environments assigned to a processor, a cyclic distribution of the local environments to the processors balance evenly the load.

5 Conclusion

As shown in this paper, parallelisation of rendering algorithms have been largely investigated for now more than ten years. Proposed solutions are both complex and often not independent from the underlying architecture. Even if such solutions have proven their efficiencies, parallel rendering algorithms are not widely used in production, especially for the production of movies based on image synthesis techniques. Such fact can be explained as follow. A movie is a set of image frames that can be computed in parallel. Such trivial exploitation of parallelism can be illustrated by the production of the *Toy Story* movie, which was made entirely with computer generated images. The making of such movie was performed using a network of dozens of workstations. Each of them was in charge of rendering one image frame that took an average of 1.23 hours. Therefore, one has to raise the following question: why do we have still to contribute to new parallelisation techniques for rendering algorithms? Three answers can be given to this cumbersome question. The first one is obvious, there are still a need to produce an image in the shortest time (for lighting simulation application). The second answer comes from the fact that both radiosity and ray-tracing techniques can be applied to other kinds of applications (wave propagation, sound simulation, ...). For such applications, it is an iterative process that consists in analysing the results of a simulation before starting another simulation with different parameters. In such cases, parallelisation of ray-tracing and radiosity algorithms is required to speedup the whole simulation process. The last answer comes from the resource management problem. Image frame parallelism is the most efficient technique if the geometric (objects) and photometric (textures) databases fit in memory. If such databases exceed the size of the physical memory, frequent accesses to the disk will slow down the computation times. By using several computers to put their resources (both processors and memory) together, a single image can be computed more efficiently. The challenge for the next decade will be, no doubt, the design of new parallel rendering algorithms capable of rendering large complex scenes having several millions of objects. Parallel computers have not always to be seen as a mean to reduce computing times but also as a mean to compute larger problem size which cannot afford by sequential computer.

Acknowledgments

I would like to thank B. Arnaldi, D. Badouel, K. Bouatouch, X. Pueyo and L. Renambot who contributed to this work.

References

[1] Bruno Arnaldi, Thierry Priol, Luc Renambot, and Xavier Pueyo. Visibility masks for solving complex radiosity computations on multiprocessors. *Parallel Computing*, 23(7):887–897, July 1997.

[2] Bruno Arnaldi, Xavier Pueyo, and Josep Vilaplana. On the Division of Environments by Virtual Walls for Radiosity Computation. In *Proc. of the Second Eurographics Workshop on Rendering*, pages 198–205, Barcelona, 1991. Springer-Verlag.

[3] D. Badouel, K. Bouatouch, and T. Priol. Ray tracing on distributed memory parallel computers: Strategies for distributing computation and data. *IEEE Computer Graphics and Application*, 14(4):69–77, July 1994.

[4] D. Badouel and T. Priol. An Efficient Parallel Ray Tracing Scheme for Highly Parallel Architectures. In *Eurographics Hardware Workshop*, Lausanne, Switzerland, September 1990.

[5] Daniel R. Baum and James M. Winget. Real time radiosity through parallel processing and hardware acceleration. *ACM Workshop on interactive 3D Graphics*, pages 67–75, March 1990.

[6] K. Bouatouch and T. Priol. Data Management Scheme for Parallel Radiosity. *Computer-Aided Design*, 26(12):876–882, December 1994.

[7] C. Bouville, R. Brusq, J.L. Dubois, and I. Marchal. Synthèse d'images par lancer de rayons: algorithmes et architecture. *ACTA ELECTRONICA*, 26(3-4):249–259, 1984.

[8] E. Caspary and I.D. Scherson. A self balanced parallel ray tracing algorithm. In *Parallel Processing for Computer Vision and Display*, UK, january 1988. University of Leeds.

[9] R. Caubet, Y. Duthen, and V. Gaildrat-Inguimbert. Voxar: A tridimentional architecture for fast realistic image synthesis. In *Computer Graphics 1988 (Proceedings of CGI'88)*, pages 135–149, May 1988.

[10] Alan G. Chalmers and Derek J. Paddon. Parallel processing of progressive refinement radiosity methods. In *Second Eurographics Workshop on Rendering*, volume 4, Barcelone, May 1991.

[11] Shenchang Eric Chen. A progressive radiosity method and its implementation in a distributed processing environment. Master's thesis, Cornell University, 1989.

[12] J.G. Cleary, B. Wyvill, G.M. Birtwistle, and R. Vatti. Multiprocessor ray tracing. *Computer Graphics Forum*, 5(1):3–12, March 1986.

[13] Michael F. Cohen, Shenchang E. Chen, John R. Wallace, and Donald P. Greenberg. A progressive refinement approach to fast radiosity image generation. *SIGGRAPH'88 Conference proceedings*, pages 75–84, August 1988.

[14] Michael F. Cohen and John R. Wallace. *Radiosity and Realistic Image Synthesis.* Academic Press Professional, Boston, MA, 1993.

[15] E.W. Dijkstra, W.H.J. Feijen, and A.J.M. Van Gasteren. Derivation of a Termination Detection Algorithm for Distributed Computation. *Inf. Proc. Letters*, 16:217–219, June 1983.

[16] M. Dippé and J. Swensen. An adaptative subdivision algorithm and parallel architecture for realistic image synthesis. In *SIGGRAPH'84*, pages 149–157, New York, 1984.

[17] Steven M. Drucker and Peter Shröder. Fast radiosity using a data parallel architecture. *Third Eurographics Workshop on rendering*, 1992.

[18] A. Fujimoto, T. Tanaka, and K. Iawata. ARTS : Accelerated Ray Tracing System. *IEEE Computer Graphics and Applications*, 6(4):16–26, April 1986.

[19] T. Funkhouser, S. Teller, and D. Khorramabadi. The UC Berkeley System for Interactive Visualization of Large Architectural Models. *Presence, Teleoperators and Virtual environments*, 5(1):13–44, Winter 1996.

[20] S. Green. *Parallel Processing for Computer Graphics.* MIT Press, 1991.

[21] S.A. Green and D.J. Paddon. A highly flexible multiprocessor solution for ray tracing. *The Visual Computer*, 5(6):62–73, March 1990.
[22] Pascal Guitton, Jean Roman, and Gilles Subrenat. Implementation Results and Analysis of a Parallel Progressive Radiosity. In *IEEE/ACM 1995 Parallel Rendering Symposium (PRS '95)*, pages 31–38,101, Atlanta, Georgia, October 1995.
[23] V. İşler, C. Aykanat, and B. Ozguç. Subdivision of 3d space based on the graph partitioning for parallel ray tracing. In *2nd Eurographics Workshop on Rendering*. Polytechnic University of Catalogna, May 1991.
[24] M. J. Keates and R. J. Hubbold. Interactive ray tracing on a virtual shared-memory parallel computer. *Computer Graphics Forum*, 14(4):189–202, October 1995.
[25] H. Kobayashi, T. Nakamura, and Y. Shigei. A strategy for mapping parallel ray-tracing into a hypercube multiprocessor system. In *Computer Graphics International'88*, pages 160–169. Computer Graphics Society, May 1988.
[26] Eric Lepretre, Christophe Renaud, and Michel Meriaux. La radiosité sur tranputers. *La lettre du tranputer et des calculateurs distribués*, pages 49–66, December 1991.
[27] T. Naruse, M. Yoshida, T. Takahashi, and S. Naito. Sight : A dedicated computer graphics machine. *Computer Graphics Forum*, 6(4):327–334, 1987.
[28] K. Nemoto and T. Omachi. An adaptative subdivision by sliding boundary surfaces for fast ray tracing. In *Graphics Interface'86*, pages 43–48, May 1986.
[29] H. Nishimura, H. Ohno, T. Kawata, I. Shirakawa, and K. Omuira. Links-1: A parallel pipelined multimicrocomputer system for image creation. In *Proc. of the 10th Symp. on Computer Architecture*, pages 387–394, 1983.
[30] M. Potmesil and E.M. Hoffert. The pixel machine : A parallel image computer. In *SIGGRAPH'89*, Boston, 1989. ACM.
[31] T. Priol and K. Bouatouch. Static load balancing for a parallel ray tracing on a mimd hypercube. *The Visual Computer*, 5:109–119, March 1989.
[32] Werner Purgathofer and Michael Zeiller. Fast radiosity by parallelization. *Eurographics Workshop on Photosimulation, Realism and Physics in Computer Graphics*, pages 173–184, June 1990.
[33] Rodney J. Recker, David W. George, and Donald P. Greenberg. Acceleration techniques for progressive refinement techniques. *ACM Workshop on Interactive 3D graphics*, pages 59–66, March 1990.
[34] Erik Reinhard and Frederik W. Jansen. Rendering large scenes using parallel ray tracing. *Parallel Computing*, 23(7):873–885, July 1997.
[35] Luc Renambot, Bruno Arnaldi, Thierry Priol, and Xavier Pueyo. Towards efficient parallel radiosity for dsm-based parallel computers using virtual interfaces. In *Proceedings of the Third Parallel Rendering Symposium (PRS '97)*, Phoenix, AZ, October 1997. IEEE Computer Society.
[36] John Salmon and Jeff Goldsmith. A hypercube ray-tracer. *The 3rd Conference on Hypercube Concurrent Computers and Applications*, 2, Applications:1194–1206, January 1988.
[37] R. van Liere. Divide and Conquer Radiosity. In *Proc. of the Second Eurographics Workshop on Rendering*, pages 191–197, Barcelona, 1991. Springer-Verlag.
[38] Amitabh Varshney and Jan F. Prins. An environment projection approach to radiosity for mesh-connected computers. *Third Eurographics Workshop on rendering*, 1992.
[39] Hau Xu, Qun-Sheng Peng, and You-Dong Liang. Accelerated Radiosity Method for Complex Environments. In *Eurographics '89*, pages 51–61. Elsevier Science Publishers, Amsterdam, September 1989.

Parallel Implementations of Morphological Connected Operators Based on Irregular Data Structures

Christophe Laurent[1,2] and Jean Roman[2]

[1] France Telecom - CNET/DIH/HDM, 4, Rue du Clos Courtel
35 512 Cesson Sévigné Cedex, France
[2] LaBRI UMR CNRS 5800, 351, Cours de la Libération
33 405 Talence Cedex, France

Abstract. In this paper, we present parallel implementations of connected operators. These kind of operators have been recently defined in mathematical morphology and have attracted a large amount of research due to their efficient use in image processing applicati ons where contour information is essential (image segmentation, pattern recognition ...). In this work, we focus on connected transformations based on geodesic reconstruction process and we present parallel algorithms based on irregular data structures. We show that the parallelization poses several problems which are solved by using appropria te communication schemes as well as advanced data structures.

1 Introduction

In the area of image processing, mathematical morphology [1] has always proved its efficiency by providing a geometrical approach to image interpretation. Thus, contrary to usual approaches, image objects are not described by their frequential spectrum but by more visual attributes such as size, shape, contrast ....

Recently, complex morphological operators, known as *connected operators* [2], have been defined and become today increasingly popular in image processing because they have the fundamental property of simplifying an image without corrupting contour information. Through this property, this class of operators can be used for all applications where contour information is essential (image segmentation, pattern recognition ...).

In this paper, we focus on the parallelisation of connected operators based on a geodesic reconstruction process [3] aimed at reconstructing the contours of a reference image from a simplified one. In spite of their efficiency in the sequential case, these transformations are difficult to parallelise efficiently, due to complex propagation and re-computation phenomena, and thus, advanced communication schemes and irregular data structures have to be used in order to solve these problems.

The proposed parallel algorithms are designed for MIMD (*Multiple Instruction Multiple Data*) architectures with distributed memory and use a message passing programming model.

J. Palma, J. Dongarra, and V. Hernández (Eds.): VECPAR'98, LNCS 1573, pp. 579–592, 1999.

This paper is organised as follows : section 2 introduces the theoretical foundations of morphological connected operators and presents the connected transformations based on geodesic reconstruction process. Section 3 makes a survey of the most efficient sequential implementations which are used as starting point of the parallelisation. In section 4, we propose to detail all parallel algorithms for morphological reconstruction and the experimental results obtained on a IBM SP2 machine are presented in section 5. Finally, we conclude in section 6.

2 Overview of Morphological Connected Operators

This section presents the concept of morphological connected operators and we invite the reader to refer to [2, 4, 5] for a more theoretical study.

In the framework of mathematical morphology [1], the basic working structure is the *complete lattice*. Let us recall that a complete lattice is composed of a set equipped with a total or partial order such that each family of elements $\{x_i\}$ possesses a supremum $\vee\{x_i\}$ and an infinum $\wedge\{x_i\}$. In the area of grey-level image processing, the lattice of functions (where the order, $\vee$ and $\wedge$ are respectively defined by $\leq$, Max and Min) is used.

Following this structure, the definition of grey-level connected operators has been given in [2, 4] by using the notion of *partition of flat zones*. Let us recall that a partition of a space E is a set of connected component $\{A_i\}$ which are disjoints ($A_i \cap A_j = \emptyset \; \forall i \neq j$) and the union of which is the entire space ($\bigcup_i A_i = E$). Each connected component A_i is then called a *partition class*. Moreover, a partition $\{A_i\}$ is said to be *finer* than a partition $\{B_i\}$ if any pair of points belonging to the same class A_i also belongs to a unique partition class B_j. Finally, the flat zones of a grey-level function f are defined by the set of connected components of the space where f is constant. In [2], the authors have shown that the set of flat zones of a function defines a partition of the space, called the *partition of flat zones of the function.*

From these notions, a grey-level connected operator can be formally defined as follows :

Definition 1. *An operator Ψ acting on grey-level functions is said to be connected if, for any function f, the partition of flat zones of $\Psi(f)$ is less fine than the partition of flat zones of f.*

This last definition shows that connected operators have the fundamental property of simplifying an image while preserving contour information. Indeed, since the associated partition of $\Psi(f)$ is less fine than the associated partition of f, each flat zone of f is either preserved or merged in a neighbour flat zone. Thus, no new contour is created.

Following this principle, we can decompose a grey-level connected operator in two steps : the first one, called *selection step*, assesses a given criterion (size, contrast, complexity ...) for each flat zone and from this criterion, the second step called *decision step* decides to preserve or merge the flat zone. A large number of grey-level connected operators can thus be defined, only differencing in the criterion used by the selection step. For a presentation of different connected operators, refer to [6].

In this paper, we focus on connected operators based on geodesic reconstruction process [3]. This kind of transformation is actively used in applications such as grey-level and calor image and video segmentation [7–9], default detection [10], texture classification [11] and so forth. The reconstruction process is base d on the definition of the *elementary geodesic dilation* $\delta^{(1)}(f,g)$ of the grey-level image $g \leq f$ "under" f defined by $\delta^{(1)}(f,g) = Min\{\delta_1(g), f\}$ where $\delta_1(g)$ defines the elementary numerical dilation of g given by $\delta_1(g)(p) = Max\{g(q), q \in N(p) \cup \{p\}\}$ with $N(p)$ denoting the neighbourhood of the pixel p in the image g.

The geodesic reconstruction of a *marker image* g with reference to a *mask image* f with $g \leq f$ is obtained by iterating elementary gray-level geodesic dilation of g "under" f until stability :

$$\rho(g \mid f) = \delta^{(\infty)}(f,g) = \delta^{(1)}(\delta^{(1)}(\ldots(\delta^{(1)}(f,g))\ldots)) \qquad (1)$$

Figure 1 shows a 1D example of the reconstruction process $\rho(g \mid f)$ for which we can note that all contours of the function f are perfectly preserved after the reconstruction.

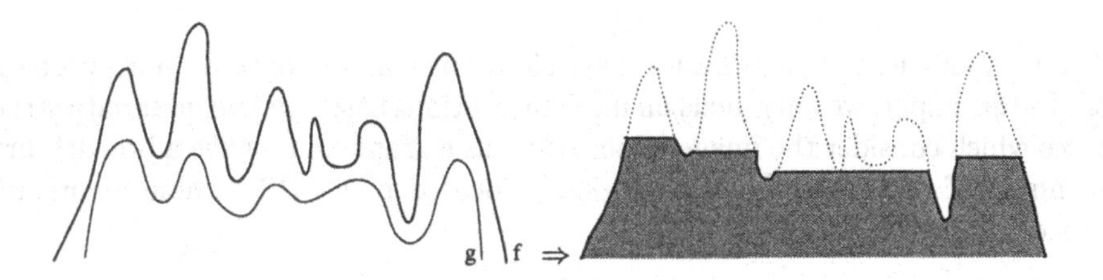

Fig. 1. 1D example of gray-scale reconstruction of mask f from marker g

Following the technique used to obtain the marker image g, it is straightforward that a large number of connected operators can be defined . Two connected operators, based on this reconstruction process, are intensively used in literature. The first one, known as *opening by reconstruction*, has a size oriented simplification effect since the marker image is obtained by a morphological opening removing all bright objects smaller than the size of the structuring element. The second one, known as $\lambda - Max$ operator, has a contrast oriented simplification effect since the marker image is obtained by subtracting a constant λ to the mask image ($g = f - \lambda$). Figure 2 shows these two transformations. From the *Cameraman* test image (see Figure 2(a)), Figure 2(b) shows the result of an opening by reconstruction $\rho(\gamma_{15}(f) \mid f)$ where $\gamma_n(f)$ denotes the morphological opening of the function f using a structuring element of size n. Figure 2(c) shows a $\lambda - Max$ operator with $\lambda = 40$ denoted by $\rho(f - 40 \mid f)$.

Note that for each defined connected operator, a dual transformation can be obtained by reversing the Max and Min operator. We thus obtain a *closing by*

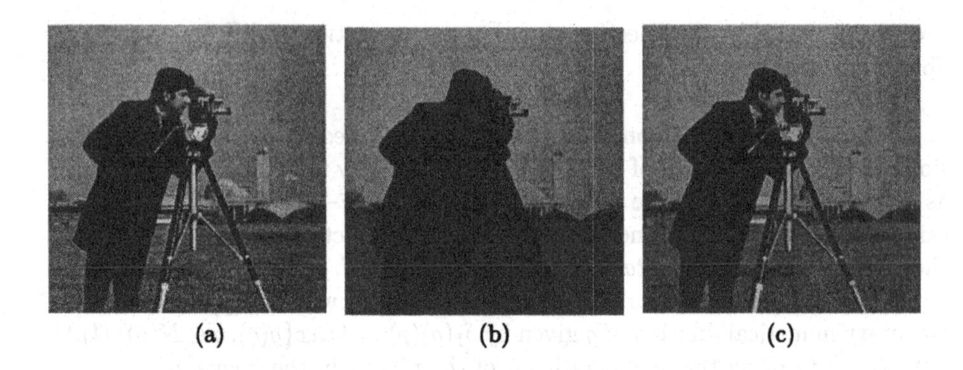

(a) (b) (c)

Fig. 2. Example of geodesic reconstruction on grey-level images

reconstruction transformation, that removes all dark objects smaller than the structuring element, and a $\lambda - Min$ transformation, that removes all objects which have a contrast higher than λ. Due to this duality relation, we only study in this paper the reconstruction process by geodesic dilation (see equation (1)).

3 Efficient Sequential Implementations of Geodesic Reconstruction

The most efficient algorithms for geodesic reconstruction have been proposed in [3]. In this paper, we only focus on implementations based on irregular data structures which consider the image under study as a graph and realize a breadth-first scanning of the graph from strategically located pixels [12]. These algorithms proceed in two main steps :

- detection of the pixels which can initiate the reconstruction process,
- propagation of the information only in the relevant image parts.

In the case of reconstruction by geodesic dilation (see equation (1)), we can easily show that a pixel p can initiate the reconstruction of a mask image f from a marker image g if it has in its neighborhood at least one pixel q such that $g(q) < g(p)$ and $g(q) < f(q)$.

Proposition 1. *In the reconstruction of a mask image f from a marker image g with $g \leq f$, the only pixels p which can propagate their grey-level value in their neighborhood verify : $\exists q \in N(p), g(q) < g(p)$ and $g(q) < f(q)$.*

Proof. Let p be a pixel such that $\exists q \in N(p), g(q) < g(p)$ and $g(q) < f(q)$.

We have $\delta^{(1)}(f,g)(q) = Min\{\delta_1(g)(q), f(q)\}$ with $\delta_1(g)(q) = Max\{g(t), t \in N(q) \cup \{q\}\}$.

Since $q \in N(p)$, we have $p \in N(q)$. Moreover, we know that $g(q) < g(p)$ and thus, q is not a fixed point of the transformation δ_1 eg $\delta_1(g)(q) > g(q)$.

On the other hand, since $g(q) < f(q)$, the transformation $\delta^{(1)}(f, g)$ has not reached stability at the location of q. As a result, the pixel q can receive a propagation from the pixel p.

To conclude the proof, it is straightforward that if $\forall q \in N(p), g(q) > g(p)$, the pixel p cannot propagate its grey-level value but receives a propagation from its neighborhood. □

Note that a same pixel can propagate its grey-level value on a neighbor pixel and receive a propagation from another neighbor pixel.

Following this principle, two methods have been proposed in [3] to detect initiator pixels. The first one consists in computing the regional maxima of the marker image g which designate the set of connected components $\mathcal{M}$ with a grey-level value h in g such that every pixel in the neighborhood of $\mathcal{M}$ has a strictly lower grey-level value than h. In this case, initiator pixels are those located in the interior boundaries of regional maxima of g.

The second method is based on two scanning of the marker image g. The first scanning is done in a raster order (from top to bottom and left to right) and each grey-level value $g(p)$ becomes $g(p) = Min\{Max\{g(q), q \in N^+(p) \cup \{p\}\}, f(p)\}$ where $N^+(p)$ is the set of neighbors of p which are reached before p in a raster scanning (see Figure 3(a)). The second scanning is done in the anti-raster order and each grey-level value $g(p)$ becomes $g(p) = Min\{Max\{g(q), q \in N^-(p) \cup \{p\}\}, f(p)\}$ where $N^-(p)$ designates the neighbors of p reached after p in the raster scanning order (see Figure 3(b)). In this case, the initiator pixels are detected in the second scanning and are those which could propagate their grey-level value during the next raster scanning. These pixels p verify : $\exists q \in N^-(p), g(q) < g(p)$ and $g(q) < f(q)$.

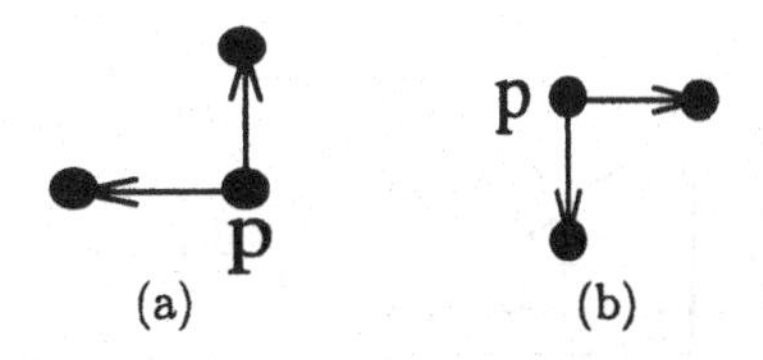

Fig. 3. Definition of $N^+(p)$ (a) and $N^-(p)$ (b) in the case of 4-connectivity

One can easily see that these two methods verify the general principle exposed in Proposition 1.

Once the initiator pixels are detected, the information has to be propagated by a breadth-first scanning. For this purpose, the breadth-first scanning is implemented in [3] by a queue of pixels represented by a FIFO (*First In First Out*) data structure. The initiator pixels are first inserted in the queue and the propagation consists in extracting the first pixel p from the queue and propagating its grey-level value $g(p)$ to all of its neighbors q such that $g(q) < g(p)$ and $g(q) < f(q)$. The grey-level of these pixels becomes then $g(q) = Min\{g(p), f(q)\}$

and the last operation consists in inserting these pixels q in the queue in order to continue the propagation. The reconstruction stops when the queue is empty.

4 Parallel Implementations of Geodesic Reconstruction

4.1 Preliminaries

In this paper, we are interested in the parallelization of morphological reconstruction based on geodesic dilation since by duality, reconstruction by geodesic erosion can be obtained by reversing the Min and Max operators in equation (1).

All proposed parallel algorithms are designed for MIMD (*Multiple Instruction Multiple Data*) architectures with distributed memory and use a message passing programming model. Based on this parallel context, we denote by p the number of processors and by P_i the processor indexed by i $(0 \leq i < p)$.

The mask image f and the marker image g, for which the domain of definition is denoted by D, are splitted into p sub-images f_i and g_i defined on disjoint domains D_i $(0 \leq i < p)$. For the sake of simplicity, we assume that the partition ning is made in an horizontal 1D-rectilinear fashion. Thus, if we suppose that f and g are of size $n \times n$, each processor P_i owns a mask sub-image f_i and a marker sub-image g_i, each of size $\frac{n}{p} \times n$. Note that all proposed algorithms can be extended to a 2D-rectilinear partitionning scheme without difficulty.

Moreover, in order to study the propagation property of its local pixels on non local pixels located in neighbor sub-images, each processor P_i owns two 1-pixel wide overlapping zones on its neigbhor sub-images f_{i-1} and g_{i-1} (for $i > 0$), and f_{i+1} and g_{i+1} (for $i < p-1$). From this extended sub-domain, denoted by ED_i, we denote by $LB(i,j)$ the local pixels located on the frontier between P_i and P_j that is $LB(i,j) = D_i \cap ED_j$, and by $NB(i,j)$ the non local pixels located on the frontier between P_i and P_j that is $NB(i,j) = ED_i \cap D_j$. Figure 4 shows all of these notations.

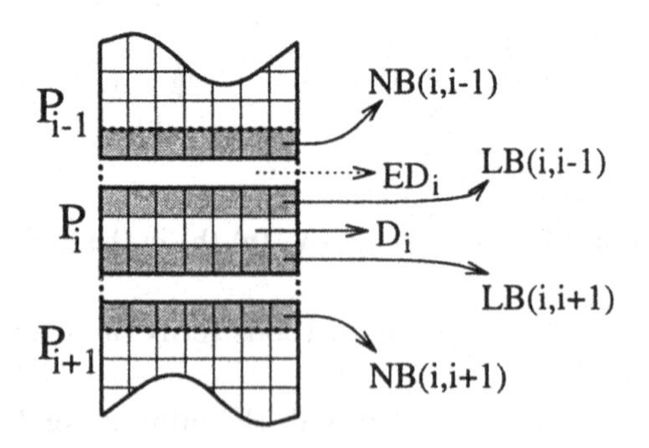

Fig. 4. 1D rectilinear data partitionning

On the other hand, we denote by $N_D(p)$ the neighborhood of the pixel p in the domain D and by $N_{D_i}(p)$ the neighborhood of the pixel p restricted to the sub-domain D_i.

Before presenting our parallel geodesic reconstruction algorithms, we would like to underline the difficulty of this parallelization and show the motivation of this work. As mentioned in section 2, geodesic reconstruction is not a locally defined transformation resulting in a complex propagation phenomenon, that makes it difficult to parallelize efficiently. If we consider the example of reconstruction by geodesic dilation (see equation (1)), we remark that a bright pixel can propagate its grey-level value to a large part of the image and no technique allows us to predict this phenomenon a priori. Now, in parallel implementations where initial images are splitted and distributed among processors, each proce ssor proceeds reconstruction from its local data and thus, when a propagation phenomenon appears and crosses inter-processors frontier, it can generate the re-computation of a large number of pixels located in the sub-image that have received the propagat ion. Figure 5 illustrates this phenomenon with a 1D distribution in which the mask image f and the marker image g are distributed on two processors P_0 and P_1. On this example, the processor P_1 owns a dark marker sub-image w hereas the processor P_0 owns a bright marker sub-image (see Figure 5(a)). In a computational point of view, if P_1 entirely reconstructs its local image before taking into account the propagation messages sent by P_0, it is strai ghtforward that P_1 will have to re-compute a large number of pixels and the resulting total execution time will be slowed down (see Figures 5(b,c)). The motivation of this work is thus to adopt communication schemes adapted to this i nter-processors propagation phenomenon and to propose efficient data structures limiting the re-computation phenomenon.

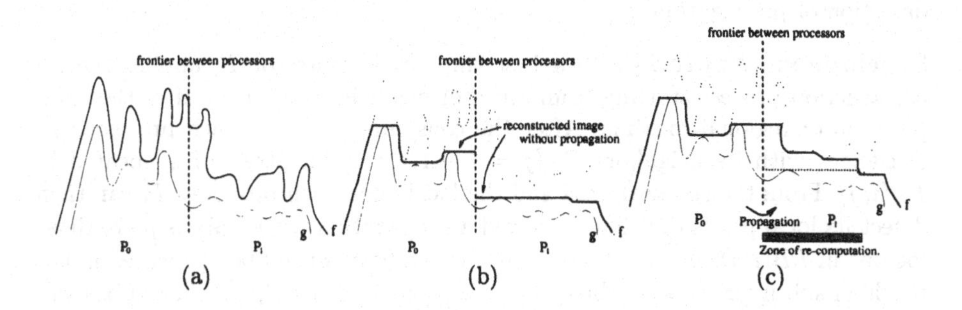

Fig. 5. Propagation and re-computation phenomenons

In this paper, we only focus on parallel geodesic reconstruction algorithms based on irregular data structures represented by pixel queues because of their efficiency in the sequential case. As mentionned in section 3, sequential algorithm s proceed in two steps : detection of initiator pixels and propagation of the information. One can easily remark that parallel algorithms which follow this principle are very irregular since on one hand, the number of detected initiator pixels can differ on each processor and on the other hand, the propagation of information can differently evolve on each processor. Moreover, from this irregular nature, communication primitives, that have to take place in order to propagate information between processors, will be irregular since the number of sent and received propagations can differ on each processor.

4.2 Parallel Reconstruction Based on Regional Maxima

The parallel algorithm based on the use of marker image regional maxima proceeds in two steps. In the first one, all processors compute in parallel the regional maxima of the marker image g (for this step, refer to [13]) and product a temporary image $\mathcal{R}$ defined by :

$$\mathcal{R}(p) = \begin{cases} g(p) & \text{if } p \text{ belongs to a regional maxima} \\ 0 & \text{otherwise.} \end{cases}$$

From the image $\mathcal{R}$, each processor P_i initializes its local pixel queue F_i by inserting all pixels $p \in D_i$ located on internal boundaries of regional maxima :

$$F_i = \{p \in D_i, \mathcal{R}(p) \neq 0 \text{ and } \exists q \in N_D(p) \text{ such that } \mathcal{R}(q) = 0\}.$$

During the last step, the information is propagated through the image and each processor can then receive a propagation from its neighbor processors. These interactions are implemented by communication primitives and two methods can be proposed for this s tep : the *synchronous approach* in which communications are considered as synchronization points which regularly appear, and the *asynchronous approach* in which communications are exchanged immediatly after the detection of propagation.

Synchronous Approach. In a first time, each processor P_i applies the reconstruction process starting from initiator pixels inserted in F_i after the computation of regional maxima. After the consumption of F_i, each processor P_i exchanges with its neighbors P_j ($j = i-1, i+1$) all pixels of g_i located in $LB(i,j)$. From the received data and the local ones, each processor P_i can then detect its local pixels that have received a propagation. These pixels p are those located in $LB(i,j)$ ($j = i-1, i+1$) and which have a neighbor q located in the neighbor sub-image g_j such that $g_j(q) > g_i(p)$ and $g_i(p) < f_i(p)$. These pixel values become then $g_i(p) = Min\{g_j(p), f_i(p)\}$ (see Figure 6). These pixels form a new set of initiator pi xels and are thus inserted in turn in F_i. The reconstruction process can then be iterated with this new set.

This general scheme is iterated n_{it} times until a global stabilization, which is detected when no processor receives a propagation from its neighbors.

This approach follows a "divide and conquer" programming paradigm and can be qualified as semi-irregular since the local propagation step remains irregular because based on FIFO data structure but communications are performed in a regular and synchroniz ed fashion.

The main disadvantage of this technique is to take into account the interprocessor propagations at latest, resulting in an important re-computation phenomenon. Moreover, a propagation starting on a processor P_i and going up to

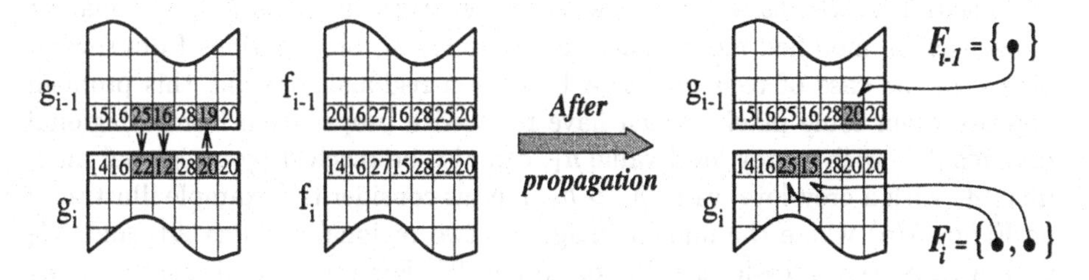

Fig. 6. Example of inter-processor propagation

P_{i+k} ($k \in [-i, p-i-1]$) will be take into account by P_{i+k} after k communication steps. Finally, due to the regular nature of the communications, no overlapping of computation and communications are proposed by this approach.

Asynchronous Approach. Contrary to the previous approach, inter-processor propagation messages are here exchanged as one goes along their detection in order to take them into account as soon as possible. This technique solves the problems posed by the synchronous approach since it reduces the amount of communications crossing the network and limits the re-computation phenomenon.

In this approach, the reconstruction stops when all processors have cleaned their local pixel queue and when all sent propagation messages have been taken into account by their recipients. In order to detect this stabilization time, a token is used reporting all sent propagation messages which have not yet been taken into account. This token is implemented by a vector T of size p in which $T[i]$ ($0 \leq i < p$) designates the number of messages sent to P_i and not yet received. Thus, the global stabilization is reached when all entries of T become nil.

In the same manner of the synchronous approach, when the grey-level value of a boundary pixel p is modified by a processor P_i, a message is sent to the corresponding neighbor processor P_j indicating the position x and the new grey-level value $g_i(p)$ of the modified pixel p. The processor P_j receiving a message $(x, g_i(p))$ of this kind takes into account the propagation effect of the pixel p on its own marker sub-image g_j. For this purpose, it scans all pixels q located in g_j and in the neighborhood of p ($q \in N_{D_j}(p)$) and it reports propagation on all of these pixels verifying $g_j(q) < g_i(p)$ and $g_j(q) < f_j(q)$. The grey-level value of these pixels becomes then $g_j(q) = Min\{g_i(p), f_j(q)\}$. The last step consists in inserting the pixel q in the queue F_j.

Finally, once a processor P_i has finished its local reconstruction ($F_i = \emptyset$), it enters in a blocking state until the global stabilization is reached. In this state, it can either receive a propagation message from a neighbor processor that hav e not yet terminated its reconstruction or a message for the token management.

In these synchronous and asynchronous irregular algorithms, the consumption order of the pixels during the propagation step directly depends on their insertion order in the FIFO, which depends in turn on the image scanning order in the initialization phas e. Following this observation, we can remark that for a sequential as well as parallel geodesic reconstruction, some pixels of marker image can be modified more than once by receiving propagations from several pixels. In the case of reconstruction based o n regional maxima, this problem appears when some pixels, which have received a propagation from a regional maxima $\mathcal{M}_1$ with a grey-level value h_1, can also be reached by a regional maxima $\mathcal{M}_2$ with a grey-level value $h_2 > h_1$. L et us consider the example illustrated on Figure 7(a) where the marker image has two regional maxima $\mathcal{M}_1$ and $\mathcal{M}_2$ with a respective grey-level value of 10 and 20. For the sake of simplicity, we assume that the mask image has a constant grey-level value equal to 25. Suppose now that during the initiator pixels detection step, the pixels located in internal boundaries of $\mathcal{M}_1$ are inserted in the FIFO before the pixels located in internal boundaries of $\mathcal{M}_2$. On this Figure, where arrows designate the propagations sense, each maximum is extended to its neighborhood. At the second step of propagation (see Figure 7(b)), the maximum $\mathcal{M}_2$, that has a grey-level value higher than h_1 , begins to generate the re-computation of some pixels previously modified by the propagation phase initialized from $\mathcal{M}_1$. As one goes along of iterations, each of maximum is spatially extended and all pixels located in the intersection of thes e extensions are re-computed (see Figure 7(c,d) where re-computation are designed by dark grey dashed square). At the end of the reconstruction, all pixels of marker image have a grey-value of 20 and thus, all pixels modified during the extension of $\mathcal{M}_1$ have been re-computed by receiving a propagation from $\mathcal{M}_2$.

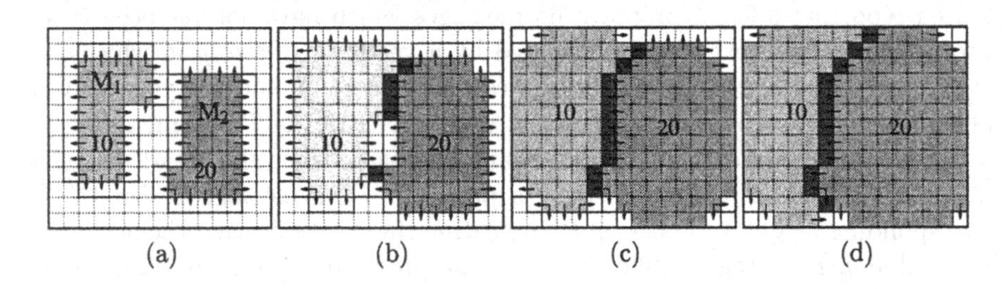

Fig. 7. Re-computation phenomenon for reconstruction based on regional maxima

It would be thus interesting to propose a data structure ensuring the modification unicity of each pixel. For this purpose, it is important to observe that in the case of reconstruction by geodesic dilation (see equation (1)), the pr opagation sense is always from bright pixels to dark ones, and thus we can affect a priority of reconstruction to each pixel, attached with its grey-level value. In the case of reconstruction by geodesic dilation of a mask image f from a marker image g, each pixel p will thus be inserted in the queue with a priority given by $g(p)$. One can easily see that this technique cancels the re-computation phenomenon in the sequential case and limits it as far as possible in the parallel case. Moreover, the priority mechanism can be efficiently implemented by using hierarchical FIFO [14] with as much priority as grey-level values in the marker image.

From this new data structure, all proposed algorithms can be rewritten by only modifying the calls to FIFO management.

4.3 Hybrid Parallel Reconstruction

In the previous implementations, a large part of the reconstruction time is dedicated to the computation of regional maxima. To solve this problem, it has been proposed in [3] to detect initiator pixels from two scanning of marker image as explained in section 3. Following this principle, all techniques presented in the previous section and devoted to the propagation step can be applied here since the only modified step concerns with the detection of initiator pixels. Thus, the parallel algorithms based on this technique proceed in two steps : the initiator pixels are first detected from two scannings of marker sub-image, and the propagation of information is then executed in a synchronous or asynchronous way by using a classical or hierarchical FIFO.

5 Experimental Results

In this section, we present some experimentations of our parallel algorithms obtained on a IBM SP2 machine with 16 processing nodes by using the MPI (*Message Passing Interface*) [15] communication library. For these measures, four test images (*Cameraman*, *Lenna*, *Landsat* and *Peppers*) have been used, each of size 256 × 256. For *Cameraman* and *Lenna* test images, a parallel opening by reconstruction has been measured for which the marker image has been obtained by a morphological opening of size 5 and 15 respectively. For *Landsat* and *Peppers* test images, a $\lambda - Max$ operator has been measured for which the marker image has been obtained by substracting the constant values 20 and 40 respectively to the mask image.

As explained in section 4.3, the most promising parallel algorithms uses two marker image scannings in order to detect initiator pixels and thus, we only analyse in this section the four algorithms based on this technique :

- **algorithm 1** : synchronous propagation step and use of classical FIFO,
- **algorithm 2** : asynchronous propagation step and use of classical FIFO,
- **algorithm 3** : synchronous propagation step and use of hierarchical FIFO,
- **algorithm 4** : asynchronous propagation step and use of hierarchical FIFO.

Figure 8 shows the speedup coefficients obtained by all proposed algorithms by using the four test images. First of all, we can note that these coefficients are limited because of, on one hand, the irregularity of the proposed algorithms, and on the other hand, the small size of the processed images. However, these results well show the behavior of our algorithms. We can thus observe that synchronous approaches (algorithms 1 and 3) are not adapted to the marked irregularity of the algorithms because the inter-processor propagations are taken into account at latest resulting in a marked load imbalance in some cases. We can thus observe

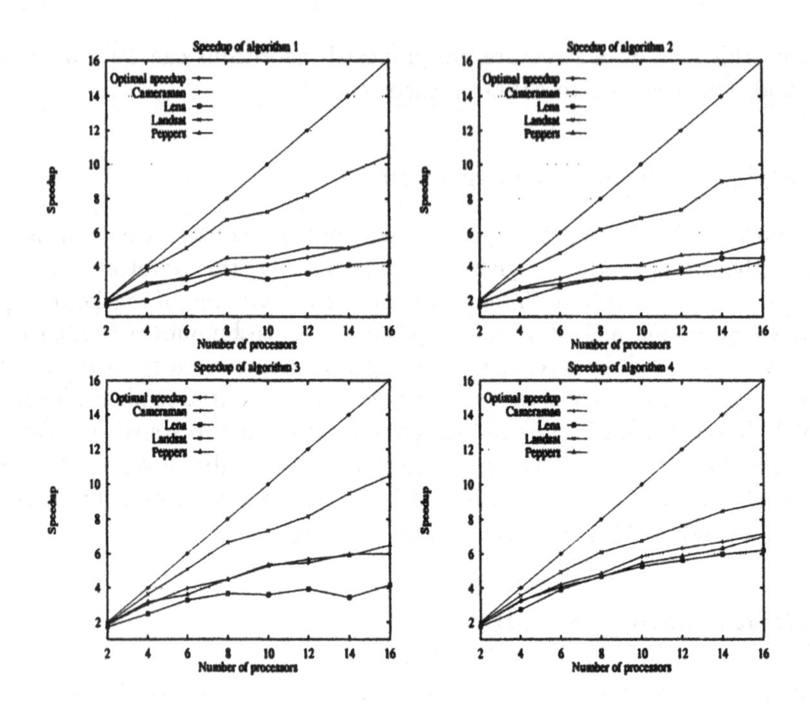

Fig. 8. Speedup coefficients of all proposed algorithms

a chaotic behavior with some images showing a high reconstruction complexity (see, for example, the obtained measures on *Lenna* image). Moreover, we can remark that these synchronous algorithms can also bring a totally sequential behavior for some images where the evolution of the grey-level values is continuous from top to bottom. As a result, we can conclude that parallel geodesic reconstruction algorithms based on synchronous approach are not scalable.

The algorithm 2 presents very limited speedup caused by the use of a classical FIFO data structure. Indeed, as explained in section 4, this data structure does not ensure the unicity of modification of the pixels and thus, a communication is performed each time a pixel located on an inter-processor frontier is modified. As a result, the number of communication is higher than for synchronous approaches since for these approaches, a communication is performed only after the t ermination of all local reconstruction.

Finally, the algorithm 4 gives the best results and shows the most regular behavior for all test images. For this small image size, the relative efficiency is always superior to 40% and we can note that this algorithm shows the best scalability.

In order to further study the scalability of our algorithms, we have tested them on a variable problem size. For this purpose, we have used test images of size 512×512. Figure 9 shows speedup coefficients of the four proposed algorithms on the *Lenna* test image and for the opening by reconstruction $\rho(\gamma_{15}(f) \mid f)$. In this Figure, we can note that algorithm 4 shows again the best behavior and

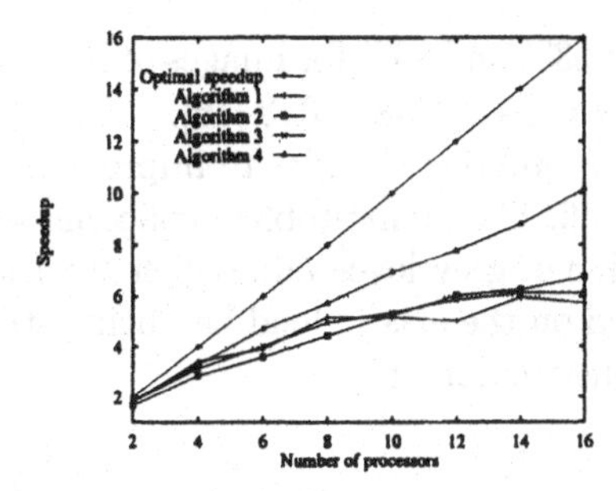

Fig. 9. Speedup of algorithms with a variable problem size

the efficiency is here very satisfying since it varies from 63% to 91% whereas the efficiency of all other algorithms is inferior to 38% with 16 processors.

From all of these experimentations, we can conclude that algorithm 4 is the best suited for the implementation of connected operators based on reconstruction process since the obtained results are improved with the use of larger images whereas the other algorithms show poor parallel behaviors for all image sizes. On the other hand, these experimentations and those presented in [16] show that the limited speedup obtained by the algorithm 4 for small image sizes is only caused by the thin computation grain. Indeed, as mentionned previously, the use of hierarchical FIFO limits as far as possible the re-computation rate. However, for small images, the computation grain becomes then very thin which explains a fair parallel behavior that is improved with the use of larger images presenting a thicker computation grain. Moreover, the irregularity in a small problem size also brings a limitation of performances.

Finally, the increase of problem size have also shown an improvement of the load imbalance problem [16]. Indeed, while this problem is visible for images smaller than 256×256, it becomes slightly perceptible for images of size 512×512 and disappears for higher image sizes. This behavior is a direct consequence of the increase of computation grain since with high problem size, the processors do not terminate their local reconstruction before they receive the first propagation messages and thus, the re-computation rate can be improved.

6 Conclusion and Perspectives

In this paper, we have presented several parallel algorithms for geodesic reconstruction based on irregular data structures. We have shown that these transformations present complex propagation and re-computation phenomena that have been solved in this paper by using an asynchronous approach as well as a hierarchical data structure. The resulting algorithm presents a marked irregularity and the experimental results have shown that for small images, this approach suffers from a thin computation grain and obtains relative efficiency of about 50% on 16 processors. The measures obtained on a variable problem size have shown up this computation grain problem and have proved that with a higher problem size, the algorithm shows a good parallel behaviour since the obtained efficiency varies between 63% and 85% for images of size 1024×1024 and between 65% and 95% for images of size 2048×2048 [16].

However, works are in progress in order to improve the performances for images smaller than 256×256. The main problem concerns with the load imbalance caused by the distribution of grey level values over the image. Presently, the retained technique is based on the elastic load balancing strategy proposed in [17] and on progressive reconstruction [16].

References

1. Jean Serra. *Image Analysis and Mathematical Morphology.* Academic Press, 1982.
2. Serra J. and Salembier P. Connected Operators and Pyramids. In *Proceedings SPIE Image Algebra and Mathematical Morphology*, volume 2030, pages 65–76, San Diego, 1993.
3. Vincent L. Morphological Grayscale Reconstruction in Image Analysis : Applications and Efficient Algorithms. *IEEE Trans. on Image Proc.*, 2(2):176–201, April 1993.
4. Salembier P. and Serra J. Flat Zones Filtering,Connected Operators and Filters by Reconstruction. *IEEE Trans. on Image Proc.*, 4(8):1153–1160, August 1995.
5. Crespo J., Serra J. and Schafer R.W. Theoretical Aspects of Morphological Filters by Reconstruction. *Signal Processing*, 47:201–225, 1995.
6. Meyer F., Oliveras A., Salembier P. and Vachier C. Morphological Tools for Segmentation : Connected Filters and Watersheds. *Annales des Télécommunications*, 52(7-8):367–379, July 1997.
7. Salembier P. Morphological Multiscale Segmentation for Image Coding. *Signal Processing*, 38:359–386, 1994.
8. Salembier P. and Pardàs M. Hierarchical Morphological Segmentation for Image Sequence Coding. *IEEE Trans. on Image Proc.*, 3(5):639–651, September 1994.
9. Gu C. *Multivalued Morphology and Segmentation-Based Coding.* PhD thesis, Ecole Polytechnique Fédérale de Lausanne, 1995.
10. Salembier P. and Kunt M. Size-sensitive Multiresolution Decomposition of Images with Rank Order Filters. *Signal Processing*, 27:205–241, 1992.
11. Li W., Haese-Coat V. and Ronsin J. Residues of Morphological Filtering by Reconstruction for Texture Classification. *Pattern Recognition*, 30(7):1081–1093, 1997.
12. Vincent L. *Algorithmes Morphologiques à Base de Files d'Attente et de Lacets. Extension aux Graphes.* PhD thesis, Ecole Nationale Supérieure des Mines de Paris, France, May 1990.
13. Moga A. *Parallel Waterhed Algorithms for Image Segmentation.* PhD thesis, Tampere University of Technology, Finland, February 1997.
14. Breen E. and Monro D.H. An Evaluation of Priority Queues for Mathematical Morphology. In Jean Serra and Pierre Soille, editors, *Mathematical Morphology and Its Applications to Image Processing*, pages 249–256. Kluwer Acecdemics, 1994.
15. Message Passing Interface Forum. *MPI : A Message Passing Interface Standard*, May 1994.
16. Laurent C. *Conception d'algorithmes parallèles pour le traitement d'images utilsant la morphologie mathématique. Application à la segmentation d'images.* PhD thesis, Université de Bordeaux I, September 1998.
17. Pierson J.M. *Equilibrage de charge dirigé par les données. Applications à la synthèse d'images.* PhD thesis, Ecole normale supérieure de Lyon, October 1996.

Chapter 6:

Parallel Database Servers

The Design of an ODMG Compatible Parallel Object Database Server (Invited Talk)

Paul Watson

Department of Computing Science, University of Newcastle-upon-Tyne, NE1 7RU, UK
Paul.Watson@newcastle.ac.uk

Abstract. The *Polar* project has the aim of designing a parallel, ODMG compatible object database server. This paper describes the server requirements and investigates issues in designing a system to achieve them. We believe that it is important to build on experience gained in the design and usage of parallel relational database systems over the last ten years, as much is also relevant to parallel object database systems. Therefore we present an overview of the design of parallel relational database servers and investigate how their design choices could be adopted for a parallel object database server. We conclude that while there are many similarities in the requirements and design options for these two types of parallel database servers, there are a number of significant differences, particularly in the areas of object access and method execution.

1 Introduction

The parallel database server has become the "killer app" of parallel computing. The commercial market for these systems is now significantly larger than that for parallel systems running numeric applications, making them mainstream IT system components offered by a number of major computer vendors. They can provide high performance, high availability, and high storage capacity, and it is this combination of attributes which has allowed them to meet the growing requirements of the increasing number of computer system users who need to store and access large amounts of information.

There are a number of reasons to explain the rapid rise of parallel database servers, including:

J. Palma, J. Dongarra, and V. Hernández (Eds.): VECPAR'98, LNCS 1573, pp. 593-621, 1999.

- they offer higher performance at a better cost-performance ratio than do the previous dominant systems in their market - mainframe computers.
- designers have been able to produce highly available systems by exploiting the natural redundancy of components in parallel systems. High availability is important because many of these systems are used for business-critical applications in which the financial performance of the business is compromised if the data becomes inaccessible.
- the 1990s have seen a process of re-centralisation of computer systems following the trend towards downsizing in the 1980s. The reasons for this include: cost savings (particularly in software and system management), regaining central control over information, improving data integrity by reducing duplication, and increasing access to information. This process has created the demand for powerful information servers.
- there has been a realisation that many organisations can derive and infer valuable information from the data held in their databases. This has led to the use of techniques, such as data-mining, which can place additional load on the database server from which the base data on which they operate must be accessed.
- the growing use of the Internet and intranets as ways of making information available both outside and inside organisations has increased the need for systems which can make large quantities of data available to large numbers of simultaneous users.

The growing importance of parallel database servers is reflected in the design of commercial parallel platforms. Efficient support for parallel database servers is now a key design requirement for the majority of parallel systems.

To date, almost all parallel database servers have been designed to support relational database management systems (RDBMS) [1]. A major factor which has simplified, and so encouraged, the deployment of parallel RDBMS by organisations is their structure. Relational database systems have a client-server architecture in which client applications can only access the server through a single restricted and well defined query interface. To access data, clients must send an SQL (Structured Query Language) query to the server where it is compiled and executed. This architecture allows a serial server, which is not able to handle the workload generated by a set of clients, to be replaced by a parallel server with higher performance. The client applications are unchanged: they still send the same SQL to the parallel server as they did to the serial server because the exploitation of parallelism is completely internal to the server.

Existing parallel relational database servers exploit two major types of parallelism. Inter-query parallelism is concerned with the simultaneous execution of a set of queries. It is typically used for On-Line Transaction Processing (OLTP) workloads in which the server processes a continuous stream of small transactions generated by a set of clients. Intra-Query parallelism is concerned with exploiting parallelism within the execution of single queries so as to reduce their response time.

Despite the current market dominance of relational database servers, there is a growing belief that relational databases are not ideal for a number of types of applications, and in recent years there has been a growth of interest in object oriented databases which are able to overcome many of the problems inherent in relational systems [2]. In particular, the growth in the use of object oriented programming languages, such as C++ and Java, coupled with the increasing importance of object-based distributed systems has promoted the use of object database management systems (ODBMS) as key system components for object storage and access. A consequence of the interest in object databases has been an attempt to define a standard specification for all the key ODBMS interfaces - the Object Database Management Group ODMG 2.0 standard [3].

The *Polar* project has the aim of designing and implementing a prototype ODMG compatible parallel object database server. Restricting the scope of the project to this standard allows us to ignore many issues, such as query language design and programming language interfaces, and instead focus directly on methods for exploiting parallelism within the framework imposed by the standard.

In this paper we describe the requirements that the *Polar* server must meet and investigate issues in designing a system to achieve them. Rather than design the system in isolation, we believe that it is important to build, where possible, on the extensive experience gained over the last ten years in the design and usage of parallel relational database systems. However, as we describe in this paper, differences between the object and relational database paradigms result in significant differences in some areas of the design of parallel servers to support them. Those differences in the paradigms, which have most impact on the design, are:

- objects in an object database can be referenced by a unique identifier, or (indirectly) as members of a collection. In contrast, tables (collections of rows) are the only entities which can be referenced by a client of a relational database (i.e. individual table rows cannot be directly referenced).
- there are two ways to access data held in an object database: through a query language (OQL), and by directly mapping database objects into client application program objects. In a relational database, the query language (SQL) is the only way to access data.
- objects in an object database can have associated user-defined methods which may be called within queries, and by client applications which have mapped database objects into program objects. In a relational database, there is no equivalent of user-defined methods: only a fixed set of operations is provided by SQL.

The structure of the rest of this paper is as follows. In Section 2 we give an overview of the design of parallel relational database servers, based on our experiences in two previous parallel database server projects: EDS [4], and Goldrush [1]. Next, in Section 3 we define our requirements for the *Polar* parallel ODBMS. Some of these are identical to those of parallel RDBMS; others have emerged from experience of the limitations of existing parallel servers; while others are derived from our view of the potential use of parallel ODBMS as components in distributed systems. Based on these requirements, in Section 4, we present an overview of issues in the design of a parallel ODBMS. This allows us to highlight those areas in the design of a parallel object database server where it is possible to adopt solutions based on parallel RDBMS or serial ODBMS, and, in contrast, those areas where new solutions are required. Finally, in Section 5 we draw conclusions from our investigations, and point to further work.

1.1 Related Work

There have been a number of parallel relational database systems described in the literature. Those which have most influenced this paper are the two on which we previously worked: EDS and Goldrush.

The EDS project [4] designed and implemented a complete parallel database server, including hardware, operating system and database. The database itself was basically relational though there were some extensions to provide support for objects.

The Goldrush project within ICL High Performance Systems [1, 5, 6] designed a parallel relational database server product running Parallel Oracle.

There is extensive coverage in the literature of research into the design of serial ODBMS. One of the most complete is the description of the O2 system [7].

Recently, there has been some research into the design of parallel ODBMS. For example, Goblin [8] is a parallel ODBMS. However, unlike the system described in this paper, it is limited to a main-memory database.

Work on object servers such as Shore [9] and Thor [10] is also of relevance as this is a key component of any parallel ODBMS. We will refer to this work at appropriate points in the body of the paper.

2 The Design of Parallel Relational Database Servers

Parallel relational database servers have been designed, implemented and utilised for over ten years, and it is important that this experience is used to inform the design of parallel object database servers. Therefore, in this section we give an overview of the design of parallel relational database servers. This will then allow us to highlight commonalties and differences in the requirements (Section 3) and design options (Section 4) between the two types of parallel database servers. The rest of this section is structured as follows. We begin by describing the architectures of parallel platforms (hardware and operating system) designed to support parallel database servers. Next, we describe methods for exploiting parallelism found in relational database workloads. Throughout this section we will draw on examples from our experience in the design and use of the ICL Goldrush MegaServer [1].

2.1 Parallel Platforms

In this section, we describe the design of parallel platforms (which we define as comprising the hardware and operating system) to meet the requirements of database servers.

We are interested in systems utilising the highly scaleable distributed memory parallel hardware architecture [5] in which a set of computing nodes are connected by a high performance network (Fig. 1). Each node has the architecture of a uniprocessor or shared store multiprocessor - one or more CPUs share the local main memory over a bus. Typically, each node also has a set of locally connected disks for the persistent storage of data. Connecting disks to each node allows the system to provide both high IO performance, by supporting parallel disk access, and high storage capacity. In most current systems the disks are arranged in a share nothing configuration - each disk is physically connected to only one node. As will be seen, this has major implications for the design of the parallel database server software. It is likely that in future the availability of high bandwidth peripheral interconnects will lead to the design of platforms in which each disk is physically connected to more than one node [11], however this `shared disk' configuration is not discussed further in this paper as we focus on currently prevalent hardware platform technology. Database servers require large main memory caches for efficient performance (so as to reduce the number of disk accesses) and so the main memories tend to be large (currently 0.25-4GB is typical). Some of the nodes in a parallel platform will have external network connections to which clients are connected. External database clients send database queries to the server through these connections and later receive the results via them. The number of these nodes (which we term Communications Nodes) varies depending on the required performance and availability. In terms of performance, it is important that there are enough Communications Nodes to perform client communication without it becoming a bottleneck, while for availability it is important to have more than one route from a client to a parallel server so that if one fails, another is available.

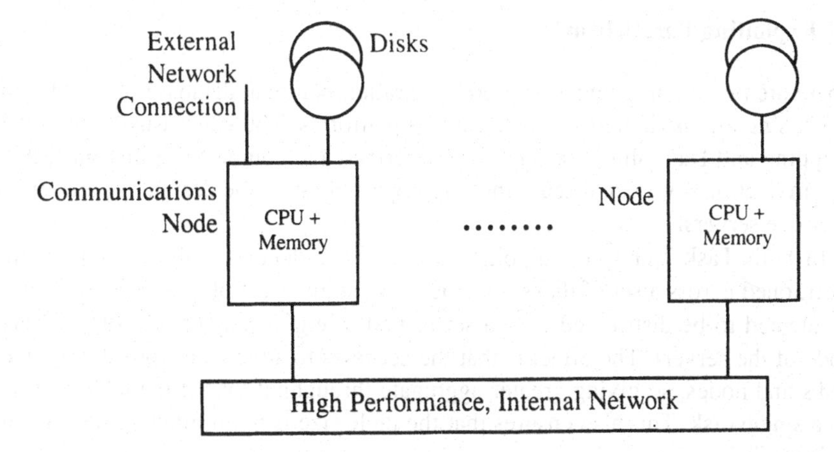

Fig. 1. A Distributed Memory Parallel Architecture

The distributed memory parallel hardware architecture is highly scaleable as adding a node to a system increases all the key performance parameters including: processing power, disk throughput and capacity, and main memory bandwidth and capacity. For a database server, this translates into increased query processing power, greater database capacity, higher disk throughput and a larger database cache.

The design of the hardware must also contribute to the creation of a high availability system. Methods of achieving this include providing component redundancy and the ability to replace failed components through hot-pull & push techniques without having to take the system down [5].

The requirement to support a database server also influences the design of the operating system in a number of ways. Firstly, commercial database server software depends on the availability of a relatively complete set of standard operating system facilities, including support for: file accessing, processes (with the associated inter-process communications) and external communications to clients through standard protocols. Secondly, it is important that the cost of inter-node communications is minimised as this directly affects the performance of a set of key functions including remote data access and query execution. Finally, the operating system must be designed to contribute to the construction of a high availability system. This includes ensuring that the failure of one node does not cause other nodes to fail, or unduly affect their performance (for example through having key services such as distributed file systems delayed for a significant time waiting, until a long time-out occurs, for a response from the failed node).

Experience in a number of projects suggests that micro-kernel based operating systems provide better support than do monolithic kernels for adding and modifying the functionality of the kernel to meet the requirements of parallel database servers [1]. This is for two reasons: they provide a structured way in which new services can be added, and their modularity simplifies the task of modifying existing services, for example to support high-availability [4].

2.2 Exploiting Parallelism

There are two main schemes utilised by parallel relational database servers for query processing on distributed memory parallel platforms. These are usually termed Task Shipping and Data Shipping. They are described and compared in this subsection so that in Section 4 we can discuss their appropriateness to the design of parallel object database servers.

In both Task and Data shipping schemes, the database tables are horizontally partitioned across a set of disks (i.e. each disk stores a set of rows). The set of disks is selected to be distributed over a set of nodes (e.g. it may consist of one disk per node of the server). The effect is that the accesses to a table are spread over a set of disks and nodes, so giving greater aggregate throughput than if the table was stored on a single disk. This also ensures that the table size is not limited to the capacity of a single disk.

In both schemes, each node runs a database server which can compile and execute a query or, as will be described, execute part of a parallelised query. When a client sends an SQL query to one of the Communications Nodes of the parallel server it is directed to a node where it is compiled and executed either serially, on a single node, or in parallel, across a set of nodes. The difference between the two schemes is in the method of accessing database tables as is now described.

2.2.1 Data Shipping

In the data shipping scheme the parallel server runs a distributed filesystem which allows each node to access data from any disk in the system. Therefore, a query running on a node can access any row of any table, irrespective of the physical location of the disk on which the row is stored.

When inter-query parallelism is exploited, a set of external clients generate queries which are sent to the parallel server for execution. On arrival, the Communications Node forwards them to another node for compilation and execution. Therefore the Communication Nodes need to use a load balancing algorithm to select a node for the compilation and execution of each query. Usually, queries in OLTP workloads are too small to justify parallelisation and so each query is executed only on one node. When a query is executed, the database server accesses the required data in the form of table rows. Each node holds a database cache in main memory and so if a row is already in the local cache it can be accessed immediately. However, if it is not in the cache then the distributed filesystem is used to fetch it from disk. The unit of transfer between the disk and cache is a page of rows. When query execution is complete the result is returned to the client.

When a single complex query is to be executed in parallel (intra-query parallelism) then parallelism is exploited within the standard operators used to execute queries: scan, join and sort [12]. For example, if a table has to be scanned to select rows which meet a particular criteria then this can be done in parallel by having each of a set of nodes scan a part of the table. The results from each node are appended to produce the final result. Another type of parallelism, pipeline parallelism, can be exploited in queries which require multiple levels of operators by streaming the results from one operator to the inputs of others.

Because, in the Data Shipping model, any node can access any row of any table the compiler is free to decide for each query both how much parallelism it is sensible to exploit, and the nodes on which it should be executed. Criteria for making these decisions include the granularity of parallelism and the current loading of the server, but it is important to note that these decisions are not constrained by the location of the data on which the query operates. For example, even if a table is only partitioned over the disks of three nodes, a compiler can still choose to execute a scan of that table over eight nodes.

The fact that any node can access data from disks located anywhere in the parallel system has a number of major implications for the design of the parallel database server:

- the distributed filesystem must meet a number of requirements which are not found in typical, conventional distributed filesystems such as NFS. These include high performance access to both local and remote disks, continuous access to data even in the presence of disk and node failures, and the ability to perform synchronous writes to remote disks. Consequently, considerable effort has been expended in this area by developers of parallel database platforms.
- a global lock manager is required as a resource shared by the entire parallel system. A table row can be accessed by a query running on any node of the system. Therefore the lock which protects that row must be managed by a single component in the system so that queries running on different nodes of the system see the same lock state. Consequently, any node which wishes to take a lock must communicate with the single component that manages that lock. If all locks in a system were to be managed by a single component - a centralised lock manager - then the node on which that component ran would become a bottleneck, reducing the system's scalability. Therefore, parallel database systems usually implement the lock manager in a distributed manner. Each node runs an instance of the lock manager which manages a subset of the locks [1]. When a query needs to acquire or drop a lock it sends a message to the node whose lock manager instance manages that lock. One method of determining the node which manages a lock is to use a hash function to map a lock identifier to a node. All the lock manager instances need to co-operate and communicate to determine deadlocks, as circles of dependencies can contain a set of locks managed by more than one lock manager instance [13].
- distributed cache management is required. A row can be accessed by more than one node and this raises the issue that a row could be cached in more than one node at the same time. It is therefore necessary to have a scheme for maintaining cache coherency. One solution is the use of cache locks managed by the lock manager.

The efficiency of database query execution in a Data Shipping system is highly dependent on the cache hit rate. A cache miss requires a disk access and the execution of filesystem code on two nodes (assuming the disk is remote), which increases response time and reduces throughput. Two techniques can be used to reduce the resulting performance degradation. The first is to, where possible, direct queries which access the same data to the same node. For example all banking queries from one branch could be directed to the same node. This increases cache hit rates and reduces pinging: the process by which cached pages have to be moved around the caches of different nodes of the system because the data contained in them is updated by queries running on more than one node. The second technique extends the first technique so that not only are queries operating on the same data sent to the same node, but that node is selected as the one which holds the data on its local disks. The result is that if there is a cache miss then only a local disk access is required. This not only reduces latency, but also increases throughput as less code is executed in the filesystem for a local access than a remote access. This second technique is closely related to Task Shipping, which is now described.

2.2.2 Task Shipping

The key difference between the two Shipping schemes is that in Task Shipping only those parts of tables stored on local disks can be accessed during query evaluation. Consequently, each query must be decomposed into sub-queries, each of which only requires access to the table rows stored on a single node. These sub-queries are then shipped to the appropriate nodes for execution. For a parallel scan operation, this is straightforward to organise, though it does mean that, unlike in the case of Data Shipping, the parallel speed-up is limited by the number of nodes across which the table is partitioned, unless the data is temporarily re-partitioned by the query. When a join is performed, if the two tables to be joined have been partitioned across the nodes such that rows with matching values of the join attribute are stored on disks of the same node, then the join can straightforwardly be carried out as a parallel set of local joins, one on each node. However, if this is not the case, then at least one of the tables will have to be re-partitioned before the join can be carried out. This is achieved by sending each row of the table(s) to the node in which it will participate in a local join. Parallel sorting algorithms also require this type of redistribution of rows, and so it may be said that the term Task Shipping is misleading because in common situations it is necessary to ship data around the parallel system. However because data is only ever accessed from disk on one node - the node connected to the disk on which it is stored, then some aspects of the system design are simplified when compared to the Data Shipping scheme, viz.:

- there is no requirement for a distributed filesystem because data is always accessed from local disk (however, in order that the system be resilient to node failure, it will still be necessary to duplicate data on a remote node or nodes - see Section 2.3).
- there is no requirement for global cache coherency because data is only cached on one node - that connected to the disk on which it is stored.
- lock management is localised. All the accesses to a particular piece of data are made only from one node - the node connected to the disk on which it is stored. This can therefore run a local lock manager responsible only for the local table rows. However, a 2-phase commitment protocol is still required across the server to allow transactions which have been fragmented across a set of nodes to commit in a co-ordinated manner. Further, global deadlock detection is still required.

An implication of the Task Shipping scheme is that even small queries which contain little or no parallelism still have to be divided into sub-queries if they access a set of table rows which are not all stored on one node. This will be inefficient if the sub-queries have low granularity.

2.3 Availability

If a system is to be highly available then its components must be designed to contribute to this goal. In this section we discuss how the database server software can be designed to continue to operate when key components of a parallel platform fail.

A large parallel database server is likely to have many disks, and so disk failure will be a relatively common occurrence. In order to avoid a break of service for recovery when a disk fails it is necessary to duplicate data on more than one disk. Plexing and RAID schemes can be used for this. In Goldrush, data is partitioned over a set of disk volumes, each of which is duplexed. The two plexes are always chosen to be remote from each other, i.e. held on disks not connected to the same node, so that even if a node fails then at least one plex is still accessible.

If a node fails then the database server running on that node will be lost, but this need not prevent other nodes from continuing to provide a service, all be it with reduced aggregate performance. Any lock management data held in memory on the failed node becomes inaccessible and so, as this may be required by transactions running on other nodes, the lock manager design must ensure that the information is still available from another node. This can be achieved by plexing this information across the memories of two nodes.

Failure of an external network link may cause the parallel server to loose contact with an external client. In the Goldrush system, the parallel server holds information on alternative routes to clients and monitors the external network connections so that if there is a failure then an alternative route can be automatically chosen.

3 Parallel ODBMS Requirements

In this section we describe and justify the requirements which we believe a parallel ODBMS server should meet. A major source of requirements is derived from our experience with the design and usage of the ICL Goldrush MegaServer [1, 5, 6] which embodies a number of the functions and attributes that will also be required in a parallel ODBMS. However, it is also the case that there are interesting differences between the requirements for a parallel RDBMS, such as Goldrush, and a parallel ODBMS. In particular, we envisage a parallel ODBMS as a key component for building high performance distributed applications as it has attributes not found in alternative options including: performance, availability, rich interfaces for accessing information (including querying), and transactional capability to preserve database integrity. However, if a parallel ODBMS is to fulfill its potential in this area then we believe that there are a set of requirements which it must meet, and these are discussed in this section.

The rest of this section is structured as follows. We first describe the overall system requirements, before examining the non-functional requirements of performance and availability.

3.1 Systems Architecture

Our main requirement is for a server which provides high performance to ODMG compatible database clients by exploiting both inter-query and intra-query parallelism. This requires a system architecture similar to that described for RDBMS in Section 2. Client applications may generate OQL queries which are sent for execution to the parallel server (via the Communications Nodes). To support intra-query parallelism the server must be capable of executing these queries in parallel across a set of nodes. The server must also support inter-query parallelism by simultaneously executing a set of OQL queries sent by a set of clients. Clients may also map database objects into program objects in order to (locally) access their properties and call methods on them. Therefore, the server must also satisfy requests from clients for copies of objects in the database. The roles of the parallel system, serving clients, are summarised by Figure 2.

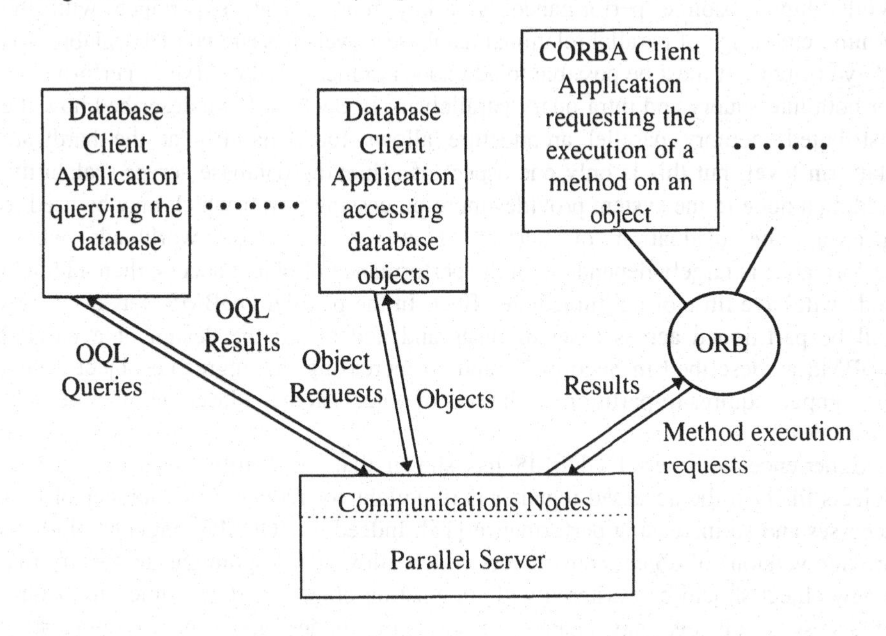

Fig. 2. Clients of the Parallel ODBMS Server

The architecture does not preclude certain clients from choosing to execute OQL queries locally, rather than on the parallel server. This would require the client to have local OQL execution capability, but the client would still need to access the objects required during query execution from the server. Reasons for doing this might be to exploit a particularly high performance client, or to remove load from the server so that it can spend more of its computational resources on other, higher priority queries.

Where the ODBMS is to act as a component in a distributed system, we require that it can be accessed through standard CORBA interfaces [14]. The ODMG standard proposes that a special component - an Object Database Adapter (ODA) - is used to connect the ODBMS to a CORBA Object Request Broker (ORB) [3]. This allows the ODBMS to register a set of objects as being available for external access via the ORB. CORBA clients can then generate requests for the execution of methods on these objects. The parallel ODBMS services these requests, returning the results to the clients. In order to support high throughput, the design of the parallel ODBMS must ensure that the stream of requests through the ODA is executed in parallel.

3.2 Performance

Whilst high absolute performance is a key requirement, experience with the commercial usage of parallel relational database servers has shown that scalability is a key attribute. It must be possible to add nodes to increase the server's performance for both inter-query and intra-query parallelism. In Section 2, we described how the distributed memory parallel architecture allows for scalability at the hardware platform level, but this is only one aspect of achieving database server scalability. Adding a node to the system provides more processing power which may be used to speed-up the evaluation of queries. However for those workloads whose performance is largely dependent on the performance of object access, then adding a node will have little or no immediate effect. In the parallel ODBMS, sets of objects will be partitioned across a set of disks and nodes (c.f. the design of a parallel RDBMS as described in Section 2), and so increasing the aggregate object access throughput requires re-partitioning the sets of objects across a larger number of disks and nodes.

Experience with serial ODBMS has shown that the ability to cluster on disk objects likely to be accessed together is important for reducing the number of disk accesses and so increasing performance [15]. Indeed, the ODMG language bindings provide versions of object constructors which allow the programmer to specify that a new object should be clustered with an existing object [3]. Over time, the pattern of access to objects may change, or be better understood (due to information provided by performance monitoring), and so support for re-clustering is also required to allow performance tuning.

As will be seen in Section 4, the requirement to support the re-partitioning and re-clustering of objects has major implications for the design of a parallel ODBMS.

3.3 Performance Management

This section describes requirements which relate to the use of a database as a resource shared among a set of services with different performance requirements.

The computational resources (CPU, disk IOs, etc.) made available by high performance systems are expensive - customers pay more per unit resource on a high performance system than they do on a commodity computer system. In many cases, resources will not be completely exhausted by a single service but if the resources are to be shared then mechanisms are needed to control this sharing.

An example of the need for controlled resource sharing is where a database runs a business-critical OLTP workload which does not utilise all the server's resources. It may be desirable to run other services against the same database, for example in order to perform data mining, but it is vital that the performance of the business-critical OLTP service is not affected by the loss of system resources utilised by other services. In a distributed memory parallel server, the two services may be kept apart on non-intersecting sets of nodes in order to avoid conflicts in the sharing of CPU resources. However, they will still have to share the disks on which the database is held. A solution might be to analyse the disk access requirements of the two services and try to ensure that the data is partitioned over a sufficiently large set of nodes and disks so as to be able to meet the combined performance requirements of both services. However this is only possible where the performance requirements of the two services are relatively static and therefore predictable. This may be true of the OLTP service, but the performance of a complex query or data mining service may not be predictable. A better solution would be to ensure that the OLTP service was allocated the share of the resources that it required to meet its performance requirements, and only allow the other service the remainder.

A further example of the importance of controlled resource sharing is where parallel ODBMS act as repositories for persistent information on the Internet and intranets. If information is made available in this way then it is important that the limited resources of the ODBMS are shared in a controlled way among clients. For example there is a danger of denial of service attacks in which all the resources of system are used by a malicious client. Similarly, it would be easy for a non-malicious user to generate a query which required large computational resources, and so reduced the performance available to other clients below an acceptable level. Finally, it may be desirable for the database provider to offer different levels of performance to different classes of clients depending on their importance or payment.

The need for controlled sharing of resources has long been recognised in the mainstream computing world. For example, mainframe computer systems have for several decades provided mechanisms to control the sharing of their computational resources among a set of users or services. Some systems allow CPU time and disk IOs to be divided among a set of services in any desired ratio. Priority mechanisms for allocating resources to users and services may also be offered. Such mechanisms are now also provided for some shared memory parallel systems.

Unfortunately, the finest granularity at which these systems control the sharing of resources tends to be at an Operating System process level. This is too crude for database servers, which usually have at their heart a single, multi-threaded process which executes the queries generated by a set of clients (so reducing switching costs when compared to an implementation in which each client has its own process running on the server). Therefore, in conclusion, we believe that the database server must provide its own mechanisms to share resources among a set of services in a controlled manner if expensive parallel system resources are to be fully utilised, and if object database servers are to fulfil their potential as key components in distributed systems.

3.4 Availability

The requirements here are identical to those of a parallel RDBMS. The parallel ODBMS must be able to be continue to operate in the presence of node, disk and network failures. It is also important that availability is considered when designing database management operations which may effect the availability of the database service. These include: archiving, upgrading the server software and hardware, and re-partitioning and re-clustering data to increase performance. Ideally these should all be achievable on-line, without the need to shut-down the database service, but if this is not possible then the time for which the database service is down should be minimised.

A common problem causing loss of availability during management operations is human error. Manually managing a system with up to tens of nodes, and hundreds of disks is very error prone. Therefore tools to automate management tasks are important for reducing errors and increasing availability.

3.5 Hardware Platform

The *Polar* project has access to a parallel database platform, an ICL Goldrush MegaServer [1], however we are also investigating alternatives. Current commercial parallel database platforms use custom designed components (including internal networks, processor-network interfaces and cabinetry), which leads to high design, development and manufacturing costs, which are then passed on to users. Such systems generally have a cost-performance ratio that is significantly higher than that of commodity, uniprocessor systems. We are investigating a solution to this problem and have built a parallel machine, the Affordable Parallel Platform (APP), entirely from standard, low-cost, commodity components - high-performance PCs interconnected by a high throughput, scaleable ATM network. Our current system consists of 13 nodes interconnected by 155Mbps ATM. The network is scaleable as each PC has a full 155Mpbs connection to the other PCs. This architecture has the interesting property that high-performance clients can be connected directly to the ATM switch giving them a very high bandwidth connection to the parallel platform. This may, for example, be advantageous if large multimedia objects must be shipped to clients.

Our overall aim in this area is to determine whether commodity systems such as the APP can compete with custom parallel systems in terms of database performance. Therefore we require that the database server design and implementation is portable across, and tuneable for, both types of parallel platform.

4 Parallel ODBMS Design

In this section we consider issues in the design of a parallel ODBMS to meet the requirements discussed in the last section. We cover object access in the most detail as it is a key area in which existing parallel RDBMS and serial ODBMS designs do not offer a solution. Cache coherency, performance management and query processing are also discussed.

4.1 Object Access

Objects in an object database can be individually accessed through their unique identifier. This is a property not found in relational databases and so we first discuss the issues in achieving this on the parallel server (we leave discussion of accessing collections of objects until Section 4.3).

As described earlier, the sets of objects stored in the parallel server are partitioned across a set of disks and nodes in order to provide a high aggregate access throughput to the set of objects. When an object is created, a decision must be made as to the node on which it will be stored. When a client needs to access a persistent object then it must send the request to the node holding the object.

In the rest of this section we discuss in more detail the various issues in the storage and accessing of objects in a parallel ODBMS as these differ significantly from the design of a serial ODBMS or a parallel RDBMS. Firstly we discuss how the Task Shipping and Data Shipping mechanisms described in Section 2 apply to a parallel ODBMS. We then discuss possible options for locating and accessing persistent objects.

4.1.1 Task Shipping vs. Data Shipping

In a serial ODBMS, the database server stores the objects and provides a run-time system which allows applications running on external clients to transparently access objects. Usually a page server interface is offered to clients, i.e. a client requests an object from the server which responds with a page of objects [7]. The page is cached in the client because it is likely that the requested object and others in the same page will be accessed in the near future. This is a Data Shipping architecture, and has the implication that the same object may be cached in more than one client, so necessitating a cache coherency mechanism. If Data Shipping was also adopted within the parallel server then nodes executing OQL would also access and cache objects from remote nodes.

We now consider if the alternative used in some parallel RDBMS - a Task Shipping architecture - is a viable option for a parallel ODBMS. There are two computations which can be carried out on objects: executing a method on an object and accessing a property of an object. These computations can occur either in an application running on an external client, or during OQL execution on a node of the parallel server.

In order to implement task shipping, the external clients, and nodes executing OQL would not cache remote objects and process them locally. Instead, they would have to send a message to the node which stored the object requesting that it perform the computation (method execution or property access) locally and return the result. In this way, the client does not cache objects, and so cache coherency is not an issue. However Task Shipping in object database servers does have two major drawbacks. Firstly, the load on the server is increased when compared to the data shipping scheme. This means that a server is likely to be able to support fewer clients. Secondly, the latency of task shipping - the time between sending the task to the server and receiving the response - is likely to result in performance degradation at the client when compared to the data shipping scheme in which the client can operate directly on the object without any additional latency once it has been cached. Many database clients are likely to be single application systems, and so there will not be other processes to run while a computation on an object is being carried out on the parallel server - consequently the CPU will be idle during this time. Further, the user of an interactive client application will be affected by the increase in response time caused by the latency.

The situation is slightly different for the case of OQL query execution on the nodes of the server under a Task Shipping scheme. When query execution is suspended while a computation on an object is executed remotely, it is very likely that there will be other work to do on the node (for example, servicing requests for operations on objects and executing other queries) and so the CPU will not be idle. However, it is likely that the granularity of parallel processing (the ratio of computation to communication) will be low due to the frequent need to generate requests for computations on remote objects. Therefore, although the node may not be idle, it will be executing frequent task switches and sending/receiving requests for computations on objects. The effect of this is likely to be a reduction in throughput when compared to a data shipping scheme with high cache hit rates which will allow a greater granularity of parallelism.

For these reasons, we do not further consider a task shipping scheme, and so must address the cache coherence issue (Section 4.2). Figure 3 shows the resulting parallel database server architecture. The objects in the database are partitioned across disks connected to the nodes of the system, taking into account clustering. Requests to access objects will come from both: applications running on external database clients; and the nodes of the parallel server which are executing queries or processing requests for method execution from CORBA clients. The database client applications are supported by a run-time system (RTS) which maintains an object cache. If the application accesses an object which is not cached then an object request is generated and sent to a Communication Node (CN) of the parallel server. The CN uses a Global Object Access module to determine the node which has the object stored on one of its local disks. The Remote Object Access component forwards the request to this node where it is serviced by the Local Object Access component which returns the page containing the object back to the client application RTS via the CN.

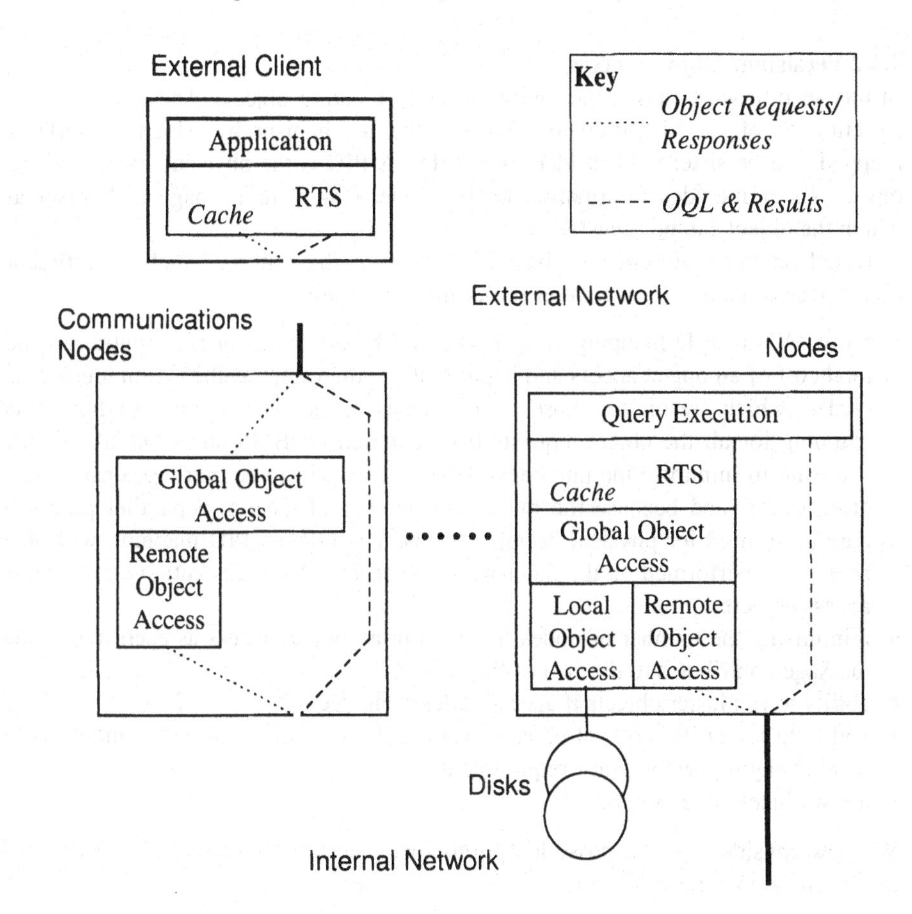

Fig. 3. Database Server System Architecture

If the client application issues an OQL query then the client's database RTS sends it to a CN of the parallel server. From there it is forwarded to a node for compilation. This generates an execution plan which is sent to the Query Execution components on one or more nodes for execution. The Query Execution components on each node are supported by a local run time system, identical to that found on the external database clients, which accesses and caches objects as they are required, making use of the Global Object Access component to locate the objects. The question of how objects are located is key, and is discussed in the next section.

If there are CORBA based clients generating requests for the execution of object methods then the requests will be directed (by the ORB) to a CN. In order to maximise the proportion of local object accesses (which are more efficient than remote object accesses) then the CN will route the request to the node which stores the object for execution. However, there will be circumstances in which this will lead to an imbalance in the loading on the nodes, for example if one object frequently occurs in the requests. In these circumstances, the CN may choose to direct the request to another, more lightly loaded, node.

4.1.2 Persistent Object Access

In this section we describe the major issues in locating objects. These include the structure of object identifiers (OIDs) and the mechanism by which an OID is mapped to a persistent object identifier (PID). A PID is the physical address of the object, i.e. it includes information on the node, disk volume, page and offset at which the object can be accessed.

Based on the requirements given in Section 3, the criteria which an efficient object access scheme for a parallel server must meet are:

- a low OID to PID mapping cost. This could be a significant contribution to the total cost of an object access and is particularly important as the Communications Nodes (which are only a subset of the nodes in the system) must perform this mapping for all the object requests from external ODBMS and CORBA clients. We want to minimise the number of CNs as they will to be more expensive than other nodes, and because the maximum number of CNs in a parallel platform may be limited for physical design reasons. The OID to PID mapping will also have to be performed on the other nodes when the Query Execution components access objects.
- minimising the number of nodes involved in an object access as each inter-node message contributes to the cost of the access.
- ability to re-cluster objects if access patterns change.
- ability to re-partition sets of objects over a different number of nodes and disks to meet changing performance requirements.
- a low object allocation cost.

We now consider a set of possible options for object access and consider how they match up against these criteria.

4.1.2.1 Physical OID schemes

In this option, the PID is directly encoded in the OID. For example, the OID could have the structure:

Node	Disk Volume	Page	Offset

The major advantage of this scheme is that the cost of mapping the OID to the PID is zero. The Communications Nodes (CNs) only have to examine the OIDs within incoming object requests from external clients and forward the request to the node contained in the OID. This is just a low-level routing function and could probably be most efficiently performed at a low level in the communications stack. Therefore, the load on a CN incurred by each external object access will be low. Similarly, object accesses generated on a node by the Query Execution component can be directed straight to the node which owns the object. Therefore this scheme does minimise the number of nodes involved in an access. However, because the OID contains the exact physical location of the object, it is not possible to move an object to a different page, disk volume or node. This rules out re-clustering or re-partitioning.

When an object is to be created then the first step is to select a node. This could be done by the CN using a round-robin selection scheme to evenly spread out the set of objects. The object is then created on the chosen node. Therefore object creation is a low cost operation.

Therefore, this scheme will perform well in a stable system but as it does not allow re-partitioning or re-clustering, it is likely that over time the system will become de-tuned as the server's workload changes.

We now consider three possible variations of the scheme which attempt to overcome this problem:

Indirections. The OID scheme used by O2 [7] includes a physical volume number in the OID, and so runs into similar problems if an Object is to be moved to a different volume. The O2 solution is to place an indirection at the original location of the object pointing to the new location of the object. Therefore, potentially two disk accesses might be incurred and (in a parallel server) two nodes might be involved in an object access. Also, this is not a viable solution for re-partitioning a set of objects across a larger number of disks and nodes in order to increase the aggregate throughput to those objects. The initial accesses to all the objects will still have to go to the indirections on the original disks/nodes and so they will still be a bottleneck. For this reason we disregard this variant as a solution.

Only include the Physical Node in the OID. In this variant, the OID structure would contain the physical node address, and then a logical address within the node, i.e. :

Node	Logical address within node

The logical address would be mapped to the physical address at the node; methods for this include hash tables and B-trees (the Direct Mapping scheme[16] is however not an option as it is not scaleable: once allocated, the handles which act as indirections to the objects cannot be moved). This would allow re-clustering within a node by updating the mapping function, and re-clustering across nodes using indirections, but it would not support re-partitioning across more nodes as the Node address is still fixed by the OID. Therefore we disregard this variant as a solution.

Update OIDs when an Object is Moved. If this could be implemented efficiently then it would have a number of advantages. Objects could be freely moved within and between nodes for re-clustering and re-partitioning, but because their OID would be updated to contain their new physical address then the mapping function would continue to have zero cost, and no indirections would be required. However, such a scheme would place restrictions on the external use of OIDs. Imagine a scenario in which a client application acquires an OID and stores it. The parallel ODBMS then undergoes a re-organisation which changes the location of some objects, including the one whose OID was stored by the client application. That OID is now incorrect and points to the wrong physical object location. A solution to this problem is to limit the temporal scope of OIDs to a client's database session. The ODMG standard allows objects to be named, and provides a mapping function to derive the object's OID from its name. It could be made mandatory for external database clients to only refer to objects by name outside of a database session. Within the session the client would use the mapping function to acquire the OID of an object. During the session, this OID could be used, but when the session ended then the OID would cease to have meaning and at the beginning of subsequent sessions OIDs would have to re-acquired using the name mapping function. This protocol leaves the database system free to change OIDs while there are no live database sessions running. When an object is moved then the name mapping tables would be updated. Any references to the object from other objects in the database would also have to be updated. If each reference in a database is bi-directional then this process is simplified, as all references to an object could be directly accessed and updated. The ODMG standard does permit unidirectional references but these can still be implemented bi-directionally. Another option is to scan the whole database to locate any references to the object to be moved. In a large database then this could be prohibitively time-consuming. An alternative is the Thor indirection based scheme [10], in which, over a period of time, references are updated to remove indirections.

A name based external access scheme could be used for CORBA client access to objects. Objects could be registered with an ORB by name. Therefore, incoming requests from CORBA based clients would require the object name to be mapped to an OID at a CN. A CORBA client could also navigate around a database so long as it started from a named object, and provided that the OIDs of any other objects that it reached were registered with the ORB. However, if the client wished to store a reference to an object for later access (outside the current session) then it would have to ask the ODBMS to create a name for the object. The name would be stored in the mapping tables on the CNs, ready for subsequent accesses. This procedure can, for example, be used to create interlinked, distributed databases in which references in one database can refer to objects in another.

4.1.2.2 Logical OID schemes

At the other extreme from the last scheme is one in which the OID is completely logical and so contains no information about the physical location of the object on disk. Consequently, each object access requires a full logical to physical mapping to determine the PID of the object. Methods of performing this type of mapping have been extensively studied with respect to a serial ODBMS [16], however their appropriateness for a parallel system requires further investigation.

As stated above, the logical to physical mapping itself can be achieved by a number of methods including hash tables and B-trees. Whatever method is used for the mapping, the Communication Nodes will have to perform the mapping for requests received by external clients. This may require disk accesses as the mapping table for a large database will not fit into main memory. Therefore there will be a significant load on the CNs, and it will be necessary to ensure that there are enough CNs configured in a system so that they are not bottlenecks. Similarly, the nodes executing OQL will also have to perform the expensive OID to PID mapping. However, once the mapping has been carried out then the request can be routed directly to the node on which the object is located.

As all nodes will need to perform the OID to PID mapping, then each will need a complete copy of the mapping information. We rule out the alternative solution of having only a subset of the nodes able to perform this mapping due to the extra message passing latency that this would add to each object access. This has two implications: each node needs to allocate storage space to hold the mapping; and whenever a new object is created then the mapping information in all nodes needs to be updated, which results in a greater response time, and worse throughput for object creation than was the case for the Physical OID scheme described in the last section.

With this scheme, unlike the Physical OID scheme, names are not mandatory for storing external references to objects. As OIDs are unchanging, they can be used both internally as well as externally. This removes the need for CNs to map names to OIDs for incoming requests from CORBA clients.

4.1.2.3 Virtual Address Schemes

There have been proposals to exploit the large virtual address spaces available in some modern processors, in object database servers, both parallel [17] and serial [18]. Persistent objects are allocated into the virtual address space and their virtual address is used as their OID. This has the benefit of simplifying the task of locating a cached object in main memory. However an OID to PID mapping mechanism is still required when a page fault occurs and the page containing an object must be retrieved from disk. The structure of a typical virtual address is:

Segment	Page	Offset

In a parallel ODBMS, when a page fault occurs then the segment and page part of the address are mapped to a physical page address. The page is fetched, installed in real memory and the CPU's memory management unit tables are configured so that any accesses to OIDs in that page are directed to the correct real memory address.

The key characteristic of this scheme is that the OID to PID mapping is done at the page level, i.e. in the [Segment, Page, Offset] structure shown above only the Segment and Page fields are used in determining the physical location of the object. Therefore it is possible to re-partition sets of objects by changing the mapping, but it is not possible to re-cluster objects within pages unless changing OIDs, or indirections are supported (as described in Section 4.1.2.1).

The OID to PID mapping can be done using the logical or physical schemes described earlier, with all their associated advantages and disadvantages. However, it is worth noting that if a logical mapping is used then the amount of information stored will be smaller than that required for the logical OID scheme of Section 4.1.2.2 as only pages, rather than individual objects must be mapped.

4.1.2.4 Logical Volume Scheme

None of the above schemes is ideal, and so we have also investigated the design of an alternative scheme which allows the re-partitioning of data without the need to update addresses nor incur the cost of an expensive logical to physical mapping in order to determine the node on which an object is located. In describing the scheme we will use the term *volume set* to denote the set of disk volumes across which the objects in a class are partitioned. OIDs have the following structure:

Class	Logical Volume	Body

The *Class* field uniquely identifies the class of the object. As will be described in detail later, the *Logical Volume* field is used to ensure that objects which must be clustered are always located in the same disk volume, even after re-partitioning. The size of this field is chosen so that its maximum value is much greater than the maximum number of physical volumes that might exist in a parallel system.

Each node of the server has access to a definition of the Volume Set for each class. This allows each node, when presented with an OID, to use the *Class* field of the OID to determine the number and location of the physical volumes in the Volume Set of the class to which the object belongs.

If we denote the cardinality of Volume Set V as |V|, then the physical volume on which the object is stored is calculated cheaply as:

$$\text{Physical Volume} = V[\ \text{LogicalVolume modulus } |V|\ ]$$

where the notation V[i] denotes the i'th Volume in the Volume Set V. A node wishing to access the object uses this information to forward the request to the node on which the object is held. On that node, the unique serial number for the object within the physical volume can be calculated as:

$$\text{Serial} = (\text{Body div } |V|) + |\text{Body}| * (\text{LogicalVolume div } |V|)$$

where: a div b is the whole number part of the division of a by b, and |Body| is the number of unique values that the body field can take.

The Serial number can then be mapped to a PID using a local logical to physical mapping scheme.

If it is necessary to re-partition a class then this can be done by creating a new volume set with a different cardinality and copying each object from the old volume set into the correct volume of the new volume set using the above formula. Therefore a class of objects can be re-partitioned over an arbitrary set of volumes without having to change the OIDs of those objects.

The main benefit of the scheme is that objects in the same class with the same *Logical Volume* field will always be mapped to the same physical volume. This is still true even after re-partitioning, and so ensures that objects clustered before partitioning can still be clustered afterwards. Therefore, when objects which must be clustered together are created they should be given OIDs with the same value in the *Logical Volume* field. To ensure that the OID is unique they must have different values for their *Body* fields.

Where it is desirable to cluster objects of different classes then this can be achieved by ensuring that both classes share the same volume set, and that objects which are to be clustered share the same *Logical Volume* value. This ensures that they are always stored on the same physical volume.

When an external client application needs to create a new object then it sends the request, including the class of the object, to a Communication Node. These hold information on all the volume sets in the server, and so can select one physical volume from that set (using a round-robin or random algorithm). The CN then forwards the 'create object' request to the node that contains the chosen physical volume, and the OID is then generated on that node as follows. Each node contains the definition of the volume set for each class. The required values of the Class and Logical Volume fields of the object will have been sent to the node by the CN. The node also keeps information on the free *Body* values for each of the logical volumes which map to a physical volume connected to that node. The node can therefore choose a free *Body* value for the Logical Volume chosen by the CN. This completes the OID. The object is then allocated to a physical address and the local OID mapping information updated to reflect this new OID to PID mapping.

The creation algorithm is different in cases where it is desirable to create a new object which is to be clustered with an existing object. The new object must be created with the same *Logical Volume* field setting as the existing object so that it will always by stored on the same Physical Volume as the existing object, even after a re-partition. Therefore it is this Logical Volume value which is used to determine the node on which the object is created. On that node the Body field of the new object will be selected, and the object will be allocated into physical disk storage clustered with the existing object (assuming that space available).

This scheme therefore allows us to define a scaleable, object manager in which it is possible to re-partition a set of objects in order to meet increased performance requirements. Existing clustering can be preserved over re-partitioning of the data, however if re-clustering is required then it would have to be done by either using indirections or by updating OIDs. Determining the node on which an object resides is cheap and this should reduce the time required on the CNs to process incoming object access requests. The vast bulk of the information required to map an OID to a PID is local to the node on which the object is stored, so reducing the total amount of storage in the system required for OID mapping, and reducing the response time and execution cost of object creation.

4.1.2.5 Summary

The above schemes all have advantages and disadvantages:

- The Physical OID scheme has a low run time cost but our requirements for re-clustering and re-partitioning can only be met if OIDs can be changed. However, the use of names as permanent identifiers does permit this.
- The Logical OID system supports both re-partitioning and re-clustering. However it is expensive both for object creation and in terms of the load it places on the CNs.
- The Virtual Address scheme supports re-partitioning, but re-clustering is only possible through indirections or by changing OIDs. Creating an object is costly

when a new page has to be allocated, as it requires an updating of the OID to PID mapping on all nodes. Also, the cost of mapping OIDs to PIDs for accesses from external clients may place a significant load on the CNs.
- The Logical Volume scheme may be a reasonable compromise between completely logical and completely physical schemes. However indirections or changeable OIDs are required for re-clustering.

We are therefore investigating these schemes further in order to perform a quantitative comparison between them.

4.2 Concurrency Control and Cache Management

Concurrency control is important in any system in which multiple clients can be accessing an object simultaneously. In our system, each node runs an Object Manager which is responsible for the set of local objects including: concurrency control when those objects are accessed; and the logging mechanisms required to ensure that it is possible to recover of the state of those objects after a node or system failure.

In the system we have described there are three types of clients of the Object Managers: database applications running on external clients; Query Execution components on the parallel server nodes; and, also on the nodes, object method calls instigated by CORBA based clients. Therefore there are caches both on the nodes of the server, and in the external database clients. More than one client may require access to a particular object and so concurrency control and cache coherency mechanisms must manage the fact that an object may be held in more than one cache. This situation is not unique to parallel systems, and occurs on any client-server object database system which supports multiple clients. A number of concurrency control/cache coherency mechanisms have been proposed for this type of system [19], and there appears to be nothing special about parallel systems which prevents one of these from being adopted.

4.3 Query Execution

In a parallel system we wish to be able to improve the response time of individual OQL queries by exploiting intra-query parallelism. In this section we discuss some of the issues in achieving this.

An OQL query generated by a client will be received first by a Communications Node which will use a load balancing scheme to choose a node on which to compile the query. Compilation generates an execution plan: a graph whose nodes are operations on objects, and whose arcs represent the flow of objects from the output of one operation to the input of another. As described earlier, the system we propose supports data shipping but for the reasons discussed in Section 2.2.1, it will give performance benefits if operations are carried out on local rather than remote objects where possible. To achieve this it is necessary to structure collections which are used to access objects (e.g. extents) so as to allow computations to be parallelised on the basis of location. One option [17] is to represent collections of objects through two-level structures. At the top level is a collection of sets of objects with one set per node of the parallel system. Each set contains only the objects held on one node. This allows computations on collections of objects to be parallelised in a straightforward manner: each node will run an operation which processes those objects stored locally. However, in our data shipping system if the objects in a collection are not evenly partitioned over the nodes then it is still possible to partition the work of processing the objects among the nodes in a way that is not based on the locations of the objects. One advantage of the Logical Volume OID mapping scheme described in Section 4.1.2.4 is that it removes the need to explicitly maintain these structures for class extents as the existing OID to PID mapping information makes it possible to locate all the objects in a given class stored on the local node.

A major difference between object and relational databases is that object database queries can include method calls. This has two major implications.

Firstly, as the methods must be executed on the nodes of the parallel server a mechanism is required to allow methods to be compiled into executable code that runs on the server. This is complicated by the fact that the ODMG standard allows methods to be written in a number of languages, and so either the server will have to provide environments to compile and execute each of these languages, or alternatively client applications using the system will have to be forced to specify methods only in those languages which the server can support. Further, the architecture that we have proposed in this paper has client applications executing method calls in navigational code on the client. Therefore executable versions of the code must be available both in the server (for OQL) and on the clients. This introduces risks that the two versions of the code will get out of step. This may, for example, occur if a developer modifies and recompiles the code on a client but forgets to do the same on the server. Also a malicious user might develop, compile and execute methods on the client specifically so as to access object properties not available through the methods installed in the server. Standardising on Java or another interpreted language for writing methods might appear to be a solution to these problems because they have a standard, portable executable representation to which a method could be compiled once and stored in the database before execution on either a client or server. However, the relatively poor performance of interpreted languages when compared to a compiled language such as C++ is a deterrent because the main reason for utilising a parallel server is to provide high performance.

Secondly, the cost of executing a method may be a significant component of the overall performance of a database workload, and may greatly vary in cost depending on its arguments. This could make it difficult to statically balance the work of executing a query over a set of nodes. For example, consider an extent of objects that is perfectly partitioned over a set of nodes such that each node contains the same number of objects. A query is executed which applies a method to each object but not all method calls take the same time, and so, if the computation is partitioned on the basis of object locality, some nodes may have completed their part of the processing and be idle while others are still busy. Therefore it may be necessary to adopt dynamic load balancing schemes which delay decisions on where to execute work for as long as possible (at run-time) so as to try to evenly spread the work over the available nodes. For workloads in which method execution time dominates and not all nodes are fully utilised (perhaps because the method calls are on a set of objects whose cardinality is less than the number of nodes in the parallel system) then it may be necessary to support intra-method parallelism, in which parallelism is exploited within the execution of a single method. This would require methods to be written in a language which was amenable to parallelisation, rather than a serial language such as C++ or Java. One promising candidate is UFO [20], an implicit parallel language with support for objects.

4.4 Performance Management

Section 3 outlined the requirements for performance management in a parallel database server. In this section we describe how we intend to meet these requirements.

In recent years, there has been an interest in the performance management of multimedia systems [21]. When a client wishes to establish a session with a multimedia server, the client specifies the required Quality of Service and the server decides if it can meet this requirement, given its own performance characteristics and current workload. If it can, then it schedules its own CPU and disk resources to ensure that the requirements are met. Ideally, we would like this type of solution for a parallel database server, but unfortunately it is significantly more difficult due to the greater complexity of database workloads. The Quality of Service requirements of the client of a multimedia server can be expressed in simple terms (e.g. throughput and jitter), and the mapping of these requirements onto the usages of the components of the server is tractable. In contrast, whilst the requirements of a database workload may be specified simply, e.g. response time and throughput, in many cases the mapping of the workload onto the usage of the resources of the system cannot be easily predicted. For example, the CPU and disk utilisation of a complex query are likely to depend on the state of the data on which the query operates.

We have therefore decided to investigate a more pragmatic approach to meeting the requirements for performance management in a parallel database server which is based on priorities [11]. When an application running on an external client connects to the database then that database session will be assigned a priority by the server. For example, considering the examples given in Section 3, the sessions of clients in an OLTP workload will have high priority, the sessions of clients generating complex queries will have medium priority while the sessions of agents will have low priority. Each unit of work - an object access or an OQL query - sent to the server from a client will be tagged with the priority and this will be used by the disk and CPU schedulers. If there are multiple units of work, either object accesses or fragments of OQL available for execution, then they will be serviced in the order of their priority, while pre-emptive round-robin scheduling will be used for work of the same priority.

5 Conclusions

The aim of the *Polar* project is to investigate the design of a parallel, ODMG compatible ODBMS. In this paper we have highlighted the system requirements and key design issues, using as a starting point our previous experience in the design and usage of parallel RDBMS. We have shown that differences between the two types of database paradigms lead to a number of significant differences in the design of parallel servers. The main differences are:

- rows in RDBMS tables are never accessed individually by applications, whereas objects in an ODBMS can be accessed individually by their unique OIDs. The choice of OID to PID mapping is therefore very important. A number of schemes for structuring OIDs and mapping them to PIDs were presented and compared. None is ideal, and further work is needed to quantify the differences between them.
- there are two ways to access data held in an object database: through a query language (OQL), and by directly mapping database objects into client application program objects. In a relational database, the query language (SQL) is the only way to access data. A major consequence of this difference is that parallel RDBMS can utilise either task shipping or data shipping, but task shipping is not a viable option for a parallel ODBMS with external clients.
- object database queries written in OQL can contain arbitrary methods (unlike RDBMS queries) written in one of several high level programming languages. Mechanisms are therefore required in a parallel ODBMS to make the code of a method executable on the nodes of the parallel server. This will complicate the process of replacing an existing serial ODBMS, in which methods are only executed on clients, with a parallel server. It also raises potential issues of security and the need to co-ordinate the introduction of method software releases. Further, as methods can contain arbitrary code, estimating execution costs is difficult and dynamic load-balancing, to spread work evenly over the set of nodes, may be required.

We have also highlighted the potential role for parallel ODBMS in distributed systems and described the corresponding features that we believe will be required. These include performance management to share the system resources in an appropriate manner, and the ability to accept and load-balance requests from CORBA clients for method execution on objects.

5.1 Future Work

Having carried out the initial investigations described in this paper we are now in a position to explore the design options in more detail through simulation and the building of a prototype system. We are building the prototype in a portable manner so that, as described in Section 3 we can compare the relative merits of a custom parallel machine and one constructed entirely from commodity hardware.

Our investigations so far have also raised a number of areas for further exploration, and these may influence our design in the longer term:

- More sophisticated methods of controlling resource usage in the parallel server are needed so as to make it possible to guarantee to meet the performance requirements of a workload. The solution based on priorities, described in Section 3, does not give as much control over the scheduling of resources as is required to be able to guarantee that the individual performance requirements of a set of workloads will all be met. Therefore we are pursuing other options based on building models of the behaviour of the parallel server and the workloads which run on it. Each workload is given the proportion of the systems resources that it needs to meet its performance targets.
- If it is possible to control the resource usage of database workloads then, in the longer term, it would be desirable to extend the system to support continuous media. Currently multimedia servers (which support continuous media) and object database servers have different architectures and functionality, but there would be many advantages in unifying them, for example to support the high-performance, content-based searching of multimedia libraries.
- The ODMG standard does not define methods of protecting access to objects equivalent to that found in RDBMS. Without such a scheme, there is a danger that clients will be able to access and update information which should not be available to them. This will be especially a problem if the database is widely accessible, for example via the Internet. Work is needed in this area to make information more secure from both malicious attack and programmer error.

Acknowledgements

The *Polar* project is a collaboration between research groups in the Department of Computing Science at the University of Newcastle, and the Department of Computer Science at the University of Manchester. We would like to thank Francisco Pereira and Jim Smith (Newcastle), and Norman Paton (Manchester) for discussions which have contributed to the contents of this paper. The *Polar* project is supported by the UK Engineering and Physical Science Research Council through grant GR/L89655.

References

1. Watson, P. and G.W. Catlow. *The Architecture of the ICL Goldrush MegaServer*. in *BNCOD13*. 1995. Manchester, LNCS 940, Springer-Verlag.
2. Loomis, M.E.S., *Object Databases, The Essentials*. 1995: Addison-Wesley.
3. Cattell, R.G.G., ed. *The Object Database Standard:ODMG 2.0.* , Morgan Kaufman.

4. Watson, P. and P. Townsend, *The EDS Parallel Relational Database System*, in *Parallel Database Systems*, P. America, Editor. 1991, LNCS 503, Springer-Verlag.
5. Watson, P. and E.H. Robinson, *The Hardware Architecture of the ICL Goldrush MegaServer.* Ingenuity- The ICL Technical Journal, 1995. **10**(2): p. 206-219.
6. Watson, P., M. Ward, and K. Hoyle. *The System Management of the ICL Goldrush Parallel Database Server.* in *HPCN Europe*. 1996: Springer-Verlag.
7. Bancilhon, F., C. Delobel, and P. Kanellakis, eds. *Building an Object-Oriented Database System : The Story of O2*. 1992, Morgan Kaufmann.
8. van den Berg, C.A., *Dynamic Query Processing in a Parallel Object-Oriented Database System*, 1994, CWI, Amsterdam.
9. Carey, M. et al., *Shoring up persistent applications.* in *1994 ACM SIGMOD Conf.* 1994. Mineapolis MN.
10. Day, M., *et al.*, *References to Remote Mobile Objects in Thor.* ACM Letters on Programming Languages and Systems, 1994.
11. Rahm, E. *Dynamic Load Balancing in Parallel Database Systems.* in *EURO-PAR 96*. 1996. Lyon: Springer-Verlag.
12. Tamer Ozsu, M. and P. Valduriez, *Distributed and parallel database systems.* ACM Computing Surveys,, 1996. **28**(1).
13. Hilditch, S. and C.M. Thomson, *Distributed Deadlock Detection: Algorithms and Proofs*, UMCS-89-6-1, 1989, Dept. of Computer Science, University of Manchester.
14. OMG, *The Common Object Request Broker: Architecture and Specification*. 1991: Object Management Group and X/Open.
15. Gerlhof, C.A., *et al.*, *Clustering in Object Bases*, TR 6/92. 1992, Fakuly for Informatik, University Karlsruhe.
16. Eickler, A., C.A. Gerlhof, and D. Kossmann. *A Performance Evaluation of OID Mapping Techniques*. in *VLDB*. 1995.
17. Gruber, O. and P. Valduriez, *Object management in parallel database servers*, in *Parallel Processing and Data Management*, P. Valduriez, Editor. 1992, Chapman & Hall. p. 275-291.
18. Singhal, V., S. Kakkad, and P. Wilson, *Texas: An Efficient, Portable Persistent Store*, in *Proc. 5th International Workshop on Persistent Object Stores*. 1992. p. 11-13.
19. Franklin, M.J., M.J. Carey, and M. Livny, *Transactional client-server cache consistency: alternatives and performance.* ACM Transactions on Database Systems, 1997. **22**(3): p. 315-363.
20. Sargeant, J., *Unified Functions and Objects: an Overview*, UMCS-93-1-4, 1993, University of Manchester, Department of Computer Science.
21. Adjeroh, D.A. and K.C. Nwosu, *Multimedia Database Management - Requirements and Issues.* IEEE Multimedia, 1997. **4**(3): p. 24-33.

Chapter 7: Nonlinear Problems

Introduction

Heather Ruskin[1] and José A.M.S. Duarte[2]

[1] School of Computer Applications
Dublin City University
Dublin 9, Ireland,
heather@dcs.qmw.ac.uk

[2] Faculdade de Ciências da Universidade do Porto
Departamento de Física
Rua do Campo Alegre 687, 4169-007 Porto, Portugal
jduarte@fc.up.pt

The diversity of problems which attract a parallel solution is the theme of the concluding chapter of these proceedings and was strongly apparent in the conference presentations, with some outstanding student contributions. Papers included applications as varied as the study of the astrophysical N-body problem, parallel grid manipulation in Earth science calculations, simulations of magnetised plasma and sensitive use of techniques required for a range of industrial problems. A number also considered general questions of algorithm design, coding features and cost implications for implementation. Efficiency and scalability of the developed code, on a variety of parallel platforms, were discussed by several authors and performance analyses provided. Parallelisation strategies were evaluated in the context of architecture availability for portability, speed and communication features.

A Parallel N-body Integrator using MPI, by N. Pereira, presents results of a parallel integrator, using the Message Passing Interface (MPI) library for the astrophysical N-body problem. The initial motivation of the work was described as the study of the exponential instability in self-gravitating N-body systems and the relationship between collisions and the growth of perturbations in system evolution was considered via particle methods. A serial version of the particle-particle method was given, from which a parallel version was derived. Load-balancing features were handled in a straightforward manner, with each processor being allocated the same number of particles. The MPI framework was chosen in order to avail of features such as source-code portability and efficiency of implementation across a range of architectures, including heterogeneous configurations. Performance metrics, including execution time, relative efficiency and relative speedup were used to analyse scalability on an exploratory basis, with a view to investigating their sensitivity to increasing number of processors

J. Palma, J. Dongarra, and V. Hernández (Eds.): VECPAR'98, LNCS 1573, pp. 623–626, 1999.

for a fixed problem size and also to exploring the behaviour of the algorithm for variable problem size. The design for the parallelisation of the algorithm was found to be appropriate with code portability achieving high efficiency and good scalability.

Impressive speedup and high efficiency on a 4 node RS/6000 computer with distributed memory were reported by Borges and Falcão in their work entitled *A Parallelisation Strategy for Power Systems Composite Reliability Evaluation.* They presented a methodology for evaluation of power systems performance based on a parallel Monte Carlo simulation strategy. A coarse-grained parallelism was adopted for the distributed environment, with communications focused on initial data distribution, collation of results and control of the global parallel convergence. Spurious correlations between processor outputs for adequacy analysis on sampled states was avoided by a distribution policy which generated system states directly at the processors and by using a common random number seed, with initiator based on the rank of the processor in the parallel computation. An asynchronous parallel strategy was found to produce best results in terms of performance, which was evaluated on five different systems, producing practically linear speedup and efficiencies close to 100%. Migration of the methodology to a network of workstations might be expected to provide considerable economic potential, but is yet to be realised.

Practical applications in Tribology and Earth Science were considered by contributors, Arenaz et al. and Sawyer et al., respectively. The former (*High Performance Computing of a New Numerical Algorithm for an Industrial Problem in Tribology*) discussed the advantages of HPC for implementation of a numerical algorithm, which enables simulation of lubricant behaviour. Mathematical models of thin film displacement of a fluid, based on ball-plane geometry, underpinned initial vectorisation of the algorithm. Functional program blocks, associated with the mathematical steps, provided the basis for identification of the most time-consuming elements of the computation. The accuracy of the approximation was increased with the introduction of finer grids and a first parallel version presented in outline. Some loss of generality and architecture dependence compromised the improved efficiency.

Generalisation of unstructured grids, for which the mesh can then be adaptively refined, was discussed by Sawyer et al. (*Parallel Grid Manipulations in Earth Science Calculations*), who described the design of a parallel library, applicable to multiple grid situations. At least three distinct grids are generally employed in the Goddard Earth Observing System (GEOS), with the Data Assimilation Office (DAS) using observational data, which features both systematic and random errors, to estimate parameters of the global earth system. These include an unstructured observation grid, a structured geophysical grid and a block-structured computational grid, with numerous interactions between the grids. Unstructured grids are useful for resolution trade-offs in terms of key regions of the domain and are employed in a variety of problems, with some considerable recent research efforts focused on their generation. The PILGRIM parallel library, described by the Sawyer et al., is based on a set of assumptions, which include manipulation of the local data at local level, (also a feature of the data decomposition). Communication and decomposition utilities are implemented using MPI, with all message - passing nevertheless provided in a single Fortran 90 module, to facilitate other options. An outline of the code for definition of the grid rotation matrix is given here, and the flexible design features, (e.g. modularity and extensibility) of PILGRIM are emphasised. Plans include interfacing forecasting and statistical analysis options, before offering the library to the public domain.

The Versatile Advection Code (VAC), as a vehicle for solving the equations of magnetohydrodynamics through simulations of the evolution of magnetised plasma, was described by Keppens and Toth (*Simulating Magnetised Plasma with the Versatile Advection Code*). The versatility of VAC was demonstrated for calculations on the Rayleigh-Taylor instability, with example results from a Cray J90 implementation. Simulations in two and three-spatial dimensions, with and without magnetic fields, can be represented in VAC as a single problem setup, with dimensionality specified at a pre-processing stage. This is one of the most tested and flexible algorithms reported at the conference, matched by performance and scaling results on a variety of an almost exhaustive set of architectures, by the authors, who also emphasised the suitability of VAC for distributed memory architectures, after automatic translation to HPF.

The design of highly parallel genetic algorithms was discussed by Baraglia and Perego, (in *Parallel Genetic Algorithms for Hypercube Machines*), with the well-known Travelling Salesman Problem (TSP) used as a case study to evaluate and compare different implementations. The authors briefly considered computational models used to design parallel GA's and discussed the implementation issues and results achieved. Examples included a centralised model, for parallel processing of a single *panmitic* population under a master-slave paradigm, a fine-grained model operating on a single-structured population and which exploited spatiality and neighbourhood concepts and a coarse-grained model, with sub-populations or islands which evolve in parallel. The second is, of course, particularly suited to massively parallel implementations, due to its scalable communication pattern. Different replacement criteria and crossover operators were used to study the sensitivity of the Holland's GA to the setting of the genetic operators, with every test run 32 times on different random populations, to obtain average, best and worst solutions. Comparison of coarse and fine-grained algorithms was made on the basis of fitness values obtained, after evaluation of 5.12×10^5 solutions, (allowing for convergence and comparable execution times). Coarse-grained performance showed super-linear speedup, due to the sorting algorithm, but the quality of solutions declined for increased number of nodes. The fine-grained algorithm, however, showed sensible improvement and good scalability, providing a more satisfactory compromise between quality of solution and execution time.

The papers in this chapter are sufficiently diverse, detailed and accessible to be of value to the practitioner in their respective areas. The authors and their referees have struck a happy level of legibility for specialists in other domains.

Overall, this set of papers are as representative of the increasing role of simulations on vector and parallel machines as we have witnessed in three successive editions of *VECPAR*.

A Parallel N-Body Integrator Using MPI

Nuno Sidónio Andrade Pereira *

Politechnical Institute of Beja, School of Technology and Management
Largo de São João, nº 16, 1º Esq. C e D, 7800 Beja
Portugal

Abstract. The study of the astrophysical N-body problem requires the use of numerical integration to solve a system of 6N first-order differential equations. The particle-particle codes (PP) using direct summation methods are a good example of algorithms where parallelisation can speed up the computation in an efficient way. For this purpose, a serial version of the PP code *NNEWTON* developed by the author was parallelised using the MPI library and tested on the CRAY-T3D at the EPCC. The results of the parallel code here presented show very good efficiency and scaling, up to 128 processors and for systems up to 16384 particles.

1 Introduction

We begin by an introduction to the Astrophysical N-body problem and the mathematical models used in our work. We also present an overview of particle simulation methods, and discuss the implementation of a direct summation method: the PP algorithm. A parallel version of this algorithm as well as the performance analysis are presented. Finally, the conclusions regarding the discussion of results are offered.

2 The Astrophysical N-Body Problem

The gravitational N-body problem refers to a system of interacting bodies through their mutual gravitational attraction, confined to a delimited region of space. In the universe we can select systems of bodies according to the observation scale. For instance, we can consider the Solar System with $N = 10$ (a restricted model: Sun + 9 planets). Increasing the observation scale, we have systems like open clusters (systems of young stars with typical ages of the order of 10^8 years, and $N \sim 10^2 - 10^3$), globular clusters (systems of old stars with ages of 12-15 billion years, extremely compact and spherically symmetric with $N \sim 10^4 - 10^6$), and galaxies ($N \sim 10^{10} - 10^{12}$). On the other extreme of our scale, on a cosmological scale, we have clusters of galaxies and superclusters.

* This work was supported by EPCC/TRACS under Grant ERB-FMGE-CT95-0051 and partly supported by PRAXIS XXI under GRANT BM/594/94.

J. Palma, J. Dongarra, and V. Hernández (Eds.): VECPAR'98, LNCS 1573, pp. 627–639, 1999.

If we want to consider the whole universe, the total number of galaxies in the observable part is estimated to be of the order of 10^9 (see [2], [9] and [18]).

In our work we are interested in the dynamics of systems with N up to the order of 10^4 (open clusters and small globular clusters).

2.1 The Mathematical Model

In our mathematical model of the physical system each body is considered as a mass point (hereafter referred to as particle) characterised by a mass, a position, and a velocity. We also define an inertial Cartesian coordinate system, suitably chosen in three-dimensional Euclidean space, and an independent variable t, the *absolute time* of Newtonian mechanics.

The state of the system is defined by the set S_N of $3N$ parameters: the masses, positions, and velocities of all particles. Hence:

$$S_N = \{(m_i, \mathbf{r}_i, \dot{\mathbf{r}}_i), i = 1, \ldots, N\}, \tag{1}$$

where $\mathbf{r}_i$ and $\dot{\mathbf{r}}_i$ are the position and velocity vector of particle i, respectively.

Comments. The physical state of the system can be represented as a point in a $6N$-dimensional phase-space with coordinates $(\mathbf{r}_1, \ldots, \mathbf{r}_N, \dot{\mathbf{r}}_1, \ldots, \dot{\mathbf{r}}_N)$ (see [3]). However, we will use this representation of the system which is more suitable for the discussion of the parallelisation of the N-body integrator, on Sect. 3.3.

The force exerted by particle j on particle i is given by Newton's Law of Gravity:

$$\mathbf{F}_{ij} = -Gm_i m_j \frac{\mathbf{r}_i - \mathbf{r}_j}{\| \mathbf{r}_i - \mathbf{r}_j \|^3}, \tag{2}$$

and the total force acting on particle i is

$$\mathbf{F}_i = \sum_{j=1, j \neq i}^{N} \mathbf{F}_{ij}. \tag{3}$$

The right-hand side of equation (3) represents the contribution of the other $N-1$ particles to the total force.

We can now write the equations of motion of particle i:

$$\ddot{\mathbf{r}}_i = \frac{1}{m_i} \mathbf{F}_i. \tag{4}$$

Defining $\mathbf{v}_k = \dot{\mathbf{r}}_i$ we can write the system of $6N$ first-order differential equations:

$$\dot{\mathbf{r}}_i = \mathbf{v}_i, \, \dot{\mathbf{v}}_i = \frac{1}{m_i} \mathbf{F}_i \tag{5}$$

with $i = 1, \ldots, N$. The evolution of the N-body system is determined by the solution of this system of differential equations with initial conditions (1).

For systems with $N = 2$, the two-body problem known as the Kepler problem, (e.g. the Earth-Moon system) the equations of motion (5) can be solved analytically. However, for the general N(>2)-body problem that is not the case (see [3]), and we must use numerical methods to solve the system of differential equations. In Sect. 3 we will discuss the problem of numerical integration of N-body systems.

In every mathematical model of a physical system there is always the problem of the validity of the model, that is, how suitable the model is to describe the physics of the system. In our case we are representing bodies with finite and, in general, different sizes by material points: bodies endowed with mass, but no extension. The physics of the interior of the bodies is not taken into account. However, for dynamical studies this model has proven to be suitable, and has been used to study the evolution of clusters of stars, galaxies, and the development of structures in single galaxies (see [9] and [3]).

2.2 Exponential Instabilities in N-body Systems

The initial motivation of this work was the study of the exponential instability is self-gravitating N-body systems (see [16]). In this problem we are interested in the growth of a perturbation in one or more components of the system. For a given system of N particles we consider the set

$$\mathcal{S}_N^o = \{(m_i, \mathbf{r}_i^o, \dot{\mathbf{r}}_i^o), i = 1, \ldots, N\} \tag{6}$$

of initial conditions (at time $t = t_o$), and define the set of perturbed initial conditions:

$$\Delta\mathcal{S}_N^o = \{(m_i, \Delta\mathbf{r}_i^o, \Delta\dot{\mathbf{r}}_i^o), i = 1, \ldots, N\} \tag{7}$$

where $\Delta\mathbf{r}_i^o$ and $\Delta\dot{\mathbf{r}}_i^o$ are the position and the velocity perturbation vectors for the initial conditions. To evaluate the growth of the perturbations we must solve the system of $3N$ second-order differential equations (see [6] and [16]):

$$\Delta\ddot{\mathbf{r}}_i = - \sum_{j=1, j\neq i}^{N} \mathbf{f}(\Delta\mathbf{r}_i, \Delta\mathbf{r}_j, \mathbf{r}_i, \mathbf{r}_i) \frac{m_j}{\| \mathbf{r}_i - \mathbf{r}_j \|^3} \tag{8}$$

with $i = 1, \ldots, N$, and

$$\mathbf{f}(\Delta\mathbf{r}_i, \Delta\mathbf{r}_j, \mathbf{r}_i, \mathbf{r}_j) = \Delta\mathbf{r}_i - \Delta\mathbf{r}_j - 3(\Delta\mathbf{r}_i - \Delta\mathbf{r}_j).(\mathbf{r}_i - \mathbf{r}_j)\frac{\mathbf{r}_i - \mathbf{r}_j}{\| \mathbf{r}_i - \mathbf{r}_j \|^2}. \tag{9}$$

Defining $\Delta\mathbf{v}_i = \Delta\dot{\mathbf{r}}_i$ we can rewrite (8) in the form:

$$\Delta\dot{\mathbf{r}}_i = \Delta\mathbf{v}_i, \;\; \Delta\dot{\mathbf{v}}_i = - \sum_{j=1, j\neq i}^{N} \mathbf{f}(\Delta\mathbf{r}_i, \Delta\mathbf{r}_j, \mathbf{r}_i, \mathbf{r}_i) \frac{m_j}{\| \mathbf{r}_i - \mathbf{r}_j \|^3} \tag{10}$$

with $i = 1, \ldots, N$, $\Delta\mathbf{r}_i = (\Delta x_i, \Delta y_i, \Delta z_i)$, and $\Delta\mathbf{v}_i = (\Delta\dot{x}_i, \Delta\dot{y}_i, \Delta\dot{z}_i)$. This system of $6N$ first-order differential equations, the variational equations, must be solved together with equations (5).

We now define several metrics as functions of the components of the perturbation vectors (see [6] and [16]):

$$\Delta R = \max_{i=1,\dots,N}(|\Delta x_i| + |\Delta y_i| + |\Delta z_i|) \tag{11}$$

$$< \Delta R >= \frac{1}{N}\sum_{i=1}^{N}(|\Delta x_i| + |\Delta y_i| + |\Delta z_i|) \tag{12}$$

for the perturbations in the position vectors, and

$$\Delta V = max_{i=1,\dots,N}(|\Delta \dot{x}_i| + |\Delta \dot{y}_i| + |\Delta \dot{z}_i|) \tag{13}$$

$$< \Delta V >= \frac{1}{N}\sum_{i=1}^{N}(|\Delta \dot{x}_i| + |\Delta \dot{y}_i| + |\Delta \dot{z}_i|) \tag{14}$$

for the perturbations in the velocity vectors. Each metric is evaluated for each time step of the numerical integration of equations (5) and (10).

The analysis of the quantities given by equations (11), (12), (13), and (14) is very important to understand some aspects of the dynamical behaviour of N-body systems (see [8], [10], [11], and [13]). In particular, we are interested in the relation between collisions and the growth of perturbations. The collisions between bodies are an important mechanism in the evolution of systems like open clusters and globular clusters (see [2] and [9]).

3 Numerical Simulation of N-Body Systems

In this section, we will briefly discuss the use of particle methods to solve the N-body problem with special attention to the direct summation method: the PP method (see [9], for an excellent and detailed presentation of these methods). We present a serial version of the PP method and discuss a parallel version of that method.

3.1 Overview of Particle Simulation Methods

Particle methods is the designation of a class of simulation methods in which the physical phenomena are represented by particles with certain attributes (such as mass, position, and velocity), interacting according to some physical law that determines the evolution of the system. In most cases we can establish a direct relation between the computational particles and the physical particles. In our work each computational particle is the numerical representation of one physical particle. However, in simulations of physical systems with large N, such as galaxies of 10^{11} to 10^{12} stars, each computational particle is a super-particle with the mass of approximately 10^6 stars.

We will now discuss the three principal types of particle simulation methods: a direct summation method, a particle-in-cell (PIC) method, and a hybrid method.

The Particle-Particle Method (PP). This is a direct summation method: the total force on the ith particle is the sum of the interactions with each other particles of the system. To determined the evolution of a N-body system we consider the interaction of every pair of particles, that is, $N(N-1)$ pairs (i,j), with $i,j = 1,\ldots,N \wedge i \neq j$. The numerical effort (number of floating-point operations) is observed to be proportional to N^2.

The Particle-Mesh Method (PM). This is a particle-in-cell method: the physical space is discretized by a regular mesh where a density function is defined according to the attributes of the particles (e.g. mass density for a self-gravitating N-body system). Solving a Poisson equation on the mesh, the forces at particle positions are then determined by interpolation on the array of mesh-defined values. The numerical effort is observed to be proportional to N. The gain in speed is obtained at the cost of loss of spatial resolution. This is particularly important for the simulation of N-body systems if we are interested in exact orbits.

The Particle-Particle-Particle-Mesh Method (P³M). This is a hybrid method: the interaction between one particle and the rest of the system is determined considering a short-range contribution (evaluated by the PP method) and a long-range contribution (evaluated by the PM method). The numerical effort is observed to be also proportional to N, as in the PM method. The advantage of this method over the PM method is that it can represent close encounters as accurately as the PP method. On the other hand the P³M method calculates long-range forces as fast as the PM method.

Comments. We base the choice of method according to the physics of the system under investigation. For our work we use the PP method: we are interested in simulating clusters of stars where collisions are important and, therefore, spatial resolution is important. On the other hand, for the values of N used in some of our simulations ($N \sim 16 - 1024$) the use of a direct summation method has the advantage of providing forces that are as accurate as the arithmetic precision of the computer.

3.2 The PP Serial Algorithm

In our previous work (see [16]) we have implemented the PP method using FORTRAN 77. Several programs were written (the *NNEWTON* codes) but only two versions are considered here: a PP integrator of the equations of motion, and a PP integrator of the equations of motion + variational equations. These two versions use a softened point-mass potential, that is, the force of interaction between two particles i and j is defined as (see [1], [2], and [9]):

$$\mathbf{F}_{ij} = -Gm_i m_j \frac{\mathbf{r}_i - \mathbf{r}_j}{\| (\mathbf{r}_i - \mathbf{r}_j)^2 + \epsilon^2 \|^{3/2}}. \tag{15}$$

The parameter ϵ is often called the softening parameter and is introduced to avoid numerical problems during the integration of close encounters between particles: as the distance between particles becomes smaller the force changes as $1/ \parallel \mathbf{r}_i - \mathbf{r}_j \parallel^2$ in equation (2) and extremely small time steps must be used in order to control the local error of truncation of the numerical integrator. The softening parameter will prevent the force to go to infinity for zero distance causing overflow errors.

3.3 The PP Parallel Algorithm (P-PP)

The PP method has been used to implement parallel versions of N-body integrators by several authors (see [14], for instance). Having this in mind, our first goal was to write a simple algorithm with good load-balance: each processor should perform the same amount of computations. On the other hand, the algorithm should be able to take advantage of an increased number of processors (scalability).

In our algorithm the global task is the integration of the system of equations (5), for N particles, and the sub-tasks are the integration of sub-sets $\mathcal{S}_{N_k}$ of N_k particles, with $k = 0, ..., p$, where $P = p + 1$ is the number of available processors. The parallel algorithm implements a *single program multiple data* (SPMD) programming model: each sub-task is executed by the same program operating on different data (the sub-sets $\mathcal{S}_{N_k}$ of particles).

The diagram in figure 1 shows the structure of the parallel algorithm and the main communication operations. The data are initially read from a file by one processor and a broadcast communication operation is performed to share the initial configuration of the system between every available processor. To each processor (k) is then assigned the integration of a sub-set $\mathcal{S}_{N_k}$ of particles. The global time step is also determined by a global communication operation, and at the end of each time iteration the new configuration of the particles (in each sub-set $\mathcal{S}_{N_k}$) is shared between all processors.

The load-balance problem is completely avoided in this algorithm since each processor is responsible for the same number of particles. The defined sub-sets of particles are such that

$$\sum_{k=0}^{p} \#(\mathcal{S}_{N_k}) = \sum_{k=0}^{p} N_k = N \tag{16}$$

and

$$N_i = N_j, \;\; i, j = 0, ..., p.$$

4 Implementation of the Parallel Algorithm

4.1 The Message Passing Model

The implementation of the P-PP algorithm was done in the framework of the message passing model (see [5] and [7]). In this model we consider a set of

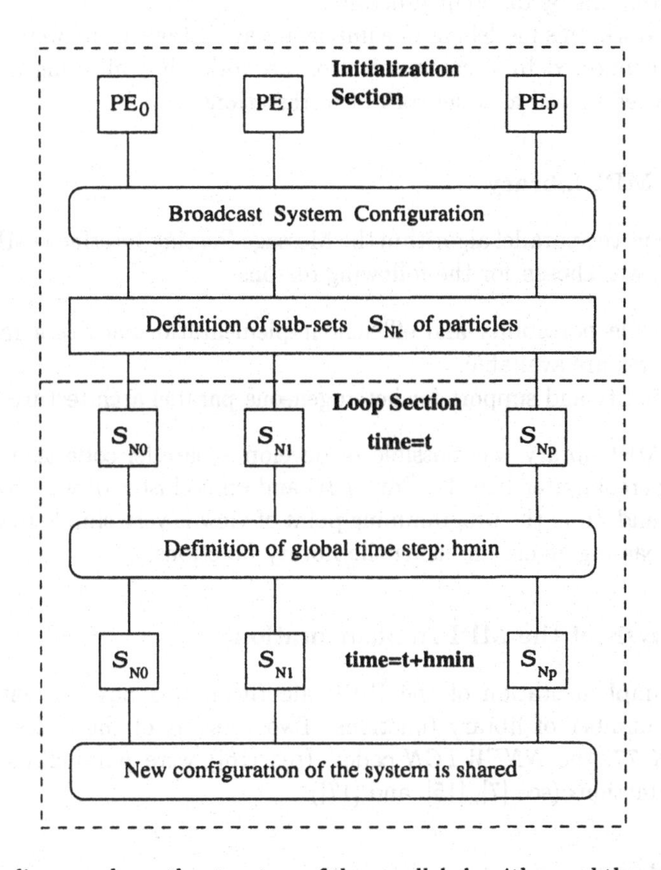

Fig. 1. The diagram shows the structure of the parallel algorithm and the main communication operations: broadcasting the initial configuration of the system to all processors, determination of the global time step and the global communication between processors to share the new configuration of the system after one time step. Each processor $PE_k, (k = 0, ...p)$ is responsible for the integration of its sub-set S_{N_k} of particles.

processes (each identified with a unique name) that have only local memory but are able to communicate with other processes by sending and receiving messages.

Most of the message passing systems implement a SPMD programming model: each process executes the same program but operates on different data. However, the message passing model does not preclude the dynamic creation of processes, the execution of multiple processes per processor, or the execution of different programs by different processes.

For our work, this model has one important advantage: it fits well on separate processors connected by a communication network, thus allowing the use of a supercomputer as well as a network of workstations.

4.2 The MPI Library

To implement the parallel algorithm the Message Passing Interface (MPI) library (see [5][12]) was chosen for the following reasons:

- source-code portability and efficient implementations across a range of architectures are available,
- functionality and support for heterogeneous parallel architectures.

Using the MPI library was possible to develop a parallel code that runs on a parallel supercomputer like the Cray-T3D and on a cluster of workstations. On the other hand, from the programming point of view is very simple to implement a message passing algorithm using the library functions.

4.3 Analysis of the MPI Implementation

The MPI implementation of the P-PP algorithm was possible with the use of a small number of library functions. Two versions of the codes written in FORTRAN 77, the *NNEWTON* codes, (see [16]) were parallelised using the following functions (see [7], [15], and [17]):

Initialisation

1. `MPI_INIT`: Initialises the MPI execution environment.
2. `MPI_COM_SIZE`: Determines the number of processors.
3. `MPI_COMM_RANK`: Determines the identifier of a processor.

Data Structures: Special data structures were defined containing the system configuration.

4. `MPI_TYPE_EXTENT`: Returns the size of a datatype.
5. `MPI_TYPE_STRUCT`: Creates a structure datatype.
6. `MPI_TYPE_COMMIT`: Commits a new datatype to the system.
7. `MPI_TYPE_FREE`: Frees a no longer needed datatype.

Communication: One of the processors broadcasts the system configuration to all other processors.

8. `MPI_BCAST`: Broadcasts a message from processor with rank "root" to all other processors of the group.

Global Operations: Used to compute the global time step, and to share the system configuration between processors after one iteration.

9. `MPI_ALLREDUCE`: Combines values from all processors and distribute the result back to all processors.
10. `MPI_ALL_GATHERV`: Gathers data from all processors and deliver it to all.

Finalisation

11. `MPI_FINALIZE`: Terminates MPI execution environment.

5 Performance Analysis

To analyse the performance of a parallel program several metrics can be considered depending on what characteristic we want to evaluate. In this work we are interested in studying the scalability of the P-PP algorithm, that is, how effectively it can use an increased number of processors. The metrics we used to evaluate the performance are functions of the program execution time (T), the problem size (N, number of particles), and processor count (P). In this section we will define the metrics (as in [5] and [14]) and discuss their application.

5.1 Metrics of Performance

We will consider three metrics for performance evaluation: execution time, relative efficiency, and relative speedup.

Definition 1. *The* execution time *of a parallel program is the time that elapses from when the first processor starts executing on the program to when the last completes execution.*

The execution time is actually the sum over the number of processors of three distinct times: computation time (during which the processor is performing calculations), communication time (time spent sending and receiving messages), and idle time (the processor is idle due to lack of computation or lack of data).

In this study the program is allowed to run for 10 iterations and the execution time is measured by the time of one iteration ($T_{one} = T_{ten}/10$).

Definition 2. *The* relative efficiency *(E_r) is the ratio between time T_1 of execution on one processor and time PT_P,*

$$E_r = \frac{T_1}{PT_P}, \tag{17}$$

where T_P is the time of execution on P processors.

The relative efficiency represents the fraction of time that processors spend doing useful work. The time each processor spends communicating with other processors or just waiting for data or tasks (idle time) will make efficiency always less than 100% (this may not be true is some cases where we have a superlinear regime due to cache effects but we will not discuss it in this work).

Definition 3. *The* relative speedup *(S_r) is defined as the ratio between time T_1 of execution on one processor and time T_P of execution on P processors,*

$$S_r = \frac{T_1}{T_P}. \tag{18}$$

The relative speedup is the factor by which execution time is reduced on P processors. Ideally, a parallel program running on P processors would be P times faster than on one processor and we would get $S_r = P$. However, communication time and idle time on each processor will make S_r always smaller than P (except on the superlinear regime).

These quantities are very useful to analyse the scalability of a parallel program however, efficiency an speedup as defined above do not constitute an absolute figure of merit since the time of execution on a single processor is used as the baseline.

5.2 Performance Results of the PNNEWTON Code

For the performance analysis of the algorithm we measured the time of one iteration for a range of values of two parameters: problem size, and number of processors. The relative efficiency and relative speedup were then evaluated using equations (17) and (18).

The objectives of this analysis are two-fold. First, we want to investigate how the metrics vary with increasing number of processors for a fixed problem size. Second, we want to investigate the behaviour of the algorithm for different problem sizes within the range of interest for our N-body simulations. For that purpose the parallel code (P*NNEWTON*) was tested on the Cray-T3D system at the Edinburgh Parallel Computer Centre (EPCC). The system consists of 512 DEC Alpha 21064 processors arranged on a tridimensional torus and running at 150 MHz. The peak performance of the T3D array itself is 76.8 Gflop/s (see [4]).

The next figures show the results of the tests for systems with $N = 2^6, \ldots, 2^{14}$. The code was integrating equations (5). Similar tests were performed for another version of the P*NNEWTON* code which integrates equations (5) and (10), and identical results were obtained.

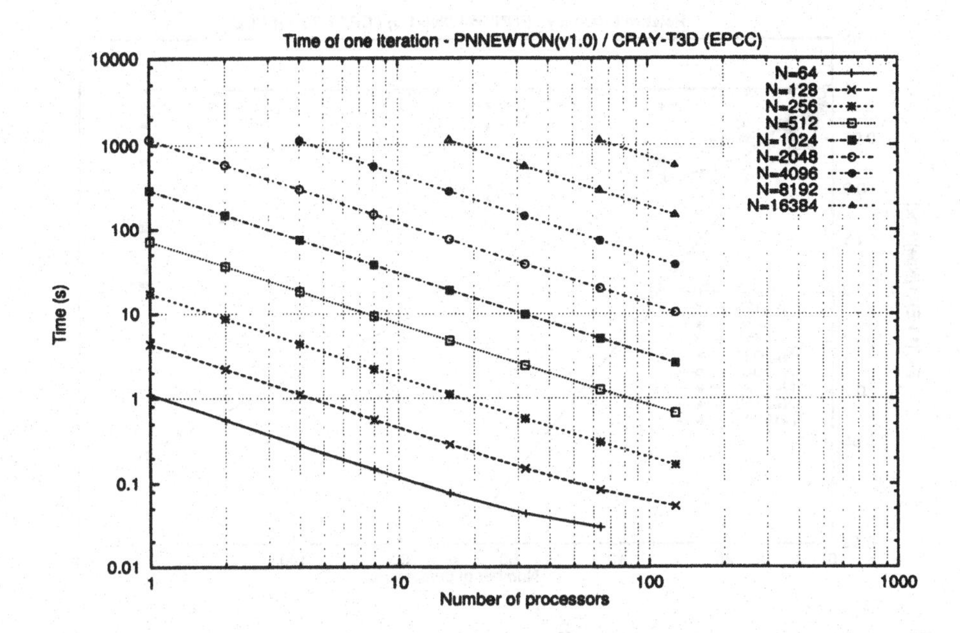

Fig. 2. For each value of $N=2^k, (k = 6, ..., 14)$ the system is allowed to evolve during ten time steps. The computation was performed on a different number of processors. The variation of the time of one iteration with the number of processors for the tested systems shows a good scaling.

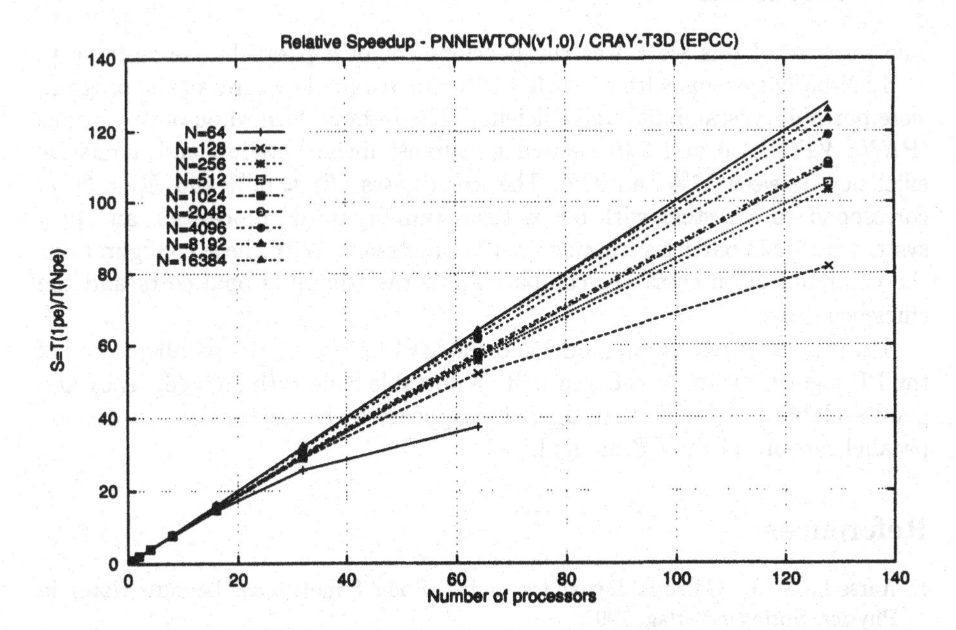

Fig. 3. The program is showing a good scalability for the tested configurations. The speed up is almost linear.

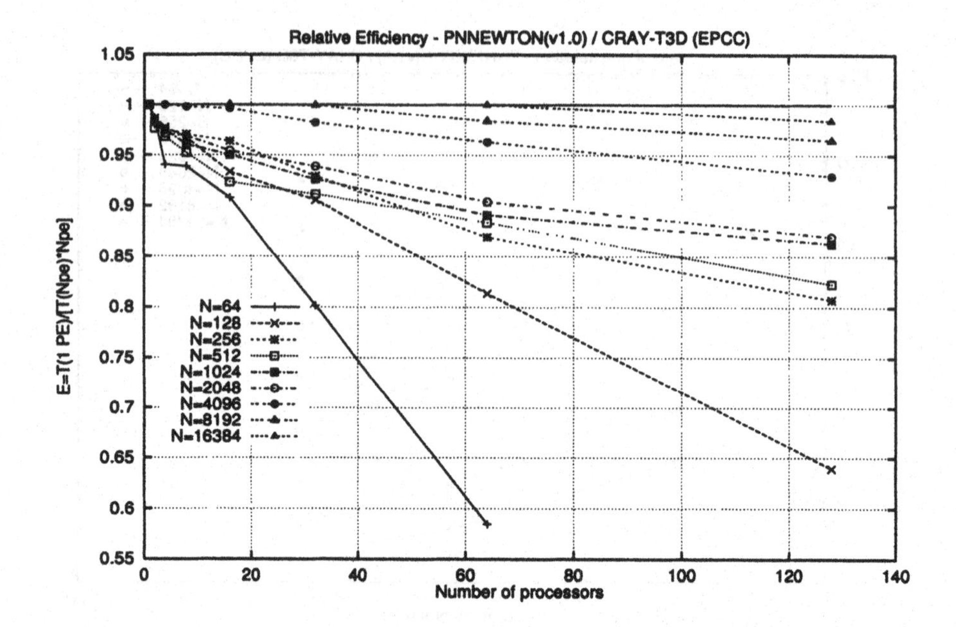

Fig. 4. The program shows high efficiency for most of the configurations tested. The lowest efficiencies correspond to cases where the cost of communications is relevant (the number of particles is the same as the number of processors).

6 Conclusions

The purpose of this work was the development of a parallel code suitable to study N-body systems with $N \sim 10 - 10^4$. The required features of the program were portability, scalability and efficiency. The tests performed on both versions (P*NNEWTON* 1.0 and 2.0) showed an almost linear speedup and a relative efficiency between 57% and 98%. The worst cases ($E_r \approx 60\%$ and $E_r \approx 65\%$) correspond to a system with 64 particles running on 64 processors, and to a system with 128 particles running on 128 processors. With those configurations the communication costs are comparable to the computational costs and the efficiency drops.

Using a message passing model and the MPI library for the parallelisation of the PP algorithm was possible to write a portable code with high efficiency and good scalability. Our parallel algorithm appears to be appropriate to develop parallel versions of the PP method.

References

1. Aarseth, S. J.: Galactic Dynamics and N-Body Simulations, Lecture Notes in Physics, Springer-Verlag, 1993.
2. Binney, J., Tremaine, S.: Galactic Dynamics, Princeton Series in Astrophysics, 1987.
3. Boccaletti, D., Pucacco, G.: Theory of Orbits, 1: Integrable Systems and Non-perturbative Methods. A&A Library, Springer-Verlag, 1996.

4. Booth, S., Fisher, J., MacDonald, N., Maccallum, P., Malard, J., Ewing, A., Minty, E., Simpson, A., Paton, S., Breuer, S.: Introduction to the Cray T3D, Edinburgh Parallel Computer Centre, The University of Edinburgh, 1997.
5. Foster, Ian.: Designing and Building Parallel Programs, Addison-Wesley, 1995.
6. Goodman, J., Heggie D. C., & Hut P.: The Astrophysical Journal, **515**:715-733, 1993.
7. Gropp, W., Lusk E., Skjellum, A.: USING MPI Portable Parallel Programming with the Message-Passing Interface, The MIT Press London, England, 1996.
8. Heggie, D. C.: Chaos in the N-Body Problem of Stellar Dynamics. Predictability, Stability, and Chaos in N-Body Dynamical Systems, Plenum Press, 1991.
9. Hockney, R. W., & Eastwood, J. W.:Computer Simulation Using Particles, Institute of Physics Publishing, Bristol and Philadelphia, 1992.
10. Kandrup, E. H., Smith, H. JR.: The Astrophysical Journal, **347**:255-265, 1991 June 10.
11. Kandrup, E. H., Smith, H. JR.: The Astrophysical Journal, **386**:635-645, 1992 February 20.
12. MacDonald, N., Minty, E., Malard, J., Harding, T., Brown, S., Antonioletti, M.: MPI Programming on the Cray T3D, Edinburgh Parallel Computer Centre, The University of Edinburgh, 1997.
13. Miller, R. H.: The Astrophysical Journal, 140,250, 1964.
14. Velde, Eric F. Van de.: Concurrent Scientific Computing, Springer-Verlag, 1994.
15. MPI: A Message-Passing Interface Standard, Message Passing Interface Forum, June 12, 1995.
16. Pereira, N. S. A.: Master Thesis, University of Lisbon, 1998.
17. Snir, M., Otto, S., Huss-Lederman, S., Walker, D., Dongarra, J.: MPI: The Complete Reference, The MIT Press London, England, 1996.
18. Zeilik, M., Gregory, S. A., Smith, E. v. P.: Introductory Astronomy and Astrophysics, Saunders, 1992.

A Parallelisation Strategy for Power Systems Composite Reliability Evaluation
(Best Student Paper Award: Honourable Mention)

Carmen L.T. Borges and Djalma M. Falcão

Federal University of Rio de Janeiro
C.P. 68504, 21945-970, Rio de Janeiro - RJ Brazil
{carmen,falcao}@coep.ufrj.br

Abstract. This paper presents a methodology for parallelisation of the power systems composite reliability evaluation using Monte Carlo simulation. A coarse grain parallelisation strategy is adopted, where the adequacy analyses of the sampled system states are distributed among the processors. An asynchronous parallel algorithm is described and tested on 5 different electric systems. The paper presents results of almost linear speedup and high efficiency obtained on a 4 node IBM RS/6000 SP distributed memory parallel computer.

1 Introduction

The power generation, transmission and distribution systems constitute a basic element in the economic and social development of the modern societies. For technical and economic reasons, these systems have evolved from a group of small and isolated system s to large and complex interconnected systems with national or, even, continental dimensions. For this reason, failure of certain components of the system can produce disturbances capable of affecting a great number of consumers. On the other hand, due to the sophistication of the electric and electronic equipments used by the consumers, the demand in terms of the power supply reliability has been increasing considerably. More recently, institutional changes in the electric energy sector, such as those origina ted by deregulation policies, privatizations, environmental restrictions, etc., have been forcing the operation of such systems closer to its limits, increasing the need to evaluate the power supply interruption risks and quality degradation in a more pre cisely form.

Probabilistic models have been largely used in the evaluation of the power systems performance. Based on information about components failures, these models allow to establish system performance indexes which can be used to aid decision making relative t o new investments, operative policies and to evaluate transactions in the electric energy market. This type of study receives the generic name of Reliability Evaluation [1] and can be accomplished in the generation, transmission and distribution levels or, still, combining the several levels. The composite system reliability evaluation refers to the reliability evaluation of

J. Palma, J. Dongarra, and V. Hernández (Eds.): VECPAR'98, LNCS 1573, pp. 640–651, 1999.

power systems composed of generation and transmission sub-sys tems, object of this work.

The basic objective of the composite generation and transmission system reliability evaluation is to assess the capacity of the system to satisfy the power demand at its main points of consumption. For this purpose, it is considered the possibility of occ urrence of failures in both generation and transmission system components and the impact of these failures in the power supply is evaluated. There are two possible approaches for reliability evaluation: analytic techniques and stochastic simulation (Monte Carlo simulation [2, 3], for example). In the case of large systems, with complex operating conditions and a high number of severe events, the Monte Carlo simulation is, generally, preferable due to the easiness of modelling complex phenomena.

For the reliability evaluation based on Monte Carlo simulation, it is necessary to analyse a very large number of system operating states. This analysis, in general, includes load flow calculations, static contingencies analysis, generation re-scheduling, load shedding, etc. In several cases, this analysis must be repeated for several different load levels and network topology scenarios.

The reliability evaluation of large power systems may demand hours of computation on high performance workstations. The majority of the computational effort is concentrated in the system states analysis phase. This analysis may be performed independently for each system state. The combination of elevated processing requirements with the concurrent events characteristic suggests the application of parallel processing for the reduction of the total computation time.

This paper describes some results obtained through a methodology under development for power system composite reliability evaluation on parallel computers with distributed memory architecture and communication via message passing. The methodology is being developed using as reference element a sequential program for reliabil ity evaluation used by the Brazilian electric energy utilities [4] .

2 Composite Reliability Evaluation

The composite generation and transmission system reliability evaluation consists of the calculation of several performance indexes, such as the Lost of Load Probability (LOLP), Expected Power Not Supplied (EPNS), Lost of Load Frequency (LOLF), etc., using a stochastic model for the electric system operation. The conceptual algorithm for this evaluation is as follows:

1. *Select an operating scenario $\underline{x}$ corresponding to a load level, components availability, operating conditions, etc.*
2. *Calculate the value of an evaluation function $F(\underline{x})$ which quantifies the effect of violations in the operating limits in this specific scenario. Corrective actions such as generation re-scheduling, load shedding minimisation, etc., can be included in this evaluation.*
3. *Update the expected value of the reliability indexes based on the result obtained in step 2.*

4. If the accuracy of the estimates is acceptable, terminate the process. Otherwise, return to step 1.

Consider a system with n components (transmission lines, transformers, electric loads, etc.). An operating scenario for this system is given by the random vector:

$$\underline{x} = (x_1, x_2, \ldots, x_m) \ , \tag{1}$$

where x_i is the state of the i-th component. The group of all possible operating states, obtained by all the possible combinations of components states, is called state space and represented by X. In Monte Carlo simulation, step 1 of the previous algorithm consists of obtaining a sample of vector $\underline{x} \in X$, by sampling the random variables probability distributions corresponding to the components operating states, using a random number generator algorithm.

In step 2 of the previous algorithm, it is necessary to simulate the operating condition of the system in the respective sampled states, in order to determine if the demand can be satisfied without operation restrictions violation. This simulation demands the solution of a contingency analysis problem [5] and, eventually, of an optimal load flow problem [6] to simulate the generation re-scheduling and the minimum load shedding. In the case of large electric systems, these simulation s require high computational effort in relation to that necessary for the other steps of the algorithm [7].

The reliability indexes calculated at step 3 correspond to estimates of the expectation of different evaluation functions $F(\underline{x})$, obtained for N system state samples by:

$$\bar{E}[F] = \frac{1}{N} \sum_{k=1}^{N} F(\underline{x}_k) \ . \tag{2}$$

The Monte Carlo simulation accuracy may be expressed by the coefficient of variation α, which is a measure of the uncertainty around the estimate, and defined as:

$$\alpha = \frac{\sqrt{V(\bar{E}[F])}}{\bar{E}[F]} \ . \tag{3}$$

The convergence criterion usually adopted is the coefficient of variation of EPNS index, which has the worst convergence rate of all reliability indexes [2].

3 Parallelisation Strategy

The composite reliability evaluation problem can be summarised in three main functions: the system states selection, the adequacy analysis of the selected system states and the reliability indexes calculation. As described before, an adequacy analysis is performed for each state selected by sampling in the Monte

Carlo simulation, i.e., the system capacity to satisfy the demand without violating operation and security limits is verified.

The algorithm is inherently parallel with a high degree of task decoupling [8]. The system states analyses are completely independent from each other and, in a coarse grain parallelisation strategy, it is only necessary to communicate at three diff erent situations:

1. For the initial distribution of data, identical for all processors and executed once during the whole computation;
2. For the final grouping of the partial results calculated in each processor, also executed only once ; and
3. For control of the global parallel convergence, which needs to be executed several times during the simulation process, with frequency that obeys some convergence control criteria.

The basic configuration for parallelisation of this problem is the master-slaves paradigm, in which a process, denominated master, is responsible for acquiring the data, distributing them to the slaves, controlling the global convergence, receiving the pa rtial results from each slave, calculating the reliability indexes and generating reports. The slaves processes are responsible for analysing the system states allocated to them, sending their local convergence data to the master and, at the end of the it erative process, also sending their partial results. It is important to point out that, in architectures where the processors have equivalent processing capacities, the master process should also analyse system states in order to improve the algorithm p erformance. For purposes of this work, each process is allocated to a different processor and is referred to simply as processor from now on.

The main points that had to be solved in the algorithm parallelisation were the system states distribution philosophy and the parallel convergence control criteria, in order to have a good load balancing. These questions are dealt with in the next subsec tions.

3.1 System States Distribution Philosophy

The most important problem to be treated in parallel implementations of Monte Carlo simulation is to avoid the existence of correlation between the sequences of random numbers generated in different processors. If the sequences are correlated in some way, the information produced by different processors will be redundant and they will not contribute to increase the statistical accuracy of the computation, degrading the algorithm performance. In some Monte Carlo applications, correlation introduces interfe rences that produce incorrect results. To initialise the random numbers generator with different seeds for each processor is not a good practice, because it can generate sequences that are correlated to each other [9].

In the system states distribution philosophy adopted, the system states are generated directly at the processors in which they will be analysed. For this purpose, all processors receive the same seed and execute the same random numbers sampling, generating the same system states. Each processor, however, starts to analyse the state with a number equal to its rank in the parallel computation and analyses the next states using as step the number of processors involved in the computation. Supposing that the number of available processors is 4, then processor 1 analyses states numbered 1, 5, 9, ..., processor 2 analyses states numbered 2, 6, 10, ..., and so on.

3.2 Parallel Convergence Control

Three different parallelisation strategies for the composite reliability evaluation problem were tried [10] with variations over the task allocation graph, the load distribution criteria and the convergence control method. The strategy that produ ces best results in terms of speedup and scalability is an asynchronous one that will be described in detail in this section. The task allocation graph for this asynchronous parallel strategy is shown in Fig. 1, where p is the number of sc heduled processors. Each processor has a rank in the parallel computation which varies from 0 to (p-1), 0 referring to the master process. The basic tasks involved in the solution of this problem can be classified in five types: **I** - Init ialization, **A** - States Aanalysis, **C** - Convergence Control, **P** - Iterative Process Termination and **F** - Finalization (calculation of reliability indexes and generation of reports). A subindex i associate d with task T means it is allocated to processor i. A superindex j associated with T_i means the j-th execution of task T by processor i.

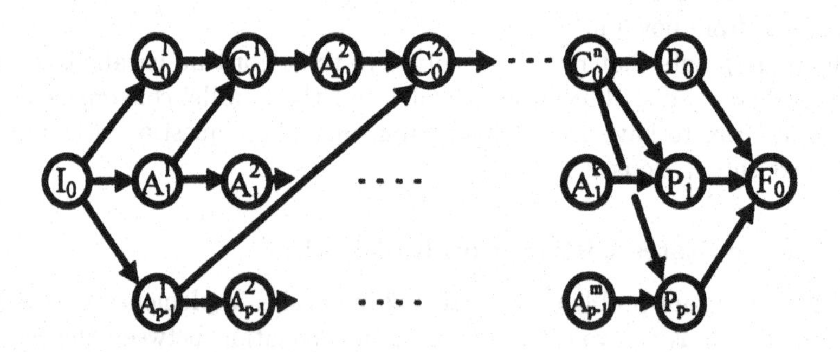

Fig. 1. Task Allocation Graph

The problem initialisation, which consists of the data acquisition and some initial computation, is executed by the master processor, followed by a broadcast of the data to all slaves. All processors, including the master, pass to the phase of states anal ysis, each one analysing different states. After a time interval Δt, each slave sends to the master the data relative to its own local convergence, independently of how many states it has analysed so far, and then continues to analyse other states . At the end of another Δt time interval, a new message is sent to the master, and this process is periodically repeated.

When the master receives a message from a slave, it verifies the status of the global parallel convergence. If it has not been achieved, the master goes back to the system states analysis task until a new message arrives and the parallel convergence need s to be checked again. When the parallel convergence is detected or the maximum number of state samples is reached, the master broadcasts a message telling the slaves to terminate the iterative process, upon what the slaves send back their partial results to the master. The master then calculates the reliability indexes, generate reports and terminate.

In this parallelisation strategy, there is no kind of synchronization during the iterative process and the load balancing is established by the processors capacities and the system states complexity. The precedence graph [11] is shown in Fig. 2, where the horizontal lines are the local time axis of the processors. In this figure only the master and one slave are represented. The horizontal arches represent the successive states by which a processor passes. The vertices represent the messages sending and receiving events. The messages are represented by the arches linking horizontal lines associated with different processors.

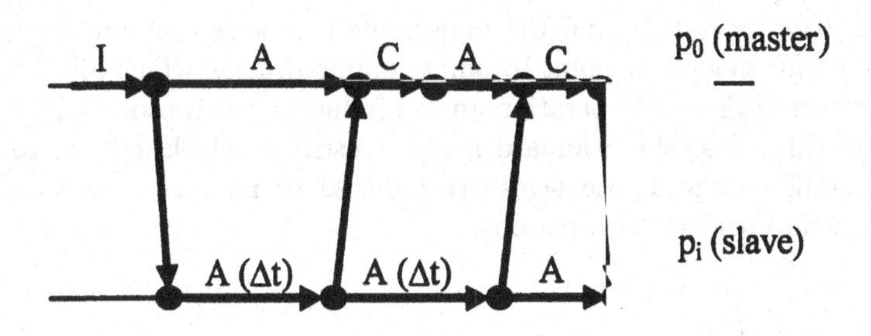

Fig. 2. Precedence Graph

An additional consideration introduced by the parallelisation strategy is the redundant simulation. During the last stage of the iterative process, the slaves execute some analyses that are beyond the minimum necessary to reach convergence. The convergenc e is detected based on the information contained in the last message sent by the slaves and, between the shipping of the last message and the reception of the message informing of the convergence, the slaves keep analysing more states. This, however, does not imply in loss of time for the computation as a whole, since no processor gets idle any time. The redundant simulation is used for the final calculation of the reliability indexes, generating indexes still more accurate than the sequential solution, s ince in Monte Carlo methods the uncertainty of the estimate is inversely proportional to the number of analysed states samples.

4 Implementation

4.1 Message Passing Interface (MPI)

MPI is a standard and portable message passing system designed to operate on a wide variety of parallel computers [12]. The standard defines the syntax and semantics of the group of subroutines that integrate the library. The main goals of MPI are portability and efficiency. Several efficient implementations of MPI already exist for different computer architectures. The MPI implementation used in this work is the one developed by IBM for AIX that complies with MPI standard version 1.1.

Message passing is a programming paradigm broadly used in parallel computers, especially scalable parallel computers with distributed memory and networks of workstations (NOWs). The basic communication mechanism is the transmittal of data between a pair o f processes, one sending and the other receiving. There are two types of communication functions in MPI: blocking and non-blocking. In this work, the non-blocking send and receive functions are used, what allow the possible overlap of message transmittal with computation and tend to improve the performance.

Other important aspects of communication are related to the semantics of the communication primitives and the underlying protocols that implement them. MPI offers four modes for point to point communication which allow to choose the semantics of the send operation and to influence the protocol of data transferring. In this work, the Standard mode is used, in which it is up to MPI to decide whether outgoing messages are buffered or not based on buffer space availability and performance reasons.

4.2 Parallel Computer

The IBM RS/6000 SP Scalable POWERparallel System [13] is a scalable parallel computer with distributed memory. Each node of this parallel machine is a complete workstation with its own CPU, memory, hard disk and net interface. The architecture of the processors may be POWER2/PowerPC 604 or Symmetrical Processor (SMP) PowerPC. The nodes are interconnected by a high performance switch dedicated exclusively for the execution of parallel programs. This switch can establish direct connection between any pair of nodes and one full-duplex connection can exist simultaneously for each node.

The parallel platform where this work was implemented is an IBM RS/6000 SP with 4 POWER2 processors interconnected by a high performance switch of 40 MB/s full-duplex bandwidth. Although the pick performance for floating point operations is 266 MFLOPS for all nodes of this machine, there are differences related to processor type, memory and processor buses, cache and RAM memory that can be more or less significant depending on the characteristics of the program in execution.

5 Results

5.1 Test Systems

Five different electric systems were used to verify the performance and scalability of the parallel implementation. The first one is the IEEE-RTS standard reliability test system for reliability evaluation [14]. The second one, CIGRÉ-NBS is a representation of the New Brunswick Power System proposed by CIGRÉ as a standard for reliability evaluation methodology comparison [15]. The third, fourth and fifth ones are representations of the actual Brazilian power system, with actual electric characteristics and dimensions, for region North-Northeastern (NNE), Southern (SUL) and Southeastern (SE), respectively. The main data for the test systems are shown in Table 1. It was adopted a convergence tolerance of 5% in the EPNS index for all test systems reliability evaluation.

Table 1. Test Systems

System	Nodes	Circuits	Areas
RTS	24	38	2
NBS	89	126	4
NNE	89	170	6
SUL	660	1072	18
SE	1389	2295	48

5.2 Results Analysis

The speedups and efficiencies achieved by the parallel code for the five test systems on 2, 3 and 4 processors, together with the CPU time of the sequential code (1 processor), are summarised in Table 2.

Table 2. Results

System	CPU time	Speedup			Efficiency		
	p=1	p=2	p=3	p=4	p=2	p=3	p=4
RTS	35.03 sec	1.87	2.70	3.52	93.68	90.07	88.07
NBS	24.36 min	1.96	2.94	3.89	98.10	97.90	97.28
NNE	17.14 min	1.99	2.99	3.98	99.75	99.63	99.44
SUL	25.00 min	1.95	2.91	3.85	97.71	97.00	96.28
SE	8.52 hour	1.98	2.93	3.88	99.12	97.70	97.04

As it can be seen, the results obtained are very good, with speedup almost linear and efficiency close to 100 % for the larger systems. The asynchronous parallel implementation provides practically ideal load balancing and negligible synchronization time. The communication time for broadcasting of the initial data and grouping of the partial results is negligible compared to the total tim e of the simulation. The communication time that has significant effect in the parallel performance is the time consumed in exchanging messages for controlling the parallel convergence. The time spent by the processors sampling states that are not analyse d by them, due to the states distribution philosophy adopted, is also negligible compared to the total computation time.

The algorithm also presented a good scalability in terms of number of processors and dimension of the test systems, with almost constant efficiency for different numbers of processors. This good behaviour of the parallel solution is due to the combination of three main aspects:

1. The high degree of parallelism inherent to the problem,
2. The coarse grain parallelisation strategy adopted and
3. The asynchronous implementation developed.

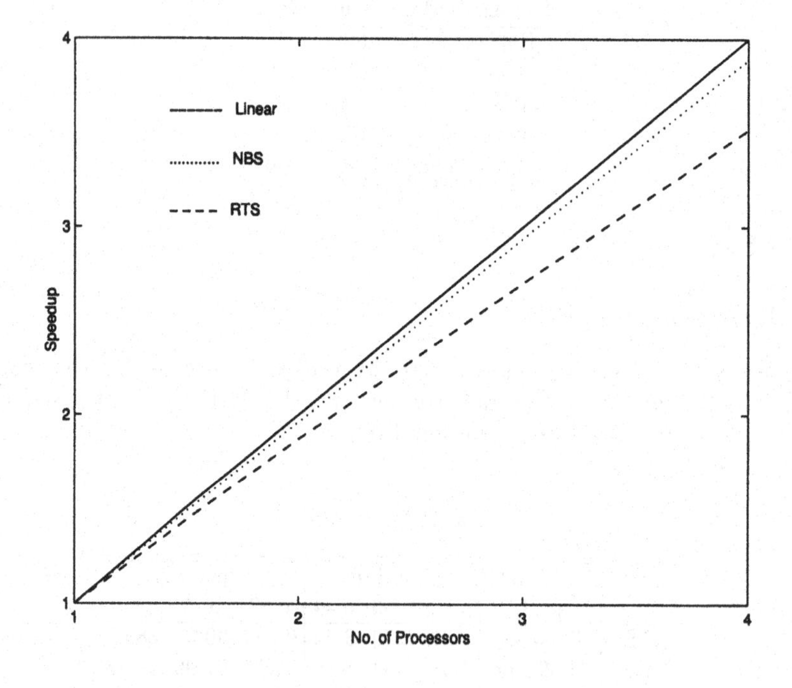

Fig. 3. Speedups - RTS and NBS

Figure 3 shows the speedup curve for the RTS and NBS test systems and Fig. 4 for the three actual Brazilian systems.

A comparison of the main indexes calculated in both sequential and parallel (4 processors) implementations can be done based on Table 3, where the

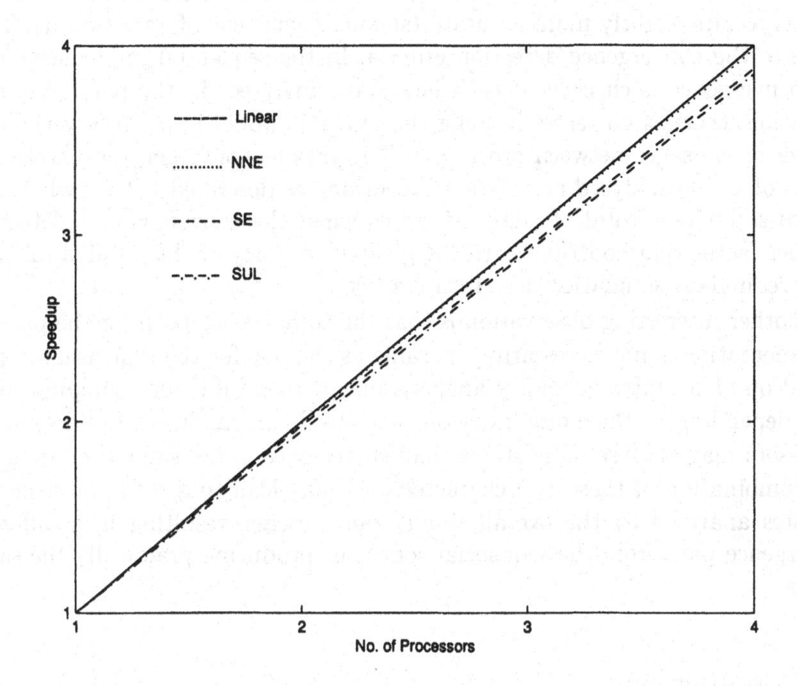

Fig. 4. Speedups - NNE, SUL and SE

Table 3. Reliability Indexes

System	LOLP		EPNS		LOLF		$\alpha(\%)$	
	p=1	p=4	p=1	p=4	p=1	p=4	p=1	p=4
RTS	0.144	0.142	25.16	24.98	26.90	26.28	4.998	4.883
NBS	0.014	0.014	2.10	2.10	10.89	10.88	4.999	4.999
NNE	0.027	0.027	1.99	2.01	23.97	24.23	5.000	4.976
SUL	0.188	0.188	3.82	3.82	69.76	69.61	4.994	4.964
SE	0.037	0.037	1.75	1.75	114.22	114.56	5.000	4.998

Expected Power Not Supplied (EPNS) is given in MW and the Lost of Load Frequency (LOLF) is given in occurrences/year.

As it can be seen, the results obtained are statistically equivalent, with the parallel results slightly more accurate (smaller coefficient of variation α). This is due to the convergence detection criteria. In the sequential implementation, the co nvergence is checked at each new state analysed. In the parallel implementation, there is no sense in doing this, since it would imply in a very large number of messages between processors. The parallel convergence is checked in chunks of states analysed at different processors, as described before, what may lead to a different total of analysed states when the convergence is detected. Another factor that contributes to the greater accuracy of the parallel solution is the redundant simulation described e arlier.

Another interesting observation is that the convergence path for the parallel implementation is not necessarily the same as the one for the sequential implementation. The states adequacy analyses may demand different computational times depending on the c omplexity of the states being analysed. Besides, faster processors may analyse more states than slower ones in the same time interval. The combination of these two characteristics may lead to a different sequence of states analysed by the overall simulat ion process, resulting in a different convergence path from the sequential code, but producing practically the same results.

6 Conclusions

The power systems composite reliability evaluation using Monte Carlo simulation demands high computation effort due to the large number of states that need to be analysed and to the complexity of these states analyses.

Since the adequacy analysis of the system states can be performed independently from each other, the use of parallel processing is a powerful tool for the reduction of the total computation time.

This paper presented a composite reliability evaluation methodology for a distributed memory parallel processing environment. A coarse grain parallelism was adopted, where the processing grain is composed by the adequacy analysis of several states. In t his methodology, the states to be analysed are generated directly at the different processors, using a distribution algorithm that is based on the rank of the processor in the parallel computation.

A parallel implementation was developed where the Monte Carlo simulation convergence is controlled in a totally asynchronous way. This asynchronous implementation has a practically ideal load balancing and worthless synchronization time.

The results obtained in a 4 nodes IBM RS/6000 SP parallel computer for five electric systems are very good, with practically linear speedup, close to the theoretical, and efficiencies also close to 100%. The reliability indexes calculated in both the se quential and parallel implementations are statistically equivalent.

A point being explored in continuation to this work is the migration of the methodology developed from a distributed memory parallel computer to a network of workstations. To ally the computation time reduction achieved by this parallel methodology with the use of networks of workstations already available at the electric energy companies is of great interest from the economic point of view.

References

1. R. Billinton and R.N. Allan, "*Reliability Assessment of Large Electric Power Systems*", Kluwer, Boston, 1988.
2. R. Billinton and W. Li, "*Reliability Assessment of Electric Power Systems Using Monte Carlo Methods*", Plenum Press, New York, 1994.
3. M.V.F. Pereira and N.J. Balu, "Composite Generation/Transmission System Reliability Evaluation", *Proceedings of IEEE*, vol. 80, no. 4, pp. 470-491, April 1992.
4. "*NH2 Computational System for Generation and Transmission Systems Reliability Evaluation - Methodology Manual*", (in Portuguese), CEPEL Report no. DPP/POL-137/93, 1993.
5. B. Sttot, O. Alsac, A.J. Monticelli, "Security Analysis and Optimization", *Proceedings of IEEE*, Vol. 75, no. 12, pp. 1623-1644, Dec 1987.
6. H.W. Dommel, W.F. Tinney, "Optimum Power Flow Solutions", *IEEE Transactions on Power Apparatus and Systems*, pp. 1866-1876, Oct 1968.
7. A.C.G. Melo, J.C.O. Mello, S.P. Roméro, G.C. Oliveira, R.N. Fontoura, "Probabilistic Evaluation of the Brazilian System Performance", *Proceedings of the IV Simpósio de Especialistas em Planejamento da Operação e Expans ão Elétrica*, (in Portuguese), 1994.
8. N. Gubbala, C. Singh, "Models and Considerations for Parallel Implementation of Monte Carlo Simulation Methods for Power System Reliability Evaluation", *IEEE Transactions on Power Systems*, Vol. 10, no. 2, pp. 779-787, May 1995.
9. G. Fox, M. Johnson, G. Lyzenga, S. Otto, J. Salmon, D. Walker, "*Solving Problems on Concurrent Processors*", Vol.1, Prentice Hall, New Jersey, 1988.
10. C.L.T. Borges, D.M. Falcão, J.C.O. Mello, A.C.G. Melo, "Reliability Evaluation of Electric Energy Systems composed of Generation and Transmission in Parallel Computers", *Proceedings of the IX Simpósio Brasileiro de Arquitet ura de Computadores / Processamento de Alto Desempenho*, (in Portuguese), pp. 211-223, Oct 1997.
11. V.C. Barbosa, "*An Introduction to Distributed Algorithms*", The MIT Press, Cambridge, Massachusetts, 1996.
12. M. Snir, S. Otto, S. Huss-Lederman, D. Walker, J. Dongarra, "*MPI: The Complete Reference*", The MIT Press, Cambrige, Massachusetts, 1996.
13. "*IBM POWERparallel Technology Briefing: Interconnection Technologies for High-Performance Computing (RS/6000 SP)*", RS/6000 Division, IBM Corporation, 1997.
14. IEEE APM Subcommittee, "IEEE Reliability Test System", *IEEE Transaction on Power Apparatus and Systems*, Vol. PAS-98, pp. 2047-2054, Nov/Dec 1979.
15. CIGRÉ Task Force 38-03-10, "*Power System Reliability Analysis - Volume 2 - Composite Power Reliability Evaluation*", 1992.

High Performance Computing of an Industrial Problem in Tribology
(Best Student Paper Award: First Prize)

M. Arenaz[1], R. Doallo[1], G. García[2], and C. Vázquez[3]

[1] Department of Electronics and Systems, University A Coruña, 15071 A Coruña, Spain
arenaz@des.fi.udc.es, doallo@udc.es

[2] Department of Applied Mathematics, University of Vigo, 36280 Vigo, Spain
guille@dma.uvigo.es

[3] Department of Mathematics, University A Coruña, 15071 A Coruña, Spain
carlosv@udc.es

Abstract. In this work we present the vectorisation of a new complex numerical algorithm to simulate the lubricant behaviour in an industrial device issued from tribology. This real technological problem leads to the mathematical model of the thin film displacement of a fluid between a rigid plane and an elastic and loaded sphere. The mathematical study and a numerical algorithm to solve the model has been proposed in the previous work [9]. This numerical algorithm mainly combines fixed point techniques, finite elements and duality methods. Nevertheless, in order to obtain a more accurate approach of different real magnitudes, it is interesting to be able to handle finer meshes. As it is well-known in finite element methods, mesh refinement carries out a great increase in the storage cost and the execution time of the algorithm. It is precisely this computational cost problem what has motivated the authors to try to increase the performance of the numerical algorithm by using high performance computing techniques. In this work we mainly apply vectorisation techniques but also we present some preliminary partial results from the design of a parallel version of the algorithm.

1 The Industrial Problem in Tribology

In a wide range of lubricated industrial devices studied in Tribology the main task is the determination of the fluid pressure distribution and the gap between the elastic surfaces which correspond to a given imposed load [4]. Most of these devic es can be represented by a ball–plane geometry (see Fig. 1). So, in order to perform a realistic numerical simulation of the device, an appropriate mathematical model must be considered. From the mathematical point of view, the lu bricant pressure is governed by the well–known Reynolds equation [4]. In the case of elastic surfaces, the computation of the lubricant pressure is coupled with

J. Palma, J. Dongarra, and V. Hernández (Eds.): VECPAR'98, LNCS 1573, pp. 652–665, 1999.

the determination of the gap. Thus, in Reynolds equation the gap depends on the pres sure. In the particular ball–bearing geometry, the local contact aspect allows us to introduce the Hertz contact theory to express this gap–pressure dependence. The inclusion of cavitation (the presence of air bubbles) and piezo-viscosity (pressure–vis cosity dependence) phenomena is modelled by a more complex set of equations, see [5]. Moreover, the balance between imposed and hydrodynamic loads is formulated as a nonlocal constraint on the fluid pressure, see [9] for the det ails on the mathematical formulation.

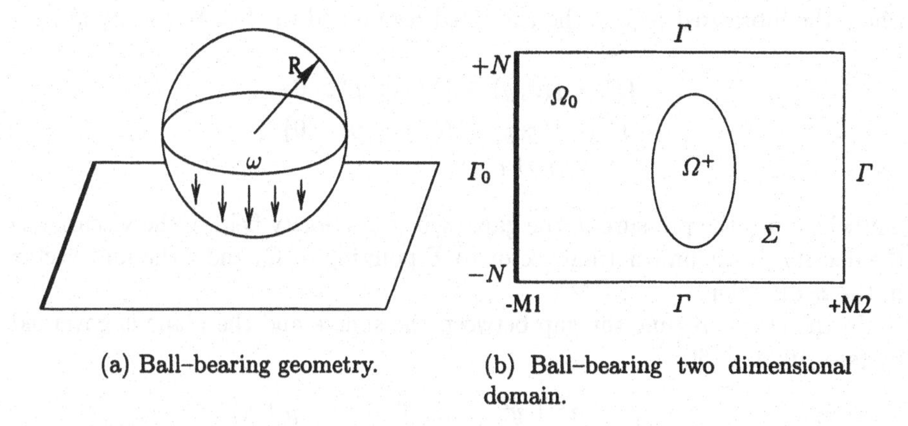

(a) Ball–bearing geometry.

(b) Ball–bearing two dimensional domain.

Fig. 1. Ball–bearing device

In Sections 2 and 3 we briefly describe the model problem and the numerical algorithm presented in [9], respectively. In Section 4 we make reference to the vect orisation and parallelisation techniques that we have applied in order to improve the performance of the algorithm. In Section 5 we present the numerical results and the execution times corresponding to several different meshes. F inally, in Section 6 we present the conclusions we have come to as a result of our work.

2 The Model Problem

As a preliminary step to describing the numerical algorithm, we introduce the notations and the equations of the mathematical model.

Thus, let Ω be given by $\Omega = (-M_1, M_2) \times (-N, N)$, with M_1, M_2, N positive constants, which represents a small neighbourhood of the *contact point*. Let $\partial\Omega$ be divided in two parts: the supply boundary $\Gamma_0 = \{(x, y) \in \partial\Omega \,/\, x = -M_1\}$ and the boundary at atmospheric pressure $\Gamma = \partial\Omega \setminus \Gamma_0$.

In order to consider the cavitation phenomenon, we introduce a new unknown θ which represents the saturation of fluid ($\theta = 1$ in the fluid region where $p > 0$ and $0 \leq \theta < 1$ in the cavitation region where $p = 0$). The mathematical formulation of the problem consists of the set of nonlinear partial differential equations (see [2] and [6] for details) verified by (p, θ):

$$\frac{\partial}{\partial x}\left(\frac{\rho}{\nu} h^3 \frac{\partial p}{\partial x}\right) + \frac{\partial}{\partial y}\left(\frac{\rho}{\nu} h^3 \frac{\partial p}{\partial y}\right) = 12s \frac{\partial}{\partial x}(\rho h)\ , \ p > 0\ , \ \theta = 1 \ \text{ in } \Omega^+ \quad (1)$$

$$\frac{\partial}{\partial x}(\rho \theta h) = 0\ , \ p = 0\ , \ 0 \leq \theta \leq 1 \ \text{ in } \Omega_0 \quad (2)$$

$$\frac{h^3}{\nu}\frac{\partial p}{\partial n} = 12s(1-\theta)\, h \cos(n, \imath)\ , \ p = 0 \ \text{ in } \Sigma \quad (3)$$

where the lubricated region, the cavitated region and the free boundary are

$$\Omega^+ = \{(x, y) \in \Omega \,/\, p(x, y) > 0\}$$
$$\Omega_0 = \{(x, y) \in \Omega \,/\, p(x, y) = 0\}$$
$$\Sigma = \partial\Omega^+ \cap \Omega$$

and where p is the pressure, h the gap, $(s, 0)$ the velocity field, ν the viscosity, ρ the density, n the unit normal vector to Σ pointing to Ω_0 and $\imath$ the unit vector in the x–direction.

In the elastic regime the gap between the sphere and the plane is governed by (see [10] and [6]):

$$h = h(x, y, p) = h_0 + \frac{x^2 + y^2}{2R} + \frac{2}{\pi E}\int_\Omega \frac{p(t, u)}{\sqrt{(x-t)^2 + (y-u)^2}}\, dtdu \quad (4)$$

where h_0 is the minimum reference gap, E is the Young equivalent modulus and R is the sphere radius. The Equation (4) arises from Hertzian contact theory for local contacts. The relation between pressure and viscosity is:

$$\nu(p) = \nu_0\, e^{\alpha p} \quad (5)$$

where α and ν_0 denote the piezo-viscosity constant and the zero pressure viscosity, respectively. Moreover, the boundary conditions are:

$$\theta = \theta_0 \text{ in } \Gamma_0 \quad (6)$$

$$p = 0 \text{ in } \Gamma\,. \quad (7)$$

The above conditions correspond to a drip feed device where the lubricant is supplied from the boundary Γ_0. Finally, the hydrodynamic load generated by fluid pressure must balance the load ω, imposed on the device, in a normal direction t o the plane (see Fig. 1(a)):

$$\omega = \int_\Omega p(x, y)\, dxdy \quad (8)$$

In this model, the parameter h_0 appearing in Equation (4) is an unknown of the problem related to this condition. The numerical solution of the problem consists of the approximation of (p, θ, h, h_0) verifying (1)-(8) .

3 The Numerical Algorithm

In order to perform the numerical solution of (1)-(8), with real industrial data, we initially proceed to its scaling in terms of the load, radius and material data. This process leads to a dimensionless fixed domain $\Omega = (-4,2) \times (-2,2)$ by introducing the Hertzian contact radius and the maximum Hertzian pressure,

$$b = \left(\frac{3\omega R}{2E}\right)^{\frac{1}{3}} \ , \quad P_h = \frac{3\omega}{2\pi b^2} \ ,$$

respectively, which only depends on the imposed load ω, the Young modulus E and the sphere radius R. Thus, the new dimensionless variables to be considered are:

$$\tilde{p} = \frac{p}{P_h} \ , \quad \tilde{h} = \frac{hR}{b^2} \ , \quad X = \frac{x}{b} \ , \quad Y = \frac{y}{b} \ ,$$

$$\overline{\alpha} = \alpha P_h \ , \quad \overline{\nu} = \frac{\nu}{\nu_0} \ , \quad \lambda = \frac{8\pi\nu_0 s R^2}{\omega b} \ , \quad \overline{\omega} = \frac{\omega}{P_h \, b^2} = \frac{2\pi}{3} \ .$$

After this scaling, the substitution in (1)-(8) leads to equations of the same type in the dimensionless variables. In [9] a new numerical algorithm is proposed to solve this new set of equations. In this paper, high perfo rmance computing techniques are applied to this algorithm, which is briefly described hereafter. In order to clarify the structure of the code, Fig. 2 shows a flowchart diagram of the numerical algorithm.

The first idea is to compute the hydrodynamic load (Eqn. (8)) for different gap parameters $\tilde{h}_0$ in order to state a monotonicity between $\tilde{h}_0$ and this hydrodynamic load. Then, the numerical solution of the problem for each $\tilde{h}_0$ is decomposed in the numerical solution of:

1. The hydrodynamic problem which is concerned with the computation of the pressure and the saturation of the fluid for a given gap. A previous reduction to an iso-viscous case is performed in order to eliminate the nonlinearity introduced by (5). In this step, the characteristics algorithm, finite elements and duality methods are the main tools. This problem is solved in the two innermost loops shown in Fig. 2, namely, characteristics and multipliers loops.
2. The elastic problem which computes the gap for a given pressure by means of numerical quadrature formulae to approximate the expression (4). This computation can be expressed by means of the loop shown below:

   LOOP in n

$$h(p^{n+1}) = h_0 + \frac{x^2+y^2}{\pi E} \sum_{k \in \tau_h} \int_k \frac{p^{n+1}(t,u)}{\sqrt{(x-t)^2+(y-u)^2}} dt du \qquad (9)$$

   which corresponds with the gap loop depicted in Fig. 2. In the above formula the updating of the gap requires the sum of the integrals over each triangle k of the finite element mesh τ_h.

The complex expression of the gap at each mesh point motivates the high computational cost at this step of the algorithm to obtain $h(p^{n+1})$.

An outer loop (load loop in Fig. 2) for the gap and pressure computations for each value of h_0 determines the value of $\tilde{h}_0$ which balances the imposed load (Eqn. (8)). This design of the algorithm is based on monotonicity arguments and *regula falsi* convergence. In a first term, an interval $[h_a, h_b]$ which contains the final solution $\tilde{h}_0$ is obtained. Then, this value is computed by using a *regula falsi* method.

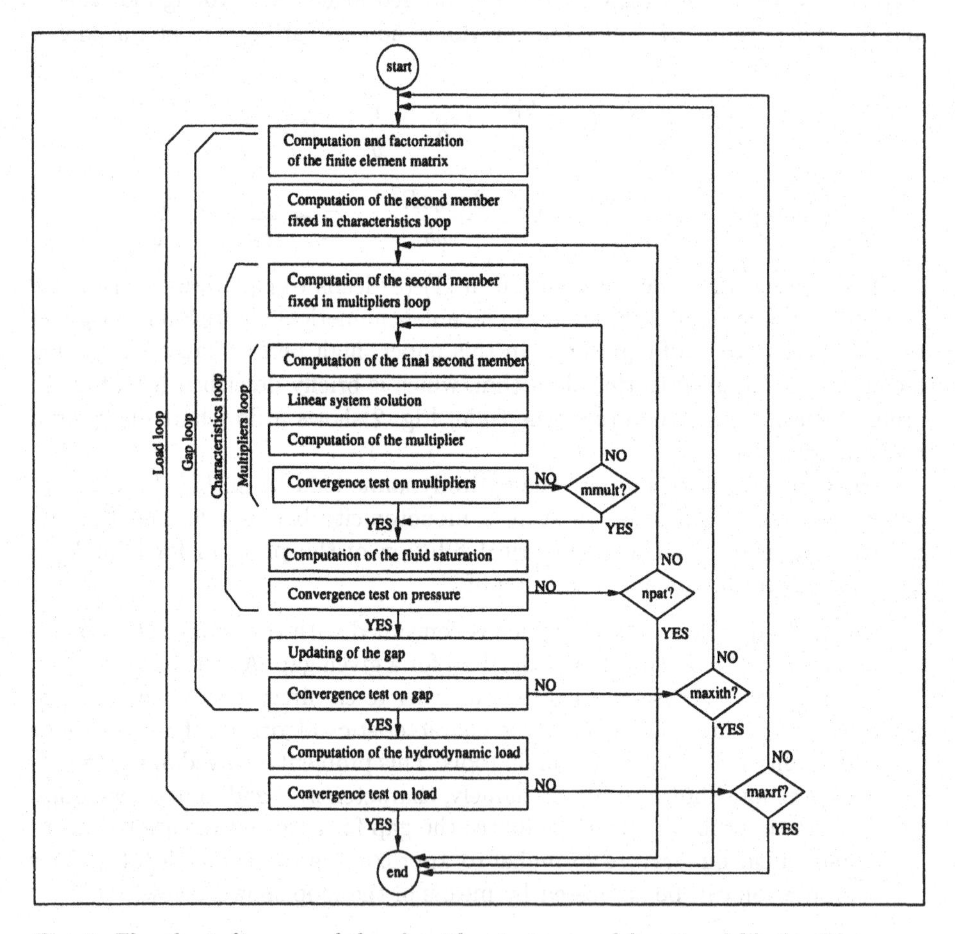

Fig. 2. Flowchart diagram of the algorithm in terms of functional blocks. The parameters *mmult*, *npat*, *maxith* and *maxrf* represent the maximum number of iterations for multipliers, characteristics, gap and load loops, respectively

A numerical flux conservation test shows that better results correspond to finer grids. Nevertheless, this mesh refinement greatly increases the computing times and motivates the interest of using high performance computing.

4 Improving Algorithm Performance

The main goal of our work is to increase the performance of the new numerical algorithm [9] briefly explained above. In order to carry out our work, we have considered two different approaches: firstly, the use of vectorisation techniques and, se condly, the use of parallelisation techniques.

The first part of our work consisted of modifying the original source code of the programme in order to get the maximum possible performance from a vector computer architecture. The target machine was the Fujitsu VP2400/10 available at the CESGA laborator ies [1].

The second step of our work consisted of developing a parallel version of the original sequential source code, using the PVM message-passing libraries. In order to check the performance of the parallelisation process, we have executed our parallel program me over a cluster of workstations based on a SPARC microprocessor architecture at 85 MHz and Solaris operating system interconnected via an Ethernet local area network.

4.1 Vectorisation Techniques

The first step that must be done when trying to vectorise a sequential programme is to analyse its source code and identify the most costly parts of the algorithm in terms of execution time. We will focus most of our efforts on these parts. We used the em ANALYZER tool [7] in order to carry out this task.

The analysis of the programme's source code lead us to identify nine functional blocks corresponding to their relationship with the mathematical methods used in the numerical algorithm, namely, finite elements, characteristics and duality methods. These b locks, shown in Fig. 2, are the following: finite element matrix, initial second member of the system, updating of the second member due to characteristics, final second member, computation of the multiplier, convergence test on multi pliers, stopping test in pressure, updating of the gap and stopping test in the gap.

We have done a study of the distribution of the total execution time of the programme, and we have come to the conclusion that 90% of that time is taken up by the blocks included in the multipliers loop (final second member and computation of the multi plier) and by the block that updates the gap. For this reason, in this paper we explain the vectorisation process corresponding to the three blocks that we have just mentioned. More detailed information about the whole vectorisation process can be found in [1].

Final Second Member. The first block inside the multipliers loop computes the final value of the second member of the linear equations system that is solved in the current iteration of the loop.

[1] CESGA (Centro de Supercomputación de Galicia): Supercomputing Center of Galicia placed at Santiago de Compostela (Spain).

This block is divided in two parts, the first one is only partially vectorisable by the compiler because of the presence of a recursive reference. Table 1 shows the execution times obtained for this part before and after the vectorisatio n process according to the format $h : m : s.\mu s$, where h, m, s and μs represent the number of hours, minutes, seconds and microseconds, respectively.

Table 1. Execution times corresponding to the routine *modsm*().

	T_Orig	*T_AV*	*iprv_AV*	*T_MV*	*iprv_MV*
*mesh*1	0.001214	0.000526	56.67%	0.000325	73.23%
*mesh*3	0.015122	0.005188	65.69%	0.004595	69.61%
*mesh*4	0.060288	0.020446	66.09%	0.018160	69.88%
*mesh*5	0.120261	0.040887	66.00%	0.036354	69.77%

The first column points out which mesh has been used for the discretisation of the domain of the problem, where *mesh*1, *mesh*3, *mesh*4 and *mesh*5 consist of 768, 9600, 38400 and 76800 finite elements, respectively. The column labelled as *T_Orig* shows the time measured when executing the programme after a scalar compilation. The columns labelled as *T_AV* and *T_MV* show, respectively, the times after setting the automatic vectorisation compiler option and after applying this last op tion over the modified source code we have implemented in order to optimise the vectorisation process. The columns *iprv_AV* and *iprv_MV* reveal the improvement with regard to the original scalar execution time. The percentages shown have been calculat ed by means of the following expressions:

$$iprv\_AV = 100 * (1 - (T\_AV/T\_Orig)) \quad (10)$$

$$iprv\_MV = 100 * (1 - (T\_MV/T\_Orig)) \quad (11)$$

The second part of this block has been vectorised by making use, on the one hand, of the loop coalescing technique [11] and, on the other, of the *in-lining* transformation technique. The last one has been applied manually so as to be abl e to optimise the source code of the particular case we are interested in. This resource introduces a generality loss in the vector version of the programme with regard to its original source code. We consider this fact acceptable because our goal was to reduce the execution time as much as possible.

Table 2 shows the execution times corresponding to this fragment of code. It is important to highlight the high improvement we have obtained with respect to the simple automatic vectorisation.

Computation of the Multiplier. The third block inside the multipliers loop mainly copes with the updating of the multiplier introduced by the duality type method proposed in [3]. It also makes the necessary computations so as to implement the convergence test on multipliers.

In the first case, we are faced with the fact that there was a loop that was not automatically vectorisable by the compiler because of the presence of, on the

Table 2. Execution times corresponding to the routine *bglfsm*()

	T_Orig	*T_AV*	*iprv_AV*	*T_MV*	*iprv_MV*
*mesh*1	0.001231	0.001255	-1.95%	0.000508	58.73%
*mesh*3	0.015383	0.015339	0.29%	0.005918	61.53%
*mesh*4	0.059497	0.057986	2.54%	0.023546	60.42%
*mesh*5	0.118632	0.115769	2.41%	0.047059	60.33%

one hand, a call to an external function and, on the other, a recursive reference on the vector whose value is computed, namely, *alfa*.

The first of these problems was solved by applying manually the *in-lining* technique. Again the resultant source code was optimised for the particular case we are interested in, with corresponding loss of generality.

The recursive reference present in the source code is not vectorisable because when computing the i-th element of the vector *alfa*, the $(i-k)$-th element is referenced, with k a positive constant with a different value for each mesh. The use of the compiler directive **VOCL LOOP,NOVREC(alfa)* (see [8]), which explicitly tells the compiler to vectorise the recursive reference, has allowed us to overcome the problem. We have checked that the numerical results remain correct , improving the percentage of vectorisation of our code.

Table 3 shows the execution times, expressed in microseconds, corresponding to the process described in the previous paragraphs.

Table 3. Execution times corresponding to the updating of the multiplier

	T_Orig	*T_AV*	*iprv_AV*	*T_MV*	*iprv_MV*
*mesh*1	0.001115	0.001113	0.18%	0.000043	96.14%
*mesh*3	0.012785	0.012903	-0.92%	0.000162	98.73%
*mesh*4	0.050319	0.049284	2.06%	0.000489	99.03%
*mesh*5	0.100197	0.098290	1.90%	0.000930	99.07%

As can be seen, improvement percentages are close to 100% (moreover, see the great improvement obtained with respect to the one obtained enabling automatic vectorisation without providing any extra information to the compiler). If we take into account that this functional block is placed in the body of the innermost loop of the algorithm (multipliers loop), which is the section of the programme that is executed the greatest number of times, we can conclude that the reduction of the execution time of th e whole programme will be very important.

Updating of the Gap. The third and last block whose vectorisation process is presented in this paper copes with the computation of the gap between the two surfaces in contact in the ball-bearing geometry of Fig. 1(a). The source code corresponding to this bloc k was fully vectorisable by the compiler. Nevertheless, we have carried out some changes by using source code optimisation techniques and the technique called use of scalar variables in reduction operations.

Table 4 shows execution times corresponding to the updating of the gap.

Table 4. Execution times corresponding to the updating of the gap

	T_Orig	*T_AV*	*iprv_AV*	*T_MV*	*iprv_MV*
*mesh*1	0:00:01.351421	0:00:00.060134	95.55%	0:00:00.052062	96.15%
*mesh*3	0:03:28.791518	0:00:08.546158	95.91%	0:00:07.511769	96.20%
*mesh*4	0:51:53.848528	0:02:17.901371	95.57%	0:02:01.627059	96.09%
*mesh*5	3:26:54.230843	0:09:09.402314	95.57%	0:08:05.177401	96.09%

As can be seen, execution time improvement percentage is slightly higher than 96% independently of the size of the mesh.

Although this part of the algorithm is executed a relatively low number of times, it is by far the most costly part of the programme (approximately, 14% of the total scalar execution time). This is due to the fact that for each node of the mesh (the nu mber of nodes is 19521 in *mesh*4, for example) it is necessary to compute an integral over the whole domain of the problem. That integral is decomposed in a sum of integrals over each finite element (the number of elements is 38400 in *mesh*4) which are solved with numerical integration.

In Section 5 we present the times corresponding to the execution of the whole programme.

4.2 Parallelisation Techniques

The parallel algorithm version is at this moment being refined. In this paper we will focus our attention on the block corresponding to the updating of the gap between the two surfaces in lubricated contact (see corresponding part of Subsection 4.1).

For each node of the mesh, the updating of the gap is done according to (9). If we analyse the source code corresponding to this computation, we can see that we are faced with two nested loops where the outermost makes one iteration for ea ch node of the mesh and the innermost makes one iteration for each finite element.

Each iteration of the outer loop calculates the gap corresponding to one node of the mesh. This computation is cross–iteration independent. This characteristic makes the efficiency of the process of parallelisation independent of the data distribution am ong the workstations, which lead us to choose one simple data distribution such as the standard block distribution.

Table 5 shows the execution times obtained for this functional block, after the parallelisation process. Fig. 3 represents, for each mesh, the *speedups* corresponding to the execution times shown in Table 5.

Table 5. Execution times corresponding to the updating of the gap

N	*mesh*1	*mesh*3	*mesh*4	*mesh*5
1	3.961	9:29.688	2:29:35.406	9:52:36.750
2	1.872	4:21.219	1:08:52.500	
4	0.953	2:19.834	0:34:35.219	
8	0.598	2:27.909	0:17:24.781	1:08:56.719
12	0.542	1:18.613	0:11:40.469	0:46:11.344
16	0.577	1:00.074	0:08:54.438	0:34:57.156

The *speedup* is usually defined according to the following formula:

$$speedup = T_1/T_N \tag{12}$$

where T_N represents the time when using N processors and T_1 the time obtained when executing the parallel programme over only one processor. In almost all cases a *superspeedup* is obtained, that is to say, the value of the *speedup* i s greater than the number of processors used. This effect may be due to the fact that we have distributed the vector of the gaps corresponding to each node of the mesh among the set of workstations. In consequence, we have reduced the size of that vector, which has decreased the number of cache misses.

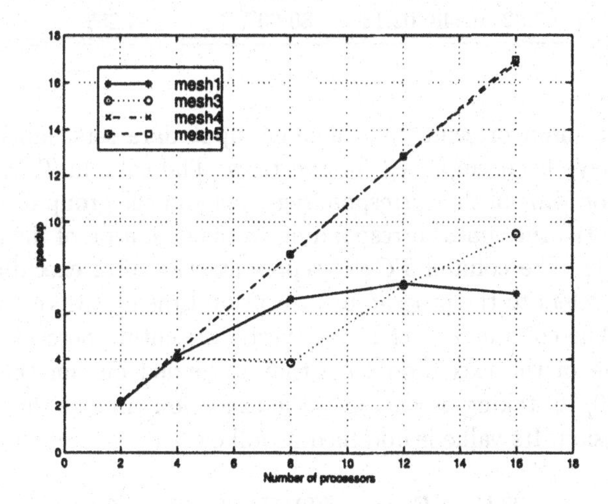

Fig. 3. Speedups corresponding to data shown in Table 5.

Note also that the performance of the programme decreases when using small meshes (*mesh*1 and *mesh*3) with a high number of workstations. This effect is due to the communications overhead.

5 Numerical Results

The results obtained from the vectorised code were validated by comparing them with those calculated by execution of the scalar code.

In Table 6 different execution time measures illustrate the good performance achieved after applying code and compiler optimisation techniques with respect to the one obtained by the vectoriser without any extra code information. The table also shows the scalar execution time. In these cases the coarser grid, namely *mesh*1 (768 triangular finite elements), is used for the spatial discretisation.

Table 6. Execution times for *mesh*1 on the VP2400/10

	T_Orig	T_AV			
		TVU	TCPU	% Vect.Exec.	% Vectorization
b	11:24	01:45	07:03	24.82%	53.51%
l	05:03	00:42	02:47	25.15%	58.75%
g	15:14	02:38	10:53	24.20%	45.84%
j	07:32	01:14	05:03	24.42%	49.34%

	T_Orig	T_MV			
		TVU	TCPU	% Vect.Exec.	% Vectorisation
b	11:24	02:31	03:08	80.32%	94.59%
l	05:03	01:00	01:16	78.95%	94.72%
g	15:14	03:50	04:47	80.14%	94.76%
j	07:32	01:48	02:14	80.60%	94.25%

The first column presents the different input data sets, labelled as b, l, g and j, that have been employed for the tests. The column T_Orig shows the scalar execution time of the corresponding tests and the group of three columns labelled as T_AV, the times corresponding to the execution of the automatically vectorised code. The column $TCPU$ represents the total execution time. The column TVU refers to the programme execution time on the vector units of the VP2400/10. The column $\%Vect.Exec.$ (vector execution percentage) measures the percentage of the execution time that is carried out on the vector units of the VP2400/10, taking as reference the execution time of the automatically vectorised version. Its value is obtained as foll ows:

$$\%Vect.Exec. = (TVU/TCPU) * 100 \tag{13}$$

Finally, the column $\%Vectorization$ provides an idea of the characteristics of the code so as to be executed on a vector machine. Its value is computed according to the formula

$$\%Vectorization = (T\_Orig\_Vect/T\_Orig) * 100 \tag{14}$$

where T_Orig_Vect is the execution time in scalar mode corresponding to the vectorisable part of the original source code, that is to say,

$$T\_Orig\_Vect = T\_Orig - (TCPU - TVU) \tag{15}$$

The group of columns labelled as T_MV show the times corresponding to the execution of the optimised version of the code that we have implemented.

Note that we have been able to increase the $\%Vectorization$ from 50% up to 94%, approximately. This represents a significant decrease of the computational cost of the algorithm. The execution times of the vectorised version for *mesh*4 and *mesh*5 are presented in Table 7.

Table 7. Execution times for *mesh*4 and *mesh*5 on the VP2400/10

	T_MV					
	*mesh*4			*mesh*5		
	TVU	TCPU	% Vect.Exec.	TVU	TCPU	% Vect.Exec.
b	08:44:36	11:11:55	72.82%			
l	17:14:28	22:08:59	77.28%	60:38:56	77:36:37	77.93%
g	20:25:24	26:17:54	76.94%			
j	33:45:50	43:38:38	76.75%	68:54:51	88:37:47	77.28%

The aim of the work was not only to reduce computing time but also to test in practice the convergence of the finite element space discretization. So, the finer grids *mesh*4 (38400 triangular finite elements) and *mesh*5 (76800 triangular finite elem ents) were considered. For both new meshes, the analysis of huge partial computing times prevented us from executing the scalar code. In Figs. 4 and 5 the pressure and gap approximation profiles are presented for *mesh*4 and *mesh*5, respectively, in an appropriate scale. In both figures, the relevant real parameter is the imposed load which is taken to be $\omega = 3$.

Graphics included in Figs. 4 and 5 represent the low boundary of the domain of the problem (see Fig. 1) on the x axis. There are two sets of curves. The mountain–like one represents the pressure o f the lubricant in the contact zone. The parabola–like one represents the gap between the two surfaces in contact.

Next, in order to verify the expected qualitative behaviour already observed in previous tests with *mesh*3 when increasing the charge, we considered *mesh*4 for the datum $\omega = 5$. In Fig. 6 pressure and gap profiles were computed for $\omega = 5$.

6 Conclusions

In this paper we have mainly concentrated on increasing the performance of a new numerical algorithm that simulates the behaviour of a lubricated industrial device studied in Tribology.

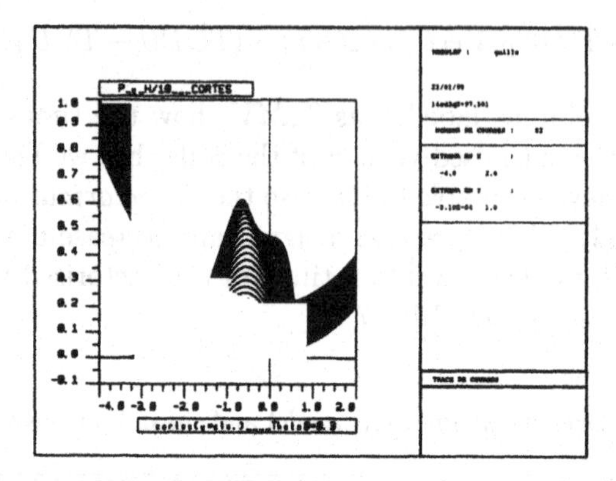

Fig. 4. Pressure and gap profiles for $\omega = 3$ and $\theta_0 = 0.3$ with *mesh4*.

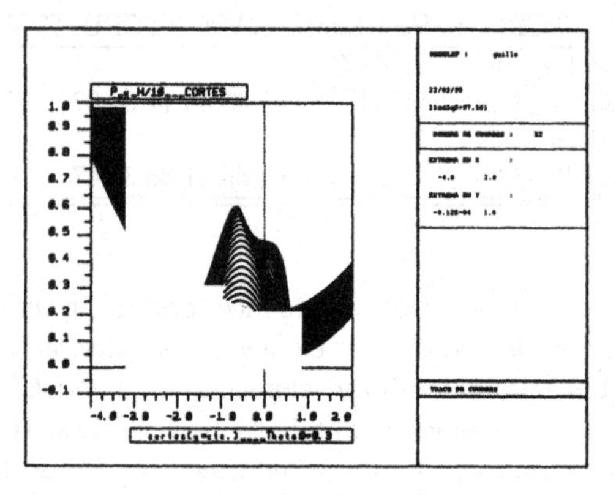

Fig. 5. Pressure and gap profiles for $\omega = 3$ and $\theta_0 = 0.3$ with *mesh5*.

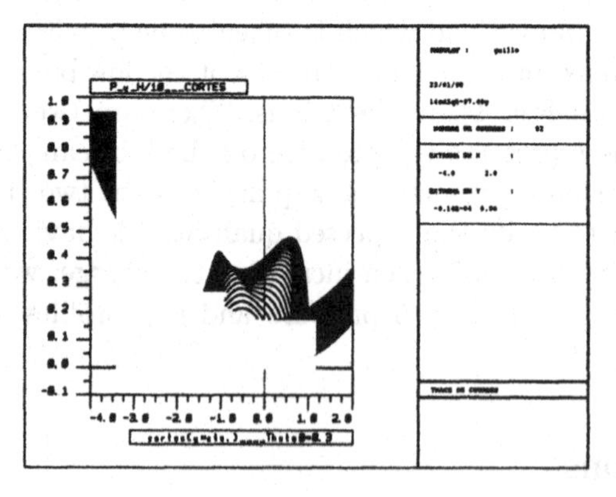

Fig. 6. Pressure and gap profiles for $\omega = 5$ and $\theta_0 = 0.3$ with *mesh4*.

Firstly, we have developed a vector version where about 94% of the original source code has been vectorised. As a drawback, our vector version has partially lost the generality of the original programme and has been implemented in a manner that is depe ndent on the architecture of the machine.

From the numerical viewpoint, the efficiency of the vector version illustrates the theoretical convergence of the algorithm in practice. Moreover, the accuracy of the approximation has been increased with the introduction of finer grids ($mesh4$ and $mesh5$) that confirm the expected results for the numerical method implemented by the algorithm.

On the other hand, a first approach to a parallel version of the algorithm is being developed. As an example, we have presented a parallel implementation of one functional block of the algorithm, notable for the *superspeedup* obtained in some cases.

We are currently refining the parallel version so as to reduce the execution time corresponding to the multipliers loop and so as to execute a final version on a multiprocessor architecture like the Cray T3E or the Fujitsu AP3000.

References

1. Arenaz M.: *Improving the performance of the finite element software for an elastohydrodynamic punctual contact problem under imposed load.* (in spanish) Master Thesis, Department of Electronics and Systems, University of A Coru na (1998).
2. Bayada G., Chambat M.: *Sur Quelques Modelisations de la Zone de Cavitation en Lubrification Hydrodynamique.* J. of Theor. and Appl. Mech., **5(5)** (1986) 703–729.
3. Bermúdez A., Moreno C.: *Duality methods for solving variational inequalities.* Comp. Math. with Appl., **7** (1981) 43–58.
4. Cameron A.: *Basic Lubrication Theory.* John Wiley and Sons, Chichester (1981).
5. Durany J., García G., Vázquez C.: *Numerical Computation of Free Boundary Problems in Elastohydrodynamic Lubrication.* Appl. Math. Modelling, **20** (1996) 104–113.
6. Durany J., García G., Vázquez C.: *A Mixed Dirichlet-Neumann Problem for a Nonlinear Reynolds Equation in Elastohydrodynamic Piezoviscous Lubrication.* Proceedings of the Edinburgh Mathematical Society, **39** (1996) 151–162.
7. Fujitsu España: *FVPLEB user's guide.* (1992).
8. Fujitsu España: *FORTRAN77 EX/VP user's guide.* (1992).
9. García G.: *Mathematical study of cavitation phenomena in elastohydrodynamic piezoviscous lubrication.* (in spanish) Ph.D. Thesis, University of Santiago (1996).
10. Lubrecht A.A.: *The Numerical Solution of the Elastohydrodynamic Lubricated Line and Point Contact Problems using Multigrid Techniques.* Ph.D. Thesis, Twente University (1987).
11. Wolfe M.: *High Performance Compilers for Parallel Computing.* Addison–Wesley Publishing Company (1996).

Parallel Grid Manipulations in Earth Science Calculations

William Sawyer[1,2], Lawrence Takacs[1], Arlindo da Silva[1], and Peter Lyster[1,2]

[1] NASA Goddard Space Flight Center, Data Assimilation Office
Code 910.3, Greenbelt MD, 20771, USA
{sawyer,takacs,dasilva,lys}@dao.gsfc.nasa.gov
http://dao.gsfc.nasa.gov

[2] Department of Meteorology, University of Maryland at College Park
College Park MD, 20742-2425, USA
{sawyer,lys}@atmos.umd.edu

Abstract. We introduce the parallel grid manipulations needed in the Earth Science applications currently being implemented at the Data Assimilation Office (DAO) of the National Aeronautics and Space Administration (NASA). Due to real-time constraints the DAO software must run efficiently on parallel computers. Numerous grids, structured and unstructured are employed in the software.
The DAO has implemented the PILGRIM library to support multiple grids and the various grid transformations between them, e.g., interpolations, rotations, prolongations and restrictions. It allows grids to be distributed over an array of processing elements (PEs) and manipulated with high parallel efficiency. The design of PILGRIM closely follows the DAO's requirements, but it can support other applications which employ certain types of grids. New grid definitions can be written to support still others. Results illustrate that PILGRIM can solve grid manipulation problems efficiently on parallel platforms such as the Cray T3E.

1 Introduction

The need to discretise continuous models in order to solve scientific problems gives rise to finite *grids* — sets of points at which prognostic variables are sought. So prevalent is the use of grids in science that it is possible to forget that a computer-calculated solution is not the solution to the original problem but rather of a discretized representation of the original problem, and moreover is only an approximate solution, due to finite precision arithmetic. Grids are ubiquitous where analytical solutions to continuous problems are not obtainable, e.g., the solution of many differential equations.

Classically a structured grid is chosen a priori for a given problem. If the quality of the solution is not acceptable, then the grid is made finer, in order to better approximate the continuous problem.

For some time the practicality of *unstructured* grids has also been recognised. In such grids it is possible to cluster points in regions of the domain which require

J. Palma, J. Dongarra, and V. Hernández (Eds.): VECPAR'98, LNCS 1573, pp. 666–679, 1999.

higher resolution, while retaining coarse resolution in other parts of the domain. Unstructured grids are often employed in device simulation [1], computational fluid dynamics [2], and even in oceanographic models [3]. Although these grids are more difficult to lay out than structured grids, much research has been done in generating them automatically [4]. In addition, once the grid has been generated, numerous methods and libraries are available to adaptively refine the mesh [5] to provide a more precise solution.

Furthermore, the advantages of multiple grids of varying resolutions for a given domain have been recognised. This is best known in the Multigrid technique [6] in which low frequency error components of the discrete solution are eliminated if values on a given grid are restricted to a coarser grid on which a smoother is applied. But multiple grids also find application in other fields such as speeding up graph partitioning algorithms [7].

An additional level of complexity has arisen in the last few years: many contemporary scientific problems must be decomposed over an array of processing elements (or PEs) in order to obtain a solution in an expedient manner. Depending on the parallelisation technique, not only the work load but also the grid itself may be distributed over the PEs, meaning that different parts of the data reside in completely different memory areas of the parallel machine. This makes the programming of such an application much more difficult for the developer.

The Goddard Earth Observing System (GEOS) Data Assimilation System (DAS) software currently being developed at the Data Assimilation Office (DAO) is no exception to the list of modern grid applications. GEOS DAS uses observational data with systematic and random errors and incomplete global coverage to estimate the complete, dynamic and constituent state of the global earth system. The GEOS DAS consists of two main components, an atmospheric General Circulation Model (GCM) [8] to predict the time evolution of the global earth system and a Physical-space Statistical Analysis Scheme (PSAS) [9] to periodically incorporate observational data.

At least three distinct grids are being employed in GEOS DAS: an *observation grid* — an unstructured grid of points where physical quantities measured by instruments or satellites are associated — a structured *geophysical grid* of points spanning the earth at uniform latitude and longitude locations where prognostic quantities are determined, and a block-structured *computational grid* which may be stretched in latitude and longitude. Each of these grids has a different structure and number of constituent points, but there are numerous interactions between them. Finally the GEOS DAS application is targeted for distributed memory architectures and employs a message-passing paradigm for the communication between PEs.

In this document we describe the design of PILGRIM (Fig. 1), a parallel library for grid manipulations, which fulfils the requirements of GEOS DAS. The design of PILGRIM is *object-oriented* [10] in the sense that it is modular, data is encapsulated in each design layer, operations are overloaded, and different instantiations of grids can coexist simultaneously. The library is realized in Fortran 90, which allows the necessary software engineering techniques such as modules

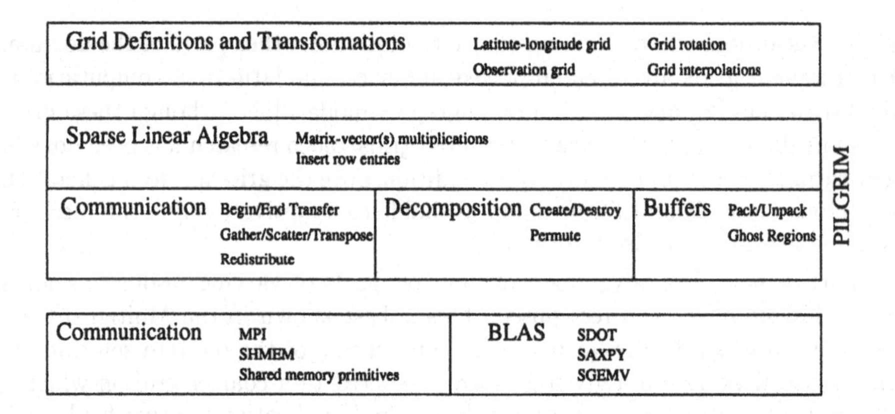

Fig. 1. PILGRIM assumes the existence of fundamental communication primitives such as the Message-Passing Interface (MPI) and optimised Basic Linear Algebra Subroutines (BLAS). PILGRIM's first layer contains routines for communication as well as for decomposing the domain and packing and unpacking sub-regions of the local domain. Above this is a sparse linear algebra layer which performs basic sparse matrix operations for grid transformations. Above PILGRIM, modules define and support different grids. Currently only the grids needed in GEOS DAS are implemented, but the further modules could be designed to support yet other grids.

and derived data types, while keeping in line with other Fortran developments at the DAO. The communication layer is implemented using MPI [11]; however, the communication interfaces defined in PILGRIM's primary layer could conceivably be implemented with other message-passing libraries such as PVM [12], or with other paradigms, e.g., Cray SHMEM [13] or with shared-memory primitives which are available on shared-memory machines like the SGI Origin or SUN Enterprise.

This document is structured in a bottom-up fashion. Design assumptions are made in Sect. 2 in order to ease the implementation. The layer for communication, decompositions, and buffer packaging is discussed in Sect. 3. The sparse linear algebra layer is specified in Sect. 4. The plug-in grid modules are defined in Sect. 5 to the degree necessary to meet the requirements of GEOS DAS. In Sect. 6 some examples and prototype benchmarks are presented for the interaction of all the components. Finally we summarise our work in Sect. 7.

2 Design Assumptions

A literature search was the first step taken in the PILGRIM design process in order to find public domain libraries which might be sufficient for the DAO's requirements [14]. Surprisingly, none of the common parallel libraries for the solution of sparse matrix problems, e.g., PETSc [15], Aztec [16], PLUMP [17], et al., was sufficient for our purposes. These libraries all try to make the parallel implementation transparent to the application. In particular, the application is not supposed to know how the data are actually distributed over the PEs.

This trend in libraries is not universally applicable for the simple reason that if an application is to be parallelised, the developers generally have a good idea of how the underlying data should be distributed and manipulated. Experience has shown us that hiding complexity often leads to poor performance, and the developer often resorts to workarounds to make the system perform in the manner she or he envisions. If the developer of a parallel program is capable of deciding on the proper data distribution and manipulation of local data, then those decisions need to be supported.

In order to minimise the scope of PILGRIM, several simplifying assumptions were made about the way the library will be used:

1. The local portion of the distributed grid array is assumed to be a contiguous section of memory. The local array can have any rank, but if the rank is greater than one, the developer must assure that no gaps are introduced into the actual data representation, for example, by packing it into a 1-D array if necessary.
2. Grid transformations are assumed to be a *sparse linear* transformations. That is, a value on one grid is derived from only a few values on the other grid. The linear transformation therefore corresponds to a sparse matrix with a predictable number of non-zero entries per row. This assumption is realistic for the localised interpolations used in GEOS DAS.
3. At a high level, the application can access data through global indices, i.e., the indices of the original undistributed problem. However, at the level where most computation is performed, the application needs to work with local indices (ranging from one to the total number of entries in the local contiguous array). The information to perform global-to-local and local-to-global mappings must be contained in the data structure defining the grid. However, it is assumed that these mappings are seldom performed, e.g., at the beginning and end of execution, and these mappings need not be efficient.
4. All decomposition-related information is replicated on all PEs.

These assumptions are significant. The first avoids the introduction of an opaque type for data and allows the application to manipulate the local data as it sees fit. The fact that the data are contained in a simple data structure generally allows higher performance than an implementation which buries the data inside a derived type. The second assumption ensures that the matrix transformation are not memory limited. The third implies that most of the calculation is performed on the data in a *local* fashion. In GEOS DAS it is fairly straightforward to run in this mode; however, it might not be the case in other applications. The last assumption assures that every PE knows about the entire data decomposition.

3 Communication and Decomposition Utilities

In this layer communication routines are isolated, and basic functionality is provided for defining and using data decompositions as well as for moving sections of data arrays to and from buffers.

The operations on data decompositions are embedded in a Fortran 90 module which also supplies a generic `DecompType` to describe a decomposition. Any instance of `DecompType` is replicated on all PEs such that every PE has access to information about the entire decomposition. The decomposition utilities consist of the following:

DecompRegular1d	Create a 1-D blockwise data decomposition
DecompRegular2d	Create a 2-D block-block data decomposition
DecompRegular3d	Create a 3-D block-block-block data decomposition
DecompIrregular	Create an irregular data decomposition
DecompCopy	Create new decomposition with contents of another
DecompPermute	Permute PE assignment in a given decomposition
DecompFree	Free a decomposition and the related memory
DecompGlobalToLocal1d	Map global 1-D index to local (pe,index)
DecompGlobalToLocal2d	Map global 2-D index to local (pe,index)
DecompLocalToGlobal1d	Map local (pe,index) to global 1-D index
DecompLocalToGlobal2d	Map local (pe,index) to global 2-D index

Using the Fortran 90 overloading feature, the routines which create new decompositions are denoted by `DecompCreate`. Similarly, the 1-D and 2-D global-to-local and local-to-global mappings are denoted by `DecompGlobalToLocal` and `DecompLocalToGlobal`, resulting in a total of five fundamental operations.

Communication primitives are confined to this layer because it may be necessary at some point to implement them with a message-passing library other than MPI such as PVM or SHMEM, or even with shared-memory primitives such as those on the SGI Origin (the principle platform at the DAO). Thus it is wise to encapsulate all message-passing into one Fortran 90 module. For brevity, only the overloaded functionality is presented:

ParInit	Initialise the parallel code segment
ParExit	Exit from the parallel code segment
ParSplit	Split parallel code segment into two groups
ParMerge	Merge two code segments
ParScatter	Scatter global array to given data decomposition
ParGather	Gather from data decomposition to global array
ParBeginTransfer	Begin asynchronous data transfer
ParEndTransfer	End asynchronous data transfer
ParExchangeVector	Transpose block-distributed vector over all PEs
ParRedistribute	Redistribute one data decomposition to another

In order to perform calculations locally on a given PE it is often necessary to "ghost" adjacent regions, that is, send boundary regions of the local domain to adjacent PEs. To this end a module has been constructed to move ghost regions to and from buffers. The buffers can be transferred to other PEs with the communication primitives such as `ParBeginTransfer` and `ParEndTransfer`. Currently the buffer module contains the following non-overloaded functionality:

BufferPackGhost2dReal	Pack a 2-D array sub-region into buffer
BufferUnpackGhost2dReal	Unpack buffer into 2-D array sub-region
BufferPackGhost3dReal	Pack a 3-D array sub-region into buffer
BufferUnpackGhost3dReal	Unpack buffer into 3-D array sub-region
BufferPackSparseReal	Pack specified entries of vector into buffer
BufferUnpackSparseReal	Unpack buffer into specified entries of vector

In this module, as in most others, the local coordinate indices are used instead of global indices. Clearly this puts responsibility on the developer to keep track of the indices which correspond to the ghost regions. In GEOS DAS this turns out to be fairly straightforward.

4 Sparse Linear Algebra

The concept of transforming one grid to another involves interpolating the values defined on one grid to grid-points on another. These values are stored as contiguous vectors with a given length, $1 \ldots N_{local}$, and distribution defined by the grid decomposition (although the vector might actually represent a multi-dimensional array at a higher level). Thus the sparse linear algebra layer fundamentally consists of a facility to perform linear transformations on distributed vectors.

As in other parallel sparse linear algebra packages, e.g., PETSc [15] and Aztec [16], the linear transformation is stored in a distributed sparse matrix format. Unlike those libraries, however, local indices are used when referring to individual matrix entries, although the mapping `DecompGlobalToLocal` can be used to translate from global to local indices. In addition, the application of the linear transformation is a matrix-vector multiplication where the matrix is not necessarily square, and the resulting vector may be distributed differently than the original.

There are many approaches to storing distributed sparse matrices and performing a the matrix-vector product. PILGRIM uses a format similar to that described in [17], which is optimal if the number of non-zero entries per row is constant.

Assumption 3 in Sect. 2 implies that the matrix definition is not time-consuming. In GEOS DAS the template of any given interpolation is initialised once, but the interpolation itself is performed repeatedly. Thus relatively little attention has been paid to the optimisation of the matrix creation and definition. The basic operations for creating and storing matrix entries are:

SparseMatCreate	Create a sparse matrix
SparseMatDestroy	Destroy a sparse matrix
SparseInsertEntries	Insert entries replicated on all PEs
SparseInsertLocalEntries	Insert entries replicated on all PEs

Two scenarios for inserting entries are supported. In the first scenario, every PE inserts all matrix entries. Thus every argument of the corresponding routine,

`SparseInsertEntries`, is replicated. The local PE picks up only the data which it needs, leaving other data to the appropriate PEs. This scenario is the easiest to program if the sequential code version is used as the code base.

In the second scenario the domain is partitioned over the PEs, meaning that each PE is responsible for a disjoint subset of the matrix entries, and the matrix generation is performed in parallel. Clearly this is the more efficient scenario. The corresponding routine, `SparseInsertLocalEntries` assumes that no two PEs try to add the same matrix entry. However, it does not assume that the all matrix entries reside on the local PE, and it will perform the necessary communication to put the matrix entries in their correct locations.

The efficient application of the matrix to a vector or group of vectors is crucial to the overall performance of GEOS DAS, since the linear transformations are performed continually on assimilation runs for days or weeks at a time. The most common transformation is between three-dimensional arrays of two different grids which describe global atmospheric quantities such as wind velocity or temperature. One 3-D array might be correspond to the geophysical grid which covers the globe, while another might be the computational grid which is more appropriate for the dynamical calculation. The explicit description of such a 3-D transformation would be prohibitive in terms of memory. But fortunately, this transformation only has dependencies in two of the three dimensions as it acts on 2-D horizontal cross-sections independently.

To fulfil the assumptions in Sect. 2, a 2-D array is considered a vector x. Using this representation the transformations become parallel matrix-vector multiplications, which can be performed with one of the following two operations:

SparseMatVecMult	Perform $y \leftarrow \alpha Ax + \beta y$
SparseMatTransVecMult	Perform $y \leftarrow \alpha A^T x + \beta y$

In order to transform several arrays simultaneously, the arrays are grouped into multiple vectors, that is, into a $n \times m$ matrix where n is the length of the vector (e.g., the number of values in the 2-D array), and m is the number of vectors. The following matrix-matrix and matrix-transpose-matrix multiplications can group messages in such a way as to drastically minimise latencies and utilise BLAS-2 operations instead of BLAS-1:

SparseMatMatMult	Perform $Y \leftarrow \alpha AX + \beta Y$
SparseMatTransMatMult	Perform $Y \leftarrow \alpha A^T X + \beta Y$

The distributed representation of the matrix contains, in addition to the matrix information itself, space for the *communication pattern.* Upon entering any one of the four matrix operations, the the matrix is checked for new entries which may have been added since its last application. If the matrix has been modified, the operation first generates the communication pattern — an optimal map of the information which has to be exchanged between PEs — before performing the matrix multiplication. This is a fairly expensive operation, but in GEOS DAS it only needs to be done occasionally. Subsequently, the matrix multiplication can be performed repetitively in the most efficient manner possible.

5 Supported Grids

The *grid data structure* describes a set of *grid-points* and their decomposition over a group of PEs as well as other information, such as the size of the domain. The grid data structure itself does not contain actual data and can be replicated on all PEs due to its minimal memory requirements. The data reside in arrays distributed over the PEs and given meaning by the information in the grid data structure. There is no limitation on how the application accesses and manipulates the local data arrays. Two types of grids employed in GEOS DAS are described here, but others are conceivable and could be supported by PILGRIM without modifications to the library.

The latitude-longitude grid defines a *lat-lon* coordinate system — a regular grid spanning the earth with all points in one row having a given latitude and all points in a column a given longitude. The grid encompasses the entire earth from $-\pi$ to π longitudinally and from $-\pi/2$ to $\pi/2$ in latitude.

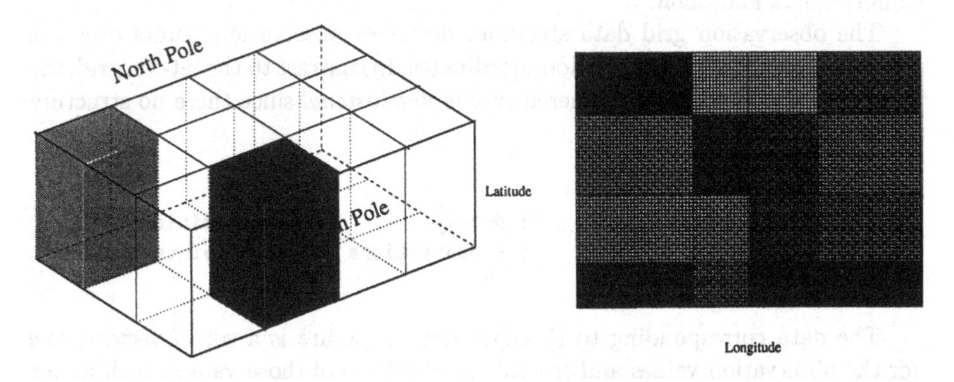

Fig. 2. GEOS DAS uses a column decomposition of data (left), also termed a "checkerboard" decomposition due to the distribution of any given horizontal cross-section (right). The width and breadth of a column can be variable, although generally an approximately equal number of points are assigned to every PE.

The decomposition of this grid is a "checkerboard" (Fig. 2), because the underlying three-dimensional data array comprises all levels of the column designated by the 2-D decomposition of the horizontal cross-section. This decomposition can contain a variable-sized rectangle of points — it is not necessary for each PE to be assigned an equal number — and thus some freedom for load balancing exists. The basic data structure for the lat-lon grid is defined below in its Fortran 90 syntax:

```
TYPE LatLonGridType
  TYPE (DecompType)    :: Decomp       ! Decomposition
  INTEGER              :: ImGlobal     ! Global Size in X
  INTEGER              :: JmGlobal     ! Global Size in Y
  REAL                 :: Tilt         ! Tilt of remapped NP
  REAL                 :: Rotation     ! Rotation of remapped NP
  REAL                 :: Precession   ! Precession of remapped NP
  REAL,POINTER         :: dLat(:)      ! Latitudes
  REAL,POINTER         :: dLon(:)      ! Longitudes
END TYPE LatLonGridType
```

This grid suffices to describe both the GEOS DAS computational grid used for dynamical calculations and the geophysical grid in which the prognostic variables are sought. The former makes use of the parameters `Tilt`, `Rotation` and `Precession` to describe its view of the earth (Fig. 3), and the `dLat` and `dLon` grid box sizes to describe the grid stretching. The latter is defined by the normal geophysical values for `Tilt`, `Rotation` and `Precession` $= (\frac{\pi}{2}, 0, 0)$ and uniform `dLat` and `dLon`.

The observation grid data structure describes observation points over the globe, as described by their lat-lon coordinates. In contrast to the lat-lon grid, the point grid decomposition is inherently *one-dimensional* since there no structure to the grid.

```
TYPE ObsGridType
  TYPE (DecompType)    :: Decomp         ! Decomposition
  INTEGER              :: Nobservations  ! Total points
END TYPE ObsGridType
```

The data corresponding to this grid data structure is a set of vectors, one for the observation values and several for *attributes* of those values, such as the latitude, longitude and level at which an observation was made.

6 Results

An example of a non-trivial transformation employed in atmospheric science applications is grid rotation [18]. Computational instabilities from finite difference schemes can arise in the polar regions of the geophysical grid when a strong cross-polar flow occurs. By placing the pole of the computational grid to the geographic equator, however, the instability near the geographic pole is removed due to the vanishing Coriolis term.

It is generally accepted that the physical processes such as those related to long- and short-wave radiation can be calculated directly on the geophysical grid. Dynamics, where the numerical instability occurs, needs to be calculated on the computational grid. An additional refinement involves calculating the dynamics on a rotated *stretched* grid, in which the grid-points are not uniform in latitude and longitude. The `LatLonGridType` allows for both variable lat-lon coordinates as well as the description of any lat-lon view of the world where the poles are assigned to a new geographical location. The grid rotation (without stretching) is depicted in Fig. 3.

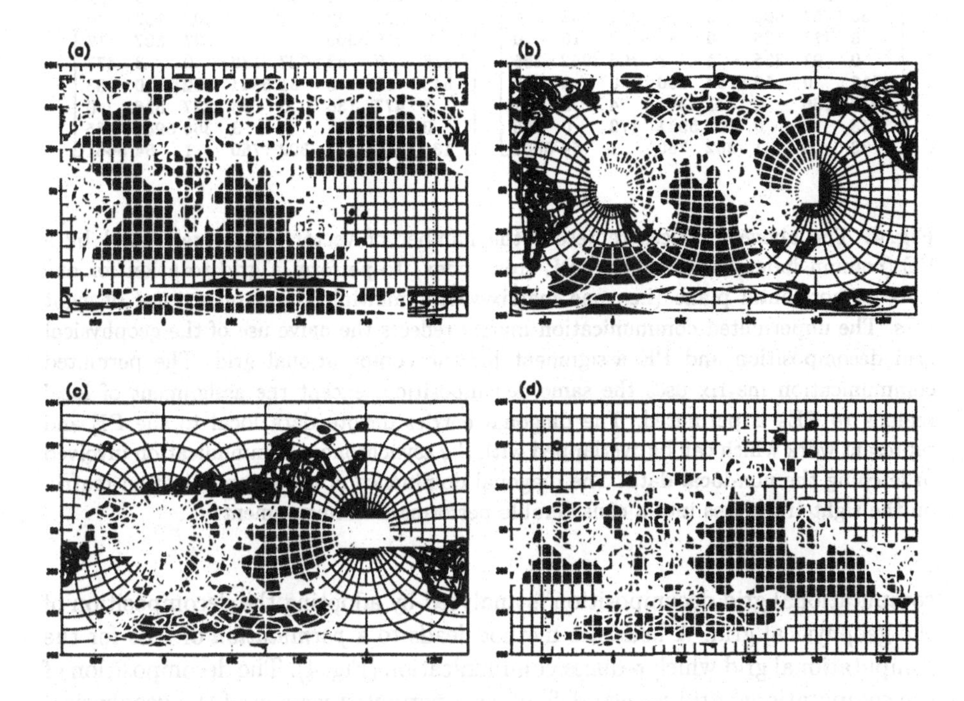

Fig. 3. The use of the latitude-longitude grid (a) and (c) as the computational grid results in instabilities at the poles due to the Coriolis term. The instabilities vanish with on a grid (b) where the pole has been rotated to the equator. The computational grid is therefore a lat-lon grid (d) where the "poles" on the top and bottom are in the Pacific and Atlantic Oceans, respectively.

It would be natural to use the same decomposition for both the geophysical and computational grids. It turns out, however, that this approach disturbs data locality inherent to this transformation (Fig. 4). If the application could have unlimited freedom to choose the decomposition of the computational grid, the forward and reverse grid rotations could exhibit excellent data locality, and the matrix application would be much more efficient.[1] Unfortunately, practicality limits the decomposition of both the geophysical and computational grids to be a checkerboard decomposition.

However, there are still several degrees of freedom in the decomposition, namely the number of points on each PE and the assignment of local regions to PEs. While an approximately uniform number of points per PE is generally best for the dynamics calculation, the assignment of PEs is arbitrary. The following optimisation is therefore applied: the potential communication pattern of a naive

[1] A simply connected region in one domain will map to at most two simply connected regions in the other.

Unpermuted Communication Matrix

$$\begin{bmatrix} 0 & 219 & 5905 & 172 & 0 & 97 & 507 & 12 \\ 53 & 5731 & 690 & 2 & 1 & 303 & 132 & 0 \\ 3 & 477 & 136 & 0 & 53 & 5727 & 516 & 0 \\ 0 & 97 & 335 & 4 & 3 & 366 & 5942 & 165 \\ 516 & 0 & 53 & 5727 & 136 & 0 & 3 & 477 \\ 5967 & 166 & 2 & 341 & 371 & 4 & 0 & 61 \\ 543 & 12 & 0 & 61 & 5941 & 172 & 0 & 183 \\ 133 & 0 & 1 & 302 & 760 & 1 & 54 & 5661 \end{bmatrix}$$

Permuted Communication Matrix

$$\begin{bmatrix} 5967 & 166 & 2 & 341 & 371 & 4 & 0 & 61 \\ 53 & 5731 & 690 & 2 & 1 & 303 & 132 & 0 \\ 0 & 219 & 5905 & 172 & 0 & 97 & 507 & 12 \\ 516 & 0 & 53 & 5727 & 136 & 0 & 3 & 477 \\ 543 & 12 & 0 & 61 & 5941 & 172 & 0 & 183 \\ 3 & 477 & 136 & 0 & 53 & 5727 & 516 & 0 \\ 0 & 97 & 335 & 4 & 3 & 366 & 5942 & 165 \\ 133 & 0 & 1 & 302 & 760 & 1 & 54 & 5661 \end{bmatrix}$$

Fig. 4. The above matrices represent the number of vector entries requested by a PE (column index) from another PE (row index) to perform a grid rotation for one 72×48 horizontal plane (i.e., one matrix-vector multiplication) on a total of eight PEs. The unpermuted communication matrix reflects the naive use of the geophysical grid decomposition and PE assignment for the computational grid. The permuted communication matrix uses the same decomposition, except the assignment of local regions to PEs is permuted. The diagonal entries denote data local to the PE and represent work which can be overlapped with the asynchronous communication involved in fetching the non-local data. The diagonal dominance of the communication matrix on the right translates into a considerable performance improvement.

computational grid decomposition is analysed by adopting the decomposition of the geophysical grid. A heuristic method leads to a *permutation* of PEs for the computational grid which reduces communication (Fig. 4). The decomposition of the computational grid is then defined as a permuted version of the geophysical grid. Only then is the grid rotation matrix defined. An outline of the code is as given in Algorithm 1.

Algorithm 1 (Optimised Grid Rotation) *Given the geophysical grid decomposition, find a permutation of the PEs which will maximise the data locality of the geophysical-to-computational grid transformation, create and permute the computation grid decomposition, and define the transformation in both directions.*

```
SparseMatrixCreate( ..., GeoToComp )
SparseMatrixCreate( ..., CompToGeo )
DecompCreate( ..., GeoPhysDecomp )
LatLonCreate( GeoPhysDecomp, ...., GeoPhysGrid )
AnalyzeGridTransform( GeoPhysDecomp, ...., Permutation )
DecompCopy( GeoPhysDecomp, CompDecomp )
DecompPermute( Permutation, CompDecomp )
LatLonCreate( CompDecomp, ...., CompGrid )
GridTransform( GeoPhysGrid, CompGrid, GeoToComp )
GridTransform( CompGrid, GeoPhysGrid, CompToGeo )
```

In `GridTransform` the coordinates of one lat-lon grid are mapped to another. Interpolation coefficients are determined by the proximity of rotated grid-points to grid-points on the other grid (Fig. 3). Various interpolation schemes can be employed including bi-linear or bi-cubic; the latter is employed in GEOS DAS.

The transformation matrix can be completely defined by the two grids — the values on those grids are not necessary.

Once the transformation matrix is defined, sets of grid values, such as individual levels or planes of atmospheric data, can be transformed ad infinitum using a matrix-vector multiplication.

```
DO L = 1, GLOBAL_Z
  CALL SparseMatVecMult(GeoToComp, 1.0, In(1,1,L), 0.0, Out1(1,1,L))
END DO
```

Alternatively, the transformation of the entire 3-D data set can be performed with one matrix-matrix product:

```
CALL SparseMatMatMult( GeoToComp, GLOBAL_Z, 1.0, In, 0.0, Out2 )
```

Note that the pole rotation is trivial (embarrassingly parallel) if any given plane resides entirely on one PE, i.e., if the 3-D array is decomposed in the z-dimension. Unfortunately, there are compelling reasons to distribute the data in vertical columns with the checkerboard decomposition.

Fig. 5 compares the performance of the unpermuted rotation with that of the permuted rotation on the Cray T3E. A further optimisation is performed by replacing the non-blocking MPI primitives used in ParBeginTransform by faster Cray SHMEM primitives. The result of these optimisations is the improvement in scalability from tens of PEs to hundreds of PEs. The absolute performance in GFlop/s is presented in Fig. 6.

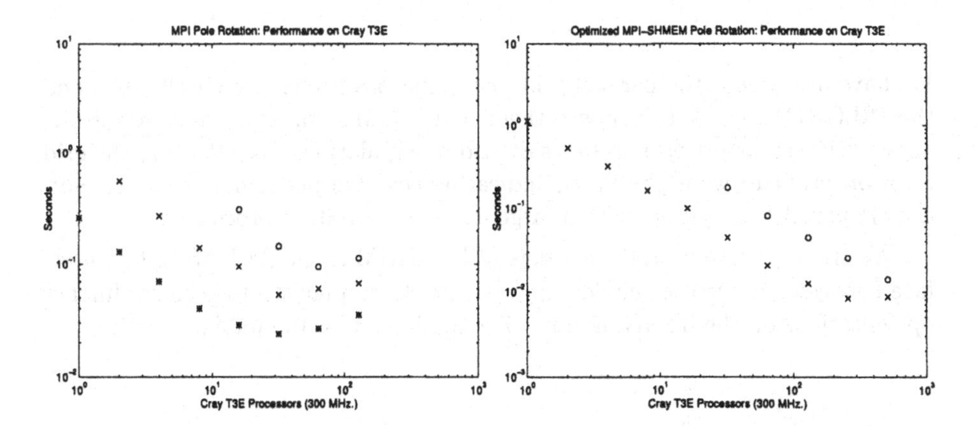

Fig. 5. With a naive decomposition of both the geophysical and computational grids and a straightforward MPI implementation, the performances at the left for the $72 \times 46 \times 70$ (*), $144 \times 91 \times 70$ (**x**), and $288 \times 181 \times 70$ (**o**) resolutions yield good scalability only to 10-50 processors. The optimised MPI-SHMEM hybrid version on the right scales to nearly the entire extent of the machine (512 processors).

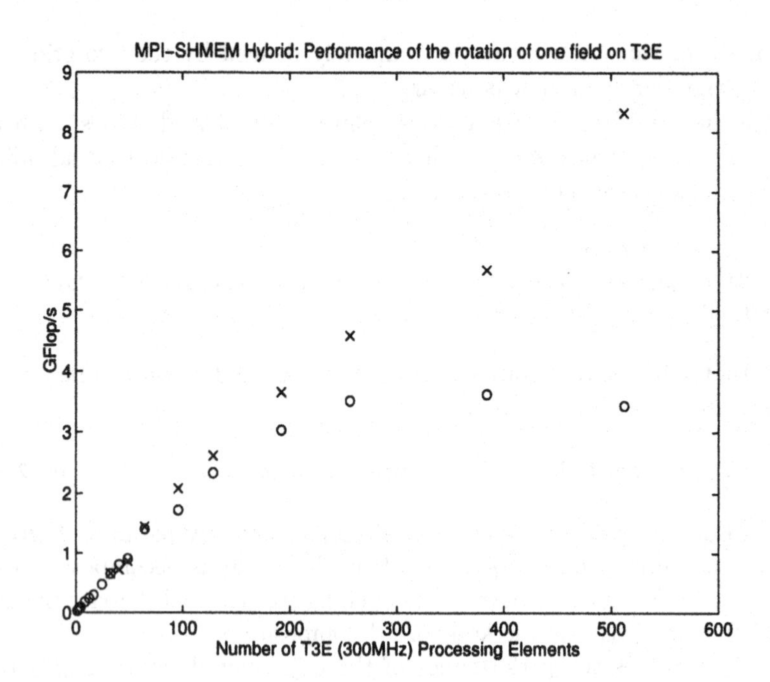

Fig. 6. The GFlop/s performances of the grid rotation on grids with $144 \times 91 \times 70$ (**o**), and $288 \times 181 \times 70$ (**x**) resolutions is depicted. These results are an indication that the grid rotation will not represent a bottleneck for the overall GEOS DAS system.

7 Summary

We have introduced the parallel grid manipulations needed by GEOS DAS and the PILGRIM library to support them. PILGRIM is modular and extensible, allowing it to support various types of grid manipulations. Results from the grid rotation problem were presented, indicating scalable performance on state-of-the-art parallel computers with a large number (> 100) of processors.

We are hoping to extend the usage of PILGRIM in GEOS DAS to the interface between the forecast model and the statistical analysis, to perform further optimisations on the library, and to offer the library to the public domain.

Acknowledgments

We would like to thank Jay Larson, Rob Lucchesi, Max Suarez, and Dan Schaffer for their valuable suggestions. The work of Will Sawyer and Peter Lyster at the Data Assimilation Office was funded by the High Performance Computing and Communications Initiative (HPCC) Earth and Space Science (ESS) program.

References

[1] G. Heiser, C. Pommerell, J. Weis, and W. Fichtner. Three dimensional numerical semiconductor device simulation: Algorithms, architectures, results. *IEEE Transactions on Computer-Aided Design of Integrated Circuits*, 10(10):1218–1230, 1991.

[2] A. Ecer, J. Hauser, P. Leca, and J. Périaux. *Parallel Computational Fluid Dynamics*. Elsevier Science Publishers B.V. (North–Holland), Amsterdam, 1995.

[3] H.-P. Kersken, B. Fritzsch, O. Schenk, W. Hiller, J. Behrens, and E. Kraube. Parallelization of large scale ocean models by data decomposition. *Lecture Notes in Computer Science*, 796:323–336, 1994.

[4] P. Knupp and S. Steinberg. *Fundamentals of Grid Generation*. CRC Press, Boca Raton, FL, 1994.

[5] M. T. Jones and P. E. Plassmann. Parallel algorithms for the adaptive refinement and partitioning of unstructured meshes. In IEEE, editor, *Proceedings of the Scalable High-Performance Computing Conference, May 23–25, 1994, Knoxville, Tennessee*, pages 478–485. IEEE Computer Society Press, 1994.

[6] W. L. Briggs. *A Multigrid Tutorial*. SIAM, 1987.

[7] S. T. Barnard and H. D. Simon. A Fast Multilevel Implementation of Recursive Spectral Bisection for Partitioning Unstructured Problems. Technical Report RNR-092-033, NASA Ames Research Center, 1992.

[8] L. L. Takacs, A. Molod, and T. Wang. Documentation of the Goddard Earth Observing System (GEOS) General Circulation Model — Version 1. NASA Technical Memorandum 104606, NASA, 1994.

[9] A. da Silva and J. Guo. Documentation of the Physical-space Statistical Analysis System (PSAS), Part 1: The Conjugate Gradient Solver, Version PSAS 1.00. DAO Office Note 96-02, Data Assimilation Office, NASA, 1996.

[10] T. Budd. *Object-Oriented Programming*. Addison-Wesley, New York, N.Y., 1991.

[11] MPIF (Message Passing Interface Forum). MPI: A Message-Passing Interface Standard. *International Journal of Supercomputer Applications*, 8(3&4):157–416, 1994.

[12] A. Geist, A. Beguelin, J. Dongarra, W. Jiang, R. Manchek, and V. Sunderam. *PVM: A Users' Guide and Tutorial for Networked Parallel Computing*. MIT Press, 1994.

[13] Cray Research. CRAY T3E Applications Programming. 1997, 1997.

[14] DAO Staff. GEOS-3 Primary System Requirements Document. Internal document, available on request., 1996.

[15] L. C. McInnes and B. F. Smith. Petsc 2.0: A case study of using mpi to develop numerical software libraries. In *1995 MPI Developers' Conference*, 1995.

[16] S. A. Hutchinson, J. A. Shadid, and R.S. Tuminaro. *The Aztec User's Guide - Version 1.0*. 1995.

[17] O. Bröker, V. Deshpande, P. Messmer, and W. Sawyer. Parallel library for unstructured mesh problems. Tech. Report CSCS-TR-96-15, Centro Svizzero di Calcolo Scientifico, 1996.

[18] M. J. Suarez and L. L. Takacs. Documentation of the ARIES/GEOS Dynamical Core: Version 2. NASA Technical Memorandum 104606, NASA, 1995.

Simulating Magnetised Plasma with the Versatile Advection Code

Rony Keppens[1] and Gábor Tóth[2]

[1] FOM-Institute for Plasma-Physics Rijnhuizen,
P.O. Box 1207, 3430 BE Nieuwegein, The Netherlands,
keppens@rijnh.nl
[2] Department of Atomic Physics, Eötvös University,
Pázmány Péter sétány 2, Budapest H-1117, Hungary,
gtoth@hermes.elte.hu

Abstract. Matter in the universe mainly consists of plasma. The dynamics of plasmas is controlled by magnetic fields. To simulate the evolution of magnetised plasma, we solve the equations of magnetohydrodynamics using the Versatile Advection Code (VAC).
To demonstrate the versatility of VAC, we present calculations of the Rayleigh-Taylor instability, causing a heavy compressible gas to mix into a lighter one underneath, in an external gravitational field. Using a single source code, we can study and compare the development of this instability in two and three spatial dimensions, without and with magnetic fields. The results are visualised and analysed using IDL (Interactive Data Language) and AVS (Advanced Visual Systems).
The example calculations are performed on a Cray J90. VAC also runs on distributed memory architectures, after automatic translation to High Performance Fortran. We present performance and scaling results on a variety of architectures, including Cray T3D, Cray T3E, and IBM SP platforms.

1 Magneto-Hydrodynamics

The MHD equations describe the behaviour of a perfectly conducting fluid in the presence of a magnetic field. The eight primitive variables are the density $\rho(\mathbf{r},t)$, the three components of the velocity field $\mathbf{v}(\mathbf{r},t)$, the thermal pressure $p(\mathbf{r},t)$, and the three components of the magnetic field $\mathbf{B}(\mathbf{r},t)$. When written in conservation form, the conservative variables are density ρ, momentum $\rho\mathbf{v}$, energy density $\mathcal{E}$, and the magnetic field $\mathbf{B}$. The thermal pressure p is related to the energy density as $p=(\gamma-1)(\mathcal{E}-\frac{1}{2}\rho v^2-\frac{1}{2}B^2)$, with γ the ratio of specific heats. The eight non-linear partial differential equations express: (1) mass conservation; (2) the momentum evolution (including the Lorentz force); (3) energy conservation; and (4) the evolution of the magnetic field in an induction equation. The equations are given by

$$\frac{\partial\rho}{\partial t}+\nabla\cdot(\rho\mathbf{v})=0, \tag{1}$$

J. Palma, J. Dongarra, and V. Hernández (Eds.): VECPAR'98, LNCS 1573, pp. 680–690, 1999.

$$\frac{\partial(\rho\mathbf{v})}{\partial t} + \nabla \cdot [\rho\mathbf{v}\mathbf{v} + p_{tot}I - \mathbf{B}\mathbf{B}] = \rho\mathbf{g}, \tag{2}$$

$$\frac{\partial\mathcal{E}}{\partial t} + \nabla \cdot (\mathcal{E}\mathbf{v}) + \nabla \cdot (p_{tot}\mathbf{v}) - \nabla \cdot (\mathbf{v} \cdot \mathbf{B}\mathbf{B}) = \rho\mathbf{g} \cdot \mathbf{v} + \nabla \cdot [\mathbf{B} \times \eta(\nabla \times \mathbf{B})], \tag{3}$$

$$\frac{\partial\mathbf{B}}{\partial t} + \nabla \cdot (\mathbf{v}\mathbf{B} - \mathbf{B}\mathbf{v}) = -\nabla \times [\eta(\nabla \times \mathbf{B})]. \tag{4}$$

We introduced $p_{tot} = p + \frac{1}{2}B^2$ as the total pressure, I as the identity tensor, $\mathbf{g}$ as the external gravitational field, and defined magnetic units such that the magnetic permeability is unity.

Ideal MHD corresponds to a zero resistivity η and ensures that magnetic flux is conserved. In resistive MHD, field lines can reconnect. An extra constraint arises from the non-existence of magnetic monopoles, expressed by $\nabla \cdot \mathbf{B} = 0$. The ideal MHD equations allow for Alfvén and magneto-acoustic wave modes, while the induction equation prescribes that flow across the magnetic field entrails the field lines, so that field lines are 'frozen-in'. The field may, in turn, confine the plasma. The MHD description can be used to study both laboratory and astrophysical plasma phenomena. We refer the interested reader to [2] for a derivation of the MHD equations starting from a kinetic description of the plasma, while excellent treatments of MHD theory can be found in, e.g. [4,1].

2 The Versatile Advection Code

The Versatile Advection Code (VAC) is a general purpose software package for solving a conservative system of hyperbolic partial differential equations with additional non-hyperbolic source terms [13,14], in particular the hydrodynamic ($\mathbf{B} = 0$) and magnetohydrodynamic equations (1)-(4), with optional terms for gravity, viscosity, thermal conduction, and resistivity.

VAC is implemented in a modular way, which ensures its capacity to model several systems of conservation laws, and makes it possible to share solution algorithms among all systems. A variety of spatial and temporal discretizations are implemented for solving such systems on a finite volume structured grid. The spatial discretizations include two Flux Corrected Transport variants and four Total Variation Diminishing (TVD) algorithms (see [18]). These numerical schemes are shock-capturing and second order accurate in space and time.

Explicit time integration may exploit predictor-corrector and Runge-Kutta time stepping, while for multi-timescale problems, mixed implicit/explicit time integration is available to treat only some variables, or some terms in the governing equations implicitly [10,11]. Fully implicit time integration can be of interest when modelling steady-state problems. Typical astrophysical applications where semi-implicit and implicit methods are efficiently used can be found in [17,9].

VAC runs on personal computers (Pentium PC under Linux), on a variety of workstations (DEC, Sun, HP, IBM) and has been used on SGI Power Challenge, Cray J90 and Cray C90 platforms. To run VAC on distributed memory architectures, an automatic translation to High Performance Fortran (HPF) is done at the preprocessing phase (see [12]). We have tested the generated HPF code on several platforms, including a cluster of Sun workstations, a Cray T3D, a 16-node Connection Machine 5 (using an automatic translation to CM-Fortran), an IBM SP and a Cray T3E. Scaling and performance is discussed in section 3.

On-line manual pages, general visualisation macros (for IDL, MatLab and SM), and file format transformation programs (for AVS, DX, and Gnuplot) facilitate the use of the code and aid in the subsequent data analysis.

In this manuscript, we present calculations done in two and three spatial dimensions, for both hydrodynamic and magnetohydrodynamic problems. This serves to show how VAC allows a single problem setup to be studied under various physical conditions. We have used IDL and AVS to analyse the application presented here. Our data analysis and visualisation encompasses X-term animation, generating MPEG-movies, and video production.

3 Scaling Results

As detailed in [12], the source code uses a limited subset of the Fortran 90 language, extended with the HPF *forall* statement and the Loop Annotation SYntax (LASY) which provides a dimension independent notation. The LASY notation [15] is translated by the VAC preprocessor according to the dimensionality of the problem. Further translation to HPF involves distributing all global non-static arrays across the processors, which is accomplished in the preprocessing stage by another Perl script.

Figure 1 summaries timing results obtained on two vector (Cray J90 and C90) and three massively parallel platforms (Cray T3D, T3E and IBM SP). We solve the shallow water equations (1)-(2) with $\mathbf{B} = 0$ and $p = (g/2)\rho^2$ on a 104×104 grid on 1, 2, 4, 8, and 13 processors. This simple model problem is described in [16], and our solution method contains the full complexity of a real physics application. We used an explicit TVD scheme exploiting a Roe-type approximate Riemann solver. We plot the number of physical grid cell updates per second against the number of processors (solid lines). The dashed lines show the improved scaling for a larger problem of size 208×208, up to 16 processors. On all parallel platforms, we exploited the Portland Group *pghpf* compiler. We find an almost linear speedup on the Cray T3D and T3E architectures, which is rather encouraging for such small problem sizes. Note how the single node execution on the IBM SP platform is a factor of 2 to 3 faster than the Cray T3E, but the scaling results are poor. The figure indicates clearly that for this hydrodynamic application, on the order of 10 processors of the Cray T3E and IBM SP are needed to outperform a vectorised Fortran 90 run on one processor of the Cray C90. Detailed optimisation strategies for all architectures shown in Figure 1 (note the Pentium PC result and the DEC Alpha workstation timing in the bottom left corner) are discussed in [16].

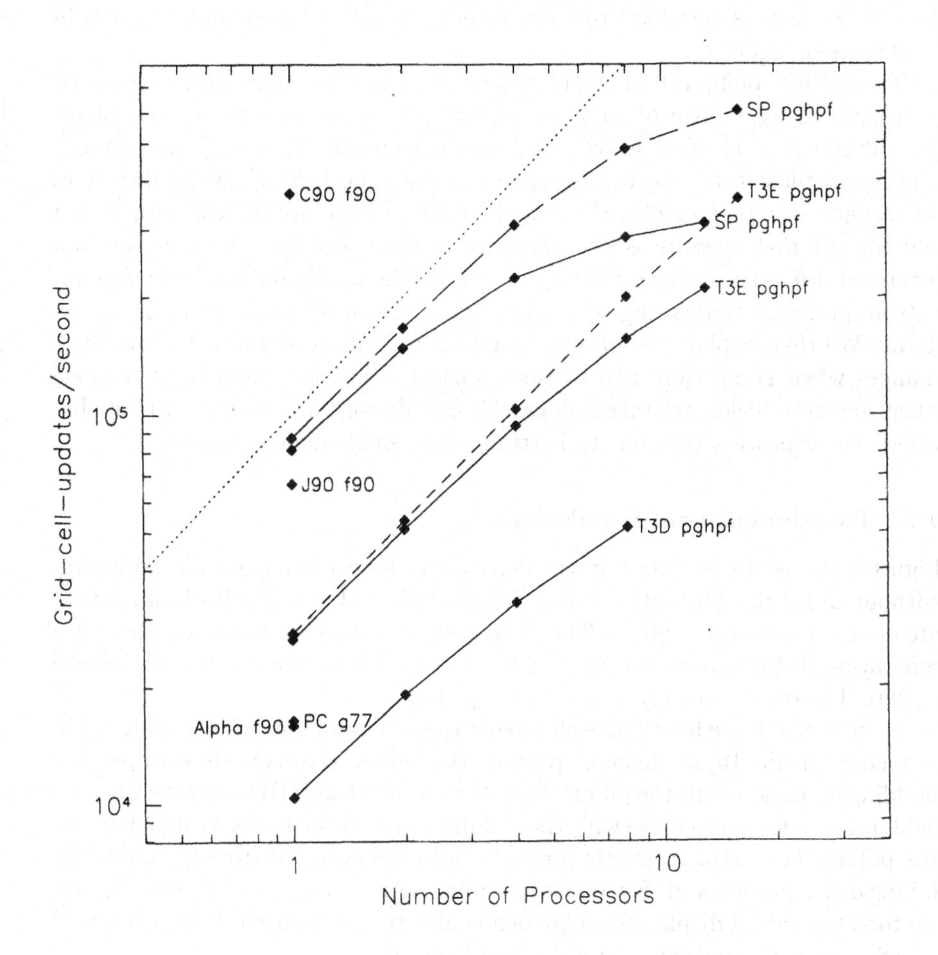

Fig. 1. Combined performance and scaling results for running the Versatile Advection Code on vector and parallel platforms. See text for details.

4 Simulating Rayleigh-Taylor Instabilities

To demonstrate the advantages of having a versatile source code for simulating fluid flow, we consider what happens when a heavy compressible plasma is sitting on top of a lighter plasma in an external gravitational field. Such a situation is unstable as soon as the interface between the two is perturbed from perfect flatness. The instability is known as the Rayleigh-Taylor instability. Early analytic investigations date back to a comprehensive and detailed analysis given by Chandrasekhar [3].

The initial configuration is one where two layers of prescribed density ratio (dense to light ratio of $\rho_d/\rho_l = 10$) are left to evolve between two planes ($y = 0$ and $y = 1$), with gravity pointing downwards ($\mathbf{g} = -\hat{e}_y$ unit vector). The heavy plasma on the top is separated from the light plasma below it by the surface $y = y_0 + \epsilon \sin(k_x x) \sin(k_z z)$. Initially, both are at rest with $\mathbf{v} = 0$, and the thermal pressure is set according to the hydrostatic balance equation (cantered differenced formula $dp/dy = -\rho$). Boundary conditions make top and bottom perfectly conducting solid walls, while the horizontal directions are periodic. We then exploit the options available in VAC to see how the evolution changes when going from two to three spatial dimensions, and what happens when magnetic fields are taken along. All calculations are done on a Cray J90, where we preprocess the code to Fortran 90 for single-node execution.

4.1 Two-dimensional Simulations

Figure 2 shows the evolution of the density in two two-dimensional simulations without and with an initial horizontal magnetic field $\mathbf{B} = 0.1\hat{e}_x$. Both simulations are done on a uniform 100×100 square grid, and the parameters for the initial separating surface are $y_0 = 0.8$, $\epsilon = 0.05$, and $k_x = 2\pi$ (there is no z dependence in 2D). The data is readily analysed using IDL.

In both cases, the heavy plasma is redistributed in falling spikes or pillars, also termed Rayleigh-Taylor 'fingers', pushing the lighter plasma aside with pressure building up underneath the pillars. However, in the ideal MHD case, the frozen-in field lines are forced to move with the sinking material, so it gets wrapped around the pillars. The extra magnetic pressure and tension forces thereby confine the falling dense plasma and slow down the sinking and mixing process. In fact, since we took the initial displacement perpendicular to the horizontal magnetic field, we effectively maximised its stabilising influence.

In [3], the linear phase of the Rayleigh-Taylor instability in both hydrodynamic and magnetohydrodynamic incompressible fluids is treated analytically. The stabilising effect of the uniform horizontal magnetic field is evident from the expression of the growthrate n as a function of the wavenumber k_x

$$n^2 = gk_x \frac{\rho_d - \rho_l}{\rho_d + \rho_l} - \frac{B^2 k_x^2}{2\pi(\rho_d + \rho_l)}. \tag{5}$$

Hence, while the shortest wavelength perturbations are the most unstable ones in hydrodynamics ($B = 0$), all wavelengths below a critical $\lambda_{crit} = B^2/g(\rho_d - \rho_l)$

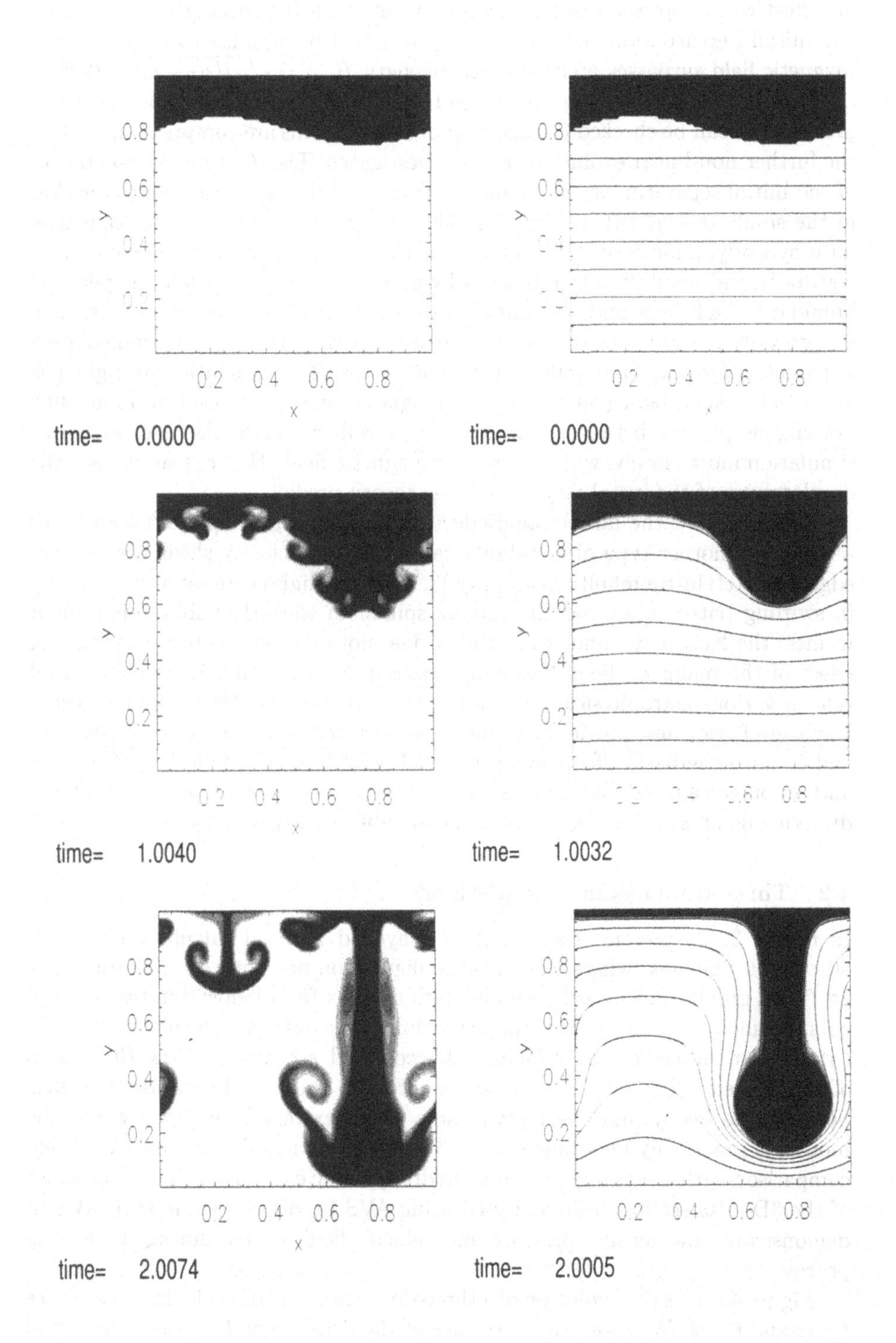

Fig. 2. Rayleigh-Taylor instability simulated in two spatial dimensions, in a hydrodynamic (left) and magnetohydrodynamic (right) case. The logarithm of the density and, in the magnetohydrodynamic case, also the magnetic field lines, are plotted.

are effectively suppressed by a horizontal magnetic field of strength B. Similarly, our initial perturbation with $\lambda = 2\pi/k_x = 1$ will be stabilised as soon as the magnetic field surpasses a critical field strength $B_{crit} = \sqrt{g\lambda(\rho_d - \rho_l)} \simeq 0.95$.

The simulations confirm and extend these analytic findings: the predicted growth-rate can be checked (noting that our simulations are compressible), while the further non-linear evolution can be investigated. The discrete representation of the initial separating surface causes intricate small-scale structure to develop in the simulation at left of Figure 2. This is consistent with the fact that in a pure hydrodynamic case, the shortest wavelengths are the most unstable ones. Naturally, the simulation is influenced by numerical diffusion, while the periodic boundary conditions and the initial state select preferred wavenumbers. The suppression of short wavelength disturbances in the MHD case is immediately apparent, since no small-scale structure develops. The simulation at right has an initial plasma beta (ratio of gas to magnetic pressure forces) of about 400. For higher plasma beta yet, the MHD case will resemble the hydrodynamic simulation more closely, while a stronger magnetic field ($\mathbf{B} = \hat{e}_x$) suppresses the development of the instability entirely, as theory predicts.

Note also how the falling pillars develop a mushroom shape (left frames) as a result of another type of instability caused by the velocity shear across their edge: the Kelvin-Helmholtz instability [3, 7, 8]. The lighter material is swept up in swirling patterns around the sinking spikes. In the MHD simulation (right frames) the Kelvin-Helmholtz instability does not develop due to the stabilising effect of the magnetic field. Typically however, *both* instabilities play a crucial role in various astrophysical situations. Two dimensional MHD simulations of Rayleigh-Taylor instabilities in young supernova remnants [5] demonstrate this, and confirm the basic effects evident from Figure 2: magnetic fields get warped and amplified around the 'fingers'. General discussions of these and other hydrodynamic and magnetohydrodynamic instabilities are found in [3].

4.2 Three-dimensional Simulations

In Figure 3, we present a snapshot of a hydrodynamical calculation in a 3D $50\times50\times50$ unit box, where the initial configuration has both $k_x = 2\pi$ and $k_z = 2\pi$. With gravity downwards, we look into the box from below. On two vertical cuts, we show at time $t = 2$ (i) the logarithm of the density in a colour scale and (ii) the streamlines of the velocity field, coloured according to the (logarithm of the) density. The cuts are chosen to intersect the initial separating surface between the heavy and the light plasma at its extremal positions where the motion is practically two-dimensional. 3D effects are readily identified by direct comparison with the two-dimensional hydrodynamic calculation. The time series of the 3D data set has been analysed using AVS (a video is made with AVS to demonstrate how density, pressure and velocity fields evolve during the mixing process).

Figure 4 shows the evolution of a three-dimensional MHD calculation at times $t = 1$ and $t = 2$. We show an iso-surface of the density (at 1% above the initial value for ρ_d), coloured according to the thermal pressure. A cutting plane also

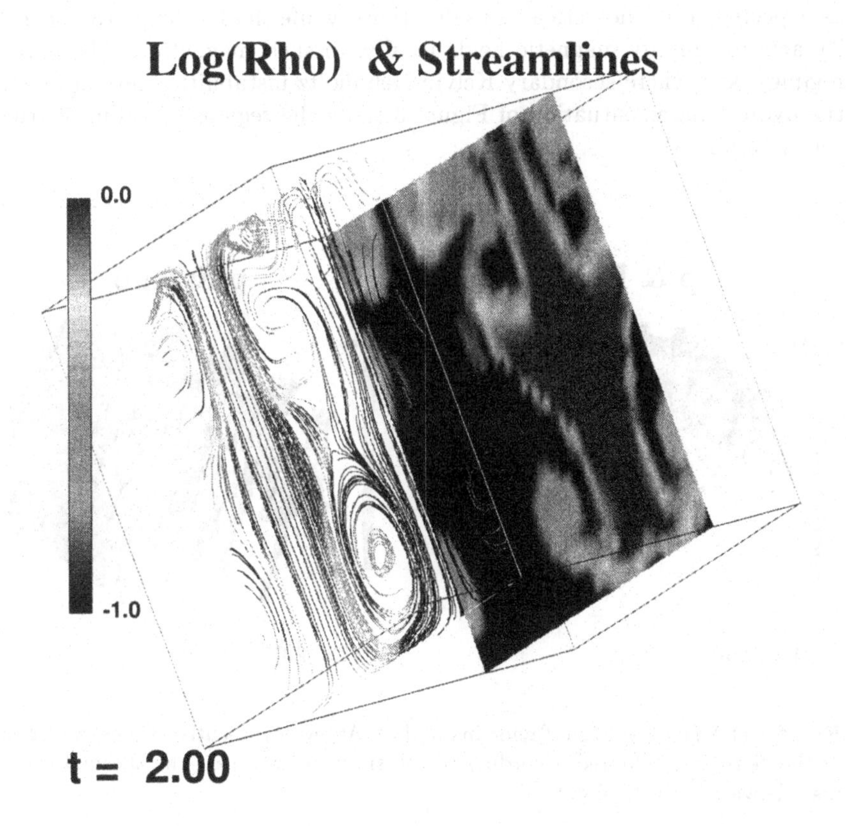

Fig. 3. Rayleigh-Taylor instability in 3D, purely hydrodynamic. We show streamlines (left) and density contours (right) in two vertical cutting planes.

shows the vertical stratification of the thermal pressure. Note the change in the initial configuration ($k_x = 6\pi$ and $k_z = 4\pi$, with $y_0 = 0.7$): more and narrower spikes are seen to grow and to split up. The AVS analysis of the full time series shows how droplets form at the tips of the falling pillars, which seem to expand horizontally to a critical size before continuing their fall. At the same time, the magnetic field gets wrapped around the falling pillars. Figure 4 nicely confirms that places where spikes branch into narrower ones correspond to places with excess pressure underneath. Similar studies of incompressible 3D ideal MHD cases are found in [6]. They confirm that strong tangential fields suppress the growth as expected from theoretical considerations, while the Rayleigh-Taylor instability acts to amplify magnetic fields locally. In such magnetic fluids, parameter regimes exist where secondary Kelvin-Helmholtz instabilities develop, just as in the hydrodynamic situation of Figure 3 (note the regions of strong vorticity in the streamlines).

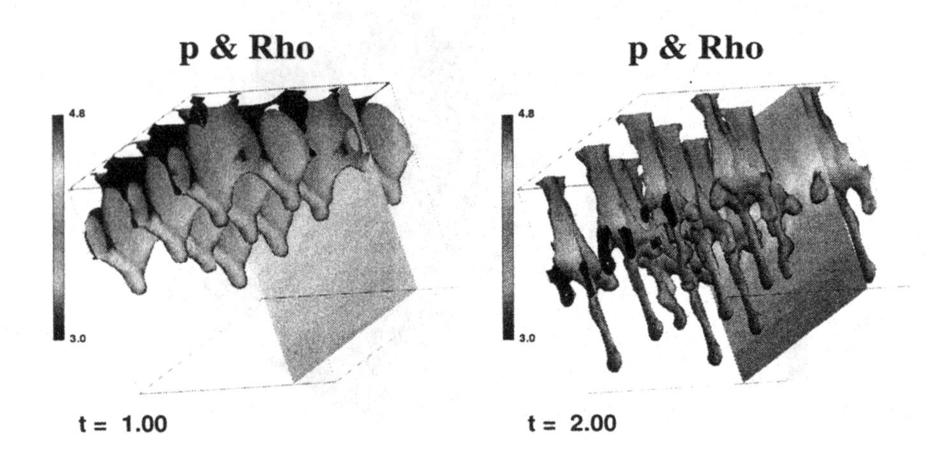

Fig. 4. 3D MHD Rayleigh-Taylor instability. At two consecutive times, an iso-surface of the density is coloured according to the thermal pressure. The thermal pressure is also shown in a vertical cut.

5 Conclusions

We have developed a powerful tool to simulate magnetised fluid dynamics. The Versatile Advection Code runs on many platforms, from PC's to supercomputers including distributed memory architectures. The rapidly maturing HPF compilers can yield scalable parallel performance for general fluid dynamical simulations. Clearly, the scaling and performance of VAC make high resolution 3D simulations possible, and detailed investigations may broaden our insight in the intricate dynamics of magneto-fluids and plasmas.

We presented simulations of the Rayleigh-Taylor instability in two and three spatial dimensions, with and without magnetic fields. VAC allows one to do all these simulations with a single problem setup, since the equations to solve and the dimensionality of the problem is simply specified in a preprocessing phase. Data analysis can be done using a variety of data visualisation packages, including IDL and AVS as demonstrated here. In the future, we plan to use VAC to investigate challenging astrophysical problems, like winds and jets emanating from stellar objects [9], magnetic loop dynamics, accretion onto black holes, etc. .

Web-site info on the code is available at `http://www.phys.uu.nl/~toth/` and at `http://www.phys.uu.nl/~mpr/`. MPEG-animations of various test problems can also be found there.

Acknowledgements

The Versatile Advection Code was developed as part of the project on 'Parallel Computational Magneto-Fluid Dynamics', funded by the Dutch Science Foundation (NWO) Priority Program on Massively Parallel Computing, and coordinated by Prof. Dr. J.P. Goedbloed. Computer time on the CM-5, the Cray T3E and IBM SP machines was sponsored by the Dutch 'Stichting Nationale Computerfaciliteiten' (NCF). R.K. performed the simulations on the Cray T3D, J90, and Sun workstation cluster at the Edinburgh Parallel Computing Centre with support from the TRACS programme as part of his research at FOM. G.T. receives a postdoctoral fellowship (D 25519) from the Hungarian Science Foundation (OTKA), and is supported by the OTKA grant F 017313.

References

1. Biskamp, D.: Nonlinear Magnetohydrodynamics. Cambridge Monographs on Plasma Physics 1, Cambridge University Press, Cambridge (1993)
2. Bittencourt, J.A.: Fundamentals of Plasma Physics. Pergamon Press, Oxford (1986)
3. Chandrasekhar, S.: Hydrodynamic and Hydromagnetic stability. Oxford University Press, New York (1961)
4. Freidberg, J.P.: Ideal Magnetohydrodynamics. Plenum Press, New York (1987)
5. Jun, B.-I., Norman, M.L.: MHD simulations of Rayleigh-Taylor instability in young supernova remnants. Astrophys. and Space Science **233** (1995) 267-272
6. Jun, B.-I., Norman, M.L., Stone, J.M.: A numerical study of Rayleigh-Taylor instability in magnetic fluids. Astrophys. J. **453** (1995) 332-349
7. Keppens, R., Tóth, G., Westermann, R.H.J., Goedbloed, J.P.: Growth and saturation of the Kelvin-Helmholtz instability with parallel and anti-parallel magnetic fields. Accepted by J. Plasma Phys. (1998)
8. Keppens, R., Tóth, G.: Non-linear dynamics of Kelvin-Helmholtz unstable magnetized jets: three-dimensional effects. Submitted for publication (1998)
9. Keppens, R., Goedbloed, J.P.: Numerical simulations of stellar winds: polytropic models. Accepted by Astron. & Astrophys. (1998)

10. Keppens, R., Tóth, G., Botchev, M.A., van der Ploeg, A.: Implicit and semi-implicit schemes in the Versatile Advection Code: algorithms. Submitted for publication (1998)
11. van der Ploeg, A., Keppens, R., Tóth, G.: Block Incomplete LU-preconditioners for Implicit Solution of Advection Dominated Problems. In: Hertzberger, B., Sloot, P. (eds.): Proceedings of High Performance Computing and Networking Europe 1997. Lecture Notes in Computer Science, Vol. 1225. Springer-Verlag, Berlin Heidelberg New York (1997) 421-430
12. Tóth, G.: Preprocessor based implementation of the Versatile Advection Code for workstations, vector and parallel computers. Proceedings of VECPAR'98 (3rd international meeting on vector and parallel processing), June 21-23 (1998), Porto, Portugal, Part II, p. 553-560.
13. Tóth, G.: Versatile Advection Code. In: Hertzberger, B., Sloot, P. (eds.): Proceedings of High Performance Computing and Networking Europe 1997. Lecture Notes in Computer Science, Vol. 1225. Springer-Verlag, Berlin Heidelberg New York (1997) 253-262
14. Tóth, G.: A general code for modeling MHD flows on parallel computers: Versatile advection code. Astrophys. Lett. & Comm. **34** (1996) 245
15. Tóth, G.: The LASY Preprocessor and Its Application to general Multidimensional Codes. J. Comput. Phys. **138** (1997) 981-990
16. Tóth, G., Keppens, R.: Comparison of Different Computer Platforms for Running the Versatile Advection Code. In: Sloot, P., Bubak, M., and Hertzberger, B. (eds.): Proceedings of High Performance Computing and Networking Europe 1998, Lecture Notes in Computer Science, Vol. 1401, Springer-Verlag, Berlin Heidelberg New York (1998) 368-376
17. Tóth, G., Keppens, R., Botchev, M.A.: Implicit and semi-implicit schemes in the Versatile Advection Code: numerical tests. Astron. & Astrophys. **332** (1998) 1159-1170
18. Tóth, G., Odstrčil, D.: Comparison of some Flux Corrected Transport and Total Variation Diminishing Numerical Schemes for Hydrodynamic and Magnetohydrodynamic Problems. J. Comput. Phys. **128** (1996) 82

Parallel Genetic Algorithms for Hypercube Machines

Ranieri Baraglia and Raffaele Perego

Istituto CNUCE, Consiglio Nazionale delle Ricerche (CNR)
via S. Maria 36, Pisa, Italy
{r.baraglia,r.perego}@cnuce.cnr.it

Abstract In this paper we investigate the design of highly parallel Genetic Algorithms. The Traveling Salesman Problem is used as a case study to evaluate and compare different implementations. To fix the various parameters of Genetic Algorithms to the case study considered, the Holland sequential Genetic Algorithm, which adopts different population replacement methods and crossover operators, has been implemented and tested. Both *fine – grained* and *coarse – grained* parallel GAs which adopt the selected genetic operators have been designed and implemented on a 128-node nCUBE 2 multicomputer. The *fine – grained* algorithm uses an innovative *mapping* strategy that makes the number of solutions managed independent of the number of processing nodes used. Complete performance results showing the behaviour of Parallel Genetic Algorithms for different population sizes, number of processors used, migration strategies are reported.

1 Introduction

Genetic Algorithms (GAs) [11, 12] are stochastic optimisation heuristics in which searches in the solution space are carried out by imitating the population genetics stated in Darwin's theory of evolution. Selection, crossover and mutation operators, directly derived by from natural evolution mechanisms are applied to a population of solutions, thus favouring the birth and survival of the best solutions. GAs have been successfully applied to many NP-hard combinatorial optimisation problems [6], in several application fields such as business, engineering, and science.

In order to apply GAs to a problem, a genetic representation of each individual (*chromosome*) that constitutes a solution of the problem has to be found. Then, we need to create an initial population, to define a cost function to measure the *fitness* of each solution, and to design the genetic operators that will allow us to produce a new population of solutions from a previous one. By iteratively applying the genetic operators to the current population, the fitness of the best individuals in the population converges to local optima.

Figure 1 reports the pseudo–code of the Holland genetic algorithm. After randomly generating the initial population $\beta(0)$, the algorithm at each iteration of the outer `repeat--until` loop generates a new population $\beta(t+1)$ from $\beta(t)$ by

J. Palma, J. Dongarra, and V. Hernández (Eds.): VECPAR'98, LNCS 1573, pp. 691–703, 1999.

selecting the best individuals of $\beta(t)$ (function SELECT()) and probabilistically applying the *crossover* and *mutation* genetic operators. The selection mechanism must ensure that the greater the fitness of an individual A_k is, the higher the probability of A_k being selected for reproduction. Once A_k has been selected, P_C is its probability of generating a son by applying the crossover operator to A_k and another individual A_i, while P_M and P_I are the probabilities of applying respectively, mutation and inversion operators to the generated individual respectively.

The crossover operator randomly selects parts of the parents' *chromosomes* and combines them to breed a new individual. The mutation operator randomly changes the value of a gene (a single bit if the binary representation scheme is used) within the chromosome of the individual to which it is applied. It is used to change the current solutions in order to avoid the convergence of the solutions to "bad" local optima.

The new individual is then inserted into population $\beta(t+1)$. Two main replacement methods can be used for this purpose. By adopting the *discrete* population model, the whole population $\beta(t)$ is replaced by new generated individuals at the end of the outer loop iteration. A variation on this model was proposed in [13] by using a parameter that controls the percentage of the population replaced at each generation. The *continuous* population model states, on the other hand, that the new individuals are soon inserted into the current population to replace older individuals with worse fitness. This replacement method allows potentially good individuals to be exploited as soon as they become available.

Irrespective of the replacement policy adopted, population $\beta(t+1)$ is expected to contain a greater number of individuals with good fitness than population $\beta(t)$. The GA end condition can be to reach a maximum number of generated populations, after which the algorithm is forced to stop or the algorithm converges to stable average fitness values.

The following are some important properties of GAs:

- they do not deal directly with problem solutions but with their genetic representation thus making GA implementation independent from the problem in question;
- they do not treat individuals but rather populations, thus increasing the probability of finding good solutions;
- they use probabilistic methods to generate new populations of solutions, thus avoiding being trapped in "bad" local optima.

On the other hand, GAs do not guarantee that global optima will be reached and their effectiveness very much depends on many parameters whose fixing may depend on the problem considered. The size of the population is particularly important. The larger the population is, the greater the possibility of reaching the optimal solution. Increasing the population clearly results in a large increase in the GA computational cost which, as we will see later, can be mitigated by exploiting parallelism.

The rest of the paper is organised as follows: Section 2 briefly describes the computational models proposed to design parallel GAs; Section 3 introduces

```
Program Holland_Genetic_Algorithm;
begin
t=0;
β (t) = INITIAL_POPULATION() ;
repeat

    for i = 1 to number_of_individuals do
        F(A_i) = COMPUTE_FITNESS(A_i);
    Average_fitness = COMPUTE_AVERAGE_FITNESS(F);
    for k = 1 to number_of_individuals do
    begin
        A_k = SELECT(β (t));
        if (P_C > random(0,1)) then
        begin
            A_i = SELECT(β (t));
            A_child = CROSSOVER( A_i,A_k);
            if (P_M > random (0,1)) then MUTATION
            (A_child);
            β (t+1)=UPDATE_POPULATION (A_child);
        end
    end;

t=t+1;
until (end_condition);
end
```

Figurel. Pseudo–code of the Holland Genetic Algorithm.

the Travelling Salesman Problem used as our case study, discusses the implementation issues and presents the results achieved on a 128–node hypercube multicomputer; finally Section 4 outlines the conclusions.

2 Parallel Genetic Algorithms

The availability of ever faster parallel computers means that parallel GAs can be exploited to reduce execution times and improve the quality of the solutions reached by increasing the sizes of populations managed.

In [5, 3] the parallelisation models adopted to implement GAs are classified. The models described are:

- **centralized model.** A single unstructured *panmitic* population is processed in parallel. A master processor manages the population and the selection strategy and requests a set of slave processors to compute the fitness function and other genetic operators on the chosen individuals. The model scales poorly and explores the solution space like a sequential algorithm which uses the same genetic operators. Several implementations of centralised parallel GAs are described in [1].

- **fine-grained model.** This model operates on a single structured population by exploiting the concepts of *spatiality* and *neighbourhood*. The first concept defines that a very small sub-population, ideally just an individual, is stored in one element (node) of the logical connection topology used, while the second specifies that the selection and crossover operators are applied only between individuals located on nearest–neighbour nodes. The neighbours of an individual determine all its possible partners, but since the neighbour sets of partner nodes overlap, this provides a way to spread good solutions across the entire population. Because of its scalable communication pattern, this model is particularly suited for massively parallel implementations. Implementations of fine–grained parallel GAs applied to different application problems can be found in [8, 9, 14, 15, 19].
- **coarse-grained model.** The whole population is partitioned into sub-populations, called *islands*, which evolve in parallel. Each island is assigned to a different processor and the evolution process takes place only among individuals belonging to the same island. This feature means that a greater genetic diversity can be maintained with respect to the exploitation of a panmitic population, thus improving the solution space exploration. Moreover, in order to improve the sub-population genotypes, a migration operator that periodically exchanges the best solutions among different islands is provided. Depending on the migration operator chosen we can distinguish between *island* and *stepping stone* implementations. In *island* implementations the migration occurs among every island, while in *stepping stone* implementations the migration occurs only between neighbouring islands. Studies have shown that there are two critical factors [10]: the number of solutions migrated each time and the interval time between two consecutive migrations. A large number of migrants leads to the behaviour of the island model similar to the behaviour of a panmitic model. A few migrants prevent the GA from mixing the genotypes, and thus reduce the possibility to bypass the local optimum value inside the islands. Implementations of coarse grained parallel GAs can be found in [10, 20, 21, 4, 18, 16].

3 Designing Parallel GAs

We implemented both $fine-grained$ and $coarse-grained$ parallel GAs applied to the classic Travelling Salesman Problem on a 128–node nCUBE 2 hypercube. Their performance was measured by varying the type and value of some genetic operators. In the following subsection the TSP case study is described and the parallel GA implementations are discussed and evaluated.

3.1 The Travelling Salesman Problem

The Travelling Salesman Problem (TSP) may be formally defined as follow: let $C = \{c_1, c_2, ..., c_n\}$ be a set of n cities and $\forall i$, $\forall j$ $d(c_i, c_j)$ the distance between city c_i and c_j with $d(c_i, c_j) = d(c_j, c_i)$. Solving the TSP entails finding a permutation $\pi\prime$ of the cities $(c_{\pi'(1)}, c_{\pi'(2)}, ..., c_{\pi'(n)})$, such that

$$\sum_{i=1}^{n} d(c_{\pi'(i)}, c_{\pi'(i+1)}) \leq \sum_{i=1}^{n} d(c_{\pi^k(i)}, c_{\pi^k(i+1)}) \qquad \forall \pi^k \neq \pi', (n+1) \equiv 1 \quad (1)$$

According to the TSP *path representation* described in [9], tours are represented by ordered sequences of integer numbers of length n, where sequence $(\pi(1), \pi(2), ..., \pi(n))$ represents a tour joining, in the order, cities $c_{\pi(1)}, c_{\pi(2)}$, ..., $c_{\pi(n)}$. The search space for the TSP is therefore the set of all permutations of n cities. The optimal solution is a permutation which yields the minimum cost of the tour.

The TSP instances used in the tests are: **GR48**, a 48-city problem that has an optimal solution equal to 5046, and **LIN105**, a 105-city problem that has a 14379 optimal solution[1].

3.2 Fixing the Genetic Operators

In order to study the sensitivity of the GAs for the TSP to the setting of the genetic operators, we used Holland's sequential GA by adopting the *discrete generation model*, one and two point crossover operators, and three different population replacement criteria.

- The *discrete generation model* separates sons' population from parents' population. Once all the sons' population has been generated, it is merged with the parents' population according to the replacement criteria adopted [8].
- One point crossover breaks the parents' tours into two parts and recombines them in the son in a way that ensures tour legality [2]. Two points crossover [7] works like the one point version but breaks the parents' tours into three different parts. A mutation operator which simply exchanges the order of two cities of the tour has been also implemented and used [9].
- The replacement criterion specifies a rule for merging current and new populations. We tested three different replacement criteria, called R1, R2 and R3. R1 replaces solutions with lower fitnesses of the current population with all the son solutions unaware of their fitness. R2 orders the sons by fitness, and replaces an individual i of the current population with son j only if the fitness of i is lower than the fitness of j. R2 has a higher control on the population than R1, and allows only the best sons to enter the new population. R3 selects the parents with a lower average fitness, and replaces them with the sons with above average fitnesses.

[1] Both the TSP instances are available at: `ftp://elib.zib-berlin.de/pub/mp-testdata/tsp/tsplib.html`

The tests for setting the genetic operators were carried out by using a 640 solution population, a 0.2 mutation parameter (to apply a mutation to 20% of the total population), 2000 generations for the 48-city TSP, and 3000 generations for the 105-city TSP. Every test was run 32 times, starting from different random populations, to obtain an average behaviour. From the results of the 32 tests we computed:

- the average solution: $AVG = \frac{\sum_{i=1}^{32} F_{E_i}}{32}$, where F_{E_i} is the best *fitness* obtained with run E_i;
- the best solution: $BST = min\{F_{E_i}, i = 1, \ldots, 32\}$;
- the worst solution: $WST = max\{F_{E_i}, i = 1, \ldots, 32\}$.

These preliminary tests allow us to choose some of the most suitable genetic operators and parameters for the TSP. Figure 2 plots the average fitness obtained by varying the crossover type on the 48-city TSP problem. The crossover was applied to 40% of the population and the R2 replacement criterion was used. As can be seen, the two point crossover converges to better average solutions than the one point operator. The one point crossover initially exhibits a better behaviour, but after 2000 generations, converges to solutions that have considerably higher costs. We obtained a similar behaviour for the other replacement criteria and for the 105-city TSP.

Table 1 reports AVG, BST and WST results for the 48-city TSP obtained by varying both the population replacement criterion and the percentage of the population to which the two point crossover has been applied. On average, the crossover parameter values in the range 0.4 – 0.6 lead to better solutions, almost irrespective of the replacement criterion adopted. Figure 3 shows the behaviour of the various replacement criteria for a 0.4 crossover value. The R2 and R3 replacement criteria resulted in a faster convergence than R1, and they converged to very near fitnesses.

		Crossover parameter			
		0.2	0.4	0.6	0.8
	AVG	6255	5632	5585	5870
R1	**BST**	5510	5315	5135	5305
	WST	7828	6079	6231	6693
	AVG	5902	5696	5735	5743
R2	**BST**	5405	5243	5323	5410
	WST	7122	6180	6225	6243
	AVG	6251	5669	5722	5773
R3	**BST**	5441	5178	5281	5200
	WST	7354	6140	6370	6594

Table1. Fitness values obtained with the execution of the sequential GA on the 48-city TSP by varying the value of the crossover parameter and the population replacement criterion.

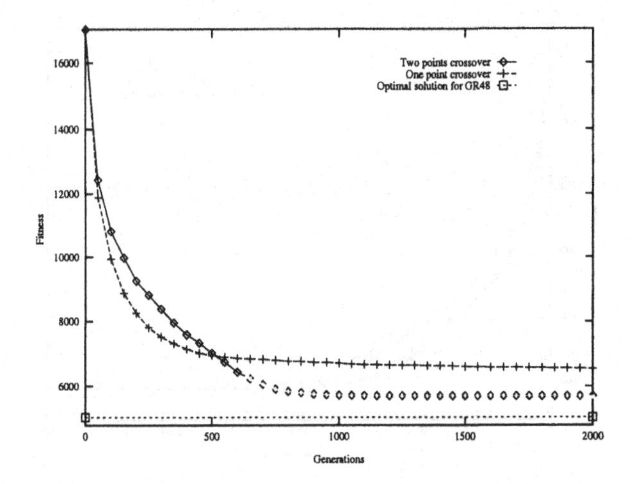

Figure2. Fitness values obtained with the execution of the sequential GA on the 48-city TSP by varying the crossover operator, and by using the R2 replacement criteria.

3.3 The Coarse Grained Implementation

The coarse grained parallel GA was designed according to the *discrete generation* and *stepping stone* models. Therefore, the new solutions are merged with the current population at the end of each generation phase, and the migration of the best individuals among sub-population is performed among ring-connected islands. Each of the P processors manages N/P individuals, with N population size (640 individuals in our case). The number of migrants is a fixed percentage of the sub-population. As in [4], migration occurs periodically in a regular time rhythm, after a fixed number of generations.

In order to include all the migrants in the current sub-populations, and to merge the sub-population with the locally generated solutions, R1 and R2 replacement criteria were used, respectively. Moreover, the two point crossover operator was adopted.

Table 2 reports some results obtained by running the coarse grained parallel GA on the 48-city TSP. **M** denotes the migration parameter. The same data for a migration parameter equal to 0.1 are plotted in Figure 4. It can be seen that AVG, BST and WST solutions get worse values by increasing the number of the nodes used. This depends on the constant population size used: with 4 nodes sub-populations of 160 solutions are exploited, while with 64 nodes the sub-populations only consists of 10 individuals. Decreasing the number of solutions that forms a sub-population worsens the search in the solution space; small sub-populations result in an insufficient exploration of the solution space. The influence of the number of migrants on the convergence is clear from Table 2. When the sub-populations are small, a higher value of the migration parameter may improve the quality of solutions through a better mix of the genetic material.

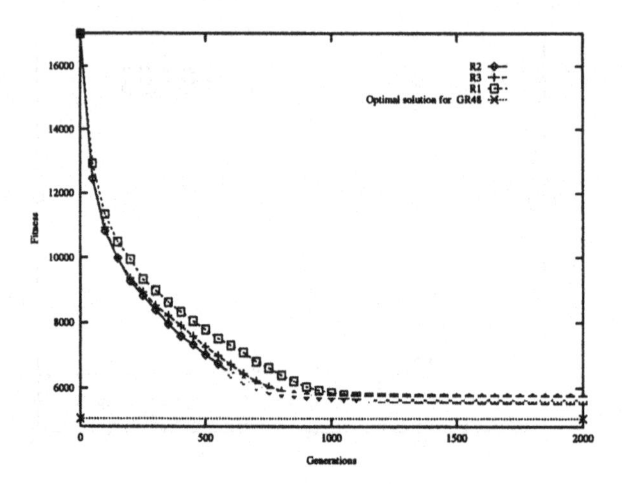

Figure3. Fitness values obtained by executing the sequential GA on the 48-city TSP with a 0.4 crossover parameter, and by varying the replacement criterion.

3.4 The Fine Grained Implementation

The fine grained parallel GA was designed according to the *continuous generation model*, which is much more suited for fine grained parallel GAs than the *discrete* one. The two point crossover operator was applied.

According to the fine grained model the population is structured in a logic topology which fixes the rules of interaction between the solution and other solutions: each solution s is placed at a vertex $v(s)$ of logic topology T. The crossover operation can only be applied among nearest neighbour solutions placed on the vertices directly connected in T. Our implementation exploits the physical topology of the target multicomputer, therefore the population of 2^N individuals is structured as a N-dimensional hypercube.

By exploiting the recursivity of the hypercube topology definition, we made the number of solutions treated independent of the number of nodes used to execute the algorithm. As can be seen in Figure 5, a $2^3 = 8$ solution population can be placed on a $2^2 = 4$ node hypercube, using a simple mapping function which masks the first (or the last) *bit* of the **Grey** code used to numerate the logical hypercube vertices [17]. Physical node $X00$ will hold solutions 000 and 100 of the logic topology, not violating the neighbourhood relationships fixed by the population structure. In fact, the solutions on the neighbourhood of each solution s will still be in the physical topology on directly connected nodes or on the node holding s itself.

We can generalise this *mapping* scheme: to determine the allocation of a 2^N solution population on a 2^M node physical hypercube, with $M < N$, we simply mask the first (the last) $N - M$ *bits* of the binary coding of each solution.

Table 3 reports the fitness values obtained with the execution of the fine grained GA by varying the population dimension from 128 solutions (a 7 di-

		Number of processing nodes					
		4	**8**	**16**	**32**	**64**	**128**
	AVG	5780	5786	5933	6080	6383	6995
M=0.1	**BST**	5438	5315	5521	5633	5880	6625
	WST	6250	6387	6516	6648	8177	8175
	AVG	5807	5877	5969	6039	6383	6623
M=0.3	**BST**	5194	5258	5467	5470	5727	6198
	WST	6288	6644	7030	6540	8250	7915
	AVG	5900	5866	5870	6067	6329	6617
M=0.5	**BST**	5419	5475	5483	5372	6017	6108
	WST	6335	6550	7029	6540	8250	7615

Table2. Fitness values obtained with the execution of the coarse grained GA on the 48-city TSP by varying the migration parameter.

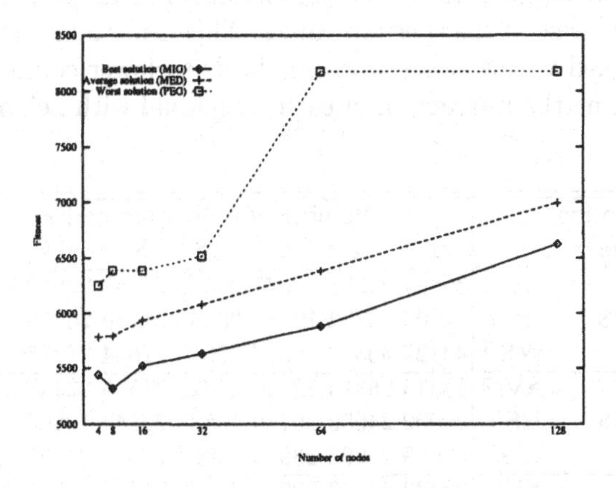

Figure4. AVG, BST and WST values obtained by executing the coarse grained GA on the 48-city TSP as a function of the number of nodes used, and 0.1 as migration parameter.

mension hypercube) to 1024 solutions (a 10 dimension hypercube). As expected, the ability to exploit population sizes larger than the number of processors used in our mapping scheme, leads to better quality solutions especially when few processing nodes are used. The improvement in the fitness values by increasing the number of nodes while maintaining the population size fixed, is due to a particular feature of the implementation, which aims to minimise the communication times to the detriment of the diversity of the same node solutions. Selection rules tend to choose partner solutions in the same node. The consequence is a greater uniformity in solutions obtained on few nodes, which worsens the exploration of the solution space. The coarse grained implementation suffered of the opposite problem which resulted in worse solutions obtained as the number of nodes was increased. This behaviour can be observed by comparing

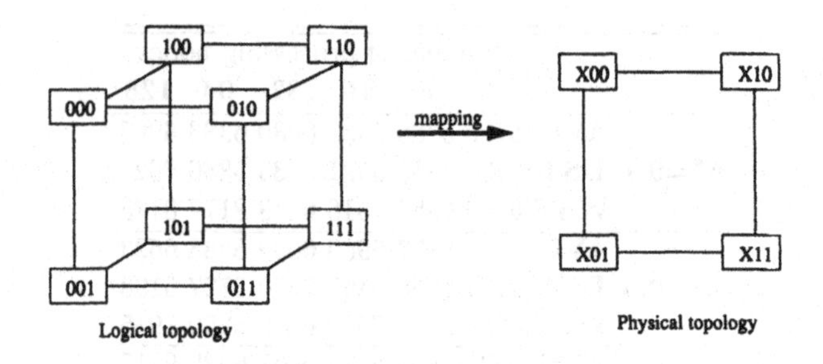

Figure5. Example of an application of the *mapping* scheme.

Figure 4, concerning the coarse grained GA, and Figure 6, concerning the fine grained algorithm with a 128 solution population applied to the 48-city TSP. Table 4 shows that an increase in the number of solutions processed results in a corresponding increase in the speedup values. This is because a larger number of individuals assigned to the same processor leads to lower communication overheads for managing the interaction of each individual with neighbour partners.

Population size		Number of processing nodes 1	4	8	16	32	64	128
	AVG	39894	24361	23271	23570	23963	22519	22567
128	**BST**	34207	20774	19830	20532	21230	20269	20593
	WST	42127	30312	26677	27610	27634	28256	25931
	AVG	33375	25313	22146	21616	21695	22247	21187
256	**BST**	29002	24059	20710	19833	20144	19660	19759
	WST	40989	26998	23980	24007	23973	24337	22256
	AVG	28422	23193	22032	21553	20677	20111	20364
512	**BST**	28987	22126	19336	20333	19093	18985	18917
	WST	41020	25684	23450	22807	22213	21696	21647
	AVG	25932	23659	22256	20366	19370	18948	19152
1024	**BST**	27010	21581	21480	18830	18256	18252	17446
	WST	40901	25307	22757	21714	20766	19525	20661

Table3. Fitness values obtained with the execution of the fine grained GA applied to the 105-city TSP after 3000 generations, by varying the number of solutions per node.

3.5 Comparisons

We compared the fine and coarse grained algorithms on the basis of the execution time required and the fitness values obtained by each one after the evaluation of 512000 solutions. This comparison criterion was chosen because it allows to overcome computational models diversity that make non comparable the fine and

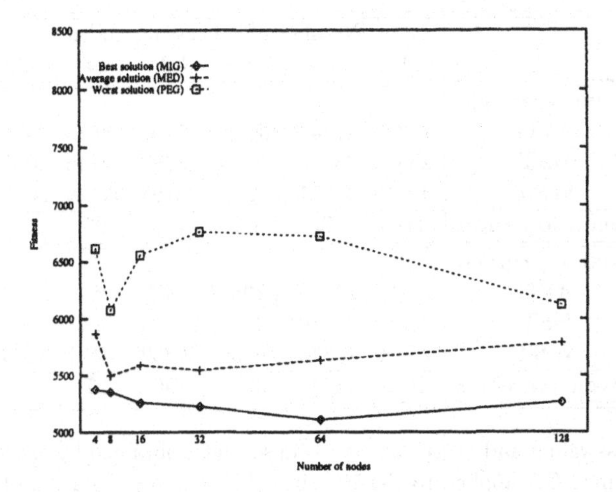

Figure6. AVG, BST and WST values obtained by executing the fine grained GA applied to the 48-city TSP as a function of the number of nodes used. The population size was fixed to 128 individuals.

Number	Number of individuals			
of nodes	128	256	512	1024
1	1	1	1	1
4	3.79	3.84	3.89	3.92
8	7.25	7.47	7.61	7.74
16	13.7	14.31	14.78	15.12
32	25.25	27.02	28.03	29.38
64	44.84	49.57	53.29	56.13
128	78.47	88.5	98.06	105.32

Table4. Speedup of the of the fine grained GA applied on the 105-city TSP for different population sizes.

coarse grained algorithms. The evaluation of 512000 new solutions allows both the algorithms to converge and requires comparable execution times. Table 5 shows the results of this comparison. It can be seen that when the number of the nodes used increases the fine grained algorithm gets sensibly better results than the coarse grained one. On the other hand, the coarse grained algorithm shows a superlinear speedup due to the *quick sort* algorithm used by each node for ordering by fitness the solutions managed. As the number of nodes is increased, the number of individuals assigned to each node decreases, thus requiring considerably less time to sort the sub-population.

	Number of processing nodes					
	4	**8**	**16**	**32**	**64**	**128**
Fine grained						
AVG	26373	26349	25922	25992	24613	23227
BST	23979	23906	23173	21992	22145	20669
WST	30140	27851	29237	29193	30176	26802
Execution times	1160	606	321	174	98	51
Coarse grained						
AVG	23860	24526	27063	29422	32542	35342
BST	20219	21120	23510	25783	30927	33015
WST	25299	29348	36795	39330	39131	41508
Execution times	1670	804	392	196	97	47

Table5. Fitness values and execution times (in seconds) obtained by executing the fine and coarse grained GA applied to the 105-city TSP with a population of 128 and 640 individuals, respectively.

4 Conclusions

We have discussed the results of the application of parallel GA algorithms to the TSP. In order to analyse the behaviour of different replacement criteria and crossover operators and values Holland's sequential GA was implemented. The tests showed that the two point crossover finds better solutions, as does a replacement criteria which replaces an individual i of the current population with son j only if the fitness of i is worse than the fitness of j. To implement the $fine-grained$ and $coarse-grained$ parallel GAs on a hypercube parallel computer the most suitable operators were adopted. For the coarse grained GA we observed that the quality of solutions gets worse if he number of nodes used is increased. Moreover, due to the sorting algorithm used to order each sub-population by fitness, the speedup of the coarse grained GA were superlinear. Our fine-grained algorithm adopts a mapping strategy that allows the number of solutions to be independent of the number of nodes used. The ability to exploit population sizes larger than the number of processors used gives better quality solutions especially when only a few processing nodes are used. Moreover, the quality of solutions does not get worse if the number of the nodes used is increased. The fine grained algorithm showed good scalability. A comparison between the fine and coarse grained algorithms highlighted that fine grained algorithms represent the better compromise between quality of the solution reached and the execution time spent on finding it.
The GAs implemented reached only "good" solutions. In order to improve the quality of solutions obtained, we are working to include a local search procedure within the GA.

References

1. R. Bianchini and C.M. Brown. Parallel genetic algorithm on distributed-memory architectures. Technical Report TR 436, Computer Sciences Department University of Rochester, 1993.
2. H. Braun. On solving travelling salesman problems by genetic algorithms. In *Parallel Problem Solving from Nature - Proceedings of 1st Workshop PPSN , volume 496 of Lecture Notes in Computer Science*, pages 129–133. Springer-Verlag, 1991.
3. E. Cantu-Paz. A summary of research on parallel genetic algoritms. Technical Report 95007, University of Illinois at Urbana-Champaign, Genetic Algoritms Lab. (IlliGAL), http://gal4.ge.uiuc.edu/illigal.home.html, July 1995.
4. S. Cohoon, J. Hedge, S. Martin, and D. Richards. Punctuated equilibria: a parallel genetic algorithm. *IEEE Transaction on CAD*, 10(4):483–491, April 1991.
5. M. Dorigo and V. Maniezzo. Parallel genetic algorithms: Introduction and overview of current research. In *Parallel Genetic Algorithms*, pages 5–42. IOS Press, 1993.
6. M. R. Garey and D.S. Jonshon. *Computers and Intractability: A Guide to the Theory of NP-Completeness.* Freeman, San Francisco, 1979.
7. D. Goldberg and R. Lingle. Alleles, loci, and the tsp. In *Proc. of the First International Conference on Genetic Algorithms*, pages 154–159, 1985.
8. M. Gorges-Schleuter. Explicit parallelism of genetic algorithms through population structures. In *Parallel Problem Solving from Nature - Proceedings of 1st Workshop PPSN*, volume 496, pages 398–406. Lecture Notes in Computer Science, 1990.
9. M. Gorges-Schleuter. *Genetic Algoritms and Population Structure.* PhD thesis, University of Dortmund, 1991.
10. P. Grosso. *Computer Simulations of Genetic Adaptation: Parallel Subcomponent Interaction in a Multilocus Model.* PhD thesis, University of Michigan, 1985.
11. J. Holland. *Adaptation in Natural and Artificial Systems.* Univ. of Mitchigan Press, 1975.
12. J.H. Holland. Algoritmi genetici. *Le Scienze*, 289:50–57, 1992.
13. K.A. De Jong. *An analysis of the behavior of a class of genetic adaptive systems.* PhD thesis, University of Michigan, 1975.
14. B. Manderick and P. Spiessens. Fine-grained parallel genetic algorithms. In *Proceedings of the Third International Conference on Genetic Algorithms*, pages 428–433. Morgan Kaufmann Publishers, 1989.
15. H. Muhlenbein. Parallel genetic algorithms, population genetic and combinatorial optimization. In *Parallel Problem Solving from Nature - Proceedings of 1st Workshop PPSN*, volume 496, pages 407–417. Lecture Notes in Computer Science, 1991.
16. H. Muhlenbein, M. Schomisch, and J. Born. The parallel genetic algorithm as function optimizer. *Parallel Computing*, 17:619–632, 1991.
17. nCUBE Corporation. ncube2 processor manual. 1990.
18. C. Pettey, M. Lenze, and J. Grefenstette. A parallel genetic algorithm. In *Proceedings of the Second International Conference on Genetic Algorithms*, pages 155–161. L. Erlbaum Associates, 1987.
19. M. Schwehm. Implementation of genetic algorithms on various interconnections networks. *Parallel Computing and Transputer applications*, pages 195–203, 1992.
20. R. Tanese. Parallel genetic algorithms for a hypercube. In *Proceedings of the Second International Conference on Genetic Algorithms*, pages 177–183. L. Erlbaum Associates, 1987.
21. R. Tanese. Distributed genetic algorithms. In *Proceedings of the Third International Conference on Genetic Algorithms*, pages 434–440. M. Kaufmann, 1989.

Author Index

Lecture Notes in Computer Science

For information about Vols. 1–1553
please contact your bookseller or Springer-Verlag

Vol. 1554: S. Nishio, F. Kishino (Eds.), Advanced Multimedia Content Processing. Proceedings, 1998. XIV, 454 pages. 1999.

Vol. 1555: J.P. Müller, M.P. Singh, A.S. Rao (Eds.), Intelligent Agents V. Proceedings, 1998. XXIV, 455 pages. 1999. (Subseries LNAI).

Vol. 1556: S. Tavares, H. Meijer (Eds.), Selected Areas in Cryptography. Proceedings, 1998. IX, 377 pages. 1999.

Vol. 1557: P. Zinterhof, M. Vajteršic, A. Uhl (Eds.), Parallel Computation. Proceedings, 1999. XV, 604 pages. 1999.

Vol. 1558: H. J.v.d. Herik, H. Iida (Eds.), Computers and Games. Proceedings, 1998. XVIII, 337 pages. 1999.

Vol. 1559: P. Flener (Ed.), Logic-Based Program Synthesis and Transformation. Proceedings, 1998. X, 331 pages. 1999.

Vol. 1560: K. Imai, Y. Zheng (Eds.), Public Key Cryptography. Proceedings, 1999. IX, 327 pages. 1999.

Vol. 1561: I. Damgård (Ed.), Lectures on Data Security. VII, 250 pages. 1999.

Vol. 1562: C.L. Nehaniv (Ed.), Computation for Metaphors, Analogy, and Agents. X, 389 pages. 1999. (Subseries LNAI).

Vol. 1563: Ch. Meinel, S. Tison (Eds.), STACS 99. Proceedings, 1999. XIV, 582 pages. 1999.

Vol. 1565: P. P. Chen, J. Akoka, H. Kangassalo, B. Thalheim (Eds.), Conceptual Modeling. XXIV, 303 pages. 1999.

Vol. 1567: P. Antsaklis, W. Kohn, M. Lemmon, A. Nerode, S. Sastry (Eds.), Hybrid Systems V. X, 445 pages. 1999.

Vol. 1568: G. Bertrand, M. Couprie, L. Perroton (Eds.), Discrete Geometry for Computer Imagery. Proceedings, 1999. XI, 459 pages. 1999.

Vol. 1569: F.W. Vaandrager, J.H. van Schuppen (Eds.), Hybrid Systems: Computation and Control. Proceedings, 1999. X, 271 pages. 1999.

Vol. 1570: F. Puppe (Ed.), XPS-99: Knowledge-Based Systems. VIII, 227 pages. 1999. (Subseries LNAI).

Vol. 1571: P. Noriega, C. Sierra (Eds.), Agent Mediated Electronic Commerce. Proceedings, 1998. IX, 207 pages. 1999. (Subseries LNAI).

Vol. 1573: J.M.L.M. Palma, J. Dongarra, V. Hernández (Eds.), Vector and Parallel Processing - VECPAR'98. Proceedings, 1998. XVI, 706 pages. 1999.

Vol. 1572: P. Fischer, H.U. Simon (Eds.), Computational Learning Theory. Proceedings, 1999. X, 301 pages. 1999. (Subseries LNAI).

Vol. 1574: N. Zhong, L. Zhou (Eds.), Methodologies for Knowledge Discovery and Data Mining. Proceedings, 1999. XV, 533 pages. 1999. (Subseries LNAI).

Vol. 1575: S. Jähnichen (Ed.), Compiler Construction. Proceedings, 1999. X, 301 pages. 1999.

Vol. 1576: S.D. Swierstra (Ed.), Programming Languages and Systems. Proceedings, 1999. X, 307 pages. 1999.

Vol. 1577: J.-P. Finance (Ed.), Fundamental Approaches to Software Engineering. Proceedings, 1999. X, 245 pages. 1999.

Vol. 1578: W. Thomas (Ed.), Foundations of Software Science and Computation Structures. Proceedings, 1999. X, 323 pages. 1999.

Vol. 1579: W.R. Cleaveland (Ed.), Tools and Algorithms for the Construction and Analysis of Systems. Proceedings, 1999. XI, 445 pages. 1999.

Vol. 1580: A. Včkovski, K.E. Brassel, H.-J. Schek (Eds.), Interoperating Geographic Information Systems. Proceedings, 1999. XI, 329 pages. 1999.

Vol. 1581: J.-Y. Girard (Ed.), Typed Lambda Calculi and Applications. Proceedings, 1999. VIII, 397 pages. 1999.

Vol. 1582: A. Lecomte, F. Lamarche, G. Perrier (Eds.), Logical Aspects of Computational Linguistics. Proceedings, 1997. XI, 251 pages. 1999. (Subseries LNAI).

Vol. 1583: D. Scharstein, View Synthesis Using Stereo Vision. XV, 163 pages. 1999.

Vol. 1584: G. Gottlob, E. Grandjean, K. Seyr (Eds.), Computer Science Logic. Proceedings, 1998. X, 431 pages. 1999.

Vol. 1585: B. McKay, X. Yao, C.S. Newton, J.-H. Kim, T. Furuhashi (Eds.), Simulated Evolution and Learning. Proceedings, 1998. XIII, 472 pages. 1999. (Subseries LNAI).

Vol. 1586: J. Rolim et al. (Eds.), Parallel and Distributed Processing. Proceedings, 1999. XVII, 1443 pages. 1999.

Vol. 1587: J. Pieprzyk, R. Safavi-Naini, J. Seberry (Eds.), Information Security and Privacy. Proceedings, 1999. XI, 327 pages. 1999.

Vol. 1590: P. Atzeni, A. Mendelzon, G. Mecca (Eds.), The World Wide Web and Databases. Proceedings, 1998. VIII, 213 pages. 1999.

Vol. 1592: J. Stern (Ed.), Advances in Cryptology – EUROCRYPT '99. Proceedings, 1999. XII, 475 pages. 1999.

Vol. 1593: P. Sloot, M. Bubak, A. Hoekstra, B. Hertzberger (Eds.), High-Performance Computing and Networking. Proceedings, 1999. XXIII, 1318 pages. 1999.

Vol. 1594: P. Ciancarini, A.L. Wolf (Eds.), Coordination Languages and Models. Proceedings, 1999. IX, 420 pages. 1999.

Vol. 1595: K. Hammond, T. Davie, C. Clack (Eds.), Implementation of Functional Languages. Proceedings, 1998. X, 247 pages. 1999.

Vol. 1596: R. Poli, H.-M. Voigt, S. Cagnoni, D. Corne, G.D. Smith, T.C. Fogarty (Eds.), Evolutionary Image Analysis, Signal Processing and Telecommunications. Proceedings, 1999. X, 225 pages. 1999.

Vol. 1597: H. Zuidweg, M. Campolargo, J. Delgado, A. Mullery (Eds.), Intelligence in Services and Networks. Proceedings, 1999. XII, 552 pages. 1999.

Vol. 1598: R. Poli, P. Nordin, W.B. Langdon, T.C. Fogarty (Eds.), Genetic Programming. Proceedings, 1999. X, 283 pages. 1999.

Vol. 1599: T. Ishida (Ed.), Multiagent Platforms. Proceedings, 1998. VIII, 187 pages. 1999. (Subseries LNAI).

Vol. 1601: J.-P. Katoen (Ed.), Formal Methods for Real-Time and Probabilistic Systems. Proceedings, 1999. X, 355 pages. 1999.

Vol. 1602: A. Sivasubramaniam, M. Lauria (Eds.), Network-Based Parallel Computing. Proceedings, 1999. VIII, 225 pages. 1999.

Vol. 1603: J. Vitek, C.D. Jensen (Eds.), Secure Internet Programming. X, 501 pages. 1999.

Vol. 1605: J. Billington, M. Diaz, G. Rozenberg (Eds.), Application of Petri Nets to Communication Networks. IX, 303 pages. 1999.

Vol. 1606: J. Mira, J.V. Sánchez-Andrés (Eds.), Foundations and Tools for Neural Modeling. Proceedings, Vol. I, 1999. XXIII, 865 pages. 1999.

Vol. 1607: J. Mira, J.V. Sánchez-Andrés (Eds.), Engineering Applications of Bio-Inspired Artificial Neural Networks. Proceedings, Vol. II, 1999. XXIII, 907 pages. 1999.

Vol. 1608: S. Doaitse Swierstra, P.R. Henriques, J.N. Oliveira (Eds.), Advanced Functional Programming. Proceedings, 1998. XII, 289 pages. 1999.

Vol. 1609: Z. W. Raś, A. Skowron (Eds.), Foundations of Intelligent Systems. Proceedings, 1999. XII, 676 pages. 1999. (Subseries LNAI).

Vol. 1610: G. Cornuéjols, R.E. Burkard, G.J. Woeginger (Eds.), Integer Programming and Combinatorial Optimization. Proceedings, 1999. IX, 453 pages. 1999.

Vol. 1611: I. Imam, Y. Kodratoff, A. El-Dessouki, M. Ali (Eds.), Multiple Approaches to Intelligent Systems. Proceedings, 1999. XIX, 899 pages. 1999. (Subseries LNAI).

Vol. 1612: R. Bergmann, S. Breen, M. Göker, M. Manago, S. Wess, Developing Industrial Case-Based Reasoning Applications. XX, 188 pages. 1999. (Subseries LNAI).

Vol. 1613: A. Kuba, M. Šámal, A. Todd-Pokropek (Eds.), Information Processing in Medical Imaging. Proceedings, 1999. XVII, 508 pages. 1999.

Vol. 1614: D.P. Huijsmans, A.W.M. Smeulders (Eds.), Visual Information and Information Systems. Proceedings, 1999. XVII, 827 pages. 1999.

Vol. 1615: C. Polychronopoulos, K. Joe, A. Fukuda, S. Tomita (Eds.), High Performance Computing. Proceedings, 1999. XIV, 408 pages. 1999.

Vol. 1617: N.V. Murray (Ed.), Automated Reasoning with Analytic Tableaux and Related Methods. Proceedings, 1999. X, 325 pages. 1999. (Subseries LNAI).

Vol. 1619: M.T. Goodrich, C.C. McGeoch (Eds.), Algorithm Engineering and Experimentation. Proceedings, 1999. VIII, 349 pages. 1999.

Vol. 1620: W. Horn, Y. Shahar, G. Lindberg, S. Andreassen, J. Wyatt (Eds.), Artificial Intelligence in Medicine. Proceedings, 1999. XIII, 454 pages. 1999. (Subseries LNAI).

Vol. 1621: D. Fensel, R. Studer (Eds.), Knowledge Acquisition Modeling and Management. Proceedings, 1999. XI, 404 pages. 1999. (Subseries LNAI).

Vol. 1622: M. González Harbour, J.A. de la Puente (Eds.), Reliable Software Technologies – Ada-Europe'99. Proceedings, 1999. XIII, 451 pages. 1999.

Vol. 1625: B. Reusch (Ed.), Computational Intelligence. Proceedings, 1999. XIV, 710 pages. 1999.

Vol. 1626: M. Jarke, A. Oberweis (Eds.), Advanced Information Systems Engineering. Proceedings, 1999. XIV, 478 pages. 1999.

Vol. 1627: T. Asano, H. Imai, D.T. Lee, S.-i. Nakano, T. Tokuyama (Eds.), Computing and Combinatorics. Proceedings, 1999. XIV, 494 pages. 1999.

Col. 1628: R. Guerraoui (Ed.), ECOOP'99 - Object-Oriented Programming. Proceedings, 1999. XIII, 529 pages. 1999.

Vol. 1629: H. Leopold, N. García (Eds.), Multimedia Applications, Services and Techniques - ECMAST'99. Proceedings, 1999. XV, 574 pages. 1999.

Vol. 1631: P. Narendran, M. Rusinowitch (Eds.), Rewriting Techniques and Applications. Proceedings, 1999. XI, 397 pages. 1999.

Vol. 1632: H. Ganzinger (Ed.), Automated Deduction – Cade-16. Proceedings, 1999. XIV, 429 pages. 1999. (Subseries LNAI).

Vol. 1633: N. Halbwachs, D. Peled (Eds.), Computer Aided Verification. Proceedings, 1999. XII, 506 pages. 1999.

Vol. 1634: S. Džeroski, P. Flach (Eds.), Inductive Logic Programming. Proceedings, 1999. VIII, 303 pages. 1999. (Subseries LNAI).

Vol. 1636: L. Knudsen (Ed.), Fast Software Encryption. Proceedings, 1999. VIII, 317 pages. 1999.

Vol. 1638: A. Hunter, S. Parsons (Eds.), Symbolic and Quantitative Approaches to Reasoning and Uncertainty. Proceedings, 1999. IX, 397 pages. 1999. (Subseries LNAI).

Vol. 1639: S. Donatelli, J. Kleijn (Eds.), Application and Theory of Petri Nets 1999. Proceedings, 1999. VIII, 425 pages. 1999.

Vol. 1640: W. Tepfenhart, W. Cyre (Eds.), Conceptual Structures: Standards and Practices. Proceedings, 1999. XII, 515 pages. 1999. (Subseries LNAI).

Vol. 1644: J. Wiedermann, P. van Emde Boas, M. Nielsen (Eds.), Automata, Languages, and Programming. Proceedings, 1999. XIII, 720 pages. 1999.

Vol. 1649: R.Y. Pinter, S. Tsur (Eds.), Next Generation Information Technologies and Systems. Proceedings, 1999. IX, 327 pages. 1999.

Vol. 1650: K.-D. Althoff, R. Bergmann, L.K. Branting (Eds.), Case-Based Reasoning Research and Development. Proceedings, 1999. XII, 598 pages. 1999. (Subseries LNAI).

Vol. 1653: S. Covaci (Ed.), Active Networks. Proceedings, 1999. XIII, 346 pages. 1999.